W9-ACJ-552

VOLUME 6
1900-1949

Science and Its Times

Understanding the Social Significance of Scientific Discovery

VOLUME 6
1900-1949

Science and Its Times

Understanding the Social Significance of Scientific Discovery

Neil Schlager, Editor
Josh Lauer, Associate Editor

Produced by Schlager Information Group

Detroit
New York
San Francisco
London
Boston
Woodbridge, CT

Science and Its Times

VOLUME 6

1900-1949

NEIL SCHLAGER, *Editor*
JOSH LAUER, *Associate Editor*

GALE GROUP STAFF

Amy Loerch Strumolo, *Project Coordinator*
Christine B. Jeryan, *Contributing Editor*

Mary K. Fyke, *Editorial Technical Specialist*

Maria Franklin, *Permissions Manager*
Margaret A. Chamberlain, *Permissions Specialist*
Shalice Shah-Caldwell, *Permissions Associate*

Mary Beth Trimper, *Production Director*
Evi Seoud, *Assistant Production Manager*
Stacy L. Melson, *Buyer*

Kenn Zorn, *Product Design Manager*
Tracey Rowens, *Senior Art Director*
Barbara Yarrow, *Graphic Services Manager*
Randy Bassett, *Image Database Supervisor*
Mike Logusz, *Imaging Specialist*
Pamela A. Reed, *Photography Coordinator*
Leitha Etheridge-Sims *Junior Image Cataloger*

The paper used in this publication meets the minimum requirements of American National Standard for Information Sciences—Permanence Paper for Printed Library Materials, ANSI Z39.48-1984.

ISBN: 0-7876-3938-9

Printed in the United States of America
10 9 8 7 6 5 4 3 2 1

Contents

Mathematics

Medicine

Physical Sciences

Technology and Invention

Preface

The interaction of science and society is increasingly a focal point of high school studies, and with good reason: by exploring the achievements of science within their historical context, students can better understand a given event, era, or culture. This cross-disciplinary approach to science is at the heart of *Science and Its Times.*

Readers of *Science and Its Times* will find a comprehensive treatment of the history of science, including specific events, issues, and trends through history as well as the scientists who set in motion—or who were influenced by—those events. From the ancient world's invention of the plowshare and development of seafaring vessels; to the Renaissance-era conflict between the Catholic Church and scientists advocating a sun-centered solar system; to the development of modern surgery in the nineteenth century; and to the mass migration of European scientists to the United States as a result of Adolf Hitler's Nazi regime in Germany during the 1930s and 1940s, science's involvement in human progress—and sometimes brutality—is indisputable.

While science has had an enormous impact on society, that impact has often worked in the opposite direction, with social norms greatly influencing the course of scientific achievement through the ages. In the same way, just as history can not be viewed as an unbroken line of ever-expanding progress, neither can science be seen as a string of ever-more amazing triumphs. *Science and Its Times* aims to present the history of science within its historical context—a context marked not only by genius and stunning invention but also by war, disease, bigotry, and persecution.

Format of the Series

Science and Its Times is divided into seven volumes, each covering a distinct time period:

Volume 1: 2000 B.C.-699 A.D.

Volume 2: 700-1449

Volume 3: 1450-1699

Volume 4: 1700-1799

Volume 5: 1800-1899

Volume 6: 1900-1949

Volume 7: 1950-present

Dividing the history of science according to such strict chronological subsets has its own drawbacks. Many scientific events—and scientists themselves—overlap two different time periods. Also, throughout history it has been common for the impact of a certain scientific advancement to fall much later than the advancement itself. Readers looking for information about a topic should begin their search by checking the index at the back of each volume. Readers perusing more than one volume may find the same scientist featured in two different volumes.

Readers should also be aware that many scientists worked in more than one discipline during their lives. In such cases, scientists may be featured in two different chapters in the same volume. To facilitate searches for a specific person or subject, main entries on a given person or subject are indicated by bold-faced page numbers in the index.

Within each volume, material is divided into chapters according to subject area. For volumes 5, 6, and 7, these areas are: Exploration and Discovery, Life Sciences, Mathematics, Medicine, Physical Sciences, and Technology and Invention. For volumes 1, 2, 3, and 4, readers will find that the Life Sciences and Medicine chapters have been combined into a single section, reflecting the historical union of these disciplines before 1800.

Arrangement of Volume 6: 1900-1949

Volume 6 begins with two notable sections in the frontmatter: a general introduction to science and society during the period, and a general chronology that presents key scientific events during the period alongside key world historical events.

The volume is then organized into six chapters, corresponding to the six subject areas listed above in "Format of the Series." Within each chapter, readers will find the following entry types:

Chronology of Key Events: Notable events in the subject area during the period are featured in this section.

Overview: This essay provides an overview of important trends, issues, and scientists in the subject area during the period.

Topical Essays: Ranging between 1,500 and 2,000 words, these essays discuss notable events, issues, and trends in a given subject area. Each essay includes a Further Reading section that points users to additional sources of information on the topic, including books, articles, and web sites.

Biographical Sketches: Key scientists during the era are featured in entries ranging between 500 and 1,000 words in length.

Biographical Mentions: Additional brief biographical entries on notable scientists during the era.

Bibliography of Primary Source Documents: These annotated bibliographic listings feature key books and articles pertaining to the subject area.

Following the final chapter are two additional sections: a general bibliography of sources related to the history of science, and a general subject index. Readers are urged to make heavy use of the index, because many scientists and topics are discussed in several different entries.

A note should be made about the arrangement of individual entries within each chapter: while the long and short biographical sketches are arranged alphabetically according to the scientist's surname, the topical essays lend themselves to no such easy arrangement. Again, readers looking for a specific topic should consult the index. Readers wanting to browse the list of essays in a given subject area can refer to the table of contents in the book's frontmatter.

Additional Features

Throughout each volume readers will find sidebars whose purpose is to feature interesting events or issues that otherwise might be overlooked. These sidebars add an engaging element to the more straightforward presentation of science and its times in the rest of the entries. In addition, the volume contains photographs, illustrations, and maps scattered throughout the chapters.

Comments and Suggestions

Your comments on this series and suggestions for future editions are welcome. Please write: The Editor, *Science and Its Times,* Gale Group, 27500 Drake Road, Farmington Hills, MI 48331.

Advisory Board

Contributors

Lloyd T. Ackert, Jr.
Graduate Student in the History of Science
Johns Hopkins University

James A. Altena
The University of Chicago

Peter J. Andrews
Freelance Writer

Kenneth E. Barber
Professor of Biology
Western Oklahoma State College

Bob Batchelor
Writer
Arter & Hadden LLP

Sherri Chasin Calvo
Freelance Writer

Thomas Drucker
Graduate Student, Department of Philosophy
University of Wisconsin

H. J. Eisenman
Professor of History
University of Missouri-Rolla

Ellen Elghobashi
Freelance Writer

Lindsay Evans
Freelance Writer

Loren Butler Feffer
Independent Scholar

Keith Ferrell
Freelance Writer

Randolph Fillmore
Freelance Science Writer

Mark R. Finlay
Associate Professor of History
Armstrong Atlantic State University

Richard Fitzgerald
Freelance Writer

Maura C. Flannery
Professor of Biology
St. John's University, New York

Donald R. Franceschetti
Distinguished Service Professor of Physics and Chemistry
The University of Memphis

Jean-François Gauvin
Historian of Science
Musée Stewart au Fort de l'ile Sainte-Hélène, Montréal

Phillip H. Gochenour
Freelance Editor and Writer

Brook Ellen Hall
Professor of Biology
California State University at Sacramento

Diane K. Hawkins
Head, Reference Services—Health Sciences Library
SUNY Upstate Medical University

Robert Hendrick
Professor of History
St. John's University, New York

James J. Hoffmann
Diablo Valley College

Mary Hrovat
Freelance Writer

Leslie Hutchinson
Freelance Writer

P. Andrew Karam
Environmental Medicine Department
University of Rochester

Evelyn B. Kelly
Professor of Education
Saint Leo University, Florida

Israel Kleiner
Professor of Mathematics
York University

Judson Knight
Freelance Writer

Lyndall Landauer
Professor of History
Lake Tahoe Community College

Mark Largent
University of Minnesota

Josh Lauer
Editor and Writer
President, Lauer InfoText Inc.

Lynn M. L. Lauerman
Freelance Writer

Dave Lee
Section for the History of Psychology
Department of Psychology
University of Groningen, The Netherlands

Adrienne Wilmoth Lerner
Division of History, Politics, and International Studies
Oglethorpe University

Brenda Wilmoth Lerner
Science Correspondent

K. Lee Lerner
Prof. Fellow (r), Science Research & Policy Institute
Advanced Physics, Chemistry and Mathematics, Shaw School

Eric v. d. Luft
Curator of Historical Collections
SUNY Upstate Medical University

Lois N. Magner
Professor Emerita
Purdue University

Marjorie C. Malley
Historian of Science

Ann T. Marsden
Writer

Kyla Maslaniec
Freelance Writer

William McPeak
Independent Scholar
Institute for Historical Study (San Francisco)

Lolly Merrell
Freelance Writer

Leslie Mertz
Biologist and Freelance Science Writer

Kelli Miller
Freelance Writer

J. William Moncrief
Professor of Chemistry
Lyon College

Stacey R. Murray
Freelance Writer

Stephen D. Norton
Committee on the History & Philosophy of Science
University of Maryland, College Park

Brian Regal
Historian
Mary Baker Eddy Library

Sue Rabbitt Roff
Cookson Senior Research Fellow
Centre for Medical Education
Dundee University Medical School

Steve Ruskin
Freelance Writer

Elizabeth D. Schafer
Independent Scholar

Neil Schlager
Editor and Writer
President, Schlager Information Group

Brian C. Shipley
Department of History
Dalhousie University

Devorah Slavin
Georgia Tech School of History, Technology, and Society
Georgia Institute of Technology

Keir B. Sterling
Historian, U.S. Army Combined Arms Support Command
Fort Lee, Virginia

Zeno G. Swijtink
Professor of Philosophy
Sonoma State University

David Tulloch
Graduate Student
Victoria University of Wellington, New Zealand

Roger Turner
Brown University

A. Bowdoin Van Riper
Adjunct Professor of History
Southern Polytechnic State University

Stephanie Watson
Freelance Writer

Richard Weikart
Associate Professor of History
California State University, Stanislaus

Giselle Weiss
Freelance Writer

A.J. Wright
Librarian
Department of Anesthesiology
School of Medicine
University of Alabama at Birmingham

Michael T. Yancey
Freelance Writer

Introduction: 1900–1949

Overview

The years between 1900 and 1949 were a time of extremes that included two world wars, a revolution in Russia, and the Depression in the United States. Scientific revolutions brought exciting new knowledge but also called into question the most basic concepts. New technologies and materials provided more deadly instruments of war but also lengthened lives and eased the demands on raw human muscle power. Exploration in this period echoed these extremes. Since many of the easier targets had already been reached, explorers headed for the poles, ventured deep beneath the ocean's surface, and high into the atmosphere.

Scientists also explored new frontiers, introducing new concepts while at the same time upsetting much of the perceived order by which the natural world worked. Revolutions in physics drew an entirely new picture of the most basic elements of physical reality. Our knowledge of the age, extent, and history of the universe, and Earth's place in it, shifted radically. The concept of evolution, introduced in the nineteenth century and developed in the twentieth, had profound implications for the nature and meaning of human life. Culturally and intellectually, these were exhilarating but deeply unsettling times.

Science provided so many answers during this period that scientists enjoyed increased prestige, especially after the creation of the first atomic weapon that effectively ended World War II. In a world that was growing increasingly secular, science offered a source of understanding and meaning; science took on questions that were once the province of religion, such as human origins, the creation of the universe, and the nature of consciousness and behavior. Scientific growth is visible in the founding of research institutes and the growth of professional scientific associations.

Looking Back to the Nineteenth Century

The rapid pace of invention that accelerated in the twentieth century began in the nineteenth. New machines and new capabilities in communication and transportation brought large-scale social changes. Many of the inventions of the nineteenth century followed advances in the understanding of electricity, thermodynamics, and the chemical properties of materials. These advances were part of a science that expressed itself by classifying and rationalizing the world around us, so as to better understand and utilize its complexities.

In addition to studying physical phenomena, scientists explored and classified the living world. This study of living things led to the emergence of the concept of evolution, in which natural variations in animals lead, through natural selection, to changes that tailor animals to specific habitats. It's no surprise that in this time when people's daily lives were changing as a result of new technologies and gadgets, the idea of change itself, and the idea of progress toward a future that humans can shape, became central to Western experience.

Scientific Revolutions and Developments

In 1900 scientists had identified only one sub-atomic particle (the electron). Over the next 50 years, physicists identified the other two major sub-atomic particles (protons and neutrons) and developed a surprising picture of the atom, based on the idea that the amount of energy an atomic particle can have is limited to certain discrete values, or quanta. Quantum mechanics, developed independently by Erwin Schrödinger and Werner Heisenberg, plus a series of developments from Albert Einstein, including general and special relativity and the equivalence between mass and energy, utterly transformed not only physics but astronomy and chemistry, explaining atomic structure, the properties of chemical elements, and the process of nuclear fusion that fuels the stars. Furthermore, exotic concepts from quantum physics and relativity (such as the dual wave-particle nature of light) called into question such everyday concepts as space, time, and reality, and the Heisenberg uncertainty principle suggested that there are inherent limits to our measurements of physical properties.

Physics was deeply affected by World War II. The Manhattan Project, an American effort that developed the first atomic bomb, did much to bring physics and physicists dramatically to the forefront of public consciousness. The use of atomic bombs over the Japanese cities of Hiroshima and Nagasaki hastened the end of World War II and set the stage for an arms race in the second half of the twentieth century. The peaceful development of nuclear power held great promise toward the end of the first half of the twentieth century (although later developments had mixed results).

Astronomers during this period resolved an important debate about spiral nebulae, determining that these were separate, distant "island universes" or galaxies. This shifted the Milky Way from being the only such system in the universe to being only one among many. Edwin Hubble discovered that other galaxies are receding from us, with speeds proportional to their distances. This was the first evidence for an expanding universe, in which galaxies rush away from each other as the result of the birth of the universe in an initial Big Bang.

Closer to home, scientists also studied the planets of the solar system. Percival Lowell's observations of Mars convinced him that he had seen evidence of life on the red planet. Although his perception of canals on the Martian surface later proved to be mistaken, his work contributed to a great interest in Martian and other extraterrestrial life. This interest is evident in the science fiction of the times, including the novels of Edgar Rice Burroughs and the novel *War of the Worlds* by H. G. Wells. When a radio script based on this novel was presented on the air in 1938, large numbers of people mistook the broadcast for a news report, and panic followed, illustrating the growing power of the media as well as the hold other worlds had on the popular imagination.

Atomic and quantum physics provided tools that chemists used to explain the chemical properties of the elements and how these relate to their places on the periodic table, continuing and refining the work done in the nineteenth century to organize and rationalize this science. Chemists developed new materials—for example, silicones—later widely used as lubricants and water repellents, and built on previous discoveries related to radioactivity. In 1927, for example, Hermann Müller demonstrated that genetic mutations are simply chemical changes that can be induced by radioactivity.

Müller's work was part of the major biological effort of this time period, understanding the mechanisms behind natural selection. The rediscovery of Gregor Mendel's laws of heredity in 1900 helped biologists begin to understand the genetic foundations of natural selection. Biologists also explored the role of chromosomes in heredity and advanced their understanding of other areas such as the ways that cells and hormones work.

Biology had a very direct impact on people's lives, as indicated by the Scopes trial in Tennessee in 1925 and the use of eugenics as a rationale for efforts to eliminate "undesirables," such as Germany's 1933 Eugenic Sterilization Law. Genetics was also applied to agriculture, with the introduction of hybrids and the improvement of animal stocks. Developments in agriculture, including dam-building and irrigation that allowed dry lands to be cultivated, resulted in shifts in occupation and population, as fewer farmers were able to feed more people.

Science also sought to bring a rational, systematic approach to the mysteries of the mind. Ivan Pavlov, B. F. Skinner, and other scientists studied behavior in animals, and some of their discoveries were applied to the developing schools of psychology attempting to explain human behavior. In addition to psychoanalysis, shock therapies, such as electroshock therapy and surgical lobotomy, were developed to treat mental illness.

New Capabilities and Conveniences

Scientists' ability to identify causes and solve problems was especially evident in public health during this period. Life-saving developments include the chlorination of drinking water to kill disease-causing micro-organisms, the discovery of the specific agents that cause infectious diseases such as yellow fever, whooping cough, measles, and scarlet fever, the realization that some diseases are caused by a vitamin deficiency and can be treated by vitamins, the development of the first antibiotics, the development of diagnostic tests such as the Pap test, and the fluoridation of water to prevent tooth decay.

New machines eased some of the burdens of maintaining a household. Women's lives began to change as a result of electric appliances such as refrigerators and electric irons; the first microwave was patented in 1945. New materials found a variety of uses and also made life easier. Bakelite, the first synthetic plastic, was patented in 1907. Advances in the understanding of polymers, an effort that began in the 1920s, led to increased use of plastics after World War II. In the 1930s nylon became the first successful synthetic fiber product, introducing clothing that was easier to maintain and less dependent on natural sources such as silk.

Transportation

The first powered flight at the beginning of the century heralded an era of rapid developments in transportation. Passenger travel in dirigibles ended due to the potential for disaster, demonstrated in the *Hindenburg* explosion in 1936. The patenting of the jet engine in 1930, however, helped make airplanes increasingly important for carrying passengers and goods, as aviation progressed from its adventurous early days marked by the daring flights of Charles Lindbergh, Amelia Earhart, and others. Aviation was in its infancy in World War I, but by World War II air power played an important role, making possible the fire-bombing of cities and providing a way to deliver the first atomic bombs.

Humanity looked to the skies in other ways as well, with the first liquid-fueled rocket, launched by Robert Goddard in 1926. Although one of the earliest applications of rocketry was warfare, the German V-2 rockets, used against England during the war, were later the basis for early experiments in space-going rockets developed in the United States.

Everyday life was overwhelmingly affected by the automobile, which became much more widely used due to Henry Ford's development of the Model T, the first car to be mass-produced on an assembly line. Relatively cheap and easily available, it launched a series of changes that transformed modern life, making the United States in particular a much more mobile country, with cities and lifestyles increasingly built around the personal car.

Mass Production and Mass Communication

Mass production and mass communication had their beginnings in assembly lines like Ford's and the introduction of radio, television, and movies. From the first regular radio broadcasts in Pittsburgh in 1920, to the first regular TV broadcasts by the BBC, to the earliest regularly scheduled U.S. TV in 1941, mass media became increasingly available in the home. Popular mass culture began to develop in this age when more and more people were exposed to the same news and entertainment content and the same consumer goods. Advances in transportation combined with these developments to make the world seem a smaller place, well on its way to the global culture of the close of the century.

The Legacy of Science: 1900-1949

Building on the science and technology of this period, and funded by post-war affluence, science introduced a dizzying array of new technologies that have wrought radical social changes. While many new technologies (for example, nuclear power and the creation of artificial materials such as plastics) were originally greeted enthusiastically, during the second half of the century the negative side of a highly technical culture became more evident, and science was both castigated as the cause of these ills and hailed as the means for dealing with increasing population, dwindling natural resources, and pollution. The physical sciences were prominent in the immediate post-war period, as the space race followed up on the rocketry work of World War II. The biological sciences attained more prominence later in the century as advances in genetics, diagnostic and surgical procedures, and drug therapies improved the length and quality of life. Mass media, combined with computing and satellite technologies, continued to develop far beyond their beginnings in the first half of the century. As the twenty-first century begins, the networked world we now know grows and changes daily.

MARY HROVAT

Chronology: 1900–1949

1900 Austrian-American physician Karl Landsteiner discovers the four blood types (A, B, AB, and O), thus making possible safe and practical transfusions.

1903 Using a craft he designed with his brother Wilbur, Orville Wright makes the first controlled and powered flight in history, at Kitty Hawk, North Carolina.

1905 Albert Einstein publishes his first papers on the special theory of relativity (including the famous equation $E = mc^2$), which he will follow up with his general theory of relativity in 1916.

1908 Henry Ford designs his Model T, which will become the first automobile produced by assembly-line techniques; by the time of its discontinuation 20 years later, Ford will have sold more than 15 million Model Ts.

1914-18 World War I claims more than 10 million lives; ends the German, Austrian, Russian, and Ottoman empires; and ushers in the atmosphere of disillusionment that characterizes the early twentieth century.

1915-22 Ottoman Turkey makes systematic efforts to kill off its Armenian minority, leaving some 1.5 million Armenians dead.

1917 Seizure of power in Russia by the Bolsheviks under V. I. Lenin leads to the establishment of Soviet Communist totalitarianism.

1918-20 A worldwide outbreak of influenza, whose spread is hastened by the movement of populations associated with the war, claims more than 20 million lives.

1920 Building on a quarter-century of technological development that includes the creation of wireless telegraphy by Guglielmo Marconi (1896) and Lee DeForest's triode amplifying tube (1907), KDKA in Pittsburgh begins the world's first regular radio broadcasts.

1920s Innovations in the arts: cubism, surrealism, and Dadaism; jazz and the Impressionist music of Debussy and Ravel; James Joyce's *Ulysses,* Marcel Proust's *Remembrance of Things Past,* and T. S. Eliot's *The Waste Land.*

1922 Benito Mussolini takes power in Italy, introducing a new form of totalitarianism to compete with Communism: Fascism.

1922 Howard Carter, a British Egyptologist, discovers the tomb of Tutankhamen, the only pharaoh's grave not looted by grave robbers.

1927 Charles A. Lindbergh becomes the first man to complete a solo, non-stop flight across the Atlantic Ocean.

1927 Werner Karl Heisenberg, a German physicist, postulates his principle of indeterminacy, which states that it is impossible to determine accurately and simultaneously two variables of an electron.

1928 Scottish bacteriologist Alexander Fleming discovers penicillin, the first antibacterial "wonder drug."

1928 In writing *The Theory of Games and Economic Behavior* (published 1944), John von Neumann and Oskar Morgenstern establish the principles of game theory, which will come to prominence in the latter half of the twentieth century.

1929 American astronomer Edwin Hubble formulates a law, named after him, which marks the beginning of the theory of the expanding universe.

1929 The crash of the U.S. stock market begins the Great Depression, which spreads from the United States to Europe with the failure of Austria's Credit-Anstalt in 1931.

1929 Josef Stalin emerges as the dictator of Soviet Russia; launches a brutal collectivization program; and begins vanquishing all opposition through extermination of "rich peasants" (1929-34), man-made famine (1932-33), and party purges (1936-38).

1931 Austrian mathematician Kurt Gödel presents his incompleteness theorem, which states that within any rigidly logical mathematical system, there are propositions that cannot be proved or disproved by the axioms within that system.

1933 Adolf Hitler becomes chancellor of Germany; Nazi persecution of Jews and others begins immediately, and culminates in the "Final Solution" during World War II, when more than 6 million are murdered in death camps.

1936-39 A civil war in Spain becomes a focal point for international tensions, pitting a regime allied with Soviet Russia against a force supplied by Nazi Germany and Fascist Italy.

1937 Japan, now under the control of a militarist regime, invades China, which is torn by civil war between Nationalist and Communist forces.

1939-45 World War II claims an incalculable death toll; ushers in the nuclear age; establishes the United States and Soviet Union as superpowers; and brings about the end of European colonial empires in Africa and Asia.

1944 Canadian scientists Oswald Theodore Avery, Maclyn McCarty, and Colin Munro Macleod discover that DNA carries a genetic "blueprint."

1945 Years of atomic research culminate in the destruction of two Japanese cities, Hiroshima and Nagasaki, by U.S. atomic bombs—the first and only use of nuclear power as a weapon—which leads to the Japanese surrender and the end of World War II.

Exploration and Discovery

Chronology

1900 Archaeologist Arthur Evans uncovers the remains of the previously unidentified Minoan civilization at Knossos in Crete.

1909 Americans Robert E. Peary and Matthew A. Henson become the first men to reach the North Pole.

1911 American Hiram Bingham discovers the Inca city of Machu Picchu high in the Andes.

1911 Norwegian Roald Amundsen becomes the first man to reach the South Pole.

1922 Howard Carter, a British Egyptologist, discovers the tomb of Tutankhamen, the only pharaoh's grave not looted by grave robbers.

1927 Charles A. Lindbergh becomes the first man to complete a solo, non-stop flight across the Atlantic Ocean.

1931 Aboard a pressurized balloon, Auguste Piccard and Paul Kipfer become the first men to enter the stratosphere.

1933 Wiley Post, an American aviator, makes the first-ever solo flight around the world.

1934 In their submersible bathysphere, Charles William Beebe and Otis Barton dive to a record-breaking depth of 3,028 feet (923 m).

1940 Prehistoric cave paintings are discovered at Lascaux, France.

1947 Chuck Yeager, an American pilot flying a Bell X-1 aircraft, becomes the first human being to break the speed of sound.

1947 At Qumran, Jordan, two Bedouin boys discover a set of first-century Jewish religious manuscripts that come to be known as the Dead Sea Scrolls.

Overview: Exploration and Discovery 1900-1949

One of the driving forces behind exploration and discovery is the overwhelming curiosity of certain men and women who are not content to sit at home with unanswered questions about Earth's mysteries. Throughout the ages, men and women have journeyed through jungles and forests, across scorched deserts and icy tundras, over mountains, along rivers, and across oceans in a quest for answers about geography, peoples, and ancient history.

In the nineteenth century, historic ocean voyages, epic adventures, and exhaustive expeditions rapidly expanded national boundaries and imperial domains as well as scientific knowledge in the fields of botany, zoology, ornithology, marine biology, geology, and cultural anthropology. By the end of the nineteenth century, few areas of the world remained undiscovered and unexplored by humans.

Two of the less explored regions of the world at the turn of the twentieth century were South America and Central Asia. In expeditions to South America from 1906-12, British army officer Percy Fawcett (1867-1925?) mapped Bolivia's boundaries with Brazil, Paraguay, and Peru, explored uncharted interior regions south of the Amazon, and discovered the source of the Rio Verde before mysteriously disappearing in the Mato Grosso region of Brazil. From 1913-14, former American president Theodore Roosevelt (1858-1919) and Brazilian Cândido Rondón (1865-1958) led a mapping expedition down a totally unknown Brazilian river that Rondón christened the Rio da Dúvida (River of Doubt), later rechristened the Rio Roosevelt.

From 1893 to 1933, Swedish geographer and explorer Sven Hedin (1865-1952) made significant expeditions to Central Asia, one of the most mysterious regions of the world—a wasteland of deserts and mountains from Afghanistan to Tibet to Mongolia and Siberia. In 1931 French automobile-manufacturer Citroën, who had sponsored expeditions to the Sahara Desert in 1922 and from Algeria to Madagascar in 1924, conducted a retracing of the ancient Silk Road followed by Marco Polo in the thirteenth century.

At the beginning of the twentieth century, the greatest ambition of many explorers was to be the first to reach the North and South Poles—literally the last places on Earth to be visited by man. The most famous of these explorers was Robert Peary (1856-1920), who had faced the icy Arctic winds eight times only to lose to the elements. In 1909, at age 52, he set out for his ninth and last try, and finally reached the North Pole on April 6th, along with Matthew Henson (1866-1955), his famous African-American companion who had been with him on all previous attempts, and four Eskimos. His accomplishment fired the race for the South Pole between Norwegian Roald Amundsen (1872-1928), who reached the pole in December 1911, and British naval officer Robert Falcon Scott (1868-1912), who reached the pole in January 1912—second by just weeks—and died tragically in a blizzard on the homeward journey.

Significant scientific exploration of the Arctic and Antarctic continued throughout the twentieth century. During the years between the two World Wars, numerous expeditions to Antarctica traveled overland onto the central plateau, mapping further areas of the coast and gathering much valuable scientific data. However, the most significant technological advance for twentieth-century exploration was the invention of the airplane (a technology whose progress was aided by the discovery of vast fields of fossil fuel in the Middle East in the early 1900s). The introduction of the airplane drastically transformed the technique of polar exploration. From the 1920s through the late 1940s, Admiral Richard E. Byrd (1888-1957) revolutionized Antarctic exploration by the use of aircraft, especially as part of Operation High Jump, a United States Naval expedition he commanded from 1946-47. Byrd was also the first to fly over both the North and South Poles in May 1926 and November 1929, respectively.

From the moment Orville (1871-1948) and Wilbur (1867-1912) Wright put their plane in the air at Kitty Hawk in 1903, the competition to be the first, the fastest, and to fly higher and farther was on. The science of flight—with its corresponding technological, scientific, and military applications—rapidly advanced in the first half of the twentieth century. World War I brought enormous advances in the field of aviation. Pilots took to the air and became heroes to a fascinated public. Men such as the Wright brothers and Charles Lindbergh (1901-1974) were celebrated in books, songs, and films. Women like Elise Deroche (1889-1919), the first licensed female pilot, and Amelia Earhart (1898-1937), the first woman to fly solo across both the Atlantic and the Pacific, were also a part of the excitement surrounding the airplane.

In addition to aviation, twentieth-century scientists and engineers designed vessels that could venture to the absolute heights of the atmosphere and the depths of the sea. In 1931 Auguste Piccard (1884-1963) and Paul Kipfer were the first to reach the stratosphere in a balloon—reaching 51,762 feet (15,781 m) before returning to the Earth's surface. Other adventurers such as William Beebe (1877-1962) and Otis Barton were the first to explore the ocean depths. In 1934 Beebe and Barton took their "bathysphere" to a record depth of 3,028 feet (923 m) to study deep-sea marine life near Bermuda. Later, in 1960, Jacques Piccard (1922-), son of Auguste Piccard, would pilot his bathyscaphe *Trieste,* a redesign of his father's balloon gondola, to a record depth of 35,800 feet (10,912 m) in the Mariana Trench, nearly 7 miles (11.3 km) down.

While some twentieth-century explorers set out to unravel Earth's mysteries, others were fascinated by the mysteries of mankind and human civilization. Beginning in 1900, several momentous archaeological expeditions returned libraries of new data and priceless artifacts to museums, universities, and collectors around the world. In 1900 Sir Arthur Evans (1851-1941) began 31 years of systematic excavations of Knossos, the Great Palace of Minos on Crete, originally discovered in 1878. A year later, in 1901, a team of French archaeologists discovered a black stone pillar in Iraq bearing the Code of Hammurabi, one of the earliest known written compilations of law. Then, in 1907, while tracing ancient caravan routes between China and the West, Sir Aurel Stein (1862-1943) uncovered frescoes, statues, and a vast collection of priceless manuscripts in Ch'ien Fo-Tung, the Cave of the Thousand Buddhas. (During the next 25 years, Stein made several more journeys through Central Asian deserts, each producing a rich harvest of archaeological treasures.) In 1911 Yale University professor Hiram Bingham (1875-1956) discovered the ruins of the great Inca city of Machu Picchu in the Peruvian Andes. In 1922 one of the most astonishing archaeological discoveries of all time was made by Howard Carter (1873-1939) and Lord George Carnarvon (1866-1923), who unearthed the tomb of King Tutankhamen, the only tomb of an Egyptian pharaoh to survive essentially intact over the centuries since its closing. A total of 3,500 artifacts were removed from the tomb.

Most of the archaeological discoveries of the early twentieth century were made by historians, anthropologists, and archaeologists who had spent years examining the past civilizations for which they were searching. However, in September 1940 four boys exploring caves in the woods near Lascaux in southern France stumbled upon Paleolithic engravings, drawings, and paintings of figures and animals that dated to about 15,000 B.C.. Then, in Jordan in 1947, two young Bedouins looking for a goat accidentally discovered a cave containing the Dead Sea Scrolls, over 800 manuscripts written between 250 B.C. and A.D. 68, perhaps the most important archaeological discovery of the century.

The first half of the twentieth century saw the last of the truly terrestrial explorers. Building on the advancements in technology up to 1950, science fiction became science fact in the later part of the twentieth century as humans finally realized the dream of visiting the Moon. Earth exploration became three-dimensional—into the oceans and skies, under the icecaps, and down into the Earth's crust. Fascinating archaeological discoveries also refocused attention on man's ancient past, the subject of intensive study during the second half of the century. Thanks to the technologies developed in the first half of the century, explorers were able to go where previously only books and dreams could take them. The quest for answers to the last of Earth's mysteries extended to the universe beyond Earth.

ANN T. MARSDEN

The Palace at Knossos: The Archaeological Discovery of Minoan Civilization

Overview

Knossos (also spelled Cnossus) is located 3.1 miles (5 km) inland from the northern coast of Crete near the present-day town center of Heraklion (Iraklion). Known in Greek mythology as the capitol of King Minos and the site of the Minotaur's labyrinth, Knossos was the center of Minoan civilization, the earliest of all Aegean civilizations.

Greek myth and epic poetry attested to the existence of an ancient city called Knossos on Crete. Later inhabitants of the region often found artifacts of previous civilizations when they tilled their fields. Originally attracted by the discovery of stones bearing an unknown script, British linguist and archaeologist Sir Arthur Evans (1851-1941) first visited Crete in 1894 hoping to decipher the script and link the Cretan tablets with similar artifacts of the recently discovered Mycenaean civilization in Greece.

When Evans published his research a year later in *Prae-Phoenician Script,* he acknowledged that the Cretan pictorial script, and a later linear script he named Linear A, were that of another culture. Only one script, Linear B, was found to have a direct link with Mycenae. Three years later, Evans began a survey of the Knossos site that was based on studies of myth, language, art, and material culture. He spent the next 31 years excavating the archaeological site, work that revealed a palace and surrounding network of buildings that were the capitol of the Bronze Age culture that dominated the Aegean from 1600 to 1400 B.C..

Background

The first person to excavate in the area was an Iraklionian merchant and amateur archaeologist, Minos Kalokairinos. He had already uncovered two storerooms by 1878, when the site's landowners forced him to stop his investigation of the ruins. In 1900 Evans began a systematic excavation of the site. After expanding upon Kalokairinos's initial dig, Evans discovered a complex network of corridors and rooms that reminded him of the legendary labyrinth of King Minos. Evans accordingly named the palace, Knossos, after that of Minos.

As the ongoing excavations yielded evidence of distinct pottery, artwork, and architecture, Evans realized that he had indeed discovered a civilization distinct from that of the recently discovered Mycenae. This Cretan civilization became known as the Minoan.

Recognizing the site's uniqueness, Evans rapidly expanded his excavations. The intricate, multistoried palace he unearthed spanned an area of 22,000 square miles (56,980 sq km). Each section had a specific use. The western area, with large rooms and a theater, was built for administrative and court functions. The eastern area consisted of smaller rooms with verandas and numerous interior frescos characteristic of Minoan art, many portraying men and women leaping bulls. These frescoes, along with several artifacts from cultures foreign to Crete, suggest that the Minoan civilization at its zenith was a highly developed society that routinely traded with its Mediterranean neighbors.

Archaeological evidence also suggests that this "Great Palace of Minos," as Evans called it, was destroyed by fire in 1400 B.C.. Around the same time, the political center of Aegean civilization shifted to Mycenae. As a result, the palace at Knossos was not rebuilt to its previous form, although small-scale inhabitation of the site continued in subsequent centuries. By 1903 Evans had unearthed nearly all of the palace structure and survey work began on the surrounding area.

Soon after the large palace had been excavated and mapped, Evans and his team discovered that it was not the only structure that had been built upon the site. Below the Palace of Minos lay the ruins of yet another, earlier palace that had a simpler form and consisted of a series of structures surrounding a rectangular court. It had been built around 2000 B.C. and destroyed by earthquake, according to Evans's hypothesis, some 300 years later.

Evans developed a chronology for the site using a complex pottery sequence, which he established by associating types of pot shards, or small fragments of pottery, with Egyptian or Mycenaean trade goods and other artifacts that had been more concretely dated. Through this work, Evans realized that the Minoan civilization had existed within the chronological context of the larger Aegean civilization. He continued to dig below the first palace structure in the

Ancient ruins: the palace of Minos at Knossos. *(Wolfgang Kaehler/Corbis. Reproduced with permission.)*

hope of finding some proof of earlier inhabitation on the site.

Excavations below the Bronze Age strata revealed that the area had been inhabited as far back as the Neolithic period (6000 B.C. and perhaps even earlier.) The excavated Neolithic levels at Knossos are still among the deepest in Europe. Archaeological survey of the upper strata of the Neolithic site revealed artifacts such as gold jewelry, glazed pottery, and bronze. A prepalace structure from 3000 B.C. was also identified, thus making the Early Minoan Period contemporary with the emergence of the Early Bronze Age in the Aegean.

Impact

Evans's discoveries in Crete were a major influence on the field of linguistics. He published the first study on the Minoan scripts, but it was not until English architect and cryptographer Michael Ventris (1922-1956) deciphered Linear B in 1952 that the approximate relationships of the different scripts were known. Ventris identified the Minoan Linear B as an early form of Greek that dated as far back as 1400 B.C.—roughly contemporary with the events depicted in the epic poems of Homer. Though the phonetic sounds depicted in the Minoan Linear A are decipherable from Linear B, the actual language represented by Linear A is unknown. How the Minoan pictorial, or hieroglyphic, script identified by Evans is related to the linear scripts as well remains unknown. This linguistic connection offered further proof that the Minoan and Mycenaean were among the ancestral civilizations of Greek and later Western civilization.

Evans's archaeological research at Knossos also provided new insight into the technology of urban planning. For example, the Palace of Minos at Knossos had an elaborate system of drains, ducts, and pipes that brought a water supply to the community and improved its sanitation, innovations once thought only to have emerged in later, more advanced Greek and Roman civilizations. The palace also contained workshops and massive storage compartments for agricultural products, and the urban area around the palace was connected to outlying towns and ports by a network of paved roads. Thus, Knossos was not just a ceremonial palace, but a vibrant working city with seemingly modern amenities.

In addition, the discovery and excavations of Knossos and Mycenae added an historical and cultural frame through which to view Greek mythological literature. These ancient myths, if not wholly reliable, could be corroborated with newly found physical evidence to offer an additional dimension of understanding and contextual insight into the history and society of the Greeks.

Several later scholars have disputed some of Evans's earlier conclusions. More recent excavations brought new evidence to light that suggests flaws in the pottery sequence that Evans developed. Advances in archaeological dating techniques, most notably the advent of radiocarbon (C-14) dating, have helped refine the established chronology of Knossos. Scholars also hesitate to rely on mythological explanations for the political structure of Knossos. Though there is evidence that Knossos was ruled by an individual, the archaeological record yields few clues about the governmental structure of Minoan civilization.

Though Evans was one of the pioneers of scientific excavation, his preservation methods remain controversial. As a greater portion of the palace grounds was excavated, questions arose of how best to protect and preserve the structure, art, and artifacts that were exposed. In the later phase of his excavations, Evans decided to reconstruct part of the palace. His reconstruction of part of the Palace at Knossos used contemporary building materials such as reinforced concrete, lead paints, and plasters. This method of preservation and reproduction received considerable criticism because it combined ancient artifacts and modern building materials that were foreign to Minoan construction practices. Among modern archaeologists, the reconstruction of archaeological ruins has largely fallen out of favor.

Regardless of modern questions surrounding his research at Knossos, Evans is still widely admired as an astute scholar and pioneering archaeologist. Knossos remains an active archaeological site and excavations still continue today.

ADRIENNE WILMOTH LERNER

Further Reading

Chadwick, John. *The Decipherment of Linear B.* Cambridge: Cambridge University Press, 1991.

Farnoux, Alexandre. *Knossos: Searching for the Legendary Palace of King Minos.* Translated by David J. Baker. New York: Harry N. Abrams, 1996.

Hamilton, Edith. *Mythology.* New York: Warner Books, 1999.

Woodard, Roger D. *Greek Writing from Knossos to Homer: A Linguistic Interpretation of the Origin of the Greek Alphabet and the Continuity of Ancient Greek Literacy.* Oxford: Oxford University Press, 1997.

Ancient Writings Shed Light on Past Civilizations

Overview

In 1901 a team of French archaeologists discovered a black stone pillar bearing the Code of Hammurabi. Inscribed in Babylon over 3,700 years ago, it is one of the earliest known compilations of law. Almost half a century after its discovery, Bedouin shepherds found a collection of ancient scrolls stored in jars and hidden away in a cave near the Dead Sea. Eventually more than 800 manuscripts were discovered in the area, most of which were fragmentary. They were written between about 250 B.C. and A.D. 68, and provide unique insight into the time and place from which both rabbinic Judaism and Christianity emerged.

Background

The civilizations of the ancient Middle East had an immense impact on world history. In fact, history itself may be said to have begun in the "fertile crescent" of Mesopotamia. It was there that writing emerged among the Sumerians, who lived in the world's first cities more than 5,000 years ago. With writing came the ability to record business transactions, codify laws and religion, and chronicle events. Writing provides a measure of immortality, as it makes known the names and exploits of people who moved through history thousands of years before us.

One of the prominent figures we know of from ancient writings is King Hammurabi, who ascended the throne of Babylon in about 1792 B.C.. By this time, Mesopotamia had experienced a number of invasions and migrations from surrounding areas, but the more advanced local culture, based on Sumerian civilization, remained dominant. Many of the newcomers belonged to nomadic tribes, speaking Semitic languages with origins in the Arabian peninsula. Hammurabi was a member of one such group, the Amorites. The language of his kingdom was Akkadian, a Semitic tongue related to Hebrew and Arabic. Hammurabi was arguably the great-

A fragment of the Dead Sea Scrolls. *(Corbis-Bettmann. Reproduced with permission.)*

est king ever to rule Babylonia. He transformed a beleaguered city-state into a large empire. Arts, sciences, commerce, and government all prospered during his reign.

In 1901 a team of French archaeologists under the direction of Jacques de Morgan (1857-1924) was excavating the town of Susa (the biblical Shushan) at the foot of the Zagros Mountains in Persia. They came upon a 7-foot (2.13 m) tall black basalt pillar with almost 4,000 lines of inscriptions in the wedge-shaped cuneiform script of Mesopotamia. When the text was translated by the Dominican friar and Orientalist Jean-Vincent Scheil, it was found to be a compilation of 282 laws. Scholars knew of older Sumerian legal codes, including the laws of Eshnunna and the even more ancient laws of Ur-Nammu, dating from around 2,100 B.C., but the Code of Hammurabi was the most complete listing that had ever been found. The column is believed to have once rested in the center of Babylon's temple of Marduk, the national god, whence it was carried off to Persia as booty by the invading Elamites. Today it is exhibited in the Louvre museum in Paris.

The 1947 discovery of the Dead Sea Scrolls was very different. Rather than a team of archaeologists, the discoverers were a pair of young Bedouins looking for a lost goat. In a cave nestled among cliffs above the Dead Sea, 13 miles (21 km) east of Jerusalem, they found seven ancient scrolls hidden away in tall clay jars. During the 1950s, archaeologists and Bedouin tribesmen alike scoured the area for more manuscripts. Eventually about 25,000 fragments were found, some no bigger than a postage stamp. Altogether more than 800 manuscripts were at least partially pieced together. Meanwhile, excavations continued at the nearby site of Qumran, as scholars attempted to understand the Jewish sect that had hidden this library away.

Impact

Scholars studying the Code of Hammurabi quickly noticed many similarities to the Mosaic laws recorded in the Hebrew scriptures, both in form and in content. In addition, a relief carving at the top of the pillar from Susa showed Shamash, the Babylonian god of justice, dictating the laws to Hammurabi. The scene is reminiscent of Moses receiving the Ten Commandments, which were also carved in stone.

Such echoes can be explained by the Sumerian influence both on later civilizations of Mesopotamia and on Hebrew culture and thought. For example, some of the Creation accounts in the Hebrew scriptures closely parallel Mesopotamian writings, such as the Epic of Gilgamesh. Abraham, whom the Jewish and Arab peoples regard as their common patriarch, is believed to have been one of a tribe of Semitic nomads living in the Mesopotamian city of Ur. It was from there that the Hebrews migrated westward to Canaan. In their travels around the Mediterranean area, and especially through the influence of the Bible, they helped make Sumerian culture one of the foundation stones of the Western world.

Like the laws of the Hebrews, the Code of Hammurabi bears the marks of both the Sumerian legal codes and the tribal customs of their Semitic adopters. From the vantage point of a society that takes the Bill of Rights for granted, even while arguing over its interpretation, these ancient legal systems seem extremely primitive. The criminal law in the Code of Hammurabi, for example, was based on the principle of *lex talionis*, or the law of retaliation. This is familiar from the Hebrew scriptures as "an eye for an eye, a tooth for a tooth." However, it is important to realize that such formulations, by setting a limit to retaliation, were a major advance over allowing feuds to continue indefinitely.

We also tend to forget how recently we arrived at our own civil rights-based concept of the rule of law. The basics of Hammurabi's criminal code, complete with trial by physical ordeal and a hair-raising catalog of execution methods—drowning, burning, hanging, impaling—for a variety of offenses, held sway in Europe until the eighteenth century. Many of the laws, criminal and otherwise, survived to influence Islamic jurisprudence as well.

The entire concept of a legal code was made necessary by cities in which people lived in close quarters. Laws allowed the existence of an ordered society. In addition to the criminal code, the laws of Hammurabi dealt with commerce, the treatment of slaves and other workers, and family relations. They assumed the division of society into three social classes, each with its own rights and responsibilities. Like the laws of Moses, they reflect the subordinate status of women in the ancient Middle East even while providing some measure of protection for widows and orphans. The inscriptions found at Susa refer to another copy, and additional fragments of a similar pillar have been discovered, as well as copies of the laws on clay tablets at Nineveh. The repetition suggests a concerted attempt to standardize the legal system across Hammurabi's kingdom. This effort may have been prompted by a desire to assimilate recently conquered territories into the empire.

The impact of the discovery of the Dead Sea scrolls was even greater than that of the Code of Hammurabi. It was perhaps the most important archaeological discovery of the twentieth century. The manuscripts include biblical texts, previously unknown psalms, commentaries, laws and liturgies, and elaborations of biblical stories. While some of the material was actually composed in the earliest days of Israelite history, the texts were written onto the scrolls between the third century B.C. and the first century A.D.. The sudden appearance of a large number of texts dating from a time when great changes were occurring in Judaism, and shortly before the emergence of Christianity, was unprecedented in Biblical scholarship.

Controversy soon erupted over who was to have access to the scrolls. The manuscripts were mostly in Hebrew, and the remainder were in Aramaic, a closely related language commonly spoken in the Jewish community 2,000 years ago and still used in some Jewish prayers and texts. So the more-or-less intact scrolls, of which there were about a dozen, could be read without much difficulty. In 1958, they were published by Israeli and American scholars. However, the bulk of the material was fragmentary, and had to be pieced together like a collection of jigsaw puzzles. In order to accomplish this, all the fragments needed to end up in the same place.

With an antiquities dealer in Bethlehem acting as a middleman, the fragments found their way from their Bedouin discoverers to what was then called the Palestine Archaeological Museum in east Jerusalem. At that time the Old City of Jerusalem was controlled by Jordan, and it was under Jordanian auspices that a team of international scholars were assembled.

Because both Jordan and the president of the museum were hostile to the state of Israel at that time, no Jews were permitted to participate in the study of these Jewish documents. Most of the team consisted of Catholic priests, who accomplished the impressive feat of arranging, deciphering, and transcribing the many thousands of fragments over the course of about seven years beginning in 1953. Since the scrolls date from 250 B.C. to A.D. 68, the Catholic scholars were naturally most interested in the extent to which the texts anticipated or depicted the events recounted in the New Testament, specifically the life and ministry of Jesus.

Unfortunately, the team refused to release the text of the fragmentary scrolls, and this caused a bitter academic scandal. The information void was filled with angry speculation by excluded scholars and sensational conjecture by the media. Many suggested that the largely Catholic team was not releasing the material because of content that in some way conflicted with the Christian faith. However, scholarly tradition held that those working on an ancient text controlled access to it until they published it, and the tradition specified no time limit. After 1987, a few Jewish researchers were finally admitted to the team, but access remained confined to this inner circle. At last, after a 30-year struggle, the "secret" texts were published in 1991, and their study has become an academic discipline in its own right.

The scrolls tell us a great deal about Judaism at a time of great social and religious upheaval. They date from a time when traditional Jews felt threatened by Hellenistic influence. The Greek way of life had been introduced to the Middle East after much of it was conquered by Alexander in 332 B.C.. When the scrolls were written, the Second Temple still stood in Jerusalem, but was soon to be destroyed by another wave of conquerors, the Romans. The synagogue-centered, rabbinical religion beginning to emerge was that which would sustain the Jewish people during their long exile. The "official," or canonical version of the Bible was not yet established, and slightly different versions of its various books were still in circulation. Jewish sects, such as the one centered at Qumran, experimented with ways of living in the world that was changing around them.

Contrary to some of the sensationalist accounts that circulated before their public release, the scrolls do not mention Jesus, nor are they the documents of an early Christian sect. Nonetheless, they remain important to Christians as well as Jews. Most Christian scholars agree that, in order to understand the life and teachings of Jesus, and the rise of the Christian movement, it is necessary to understand the religious and social context in which these events occurred. The Dead Sea scrolls, in illuminating the Jewish culture out of which Jesus and his message came, have much to tell New Testament scholars. Many are choosing to learn Hebrew and Aramaic as a result.

Ancient writings like the Code of Hammurabi and the Dead Sea Scrolls provide a unique view of the cultural developments and historical events that have shaped the world. In learning about them, we learn about the origins of our own civilization and way of life.

SHERRI CHASIN CALVO

Further Reading

Covensky, Milton. *The Ancient Near Eastern Tradition.* New York: Harper & Row, 1966.

Honour, Alan. *Cave of Riches: The Story of the Dead Sea Scrolls.* New York: McGraw-Hill, 1956.

Schiffman, Lawrence H. *Reclaiming the Dead Sea Scrolls.* Philadelphia: Jewish Publication Society, 1994.

Shanks, Hershel. *The Mystery and Meaning of the Dead Sea Scrolls.* New York: Random House, 1998.

Time-Life Books. *Mesopotamia: The Mighty Kings.* Alexandria, VA: Time-Life Books, 1995.

The Northwest Passage, Sought by Europeans for 400 Years, Is Found and Traversed by Ship by Roald Amundsen

Overview

Europeans searched for 400 years for a passage through North America that would take them to the fabled lands of the Orient. Their searches met with little success. The first man to take a sailing ship from the Atlantic to the Pacific Ocean through this fabled passage was Roald Amundsen (1872-1928) of Norway in 1903-06. By then, Europeans had found other ways to get to China and southeast Asia, and entrepreneurs and traders were no longer interested in this passage. Nonetheless, Amundsen's voyage was a culmination of human effort as well as a signal of the new understanding of the lands at the northern end of the earth.

Background

The 1492 discovery by Christopher Columbus (1451-1506) of new land masses unknown to the Europeans created a problem that was not solved for over 400 years. Columbus refused to admit that he had not reached India, and he thought that China was just on the other side of the land he did find. Subsequent explorers tried to find a way through or around these two massive continents that lay between Europe and the Orient.

It is ironic that America was discovered by accident and, when found, a great deal of effort was expended to find a way through or around it. Europeans wanted to reach India and China to trade for the gold, silver, spices, brocades, and silks that the nations of the Orient were reputed to have. Columbus's fourth voyage in 1502 focused on finding a way through the continent. He did not find it. The Spanish claimed much of the southern continent, but English and French merchants were more interested in passing through or around this obstacle. Explorations fanned out north and south with the hopes of finding a bay, a river, or an inlet that would take them all the way through the continent and shorten the trip to the Orient. Riches awaited the person who found it.

Vasco Nuñez de Balboa (1475-1519) was part of a Spanish expedition to Venezuela in 1501. He eventually helped found the new settlement of Darièn on the isthmus of what is now Panama, and became its governor. In 1513 he crossed the isthmus, climbed to the top of a mountain, and discovered a huge ocean, which he called the *Mar del Sur* (South Sea) which he claimed for Spain. (Seven years later it was renamed the Pacific Ocean by Ferdinand Magellan.) This exciting discovery, however, did nothing to help ships get from the Atlantic to the newly discovered Pacific, even though the land was only 130 miles wide. Voyages of discovery were soon sent out to find a new way to the Orient. Among these were Ferdinand Magellan (c. 1480-1521), sent by Portugal in 1520. He found a southern, very difficult passage through South America, now called the Straits of Magellan. It was the only way around the continent for centuries. Magellan was the first to sail across the Pacific Ocean and around the world. Sir Francis Drake (1540?-1596) of England was looking for a western outlet for the Northwest Passage in 1577 when he also traversed the Pacific Ocean and went around the world.

During these and subsequent voyages to South America, sailors realized that there was no easy way through the huge land mass. That left North America to explore in the hopes of finding a Northwest Passage. In 1524 Giovanni da Verrazzano (1485-1528), sailing for France, followed the North American coast as far north as Maine. In 1535 Jacques Cartier (1491-1557), also of France, reached Canada and thought he had found the passage. He sailed a thousand miles up the St. Lawrence River before he decided that it was not the passage. In 1576 Martin Frobisher (1535?-1594) of England got as far north as Baffin island without finding a passage. In 1609 Henry Hudson (1565?-1611) discovered the Hudson River in New York, and Hudson Bay in northern Canada. Both had possibilities but were disappointing. The weather and ice were often so bad, especially in the high northern latitudes that taking sailing ships into the area was hazardous and sometimes fatal.

Soon Russia, the Netherlands, and Denmark joined the frantic search for a way through the North American continent to the Pacific. While no passage was found, a great deal of territory was mapped, charted, and claimed for European countries.

Explorers continued to look for a Northwest Passage in the eighteenth century, but it was seldom the major purpose of these voyages and hopes for its discovery became more symbolic than practical. Captain James Cook (1728-1779) of England made several trips in the 1770s to look for it, but his main focus was the South Pacific. In the 1790s English sea captain George Vancouver (1757-1798) made a number of trips into the Pacific Ocean. Appended to these voyages were several trips to North America to look for the passage. While there he also discovered the Canadian island that still bears his name, as well as the Puget Sound.

It has been said that the Northwest Passage was not found until the idea of reaching gold and spices in the Orient was replaced by the passion for discovery for its own sake. A more likely answer is that more thorough exploration could not be made until ships and equipment were available to stand the bitter cold in the north and until men and supplies were prepared for it. One man came close to finding the passage but turned back. John Davis (c. 1550-1605) of England got to northern Canada in 1616. He eliminated one dead end and found another bay that was a possibility. He reached Lancaster Sound between Devon Island and the north end of the huge land mass called Baffin Island. The strait to the east, now called Davis Strait, leads into a viable passage, but Davis could go no further and did not find it. This turns out to be the eastern end and the actual beginning of a viable Northwest Passage.

In the nineteenth century expeditions were no longer sponsored by merchants, and it was government-sponsored naval expeditions that continued the search for the fabled passage. John Franklin (1786-1847) of England came close to finding the passage, but all members of his expedition perished when the ship was caught and crushed in the ice. While searching for Franklin's crew, Sir Robert John Le Mesurier McClure (1807-1873) discovered two entrances to the passage. After his ship became icebound, McClure abandoned it and began to travel over land by sledge. He was eventually rescued by Henry Kellet; when the two men made their way on foot to Beechey Island, they became the first men to completely traverse the Northwest Passage.

The passage had still not been traversed by sea, however. Credit for that feat goes to a Norwegian. Roald Amundsen grew up determined to become a Polar explorer. He wanted to be the first to reach both the North Pole and the South Pole. He also wanted to answer many questions about the magnetic position of the poles and whether they were movable or fixed. In 1906, at the age of 29, he obtained a ship called *Gjoa*, also 29 years old. With a crew of six, he was the first to make his way by water from the Atlantic Ocean to the Pacific. A few years later, he became the first to reach the South Pole, just a few days ahead of Robert F. Scott (1868-1912). He was acclaimed a hero, the first Norwegian to achieve fame since Norway gained her independence in 1906.

Impact

Despite the centuries of effort that had been expended to find the Northwest Passage, Amundsen's achievement was not of great economic importance. There was no cheering and no rush to use the passage. No countries lined up to follow it into the Pacific. In the years of the search, European nations had made other accommodations. They had set up trading centers and colonies near the source of spices and gold and created regular shipping routes to bring the material back to Europe. This was one of the reasons that no trade routes ever traveled through the Northwest Passage. Another was the weather. At 75° north latitude, 1,000 miles (1,609 km) from the North Pole, snow, ice, and fog make the route impractical for any regular operation.

As we look back on the centuries of effort to find such a passage, the significant thing is that the effort led to genuine scientific advancement and knowledge. During the search, the configuration of Earth was made clear, giving educated men an idea of its real size and shape. The search also led to the mapping of a great deal of North and South America. To reach this goal, ships, sails, and equipment had to be improved to stand the rigors of long voyages and severe weather and ice in the north. The explorers were able to test the correctness of charts and the correct positions on the globe of islands, straits, and mountains. Correct readings of longitude and latitude were taken, and new instruments, especially the new chronometer for finding longitude, were put into use. These explorers discovered the poles and their position. And they uncovered many new peoples and places that would not have come to light were it not for the need to get through this continent.

In 1942 the Royal Canadian Mounted Police vessel *St. Roch* steamed through the Northwest Passage west to east and back again. It was a symbolic voyage, still fraught with danger and hardship. After that, the passage was seldom used. In 1960 an atomic submarine traversed the passage, but this time under the ice. These later voyages are of import because it took so long to find the passage and absorbed so much energy. They represented a symbolic end to the effort and celebrated the conquest of this most difficult and elusive passage.

LYNDALL B. LANDAUER

Further Reading

Delgado, James P. *Across the Top of the World.* New York: Checkmark Books, 1999.

Huntford, Roland. *Scott & Amundsen: The Race to the Pole.* New York: G.P. Putnam's Sons, 1979.

Morison, Samuel Eliot. *The Great Explorers: The European Discovery of America.* New York: Oxford University Press, 1978.

Wilson, Derek. *The Circumnavigators.* New York: M. Evans & Co., 1989.

Sven Hedin Maps Tibet

Overview

At the beginning of the nineteenth century approximately four-fifths of the land area in the world was still virtually unknown to the Western world. In particular, there was very little knowledge about central and eastern Asia. Between 1893-1933 Swedish geographer and explorer Sven Hedin (1865-1952) made four expeditions to central Asia to gather geographic and other scientific information. During the third expedition between 1906-08 Hedin explored and mapped large tracts of previously uncharted territory in central Asia and the Himalayas. He was able to trace the Indus, Brahmaputra, and Sutlej rivers to their source as well as discover and map the Kailas mountain range. This expedition made a significant contribution to our knowledge and the geography of Tibet and central Asia.

Background

In a previous expedition into Tibet, Hedin had disguised himself as a Mongol. Hiring a Tibetan monk as an interpreter, he had led a large and well-equipped caravan across Tibet, which in the nineteenth century was officially off limits to foreigners. Jesuit missionaries had visited Lhasa, the capital city, in the seventeenth and eighteenth centuries, but otherwise almost no Westerners had traveled in Tibet. Nominally under Chinese rule, the Tibetans were very much in control of who came into this remote place. Tibet was anxious to keep their country out of the British sphere of influence. During his 1901 expedition, Hedin took leave of his caravan and, accompanied by only an interpreter and one other person, attempted to enter Lhasa. After a series of harrowing experiences, Hedin was turned back by Tibetan warriors and escorted back to his caravan. He later attempted another crossing of Tibet but was stopped by orders from the Dalai Lama and ordered to leave Tibet.

Despite these problems, Hedin returned to Tibet in 1906, aiming to explore a region that the British maps simply designated as "unexplored." Hedin set two goals for this expedition: the mapping and surveying of the great mountain range north of the Himalayas that later he was to call the Trans-Himalaya; and to find the precise definition of the sources of the great rivers Brahmaputra, Indus, and Sutlej. This area covered a mountainous area north of Tsangpo, or the upper Brahmaputra River, and as far as was known, no Westerners had ever been there before.

This time Hedin decided to enter Tibet from the south. He was able to obtain letters of introduction from the Tashi Lama, second in command to the Dalai Lama. Through Swedish contacts he received a passport from the Chinese government allowing him to travel in Sinkiang province, where he was known because he had worked on a previous expedition in the region. Although this document made no specific mention of Tibet, he hoped it would impress officials if needed and allow his party to travel overland.

When Hedin left the British summer capital of Simla, he informed officials that his destination was Leh, capital of the British-controlled western-most province of Tibet, Ladakh. Traveling northward, he entered into Sinkiang Province and once in Leh, assembled a caravan of considerable size. He engaged the services of 21 men and bought or hired 88 horses, 30 mules, 10 yaks, and many sheep and goats. The latter served a dual purpose: if the caravan were to run into Tibetan officials, Hedin planned to disguise himself as a shepherd, while at the same time some of the animals were to provide food for the expedition.

Hedin left Leh in mid-August 1906, and for the next several months was "lost." The British authorities forbade him to enter Tibet; however, he managed to slip into the Tibetan highlands before that order was received by the British in Leh. For the next six months, through the harsh winter of the highlands, Hedin and his caravan traveled eastward across Tibet. He crossed the great chain of mountains that ran parallel to the Himalayas to the north. He measured the elevations of mountain peaks and passes and the depths of remote salty lakes, as well as making maps, drawing panoramas, and sketching landscapes and people. The temperatures reached minus 50° and sheep and horses froze to death. For months the expedition saw no other human beings. However, in early February 1907, Hedin ran into Tibetan officials, who immediately ordered him to leave Tibet.

When a shipment of mail from India "found" Hedin's group, however, the Tibetan officials were so impressed that they allowed the party to continue its journey to the Tashi Lama's residence in Lhasa. Part of the journey was by riverboat on the upper course of the Brahmaputra River. Hedin managed to keep a low profile in Lhasa and was able to stay for nearly six weeks. His Chinese passport impressed the local officials, and he was cordially received by the Tibetan officials, including the Tashi Lama. Hedin's time in Lhasa coincided with the Tibetan New Year, a time of special celebrations, and he was able to photograph and sketch the spectacular ceremonies and festivities without hindrance.

During Hedin's stay in Lhasa, negotiations occurred between Lhasa, Peking, and London regarding his future travel, since under the Chinese-British convention of 1904 foreigners were not allowed to enter Tibet. After complex negotiations Hedin was allowed to return to Leh, and he and his remaining caravan departed for western Tibet. However, it was another 17 months before he returned to India, and some of his most important scientific work was conducted during this time.

Other travelers who visited western Tibet before Hedin had established the general area where the Indus, Brahmaputra, and Sutlej rivers were located. However, it was Hedin's expedition that mapped with topographic accuracy the location of the headwaters of these three great rivers: the Brahmaputra at 15,860 feet (4,834 m) above sea level where a small rivulet runs from a large glacier; the Indus at a group of springs; and the Sutlej, one of the major rivers of Pakistan, at the Ganglung Glacier in western Tibet. By April 1908 Hedin felt satisfied that the goals of the expedition had been accomplished. He then mapped, surveyed, and sketched his way back into India.

Impact

Hedin's third expedition into Tibet lasted nearly 28 months. He accomplished his goal of mapping and surveying the Trans-Himalayan mountain range and found the precise sources of the three Asian rivers. This work made a significant contribution to our knowledge and geography of Tibet and provided the first accurate maps of the unknown regions through which he traveled. He had cleared up so many problems of Tibetan geography that no major discoveries remained; the work of future explorers would be simply to fill in the details on Hedin's maps.

Hedin's other major contribution to the geography of Tibet included his series of measurements, maps, and panoramas describing the Trans-Himalaya, the great mountain system that runs parallel to the Himalayas, north of the area occupied by the Indus and Brahmaputra rivers. He crossed seven of the major passes across the Trans-Himalaya and recorded the extent, height, and boundaries of the range. Among the descriptions of the land and people of Tibet, one of the most interesting is the chapter on the pilgrims circumambulating the holy mountain of Mount Kailas. Here the mountain rises nearly 22,966 feet (7,000 m), and around its base is a 28-mi (45-km) trail that was used by pilgrims making a circuit of the mountain. Hedin himself walked the trail around Kang-Rinpoche—the Holy Ice Mountain—with a group of Buddhist pilgrims.

Hedin published the results of his expedition in a massive work entitled *Trans-Himalaya: Discoveries and Adventures in Tibet*, written for a non-scientific readership. The three volumes (in the English edition) total nearly 1,300 pages and are illustrated with photographs and sketches made by Hedin. It contains many enthusiastic descriptions of the events that he regarded as high points of the journey.

The nine volumes of *Southern Tibet: Discoveries in Former Times Compared with My Own Researches, 1906-1908*, were published between 1917 and 1922. This set of volumes is a complete survey of the exploration and mapping of Tibet up until the early twentieth century. These volumes represent Hedin's most significant contribution to geography, with the volumes containing impeccable scholarship along with lavish illustrations. The eight volumes of text together with the separate index volume total 3,547 pages. In addition to the text there are two portfolios containing 98 maps compiled from Hedin's own maps drawn in the field and 552 panoramas, also drawn by him. The volumes contain historical and ethnographic information, survey and mapping information, as well as meteorological, astronomical, geological, and botanical information about the entire central Asia region. The books were favorably reviewed and have given readers past and present an unparalleled opportunity to travel with this eminent geographer and explorer.

The scientific research and results he and the men under his leadership obtained are his lasting legacy. He was a fine geographer, mapmaker of extraordinary talent, and an artist of unusual skill. His books were translated into many languages and he was well known throughout the world. He was honored by kings and emperors and by the great geographical societies of the time, and he left a legacy of unsurpassed exploration and discovery in Asia.

LESLIE HUTCHINSON

Further Reading

Hedin, Sven. *Trans-Himalaya: Discoveries and Adventures in Tibet*. 3 vols. Leipzig, Germany: F.A. Brockhaus, 1909-13.

Hedin, Sven. *Southern Tibet: Discoveries in Former Times Compared with My Own Researches*. 9 vols. Stockholm, Sweden: Lithographic Institute of the General Staff of the Swedish Army, 1917-1922.

Kish, George. *To the Heart of Asia: The Life of Sven Hedin*. Ann Arbor, MI: University of Michigan Press, 1984.

The Exploration of the Poles

Overview

During the first part of the twentieth century, having mapped and visited nearly the whole Earth, explorers turned their attentions to the Poles. Inhospitable, dangerous, and difficult to reach, the North and South Poles represented one of the ultimate adventures left on Earth and their conquest captured the public attention. When Roald Amundsen (1872-1928) reached the South Pole in 1911, one phase of the exploration of the Earth ended and another began.

Background

From time immemorial, mankind knew of the frozen wastes to the north. The Norse, Siberians, Inuit, and others lived in it and even merchant sailors came across icebergs broken loose from the northern ice packs. Eventually, as people came to understand that the Earth was a sphere spinning in space, they realized that the Earth's axis of rotation would be found in the far north and the far south.

While the far north was known to be hiding the North Pole, little was known about it save that it was virtually uninhabitable, cold, icy, and desolate. Some were convinced a continent lay hidden beneath the ice, others felt it contained a shallow sea, and some thought it was an ice-covered ocean. However, it was all speculation, uninformed by any real information.

As little as was known of the northern polar regions, even less was known about the far south. The ancient Greeks posited the existence of a southern continent because they felt it was needed to balance the weight of the northern landmass, but it was not sighted until 1840, when the American sailor Charles Wilkes (1798-1877) made the first confirmed sighting of land. However, as with the northern polar regions, exploration on foot was to wait for some time.

The northern polar region was the first to be extensively explored, probably because it was just a short distance from the major nations engaged in journeys of exploration. Perhaps the most daring attempt to reach the North Pole was Fridtjof Nansen's (1861-1930) voyage in the *Fram*, which left Norway in 1893 and returned in 1896. Unlike previous polar explorers, Nansen chose to work with nature. Knowing that the polar ice drifted with time, he and his companions designed a ship that would become frozen into the ice purposely, letting the ice itself carry them to the Pole. While unsuccessful, Nansen did reach a higher northern latitude than any previous explorer and became famous upon his return. The North Pole was finally reached on foot by Matthew

The Amundsen expedition to the South Pole. *(Underwood & Underwood/Corbis. Reproduced with permission.)*

Henson (1866-1955) and Robert Peary (1856-1920) in 1909. The first flight over the North Pole was claimed by American Richard Byrd (1888-1957) and pilot Floyd Bennett (1890-1928) in 1926, but doubt about their actual success remains. Byrd's diary, found in 1996, indicates that they were probably about 150 miles (240 km) short of the Pole when an engine oil leak forced them to turn back. The first documented flight over the North Pole was made three days after Byrd's attempt by the team of Roald Amundsen (Norway), Lincoln Ellsworth (U.S.), and Umberto Nobile (Italy), who crossed the Pole in a dirigible.

While the assaults on the North Pole were showing fruit, serious attempts were made to visit the South Pole, too. As with the North Pole, there were many unsuccessful attempts, including Ernest Shackleton's (1874-1922) epic voyage in 1914. Plagued by frostbite, snow blindness, and dysentery, Shackleton's expedition covered over 1,700 miles, but stopped short of the continent and spent nearly two years before its rescue. The first successful expedition to the South Pole was that of Roald Amundsen in 1911, who beat the unfortunate Englishman Robert Scott and his team to the South Pole by a matter of weeks. Tragically, Scott and his entire party died on their return from the South Pole, neither the first nor the last to fall victim to the polar regions. The ubiquitous Admiral Byrd became the first to fly over the South Pole in 1929; among his most notable achievements was the establishment of the first scientific outposts on the continent.

Impact

Public interest in polar explorers and their accomplishments was avid in the era of Byrd and Amundsen, and remains high even today, despite the fact that visiting the Poles is now almost routine. Television shows on the Arctic and Antarctic are popular, as are books and movies set there. As recently as 1999, world attention focused on the plight of a woman scientist with suspected breast cancer who could not be evacuated from the south polar station for many weeks; at the same time, a news station aired a documentary on scientific expeditions that traveled by nuclear submarine beneath the Arctic ice cap.

Part of the reason for this continuing fascination is probably because the Poles were, literally, the last places on Earth to be explored. As with Mount Everest, elaborate attention to preparation and equipment is essential, making mere survival a triumph of human will, technology, or both. It may be, too, that we are drawn by such extremes because they test us as nothing else can and, in so doing, make us see what is best in ourselves. Or, as John Kennedy once said with respect to journeying to the Moon, "We choose to do this, not because it is easy, but because it is hard."

In addition, polar exploration is, in many ways, a romantic activity, full of adventure, idealism, and with little immediate practical application. It may be that, to a public living in an increasingly practical world, following the exploits of polar explorers throughout the twentieth century provided (and continues to provide) an outlet for that part of our human nature that longs for an escape from the practicalities of modern life. Whatever the reasons, polar exploration has excited public interest for over a century, an interest that shows little sign of abating.

In the political and military arena, the Poles offer much, as well. From the start, national pride and prestige were reflected in exploration of any sort, rich as they were with potential for the discovery of new sources of raw materials, new trade routes, and the like. Conquering the Poles, however, also offered military and economic advantages that many nations have been eager to explore.

While most nations have relinquished any territorial claims to the Antarctic, many have a continuing presence on the continent. The United States, Russia, Britain, New Zealand, and others maintain scientific personnel on a more-or-less continuing basis; some countries, notably Chile and Argentina, have attempted to maintain small colonies. Chile actually set up a small village at one point, encouraging pregnant women to give birth in Antarctica to bolster their claim to portions of the continent (which have never been recognized by other nations). In general, while many countries jockey for position in Antarctica, little will likely come of it unless the Antarctic Treaty fails at some point to be renewed.

International treaty also prohibits exploring for or recovery of natural resources in Antarctica. The entire continent has been declared off-limits, in spite of evidence for deposits of coal, minerals, and perhaps petroleum. This same treaty also prohibits the use of Antarctica for military purposes, although several nations maintained military personnel on the continent to assist their scientific outposts. For that reason, the primary advantage to a continuing presence in the Antarctic is political prestige. The Arctic, however, is a different story.

Although there are no mineral resources beneath the Arctic ice cap, the Arctic environment was studied for decades by both the U.S. and the Soviet Union because of its military importance. Since the mid-1950s, nuclear submarines crossed beneath the Pole regularly, perfecting under-ice navigation and warfare techniques. The Soviet Union (and, later, Russia) developed ballistic missile submarines specifically designed to break through leads in the ice to launch their missiles at the U.S. or Europe, while the U.S. developed ways to find these submarines amid the cacophony of sounds made by drifting ice. To this end, the U.S. embarked on a continuing program of subsea exploration designed to return information about the oceanography, submarine geology and geography, as well as other factors that could provide a military advantage in wartime. Both nations studied above the ice, too, with the knowledge that nuclear missiles, if launched, would travel over the Poles.

In conclusion, the scientific returns from Antarctica are impressive and appear likely to continue for some time. Antarctic science has confirmed and deepened our understanding of terrestrial phenomena such as plate tectonics, ozone depletion, past climate change, meteorites, paleontology, glaciology, biology, and more. In addition, relatively new telescopes are providing a great deal of valuable information about high-energy astrophysics, and neutrinos, and allow long-term observations of phenomena during the months-long Antarctic night.

The most significant impact of Antarctic science, however, could simply be a better appreciation of the speed with which small changes in global temperature can cause the Antarctic ice sheets and polar cap to collapse. This could lead higher sea levels, potentially flooding many major world cities such as New York, Amsterdam, London, Buenos Aires, Tokyo, and others. An ongoing debate over the fragility of parts of the ice sheet has some scientists arguing that very little change is needed to cause such a collapse. If this is the case, they suggest, it may already be too late to do more than prepare for higher sea levels in the next century or so. Although this controversy has yet to be resolved, it will profoundly affect the way we interpret our ecological influence. If the ice sheet can collapse rapidly, we may soon be faced with a crucial choice: try to lower global temperature or prepare to evacuate hundreds of millions of people from low-lying areas around the world. Either choice will reverberate for decades or centuries.

P. ANDREW KARAM

Further Reading

Books

Gerrard, Apsley Cherry. *The Worst Journey in the World.* New York: Carrol & Graf Publishers, 1998.

Green, Bill. *Water, Ice, and Stone.* Harmony Books, 1995.

Hundford, Roland. *The Last Place on Earth.* New York: The Modern Library, 2000.

Lansing, Alfred. *Endurance: Shackleton's Incredible Voyage.* New York: Carrol & Graf Publishers, 1998.

Nansen, Fridtjof. *Farthest North.* New York: The Modern Library, 1999.

The First Women Aviators

Overview

It was only six years after the Wright brothers' first flight that the first woman flew an airplane. In the next few decades, women aviators became increasingly common and attracted an increasing amount of attention, culminating with Amelia Earhart's (1898-1937) flights in the 1920s and 1930s. With Earhart's death in 1937, women aviators became less prominent, but continued to contribute greatly to aviation, especially as auxiliary pilots during the Second World War. Thanks to the early female aviators, women are now accepted as pilots in both military and commercial aircraft.

Background

The first woman took to the air in 1784, not long after the first human flight of any sort. Flying over the French countryside, Elisabeth Thible was so thrilled she burst into song as she ascended to a height of nearly a mile. In spite of this early start, women remained by and large earthbound, relinquishing the skies to men. There were, of course, exceptions, and over 20 women flew balloons during the 1800s, but not many women took to the air.

Over a century later, in 1909, women again took to the air, this time in heavier-than-air craft. Another French woman, Raymonde de Laroche (1889-1919), who referred to herself as a baroness although the legitimacy of the title was doubtful, became the world's first licensed woman pilot in 1910. In the next few years, women in Germany, Italy, and America became licensed to fly, many of them explicitly trying to prove that women were as capable as men in the air.

The first American woman to fly solo was Blanche Scott (1890-1970), hired by the Curtiss Airplane Company to demonstrate the safety of their airplanes. For the next six years, Scott flew in aerial exhibitions, performing stunts before excited crowds. She retired in 1916, citing, among other reasons, the difficulty she had in being taken seriously by both male pilots and the crowds.

Another woman, Bessie Coleman (1893-1926), attacked barriers of race as well as gender. Although she was not permitted to attend an American flight school because of her race, she eventually earned her pilot's license in France, becoming the first black woman in the world to do so. Returning to the U.S. after this accomplishment, she opened a flight school in 1921. Unfortunately, she died in a plane crash just five years later.

There were a number of other notable women pilots in the 1910s and 1920s, including Harriet Quimby (1884-1912; the first woman to fly across the English Channel), Ruth Law (who set a non-stop distance record for both men and women), and Katherine Stinson. Most famous,

Harriet Quimby. *(Bettmann/Corbis. Reproduced with permission.)*

of course, was Amelia Earhart, whose exploits are more fully described in another essay. And, during the Second World War, the Soviet Union put women pilots into combat, mostly flying antiquated bombers to attack German positions in the Crimea.

These women flew for a number of reasons, but they had some motivations in common. In this pre-Suffrage era, many women wanted simply to show that women could do the same things that men could. Some were attracted to the danger and romance of flight, and some felt this was the only way for a woman to experience any adventure in her life since so many other avenues were closed.

Women pilots faced similar obstacles, too, no matter in what nation they flew. All met with some degree of resistance from male pilots and, in many cases, from the airplane owners, their families, and the public. In general, this resistance stemmed from a few basic causes. Some believed that women were too weak or too slow to safely control aircraft moving at high altitudes and high speeds. Flying was considered "unfeminine," and women who wanted to fly were suspected of being the same. Many tried to protect women, too, in this era, and one way of doing so was by keeping them from doing things known or suspected to be dangerous. Some men simply didn't want women stepping into the spotlight with them, while other men felt that, were a woman flyer to die in a crash, the whole field of aviation would be set back by several years because of public outcry. Nonetheless, women flew, partly to prove the men wrong, but mostly because they loved to fly.

Impact

As women took to the air, several things happened, some of which continue to this day. First, after several women set altitude, speed, or distance records, many men had to grudgingly admit that women really could fly safely and skillfully. Although this grudging respect did not necessarily carry over into other areas, it was a necessary first step in the later acceptance of women in other technical professions.

Secondly, crowds flocked to see women fly and perform stunts. Part of this attraction was because of the novelty of women performing these "masculine" deeds, leading to a wider acceptance, again, of women in other technical fields.

Finally, these were the first in what was to become many steps by women in aviation, culminating (thus far) in the first woman to command a space mission, which happened in 1999.

Although women's roles in the military remain more limited than those of men, women did begin to receive larger and more technical roles in World War II and in later years. In particular, women were permitted to join the military and to fly in supporting roles for the Allies. By ferrying planes from factory to air base and across to Europe, for example, women pilots freed men up for combat missions. However, women's roles in military aviation stalled out at this point for many years in the U.S. because of continuing public, military, and governmental reluctance to place women in harm's way. In fact, it was not until after the Persian Gulf War in the early 1990s that American women were finally permitted to fly combat aircraft in potentially hostile situations.

Things were different in other countries. Driven by a severe shortage of men during World War II, the Soviet air force enlisted the talents of women almost from the start. Women bomber pilots flew over 24,000 missions for the Soviets in their war against Hitler's Germany, and other women flew combat sorties in fighter planes. In fact, some women flew up to 18 bombing sorties per day while the top Soviet woman fighter pilot, Lilya Litvyak, downed a dozen German planes in combat.

In later years, drawing on these wartime experiences, the Soviet Union continued placing women in technical roles, including the world's first women astronauts. It was not until the 1980s that the U.S. followed suit, sending women aloft in the Space Shuttle, and in 1999 the first American woman commanded a shuttle mission. In this, finally, women had accomplished virtually everything in the air as their male counterparts, proving themselves every bit as talented and skilled, and deserving of equal respect.

P. ANDREW KARAM

Further Reading

Books

Earhart, Amelia and George Putnam. *Last Flight*. Crown Books, 1996.

Moolman, Valerie. *Women Aloft*. Time-Life Books, 1981.

The Exploration of South America

Overview

By the end of the nineteenth century, South America had barely been explored by civilized man. In the first three decades of the twentieth century, however, several adventurers made extensive treks into the heavily forested and often treacherous interior of this continent. Their work offered a glimpse of South America's flora and fauna, along with a view of its human cultures and past histories. Among the many discoveries of the era, Percy Fawcett surveyed country boundaries and mapped rivers, including the Rio Verde, from 1906-10. In 1911, Hiram Bingham (1875-1956) discovered the ruins of the Incas at Machu Picchu, and three years later Theodore Roosevelt (1858-1919) and Cândido Mariano da Silva Rondon (1865-1958) traveled the River of Doubt.

Background

South America in the early 1900s was a largely unexplored continent. It held few roads to connect major cities, relying instead on waterways for travel and trade. Often, the only available route between cities was a trip down a river that snaked through the thick vegetation of tropical forests. In many places, travel was blocked by groups of hostile, indigenous people who lived along the riverbanks. In other cases, travel was made impossible by powerful, white-water rapids.

As the times changed at the turn of the century, and the governments of South American countries sought to become more economically competitive, officials realized the importance of mapping their nations, cutting roads through the wilderness, and building cross-country communications systems. At the same time, educators and explorers from other countries saw the vast continent as a treasure trove filled with amazing potential for discovery.

One of the greatest South American explorers was Cândido Mariano da Silva Rondon. His career as an explorer began in 1890 when he became a Brazilian army engineer charged with stringing a telegraph line across the state of Mato Grosso and later building a road from the state's capital in the center of the continent to Rio de Janeiro on Brazil's Atlantic coast. With those two tasks completed, from 1900-1906 he took on

Inca ruins at Machu Picchu. *(Archive Photos, Inc. Reproduced with permission.)*

the challenge of building a telegraph line across the entire country. Rondon's adventures brought him into areas previously unseen by any civilized person.

As Rondon completed the Brazilian line, British army officer Percy Fawcett made his first trip to South America to survey the boundary between Bolivia and Brazil. Local residents piqued his interest with tales of lost cities in the South American interior, and he returned on several occasions to continue his investigation of this largely unknown territory. On his next job, surveying the Rio Verde in eastern Bolivia, he embarked with a ragtag team of men of different nationalities and specialties—including a waiter, a silversmith, and a baker. Although the trip was fraught with problems, the group completed the survey, although shortly thereafter five of the porters died, apparently from the rigors of the trip.

After Fawcett's Rio Verde journey in 1908 and a return trip in 1909, Hiram Bingham struck out into the South American wilderness in 1911 in search of the lost capital of the Inca. (The Inca were South American Indians whose empire, by the sixteenth century, ran along the western coast of South America, from the northern border of modern Ecuador to the Maule River in central Chile. By the time of the Spanish conquest, the Inca numbered about 12,000,000 people.) With several South American adventures already under his belt, including his journeys retracing historical routes through the wilderness, Bingham was searching for the city of Vilcabamba, from which the Inca had fought a last, desperate, and unsuccessful rebellion against the Spanish invaders in the 1572. In 1911 he discovered Vitcos, the last Incan capital, and the architectural wonder Machu Picchu, the most famous in a series of fortifications, inns, and signal towers along the network of Incan footpaths.

While Fawcett and Bingham were making their discoveries, Rondon continued his explorations. In 1914, he embarked on a journey with Theodore Roosevelt (U.S. president 1901-09) to traverse the River of Doubt, a waterway that Rondon had named after discovering it several years earlier. As with most other ventures into the South American forests, the team faced many difficulties; some did not make it out of the wilderness alive. In fact, Roosevelt died fewer than five years later, apparently from the lingering effects of his ordeal during the trip with Rondon. Nonetheless, the group completed its task and mapped the river, which had been unknown just a few years earlier.

Impact

Rondon, Roosevelt, Fawcett, and Bingham each made important contributions to the overall geographical and scientific knowledge of the South

American wilderness. Rondon also made great strides in protecting the region's indigenous people and their cultures.

Rondon's construction work in early twentieth-century South America brought with it some of the first comprehensive studies of the continent's interior. He was able to produce some of the first biological specimens from these territories, and provided valuable insights into the geography of the expansive, unexplored lands within the massive country of Brazil. On one eight-month expedition from Mato Grasso to the Madeira River, he and his men ran out of food less than halfway through the journey and were forced to eke out a living from the land. Although weak, they eventually reached their goal on December 25, 1909. It was on this trip that Rondon discovered a river that the Brazilians did not even know existed. He named it the River of Doubt; he later renamed it the Roosevelt in honor of the former president who would accompany him as they surveyed it.

The trip along the River of Doubt brought many perils, but the team was able to map the entire length of the river, which runs north-south in approximately the center of the continent. It eventually connects with the Madeira, which meets the Amazon River farther northeast. The American public became interested in South America when in 1914 Roosevelt released his book *Through the Brazilian Wilderness*, which chronicled his adventures with Rondon.

Besides Rondon's contributions to the geography and biology of the region, his many journeys helped him learn a great deal about the different native peoples who sparsely populated the forests and river banks. Many of those he met while stringing telegraph wire or building roads were known only through legends or stories passed from village to village. Often, he found that he was the native people's first encounter with civilization, so he had not only to complete his construction work but also strike peaceable agreements with the people who laid claim to the land or the river on which he traveled.

Through the years, Rondon developed a sense of responsibility for the native people and their cultures, and became an activist on their behalf. Largely because of his efforts, the Brazilian government in 1910 formed the National Service for the Protection of the Indians, which was designed to help the native populations retain their cultures and avoid exploitation from outside businessmen and settlers. In addition, Rondon later began the national Indian Museum. Eventually his work as an explorer and protector of indigenous people brought him considerable accolades, including the honor of having a territory named for him; that territory of Rondônia is now a state.

Fawcett added to the geographical understanding of the region by mapping the Bolivia-Brazil boundary. Before his expedition, neither government was sure where one country ended and the other began. Fawcett not only mapped the area, but learned about great civilizations that were rumored to have remains hidden deep within the forests. His desire to find these lost cities brought him back to South America time and time again, yielding valuable information about the wilderness, including a comprehensive map of the Rio Verde in eastern Bolivia. Accompanied by his son and a friend, Fawcett embarked on his last South American expedition on April 20, 1925, in Mato Grosso. All members of his team disappeared a month later. Most historians believe they were killed by a group of hostile native people.

Bingham's most celebrated expedition was his 1911 journey in which he discovered the Incan ruins of Machu Picchu in the region surrounding Rio Urubamba. Local residents led him to the site, some 8,000 feet (2,438 m) above sea level. Surrounded by thick, lush forests, the ruins had survived in good condition for hundreds of years. There Bingham saw numerous finely crafted stone buildings, including palaces and a majestic three-sided temple. The area has since become a well-known tourist destination, and in 1983 was named a UNESCO World Heritage site. Throughout his life Bingham remained convinced that Machu Picchu was the elusive city of Vilcabamba; ironically, it was another of his discoveries, the Inca city of Espíritu Pampa, which American archaeologist Gene Savoy showed to be a more likely site in 1964.

In all, these four men, and the many others who helped them on their expeditions, battled repeated hardships to pave the way into a continent's interior and provide a glimpse into the mysteries of South America.

LESLIE A. MERTZ

Further Reading

Baker, Daniel B. *Explorers and Discoverers of the World*, first edition. Detroit: Gale Research, 1993.

Bingham, Hiram. *Across South America.* Boston: Houghton Mifflin, 1911.

Fawcett, Percy. *Lost Trails, Lost Cities.* New York: Funk & Wagnalls, 1953.

Hemming, John. *Red Gold: The Conquest of the Brazilian Indians.* Cambridge: Harvard University Press, 1978.

Key, Charles E. *The Story of Twentieth Century Exploration.* New York: Alfred A. Knopf, Inc., 1938.

Roosevelt, Theodore. *Through the Brazilian Wilderness.* New York: Charles Scribner's Sons, 1914.

Finding the Tomb of King Tutankhamen

Overview

Howard Carter (1873-1939) and Lord Carnavon (1866-1923) opened a window to the past when they discovered the tomb of King Tutankhamen in 1922. Dating from 1300 B.C., the tomb is the only one of an Egyptian pharaoh to survive essentially intact. Tutankhamen's reign was brief and unremarkable, during a low ebb in Egyptian culture. Nonetheless, the quantity of gold and jewels and the exquisite beauty of the artifacts found in his tomb inspired new respect for ancient cultures.

Background

When Howard Carter and Lord Carnavon discovered the tomb of Tutankhamen, it created a worldwide sensation. Such a find was considered impossible for three reasons: First, many people believed Tutankhamen's tomb had already been discovered, because artifacts with his name on them had been found in an earlier dig. Second, it had been a dozen years since the last major discovery. Archeologists had scoured the Valley of the Kings, and it was generally agreed that no major finds were left. Third, in the three millennia since Tutankhamen had been buried, robbers had been hard at work pilfering the tombs. Wherever archeologists had gone, they'd found that nearly all the treasures had long since been taken away. An intact pharaoh's tomb was unheard of.

Archeology was still a young science when Carter and Lord Carnavon made their find. Very little digging had been done until Napoleon's time, and most of that amounted to disorganized looting until the mid-1800s when Auguste Mariette (1821-1881), on behalf of the Service des Antiquités, began to establish rules to organize excavations. Even then, there was a constant competition between treasure hunting and science.

Emile Brugsch Bey experienced such a case in 1881. A local family had managed to make a good living for six years by selling artifacts from a major site they had discovered, the tomb of Queen Astemkheb. When their secret was uncovered, Bey discovered that the site contained the mummies of 40 pharaohs, including Ramses II, believed to have been the pharaoh who contended with Moses in the biblical book of Exodus.

There was another major find in 1905—not of a pharaoh, but of a high government official, Prince Yuaa, and his wife. Although the tomb had been rifled, many artifacts were untouched, still fresh from millennia before. It was the richest find to date, and provided the most complete single view of ancient Egypt.

Howard Carter was a careful man who had begun his career as an archeologist in Egypt at the age of 17. Though he had no formal training, he had studied under William Matthew Flinders Petrie (1853-1942), the man who developed the first principles of systematic excavation. When Carter began his search for King Tutankhamen's tomb, he already had two major discoveries to his credit, the tombs of Queen Hatshepsut and Thutmose IV, both found in 1902 in the Valley of the Kings.

Carter was a meticulous planner with good intuition. He believed that Tutankhamen's tomb had not yet been found, and he began a methodical excavation of a spot he had chosen in the Valley of the Kings. Because digging in one area would have inconvenienced tourists visiting the tomb of Ramses VI, he and Carnavon spent six seasons exploring other parts of the site. Finally, Carter took on the untouched area and found a staircase. Because his partner was in England, he had his workers rebury the stairs, and he waited three weeks for Lord Carnavon to arrive. Together they supervised the digging, then the opening of a door that had been shut with the seal of Tutankhamen for over 3,000 years. Though there were signs throughout the tomb of a foiled attempt by robbers, chamber after chamber held their wonders untouched. The most remarkable treasure was the sarcophagus of the boy king himself, which was rich with gold and jewels.

The gold mask of King Tutankhamen. *(Archive Photos, Inc. Reproduced with permission.)*

Impact

The discovery of Tutankhamen's tomb was an almost immediate worldwide sensation. Carter became famous. He spent 10 years supervising the removal of artifacts from the tomb (3,500 in all), and the care with which he catalogued and photographed the site became an inspiration for future archeologists. News reports and films about the discovery of Tutankhamen's tomb increased public awareness of Egyptology. People reveled in erroneous rumors of curses (which grew in the wake of Lord Carnavon's sudden death five months after the tomb's discovery) and in stories of adventure. Carter became a model for cinematic heroes from Frank Whemple in *The Mummy* to Indiana Jones.

The public also became interested in Tutankhamen himself, but the only thing remarkable about the boy king was how unremarkable he was. Tutankhamen had been a weak king, dead by the age of 18. This seemed extraordinary, given the treasures that were buried with him. The third coffin alone is a marvel, 74 inches (188 cm) long and made of 243 pounds (110 kg) of solid gold. The mask, perhaps the most famous artifact, is exquisitely shaped from 24 pounds (11 kg) of gold and decorated with inlays of turquoise, lapis lazuli, carnelian, and amazonite. If such honor and wealth was heaped upon him, then the trappings of the builder of the Great Pyramid and other important kings of Egypt at its height must have been unimaginable.

The Egyptian designs and styles found in the tomb became popular with the public. The fashion world, and the Art Deco movement in particular, adopted these motifs for jewelry, furnishings, and clothing. At the same time, an appreciation developed for history and historical sites in general. The realization of what had been lost by thieves and plunderers over the centuries led to greater protection for archeological sites.

Today, the United Nations Educational, Scientific and Cultural Organization (UNESCO) designates World Heritage sites. This designation, which results from a 1972 treaty, calls on member states to contribute the necessary financial and intellectual resources to protect humanity's common cultural heritage. Cultural heritage, according to the UN, refers to "monuments, groups of buildings and sites with historical, aesthetic, archaeological, scientific, ethnological or anthropological value." The World Heritage Committee has established a fund to support emergency action for sites in immediate danger, repair and restore sites (particularly when the local government lacks resources), provide technical assistance and training, and promote educational activities. In Cartegena, Colombia, it helped local officials create laws and building codes that protected the historic city center. In

Delphi, as part of the agreement granting the site its designation, the committee worked out an arrangement that located an aluminum plant away from the historic area. In Giza, it helped the Egyptian government find an alternative to building a highway near the Pyramids.

World Heritage designation for Thebes and its necropolis, including Tutankhamen's tomb, came in 1975, but this designation alone has not been enough to protect the site. The Valley of the Kings is subject to floods and earthquakes. Even worse, humidity from the breath and bodies of masses of tourists has caused the murals to deteriorate. In 1992, the Egyptian government teamed with the Getty Institute and the Supreme Council of Antiquities to study and preserve the tomb.

In a virtuous circle, public interest led museums to acquire Egyptian artifacts. These drew in larger audiences which, in turn, led to more archeological exhibits. The King Tutankhamen find added momentum to a trend that already had been changing museums from the musty lairs of academics to lively centers of community education and entertainment. Museums attracted financial support that allowed them to upgrade their collections and improve the presentation of objects with more narrative and context.

A big boost to museums came in 1967 when the "Treasures of Tutankhamen," a collection of 55 prized artifacts, went to Paris as part of a tour sponsored by the Egyptian government. In the end, the exhibit visited Paris, the British Museum, four cities in the Soviet Union, and seven cities in the United States. In Washington, D.C., alone, the exhibit drew an unprecedented one million visitors. The exhibit's success encouraged similar tours, which have become a mainstay of revenue and community outreach for museums. King Tutankhamen's tour also provided favorable publicity for Egypt and encouraged tourism. One Egyptologist said, "Tutankhamen has been one of Egypt's greatest ambassadors!"

Most recently, a mystery about King Tutankhamen has arisen. Was the boy king murdered? It had long been supposed that Tutankhamen had died of lung disease or a brain tumor, until a 1968 x ray indicated that he might have died from a blow to the head. Courtiers and servants were counted as suspects. Hieroglyphics and paintings were reinterpreted, but no absolute proof was found. In 1997, a second x-ray investigation found thickening of a bone in the cranium that suggested that he was murdered in his sleep and suffered a lingering death. If this weren't enough, a forensic examination done by Egyptian experts showed evidence of poisoning. The poisoning pointed to a foreign-born official, Tutankhamen's body servants (again), and possibly even the king's wife. It is unlikely that the murderer, if there was one, will ever come to light. But the controversy shows Tutankhamen's continuing public attraction, and illustrates once again that his death has always been more interesting than his life.

PETER J. ANDREWS

Further Reading

Books

Carter, Howard and Arthur C. Mace. *The Discovery of the Tomb of Tutankhamen.* New York: Dover Publications, 1977.

Ceram, C.W. *Gods, Graves, and Scholars : The Story of Archaeology.* New York: Vintage Books, 1986.

White, Anne Terry. *Lost Worlds.* New York: Random House, 1941.

Other

Tutankhamen. http://www.sis.gov.eg/tut/html/tut00.htm

Scaling the Heights: Mountaineering Advances between 1900-1949

Overview

The sport of mountain climbing is known as mountaineering. It has garnered a significant amount of public interest because mountains are both majestic and dangerous at the same time, an unpredictable combination that intrigues and fascinates. Mountaineering began as a pursuit of prestige. Later, the focus shifted to the difficulty of the route taken to the top. In recent years the technique used to reach the summit has become paramount. In addition, new technology, equipment, and materials have increased mountaineering's popularity. The sport's triumph and tragedy have captured public imagination; the two best known examples are the ill-fated attempt of George Leigh Mallory and Andrew Levine to the attain the summit of Mount Everest in 1924, and the first successful ascent of the mountain by Edmund Hillary (1919-) and Tenzing Norgay (1914-1986) in 1953.

Background

Few people attempted to climb mountains for sport prior to 1786. Mountain peaks were ascended for religious or scientific purposes, but there is little evidence that climbing for the sake of accomplishment existed. This is probably due as much to a lack of leisure time and poor record keeping as it is to a lack of desire to climb. One of the first recognized attempts to ascend a mountain for the sheer accomplishment of it was spurred by a scientist named Horace-Bénédict de Saussure (1740-1799), who offered a reward to the first man to climb France's highest mountain, Mont Blanc (15,771 feet [4,807 m]). The summit was reached in 1786 when the French doctor Michel-Gabriel Paccard (1757-1827) and Jacques Balmat, his porter, made the first triumphant climb. Saussure himself reached the peak the next year.

The next significant event in mountaineering began in 1854 when Sir Alfred Wills successfully climbed the Wetterhorn in Switzerland. His accomplishment ushered in the "golden age" of mountain climbing during which nearly all of the major peaks in the Alps were climbed within a decade, a flurry of activity that culminated in the famous ascent of the Matterhorn by Edward Whymper (1840-1911) and six others in 1865. This famous climb is also known for one of the most famous mountaineering accidents ever. During the descent from the summit one member of the party slipped, pulling three others to their deaths. Only because the rope broke during the fall were Whymper and two guides saved from a similar fate.

With many of Europe's peaks conquered, the emphasis then switched to finding more difficult routes. This shift in emphasis, known as the "silver age" of mountain climbing, was fueled primarily by an influx of British climbers. At the same time, climbers were beginning to explore other continents for additional mountains to climb. Aconcagua, the highest peak in South America, (22,835 feet [6,960 m]) was first climbed in 1897. The highest peak in North America was reached in 1913 when Hudson Stuck, an American, ascended Alaska's Mount McKinley (20,320 feet [6,194 m]). The race was on to climb higher and higher. However, World War I halted most climbing activity as the world focused on other matters.

After World War I, the British made it clear that Mount Everest, the highest peak in the world, was their objective. Although Hillary would ultimately prevail in 1953, the British made seven unsuccessful attempts to reach the top of Everest between World War I and World War II. While many got close, there is no evidence that any actually succeeded. There is some speculation that George Leigh Mallory (1886-1924) and Andrew Irvine reached the summit in 1924, but that has never been verified because they perished somewhere near the top.

Other Himalayan peaks were also assaulted. The Himalayan mountain range has 13 separate peaks over 8,000 meters (five miles or more), and while the technical difficulties associated with climbing to such great heights in extreme weather were daunting by themselves, the political obstacles that had to be overcome to get permission from the surrounding governments were often just as challenging. In addition, the British were still at the forefront of mountaineering, but they were no longer alone. Mountaineering began to acquire a much more international flavor, and accomplishments became not only a source of individual satisfaction, but national pride as well. Austria, China, France, Japan, and Russia regularly outfitted expeditions to the highest peaks in the world. In 1933 a Russian party climbed the nearly 25,000-foot (7,620 m) Communism Peak (formerly called Stalin peak). Three years later Nanda Devi (25,643 feet [7,816 m]) was climbed by the British, and Siniolchu (22,600 feet [6,888 m]) was successfully ascended by a team from Germany.

During this time, the sport of mountaineering evolved further as the emphasis for many climbers switched from the route itself to the means by which the route was navigated. This marks the birth of extreme climbing, characterized by radical advancements in equipment and a change in the climbing philosophy that emphasized the purity of the sport. Climbers began to take greater pride in their accomplishments if they were gained with greater risks. A supreme illustration of extreme climbing was the conquest of the infamous North Wall of the Eiger in 1938 by Heinrich Harrer, Fritz Kasparek, Anderl Heckmair and Ludwig Vorg. In an exhausting week-long climb, the four were nearly killed by a series of avalanches. The team became heroes in their native Germany and Austria.

Impact

"Because it is there." George Leigh Mallory's response to the question of why he risked his life to climb sums up the allure of mountaineering, a paradoxical combination of majesty and danger. Climbing has always fostered a sense of freedom and adventure in its participants, and yet, in some cases, only a thin line separates tragedy from conquest. Because of this, society has always been enamored of mountaineering, living vicariously through its heroes. In reality, except for a few well publicized tragedies, the sport is much safer than it appears, especially with the advent of new equipment and technologies. In fact, the safety and challenge of mountaineering is limited by the height of the mountains themselves. Thus, knowledgeable participants choose a level that is appropriate for them—although some choose the most difficult climbs simply to push the limits of the sport.

Mount Everest, the highest peak in the world, has captured public imagination in a way that no other mountain has, and is the standard against which all climbs are measured. In Tibet, the mountain is called *Chomolungma* (goddess mother of the world); in Nepal, *Sagarmatha* (goddess of the sky). Because of the local topology, Everest was not thought to be the highest peak in the world, and was originally designated simply Peak XV. Thus it was something of a surprise when its height was initially measured at 29,002 feet (8,840 m) above sea level in 1852, although the officially accepted current height is 29,028 feet (8,848 m). In 1865 the mountain was named in honor of Sir George Everest, the British surveyor who mapped the Indian subcontinent. A British expedition made the first attempt to climb Everest in 1921, when Tibet first allowed foreigners to cross its borders.

After a second unsuccessful British expedition in 1922, George Leigh Mallory, a famous British climber and a member the first two teams, recruited a young engineer named Andrew Irvine as a climbing partner for the next assault in 1924. Although a relatively inexperienced climber, Irvine was in remarkable physical shape, and skilled at repairing the cumbersome oxygen tanks used by climbers in that period. On June 8, Mallory and Irvine attempted to reach the summit of Everest. The two were last seen moving at a good pace towards the top, but were soon lost to view in the cloud cover. They never returned. Irvine's climbing axe was recovered in 1933, the only relic of the ill-fated attempt until May 1999, when Mallory's body was discovered by a Nova/BBC film crew. Neither photographs nor written documents indicated that Mallory and Irvine had attained the summit.

Thirteen more climbers died in 10 additional attempts to reach the top of Everest until Edmund Hillary, a beekeeper from New Zealand, and Tenzing Norgay, a Sherpa climber, reached the summit in 1953. Hillary became a hero of the British Empire, and three countries (India, Nepal, and Tibet) beamed over the accomplishments of Norgay.

As the sport of mountaineering reached new heights a greater understanding and appreciation for the physiological effects of extreme altitude emerged. Altitudes above 25,000 feet (7,620 m) became known as the death zone, because decreased atmospheric pressure allows the air to hold only a third as much oxygen as at sea level. This increases the chances of hypothermia, frostbite, pulmonary edema, and cerebral edema. Judgment can be impaired at this altitude, even when breathing bottled oxygen. Climbers can experience extreme fatigue, impaired coordination, headaches, nausea, double vision, and even suffer strokes. To combat these threats, expeditions now spend weeks acclimating to the conditions, because climbers understand the physiological adjustments that the body must make to tolerate the harsh mountain environment. Despite better knowledge and new equipment, Everest is still a dangerous place. Nearly 1,000 people climbed Everest between 1921 and 1999; more than 150 lost their lives on the mountain, 15 of them in 1996.

When Mallory climbed Everest, he had little more than hemp rope tied around his waist for fall protection. His clothes were made cotton and wool. He carried an unreliable oxygen tank that weighed close to 30 pounds (14 kg). Today, climbers have much better equipment: Their clothes insulated and waterproof, oxygen tanks weigh a fraction of what they used to, and ropes are both lighter and sturdier. They use short wave radio and satellite communications. In fact, in many cases it was the sport that provided the impetus for these innovations. Climbing, however, is still unforgiving and unpredictable, a combination of danger and excitement that only adds to the public's enchantment with mountaineering.

JAMES J. HOFFMANN

Further Reading

Books

Bonington, Chris and Audrey Salkeld, eds. *Heroic Climbs: A Celebration of World Mountaineering*. Seattle: The Mountaineers Books, 1996.

Hemmleb, Jochen, Larry Johnson, Eric Simonson, Will Northdurft, and Clare Milikan. *Ghosts of Everest: The Search for Mallory & Irvine*. Seattle: The Mountaineers Books, 1999.

Kaufman, Andrew and William Putnam. *K2: The 1939 Tragedy*. Seattle: The Mountaineers Books, 1993.

Poindexter, Joseph. *To the Summit: Fifty Mountains that Lure, Inspire and Challenge*. New York: Blackdog & Leventhal Publishers, 1998.

Unsworth, Walt. *Everest: A Mountaineering History*. Seattle: The Mountaineers Books, 2000

Finding the Pygmies: Westerners Learn More about Africa's Ituri Forest People

Overview

The first half of the twentieth century was a difficult time for the indigenous people of Africa. Beginning in the late 1880s, the entire continent was in the process of being carved up into colonies by the European countries of Britain, France, Portugal, Germany, and Belgium. The United States, although not outright colonial, enhanced its interests via corporate and scholarly endeavors—the latter through newly established museums funding scientific expeditions collecting specimens from the diverse natural and cultural landscapes of Africa. In the early years of the twentieth century, museums funded by wealthy philanthropists and industrial giants were involved in a race to fill their institutions with specimens of plants, animals, and representations of the cultures of the world. These contacts, while giving detailed and interesting information, contributed to the strain and pressures being placed on indigenous people. Information on one group, the Ituri forest people—often referred to as Pygmies—was especially sought out during the early years of the twentieth century.

Background

The Ituri forest, a dense tropical rain forest in the northern part of the Congo River Basin, covers nearly 24,000 square miles (38,624 sq km) of land in central Africa. Indigenous people of the Ituri forest, Pygmies have long been the subject of speculation, mystery, and myth because of their small stature—they typically mature at a height of four to five feet (1.2 to 1.5 m). These forest dwellers are among the oldest inhabitants of Africa and may possibly be the original human inhabitants of the vast tropical rain forest, which stretches from coast to coast. They were well established there at the beginning of recorded history.

The earliest recorded reference of the Pygmy people is a an expedition sent from Egypt in the Fourth Dynasty, 2,500 years before the Christian Era, to discover the source of the Nile River. In the tomb of the Pharaoh Nefrikare is an account of the expedition describing the discovery of a people of the forest, a "tiny" people who sang and danced to their god. The Pharaoh requested that one of these people be brought back unharmed to Egypt. Unfortunately, this is where the story ends, although later records show the ancient Egyptians had become relatively familiar with the Pygmies.

When Homer (fl. ninth or eighth century B.C.), the famous Greek poet, refers to the Pygmies in the *Iliad*, he may be relying on this information from early Egyptian sources. By Aristotle's (384-322 B.C.) time, the mythmaking trend continued, although Aristotle himself argued that the Pygmies' existence was not fable but truth and that they lived in the land "from which flows the Nile." Mosaics found in the ruins of Pompeii show that, whether the Pygmies were believed to be myth or not, the makers of the mosaics in fact knew how they lived, even the kinds of huts they built in the forest. However, from then until the turn of the twentieth century, Western knowledge of the Pygmies decreased to the point where they were thought of as mythical creatures, semi-human, flying about in tree tops, dangling by their tails, and possessing the power to make themselves invisible.

By the thirteenth century, cartographers who drew the Mappa Mundi accurately located

Two Congolese men of average height with a group of Pygmies in their Ituri Forest village. *(Hulton-Deutsch Collection/Corbis. Reproduced with permission.)*

the Pygmy home on the African continent but portrayed the Pygmies as subhuman monsters. It appears that this view held up through the seventeenth century, when English anatomist Edward Tyson (1650-1708) published a treatise on "The Anatomy of a Pigmy Compared with That of a Monkey, an Ape, and a Man." He had obtained from Africa the necessary skeletons, on which he based his conclusion that the so-called "pigmy" was not human. This "pigmy" skeleton was preserved until recently in a London museum, and has been acknowledged as the bones of a chimpanzee. Portuguese explorers in the sixteenth and seventeenth century continued these extravagant erroneous accounts, writing about the pygmies making themselves invisible and therefore able to kill elephants.

In the late nineteenth century Sir Henry Morton Stanley (1841-1904), a British explorer for the colonization effort, crossed through the Ituri forest and made contact with the Pygmy people. In *Darkest Africa*, Stanley tells the story of his 18-month journey up the Congo River from its mouth across the Ituri forest. He describes these small-statured people of the forest as "the first specimens of the tribe of dwarfs."

Impact

There are numerous groups of people living the Ituri rain forest in central Africa. Collectively now known as the Bambuti, they are not dwarfs; they are fully developed human beings both physically and mentally. They are descendants of indigenous forest-dwelling people who speak varied languages and have knowledge of hunting and gathering skills enabling them to be completely self-sufficient and self-sustaining in procuring food, clothing, medicine, and shelter from the rain forest. They are considered the earliest inhabitants of the Congo Basin and have continued their hunting and gathering lifeways since before recorded historical times. Based on their ways of living, they are broken into two groups: village-dwelling agriculturists and the nomadic hunting and gathering peoples. In many parts of the Ituri forest the villagers and hunter-gatherers practice a form of mutual interdependence, which includes the sharing of language and customs. No longer identified as Pygmies, they are called by the name they give themselves—the Bambuti people.

Several expeditions during the early 1900s informed Western knowledge about the Bambuti. In 1925 the Brooklyn Museum of Arts and Science funded and mounted an expedition into what was then the Belgian Congo of central Africa. Directed and led by an American woman, Delia Akeley (1875-1970), the expedition sought to explore the Ituri forest and bring back cultural artifacts and accurate information for the newly established museum. Akeley had gained experience during two earlier expeditions to east (1905-1906) and central Africa (1909-1911), which were led by her husband, Carl Akeley (1865-1926), for Chicago's Field Museum of Natural History. These museum-sponsored safaris collected animal specimens for the collections at the Field Museum as well as providing an excellent training for the rigors and particulars regarding travel in Africa.

At times traveling by dugout canoe during her 1925 trip, Akeley ventured deep into the homeland of the Ituri forest people. An avowed amateur anthropologist and ethnologist, Akeley lived with the Pygmy people for several months, befriending them and documenting information about their customs, diet, and physical characteristics for the Brooklyn Museum of Arts and Science. This expedition dispelled some of the myths surrounding the Ituri forest people and provided invaluable information on hunter-gatherer societies.

Akeley found the Bambuti to be fully developed human beings with specialized skills allowing them to be completely self-sufficient. They possessed hunting skills, rituals, and creative expressions that dated back thousands of years. Skilled hunters and gatherers, the forest people used the resources of the rain forest to support themselves completely. Some used bows and arrows to hunt monkeys and forest antelope, while others used spears and nets to hunt large game like buffalo and elephant. Honey, fruits, nuts, caterpillars, termites, and mushrooms were gathered and used for food and trade. Living in beehive-shaped huts constructed of twigs and leaves, they were able to construct dwellings within hours. This enabled them to move approximately every three to four weeks to take advantage of the changing conditions of edible plants and animals. In addition, Akeley's expedition also identified and documented some of the skills of the musicians, singers, dancers, and storytellers that enriched the lives of the Ituri forest people. This information collected for the Brooklyn Museum of Arts and Science contributed to deconstructing the Pygmy myth and provided information for the next generation of scientists interested in the natural, cultural, and political history of the region.

There were, however, many negative social and cultural results from these museum-collecting expeditions, not to mention the European colonization efforts, that were generally not recognized until much later. Following the European conquest and colonization of Africa in the early twentieth century, it still was automatically assumed by Westerners that these hunter-gatherers were the most primitive members of the human race or even an earlier evolutionary stage, despite information to the contrary. This phenomena of describing the Bambuti as "a race of dwarfs," "insignificant," "primitive," "of an earlier evolutionary stage," or "subhuman" was demeaning, inaccurate, and allowed for destructive policies and attitudes to be used against the Bambuti people.

These stereotypes and subsequent policies made it extremely challenging for the Ituri forest people to continue their hunting and gathering lifeways without outside interference. The Ituri forest hunter-gatherer society was significantly affected by the resulting encroaching development, political changes, sensationalism, and attempts at cultural assimilation popular during the early twentieth century. Overall, the museum-funded expeditions of the early 1900s attempted to dispel some of the harmful and incorrect ideas about Ituri forest people. However, as late as the 1920s a member of the Ituri people was put on display at the Bronx Zoo in the United States.

LESLIE HUTCHINSON

Further Reading

Books

Duffy, Kevin. *Children of the Forest*. New York: Dodd, Mead, 1984.

Friedman, Ekholm. *Catastrophe and Creation: The Transformation of an African Culture*. Harwood Academic Publishers, 1991.

McLoone, Margo. *Women Explorers in Africa*. Capstone Press, 1997.

Turnbull, Colin. *The Forest People*. New York: Simon and Schuster, 1961.

Periodical Articles

"The Efe: Archers of the Rain Forest." *National Geographic* (November, 1989): 664-86.

Flights into History and Their Effects on Technology, Politics, and Commerce

Overview

Powered flight began early in the twentieth century. Following the first flight in 1903, airplane technology made rapid advances, leading to increasingly daring flights as aviators strove to be the first, the fastest, or to fly higher and farther than anyone else. Their daring not only gained widespread attention, making the aviators heroes, but they furthered the science of flight, making technological, scientific, and military advances.

Background

Even in ancient Greece men dreamed of flying, as the legend of Icarus demonstrates, and they likely coveted the ability for far longer. Leonardo da Vinci (1452-1519) sketched flying machines in his notebooks over 500 years ago, and over the centuries other scientists, inventors, and crackpots also tried to design flying machines. German inventor Otto Lilienthal (1848-1896) had some success with gliders in the last part of the nineteenth century, and Samuel Langley (1834-1906) attempted powered flight for several years, but genuine success was earned by the Wright brothers in 1903.

Their first powered flight lasted only a few seconds and traveled a distance shorter than the length of a single wing on the Boeing 747. However, within months Wilbur (1867-1912) and Orville (1871-1948) Wright had improved their airplane to the point where they could take a passenger aloft and, within a few years, they and others were routinely flying up to several miles at a time.

Following the Wright's first flight, there was continuous competition among aviators throughout America and Europe for aviation "firsts." In 1907 Paul Cornu (1881-1914) made the first helicopter flight; in 1913 Igor Sikorsky (1889-1972) built the first multiengine airplane; in 1914 the first scheduled airline began service in Tampa, Florida; in 1915 Fokker developed an effective airplane machine gun, introducing air warfare; in 1919 John Alcock (1882-1919) and Arthur Whitten-Brown (1886-1948) made the first nonstop transatlantic flight; in 1927 Charles Lindbergh (1901-1974) made the first solo crossing of the Atlantic Ocean; in 1930 Frank Whittle (1907-1996) received patents for the jet engine; in 1932 Amelia Earhart (1898-1937) became the first woman to make a solo crossing of the Atlantic; in 1933 Wiley Post (1899-1935) completed the first solo trip around the world; in 1947 Chuck Yeager (1923-) became the first person to maintain controlled supersonic flight; and in 1949 a U.S. Air Force plane made the first nonstop flight around the world. Interestingly, this list of firsts shows that during the airplane's first dozen years of existence most of its current major applications had already been discovered: exploration, military, and commercial. What remained was setting records, expanding roles, and improving technology.

To many, the 1920s and 1930s were aviation's golden age. Technology improved rapidly, so people were constantly flying higher, faster, farther, and to different places. In today's world, when jumbo jets routinely carry hundreds of passengers across the North Pole, it's hard to comprehend the excitement generated by Lindbergh's day-and-a-half flight across the Atlantic. It's equally hard to imagine a world in which places were not reachable by air. But this was the case in the early part of the twentieth century, and every first, every new altitude and speed record, every endurance and distance record was new and exciting.

In its first few decades, aviation was fraught with both peril and promise. Airplanes were risky, short-ranged, fragile machines with a very limited carrying capacity. They could, however, open the world to fast, relatively inexpensive transportation of people and goods if their risks could be reduced. The military realized their potential for gathering intelligence and, later, for destroying enemy matérial, while daredevils sought fame, notoriety, and records. Over the years, all saw their visions come to fruition, changing society irrevocably in the process. And, in large part, these milestones helped cement aviation in the public eye as a glamorous, important endeavor.

Impact

Aviation's first flights profoundly affected the technology of the times. In order, for example, to make the first flight, all of the engineering problems of making a powered aircraft had to be mastered. Further technological advances were needed for Lindbergh to cross the Atlantic in a single-seat craft, not the least of which was designing a plane that could make the crossing rapidly enough for a lone pilot to stay awake during the entire flight. Once developed, this advanced technology was available for all successive pilots and aircraft designers, adding to the rapid progress of aviation.

And aviation's progress *was* rapid. Consider: the first major innovation in ground transportation was the development of the chariot and other wheeled, horse-drawn conveyances. Two millennia later, the only real improvement was that two additional wheels and an enclosure had been added to form a coach. Similarly, water transportation had changed little from the development of sails until the invention of steam power, and even then the speed of water travel could be limited by wind. In aviation, however, fewer than 50 years took us from barely fluttering into the air to supersonic flight; another two decades took man to the moon.

The importance of aviation in modern military campaigns can scarcely be overstated. The reconnaissance potential for airborne observers was obvious long before powered flight; during the American Civil War observers were sent aloft in tethered balloons to note enemy positions and weaponry. These gave way to observers in primitive biplanes, then to modified bombers and fighters, and the U-2 and SR-71 of the past few decades. In addition, every improvement that made the airplane more valuable for civilian use could also be adapted for military advantage. Navigation gear that helped Lindbergh reach Europe made it possible for bombers to find their targets more reliably. The long-range engines and craft used by Wiley Post and Amelia Earhart helped aeronautical engineers design long-range B-17 and B-29 bombers. And, of course, any aircraft that carried civilian passengers and cargo could be used to ferry troops and matérial to the front. Of course, the converse was also true: innovations that made military planes more deadly almost inevitably found their way into civilian aircraft, making them faster, safer, and more efficient.

Aircraft technology has also changed the nature of warfare. In the Persian Gulf War, the Allied nations gained air supremacy almost immediately and then used it to pummel the Iraqi forces relentlessly, greatly shortening the ensuing ground war. (Almost a decade later in Kosovo a campaign was waged successfully using air power alone.) The movement of troops to the theater has also changed; most soldiers flew to the Persian Gulf, arriving in days rather than the weeks or months that troop ships would have required.

Aviation progress was frequently used by nations vying for political supremacy or advantage. There was, of course, the prestige of having the fastest, highest-flying aircraft; national pride in a citizen who accomplished an aviation first (e.g. transoceanic flight); and so forth. In addition to the status, however, was an implicit threat. A plane that could take its pilot to unheard-of altitudes would also be out of reach of enemy aircraft, giving some measure of air superiority for a time. Speed record holders had a similar advantage.

Using aircraft performance for political advantage is a trend that continues. During the Cold War, for example, the Soviet Union and the United States traded altitude and speed records regularly, each trying to show the other, and the world, that they had superior aircraft and, by extension, superior technology and military might. For nearly three decades the SR-71, the MiG-25, and other top-of-the-line aircraft tried to regain titles for their nations. During the 1990s, with the collapse of the Soviet Union and Communist East Europe, these competitions died away to some extent, but it is noteworthy that both nations (and others) continue to demonstrate their aircraft at international air shows.

Rapid long-distance air travel has also affected world politics: "shuttle diplomacy" would be impossible without it. Air travel also allows diplomats from distant nations to gather at the United Nations, in Paris, or anywhere else they decide to meet. This has made rapid response to diplomatic and military crises possible, helping to resolve them more quickly than before.

Commercial aviation is a major factor in modern economics. Commercial jet sales, for example, are important exports for the U.S., Great Britain, and France. Airmail, including the now-ubiquitous overnight delivery services, depends on aircraft, and traces its roots to the mail and freight delivery businesses launched in the first few decades of flight. Passenger air travel has increased steadily for the last 60 years. All of these are multibillion-dollar businesses that show every sign of continued growth for the foreseeable future.

CIRCLING THE GLOBE BY AIR: THE FIRST AIRBORNE CIRCUMNAVIGATION

In the National Air and Space Museum's Hall of Air Transportation stands a somewhat diminutive biplane with a hand-carved wooden propeller. Originally designed as a torpedo plane for the U.S. Navy, it was modified to become the Douglas World Cruiser. On April 6, 1924, four of these craft were piloted into the air by some of the U.S. Army's best pilots in a quest to complete the first aerial circumnavigation of the globe. Two of these planes returned to Seattle on September 28 after 5 months and over 27,000 miles (43,452 km) of travel.

This journey was accomplished in aircraft considered appallingly primitive by today's standards. The cockpits were open with only a small windscreen for protection from the elements. The engine developed only 400 horsepower, less than some of today's automobiles, and giving the plane a ceiling of only 8,000 feet (2,438 m) and a low top speed. Each plane carried two people, who alternated flying on a daily basis. During their trip, the pilots not only flew the aircraft, but serviced, guarded, maintained, and repaired it as necessary. Landing in the Indochina jungle, one plane was towed upriver by natives in sampans for repair. A few months later, flying from Iceland to Greenland, the two remaining planes were separated and lost each other in the fog. The lead plane landed successfully, followed a half hour later by the second. Taking off the next day for Labrador, the fuel pump on one of the planes failed, forcing the crew to pump gasoline by hand. In spite of everything and against all predictions, the two planes landed in Seattle, completing their journey.

P. ANDREW KARAM

To understand the importance of air transportation to the economy, consider how business relies on air travel and shipping. Most businesses require face-to-face meetings to work out contracts, solve problems, and so forth. This means that documents must be signed and exchanged, and goods or parts must be moved from factory to market. Before the advent of air travel, face-to-face meetings might take weeks because of the time needed for a train to cross the country, a ship to cross the ocean, or both. Shipping goods or parts was equally time-consuming, and even documents could only move at the speed of the fastest train or ship. In some cases, a piece of machinery or a product line could be out of commission for weeks while awaiting parts or a repairman. Today, however, it is possible to send any person, any document, and most goods anywhere in the world within one or two days. Because of this, equipment downtime is reduced, companies can maintain smaller inventories and require less warehouse space, and businesses in far-flung countries can become partners.

The evolution of powered flight also had another less direct, but still significant, effect on business. In showing that a feat was possible, aviation pioneers encouraged businessmen to make use of the possibilities that were now available. Businessmen wasted little time in capitalizing on the new capabilities, creating industries that rapidly became both important and lucrative. By showing that the technology and expertise existed, for example, to cross the English Channel by air, entrepreneurs began to think about alternate ways to move people and goods from place to place. When Rodgers crossed the U.S. by plane, by implication, vital goods and mail could be transported across the nation, too. Similarly, Alcock and Brown's nonstop flight across the Atlantic demonstrated that people, goods, and mail could also be sent from the Old World to the New in less time than a surface crossing required.

Pioneering flights and aviators also captured public attention. Lindbergh was a hero for years, as were Amelia Earhart, Wiley Post, Howard Hughes (who was an avid aviator), and other famous pilots. Books, songs, movies, and artwork celebrated their accomplishments because they did something never before possible. Today, when we take safe, rapid, efficient transportation to virtually anywhere on earth for granted, it's hard to imagine a time when this was not so. By opening up the world, often at great personal risk, these aviators showed the world what was possible, not just technologically, but in human perseverance, daring, and will. By doing so, they became legends, remembered with admiration to this day. Brendan Gill, in *Lindbergh Alone*, wrote words that can be applied to virtually any aviation pioneer: "What Lindbergh was the first to do, by an act of superb intelligence and will, millions of us accomplish regularly with the expenditure of no more intelligence and will than is required to purchase a ticket and pack a bag. . . . His valor is hard to keep fresh in our minds when the most we are asked to face and outwit above the Atlantic is boredom."

But these flights and the industries they helped spawn have had more social influence than simply the admiration given to pioneering aviators. Commercial aviation has made cultural exchange possible on a scale never before envisioned. Even a short time in a major international airport brings in flights from all over the world, bringing people and cultures together that otherwise would never have had the chance to meet. The world is perhaps a more familiar place because of this contact. An unfortunate corollary, however, is that distant people and places are perhaps less exotic and less mysterious now; perhaps by making the world more accessible, some of the romance, novelty, and excitement have been lost as cultures merge and mingle.

Finally, the Wrights, Lindbergh, Earhart, and the other aviation pioneers helped establish several scientific benchmarks that have profoundly affected society. In a general sense, any advance in scientific knowledge gives us greater knowledge of how the world and universe work, and we gain a better appreciation of our place in nature. High-speed craft such as the X-1 and X-15 let us see the other side of the once-feared "sound barrier." This led to a better understanding of aerodynamics, shock waves, and fluid flow that have applications elsewhere. High-altitude flight has given us information about the atmosphere, helped us map the annual austral ozone hole, and given unique views of the earth at a fraction of the cost of lofting a satellite into orbit.

In summary, it's obvious that aviation pioneers made an indelible impression on virtually all aspects of society, including the economy, politics, science, and culture. When all is said and done, there can be no doubt that our world would be far different without the advances made possible by these early aviators, their flights, and their equipment.

P. ANDREW KARAM

Further Reading

Books

Bryan, C.D.B. *The National Air and Space Museum.* Smithsonian Institution, 1979.

Combs, Harry. *Kill Devil Hill.* Houghton Mifflin Company, 1979.

Earhart, Amelia. *Last Flight.* Crown Books, 1997 (originally published posthumously in 1937).

Rich, Doris. *Amelia Earhart.* Smithsonian Institution, 1989.

Journey Across China

Overview

In 1931 André Citroën (1878-1935), the French automobile manufacturer, sponsored an expedition that retraced the ancient Silk Road followed by Marco Polo in the thirteenth century. The goal of the expedition, led by Georges-Marie Haardt (1884-1932), was to increase international understanding through trade, science, and art.

Background

André Citroën introduced mass production to France in a munitions plant that he transformed into an automobile factory after the First World War. Citroën was as interested in marketing and adventure as he was in cars. Known as a risk taker, he sponsored three expeditions that showcased the half-track cars designed by his company for rugged cross-country travel. The first expedition, in 1922, crossed the Sahara Desert from Algeria to Sudan. The second, in 1924, was a 1,300-mi (2,092 km) trek from Algeria south to Madagascar. These expeditions demonstrated the potential of motorized travel across French Colonial Africa, and also drew attention to the African people and landscapes through books, photographs, movies, and drawings.

Both African expeditions were led by Georges-Marie Haardt, the director-general of the Citroën factories. It was he who planned every detail, choosing routes, buying equipment, and dispatching supplies along the trail. Louis Audouin-Dubreuil (1887-1960), a former Air Service officer stationed in the Sahara during World War I, was the co-leader. Looking for new challenges, Citroën chose Haardt and Audouin-Dubreuil to lead a third expedition. The proposed itinerary was from Beirut, Lebanon, north through Russian-controlled Turkestan, then across the Gobi Desert to Peking (now Beijing). From there, the expedition would proceed south to Saigon and then west, back to Beirut, linking the two French colonies.

Preparations required two years of intense work by Haardt, Audouin-Dubreuil, and others. The most difficult task was obtaining travel permits from the Russians and Chinese. To strengthen their status as a scientific mission, Haardt enlisted a scientist from the National Geographic Society of America. The group also included the French priest and geologist Pierre Teilhard de Chardin (1881-1955), a painter, a cinematographer, and a writer, Georges Le Fèvre (1892-1968), who chronicled the expedition. The backbone of the group, however, were the mechanics, 19 in all. Altogether, 40 men took part.

Three months before their departure, the Soviet government revoked the group's permit to cross Turkestan. With 50 tons of supplies already on the way to depots across China, Haardt decided to enter China through Afghanistan and the Pamir mountain range instead. To maximize the chance of success on this much more difficult route, Haardt split the group into two parts that would start from opposite directions. The Pamir group, originating in Beirut, would traverse the Pamirs. The China group, originating in Peking, would meet them just north of the Pamirs, then return with them to Peking. In the space of three months, new routes were planned, special track-cars built for the mountainous terrain, climbing equipment assembled, and new permits obtained. At the last minute, a rebellion in Afghanistan forced the group to revise their route yet again, this time from the Pamirs to the even loftier Himalayas.

The Pamir group (as they still referred to themselves) set out from the Mediterranean on April 4, 1931. Driving under the same flag he had flown in Africa, Haardt led his group to the first stop, Baghdad, Iraq, where they were welcomed by a parade. Entering Persia (modern Iran), they saw forerunners of the petroleum industry: a caravan of 500 camels, each carrying two cans of oil. They progressed across Afghanistan at a rate of 100 miles (161 km) a day, their cars separated at "dust intervals."

As they crossed the northern plains of British India (now Pakistan), temperatures rose so high (50°C or 122°F) that gasoline began to vaporize in the engines. In this sunbaked landscape they were greeted by British officers and a battalion of Highland Guards dressed in kilts and playing bagpipes. They prepared for the upcoming Himalayan climb in the luxury of a mountain resort as guests of the Maharajah of Kashmir. Haardt estimated that the climb would take 45 days and require 400 porters or 200 pack horses, four times the number available. The only solution was to split into four groups leaving at eight-day intervals, reusing the same porters and horses.

The climb began with rain that continued for five days, washing the road away in several spots. Haardt had decided to take two cars as far up the mountain as possible, which meant hauling them around hair-pin turns and across wooden bridges. At the western end of the Himalayan range, their path was covered with 20 feet (6 m) of snow, which reduced their pace to less than a mile per hour up the 45-degree slope. Finally the cars were disassembled into 60-pound loads and carried. Adding to the physical hardship of the climb were thin air, bright sunlight, constant fatigue, and dysentery. After reaching the halfway point through the mountains, Haardt received a message that the China group had been detained in Sinkiang province.

In contrast to the Pamir group, the China group's hardships were due to politics rather than nature. In 1931 the Kuomingtang, or Nationalist, Party, had a fragile hold of the Chinese government. One of its policies was to rid China of foreigners, who had enjoyed special privileges and trade advantages since the 1840s. As a result, the French were subject to suspicion and delays as they crossed China. Just before they were scheduled to meet the Pamir group, they were detained for two months by the governor-general of Sinkiang, an autonomous province in western China. He refused to let them leave until they promised to send him three cars and three radio sets, which he thought would be helpful in controlling his war-torn land. After long negotiations, permission was given to the China group to leave and for both groups to cross Sinkiang on their way east to Peking.

Once he learned of the China group's delay, Haardt decided it would be a waste of time and energy to take the track-cars the rest of the way through the mountains, and he and his party set off on horseback. Their route followed a one-foot-wide path between nearly vertical mountain walls in a landscape of unearthly severity. A bridge over a deep gorge consisted of three ropes made of twisted brushwood. One was the path, the other two the hand rails. The four parties reassembled at the western border of China, where their belongings and fresh supplies were transferred to yaks from the porters, who, bare-footed, had carried 60-lb (27 kg) loads across the mountain peaks. Messages from the China group informed them that four track-cars would meet them in China, but that the cars and radios promised to the governor-general had been seized by rebels, and their passage through Sinkiang was therefore in jeopardy. Luckily, Citroën was shipping duplicate equipment via Moscow. On September 16, after 65 days in the mountains, the Pamir group rode east, expecting to meet the China group.

It was not until October 27, however, that the entire expedition was united in Urumchi, the capital of Sinkiang. They had hoped to set off for Peking immediately, but were forced to wait one month until the cars and radio equipment arrived. The delayed departure made the journey east even more arduous. Temperatures fell as low as -33°C (-27°F). Most of the gasoline and oil that had been hidden along the route earlier had been taken by rebel troops. The cars had to be kept running constantly so the radiators would not freeze, and hot soup froze almost before it could be drunk. Along the way they encountered Catholic missionaries, marauding bandits, and Buddhist monks who offered them salted tea with butter. They were invited to Mongol New Year celebrations, marked by fireworks and attended by princes in embroidered silk caftans.

On February 12 the expedition entered Peking where they were received by cheering crowds at the French embassy. Instead of continuing their journey to the battle-torn south, they decided to go by sea to Haiphong and then by land to Saigon, both in French Indo-China (now Vietnam). From there, the plan was to complete the circle by traveling west back to Beirut. Haardt, suffering from fever and fatigue after their stay in Peking, stopped in Hong Kong for a few days' rest. There he developed pneumonia and died on March 16. Haardt had been the mainspring of the expedition. Without him, there was no desire to continue and the expedition returned to France.

Impact

During such a turbulent time and in such a turbulent region, André Citroën's goal of increasing international understanding had little chance of success. Certainly there were positive interactions between the French and the people of the lands they traversed. For example, the expedition's doctor treated people wherever they stopped. And the goods distributed along the way—cash payment to guides, bribes to bandits, gifts to monasteries and officials, even the cars and radios extorted by the governor-general of Sinkiang—undoubtedly created the impression that France was a rich and scientifically advanced country. But China, in particular, was trying to rid itself of foreign imperialism, and moving in the opposite direction from Western capitalism.

The expedition did, however, increase communication between East and West. After its arrival in Peking, newspapers in major Chinese cities ran admiring articles about the "face" or prestige the expedition had earned for France. A book describing the expedition, *La Croisière Jaune*, translated into English as *An Eastern Odyssey*, presented a verbal and photographic description of lands and people that were largely unknown in the West. The expedition also served as a publicity device for automobiles in general and Citroën in particular. Unfortunately, its enormous cost contributed to Citroën's bankruptcy in 1935.

The scientific and artistic impact of the expedition is easier to measure. The expedition included experts in the fields of archeology, geology, botany, biology, photography, film, and art. They brought back 5,000 photos, 200,000 feet (60,960 m) of film, drawings, paintings, and ethnographic documents, as well as collections of flora and fauna, all of which became the topics of exhibitions and scholarly articles that can be consulted today.

LINDSAY EVANS

Further Reading

Books

Le Fèvre, Georges. *La Croisière Jaune. Troisième Mission Haardt-Audouin-Dubreuil.* Paris: Librairie Plon, 1933.

Le Fèvre, Georges. *An Eastern Odyssey. The Third Expedition of Haardt and Audouin-Dubreuil.* Translated by E.D. Swinton. London: Victor Gollancz Ltd., 1935.

Reynolds, John. *André Citroën: the Henry Ford of France.* New York: St. Martin's Press, 1996.

Auguste Piccard and Paul Kipfer Are the First to Enter the Stratosphere

Overview

On May 27, 1931, Auguste Piccard (1884-1963) and Paul Kipfer became the first men to safely ascend into the stratosphere, riding in a pressurized gondola borne beneath a balloon designed by Piccard. This ascent was the first of many, and unmanned flights in balloons of similar design continue to this day. In addition to setting the stage for manned and unmanned exploration of the atmosphere, it was also a precursor to manned exploration of the ocean depths, which also took place initially in a craft designed by Piccard.

Background

From the earliest of times man has envied the birds in their ability to fly. Stories and myths of human flight are among the earliest and most universal in human history, although the difficulties of flight kept man out of the skies for millennia.

The first human flight occurred in 1783, when two men flew over Paris, France, in a balloon designed and built by the Montgolfier brothers. They had been preceded into flight by a sheep, a rooster, and a duck that flew in an earlier Montgolfier balloon.

Balloons changed little over the next century, continuing to be filled with hot air, which rises because it is less dense than the colder air of the atmosphere. They saw military action as observation posts during the American Civil War, and Napoleon used them to observe enemy troop positions during his many wars. World War I saw the use of balloons for observations, too, as well as the extension of balloons into blimps and dirigibles.

In addition to their wartime roles, balloons were quickly pressed into service by scientists looking for a stable platform, indeed, any platform, for scientific studies at high altitudes. From the heights, balloons could look back at the Earth, taking photos for later study. In addition, balloons provided scientists the opportunity to obtain air samples and instrument readings at a variety of altitudes. The lowest region of atmosphere is called troposphere, and the stratosphere is the region right above the troposphere. Early balloons collected most of their data from the troposphere. A handful of balloons were able to loft instruments into the stratosphere, obtaining interesting scientific readings. Some of these readings were among the first to show the existence of cosmic rays, from which a number of important scientific discoveries were made.

In 1931, Swiss physicist Auguste Piccard and his colleague, Paul Kipfer, became the first humans to reach the stratosphere in Piccard's balloon, achieving an altitude of 51,762 feet (15,777 m). During this flight, in addition to making some scientific observations, Piccard and Kipfer were also able to demonstrate that Piccard's design worked. To allow people to survive in the stratosphere, Piccard designed the first pressurized gondola, intended to keep air pressure within the gondola at a comfortable level even in the rarefied upper atmosphere. Another Piccard innovation was to design a huge balloon that could lift the entire gondola while remaining only partially inflated. This let the gas within the balloon expand as it ascended, giving steadily increasing lift as the balloon rose.

The following year, Piccard broke his record with an ascent to nearly 55,000 feet (16,764 m), and within a few years, others had risen to nearly 61,000 feet (18,593 m). Since then, manned balloons have risen to over 113,000 feet (34,442 m), although high-altitude research for its own sake has largely ended. Recent record-breaking balloon flights, including crossings of the Atlantic and Pacific Oceans and the 1999 circumnavigation of the Earth, have all taken place, at least in part, in pressurized gondolas riding beneath balloons in the stratosphere.

Impact

The most immediate impact of Piccard's flight was to show that people could survive in the stratosphere. An earlier high-altitude balloonist, American Hawthorne Gray, had died in the 1920s because he lost consciousness from a lack of oxygen at great heights (he rose to about 40,000 feet [12,192 m] in an open gondola). By designing and ascending in a pressurized gondola, Piccard showed that the stratosphere was survivable. This same principle was, in turn, used for all subsequent high-altitude aircraft, including passenger airliners. In addition, Piccard's

Auguste Piccard in the pressurized gondola of his high-altitude balloon. *(Corbis Corporation, (New York). Reproduced with permission.)*

flight opened the door for high-altitude research into cosmic rays, the properties of the atmosphere at such altitudes, and other areas of inquiry. Finally, recent record-setting balloon flights have all used balloons very similar in design to Piccard's, proving the soundness of his original design.

From an engineering standpoint, Piccard's gondola design was, if not revolutionary, at least very significant. Others had constructed vessels designed to withstand pressure differences, but none had been previously built solely for the purpose of travel at high altitudes. Other design features made the gondola even more innovative. For example, Piccard painted one half of the gondola white and the other black, and then added a motor to spin it at a slow rate to help control temperatures. Unfortunately, the gondola stuck with the black side facing the sun and it became uncomfortably warm inside. Another mishap, corrected on a second flight, resulted in a valve sticking shut that was to be used to vent gas when descending. Instead, Piccard and Kipfer had to wait until dark, when the gas cooled enough to lower them back to earth. However, these were not significant problems and should not detract from Piccard's overall sound design.

More revolutionary was the design of the balloon itself. This was the first balloon to be designed for high altitudes and low air pressures. Instead of being completely filled when it took off, the balloon looked nearly empty, with just a small bubble of hydrogen in the top and a long, slack balloon beneath. However, as the balloon ascended and atmospheric pressure dropped, the gas bubble at the top expanded, filling the entire envelope at altitude. Previous balloons, not designed to do this, could generate large pressure differentials between the atmosphere and the internal gas, threatening to rupture the envelope. Realizing this, Piccard purposely designed a balloon with enough lifting capacity to penetrate into the stratosphere and enough extra volume to accommodate the expansion of gas at high altitudes. This same basic design has been used in virtually all high-altitude balloons, manned and unmanned, for the past 70 years.

The other important impact of Piccard's flight was the scientific knowledge returned by him and those who followed. During his flights in 1930 and 1931, Piccard conducted research on cosmic rays, which had first been identified during earlier balloon flights at lower elevations. In future flights, many more cosmic ray experiments were conducted, providing a wealth of information about this then-unknown phenomenon. In fact, this method of data collection remains important to this day, and scientific researchers in many countries routinely send experiments aloft in high-altitude balloons to study cosmic rays, extra-solar x ray and ultraviolet sources, and to search for other astronomical information.

High-altitude balloon flights have also returned an impressive amount of meteorological data over the years. Balloons designed along the lines of Piccard's ascend into the upper atmosphere on an almost daily basis, gathering information that supplements satellite data and helps scientists better understand our atmosphere and weather. This information is used for weather forecasting as well as for scientific research.

The other significant impact resulting from this flight was the effect it had on subsequent high-altitude, manned balloon flights, including recent record-setting flights and attempts. For nearly three decades after Piccard and Kipfer's flight, many nations pursued manned, high-altitude balloon flight. Two of these programs were the U.S. Navy's Skyhook program and the U.S. Air Force's Manhigh program. These programs, designed to test pilots' abilities to function at high altitudes and, later, to test space suit designs, culminated in several ascents to over 100,000 feet (30,480 m). On one of these flights, on August 16, 1960, Air Force Captain Joseph Kittinger parachuted from over 102,000 feet (31,090 m) in what is still the highest jump ever made.

Interestingly enough, Piccard's design was also adapted to undersea exploration with little difficulty. While considering the problem of constructing a vehicle to explore the oceans' depths, Piccard settled on a design in which a metal sphere, packed with instruments, was suspended beneath a large metal envelope filled with gasoline. Because gasoline is less dense than water, this design was a nearly perfect analogue of his balloon design (gasoline was used because, unlike air, it compresses very little with increasing sea pressure, thus giving more constant buoyancy with changes in depth). In this vehicle, named the *Trieste*, Piccard's son Jacques descended to over 10,000 feet (3,048 m) in 1953. The *Trieste* later dove to the bottom of the Mariana Trench, the deepest spot on Earth.

P. ANDREW KARAM

Further Reading

Books

Briggs, Carole. *Research Balloons: Exploring Hidden Worlds.* 1988.

Devorkin, David. *Race to the Stratosphere: Manned Scientific Ballooning in America.* Springer Verlag Books, 1989.

Jackson, Donald. *The Aeronauts.* Time-Life Books, 1980.

William Beebe and Otis Barton Set Depth Record

Overview

In the summer of 1930, William Beebe (1877-1962) and Otis Barton climbed into a cramped steel sphere attached by cable to a crane and were lowered over the side of a ship to a depth of over 1,000 feet (305 m)—at that time, the deepest dive ever made. Later dives brought them to a record depth of over 3,000 feet (914 m) a few years later. Beebe's vivid descriptions of the exotic life he saw sparked public interest, and, over the years, more deep-sea vessels were built, including free-moving vessels such as *Trieste* and *Alvin*. These vessels, and others that have explored the ocean depths, have returned enormous amounts of information to the surface, completely changing our view of life at great depths in the sea.

Background

The earliest record we have of any sort of underwater exploration dates to Alexander the Great, the Greek soldier and king who lived in the fourth century, B.C. Alexander is said to have descended into the ocean in a glass barrel covered with asses' skins. Between Alexander's time and the early twentieth century, manned underwater exploration scarcely changed, consisting primarily of short-term observations at shallow depths, primarily for lack of adequate systems to maintain a breathable atmosphere or to control the sea pressure. In fact, virtually all that was known of life below the depth of human dives came from nets and dredging the seafloor, neither of which (as we now know) return representative samples of the life that is present.

William Beebe (left) and Otis Barton with their bathysphere. *(Ralph White/Corbis. Reproduced with permission.)*

In the late 1920s, William Beebe, an ornithologist by training, became interested in the problems of deep-sea scientific exploration. He managed to interest an engineer, Otis Barton, in this problem. Together, they designed and constructed a steel sphere with three ports made of thick, fused quartz, the strongest known transparent substance. This sphere contained bottles of oxygen, canisters of chemicals to absorb water vapor and carbon dioxide, some scientific gear, and a telephone to the surface. This gear, and the two men were crammed into a sphere less than 5 feet (152 cm) in diameter.

To make their descent, Beebe and Barton climbed into the bathysphere, now hooked to a steel cable, and were lowered into the ocean. Following several manned and unmanned test dives, they finally reached a depth of 1,368 feet (417 m) on June 11, 1930. In constant telephone communication with their support ship, Beebe gave nonstop descriptions of changes in light levels and the strange new creatures visible beyond the portholes during their descent, stay, and subsequent recovery.

In subsequent dives, Beebe was lowered into the depths in many locations, all the while continuing to record his observations. In 1934, he and Barton made another record-breaking descent, this time to 3,028 feet (923 m), a record that was to stand for 15 years. These dives were not without risk. On several occasions, the bathysphere returned to the surface filled with water after a window seal failed and, on one manned dive, water began streaming in. In that instance, Beebe called the surface, asking to be lowered more quickly in the hopes that increasing sea pressure would force the window into its seals, a bet that paid off. In spite of the risk, however, Beebe continued to dive and to bring back a fantastic amount of information about life in the deep ocean.

Impact

There were three major areas in which the adventures of Beebe and Barton impacted society: 1. Scientists and engineers were encouraged to construct more advanced vessels with which to explore the ocean floor; 2. Science and society developed a better understanding and appreciation of the abundance and diversity of deep-sea life; 3. From this appreciation came a public fascination with both the deep sea and its denizens that made its way into popular culture in many ways.

For 15 years after Barton and Beebe's record-setting dive in 1934, their record remained unchallenged, primarily because World War II intervened. However, lessons learned in submarine construction were combined with research in ballooning by Swiss scientist Auguste Piccard (1884-1963), leading to construction of the bathyscaph, a vehicle designed to move independently. By severing the cable that tethered the bathysphere to the surface, Piccard was able to overcome the single greatest weakness of the bathysphere—lack of mobility. The bathysphere was completely dependent on support from the surface to control its depth, and it could not be readily moved laterally through the sea to follow intriguing animals. This frustrating lack of mobility was overcome by designing a vessel in which the bathysphere's steel sphere was suspended beneath a dirigible-like envelope filled with gasoline (which, unlike air, does not compress significantly with increasing sea pressure). Propellers were attached to the vessel, giving it lateral mobility, and a combination of venting gasoline or dropping ballast let it move vertically. Finally, an improved design gave this new vessel, the *Trieste*, the ability to descend to any depth in the oceans. In 1960 *Trieste* descended to the deepest point on Earth, 35,800 feet (10,912 m) in the Challenger Deep portion of the Mariana Trench, near the island of Guam.

Trieste and *Trieste II* were followed by *Alvin* and other deep-sea scientific submersibles, most of which were dedicated to the study of underwater geology and marine biology. Each successive generation of submersible was more maneuverable and more capable than the last, culminating in a number of small, agile, and relatively easy-to-use craft that debuted in the 1990s. While these vessels cannot dive as deeply as *Trieste*, they continue to return with enormous amounts of valuable scientific information. Among the findings made by these vehicles are continuing observations of deep-sea hydrothermal vents ("black smokers") and their unique colonies of life, studies of deep-water fish such as the Greenland shark, and searches for underwater archeological artifacts ranging from Roman galleys to gold-carrying steamships.

Before we explored the depths of the sea, most scientists had hypothesized that life either did not exist in deep waters or that the life there was very sparse and primitive. In addition, most scientists made the assumption that whatever life did exist probably survived on the occasional fish carcass that dropped through the water column to the sea floor. The actuality showed how naïve these assumptions were.

Barton described in breathless detail all the strange and wonderful fish he saw during his dives. Once below the reach of sunlight, he saw that many fish and other organisms were luminescent, showing up quite nicely even in the absence of external lights. He also noted that, in spite of the crushing pressures at 1,000 feet (305 m) (440 lb [200 kg] per square inch, or over 30 tons per square foot), life was not only present, but plentiful. Later dives, even those to the floor of the abyssal plains, showed that abundant life has colonized virtually every available part of every ocean, including the polar oceans.

More amazingly, much of this life is now known to have no connection at all with the surface. Until this discovery, the standard paradigm was that all life, however indirectly, depended on sunlight for its food. The discovery of tubeworm colonies, algal mats, and other forms of life surrounding hydrothermal vents that have absolutely no need for outside sources of energy or nutrition was a profound surprise to scientists. In addition to causing them to rethink their ideas of life on Earth, this also forced them to reconsider possibilities of life on other planets. In fact, current speculation about life on Europa, one of Jupiter's largest moons, would likely be purely in the realm of science fiction were it not for our knowledge of these deep-sea colonies found by deep-diving submersibles.

All of this has captured the public's imagination, perhaps more than Beebe would have guessed before his famous dive. Part of this fascination was sparked by Beebe, who wrote about his adventures in *Half Mile Down*. People were familiar with aquarium fish and those that showed up on their dinner plates, but Beebe described fish 20 feet (6 m) long that glowed while they swam through pitch-black waters. This was completely outside the realm of anyone's experience and sounded more like creatures from Jules Verne. As later expeditions brought back descriptions, photos, and specimens of even more bizarre creatures, the public's fascination continued to grow.

This interest manifested itself in several ways. Popular television shows like "Voyage to the Bottom of the Sea" may have lacked scientific and technical accuracy, but garnered large audiences week after week. Television specials, mostly by Jacques Cousteau (1910-1997) and the National Geographic Society, often focused on life in the sea, again gaining large audiences on a regular basis. Popular books were written (including Barton's), and many magazines such as *Smithsonian* and *National Geographic* carried story after story about deep-sea life, geology, and other discoveries that were made in the sea. Many movies, such as *The Abyss,* focus on adventures in the deep sea.

P. ANDREW KARAM

Further Reading

Books

Ballard, Robert D. *The Eternal Darkness : A Personal History of Deep-Sea Exploration.* Princeton University Press, 2000.

Cousteau, Jacques-Yves. *The Cousteau Almanac.* New York: Doubleday, 1981.

Cox, Donald. *Explorers of the Deep.* Hammond, 1968.

Prehistoric Cave Art Found at Lascaux

Overview

On September 12, 1940, four boys formed a small expedition team to explore a shaft they found while hiking through the sloping woods above Lascaux manor. Armed with shovels and picks, they expanded the restricted opening enough to enable them to descend into the unexplored chambers below. As they made their way through the narrow entrance shaft into the largest room of the cave, the young explorers noticed the walls and ceiling were adorned with brightly colored renderings of bulls and other animals. The boys had stumbled upon one of the greatest archaeological discoveries of the twentieth century—the Paleolithic paintings at Lascaux Cave.

The Cave at Lascaux, or Lascaux Grotto, is located in hills surrounding the Vézère River valley near the village of Montignac, Dordogne, in southwest France. It is one of 150 prehistoric settlements, and nearly two dozen painted caves, dating back to the Stone Age in the Vézère valley. Lascaux Grotto contains perhaps the most unprecedented exhibit of prehistoric art discovered to date.

Background

The first person to conduct scientific research at Lascaux was French anthropologist Henri-Edouard-Prosper Breuil (1877-1961). Breuil concluded that the cave was most likely a ceremonial site that was probably not constantly or even seasonally inhabited by humans. By analyzing the content of the murals, he theorized that the cave was used for ritualistic purposes, most likely connected with hunting practices. Though he did not conduct in-depth archaeological research, more recent excavations inside the cave have yielded no evidence of constant human habitation. Large fire pits for cooking, lithic debitage (the flakes produced when making stone tools), and abundant animal and plant remains from human subsistence are for the most part absent from Lascaux—all evidence that supports Breuil's initial conclusions.

The cave consists of a main chamber that is 66 feet (20 m) wide and 16 feet (5 m) high. The walls and ceiling of the main room and several branching chambers create steep galleries, all of which were magnificently decorated with engraved, drawn, and painted figures dating from about 15,000 B.C.. Based on carbon-14 dating, as well as the fossil record of the animal species portrayed in the paintings, the Lascaux artwork dates from the Upper Paleolithic period. The type of lithic industry, or stone tools, found and depicted further identifies Lascaux as part of the Aurignacian (Perigordian) culture present in Europe from 15,000 to 13,000 B.C.

The paintings were done against the stark contrast of the limestone, sometimes smeared

Prehistoric cave painting at Lascaux. *(Francis G.Mayer/Corbis. Reproduced with permission.)*

with a pale pigment, in various shades of red, yellow, brown, and black. Among the most captivating paintings are those of four huge auroch bulls, some 17 feet (5.25 m) long, whose horns are depicted in a stylistically twisted manner characteristic of the artworks of Lascaux and neighboring caves. These bulls are located in the main chamber of the cave and earned the room the name the "Hall of Bulls." The ceiling of the main chamber and surrounding galleries are marked with depictions of more common animals such as red deer, various felines (many now extinct), horses, and bovids.

Of these murals, there are two that are most impressive. One portrays only the heads of several large stags, 3.3 feet (1 m) tall, as if to suggest that the horses are fording a river, the other is a rare narrative scene that depicts the killing of a bison and bird-headed man with an erect phallus. The latter of these murals is perhaps an insight into the shamanistic beliefs of the Stone Age people who created the works at Lascaux. The graphic murals are some of the earliest representations of death and fertility in the archaeological record.

Impact

The creation of the murals at Lascaux required technological skill that archaeologists originally thought did not exist in the Paleolithic period. Painting in a dark cave meant that the artists needed some type of portable, long-burning light source—in other words, a lamp—and archaeologists have unearthed small stones with smoothed out depressions containing thickly smeared charcoal, consistent with a "lamp" fuel mixture of botanicals and fats. In addition, the height at which some of the paintings appear indicate that the people of Lascaux were more technologically adept than had been previously assumed. While the artists frequently stood on natural ledges while painting, round, worn notches and natural grooves in the cave walls also suggest that they constructed some sort of scaffolding. Finally, some of the pigments used in the paintings came from metal ores and oxides imbedded in rock that had to be identified, mined, and crushed to obtain the desired color. Others tints required the skillful processing of plants to extract dyes that were combined with other powder pigments and animal fats.

Though Lascaux is perhaps more famous for its contributions to art history, the cave also provides unique insight into the daily life and practices of prehistoric man. Lascaux and other caves in the region contain what may be considered the closest approximation of an historical or literary account of prehistoric Europe. Instead of interpreting only their material cultural (e.g., stone tools, botanical remains, and other artifacts), Lascaux enabled archaeologists to conduct rare cognitive studies, providing a perspective on how man reacted to his environment and what that environment actually looked like.

Though Lascaux is thought to be a ceremonial dwelling and not a site of continual human habitation, the interior paintings nonetheless depict everyday life in a hunter-gatherer, or nonagrarian, society. Scenes that depict hunting practices also depict men using tools similar to those that archaeologists excavated from the cave floor. Thus, archaeologists were able not only to evaluate the hunting implements themselves, but also to have an idea of the context in which men used such tools. For example, a painting of a bull kill shows that men hunted in groups, not just individually. This had long been speculated by archaeologists, but Lascaux and similar caves lent credence to this scientific hypothesis.

Along with deer and horses, paintings at the cave also portray several auroch bulls, a wild ox that is now extinct. The cave murals reflect not only a culture, but also an environment that was on the verge of change. The paintings at Lascaux are contemporary with the earliest stages of a slow transition from glaciation to a more temperate climate—a change characteristic of the late Pleistocene epoch.

One of the greatest lessons learned from Lascaux is the importance of historic preservation, when efforts to balance the often contradictory interests of public access and site conservation became a four-decade battle at Lascaux. The French Ministry of Culture opened the cave to the public in 1948. Though measures were taken to preserve the paintings, there was little regard for the potential archaeological discoveries that lay embedded in the sediments on the cave floor.

The narrow cave entrance was hastily enlarged to accommodate a walkway for the visitors, thus destroying archaeological information. Artificial lighting was installed to illuminate the artwork, and the feet of nearly 100,000 people a year damaged the archaeological deposits in the cave floor. The paintings themselves were damaged by the breath of so many tourists, which elevated carbon dioxide levels and introduced foreign bacteria into the cave; crowds also caused the natural temperature of the cave to rise slightly. This combination facilitated the growth of destructive algae on the cave walls, and caused the pigments used in the murals to crack and fade. To prevent further deterioration, the cave was closed in 1963.

In 1980 an ambitious project was undertaken that allowed the public to once again view the paintings at Lascaux and permit scientists to more carefully study the cave. A partial reproduction of the cave was commissioned for public visitation. The creation of the exact replica of the Hall of Bulls and the Picture Gallery demanded that scientists and artists take exact measurements of the cave surface and endeavor to copy the paintings. The cement structure mimicked the exact contours of the cave walls, and geologists even conducted studies on the wall surface to better approximate the texture of the limestone.

Artists, archaeologists, and art historians used projections of the original paintings to copy the murals onto the walls of the facsimile. Attempts were made to use some of the same techniques and materials employed by the prehistoric artists at Lascaux, but the concrete reproduction did not bond with the crushed ore pigments in the same manner as limestone, so synthetic ingredients and paints had to be used also. Inside the original cave, a computerized monitoring system was set up to allow scientists to more carefully monitor climatic conditions in the cave, such as temperature, moisture and carbon dioxide levels, and gas pressure. Lascaux II, the replica, opened to the public in 1983. It now welcomes 300,000 visitors per year.

ADRIENNE WILMOTH LERNER

Further Reading

Conkey, Margaret W., Olga Soffer, and Deborah Stratmann, eds. *Beyond Art: Pleistocene Image and Symbol.* Wattis Symposium Series in Anthropology. Los Angeles: University of California Press, 1997.

Gowlett, John. *Ascent to Civilization: The Archaeology of Early Humans.* New York: McGraw-Hill, 1992.

Ruspoli, Mario. *The Cave of Lascaux: The Final Photographs.* New York: Abrams, 1987.

Breaking the Sound Barrier

Overview

In 1947, Chuck Yeager (1923-) became the first person to fly faster than the speed of sound. Not only did he prove that the "sound barrier" could be broken, but he set in motion development that has led to the high-speed fighters that are critical to military defense. Commercial use of high-speed aircraft has been limited primarily to the Concorde, but plans exist for hypersonic flight (Mach 5 or higher).

Background

Aircraft went through a period of rapid development and commercialization after World War I, with the development of airmail and regular passenger flights. In addition, competitions, most notably the race to cross the Atlantic solo, captured public imagination and attracted investment. Airplanes became an essential weapon during World War II. The need for speed and maneuverability, particularly among fighters, spurred advances in design, including the development of jet engines.

Air superiority, particularly with a nuclear payload, became a life and death concern after the war. With the Eastern Bloc and Western Alliance locked in the Cold War, massive investments went into the design, construction, and testing of new aircraft. During what is now called the golden age of flying, test pilots regularly pushed new aircraft to their limits. One limit that many considered absolute was the speed of sound, Mach 1 (760 mph [1,223 kph] at sea level).

During World War II research into supersonic flight, pilots noticed that when aircraft velocity approached the speed of sound, the planes started to become uncontrollable. This happened when sound waves actually began to get ahead of their source, creating localized shockwaves that disrupted airflow around the aircraft. In particular, this phenomenon, called "compressibility," caused tremendous air turbulence around the wings. Compressibility was so difficult to handle that the British, after losing one of their best pilots, terminated their supersonic research flights.

The U.S. team, however, saw an engineering challenge and began further testing. The vehicle they chose was Bell's X-1, a solidly built, bullet-shaped craft that was launched from a bomber. The pilot was an Air Force captain named Chuck Yeager, who took over the mission after the commercial test pilot demanded more money. After series of powerless glides and incrementally faster tests of the X-1's rockets, Yeager hit a problem when he brought the vehicle up to 0.94 Mach. He lost pitch control, the ability to keep the nose from rising or falling. The control for the elevator on the wing refused to move.

The X-1 had hit the "barrier" that had led other aircraft to crash or disintegrate. The shock waves caused the airplane to shake violently and changed the aerodynamics so that it lost lift and began to drag. Sensor data showed that turbulence near the wing actually froze the hinge on the wing's elevator, which was why Yeager couldn't move it. The team was faced with a decision—quit or risk having the X-1 disintegrate with Yeager inside. The team's engineer, Jack Ridley, thought there was an alternative way to maintain control as Mach 1 was approached. He suggested Yeager use the horizontal stabilizer (a wing-like structure high on the tail, some distance from the air turbulence) to control the pitch. On October 14, 1947, Yeager ignited all four rockets and brought the X-1 up to speed. Again, he lost normal pitch control at Mach 0.94, but he was able to use the stabilizer as planned and regained command of the vehicle. Passing through Mach 0.96, the normal controls became operative again. Yeager took the X-1 up to Mach 1.07 (700 mph [1,126 kph]), and he was able to land the vehicle safely after having created the first man-made sonic boom.

Impact

Breaking the sound barrier meant breaking a psychological barrier as well as an engineering one. Once the possibility was proven, research was able to proceed, moving up to higher speeds and higher altitudes.

Public reaction to the breaking of the sound barrier was muted because security concerns kept it secret for eight months. The American military wanted as much of a lead in aircraft development over the Soviets as they could get. High-speed aircraft represented a significant military advantage for four reasons: First, in a dogfight, a speedier aircraft could provide the ability to acquire targets and, with the right design,

The first supersonic aircraft. *(Bettmann/Corbis. Reproduced with permission.)*

outmaneuver them. Second, faster aircraft that could outrun the enemy could retreat from dangerous situations where they did not have the advantage. Third, at high speeds, the probability of being shot down was reduced, making these jets good platforms for surveillance. Fourth, supersonic aircraft could be stationed at a safe distance from a target and still be accessible in a short period of time.

Each of these benefits has played a role in defense operations, but today some limits of supersonic flight are recognized. According to Yeager, most dogfights are done at Mach 0.9 to 1.2. Heat-seeking missiles eliminate the need to get on the enemy's tail, so anything over Mach 2.2 just uses more fuel. Since missiles travel faster than jets, running away is no longer an option, even for supersonic aircraft. Missiles have also helped neutralize the surveillance potential of supersonic aircraft, particularly when satellites offer an alternative. Distance to target remains a vital capability, but not for supersonic bombers, which are not a factor in defense due to their inability to meet cost and performance specifications.

Still, supersonic aircraft offered so much of a military advantage that they became one of the three legs of the U.S. nuclear defense (along with missiles and submarines). In addition, dozens of nations now include supersonic fighters, armed with conventional weapons, in their arsenals. For instance, there are over 1,000 F-15s currently in use. The production and development of supersonic fighters represent a multi-billion dollar industry with significant foreign exchange. (A single U.S. sale of 160 F-16s to Turkey in 1987 was valued at $4 billion.)

The economic impact of commercial supersonic flight has been minimal. With the exception of the British-French alliance that created the Concorde (which first flew in 1976), major development projects were scrapped in the 1970s due to environmental concerns and high costs. The Soviet Tu-144, which made its first flight in 1968, suffered a very public reversal when an aircraft crashed at the Paris Air Show in 1973. The model was grounded in 1978 after only 102 passenger flights. The American SST became a focus of concern because of sonic booms and ozone-destroying pollutants. (The latter led to worldwide action protecting the Earth's ozone layer, based on work for which Paul Crutzen [1933-] shared the 1995 Nobel Prize in chemistry.) Environmental groups successfully stopped government funding, and the aircraft industry put its development efforts into other subsonic projects, most notably jumbo jets like the Boeing 747. Further, the environmental groups translated the momentum for their cause into a ban on commercial supersonic flight over the continental United States and tight restrictions on flights in and out of coastal cities. This severely hampered the economics of commercial applications. In fact, only 14 of the 70 Concordes planned were built, and the program isn't likely to ever recover its development costs.

Pilots have had a place in the popular imagination almost from the time of the Wright brothers, and test pilots had their heyday, becoming popular heroes in the 1950s. Yeager never got a ticker-tape parade, but he has never lacked for speaking engagements and even helped make a John Wayne movie. Many test pilots, impressed by Yeager's accomplishments and his desire to "push the envelope," went on to become astronauts, who in turn attracted a whole generation into science. Yeager himself became one of the key figures in Tom Wolfe's book *The Right Stuff* (and the movie by the same name), which inspired renewed interest in his accomplishments.

THE TEST PILOT MYSTIQUE

The jet test pilot emerged, in the late 1940s, as a new kind of American hero. There had, of course, been pilot-heroes before: barnstormers, air racers, explorers, airmail pioneers, and the aces of two world wars. The older pilot-heroes were defined, in the public's mind, by their reckless daring, raw courage, and flamboyant behavior. The new heroes were something else altogether. Their job was to gather precise data on the handling of experimental aircraft and determine where the limits of their performance lay. Test flying demanded the ability to fly routine maneuvers time after time with absolute precision while staying alert and ready to deal, coolly and quickly, with any unexpected crisis. Test pilots judged themselves by their ability to push a plane to the very edge of its performance "envelope" and pull it back at the last instant before control was lost. The public's image of test pilots revolved around their supreme grace under pressure—their ability to deal matter-of-factly with life-threatening situations. This image of the test pilot was introduced by Chuck Yeager, who broke the sound barrier in 1947, but it was made famous by the test-pilot astronauts who went to the Moon in the 1960s and 70s.

A. BOWDOIN VAN RIPER

The dreams of supersonic flight persist. New research is underway, partly aimed at creating a successor to the Concorde fleet, which retires in 2005. Curiously, there is an American-Russian initiative that uses a Tu-144 as part of its test program. More ambitious is the work toward hypersonic flight (Mach 5 or better). The speed record for manned flight is Mach 6.7, set in 1967, but the current target is Mach 10. A radio-controlled prototype vehicle, the A43-HyperX, underwent ground testing by NASA in 2000, and many believe that the military already has a hypersonic vehicle, popularly known as Aurora.

Renewed interest in faster vehicles came in part as a result of the development of new materials, encouraging tests of a new engine concept called scramjet and new design options that may mitigate the intensity of sonic booms. The Lawrence Livermore National Laboratory's proposal for the HyperSoar vehicle would skip the aircraft off the Earth's atmosphere in a way that would allow it to provide air to the engines (inherently more efficient than using rockets) while shedding excess heat in space. Such an aircraft could take passengers from San Francisco to Tokyo in less than two hours. A HyperSoar bomber could reach any target on the planet without refueling. HyperSoar could even provide the first stage for launching payloads into space.

PETER J. ANDREWS

Further Reading

Books

Wolfe, Tom. *The Right Stuff.* New York: Bantam Books, 1983.

Yeager, Chuck, and Leo Janos. *Yeager.* New York: Bantam Books, 1985.

Other

The Boeing Company. "Faster and Faster." http://www.boeing.com/defense-space/military/f15/barrier/faster.htm

PBS Nova Online. "Supersonic Spies." http://www.pbs.org/wgbh/nova/supersonic/

Oil Is Discovered in the Middle East

Overview

Petroleum has become steadily more useful and valuable since the first oil well was drilled by Edwin Drake (1819-1880) in 1859. The first major oil fields were discovered in Pennsylvania and Ohio, with major strikes in Texas and Oklahoma to follow in 1901. Shortly after, the first oil concessions in Persia (now Iran) were granted, and the race for Middle East oil was on. Since that time, the discovery and exploitation of oil in the Middle East has had a profound influence on modern society and politics. Oil created vast fortunes and industrial empires, launched at least one war, promoted the widespread use of petroleum, gave birth to OPEC, realigned twentieth-century politics, and much more. It's safe to say that the huge reserves found in the Middle East played a tremendously important role in shaping the world we live in.

Background

Since antiquity, natural petroleum seeps were known, and the petroleum collected there was used for a variety of purposes. Not until the mid-nineteenth century, however, was any formal effort made to extract the oil for commercial use. When this finally happened, the first place people drilled for oil was near these seeps, assuming that they portended oil below. Oil seeps had been found for thousands of years in Persia.

In 1901, British businessman William D'Arcy convinced the Persian government to award him a concession for oil exploration, extraction, and sales in exchange for £20,000 and 16% of profits over the next 60 years. At one point when he was on the verge of bankruptcy, D'Arcy appealed to the British government for help; they agreed to assist him, fearing he might otherwise sell his concession to a foreign country such as Russia. Britain, still a great power at that time, also wanted to maintain a political presence in the Middle East. The British government pressured an existing British oil company, Burmah Oil, to give D'Arcy the financial assistance he needed in 1905; shortly thereafter, large amounts of oil were found.

In the following years, oil was discovered in a great many places in the Middle East: the Arabian Peninsula, beneath the Caspian Sea, beneath what would become the nations of Iraq, Kuwait, the United Arab Emirates, and others. In 1944, a prominent petroleum geologist named Everette DeGolyer reported to the U.S. government that he was certain the Middle East nations were sitting atop at least 25 billion barrels of crude oil, at least 5 billion of which were in Saudi Arabia. Not reported at that time were his unofficial estimates of up to 300 billion barrels of oil—a third of which he thought underlay Saudi Arabia. In a report to the State Department, DeGolyer's team commented that "The oil in this region is the greatest single prize in all history."

At that time, the Middle East produced slightly less than 5% of the world oil supply; over 60% came from the U.S., which was providing virtually all oil for the Allied armies in World War II. Concerns about the longevity of America's domestic petroleum reserves began to emerge at the same time that the Saudi Arabian economy began to suffer—the war kept many Muslims from making their required pilgrimage to Mecca. Saudi economic troubles and American fears complemented each other, and the U.S. began to take an active part in finding and extracting Arabian oil. This marked the beginning of Middle Eastern petroleum's ascent to its current domination of the global petroleum market.

Impact

Middle East oil was discovered during the first rush to look for oil outside the U.S., when governments and industrialists were attempting to find out how much petroleum was available for further industrial expansion. In addition, with the advent of mechanized warfare on land, the increasing use of aircraft, and the transition of naval propulsion from coal to oil, petroleum became a vital strategic commodity. The U.S. and most of Europe found themselves with a seemingly inexhaustible source of energy to power their growth. This abundant, cheap energy encouraged them to increase their reliance on power-hungry machines and internal combustion engines. This industrialization ensured a high standard of living, but its almost total dependence on access to cheap energy became the Achilles heel of the developed world.

The Organization of Petroleum Exporting Countries (OPEC) was founded in 1960 by Iran, Iraq, Kuwait, Saudi Arabia, and Venezuela primarily in retaliation for price cuts made by the oil companies. (Current membership also includes Qatar, Indonesia, Libya, the United Arab Emirates, Algeria, and Nigeria.) At that time, most oil wells were owned by petroleum companies who had been granted concessions by the nations on whose territory the wells lay. These companies paid only a fraction of their proceeds to the countries. When prices dropped, the oil-producing nations lost a great deal of money. OPEC was organized to raise and stabilize the price of crude oil. By regulating the amount of oil produced, the price could, theoretically, be maintained artificially high, increasing revenues for these nations. Of course, raising the price too high would be counter-productive because it would encourage less energy consumption, the recovery of otherwise marginal reserves, or both. So setting oil production and pricing became an intricate balancing act.

OPEC flexed its economic muscles in 1973 when, in retaliation for American support of Israel during the Yom Kippur war, it raised the price of oil from $3 per barrel to $12 per barrel and, for a short time it even stopped selling oil to the U.S. Other price increases followed; by 1980 oil was $30 per barrel. This shocked the U.S. into realizing its dependence on foreign oil, encouraged energy conservation, research into alternate forms of energy, and increased development of domestic reserves. Although oil produced by non-OPEC nations surrounding the North Sea oil fields and in Southeast Asia has diminished OPEC's power somewhat, OPEC nations still control a disproportionate share of world oil production, and retain a great deal of power as a result.

In addition to setting prices, countries like Saudi Arabia, Qatar, Iraq, and Venezuela nationalized their oil production in the 1970s, when they realized they could bank all the profits of oil production rather than just taking their concession fee. The government simply informed the oil companies that the government was buying their oil fields. The companies were paid off, asked to leave the country, and the government began operating the oilfields instead.

The influx of oil dollars, in turn, has made many OPEC nations dependent on petroleum to maintain their economy. Venezuela is an excellent example of the perils of over-dependence on a single commodity for a nation's economic well-being. Venezuela suffered in two ways: Much of the oil revenue was siphoned off from the economy by a corrupt government, which kept the revenue from benefiting the nation as a whole. In addition, the Venezuelan government counted on an unending and ever-increasing cash flow from oil. When prices dropped in the 1980s, Venezuela lost this revenue and much of their hard-won economic prosperity vanished with it. Plus, with little money saved because of rampant corruption, much of the infrastructure built with petro-dollars began to crumble.

Finally, since oil has a definite military value, protecting oil reserves, even in other countries, becomes a high priority for industrial nations. In World War II this led the Allies to bomb the Ploesti oil fields in Romania to deprive the Nazis in Germany of this energy source. In more recent years, Saddam Hussein invaded Kuwait and threatened to do the same to Saudi Arabia. This would have placed him in direct control of over 20% of total global oil production, which was considered an intolerable situation. For this, and other reasons, a coalition of forces waged war against Iraq to protect Saudi Arabia, restore Kuwait, and maintain unfettered access to Middle East petroleum. This, more than anything else, illustrates the importance of Middle East oil in today's world.

P. ANDREW KARAM

Further Reading

Books

Yergin, Daniel. *The Prize*. New York: Simon & Schuster, 1991.

Delia Julia Akeley

1875-1970

American Explorer

Although Delia Akeley was not the first woman to explore the African continent alone, her solo expeditions were remarkable. She traveled through areas where few non-Africans had journeyed. Initially accompanying her husband on safaris, Akeley acquired skills and experience, hunting wildlife for museum specimens. She sublimated this knowledge to investigate African animals and their habitats and to study natives and their cultures. Her observations provided zoologists and anthropologists with frameworks for future scientific research.

Information about Akeley's childhood is vague, primarily because Akeley disliked her family and was ashamed of her impoverished origins. Most sources state that she was born on December 5, 1875, on her parents' farm, near Beaver Dam, Wisconsin. The daughter of Irish immigrants Patrick and Margaret Denning, Akeley resented their incessant demands for her to perform domestic chores and ran away to Milwaukee at age thirteen. In 1889 she married Arthur J. Reiss, a barber who had helped her find work. Possibly during hunting trips with Reiss, she met Carl Ethan Akeley, a taxidermist and sculptor employed by the Milwaukee Public Museum.

Carl Akeley was an experimenter of more realistic taxidermy mounting methods. He believed that museum displays should depict animals in their natural habitats. Delia assisted him with this work and relocated to Chicago when Akeley accepted a position at the Field Museum of Natural History in 1895. She made plaster casts of plants, then poured wax into molds to make thousands of leaves, lichens, and flowers for museum habitats to complement animals Carl Akeley designed. After she divorced Reiss, Delia married Akeley in 1902.

The couple undertook two major African expeditions. In 1905 the Field Museum asked Carl to hunt for elephant specimens in East Africa. Delia assisted her husband and was also directed to procure mammals, insects, birds, and other interesting animals. Through 1906 the Akeleys pursued game in Kenya, and Delia Akeley shot a record-setting-size bull elephant,

Delia Akeley. *(Corbis Corporation. Reproduced with permission.)*

which the Field Museum displayed. In 1909 the American Museum of Natural History in New York City commissioned Carl to secure a family of elephants for an exhibit. Delia returned to Africa, but this trip proved more challenging than her previous adventure. She nursed her husband through bouts of malaria and wounds from an elephant attack. She became interested in studying primates and their habitats, preceding the scientific study of primate behavior by such notable primatologists as Dian Fossey (1932-1985) and Jane Goodall (1934-). Akeley transported a monkey to her Manhattan home and doted on it, even after it bit her severely. Her husband shipped the monkey to a zoo. This action strained the Akeleys' marriage and they divorced. Colleagues credited her with advancing her husband's career.

Delia Akeley returned to Africa alone on two expeditions financed by the Brooklyn Museum of Arts and Sciences from 1924 to 1925 and from 1929 to 1930. She was the first woman to direct a collecting expedition sponsored by a museum. Akeley capably trekked through Africa, gathering native crafts and photographing indigenous peoples, particularly the Pygmies of the Ituri Forest. She discovered several new species of animals. Akeley was the first non-African woman to cross the continent alone, moving from the east to west coasts. Unaware of her whereabouts, newspapers reported Akeley missing during her journeys through remote Africa. She lived with the Pygmies, recording her impressions of their culture and criticizing how outsiders treated natives and decimated wildlife populations. The Brooklyn Museum Bulletin later published her reports.

Akeley wrote two books, *J.T., Jr.: The Biography of an African Monkey* (1929) and *Jungle Portraits* (1930), and contributed articles to the *Saturday Evening Post, Collier's,* and *Century* magazines. Her artifacts were also displayed at New Jersey's Newark Museum. She married Warren D. Howe in 1939. Akeley died on May 22, 1970, in Daytona Beach, Florida. Her legacy was providing scientific institutions with difficult-to-obtain specimens from eastern and central Africa. Her years devoted to understanding African peoples and animals contributed to increasing westerners' awareness of that continent.

ELIZABETH D. SCHAFER

Roald Amundsen. *(Norwegian Information Service. Reproduced with permission.)*

Roald Amundsen
1872-1928
Norwegian Explorer

As an Arctic and Antarctic explorer, Roald Amundsen accumulated a number of titles. After he became the first human to sail through the Northwest Passage, he struck out for the never-before-reached South Pole, narrowly edging out the British team of Robert F. Scott (1868-1912), which arrived a month later. Afterward, Amundsen went on to become the second person to sail the Northeast Passage, and the first to reach the North Pole by dirigible.

Amundsen was born in Borge near Oslo, Norway, in 1872. When he was a teenager, the young man took an interest in Arctic exploration. Despite his enthusiasm for the far north, he trained to become a doctor, the wish of his recently widowed mother. Seven years after his father's death, his mother also died, and Amundsen pursued his dream of becoming an explorer.

He believed that the best route to success in his chosen field was through attainment of navigational skills. He took assorted maritime jobs, and in 1897 landed the position of first mate on the *Belgica*, which was carrying a Belgian Antarctic expedition as well as American explorer Frederick A. Cook (1865-1940). The voyage lasted longer than expected when the ship became icebound, and all aboard were forced to spend the winter in the Antarctic. After the ship finally landed in Norway in 1899, the now 27-year-old Amundsen attained his skipper's license and began arranging a voyage on his ship, the *Gjöa* through the Northwest Passage.

The *Gjöa* set out on this ambitious journey on June 16, 1903. By September, the ship had withstood a fire and some reef damage, but the crew was able to anchor on the south shore of King William Island. There, Amundsen's men spent the next two winters making magnetic and astronomical observations, mapping nearby areas, and determining the location of the magnetic North Pole. They finally set sail again in August 1905 and two weeks later completed their crossing of the Northwest Passage.

After this success, Amundsen hoped to be the first to the North Pole, and even had an expedition planned when he learned that Robert E. Peary (1856-1920) team's had beat him to it. Amundsen quickly shifted his sights to the South Pole, which humans had still not reached. Another explorer, Robert F. Scott, had the same idea, and the two men began a race to the pole. Amundsen arrived in Antarctica in January 1911 about 60 miles (96 km) closer to the pole than Scott. There, he waited out the winter. In October, his team began a two-month traverse to the South Pole and planted the Norwegian flag on the site on December 14. Scott's expedition arrived at the South Pole on January 18, 1912.

Amundsen next took the *Maud*, a ship he designed, through the Northeast Passage from 1918-1921. With his former travel through the Northwest Passage, this trip completed his global circumnavigation within the Arctic Circle. From 1925-26, he hoped to be the first to reach the North Pole by air. With financial support from American Lincoln Ellsworth, Amundsen made a nearly disastrous attempt by plane. His aircraft, an N-25, had to make an emergency landing on an ice floe and nearly collided with an iceberg. The team spent more than three weeks on the floe before finally digging out and building a sufficient snow runway to launch the plane again. The following year, Amundsen switched his means of transportation to the North Pole from a plane to a dirigible. He reached the pole just two days after Richard E. Byrd (1888-1957) arrived by plane.

Amundsen's last Arctic trip was prompted by the disappearance of Umberto Nobile (1885-1978), the man who had designed and piloted the dirigible that took Amundsen to the North Pole. Nobile's party had vanished during an expedition in 1928 to reach the North Pole by a newly designed dirigible. Learning that the party was considered lost, Amundsen struck out immediately to join one of two search parties. His plane took off on June 8, 1928, but crashed. All perished. In the meantime, the other search party found Nobile and other survivors from his expedition.

LESLIE A. MERTZ

Charles William Beebe

1877-1962

American Naturalist and Explorer

Despite an early, brilliant career in ornithology that took him on scientific expeditions from Nova Scotia to South America to the Himalayas, Dr. William Beebe is best remembered for his exploration and observations of the deep sea. With engineer Otis Barton, Beebe was the first to make direct study of the ocean and marine life within a tethered vessel a half mile down. In addition to his scientific pursuits, Beebe wrote nearly 300 articles and books, including *Half Mile Down* (1943) about his trips into the depths of the sea.

Born on July 29, 1877, to a paper dealer and a homemaker, Charles William Beebe grew up in East Orange, New Jersey. Young Will's family encouraged an interest in nature, taking trips to the American Museum of Natural History during which the boy would bring specimens to be identified. Because his family was not wealthy and thus couldn't afford an expensive education for their only son, Beebe's friendships with museum staff helped increase his knowledge of natural history as well as his skill in taxidermy and eventually led to his acceptance as a special student at Columbia University, where he studied from 1896 to 1899. He never officially graduated, although he was later awarded honorary Doctorate in Science degrees from Colgate University and Tufts College (1928).

In October 1899 the president of the American Museum of Natural History, also a professor at Columbia and the first president of the New York Zoological Society, offered Beebe the post of Assistant Curator of Birds at the new zoological park, which would open in the Bronx in November 8 of that year. As assistant curator, and eventually full curator (1902), Beebe helped stock the zoo's huge flying cage and bird house, making it one of the finest such collections in the world. With his first wife, Mary, whom he wed in 1902, Beebe traveled to study and collect specimens—to Mexico in 1903 and to Trinidad and Venezuela in 1908. In December 1909 Beebe headed an expedition to the Far East to study the pheasants of the world. The expedition traveled nearly 52,000 miles (83,686 km) and visited 22 countries before returning in May 1911. Beginning in 1918, Beebe published a four-volume set of his research from the journey, entitled *A Monograph of the Pheasants.*

Even before his monograph on pheasants was published, Beebe's focus had turned to the tropical jungles when he became director of the zoo's new Department of Tropical Research. In early 1916 he opened the Bronx Zoo's first research station in British Guiana. In 1923 Beebe was in charge of a research expedition to the Galapagos Islands, where he studied land and marine life, making oceanographic studies in the Sargasso Sea during the journey. His studies sparked a desire to explore more of the earth's oceans, especially their depths, and in the late 1920s Beebe began helmet diving and dreaming of diving deeper.

After several years of traditional diving with the helmet breathing apparatus, Beebe wanted more—and began designing a depth chamber to allow humans to dive to deeper ocean depths.

Around that time, he met a fellow diving enthusiast, Otis Barton, who had designed a spherical steel vessel that Beebe named the bathysphere. In June 1930 the duo made their first test dive, and after extensive testing and modifications to the bathysphere, Beebe and Barton descended to a record depth of 3,028 feet (923 m) in 1934. Beebe wrote about their experiences in his book *Half Mile Down* in 1943.

After the deep ocean adventures of the bathysphere, Beebe settled down to more mundane pursuits, including the editing of *The Book of Naturalists* (1944). In July 1952 he retired from his position as director of the Department of Tropical Research at the Bronx Zoo. He returned to his estate called Simla in Trinidad, where he died in June 1962. His will proved that the New York Zoological Society, the guiding force behind the Bronx Zoo, remained close to his heart when he donated Simla and its 228 acres of land to the society.

ANN T. MARSDEN

Hiram Bingham. *(Corbis Corporation. Reproduced with permission.)*

Hiram Bingham

1875-1956

American Explorer and Educator

Hiram Bingham spent several years traveling throughout South America. On his most noteworthy expedition, he set out to find the capital city of the Incan descendants, but instead discovered on July 24, 1911, Machu Picchu, an area deep in the South American interior that is filled with the ruins of one-time Incan palaces, a temple, and other buildings.

Bingham was born on Nov. 19, 1875, in Honolulu, Hawaii, into a long-time family of the islands. The son of retired missionaries, he went to school at Punahou School and Oahu College in Hawaii before continuing his education on the mainland. From 1892-1905 he studied successively at Phillips Academy in Massachusetts, Yale University, the University of California at Berkeley, and finally Harvard University, where he earned his doctorate in Latin American history. He ultimately took positions teaching history at Harvard, Yale, and Princeton universities.

Fresh out of school, Bingham ventured to South America in 1905 to attempt to retrace the path taken by Venezuelan Simón Bolívar in the early 1800s. He recounted the trip in *The Journal of an Expedition Across Venezuela and Colombia*. He continued his adventures in 1908 and 1909 by following a one-time trade route that connected Rio de Janeiro on the Atlantic coast of Brazil to the city of Lima, near the Pacific coast of Peru.

He is most remembered for his expedition to South America in 1911, however. He had hopes of finding the legendary capital city of the Incan descendants in the Rio Urubamba valley. This city, called Vilcabamba, was said to be the site of a standoff between the Spanish invaders and the Incans in 1572. Bingham's Yale University expedition set out in July. Not long into the trip, Bingham met a local man who started the expedition team on a several-hour hike to Machu Picchu, which means old peak. There, 8,000 feet (2,438 m) above sea level and set in the middle of a thick, nearly impenetrable forest, rested the well-preserved remains of impressive Incan palaces. The site held examples of fine Incan stonework, including a spectacular three-sided temple. Of the site, Bingham noted in his diary, "Would anyone believe what I have found?"

Although the buildings of Machu Picchu have since been heralded as architectural masterpieces and archaeological treasures, Bingham quickly pushed on in search of the elusive Vilcabamba and Vitcos, the Inca's last capital city. In about two weeks—and after trips to several other ruins along the way—Bingham's expedition was led by local residents to a fortress containing more than a dozen buildings and a palace. The fortress, plus a nearby pool marked with a large, carved boulder, provided enough evidence for Bingham to declare that he had finally found Vitcos. Continuing its explorations, Bingham's team a week later discovered ruins from a large Incan town in Espíritu Pampa.

He returned to South America in 1912 and 1915, and eventually came to believe that Machu Picchu, and not the fortress, was actually the capital city. The debate over the site of Vitcos continues today.

Following his last adventure in South America in 1917, Bingham returned to the United States and eventually shifted from an educational career to one in the military and later one in politics. In 1916 he became a captain in the Connecticut National Guard. With the country's involvement in World War I, Bingham became an aviator in the spring of 1917, eventually advancing to the position of lieutenant colonel and commanding a flight school in France until 1918.

His political career began in 1922 with his election to lieutenant governor of Connecticut. Two years later he became the state's governor, but served only a month before moving to the U.S. Senate to fill a vacant seat. He remained a senator until 1933, when he lost the re-election following his censure on charges of making a lobbyist a paid staff member. He remained active in politics, however, and from 1951-53 became chair of the controversial Civil Service Commission's Loyalty Review Board.

Bingham died in Washington, D.C., on June 6, 1956, and is interred in Arlington National Cemetery.

LESLIE A. MERTZ

Louis Bléirot
1872-1936
French Aviator

French aviation pioneer Louis Bléirot was the first person to fly a plane over the ocean in a heavier-than-air craft, a feat of great daring for the early 1900s.

Bléirot was born July 1, 1872, in Cambrai, France. While he amassed a great fortune at an early age by inventing automobile headlights, Bléirot's lifelong devotion was focused on aviation. His earliest flight experiments involved towed gliders on the Seine River. While in his thirties, Bléirot used the newly available lightweight engines to design his own aircraft. He taught himself to fly while continuing to improve his initial design by trial and error. His series of airplane configurations ranged from box-kite biplanes to a tail-first monoplane. In 1907, at the age of 35, he made his first flight at Bagatelle, France.

But it was Bléirot's incredibly dangerous 40-minute flight across the English Channel that launched him to fame. The English Channel posed a challenge to aviation that many could not resist. In 1909, three European pilots, Hubert Latham, Count de Lambert, and Bleriot, prepared for what many called a deadly attempt.

Aircraft damage delayed de Lambert, and Lantham's plane crashed into the water halfway through the course. Braving rough weather and more than 20 miles (32 km) of dangerous sea, Bléirot shocked the world six days later when he flew his Bleriot Model XI, 28-horsepower monoplane successfully across the Channel on July 25, 1909.

The young aviator traveled from Les Barraques, France, to Dover, England. Bléirot achieved lateral control by slightly warping the wings and castering the main undercarriage wheels in a way that would allow the aircraft to "crab" in a crosswind on the ground. His design revolutionized all operations on the ground. In the end, his triumphant flight won for him the much sought after London *Daily Mail* price of 1,000 pounds sterling. Bléirot's success sparked a wave of excitement across Europe.

After his launch to fame in 1909, Bléirot's company produced fighter aircraft for the military. His new aviation company was soon producing a line of aircraft known for their high quality and performance. The Bléirot XI was the first aircraft put to use by France and Italy in 1910; Britain followed suit two years later. By World War I, his famous S.P.A.D. fighter planes were in use by all the Allied Nations.

Bléirot's exceptional skill and ingenuity contributed significantly to the advance of aero science in his time, and popularized aviation as a sport. He remained active in the aero industry until his death on August 2, 1936.

KELLI A. MILLER

Admiral Richard Evelyn Byrd
1888-1957
American Naval Officer, Polar Explorer, and Aviator

Richard Evelyn Byrd was an American naval officer, polar explorer, and aviator who is best known for his contributions to polar exploration. He is credited for being the first to fly over the North Pole and received the Congressional Medal of Honor for this feat. However, there is considerable debate whether he actually reached the North Pole on that flight. Later, he participated in the first airplane flight over the South Pole. Throughout his life, Byrd made significant contributions to the areas of science and exploration. He had a tremendous ability for organization and made extensive use of technology on his explorations that served as a model for future endeavors. Byrd invented instruments that had application to his journeys and these greatly advanced the area of aerial navigation. Despite his controversial flight over the North Pole, Richard Byrd is considered to be an American hero.

Richard Evelyn Byrd was born in Virginia on October 25, 1888. He was part of a high-achieving family, as his father was a successful lawyer and his brother became a United States Senator. Byrd was often described as an intense person who wanted to distinguish himself through his actions. As an early example, at the age of 11, Byrd was allowed to travel alone to the Philippines to visit a relative. This sense of adventure seems to have followed him throughout his life. Byrd graduated from the United States Naval Academy in 1912 and learned to fly in 1916. He commanded a United States air station in Canada during the latter parts of World War I.

Byrd's fascination with the polar regions was certainly stimulated when he commanded an aviation detachment in 1924 during an expedition to Western Greenland. After several unsuccessful attempts to fly over the North Pole with the Navy, Byrd raised funds to embark on a private mission in 1926. He was racing another explorer, Roald Amundsen (1872-1928?), for the honor of being the first to fly over the North Pole. An unsuccessful first try caused Byrd to make a furious effort for the North Pole. A triumphant Byrd claimed that on his next try, May 9, 1926, he and his pilot, Floyd Bennett (1890-1928), were the first to fly over the North Pole. Skeptics immediately contended that Byrd's flight time was not long enough to cover the distance to the Pole and back. In addition, the plane had developed an oil leak that would have compromised their chances of attaining their goal. Despite this criticism, Byrd was declared a national hero and awarded the Congressional Medal of Honor.

Byrd traveled next to Antarctica with intentions to fly over the South Pole and claim the

Richard Byrd. *(U.S. Department of the Navy. Reproduced with permission.)*

land for the United States. His first expedition established a base called Little America. On his explorations, he mapped and claimed a significant area of uncharted territory he called Marie Byrd Land after his wife. On November 29, 1929, he was part of the first airplane flight over the South Pole. When he returned to the United States in 1930, he was commissioned as a rear admiral for this feat.

On his return visit, Byrd wanted to do something noteworthy, so he spent five solitary months at a weather station 123 miles (198 km) south of Little America. He endured extreme weather and living conditions that left him ailing with frostbite and carbon monoxide poisoning when he was rescued. He published an account of this experience in *Alone* (1938).

Byrd made a total of five expeditions to the Antarctic with each successive trip more elaborate than before. Among his many achievements was his pioneering use of technological resources for polar expeditions that provided abundant information regarding the southernmost continent. He was also directly responsible for the large-scale Naval maneuvers, Operation High Jump and Operation Deep Freeze, which further extended the knowledge of the Antarctic. Byrd died at his home in Boston on March 11, 1957.

JAMES J. HOFFMANN

Howard Carter
1874-1939
English Archeologist and Egyptologist

Howard Carter discovered and excavated one of the richest finds in the history of archeology—the tomb of King Tutankhamen. The richness of the ancient treasures recovered from King Tut's tomb, and the surrounding publicity, propelled Carter to the forefront of modern archeology. The discovery also elevated King Tut from a little-known Egyptian ruler to one of ancient Egypt's most famous and studied Pharaohs.

Carter was born in the country village of Swaffham, Norfolk, England, where as his father's apprentice he studied the fundamentals of drawing and painting. Carter sought an alternative to the family business of portrait painting. At age 17, and without a formal degree, he ventured to Egypt to work as a tracer for the Egyptian Exploration Fund. Carter was charged with copying historical inscriptions and drawings for preservation and further study. Carter worked long hours at archeological sites, and was known to finish his drawings late at night and fall asleep in the tombs.

Carter's diligence caught the attention of Sir Flinders Petrie (1853-1942), a well-known archeologist and Egyptologist of the time. Carter worked with Petrie at the excavation site at el-Amarna, an ancient capital of Egypt. At el-Amarna, Petrie taught Carter the techniques of excavation. Carter unearthed several important finds at the site, and sketched many of the unusual artifacts. Later excavations with Petrie—including work at Deir el Babri, the burial place of Queen Hatshepsut, where Carter was appointed principal artist—enabled Carter to fine-tune his excavation and drawing techniques. At age 25, Carter was approached by the Egyptian Antiquities Service and asked to serve as Inspector General of Monuments, a post he held until 1905. In this position, Carter supervised archeology along the Nile Valley.

In 1907 Carter befriended the Fifth Lord Carnarvon, an Englishman with an acute interest in Egyptology, who became benefactor for Carter's Nile Valley excavations. In less than 10 years, Carnarvon held one of the most valuable private collections of Egyptian antiquities, most of which were excavated by Carter and his team. Carter was interested in unearthing the tomb of King Tutankhamen, then a little-known ruler, and was convinced his excavations were near

Howard Carter. *(Hulton-Deutsch Collection/Corbis. Reproduced with permission.)*

Tut's burial place. Carter and his team worked for another five years, finding little. In 1922 Carnarvon grew anxious about the lack of return on his investment and told Carter funding for the excavations was at an end. Carter convinced Carnarvon to fund one additional season in which to find the tomb.

In the fall of 1922, Carter resumed digging, concentrating on an area of workman's huts he had previously found. Carter had earlier seen major tombs found near the living quarters of the workmen who built them. On November 4th, one of Carter's men found a stone step in what appeared to be a stairway leading downward into the rock. Carter, sensing an imminent discovery, covered the step and summoned Lord Carnarvon from his home in England to the excavation site. Work resumed when Carnarvon arrived, and on November 24th the stairway was cleared to reveal a door, followed by an inner door bearing the name and seal of Tutankhamen. Two days later, Carnarvon and his daughter watched as Carter peered through a hole in the door to view the royal tomb and its treasures, untouched for centuries.

Carter spent the next 10 years meticulously cataloging and supervising the removal of the treasures of the tomb. Over 3,000 items were recovered, most of which are stored in the Cairo Museum. After the painstaking work of documenting the treasures of King Tutankhamen's tomb, Carter led no more excavations in the Nile Valley. He eventually retired and quietly collected Egyptian antiquities. Carter died in London at age 65.

In the 1970s the English, American, and Egyptian governments cooperated to host a traveling exhibition of the treasures of King Tut's tomb. Wildly popular, the exhibit drew larger-than-capacity crowds in the English and American cities where it toured, forcing some tickets to be awarded in lotteries or after daylong waits in long lines. A popular book available at the exhibit detailed Carter's life in Egypt and his discovery of the tomb. Fashions in clothing and elaborate jewelry were influenced by the artifacts, and a dramatic increase in Egyptian tourism ensued that still continues.

BRENDA WILMOTH LERNER

Bessie Coleman

1893-1926

African American Pilot

The first African-American female pilot, Bessie Coleman's aviation achievements demonstrated that women and blacks were capable of mastering flight. Her accomplishments also symbolized the potential of using technology for various applications. Although Coleman earned her pilot's license before Amelia Earhart (1898-1937), the latter was a better-known aviatrix in part because she was wealthy and white. During her brief career as a pilot, Coleman was a pioneer black barnstormer and stunt flyer, but her premature death in a plane crash resulted in her contributions to aviation remaining obscure.

Born on January 26, 1893, in Atlanta, Texas, Elizabeth Coleman, known as Bessie to her family and fans, was the daughter of sharecroppers George and Susan Coleman. When George Coleman abandoned his family, Bessie Coleman helped her mother by harvesting cotton and washing and ironing laundry. Determined to improve her circumstances, Coleman voraciously read books written by influential African Americans, including Booker T. Washington. She completed high school and enrolled at Langston Industrial College, which she left after one semester because it was too expensive. In 1917 Coleman moved to Chicago, where two of her brothers lived.

Coleman ambitiously sought success, hoping to help minorities overcome racism and poverty by serving as an example of how to surmount such socioeconomic and cultural barriers. When her brother told her about French female pilots serving in World War I, Coleman focused on flying as the means to attain equality and to deliver her message. After reading numerous aviation books, Coleman attempted to enroll at several flying schools in 1919 and 1920 but was refused admission because she was an African American and a woman. Persistent, Bessie sought advice from Robert Abbott, founder of the *Chicago Weekly Defender,* an African-American newspaper. He suggested that she study French to prepare for sessions at a European flight school. Supplementing her income with money from philanthropist Jesse Binga, Coleman traveled to Le Crotoy, France, in November 1920 to practice at the Caudron Aircraft Manufacturing Company's flight school. She also took private lessons from a French war veteran.

Coleman earned a pilot's license in June 1921 from the Federation Aéronautique Internationale. She was the first African-American woman to receive an international pilot's license. She took advanced flying lessons in Germany and practiced acrobatic maneuvers. Confident in her piloting skills, Coleman left Europe. In the United States, she embarked on barnstorming tours. Her debut flight was at Curtiss Field on Long Island. Coleman also lectured to African-American audiences, emphasizing the opportunities that aviation offered and denouncing segregation. Her enthusiasm inspired crowds, who watched her perform risky stunts in the sky. Fearless and committed to publicizing aviation, Coleman first jumped from an airplane when a parachutist refused to perform. She conducted aerial advertising for the Coast Firestone Rubber Company. Coleman also raced airplanes and survived several crashes. Abbot dubbed Coleman "Queen Bess" because of her spectacular showmanship, which made her a celebrity featured in newsreels worldwide.

Coleman's primary goal was to establish a flight school in Los Angeles specifically for blacks. She increased the number of performances on her tours in order to finance her future school. During a test flight at Paxon Field in Jacksonville, Florida, on April 30, 1926, Coleman, who was not wearing a seatbelt or parachute, was killed when her plane plummeted to the earth. (A loose wrench had interfered with the controls.) Fellow pilot William J. Powell established a flying school as Coleman had envisioned for African-American pilots. Bessie Coleman Aero Clubs formed nationwide and Powell published the *Bessie Coleman Aero News*. These aviation clubs organized America's first all-black air show in 1931. Other memorials to Coleman have included a U.S. postage stamp issued in her honor and a branch of the Chicago Public Library named for her.

ELIZABETH D. SCHAFER

Elise Deroche
1889-1919
French Aviator

In 1910, less than a decade after the Wright brothers soared into the skies at Kitty Hawk (1903), the members of the Aéro-Club de France granted a pilot's license, the first awarded to a woman, to Elise Deroche. Deroche, also known as the Baroness Raymonde de la Roche, also carried the distinction of being the first woman to fly solo. She had only a brief time in the spotlight, for she was tragically killed in 1919 when the experimental plane she was riding in as a passenger went down over the River Somme.

A plumber's daughter, Elise Deroche was born in 1889. By 1909 she had dabbled in sculpture, the theater, ballooning, and car racing and was looking for the next great thrill. She found it in the airplane when she asked Charles Voisin (1882-1912) to teach her to fly the plane he had designed and named for himself.

The Voisin plane was a precarious one-seater that looked as if two box kites had been strapped together. Deroche's lessons began in October 1909 at Châlons airfield. Her worried flight instructor, Charles Voisin, had forbidden Deroche to do any independent starting attempts, but after a few "dry tests" of taxiing on the runway and around the aerodrome, she became tired of the waiting and announced energetically that she was ready to start.

The surprised Voisin, as well as an English reporter and several mechanics, watched speechlessly as Deroche brought the 50-horsepower engine up to speed, rolled down the air strip and became airborne to an altitude of 15 feet (4.6 m)—becoming the first woman to solo. It was October 22, 1909. The reporter praised her in his comment: "The airplane was gliding through the air completely level for several hundred meters before it came down gently and taxied back."

On March 8, 1910, Deroche passed her qualifying test and was issued pilot's license number 36 by the Aéro-Club de France. A fearless flyer, she commented that "Most of us spread the hazards of a lifetime over a number of years. Others pack them into minutes or hours. In any case, what is to happen, will happen. It may be that I shall tempt Fate once too often, but it is to the air that I have dedicated myself and I fly always without the slightest fear." No surprise, then, that a horrific crash in July 1910, which left her with internal injuries, head wounds, a broken arm, and two broken legs, did not deter her from continued flights after her recovery.

Within two years, Deroche was fearlessly racing again. In 1913 she won the Coupe Femina, an award established for female pilots, first won in November 1910 by Marie Marvingt (1875-1963). Later that year Deroche survived another near-fatal crash, again recovering and taking to the skies until the government banned private flying at the onset of World War I.

Immediately after the war, Deroche resumed flying, and in June 1919 she set an altitude record for women pilots by climbing to 15,300 feet (4,663 m). In an ironic twist of fate that same year, she agreed to ride as a passenger in a test flight of a new experimental plane. The pilot lost control and the plane went down over the Somme, a river in northern France. Deroche died in the crash, extinguishing her pioneering career as an aviator.

ANN T. MARSDEN

Amelia Mary Earhart
1897-1937
American Aviator

Amelia Mary Earhart was the most famous American woman aviator of our time. While she was known to be an excellent pilot and was the first woman to fly solo across the Atlantic Ocean, she is best-remembered for a flight she never completed. Amelia Earhart mysteriously disappeared during the last leg of her attempted flight around the world in 1937 and was never seen again.

Amelia was born in Atchison, Kansas. She had a chaotic childhood and spent portions of her youth living with her grandparents because her parents had financial problems. She attended six high schools because her father was transferred often in his job, as a claims attorney for the railroad. After high school, she became a nurse's aid in Canada during World War I, since she was deeply moved by the suffering of wounded soldiers returning home from the war. She then enrolled as a pre-medical student at Columbia University in New York, but only stayed one year.

Although she saw her first airplane when she was 10 years old, her interest was not ignited until 1920, when after leaving Columbia University she spent her summer visiting her parents in California. While at an air show, her father arranged for her to get an airplane ride. It made quite an impression on her, as airplanes and flying became the passion of her life. She enrolled for flying lessons, purchased her first plane (*The Canary*), and took any job she could to help pay for this expensive hobby.

Amelia's big break occurred in 1928 when G.P. Putnam's Sons Publishing Company sponsored a transatlantic flight, headed by Commander Richard Byrd (1888-1957). The sponsors were interested in having a woman on board this flight, thus increasing publicity. After an interview, Amelia was selected and became the first woman to cross the Atlantic as a passenger. This made her an immediate celebrity.

From 1928 through 1935, she had much success in her aviation endeavors. Her many accomplishments included setting new altitude and speed records, becoming the first person to fly solo from Hawaii to California, and the first person to fly nonstop from Mexico City to New York. In 1931, she married newspaper publisher George Putnam, who had earlier selected her for transatlantic crossing with Commander Byrd.

The proudest moment of Amelia Earhart's life had to be her solo flight across the Atlantic Ocean in 1932. During this flight, she set several records, besides being the first woman to make that journey. She was also the first person to make the crossing twice, and set records for the longest non-stop flight distance and fastest speed.

In 1937, after years of being in demand on the lecture circuit, Amelia felt that it was time for another extraordinary feat, so she chose to fly around the world at it's longest point, the equator, a trip of 29,000 miles (46,661 km). She and her navigator, Frederick Noonan, successfully completed the first 22,000 miles (35,398 km) of the trip. On July 2, 1937, they departed from Lae, New Guinea to complete the last 7,000 miles (11,263 km) of the journey. Included in their plans was a refueling stop on Howland Island in the North Pacific. Despite favorable flying conditions, Amelia Earhart reported running low on fuel and was not able to locate the island for refueling. Her plane disappeared and no trace of them was found. What really happened to Amelia Earhart that day is a mystery, and there are many theories regarding her disappearance. However, the most probable scenario is that the plane ran out of fuel and plunged into the ocean.

Amelia Earhart. *(The Granger Collection Ltd. Reproduced with permission.)*

JAMES J. HOFFMANN

Sir Arthur John Evans
1851-1941
British Archaeologist

Sir Arthur John Evans uncovered the ruins of the ancient city of Knossos in Crete, and with it a sophisticated Bronze Age civilization he named "Minoan." Evans also found thousands of tablets bearing Minoan scripts called Linear A and Linear B. When the Linear B script was deciphered, the language it recorded was shown to be an early form of Greek. The discoveries in Crete cast light upon a period of Aegean civilization previously known mainly by its dim reflections in the mythology of classical Greece.

Arthur Evans was born on July 3, 1851, in Nash Mills, Hertfordshire, England. He was educated at Harrow School, Brasenose College at Oxford University, and the University of Göttingen. After graduating, Evans traveled to Bosnia with his brother Lewis, and was witness to a peasant uprising against Ottoman rule. Later he worked as a correspondent for the *Manchester Guardian* in the Balkans, and became secretary of the British Fund for Balkan Refugees. He identified many ancient Roman sites in Bosnia and Macedonia, including cities and roads. However, his reports for the *Guardian* on events and conditions in the contentious Balkans led authorities to accuse him of spying. He was arrested, imprisoned, and expelled in 1882.

Four years earlier, Evans had married Margaret Freeman, who was a partner in his work until her death in 1893. Between 1884 and 1908, he was curator of Oxford's Ashmolean Museum. He was a founding member of the British Academy in 1902, and became extraordinary professor of prehistoric archaeology at Oxford in 1909.

Evans became interested in Crete after seeing ancient coins and carved sealing stamps from the island. After making his first trip in 1894, he published *Cretan Pictographs and Prae-Phoenician Script*, and proposed that the pre-classical Mycenaean civilization of the Greek mainland originated in Crete. In 1899 he purchased land on the island, including the site of Knossos, for purposes of excavation.

The island of Crete is strategically located off the tip of mainland Greece. It was the major naval and trading power in the Aegean Sea beginning around 4,000 years ago. Greek legend tells of the Cretan King Minos, whose palace grounds included a labyrinth said to house a beast called a Minotaur. The Minotaur was described as a fearsome creature that was a bull from the waist up and a man from the waist down, and was finally slain by the Greek hero Theseus. This legend may have been an allegorical description of the fall of the Minoan civilization around 1450 B.C., after a volcanic eruption on a nearby island.

Within a year of beginning his excavations at Knossos, Evans uncovered the ruins of a palace. Its grounds, covering more than five acres, were laid out in a complex arrangement that suggested the labyrinth of the legendary Minotaur. This prompted Evans to coin the name "Minoan" for the ancient civilization. Beneath the Minoan layer, he found an even earlier settlement from the Neolithic period. Evans was knighted in 1911. With the exception of a hiatus for the duration of World War I, Evans continued his excavations at Knossos until 1935, when he was 84 years old. He described his work there in *The Palace of Minos* (4 volumes, 1921-1936).

Although Evans had hoped to decipher the three scripts found at Knossos—a pictographic script, and two forms of writing called Linear A and Linear B—he was unsuccessful in this. However, a lecture he gave back in England in 1936 caught the interest of Michael Ventris (1922-1956), who eventually deciphered Linear B. Evans died on July 11, 1941, in Youlbury, a town near Oxford. His research at Knossos was taken over by the British School of Archaeology, and continues to this day.

SHERRI CHASIN CALVO

Percy Harrison Fawcett
1867-1925?
British Explorer

Percy Fawcett went to South America to survey the borders of several countries and to search for the legendary City of Gold and other lost cities. He made numerous trips into South America's interior during a two-decade period. In 1925, he set out on what was to be his last expedition with his son and his son's friend. The party was never heard from again.

Fawcett was born in 1867 in Torquay, England. A young man of 19, he joined the British army for what would become a 20-year career in the military. About halfway into his service, he put his army surveying training to work by accepting a position with duties in South America.

During his first trip to South America in 1906-07 to survey the changing boundary between Bolivia and Brazil, he became intrigued with the stories he heard about the lost civilizations of South America. In one of his reports (to the Royal Geographical Society in 1910), he said: "...I have met half a dozen men who swear to a glimpse of white Indians with red hair. Such communication as there has been in certain parts with the wild Indians asserts the existence of such a race with blue eyes. Plenty of people have heard of them in the interior."

Despite the perils of travel into the forests of South America, including an encounter with a 65-foot-long (20 m) anaconda, Fawcett took to the life of adventure and accepted the job of surveying the Rio Verde in eastern Bolivia. For this trip into the unexplored wilderness, Fawcett could find few men to accompany his team and settled for a group made up of a waiter, a silversmith, a baker, and two Indians. Shortly into the trip, they abandoned their plan of navigating the river due to extreme rapids and proceeded on foot. They also left behind their store of food, assuming they could catch fish for sustenance, but the river water soon became unsuitable and the teeming fish populations dwindled. The team members survived on palm tops and hard nuts for most of the trip, and suffered innumerable bites from insects, including inch-long poisonous black ants, before reaching civilization. Five of the porters died shortly afterward. Despite the arduous journey, the team completed its task and mapped the river.

Fawcett remained undeterred and took additional commissions in South America. In 1910, he struck out to conduct a survey of a region along the border of Peru and Bolivia. Previous expeditions into the area had resulted in the deaths of most team members, including army troops, at the hands of antagonistic native people who would kill intruders by ambush or by poisoning the team's water supplies. Fawcett made his trip with a half dozen men, three of whom were soldiers. Part way into the journey, his team met the enemy warriors, who sent a flurry of 6-foot-long arrows at the men. Instead of returning fire, one of the soldiers played the accordion during the attack. When the arrows stopped, Fawcett addressed the warriors in their native tongue. The attacking men responded by welcoming the Fawcett party, and even helping to set up camp for the guests. As an added bonus, the native group relayed a message up the river requesting that the team be given safe passage.

Fawcett left his explorations during World War I, but returned in 1920 and again in 1921 to explore the western region of Brazil. Both expeditions experienced numerous problems in the field and failed. Still lured by the hope of discovering lost civilizations, he set out on another adventure in 1925, this time with his 21-year-old son Jack and Jack's friend Raleigh Rimell. Their specific goal was to find the city of "Z," which Fawcett believed lay hidden in the Amazon basin. They left Cuiabá, the capital city of the Brazilian state of Mato Grosso, on April 20. A month later, they arrived at a camp near the Xingu River, and Fawcett sent what was to be his last message. He wrote: "Our two guides go back from here. They are more and more nervous as we push further into the Indian country." The team then disappeared.

In 1928 another expedition set out in an attempt to follow Fawcett's trail. It found some items that belonged to the party, but never located any of the men. Rumors abounded of sightings of the Fawcett group, including one in which the men were alive and living with a hostile native group, but most people familiar with the region and the hostile native populations had little doubt that the men were murdered.

LESLIE A. MERTZ

Sven Anders Hedin

1865-1952

Swedish Explorer

Swedish explorer Sven Hedin enjoyed a life of adventure that many would have envied. In the late nineteenth and early twentieth centuries, at a time when to some it seemed that all the frontiers of terrestrial exploration had been crossed, he added greatly to the knowledge of the mysterious region known as Central Asia.

Born in Stockholm on February 19, 1865, Hedin showed an early interest in geography and map making. As a teen, he was commissioned by the Swedish Geographical Society to prepare a map showing the Central Asian journeys of Russian explorer Nikolai Przhevalsky (1839-1888). The map received the praise of Arctic explorer Nils Adolf Erik Nordenskjöld (1832-1901).

Hedin took his first trip east at age 20, when he accepted a job tutoring the son of a

Sven Hedin. *(Library of Congress. Reproduced with permission.)*

Swedish engineer working in the Azerbaijan oil fields. He lived in Baku, on the Caspian Sea, in 1885 and 1886, during which time he learned Farsi, or Persian, as well as Turkish—indispensable languages for the study of the region. He then traveled with a merchants' caravan in Iran (Persia at the time) and Iraq before returning to Sweden, where he wrote the first of many books about his adventures.

In 1889 Hedin attended the University of Berlin, before serving as interpreter for a diplomatic mission to Persia. Afterward, he traveled east as far as Kashgar, the westernmost city of the Chinese Empire. In 1891 he returned to Sweden, where he published another book, and the following year he earned his doctorate in geography at the University of Halle in Germany.

Plagued with a problem that caused him to lose vision in one eye, Hedin nonetheless embarked on his first scientific expedition, funded in part by Sweden's King Oscar II, in 1893. He traveled through Tashkent and western China before winding up at the forbidding Taklamakan Desert—once visited by Marco Polo—in early 1895. One of his guides died in crossing the Taklamakan, and, indeed, Hedin's entire party might have died as well had they not finally found a well at the desert's edge.

After recovering in Kashgar, he crossed another major desert, the Tarim Basin, and near the city of Khotan found the remains of ancient cities that showed the influence of faraway Persia and India. This discovery sparked a great deal of interest among archeologists and historians. He also studied the desert lake of Lop Nor, seeking to solve the mystery of why it constantly shifts in size and location. Heading on to the capital at Beijing, he crossed the Gobi Desert's eastern end and wound up at the Trans-Siberian Railroad, which he took west to St. Petersburg and an audience with the czar.

On his next journey, Hedin had the backing of the Russian monarch, who provided him with a Cossack escort. He explored the Tarim River and Lop Nor and entered Tibet, where he hoped to travel under disguise to the capital in Lhasa—off limits to Westerners. Tibetan officials discovered him and he had to turn back. He made his way to Calcutta, where he met with Lord Curzon, the viceroy, then traveled north to Russia. Upon returning to Sweden, he became the last person in that country's history to receive a title of nobility.

In 1905 Hedin set out once more, and by 1906 he was back in Tibet. He mapped much previously unexplored territory; visited the country's second-largest city, Shigatse; and discovered the source of the Brahmaputra River. Returning to India in 1908, he moved northward to Russia, heading east on the Trans-Siberian Railway. In Japan, he met with the emperor and then returned to Sweden, where the royal family and prime minister greeted his boat in the harbor.

During the 1920s, Hedin directed explorations of Asia at the behest of German aircraft-magnate Hugo Junkers, who wanted to set up weather stations in the region. Working with the Chinese government in what was dubbed the Sino-Swedish Scientific Expedition, he directed groups of scientists who explored Inner Mongolia and other vaguely defined regions. The expedition produced 54 volumes of findings, the last being published in 1982.

Hedin supported the Nazis and was the only foreigner to deliver one of the opening speeches at the 1936 Berlin Olympics. He used his influence, however, to save the lives of 13 Norwegian resistance fighters and that of a Jewish friend in Germany. A prolific writer during the 1930s and 1940s, Hedin remained active throughout his life. In later years, an operation restored the sight that he had lost in one eye six decades earlier. He died on November 26, 1952.

JUDSON KNIGHT

Matthew Alexander Henson

1866-1955

American Explorer

Matthew A. Henson went from clothing salesman to Arctic explorer following a chance meeting with American adventurer and naval engineer Robert E. Peary (1856-1920). For three decades, Henson accompanied Peary on many expeditions to the Arctic, and was a member of the team that in 1909 became the first to reach the North Pole.

Henson was born on August 8, 1866, in Charles County, Maryland. An African-American, he was the son of free parents, both of whom died when he was a boy. After attending a segregated school in Washington, D.C., he spent most of his teen years taking jobs on ships traveling to such places as Africa, the Orient, Asia, and Europe. Eventually, he traded life on the water for one on land, and took employment in a Washington, D.C., clothing store. It was there that he met Peary, who stopped in the shop to buy a pith helmet for his trip to Central America. Peary offered Henson a job as his personal valet, and Henson accepted.

Henson made his first trip with Peary in 1887, when Peary was the engineer-in-chief of a canal project in Nicaragua. After they returned to the United States and before they headed out on their next adventure, Henson passed the time working as a messenger. During this break in their quest for adventure, Peary continued to gather financial backing for future expeditions and requested an 18-month leave from the Navy. On June 6, 1891, Henson, Peary, Peary's wife, and four others set out to explore Greenland. The success of this trip, in which Peary proved that Iceland was indeed an island, brought a long line of financial backers for Peary's future Arctic explorations.

Henson developed a strong attachment to the local people. During the 1891-92 trip, Henson gained respect and admiration from the local Inuit men and women by spending the long Arctic winter learning their language and taking an interest in their culture. They gave him the nickname Maripaluk, which means "kind Matthew." When Peary and Henson returned in 1893 to attempt a crossing of the northern Greenland ice cap, Henson adopted an orphaned Inuit boy.

On this and later trips, Peary developed a keen desire to lead the first team to the North Pole. Early attempts by him and Henson were met with blinding snowstorms, open expanses of water to traverse, and many other obstacles. Nonetheless, they continued their quest. In 1908, with the threat that financial backing would soon dry up, Peary and Henson made one last attempt to reach the North Pole. They headed out on July 6, 1908, making the initial part of the trip on board the ship *Theodore Roosevelt*. At the same time, another team was preparing to make its run to the North Pole. That team was led by Dr. Frederick Albert Cook, a man who had accompanied Henson and Peary on the 1891 expedition. The 24-man Peary-Henson team embarked on the land portion of its trip in February 1909. On March 31, Peary ordered everyone to return except himself, Henson, and four Inuits, who marched on to the North Pole. According to Peary's calculations, the six explorers reached the target on April 6, 1909. Within days of Peary's public announcement of their success, Cook reported that he had reached the pole first. Peary received overwhelming support for his claim, however, and he has been roundly credited with the honor of leading the first expedition to the pole.

Matthew A. Henson. *(Library of Congress. Reproduced with permission.)*

The voyage to and from the North Pole was the last for Henson or Peary. Henson went on to become a messenger for U.S. Customs until his retirement in 1936. Racial prejudice prevented Henson from receiving much initial recognition for his Arctic explorations, but he did receive a number of honors later in his life. In 1955 Henson died at the age of 88 in New York. His remains have since been moved from their original resting place in a private cemetery to the plot next to Peary's in the Arlington National Cemetery.

LESLIE A. MERTZ

Thor Heyerdahl
1914-
Norwegian Explorer and Anthropologist

Thor Heyerdahl, nature lover and trained zoologist, made his greatest contribution to the field of anthropology, where he advanced highly debated theories on cultural diffusion—how ancient man migrated to and populated distant places. Heyerdahl often adopted facets of the lifestyle of the ancient people he was studying. He used only materials and techniques available to the ancients to construct sailing vessels, which Heyerdahl sailed on famous expeditions to help prove the possibility of transoceanic contact between ancient cultures and civilizations.

Heyerdahl was born in 1914 in Larvik, Norway. While spending time as a child in the local museum headed by his mother, Heyerdahl was intrigued by and studied natural plant and animal life. Gaining inspiration from his mother, Heyerdahl later studied zoology and geography at the University of Oslo.

In 1937 the newly married Heyerdahl ventured to Polynesia. There, after the Polynesian chief of Tahiti adopted Heyerdahl and his wife, he conducted research on the origins of animal life on an isolated island in the Marquesa Group. Heyerdahl described his life on the island as traditional Polynesian, and it was there that he first wondered how the inhabitants first came to the remote South Pacific. Current scientific theory held that the islands were populated by voyagers from Southeast Asia. Heyerdahl disputed this, after experiencing firsthand the strong easterly winds and currents while fishing. He reasoned that, instead of paddling against the currents, the first human settlers used the currents and winds to arrive from a westerly direction.

Heyerdahl first published his theory in 1941, claiming that the original Polynesian inhabitants came from the coasts of North and South America, and followed the North Pacific conveyor to reach Polynesia in two groups. The first, Heyerdahl proposed, came from Peru on balsa rafts. The second group reached Hawaii in double-canoes from British Columbia. Heyerdahl's research was met with skepticism from the scientific community. With the outbreak of World War II, Heyerdahl's research was interrupted. He returned home to Norway in 1941 to volunteer for the Norwegian Free Forces, serving in a paratrooper unit.

In 1947, with scholars still doubtful of Heyerdahl's contentions of the origin of Polynesian settlement, Heyerdahl decided to demonstrate the practicability of his hypothesis. Heyerdahl and a five-man crew built a balsa raft, named the *Kon-Tiki,* that was an exact replica of the rafts made by ancient Peruvian Indians. They then set sail from Callao, Peru, toward Polynesia in the replica craft. After a world famous three-month voyage of 4,300 miles (6,920 km), the *Kon-Tiki* arrived in the Tuamotu Islands of Polynesia. The seaworthiness of the ancient style of raft was proven, and the voyage gave credence to Heyerdahl's idea that ancient Peruvians could have reached Polynesia by this method.

Thor Heyerdahl. *(Corbis Corporation. Reproduced with permission.)*

Heyerdahl then conducted several archeological expeditions searching for remnants of South American culture in the Pacific. In 1952, in the Galapagos Islands, Heyerdahl's team found ancient South American ceramic pieces, as well as an ancient center-board used by South American sailors to navigate on their voyages. A famous expedition to Easter Island in 1955 found evidence of ancient water reeds and other South American plants once growing on the island, along with carvings in stone similar to those of ancient Peru. As Easter Island is one of the most isolated islands in the Pacific, South American remnants found there added further credibility to Heyerdahl's ideas of ancient migration. The Easter Island finds continue to be rigorously argued among scientists.

In 1969 and 1970 Heyerdahl returned to the sea aboard his *Ra* vessels, named after the Egyptian sun god. His purpose, much like that of the *Kon-Tiki* voyage, was to demonstrate the feasibility of cultural contact between early peoples. The *Ra* boats were made of reeds and constructed in Egypt by local boat builders. Hoping that the *Ra* vessels would show that communication between the ancient people of Africa and those of Central and South America was possible, he departed from Safi, Morocco, aboard the *Ra* and sailed 3,000 miles (4,828 km) before foundering due to design defects and unsuccessful cargo loading strategies. The *Ra II* was built by the Aymaro Indians and successfully made the voyage from Safi to Barbados in 57 days.

Both the *Ra* and *Ra II* flew the flag of the United Nations, and the *Ra*'s crew consisted of seven men from seven different nations. Heyerdahl's writings contributed to the evolution and popularity of the study of ethnology. Heyerdahl's books include *Kon-Tiki* (1950), *American Indians in the Pacific* (1952), *Sea Routes to Polynesia* (1968), and *The Ra Expeditions* (1971). At the end of the twentieth century Heyerdahl was studying the ancient people of Tenerife Island, and working there to create a museum.

BRENDA WILMOTH LERNER

Thomas Edward Lawrence
1888-1935
British Military Strategist

T. E. Lawrence was a British liaison officer in the Middle East during World War I, and actively promoted Arab independence. His military exploits, recounted in his book *The Seven Pillars of Wisdom* (1926), made him a legendary figure in the popular imagination, and he became famous as "Lawrence of Arabia."

Thomas Edward Lawrence was born on August 15, 1888, in Tremadoc, Caernarvonshire, Wales. He was one of five illegitimate sons of an Anglo-Irish peer, Sir Thomas Chapman, who had run off with the family governess. The couple adopted the name "Lawrence" and lived as husband and wife.

"T.E.," who preferred his initials to his name, attended Jesus College at Oxford, pursuing an interest in medieval military architecture. His 1910 honors thesis was on Crusader castles, which he studied in France, Syria, and Palestine. For the next three years, he worked on an excavation of a Hittite settlement in Iraq under the British archaeologists D. G. Hogarth and Sir Leonard Woolley (1880-1960). In 1914 he and Woolley, along with Captain S. F. Newcombe, went on a combination scientific expedition and strategic map-making reconnaissance. They traveled the Sinai between Gaza and Aqaba, and produced a valuable study entitled *The Wilderness of Zin* (1915).

When World War I started, the Ottoman Empire, which controlled what is now Syria, Lebanon, Israel, Jordan, and most of Saudi Arabia, was allied to Germany. Experts on the Arab world were sought after by the British Army for its own

T. E. Lawrence. *(Library of Congress. Reproduced with permission.)*

Middle Eastern operations centered in Egypt. By December 1914, Lawrence was a lieutenant in Cairo assigned to intelligence. Traveling to Mecca, he met with Arabian sheiks planning a rebellion against their Ottoman overlords. He urged his superiors back in Cairo to encourage the revolt as a way to undermine the Ottoman Turks.

As the British liaison officer, Lawrence soon became a key leader in the Arab guerrilla forces, known as "Emir Dynamite," for his success at bringing down bridges and railway lines. He tried to mold the individual interests of the warring sheiks into a vision of Arab nationhood. In 1917, the Arab forces seized Aqaba, at the northern tip of the Red Sea. Lawrence then attempted to coordinate their movements with General Allenby's march toward Jerusalem. Continuing north, he arrived in Damascus in October 1918.

By this time, Lawrence was physically and mentally exhausted. He had been captured by the Turks and tortured before being able to escape. He had been wounded many times, and both witnessed and participated in all the horrors of war. He was disillusioned by the actions of the British and French, who divided the conquered territories between themselves rather than offering the independence his forces had struggled for. Back in England, he stunned King George V by politely refusing two prestigious military awards at a royal audience.

While a popular series of illustrated lectures by the American journalist Lowell Thomas were making his exploits famous, Lawrence was fruitlessly lobbying for Arab independence at the Paris Peace Conference of 1919, appearing in his desert robes and headdress. In 1921 he returned to the Middle East as an advisor to Winston Churchill, who at that time was colonial minister there. During this time he was instrumental in placing Emir Feisal on the throne of Iraq, and establishing the kingdom of Trans-Jordan (later Jordan).

Still seeking adventure, a refuge from his fame, and material for another book, Lawrence enlisted in the Royal Air Force (RAF) under an assumed name, but was soon found out and released. With the help of a friend in the War Office, he then enlisted as a private in the Royal Tank Corps, under the name T.E. Shaw. This surname may have been chosen as a result of his postwar friendship with George Bernard Shaw; he adopted it legally in 1927. The playwright Shaw intervened with the Prime Minister on behalf of his new namesake, allowing him to transfer back into the RAF in 1925. He was posted to various airbases from Plymouth to Karachi, and wrote *The Mint*, a graphic account of RAF training that horrified his superiors, although it was circulated only privately until 20 years after Lawrence's death. While in the RAF he made significant contributions to the design and testing of high-speed, seaplane-tender watercraft, after watching a crash near shore in which lives were needlessly lost because of slow rescue boats.

Lawrence was discharged from the RAF in February 1935, and returned to his home in Clouds Hill, Dorset, England. He died there on May 19, 1935, at the age of 46, after a motorcycle accident.

SHERRI CHASIN CALVO

Charles Augustus Lindbergh
1902-1974
American Aviator

Charles Lindbergh flew the first non-stop solo flight in a powered flying machine over the Atlantic Ocean in 1927. He covered 3,600 miles (5,794 km) from New York to Paris in 33 hours and 30 minutes, an astonishing achievement for the time, and showed the possibilities of the airplane.

Charles Lindbergh grew up on a farm in Minnesota and was at the University of Wisconsin majoring in engineering when he became

Charles Lindbergh. *(The Library of Congress. Reproduced with permission.)*

fascinated by flying machines. After two years, he quit to learn to fly. He became a barnstormer and traveled to fairs and air shows giving demonstrations of flying. Wanting more instruction, he attended Army Flying School in Texas and became a Second Lieutenant in the Army Air Service Reserve. Then he took a job with Robertson Aircraft in St. Louis. In 1926 he flew the inaugural flight of a new airmail route between St. Louis and Chicago.

In 1919, a $25,000 prize had been offered for the first man to fly alone nonstop from New York to Paris. Still unclaimed in 1927, Lindbergh thought he could win it. He persuaded several St. Louis businessmen to put up money for a plane to be called the *Spirit of St. Louis*. When the plane was ready, Lindbergh flew it from San Diego to New York in a record 20 hours and 21 minutes.

On May 20, 1927, he took off from New York in bad weather on his historic flight, his plane loaded with extra fuel. Thirty-three hours and 30 minutes later he landed at Le Bourget Airport in Paris in the dark. He was astounded at the crowds of people who greeted him. He was given awards, celebrations and parades and became one of the world's most recognizable people. He was awarded the Distinguished Flying Cross and the Medal of Honor by U.S. President Calvin Coolidge.

Lindbergh toured the United States and the world promoting air travel and became technical advisor to several airlines. In Mexico City he met Anne Morrow, daughter of the U.S. Ambassador to Mexico, Dwight Morrow. They were married in 1929.

Lindbergh was a natural celebrity—young, tall, and good-looking with a beautiful wife and two sons—but in 1932 his life changed. His son Charles was kidnapped and found dead ten days later. Bruno Hauptmann was executed for the crime in 1934, though some still doubt he was guilty. The press sensationalized the tragedy and would not leave the bereaved parents alone. Besieged by photographers, reporters, and intruders, they fled to Europe in 1935 searching for privacy and safety.

In Germany and France, while his wife was earning stature as a writer, Lindbergh toured aircraft plants and was impressed with Adolf Hitler's industry. Lindbergh warned Europe about German air power but accepted a medal of honor from Hermann Göring, head of the German Air Force. This caused outrage in England and America. Lindbergh did himself further damage by stating that America should keep out of World War II and charging British forces and U.S. President Franklin Roosevelt with pushing America into it. He was reviled and castigated, and he resigned from the Army. When war did begin in 1941, Lindbergh tried to reenlist. Rejected, he found ways to advise American air services and increase the proficiency planes and pilots.

After the war, President Eisenhower appointed Lindbergh a Brigadier General in the new Air Force. He published a Pulitzer Prize-winning book about his historic flight and then withdrew from public life. He settled in the town of Hana on the island of Maui, Hawaii, a place that is still difficult to reach by car boat or plane. He lived there in seclusion and wrote letters and articles on the preservation of whales and opposing the development of supersonic airplanes. He died of cancer in 1974 and is buried in Hana, Maui.

LYNDALL B. LANDAUER

THE FIRST TRANSATLANTIC FLIGHTS

The early days of aviation were a time of prizes and the accomplishments necessary to claim them. Among the prizes were $25,000 for the first non-stop flight from the New World to Paris, £10,000 for the first non-stop crossing from the New World to any part of Britain, $10,000 for the first Australian to fly from London to Australia in under a month, and so forth. Among the accomplishments were those who won these prizes, the first flight across the North Pole in 1926, (although this accomplishment has recently been questioned), and others too numerous to mention.

The first flight across the Atlantic Ocean took place in May, 1919, when a U.S. Navy Curtiss Flying Boat flew from Newfoundland to England via the Azores and Portugal. A month later, John Alcock and Arthur Brown flew a Vickers bomber nonstop from Newfoundland to Ireland, where they crashed into a bog, becoming the first to cross the ocean non-stop. This flight was noteworthy because of the horrendous weather that plagued them during the entire trip and because of the incredibly primitive airplane they flew. Flying a fragile biplane with virtually no navigational equipment, Alcock and Brown flew for over 16 hours through severe storms before crash landing. By accomplishing this feat, they claimed a £10,000 prize offered by the *London Daily Mail* in 1913. In fact, although Charles Lindbergh was the first person to successfully cross the Atlantic non-stop and alone, over 30 aviators had flown this route before him in the preceding 8 years.

P. ANDREW KARAM

George Leigh Mallory
1886-1924
British Mountaineer

George Leigh Mallory was a British mountaineer who led three pioneering expeditions to Mount Everest in the Himalayas on the border between Nepal and Tibet in the 1920s. On the third expedition to Everest in 1924, Mallory and climbing partner Andrew Irvine (d. 1924) made an attempt at the summit but disappeared in stormy weather, never to return. Whether they reached the summit before they died is not known. A 1999 expedition found Mallory's frozen body at 27,000 feet (8,229 m) on Everest's north face, but the mystery of whether the summit was reached still remains unsolved.

Mallory, the son of an English clergyman, grew up in a conventional household. As a child he had an adventuresome spirit and enjoyed exploring and climbing rocks. He entered Magdalene University in 1905 to study history and train as a schoolmaster. His career as a teacher was interrupted by World War I (1914-18), during which he served at the French front as a gunner. After the war he built a formidable reputation as a mountaineer, bold rock climber, and a competent ice climber. He was also recognized for his love of adventure and his ability to lead and inspire others.

By the early 1920s the farthest corners of the earth had been explored: the North and South poles had been reached, and the sources of the world's major rivers had been mapped. The "Third Pole," the summit of Mount Everest, the highest mountain on Earth, was still unclaimed and considered to be a great prize for international explorers. Mallory and a team of British mountaineers decided to take on the "Third Pole" for Great Britain.

Though visible as a small dot on the horizon from Darjeeling in India, Everest had remained remote and unexplored because of its location on the border between Tibet and Nepal, both countries that were not welcoming to foreign travelers. Negotiating the political protocols eventually paid off, and the expedition received permission to enter Tibet. They set off on a 6-week expedition in 1921, exploring, surveying, and carrying out a "photographic offensive" on the geographical and cultural landscape of Tibet, a country unknown to most of the Western world.

With his expedition team, Mallory searched out a possible approach to the East face of Everest. Exploring great distances, climbing peaks and glaciers, and wading swollen rivers in remote valleys, they gave up the East Route because of the ever-present danger of avalanches and the geographic reality of this approach—it was in forbidden Nepal. Mallory then decided that the most probable approach lay with the North face—called the "North Col." Mallory led a small team up that side, and although conditions were unfavorable for an attempt on Mount Everest that year, they were convinced that a clear route existed all the way to the summit. The small, poorly equipped little band dressed in tweeds and clothing appropriate for less-formidable conditions vowed to return the following year.

In the following year, 1922, the climbing party was better equipped and prepared, and they approached the North Col via the East Rongbuk valley, where they reached a height of 27,000 feet (8,229 m), a record at the time, but still 2,000 feet (609 m) below the summit of Everest. Mallory decided to attempt the summit a second time, but a fresh snow triggered massive avalanches that swept away nine members of the expedition, killing seven of them, all sherpas—native Nepalese. The loss of "these brave men" was a blow to Mallory, and he abandoned the second attempt.

In 1924 the third expedition returned to the North Col of Everest. On the morning of June 6, Andrew "Sandy" Irvine and Mallory attempted the summit with the use of supplemental oxygen, considered a risky and unpredictable gas at the time. Mallory and Irvine were last seen on June 8 by expedition geologist Noel Ode, who was following behind as support and backup. Last seen at the dangerous Second Step, nearing the base of the summit pyramid, Mallory and Irvine seemed to have a terrific chance to make the summit. Shortly afterwards, however, the visibility vanished, and the climbing team was not seen again. It was unknown if they had indeed reached the summit.

Sir Edmund Hillary (1919-) became the first man to officially reach the summit in 1953. In 1960 a Chinese expedition completed Mallory's route, and in 1975 a Chinese climber spotted what was speculated to be Andrew Irvine's remains. In 1999 a BBC/Nova film crew found Mallory's remains. However, the camera that accompanied the ascent was not found, and the mystery of whether or not Mallory and Irvine made a successful first ascent of Everest remains unanswered.

LESLIE HUTCHINSON

Beryl Markham
1902-1986
English Aviator

Beryl Markham, an aviation pioneer, was the first person to make a solo, non-stop, transatlantic flight from England to the American continent. Markham's memoirs of the flight were published in her book, *West With the Night,* which also described her childhood and life in east Africa during the early twentieth century.

Markham was born Beryl Clutterbuck in 1902 in Leicester, England. Barely four years old, Markham moved to the African highlands in Kenya in 1906 (then under British rule), where her father established a farm. Markham received only basic formal education in Kenya, but delighted in learning to speak Swahili and several African dialects that she learned from the children of families her father employed. As a child Markham played with the local Murani children, and joined the young boys armed with spears hunting animals in the African bush.

When her father turned his interests to horse breeding and training, Markham became his apprentice. She soon discovered a love of horses, and a talent for handling them. At age 18 Markham was the first woman in Africa to earn a racehorse trainer's license. Europeans in colonial Kenya enjoyed gathering at the races in Nairobi, and Markham earned local celebrity when her horses won prestigious prizes. After a brief first marriage, Markham remarried in 1927 to a wealthy young Englishman, Mansfield Markham, whose name she would keep throughout her life. The couple returned to England, where their son was born in 1929. Shortly afterwards, however, the two separated and Beryl Markham returned to Kenya alone.

In 1931 Markham began flying lessons after a mesmerizing ride in the airplane of friend Dennis Finch Hatton, an English adventurer and big-game hunter living in Africa. Markham earned her pilot's license, then embarked upon a flying career after becoming the first woman in Africa to earn a commercial pilot's license. Markham flew airmail to remote regions of the Sudan, Tanganyika (now Tanzania), Kenya, and Rhodesia (now Zimbabwe). She flew missions to remote areas of the African bush to deliver supplies or rescue injured hunters or miners. Markham frequently scouted elephants and other big game from the air for wealthy hunting parties.

Markham made her mark in aviation history in 1936, when she set off to capture a prize offered for the first non-stop flight from London to New York City. Due to prevailing headwinds encountered by westbound travelers, this route was considered more difficult than that taken by American aviator Charles Lindbergh (1902-1974) during his previous Atlantic crossing. Markham believed the crossing was not only possible, but essential in order to encourage commercial transatlantic flight. Markham took off from London on September 4, 1936, in an airplane modified to carry extra fuel, but with meager instrumentation—notably, without a radio. Though successfully crossing the Atlantic, she did not make it to New York, as she was forced to crash-land her airplane in a peat bog in Nova Scotia. Markham survived the crash and achieved instant celebrity. Commercial air service between England and the United States began three years after Markham's famous flight.

Markham moved to the United States in 1939, settling in California where she began to write about her experiences in aviation and Africa. *West With the Night* was published in

Beryl Markham. *(The Granger Collection Ltd. Reproduced with permission.)*

1942. Hailed as a literary and commercial success, the book received praise from American writer Ernest Hemingway, who also spent considerable time in the African bush. Speculation that the book was written, at least in part, by Markham's third husband, American journalist Raoul Schumacher, was never confirmed.

In 1952 Markham returned to Kenya to resume raising and training horses. Markham's horses won the Kenya Derby six times during the next 20 years. Markham spent the remainder of her life in Africa, and died in Kenya at age 84.

During Markham's last years, popular interest in the history of the late colonial period in Africa and the lives and relationships between famous Europeans and Americans who lived there was rekindled. *West With the Night* was republished in 1983, and became an international bestseller.

BRENDA WILMOTH LERNER

Robert Edwin Peary

1856-1920

American Explorer and Naval Officer

Robert Edwin Peary spent a good portion of his life in the Arctic regions of the world. He made several trips through the mostly uninhabited and unknown expanses of the Arctic, including a number of attempts to reach the North Pole. The 52-year-old Peary claimed that he and five others had finally reached the elusive pole on April 6, 1909. His assertion has been disputed over the years, but Peary is generally recognized as the man who led the first expedition to the North Pole.

Peary was born on May 5, 1856, in Cresson, Pennsylvania. He lost his father when he was three years old, and his mother took her only child back to Cape Elizabeth, Maine, where the family had lived before moving to Pennsylvania. Following his education at Portland High School, Peary attended Bowdoin College and gained an education in civil engineering. After his graduation, he spent four years as a county surveyor and cartographic draftsman before joining the U.S. Navy in 1881. Nearly four years into his service, he went to Nicaragua to survey a canal route through that Central American country.

Historians believe he became interested in the Arctic either by reading about adventures there in his youth or upon his return from Nicaragua. Either way, Peary was clearly fascinated with the far northern reaches of the Earth. In the summer of 1886, he took six-months' leave from the Navy and obtained passage on a steam whaler that brought him to Godhavn in western Greenland. In the next few months, he hoped to learn about extreme-weather survival techniques along with the varied perils of the far north. He began his first Arctic expedition with a young man, Danish Lieutenant Christian Maigaard, who was stationed at Godhavn. Peary and Maigaard spent more than three weeks hiking across the wilderness in miserably cold and windy conditions before they had to turn back just 125 miles (201 km) into their journey for lack of adequate supplies.

Back in the Navy, Peary made another trip to Nicaragua in 1887 to help with the canal survey. Before leaving for Nicaragua, however, he met a salesman at a clothing store and asked him to become his personal valet. The salesman, Matthew A. Henson (1866-1955), accompanied Peary to Nicaragua and on his many future trips to the Arctic. When Peary completed his duties as engineer-in-chief in Nicaragua, he returned to the United States and married Josephine Diebetsch.

All the while, his fascination with the Arctic never wavered. His desire to return to the far north became more intense when he learned of the successful crossing of Greenland's ice cap by Norwegian Fridtjof Nansen (1861-1930). He sought out and eventually received sufficient fi-

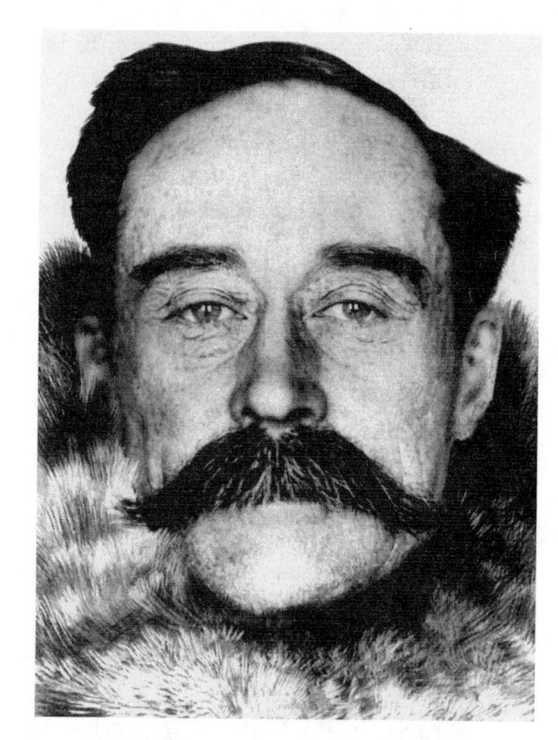

Robert Edwin Peary. *(AP/World Wide Photos, Inc. Reproduced with permission.)*

nancial backing from various scientific and geographical societies to pursue his dream. With 18-months' leave from the Navy, Peary set out for Greenland on June 6, 1891, with his wife, Henson, Dr. Frederick Albert Cook (1865-1940) (a man who would later race Peary to the Pole), and three others. The group landed in Greenland, set up a semi-permanent camp and waited out the long Arctic winter. In May, Peary went on, eventually making a month-long journey to the northeastern Atlantic shore and later to the Arctic Ocean shoreline. Peary's 1,300-mi (2,092-km) round-trip proved that Greenland was indeed an island. This successful trip also gave Peary a long line of financial backers who would assist with his next expeditions.

From 1893-1906, Peary made several attempts to reach the North Pole. During this time, he changed his tactics: instead of making the entire trip over land, he traveled by boat as far north as possible. Both water and land travel posed difficulties. Moving ice packs often threatened the boat's progress, while blizzards could cause lengthy delays on land and freezing temperatures could wreak havoc on the team members. During one of these attempts, Peary's feet froze and he lost seven or eight of his toes.

Peary made his final attempt to reach the pole during an expedition that began on July 6, 1908. This time the ship *Theodore Roosevelt* carried the explorers deep into the far north. He soon learned that Dr. Cook had embarked on a similar attempt four months earlier. The race was on. Peary's land team, consisting of two dozen men and 133 dogs, struck out in February. On March 31, Peary sent part of his team back to the *Roosevelt*, and proceeded with Henson and four Inuits. On April 6, 1909, Peary took a photograph of the other five members of the team at the point he deemed the North Pole. Within days of Peary's public announcement of his success, Cook reported that he had reached the Pole first. Peary received overwhelming support of his claim, however, and he has been roundly credited with the honor.

Barely a decade after his historic trip to the Pole—and his last—Peary died on February 20, 1920, in Washington, D.C.

LESLIE A. MERTZ

Auguste Antoine Piccard
1884-1962
Swiss Physicist, Inventor, and Explorer

Auguste Antoine Piccard was a Swiss physicist, inventor, and explorer who is famous for being the first, with a partner, to reach the stratosphere in a balloon of his own design. He also invented the bathyscaphe, a submersible capsule, making it possible to reach the lowest point in the ocean.

Auguste and his twin brother, Jean, were born in Basel, Switzerland, on January 28, 1884. Their father was a professor of chemistry at the University of Basel. The brothers followed in their father's scholarly footsteps and attended the Swiss Federal Institute of Technology; Auguste completed a degree in physics. Following graduation he remained at the Institute as a professor. In 1919, he married the daughter of a French historian.

Three years later Piccard moved to the University of Brussels, where he accepted a newly created position in the physics department. He had studied cosmic rays and was attempting to design an experiment in which he would ascend to more than 16,000 meters (52,493 ft) in order to gather more data on the subject. It was clear to him that such high altitudes would require specialized equipment so that the trip would be a success. In 1930 Piccard designed and built a round aluminum capsule, or gondola, which had the capacity to be pressurized to the air pressure at approximately sea level. The gondola was also equipped to reuse the air supply inside, and the balloon itself was large enough so that it was not necessary to have it completely filled before lifting off the ground. On May 27, 1931, Piccard and a partner became the first humans to reach the stratosphere by ascending to 51,961 feet (15,781 m) after lifting off from Augsburg, Germany; the air pressure at this altitude is approximately one-tenth that of the air pressure at sea level.

Piccard ascended to a record-breaking height of 55,563 ft (16,940 m) on August 18, 1932, this time in a balloon furnished with a radio. Following this event Piccard shifted his time and energies to conquering the depths of the sea.

Using the principles he had developed with the pressurized gondola for the stratospheric balloon trip, Piccard designed a pressurized capsule able to dive to the deepest depths of the ocean, an invention which he called the bathyscaphe. The term bathyscaphe comes from the Greek terms *bathos*, meaning "deep," and *scaphos*, meaning "ship." The gondola in this case was suspended beneath a tank filled with gasoline, which is "lighter" than water, thereby forming a virtual underwater balloon. Weights, or ballast, would cause the bathyscaphe to dive; the ballast would be released at the end of the dive so the bathyscaphe could return to the surface.

The bathyscaphe was not completed until 1948. In October of that year it was tested in an unmanned dive off the Cape Verde Islands; the dive was not a success. In 1953, Piccard and his son, Jacques, in a new and improved bathyscaphe named *Trieste*, dove to the depth of 10,330 feet (3,149 m) in the Adriatic Sea. In 1956, Auguste Piccard achieved a dive of 12,500 feet (3,810 m) in the same region. Following this success, the Piccards set their objective as the Mariana Trench. This is where the deepest known spot in the world, Challenger Deep, is located. The Mariana Trench is a 1,835-mile-long (2,953-km-long) depression along the Pacific Ocean floor; the Challenger Deep is about 200 miles (322 km) south of Guam and is approximately 36,200 feet (11,034 m) down.

On January 23, 1960, Piccard's dream was realized. His son, Jacques, with a United States Navy lieutenant, descended in the bathyscaphe *Trieste* to 35,800 feet (10,912 m) in the Mariana Trench off Guam. Auguste Piccard died in Lausanne, Switzerland, on March 24, 1962.

MICHAEL T. YANCEY

Harriet Quimby
1875-1912
American Pilot and Journalist

The first female pilot to be licensed in the United States, Harriet Quimby broke aviation gender barriers. She was also the first woman to fly solo across the English Channel. Quimby's flair for publicity resulted in more people becoming aware of aviation and its potential to improve transportation. Her beauty and sense of mystery added to the allure of her pioneering aviation efforts. Quimby's accomplishments enhanced the accessibility of technology to more people and established a foundation for future aviatrixes.

Born on May 11, 1875, near Coldwater, Michigan, Quimby was the daughter of William and Ursula Quimby. The absence of legal records, however, has resulted in a dispute over the facts concerning the date and place of her birth. The 1880 Michigan census included five-year-old Harriet in the Quimby household, living on a farm near Arcadia. The family moved to San Francisco, California, in 1887. Quimby described herself as an actress in the 1900 census, but she mostly wrote articles for San Francisco periodicals. She moved to New York City in 1903 and worked as a photojournalist and the drama critic for *Leslie's Illustrated Weekly.*

In October 1910 Quimby covered the Belmont Park International Aviation Tournament and was impressed by aviator John Moisant's flying talent. She enrolled at the Moisant School of Aviation on Long Island. Despite Moisant's death while flying, Quimby continued taking lessons in monoplanes constructed at the school. The press covered her progress. Aero Club of America officials observed Quimby perform basic aviation maneuvers and she received her pilot's license on August 1, 1911. Although other American women, including Moisant's sister Matilde, were flying at this time, Quimby was the first to earn a license.

One month later, she flew at the Richmond County Fair in the first night flight by an aviatrix. Stylish and attractive, Quimby fascinated journalists and aviation enthusiasts. Her flamboyant flying apparel, made from purple satin, contrasted with the more practical clothing most aviators wore. The dramatic Quimby created her own costume, including a cloth extension that served as a hood. Quimby penned accounts of her flying experiences for *Leslie's* and other magazines, emphasizing the commercial possibilities aviation offered for regularly scheduled passenger services, mail routes, and aerial photography and cartography. She stressed that aviators should embrace safety practices, carefully examining equipment and observing weather. Quimby also was superstitious, refusing to fly on Sundays and wearing good-luck tokens.

While flying with the Moisant International Aviators Exhibition Team at the November 1911 inauguration of Mexican President Francisco Madero, Quimby decided to become the first woman to fly across the English Channel. Worried that a European aviatrix might cross the channel first, Quimby secretly prepared to sail to England in March 1912. Louis Blériot (1872-1936), who flew across the channel in 1909, loaned her an airplane. Delayed by inclement weather, Quimby was disheartened by news that a Miss Trehawke-Davis had recently flown across the channel as a passenger in Gustave Hamel's airplane. At dawn on April 16, Quimby took off from Dover, England. Fog prevented Quimby from seeing the water, but, guided by a compass, she landed an hour later near Equihen, France. Although local residents celebrated her flight, international newspapers devoted space to coverage of the *Titanic's* sinking, barely mentioning Quimby's achievement.

Quimby agreed to fly at the Third Annual Boston Aviation Meet on July 1, 1912. Accompanied by the meet's organizer, William Willard, Quimby circled the Boston lighthouse. The plane unexpectedly lurched and Willard fell out. Quimby struggled with the controls but also was ejected. She and Willard died on impact and the plane crashed soon after. Speculation over the cause of the accident ranged from the cables becoming tangled to the airplane becoming unbalanced when Willard shifted his weight to talk to Quimby. Despite some public criticism that Quimby's accident proved women should not fly, many women considered Quimby to be a heroine and were inspired to become pilots also. Memorials to Quimby have included an American air-mail stamp featuring her image.

ELIZABETH D. SCHAFER

Knud Johan Victor Rasmussen
1879-1933
Danish Explorer and Ethnologist

Knud Johan Victor Rasmussen was a Danish explorer and ethnologist who participated in several Arctic expeditions through Greenland. He also completed an exceptionally long expedition that began in Greenland, proceeded to Canada, and ended in Alaska. During these expeditions he studied and recorded the migration and culture of the Eskimo.

Rasmussen was born to a Danish missionary father and an Eskimo mother on June 7, 1879, in Jakobshavn, Greenland, located in the west central part of the country. He attended the University of Copenhagen, Denmark, and married Dagmar Andersen; together they had one son and two daughters. From 1902-04, Rasmussen participated in expeditions to northwestern Greenland, where he encountered the Polar Eskimo. His command of the Eskimo language and his familiarity with the culture gave him a clear advantage as he studied them. In 1905, he examined the possibility of raising reindeer in western Greenland. The next two years he lived among the Polar Eskimo again, studying their way of life.

In 1910, Rasmussen built a post which he named Thule Base in Cape York, Greenland. He traded with the Eskimos from this post and used it as a point of departure for his expeditions. In 1912, Rasmussen, accompanied by three comrades, walked across the Greenland ice sheet extending from Thule to the northeast coast of the country. He surveyed the north coast of Greenland from 1916-18, then he commanded an expedition to Andmagssalik in east Greenland in 1919. His book *Greenland by the Polar Sea* was published in 1921.

Rasumussen's greatest expedition, in which he planned to visit every Eskimo tribe, began in Upernavik, Greenland, on September 7, 1921. He first explored in Greenland and then Baffin Island, then trekked across the arctic coasts of Canada to the Bering Strait. He explored northeastern Canada and reached Point Barrow, Alaska, on May 23, 1924. During this long expedition he mapped Eskimo migration routes and noted the similarities between Eskimo cultures. He also attempted to demonstrate his theory that the Eskimos as well as North American Indians had common ancestors in tribes that migrated from Asia across the Bering Strait. Rasmussen's ability to live and hunt in arctic conditions as the Eskimo did allowed him the opportunity to become close with them in order to study them. He made it his mission to study the Eskimos and examine them in their element before they made extensive contact with modern society. Rasmussen published his findings from this expedition in *Myths and Legends from Greenland.*

Rasmussen was also able to accumulate a significant amount of Eskimo poetry from the Iglulik Eskimos of the Hudson Bay area and the Musk Ox tribe of Copper Country in Canada. These poems, which were spontaneously sung or chanted during special events or accomplishments such as a bountiful hunt or a marriage, were the chief mode through which the Eskimos transmitted knowledge and traditions.

This expedition is detailed in Rasmussen's book *Across Arctic America*, which was published in two volumes in 1925, then translated into English in 1927. Rasmussen was the recipient of a number of medals from geological and anthropological societies in Scandinavia as well as other countries. He was also the honorary member of a number of geological societies.

Rasmussen died on December 21, 1933, in Copenhagen, Denmark, from complications stemming from the flu and pneumonia, as well as from a case of food poisoning which occurred on his final expedition.

MICHAEL T. YANCEY

Cândido Mariano da Silva Rondon

1865-1958

Brazilian Explorer, Military Colonel, and Engineer

Cândido Mariano da Silva Rondon was one of the foremost explorers of South America. A colonel and engineer in the Brazilian army, he mapped much of Brazil's unexplored interior. Among his most well-known expeditions was one he shared with former U.S. President Theodore Roosevelt in 1914. The men took a perilous trip down the River of Doubt, mapping the course of this previously unknown waterway, and providing information about the insects and other animals of the river basin.

Rondon was born on May 5, 1865, in Cuiabá, the capital city of the Brazilian state of Mato Grosso. In his youth, he lost both parents and went to live with an uncle. After high school, Rondon spent two years as a primary school teacher before joining the Brazilian army in 1881. He went on to the military academy in 1883, and a year after he graduated with honors in 1888, he was involved in the overthrow of the Emperor of Brazil. He also attended post-graduate school, and followed it by taking a position as substitute professor of mathematics in Rio de Janeiro's Praia Vermelha Military School. He cut short that appointment when in 1890 he joined the effort to wire Brazil's interior for telegraph communications. In this new position, he held the title of army engineer.

His post with the Telegraph Commission continued until 1895 when the telegraph wire finally spanned Mato Grasso. For the next 11 years, Rondon helped to build the only road from Rio de Janeiro to Cuiabá, an expanse that was previously traversed only via river, and to install telegraph lines through what was mostly uncharted lands throughout Brazil. Often, Rondon ran across native people who had never met a "civilized" human. Repeatedly he found himself in the position of liaison between Brazilian authorities and the native populations, and he worked hard to instill peace and cooperation into the delicate proceedings. These interactions gave him an extensive understanding of and compassion for the indigenous peoples. The understanding he gained was critical to his success during later explorations into Brazil's interior. Through the years, he became an activist in preserving the many cultures of the region. For example, he helped to encourage the Brazilian government to form the National Service for the Protection of the Indians, which was designed to help the native populations retain their cultures and avoid exploitation from outside influences, particularly businesspeople who would profit financially at the expense of the native people.

While learning about the indigenous bands of people, Rondon's work also allowed him to acquire considerable knowledge about the animals and plants of the region, and he spent many hours collecting specimens for the National Museum.

Rondon continued his explorations of Brazil's interior, and in 1909 discovered a large river he named the River of Doubt. In 1914 Roosevelt joined him to investigate the waterway. The expedition, which lasted more than three months, yielded the first map of the river, along with numerous animal specimens. Roosevelt chronicled the expedition in his book *Through the Brazilian Wilderness* (1914). This arduous journey left many men, including Roosevelt and his son, in poor health due to injuries and disease suffered during the trip. Before they completed their journey, Rondon renamed the river Roosevelt in the former president's honor. Roosevelt died less than five years after the journey, apparently from the lingering effects of his ordeal on the expedition.

Rondon went on to do further exploration of Mato Grasso until 1919 when he became chief of the army's engineering corps and head of the Telegraph Commission. In 1927, he went back to his exploratory work, and in later years began writing, helped mediate political disputes over land boundaries, continued to work on behalf of the indigenous people of Brazil, and established a national park on the Xingu River. Rondon was 93 when he died in Rio de Janeiro in 1958.

LESLIE A. MERTZ

Theodore Roosevelt
1858-1919
American President and Adventurer

In addition to his years as the 26th President of the United States, Theodore Roosevelt was an adventurer and big-game hunter who traveled to numerous remote regions of the world. He also made important discoveries when he and his son, who accompanied him on many expeditions, joined up with Brazilian Colonel Cândido Rondon (1865-1958) in 1914 to explore the River of Doubt deep in the jungles of South America. The wild trip down the River of Doubt provided a great deal of information about this previously unknown waterway, and about the insects and other animals of the area.

Roosevelt was born in New York City on October 27, 1858, as the second of four children to Theodore and Martha Roosevelt. Despite asthma and poor eyesight, the young Roosevelt became an avid sports enthusiast, taking up riding, shooting, and boxing. He also took a particular interest in the natural world. Roosevelt eventually became a student at Harvard University, and was graduated Phi Beta Kappa in 1880. During the same year, he married Alice Hathaway Lee. His interests wavered from law to history, and although he could have pursued an academic life, he turned instead to writing and politics.

After writing a history of the War of 1812, Roosevelt was elected to the New York Assembly, where he served from 1882-84. In 1883, his asthma became severe enough that he went to the Dakota Territory in an attempt to alleviate the symptoms. While there, he joined a group of ranchers who were hunting buffalo. This experience triggered a lifelong interest in big-game hunting. His last year on the New York Assembly was one of tragedy on the homefront. Both his mother and his wife died only hours apart on February 14, 1884. His wife died in childbirth, but his daughter, Alice Lee, survived.

Afterward he served as a delegate-at-large to the Chicago Republican National Convention, and continued his writing career at a ranch he had purchased in the Dakota Territory. Some of his books from this period included *Hunting Trips of a Ranchman, Ranch Life and the Hunting-Trail*, two volumes of *The Winning of the West*, and *Essays on Practical Politics*. In 1886 he tried for and lost election for mayor of New York. Following this defeat, he traveled to London, where he married Edith Kermit Carow in December of that year.

His political career continued in Washington with his appointment in 1889 as commissioner of the U.S. Civil Service. In 1894 he returned to New York where he took a post as a police commissioner, but a year later he was back in Washington, this time as assistant secretary of the U.S. Navy under President William McKinley. In May 1898, Roosevelt resigned and put together a volunteer cavalry regiment called the Rough Riders. He fought battles in Cuba—many of which were chronicled in Roosevelt's *Rough Riders* (1889)—and he gained hero's status back in the United States. When he returned, his notoriety allowed him to continue his political career as governor of New York. In 1890 President McKinley ran for reelection with Roosevelt as vice president. When McKinley was assassinated in 1901, Roosevelt—now 42 years old—took the oath of office and became the 26th president of the United States. He sought and won reelection, but turned down a third term, instead promoting William Howard Taft, who won the election. Perhaps Roosevelt's most enduring legacy as president were his efforts in the conservation arena: he set aside 125 million acres as forest reserves, established 16 national monuments and 51 wildlife refuges, and doubled the number of national parks.

Without the day-to-day demands of the presidency, Roosevelt went back to his life as an adventurer. He and his oldest son Kermit went to Kenya in 1909 for a nearly year-long hunting and scientific expedition. They traveled through Nairobi, Uganda, Sudan, and Khartoum. Roosevelt recorded their adventures in his book *African Game Trails: An Account of the African Wanderings of an American Hunter-Naturalist* (1910).

Upon his return to America, he had a fallout with President Taft, and vied unsuccessfully for the 1912 Republican presidential nomination. Undaunted, he then ran for president on the new Progressive Party ticket. During the run for office, Roosevelt faced political opposition and even an assassin's bullet. While he recovered quickly from the bullet wound, he lost the election to Woodrow Wilson.

Theodore Roosevelt. *(Library of Congress. Reproduced with permission.)*

Now, the former president set his sights on South America. He had planned a speaking tour and a steamer trip, but instead accepted an invitation from Colonel Cândido Rondon, a Brazilian army engineer and experienced explorer, to investigate the River of Doubt. Their journey began with a steamboat ride up the Paraguay River and a canoe trip along the Sepotuba River, and continued by horse and mule to a stream that would join with the River of Doubt. Other than a few miles at the entrance to the river, the waterway remained completely unknown to anyone but the few native people who dotted the forest interior.

Roosevelt, Rondon, and their party, which included Roosevelt's son Kermit, set out on their river adventure in February 1914 during the rainy season. The trip was anything but leisurely. Dangerous rapids capsized their dugout canoes time and time again, sometimes causing the loss of valuable supplies. Portage around rapids meant exhausting, time-consuming trudges through nearly impassable forests. All the while, the men were plagued by incessant, biting insects. Nonetheless, the party was able to map the river and return with specimens of many animals they had encountered along the way.

Following his return to the United States, Roosevelt continued to take an active role in politics until he died at his home on Long Island on January 6, 1919, at the age of 60.

LESLIE A. MERTZ

Sir Ernest Henry Shackleton
1874-1922
British Explorer

Sir Ernest Shackleton, British polar explorer and veteran of four Antarctic expeditions, is considered one of the greatest explorers of all time. During the early years of the twentieth century, a worldwide race was on to forge routes to the South Pole on the Antarctic continent. Extreme conditions, including treacherous seas, sub-zero temperatures, ice, and unmapped and unknown mountains and glaciers, met these expeditions on their quests for scientific and geographic information. Shackleton is respected and revered for his leadership qualities tested under conditions that were as severe as any human being had ever endured and for his scientific contributions.

Ernest Shackleton was born February 15, 1874, in Kildare, Ireland. He went to a Quaker school in Ireland until the agricultural depression in the late 1870s forced the family to move to London. At the age of 16, after attending Dulwich College, he joined the Mercantile Marine (Royal Naval Reserve) as an apprentice before signing on to the expeditions bound for Antarctica.

In 1902 Shackleton joined the British National Antarctic Expedition, sponsored by the Royal Geographic Society. Captained by Robert Falcon Scott (1868-1912), the expedition established a base camp on Ross Island in the Ross Sea. They set out with dog teams and sledges intending to be the first to reach the South Pole. Scurvy, frostbite, and a shortage of food and supplies forced them to turn back. Although the expedition contributed important geographic information, Shackleton and Scott parted on antagonistic terms.

In 1907 Shackleton and the British Antarctic Expedition set sail in the *Nimrod* for Ross Island. They intended to trek with ponies to the South Pole along the Great Beardmore Glacier. They came within 97 miles (156 km) of the Pole, turning back rather than risking the lives of the men. A Norwegian team headed by Roald Amundsen (1872-1928?) later became the first to reach the Pole (November 1911). Shackleton's

Ernest Shackleton. *(The Granger Collection Ltd. Reproduced with permission.)*

old colleague Scott reached the Pole nearly a month later and perished on the return trip.

On hearing of Amundsen's success and Scott's tragic failure, Shackleton ignored the beginnings of World War I and launched the Trans-Antarctic Expedition (1914-17). In the ship *Endurance*, Shackleton and a 28-man crew set off to cross Antarctica and reach the South Pole. A sister expedition, captained by Sir Douglas Mawson (1882-1958) in the *Aurora*, sailed to the Ross Sea and cached supply depots at intervals to the Pole. The *Endurance*, meanwhile, sailed into the unexplored Weddell Sea on the opposite side, where the team was to travel to the Pole via dog team. Once at the Pole they would rely on the supplies cached by Mawson.

The plan was derailed when the *Endurance* became trapped in the Weddell ice pack. After abandoning the *Endurance*, the expedition eventually made it to desolate Elephant Island, where most of the men remained. Shackleton and five others sailed across 850 miles (1,368 km) of rough Antarctic seas in a small open boat to a remote whaling station on South Georgia Island in order to get help. After reaching land, they made a 10-day overland trip before reaching the station at Stormness. While waiting for Shackleton to return the men on Elephant Island endured bitter cold, six months of total darkness, dangerous ice conditions, and near starvation. Almost two years later, after three unsuccessful rescue attempts, the remainder of his crew were safely brought back to England from Elephant Island. Although the expedition did not complete its goal to reach the South Pole, a vast amount of scientific data was collected. In addition, parts of the expedition were filmed by cinematographer Frank Hurley, whose film captured the courage and resourcefulness of the *Endurance* crew. The transcontinental journey would not be achieved until 1957-58, when Sir Vivian Fuchs (1908-) and Sir Edmund Hillary (1919-) completed the crossing with the aid of motorized vehicles.

On January 5, 1922, Shackleton died on his fourth Antarctic expedition to the South Georgia Island. He was buried there.

LESLIE HUTCHINSON

Dame Freya Madeline Stark
1893-1993
British Explorer and Writer

Freya Madeline Stark was a British explorer who lived during a time when explorers were regarded as heroes. She traveled to remote areas of the Middle East, where few Europeans—especially women—had traveled before. She also traveled extensively in Turkey, Greece, Italy, Nepal, and Afghanistan. She was awarded the Dame Commander of the Order of the British Empire in 1972. Through her writings, she contributed to our knowledge and understanding of the people, landscapes, and conditions unique to the world between World War I and World War II.

Stark was born in Paris in 1893 to British expatriate parents. Although she had no formal education as a child, she moved about with her artist parents and learned French, German, and Italian. She entered the University of London in 1912, but with the entry of Britain into World War I, she joined the nurse corps and was assigned to Italy. After the war she returned to London and attended the School of Oriental Studies. This work led to extensive travel in the Middle East, enabling her to eventually become fluent in Persian, Russian, and Turkish.

Stark initially made her reputation as a traveler and explorer in the Middle East while working with the British government service. During this time, most of the Middle East was controlled by the British and French, largely because the League of Nations, (forerunner to the United Nations) had given them the former Arab territories of the Ottoman Empire as "mandates" and "protectorates." After World War I ended in 1918, Britain received Palestine, Transjordan (Jordan), and Mesopotamia (now Iraq). France received Syria and Lebanon. In addition, the British occupied Egypt in the 1880s and had control of the Suez Canal, and the British Museum had extensive collections of Egyptian antiquities. The British government had converted the Royal Navy from coal to oil prior to World War I, ensuring that the British would maintain a permanent stronghold in the oil-rich Middle East. It was this world of ancient Arabian and Persian cultures, British and French intrigues, and resulting power plays that Freya Stark traveled and wrote about.

Freya Stark. *(Archive Photos, Inc. Reproduced with permission.)*

In 1927 at age 33, she saved enough money to travel to Lebanon and study the Arabic language. In 1928 she and a friend traveled by donkey to the Jebel Druze, a mountainous area in French-controlled Syria. Because the Druze resisted the French rule, the area was off limits to foreigners. She and her friend were arrested by French authorities on suspicion of espionage. During another trip, she remapped a distant region of the Elburz, a mountain range in Iran. She was searching for information about an ancient Muslim sect known as the Assassins, which during the thirteenth century had engaged in the widespread murder of Christians. She recounted this journey in *Valley of the Assassins* (1934), a classic tale that won the prestigious Gold Medal from the Royal Geographic Society. For the next 12 years she traveled and wrote, establishing her style combining travel and personal commentary on the people, places, customs, history, and politics of her travels in the Middle East.

During World War II Stark was recruited by the British government to help convince the Arab world to support the Allied cause. She worked for the British Ministry of Information in Aden, Baghdad, and Cairo. During this time she established the anti-Nazi Brotherhood of Freedom, an organization formed to educate and prepare the Arab world for its probable freedom after the war was over. The Nazi regime countered with massive propaganda to influence the Arabs to its side. Frequently under enemy fire, she traveled through the Arab world as a loyal supporter of the British Empire and influenced many individuals to the Allied cause.

After World War II, Stark authored 20 books, becoming a famous author and sought-after speaker in Europe. She was honored by the Queen of England with the order of Dame of the British Empire for her dedication and service in support of Great Britain. In 1993 she died in Italy at the age of 100.

LESLIE HUTCHINSON

Sir Wilfred Patrick Thesiger
1910-
British Explorer

Wilfred Patrick Thesiger was born in Abyssinia (present day Ethiopia) in 1910. His father was the British Minister in Addis Ababa, the capital city. The young Thesiger grew up among conditions of political turmoil. Murder and robbery were not uncommon in and around Addis Ababa, and as a child he witnessed many public floggings and hangings. For recreation he went on hunting trips with his father and brothers, and in 1918 he and his family took a short trip to India. By 1919, when he and his family moved to England, he had developed a taste for travel and adventure.

In England, Thesiger was educated in good schools. But, after their years in exotic Ethiopia, he and his brothers had difficulty adjusting to the relatively uneventful life of an English student. While at Oxford University, he was invited to return to Ethiopia to attend the coronation of the Emperor Haile Selassie. This brief trip convinced him that life in England was not for him. He returned to Oxford to finish his education, but went immediately back to Africa following his graduation.

In 1934 he began his life-long career of exploration. For six years he traveled and explored in eastern Africa. Then from 1940-44 he fought with British forces in northern Africa during the Second World War. As a member of the British Special Air Service he participated in a number of successful raids on German forces, killing and wounding an unknown number of enemy soldiers. From 1944-45 he was an advisor to the Crown Prince of Ethiopia. From 1945-50 he explored the southern deserts of the Arabian peninsula, living with the Bedu people in the region known as the "Empty Quarter." He was the first European to thoroughly explore that area. His book *Arabian Sands,* which recounts his time in the "Empty Quarter," is considered a classic account of exploration and adventure. From 1950-58 he lived with the marsh Arabs in southern Iraq, during which time he also traveled to Pakistan and Afghanistan. In 1958 he returned once more to Ethiopia and explored the central African regions in and around Kenya. In 1968 he decided to remain in that area. He lived in Kenya until 1994, when his health began to fail. He subsequently returned to England to live in a London retirement home.

Thesiger has been recognized for the numerous scientific contributions that came from his years of exploration. He expanded our knowledge of geography, ethnography, anthropology, and botany in the lands in which he traveled. Many consider him to be the last of the great explorers. Thesiger regretted, however, that his explorations helped to open many previously unknown lands and their peoples to disruptive modern technologies. The invention of automobiles, Thesiger claimed, was "the greatest disaster in human history" because it ruined the lifestyles of the nomadic peoples he had befriended and loved. These lifestyles, forged under harsh conditions and built upon strong commitments of honor and respect, were greatly disrupted by the technologies brought in from Europe and America. For this reason Thesiger long opposed global technological diffusion. Believing that modern military technology, especially nuclear weapons, could destroy all human life within the next 100 years, Thesiger was an outspoken opponent of technology as a means towards a better civilization. He preferred the simpler ways of the nomadic peoples with which he spent so many years of his life.

Thesiger has published numerous books about his travels. For his years of exploration and his many contributions to scientific knowledge, he was knighted by the British government in 1995.

STEVE RUSKIN

Sir George Hubert Wilkins
1888-1958
Australian Explorer, Pilot, and Photographer

Sir George Hubert Wilkins was an Australian explorer of the Arctic and a pilot who is regarded as a pioneer in air exploration. He also contributed to the exploration of the Arctic by submarine.

Wilkins was born in Mount Bryan East, in southern Australia on October 31, 1888. He attended the School of Mines and Industry in Adelaide, where he studied electrical engineering; he also pursued his interest in photography and took up flying in 1910. He married Suzanne Bennett in 1929. Wilkins utilized his skills in photography and flight during the Balkan War, where he served as a newsreel photographer. His work was used in newspapers as well as in movies. Following the end of the Balkan War in 1913, Wilkins began his career in Arctic exploration after he was selected by Vihjalmur Stefansson (1879-1962) to be the official photographer of the Canadian Arctic Expedition, which lasted until 1917. Wilkins was rewarded for his commitment to the expedition by being promoted to second in command beneath Stefansson.

After the expedition was completed in 1917, Wilkins joined the Australian Flying Corps as a photographer during World War I; he was stationed on the French front. While in this position he served as the official photographer of the military history department. His love of flight led him to compete, in 1919, for a $50,000 prize for a successful flight from England to Australia; he did not succeed in this effort.

Wilkins was second in command again for another expedition, this time with the British Imperial Antarctic Expedition from 1920-21. Following this effort, he returned to Antarctica in 1921 with Sir Ernest Shackleton's (1874-1922) Quest Expedition as a naturalist, or one who studies animals and plants. Wilkins was in charge of the Wilkins Australia and Islands expedition for the British Museum in Australia from 1923-25. From 1926-28 he was in charge of the *Detroit News* Arctic Expeditions. His first book, *Flying the Arctic*, was published in 1928. His second book, *Undiscovered Australia*, was published in the following year.

George Hubert Wilkins. *(Library of Congress. Reproduced with permission.)*

On April 15, 1928, Wilkins and Carl Ben Eielson flew from Point Barrow, Alaska, to Spitsbergen, Norway. The team covered a distance of 2,100 miles (3,380 km) in 20½ hours in the first east to west crossing of the Arctic; this flight was regarded as the greatest Arctic flight of the time. On June 14, 1928, Wilkins was knighted by King George V; he received the Patron's medal from the Royal Geographic Society in the same year.

In the same year of his knighting, Wilkins led the Wilkins-Hearst Antarctic Expedition. With this effort, he was the first to fly in the Antarctic and the first to fly over both poles. He was able to determine also that Graham Land was an island and not connected with a continent. Wilkins and Eielson, teamed again with Eielson as pilot, made a number of geographical observations on their 1,200-mile (1,931-km) flight along the Palmer Peninsula. Three years later, in 1931, Wilkins flew around the world in the *Graf Zeppelin*. He also made the first exploration of the Arctic by submarine, in a vessel named the *Nautilus*. In the years 1933-39, Wilkins oversaw the completion of four of Lincoln Ellsworth's (1880-1951) Antarctic expeditions. Wilkins served as a consultant to several branches of the United States government from 1942, and he consulted with the United States Army Air Force during World War II regarding Arctic clothing. He also consulted with the Weather Bureau and the Navy, and was special consultant to the Army Quartermaster Corps in 1947. He died on December 1, 1958, in Framingham, Massachusetts.

MICHAEL T. YANCEY

Charles Elwood Yeager

1923-

American Pilot

A decorated World War II fighter pilot, Charles E. ("Chuck") Yeager returned from service in Europe to become one of the leading test pilots of the jet age. He pioneered the testing of high-performance jet- and rocket-powered aircraft, setting numerous speed and distance records in the process. He is best known for becoming, in 1947, the first pilot to fly faster than sound.

Chuck Yeager. *(Bettmann/Corbis. Reproduced with permission.)*

Born in Myra, West Virginia, on February 13, 1923, Yeager enlisted in the United States Army Air Corps after graduating from high school in 1941. Originally trained as a mechanic, he began pilot training in 1942 as part of the "Flying Sergeants" program. Posted to England in November 1943, he flew a total of 64 combat missions in P-51 Mustang fighters and shot down 13 enemy aircraft—five of them in a single day.

Yeager became involved in flight testing soon after the end of the war, helping to evaluate captured German and Japanese aircraft. He also helped to evaluate the Lockheed F-80 and Republic F-84, members of the first generation of jet fighters to enter Air Force service. Yeager, by then a Captain and a graduate of the Air Force instructor and test pilot schools, was transferred to Muroc (later Edwards) Air Force Base in August 1947. Muroc, built on a dry lake bed in the California desert, was then emerging as the Air Force's principal site for high-performance flight testing and research, and Yeager was assigned as project officer for the Bell X-1.

The X-1 and its successors were not prototypes for new production aircraft, but research tools for testing new technologies and investigating the nature of flight at extreme speeds and altitudes. The X-1 program itself was a joint project of the Air Force and the National Advisory Committee on Aeronautics (NACA), designed to investigate the nature of flight just below and, if possible, just above the speed of sound. Pilots who approached the speed of sound in piston-engine planes had reported severe buffeting and sometimes loss of control, leading to speculation about an impenetrable "sound barrier" through which no aircraft could pass intact. The X-1's radical design—a bullet-shaped fuselage and thin, sharp-edged wings—was intended to reduce such control problems. On October 14, 1947, Yeager released the X-1 from its B-29 "mother ship" at 21,000 feet (6,401 m) over the desert. Firing the four liquid-fuel rockets clustered in its tail, he reached a speed of Mach 1.06 at an altitude of 45,000 feet (13,716 m), becoming the first pilot to "break the sound barrier."

Yeager made more than 49 flights in various versions of the X-1 between 1947 and 1953, when he became the first pilot to fly at two-and-a-half times the speed of sound. He went on to serve as a squadron commander and wing commander in the late 1950s and mid-to-late 1960s, flying jet fighters whose designs owed much to the data he had helped to gather in the X-1. Serving as commander of the Air Force's Aerospace Research Pilot School in the early 1960s, Yeager was directly responsible for training many of the test pilots who would be selected as astronauts for NASA's Gemini and Apollo programs. Yeager retired from active duty in 1976 with the rank of brigadier general, but remained active both as a pilot and as a consulting test pilot for the Air Force. His last official flight as an Air Force consultant took place on October 14, 1997: the 50th anniversary of his first supersonic flight in the X-1. He marked the occasion by breaking the sound barrier again, this time in an F-15 fighter.

A. BOWDOIN VAN RIPER

Biographical Mentions

Luigi Amedeo Abruzzi
1873-1933

Spanish mountaineer and explorer whose travels ranged from Africa to the Arctic. Abruzzi was the first to ascend Mt. St. Elias in Alaska (1897). His 1899 expedition to the arctic reached latitude 86° 34' N—a record for the time. In east-central Africa his expedition studied and named the geologic, topographic, and glaciers of the Ruwenzori Range. In 1909 he climbed the world's second highest mountain, K2, in the Himalayas.

Carl Ethan Akeley
1864-1926

American naturalist and explorer best known for his revolutionary taxidermic work in museum exhibition, specifically the realistic display of animals in their natural surroundings. Akeley, who traveled extensively in Africa to study, hunt, and collect specimens, developed a mounting method that resulted in unprecedented realism of displayed wildlife. His other inventions included the Akeley cement gun, used in mounting animals, and the Akeley motion-picture camera, which he was the first to use to capture motion pictures of African gorillas in their natural environment.

Mary Lee Jobe Akeley
1886-1966

American explorer, photographer, and writer who began her career as a history instructor before being commissioned by the Canadian government to study Eskimos and Indian tribes. After exploring the Canadian Northwest, she married Carl Ethan Akeley (1864-1926) in 1924 and made her first trip to Africa with him that same year to study gorillas in the Belgian Congo. Mary Akeley's photos and writings detailed the importance of preserving Africa's native cultures, wild animals, and landscapes and led to her becoming a crusader for game preserves.

John Alcock
1893-1919

English aviator who served as pilot on the first non-stop flight across the Atlantic Ocean. Alcock, guided by navigator Arthur Whitten Brown, flew a converted twin-engine Vickers Vimy bomber 2,000 miles (3,218 km) through snow, ice, and dense fog on June 15-16, 1919. The 16-hour flight began in Newfoundland and ended with an unplanned crash landing in a bog near Clifden, Ireland. Alcock received a knighthood and a share of a £10,000 prize for his achievement, but died just six months later in a crash in France.

Johan Gunnar Andersson
1874-1960

Swedish geologist, paleontologist, and archaeologist known for his study of prehistoric vertebrate fossils, especially the fossils of Peking man (*Sinanthropus pekinensis*) in southwest China. He participated in two arctic expeditions: 1898 with Alfred Nathorst and 1901-03 with Swedish explorer and geographer Otto Nordenskjold. He completed his doctorate in geology (1906) and became director of the Swedish Geological Survey. In 1928 he founded the Museum of East Asian Antiquities, a museum built around his personal collections.

Bob Bartlett
1875-1946

Canadian-American sea captain and explorer who spent his life working on and captaining sailing vessels and ships to the Arctic for exploration. He sailed with Robert Peary as first mate on the *Windward* for Peary's initial attempt to reach the North Pole in 1898-99 and again on two later attempts as the captain of Peary's ship, the *Roosevelt*. Through these and numerous other exploratory voyages to the Canadian Arctic, Bartlett helped to gather vast amounts of scientific information about this formerly little-known region of the world.

Otis Barton
b. 1899

American engineer who designed a deep-sea submersible vessel called the bathysphere—a spherical metal ball constructed of steel, with two viewing windows and capable of withstanding the pressures of deep-sea diving. Along with William Beebe, a zoologist, Barton pioneered the use of a tethered, manned submersible craft with the bathysphere. They reached a record setting depth of 1,427 feet (435 m) in 1930 and a depth of

3,028 feet (923 m) in 1934. *National Geographic* magazine carried a personally narrated account of the exploration, which gave scientists the first glimpse of the deep sea. Barton received his engineering degree from Columbia University.

Daisy May Bates
1863-1951

Irish anthropologist who studied Aborigines. Bates married an Australian cattleman in 1885. Discontented, she moved to Great Britain, where she read allegations that white Australians abused indigenous peoples. Bates returned to Australia in 1899 to investigate. She lived with Aborigines, recording information about their culture. Concerned about their health, she nursed the ill and strove to improve living conditions. The aborigines called Bates *Kabbarali,* meaning "white grandmother." She wrote *The Passing of the Aborigines* (1938), increasing anthropologists' awareness of those tribes.

Gertrude Margaret Lowthian Bell
1868-1926

English archeologist who explored the Middle East. Bell graduated from Oxford University. She visited an uncle employed in Tehran, Iran, and anonymously published a collection of essays about her experiences. Fluent in Arabic and Persian, she traveled throughout the turbulent region, meeting leaders and providing political insights to British officials prior to World War I. She wrote several books about her adventures. Bell established and served as director of the Baghdad Museum of Antiquities in Iraq.

Floyd Bennett
1890-1928

American pilot and explorer who, with Lieutenant Commander Richard E. Byrd, claimed to be the first persons to fly to the North Pole. In that historic journey, Bennett flew a three-engine Fokker monoplane, the *Josephine Ford*, to the North Pole on May 9, 1926. He and Byrd received the Medals of Honor for the feat. Bennett died three years later in the midst of planning a trip to the South Pole with Byrd.

Louise Arner Boyd
1887-1972

American explorer who expanded knowledge of the Arctic. A wealthy San Francisco heiress, Boyd invested in seven scientific Arctic expeditions. She wrote books, describing plants, animals, and topography, that were illustrated with her photographs. Boyd searched for missing explorer Roald Amundsen in 1928. She was a military consultant regarding World War II Arctic radio transmissions. Boyd received numerous awards and several geographical landmarks were named for her. In 1955 she became the first woman to fly over the North Pole.

Arthur Whitten Brown
1888-1950

Scottish-born American aviator who served as navigator on the first non-stop flight across the Atlantic Ocean. An expert aerial navigator, Brown flew from Newfoundland to Ireland on June 15-16, 1919, with pilot John Alcock at the controls. Brown acted as navigator, radio operator, and troubleshooter on the 2,000-mile (3,218 km) flight, climbing out of his open cockpit several times to service essential instruments. Brown was knighted for his achievement and received, with Alcock, a £10,000 prize.

Richard L. Burdsall
1899-1953

American mountaineer and engineer. A member of the 1932 Sikong expedition, along with A. Emmons, T. Moore, and J. Young, Burdsall traveled to the Amne Machin Range along the Chinese-Tibetan border, then thought to have peaks in excess of 30,000 feet (9,144 m). They climbed, mapped, and measured Minya Konka (24,790 feet or 7,556 m) and corrected the measurement taken by Joseph Rock in 1929, as well as collecting zoological and botanical specimens for the Chinese Academy of Sciences.

Jean-Baptiste Charcot
1867-1936

French doctor, scientist, and explorer who, after becoming a doctor, turned his hobby of sailing into a means of doing scientific research in and exploring the North Atlantic and Antarctica. Believing France should be among those countries sending exploratory and scientific expeditions to Antarctica, he purchased the *Français* in 1903 and sailed to the region. He also led other Antarctic expeditions aboard the *Pourquois-Pas* and later on, annual expeditions to the North Atlantic.

Evelyn Cheesman
1881-1969

English entomologist who collected South Pacific insects. Cheesman studied entomology because veterinary colleges refused to admit women. As the Regent's Park Zoo's insect keeper, she joined a 1924 scientific expedition to the Marquesas and Galapagos Islands. Cheesman returned to the South Pacific eight times, collecting specimens for the British Museum of Natural

History. She wrote sixteen books about her tropical adventures. As a New Guinea-gazette writer, she proved useful to Allied forces during World War II.

Frederick Albert Cook
1865-1940

American doctor who, enthralled by stories of polar exploration, joined Robert Edward Peary's 1891 expedition to Greenland as the ship doctor and ethnologist. After this trip, he accompanied other expeditions to Antarctica. Two of his most noteworthy claims—that in 1906 he was the first person to have climbed to the summit of Mount McKinley and that in 1908 he was the first man to the North Pole—were later considered untrue. Peary is roundly considered to be the first to the Pole.

Alexandra David-Neel
1868-1969

French explorer who infiltrated the Tibetan capital. David-Neel traveled throughout Asia to study philosophy and seek her spiritual home. Embracing Buddhism, she interviewed the Dalai Lama in 1912 and lived in a cave in the Himalayas. In 1923 David-Neel disguised herself as a nun to gain admittance to Lhasa, the Tibetan capital, which was closed to westerners. She wrote a book about her experiences living in Lhasa. She also published scholarly texts about Tibetan culture and religion.

Erich Dagobert von Drygalski
1865-1949

German explorer who began what was to become a lengthy expedition to Antarctica in August 1901. His ship became icebound the following February and remained there a year before the ice broke up enough for the ship to begin its return trip. While trapped in the ice, Drygalski led several land expeditions and became the first person to fly in a balloon over Antarctica.

Lincoln Ellsworth
1880-1951

American engineer and aviator who participated in expeditions to South America, the Arctic, and the Antarctic. He helped finance several of these expeditions, including a flight over the North Pole, several flights over Antarctica, and an unsuccessful effort to sail a submarine under the North Pole. Ellsworth also participated in a German expedition to fly the dirigible *Graf Zeppelin* along the Arctic coast of Siberia in 1931. This expedition helped determine the true geographical boundaries of the Arctic.

Arthur B. Emmons
1910-1962

American mountaineer and member of the 1932 Sikong expedition, along with R. Burdsall, T. Moore, and J. Young. They traveled to the Amne Machin Range along the Chinese-Tibetan border, then thought to have peaks in excess of 30,000 feet (9,144 m). They climbed, mapped, and measured Minya Konka (24,790 feet or 7,556 m) and corrected the measurement taken by Joseph Rock in 1929, as well as collecting zoological and botanical specimens for the Chinese Academy of Sciences.

Henry Farman
1874-1958

English aviator who made a variety of record-breaking flights in Europe before World War I. Farman achieved Continent-wide fame when, on January 13, 1908, he became the first European to fly a one-kilometer (0.62 miles) closed circuit. At the Rheims Air Show in August 1909, Farman won additional honors, setting a new distance-flown record of 180 kilometers (more than 100 miles) with a flight of just over three hours.

Maurice Farman
1877?-1964

French aviator who helped build the European aviation industry. The Farman brothers, Maurice and Henri, built several successful airplanes, claiming to have developed them independently of the Wright brothers. The Farman airplanes quickly became among the most popular in Europe, and, between the two brothers, over 300 airplanes rolled forth in 1912 alone.

Georges-Marie Haardt
1884-1932

French explorer who took part in a series of expeditions in the Sahara Desert during the 1920s and 1930s. Vehicles especially designed by the Citroën company were used to cross the Sahara Desert in North Africa. In 1922 Haardt, Louis Audoin-Dubreuil, and 10 others crossed the Sahara Desert from Touggourt in Algeria to Timbuktu in the Sudan in a record-breaking 20 days. The 1931-32 expedition partially followed Marco Polo's route along the Silk Road from Beirut to Beijing.

Marguerite Baker Harrison
1879-1967

American spy who was imprisoned by the Russians. A military intelligence officer in Germany, Harrison relayed information useful to diplomats during World War I peace negotiations.

Spying in Russia, she was the first American woman whom the Bolsheviks jailed. Harrison wrote several books about her ordeals, including her ten-month imprisonment. She also journeyed through Asia and assisted production of a documentary chronicling the Bakhtiari tribe's migration to Persia. Harrison helped establish the Society of Women Geographers.

Maurice Herzog
1919-

French mountaineer and engineer who in 1950 led a nine-man French expedition that explored the Annapurna massif in the north central Himalayas, encompassing fourteen 26,000 feet (8,000 m) peaks. On June 3, 1950, Herzog successfully climbed Annapurna I (26,545 feet or 8,091 m). He lost many of his fingers and toes in the climb but was able to bring back valuable information. He later served on the International Olympic Committee.

Andrew Irvine
1901-1924

British mountaineer who with George Mallory disappeared on a Mt. Everest expedition in 1924. An engineering student on leave from Merton College, he joined the Mallory team as a somewhat inexperienced mountaineer, but he was resourceful and practically minded and was able to maintain the oxygen apparatus needed for the summit of Everest. The question of whether Mallory and Irvine reached the summit still remains unanswered. A 1999 expedition on the peak discovered Mallory's body but not Irvine's.

Amy Johnson
1903-1941

English aviator who set flying records. In 1930 Johnson was the first female pilot to fly solo from England to Australia. The next year she flew from England to Moscow in one day. She also completed the first solo round trip between England and South Africa. Johnson and her husband, James Mollison, set a speed record flying from England to India. Johnson published one book, *Sky Roads of the World* (1939). She died while piloting a World War II transport aircraft.

Neil Merton Judd
1887-?

American archaeologist best known for his participation in the Pueblo Bonito excavation in Chaco Culture National Historical Park in New Mexico, a site noted not only for its great size and fine masonry architecture, but also for its wealth of unusual cultural materials, unsurpassed in the Southwest. Judd also authored several books, including one about the history of the Bureau of American Ethnology that was published in 1967.

Oakley G. Kelly
1891-1966

American aviator who, with John A. Macready, made the first nonstop flight across the United States. Kelly and Macready flew a single-engine Fokker T-2 monoplane from New York to San Diego on May 2-3, 1923, taking just under 27 hours for the 2,500-mile (4,022 km) flight. Careful planning allowed them to take advantage of a "Hudson Bay High"—a meteorological event that reversed prevailing West-to-East wind patterns over much of the country.

Alfred Vincent Kidder
1885-1963

American archaeologist and anthropologist whose pioneering work in the southwestern United States resulted in the first comprehensive regional synthesis of North American archaeology, *An Introduction to the Study of Southwestern Archaeology* (1924). Kidder's combined application of stratigraphical excavations and pottery typology revolutionized archaeology. In addition to the American Southwest, he also implemented a major multi-disciplinary study of the Maya regions of Mexico and Central America, bringing together archaeologists, ethnologists, and natural scientists.

Sir Paul Kipfer

Swiss scientist and stratospheric explorer who with Auguste Piccard (1884-1962) was the first to fly in the stratosphere, using the pressurized balloon gondola invented by Piccard (1931) and reaching a record altitude—since surpassed—of 51,793 feet (15,786 m) over the Bavarian Alps.

Henry Asbjorn Larsen
1899-1964

Norwegian-Canadian seaman who captained the first west-to-east voyage down the Northwest Passage. (Roald Amundsen made the first crossing of the Northwest Passage in 1903-05, but went from east to west.) Larsen made the trip aboard a Canadian naval ship, the *St. Roch*, in 1942. For this and other accomplishments, he received Canada's first Massey Medal, the country's highest geographical honor.

Donald Baxter MacMillan
1874-1970

American explorer, scientist, and teacher who led one of the support parties on Robert Peary's successful attempt to the North Pole in 1909. His success led to a life dedicated to the research and exploration of the peoples and lands that lay to the north. In that vane, he performed ethnological studies of the Eskimos and Indians, surveyed unexplored regions, and studied the zoology and geology of the area.

John A. Macready
1888-1979

American aviator who, with Oakley Kelly, made the first nonstop flight across the United States. Macready and Kelly flew a single-engine Fokker T-2 monoplane from New York to San Diego on May 2-3, 1923, taking just under 27 hours for the 2,500-mile (4,022 km) flight. Macready is also credited with being the first pilot to fly above 35,000 feet (10,668 m), the first to make a night parachute jump, and the first to dust crops, among other achievements.

Ella Kini Maillart
1903-1997

Swiss explorer who traveled throughout Asia. Maillart went to Moscow in 1932 and trekked to eastern Turkestan, where she lived with Kirghiz and Kazakh tribes. A correspondent for *Le Petit Parisien,* Maillart covered the Japanese invasion of Manchuria. She walked through Tibet, from China to Kashmir, and published *Forbidden Journey* (1937). She later traveled to Iran and Afghanistan and then stayed in India to explore her spirituality. She entered Nepal when its borders were opened, writing *The Land of the Sherpas* (1955).

Douglas Mawson
1882-1958

Australian researcher and Antarctic explorer who was one of the first to observe the changing location of the magnetic pole, and surveyed unknown areas to the west of the South Pole. Mawson led expeditions to Antarctica and was a member of other expeditions as well. Back at the University of Adelaide, he was a respected teacher and scientist, and became a leader in the research of uranium and other minerals related to radioactivity.

Terris Moore
1908-

American mountaineer and member of the 1932 Sikong expedition, along with R. Burdsall, T. Moore, and J. Young. They traveled to the Amne Machin Range along the Chinese-Tibetan border, then thought to have peaks in excess of 30,000 feet (9,144 m). They climbed, mapped, and measured Minya Konka (24,790 feet or 7,556 m) and corrected the measurement taken by Joseph Rock in 1929, as well as collecting zoological and botanical specimens for the Chinese Academy of Sciences.

Erik Nelson

Swedish-American aviator who participated in the first circumnavigation of the globe by air in 1924. This journey, undertaken in largely unenclosed biplanes, took about five months. At one point, after the planes had made an emergency landing in Indochina, Nelson was able to get assistance from the local Standard Oil office and local tribes, who helped tow the damaged aircraft upriver behind their sampans.

Umberto Nobile
1885-1978

Italian engineer, air force officer, and explorer who designed and built dirigibles, and made numerous expeditions to the Arctic. During one expedition with Roald Amundsen in 1926, Nobile piloted the first dirigible—and the second aircraft overall—to the North Pole. In 1928, he commanded another dirigible to the North Pole, but crashed on the return trip. Blamed for the crash and associated deaths of several crew members, Nobile resigned from the air force and ceased his Arctic explorations. In 1945, he was cleared of all charges.

Fritz von Opel
1899-1971

German automotive engineer who created and piloted one of the first rocket-propelled aircraft. No doubt influenced by his grandfather, Adam Opel, who was known for producing everything from sewing machines to bicycles to automobiles, Fritz von Opel test drove his rocket-propelled car, the Opel-Rak 1, in 1928. One year later he tested a glider that was powered by 16 rockets, designed in part with Max Valier and Friedrich Wilhelm Sander, flying the first official rocketed airplane. His work on these projects lay the foundation for modern rocketry and space exploration.

Annie Smith Peck
1850-1935

American mountaineer who achieved altitude records. Inspired to climb mountains when she saw the Matterhorn, Peck ascended that mountain in 1895. Two years later, Peck climbed Orizaba (18,700 feet or 5,697 m) in Mexico, briefly holding the women's altitude record. Frustrated by insufficient funds and unreliable guides, Peck also endured rival Fanny Workman questioning her summit measurements. Peck climbed Peru's Mount Huascaran (21,812 feet or 6,648 m) in 1908, setting a Western Hemisphere record. A peak was named for her. She wrote *A Search for the Apex of America* (1911).

Luigi Pernier
b. 1860

Italian archaeologist known for his work researching the early Bronze Age cultures of Crete. In 1900 Pernier led a team of Italian archaeologists that conducted excavations at the palace of Phaistos (3000-1100 B.C.) and uncovered the Phaistos Disc. The meaning of the terracotta disc, marked with some 45 different pictograms, has been debated since its discovery.

Wiley Post
1899-1935

American aviator who flew around the world. A corporate pilot, Post raced his employer's airplane, the *Winnie Mae,* to win the 1930 Bendix Trophy. The next year, Post and his navigator, Harold Gatty, were the first to circle the globe in an airplane. Post wrote *Around the World in Eight Days* (1931). In 1933 Post repeated the trip solo. He created a high-altitude flying suit, attaining a height of 49,000 feet. Post died in an Alaskan plane crash with humorist Will Rogers.

Albert C. Read
1887?-1967

American aviator who commanded the first heavier-than-air craft to fly across the Atlantic Ocean. Read, then a Lieutenant Commander in the U. S. Navy, commanded one of three five-man navy flying boats that left Newfoundland on May 16, 1919, bound for the Azores island chain and ultimately Portugal. The other two planes were forced to land at sea, by Read's *NC-4* remained airborne for the entire trip, flying 1,400 miles (2,253 km) in roughly 15 hours.

George Andrew Reisner
1867-1942

American Egyptologist who improved excavation methodology. After earning a Harvard University doctorate, Reisner devoted his life to field work. One of the first archeologists to apply photography to excavation, Reisner carefully recorded finds. Named the Egyptian department curator for the Boston Museum of Fine Art, Reisner excavated the Giza Pyramids. He also surveyed Nubia, carefully mapping the area before the Aswan Dam flooded the site. Reisner's books are considered classics and his techniques greatly influenced archeologists excavating other areas and cultures.

Hanna Reitsch
1913-1979

German aviator who tested military aircraft. Reitsch learned to fly gliders because engine-powered airplanes were prohibited in post-World War I Germany. She participated in a 1934 South American expedition to determine how thermal conditions affect flight. Reitsch's first distance record was for soaring 100 miles. She also won a national competition. During World War II, Luftwaffe officers asked Reitsch to serve as a military test pilot. She demonstrated helicopters and was the first pilot to fly the rocket-propelled Messerschmitt 163.

Joseph F. Rock
1884-1962

American naturalist and explorer of Tibet and China. Trained as a botanist, he also worked as an anthropologist, linguist, and philologist. Beginning in 1924, he explored northwest China and Tibet, observing local customs and collecting plant material for the Arnold Arboretum and ornithological specimens for the Museum of Comparative Zoology at Harvard University. In 1929 Minya Konka and Amne Machin, two peaks along the border of China and Tibet, were incorrectly measured by him to be 30,000 feet (9,144 m).

Alberto Santos-Dumont
1873-1932

Brazilian aviator who pioneered the development of airships and made Europe's first sustained airplane flight. Living in Paris, Santos-Dumont built and flew a series of 14 increasingly sophisticated airships between 1898 and 1905. He turned his attention to heavier-than-air craft in 1906 and, after several abortive hops, he took off and flew for 722 feet (220 m) on November 12. The flight energized Europe's aviators, though their achievements still lagged well behind those of their American rivals.

Blanche Scott
1890-1970

American aviator who was the first woman to fly an airplane solo. In 1910 Scott drove an automobile across the United States alone. That achievement emboldened her to ask Glenn Curtiss for flying lessons. Curtiss attempted to thwart Scott's efforts by placing a block on her aircraft's throttle. When the block fell off, Scott's plane rose 40 feet above the ground. The Aeronautical Society of America refused to recognize Scott's flight, claiming it was unintentional. She became a popular stunt pilot.

Robert Falcon Scott
1868-1912

English explorer and naval officer who made many attempts to be the first to the South Pole. He finally reached it on January 18, 1912, only to find a Norwegian flag and a tent with a note from explorer Roald Amundsen stating that he had reached the pole on December 14, 1911. In despair, the Scott party began their return trek but perished from exhaustion and lack of supplies.

Thomas O. Selfridge
1882?-1908

American military officer who was the first passenger to die in an airplane crash. Lieutenant Selfridge, fascinated by aviation, acted as the army's official liaison to the Aerial Experiment Association founded by Glenn Curtiss and Alexander Graham Bell. Assigned to evaluate a new Wright Brothers design for possible army use, Selfridge died of injuries suffered when the demonstration aircraft, piloted by Orville Wright, suffered a mechanical failure and crashed on September 17, 1908.

Mary (May) French Sheldon
1847-1936

American explorer, publisher, and author who was one of the first European women to explore the African interior. Sheldon, born to a wealthy family and educated in Europe, founded a London publishing house before marrying Eli Sheldon, a fellow publisher, in 1876. She left him behind to explore Africa. Sheldon narrated her travels in articles and her book *Sultan to Sultan* (1892), focusing on the women and children in the territories she visited, a notable difference from her contemporaries. Her writing made significant contributions to both ethnology and geography.

Lowell Smith

American aviator who flew during the first air circumnavigation of the globe in 1924. During this trip, Smith helped fly, performed maintenance on the aircraft, and even slept on the wings to guard the aircraft from sightseers at night. This journey, which spanned over 27,000 miles (43,500 km) and five months, began with four planes, two of which completed the circumnavigation.

Robert M. Stanley

American aviator who made the first flight in an American jet-propelled aircraft in 1942. Stanley, Bell Aircraft's chief test pilot, flew the Airacomet on October 1, 1942. This airplane, designated the XP-59A, was the direct ancestor of every subsequent American jet. During its first flights, to maintain security, the Airacomet was disguised by placing a fake propeller on its nose. Most who first saw it fly, however, refused to believe an airplane with no propeller could really fly.

Vilhjalmur Stefansson
1879-1962

Canadian anthropologist who studied the Inuit language and spent many years on the south side of Victoria Island living among and observing a group of so-called Copper Inuit. He moved to the United States when he was two, eventually attending the University of North Dakota and later studying anthropology. His interest in the Inuit people began when he joined an Arctic expedition led by Ejnar Mikkelsen in 1906, and continued with other expeditions, including the first Canadian Arctic Expedition in 1913.

Sir Mark Aurel Stein
1862-1943

Hungarian-born British archaeologist and geographer who traveled extensively in central and southern Asia in the 1920s and 1930s. Stein investigated and recorded a large number of prehistoric, classical, and medieval sites in Iran, India, and particularly Chinese Turkistan. His most significant discovery (1906-08) was the uncovering of frescoes (and a vast collection of priceless manuscripts) in Ch'ien Fo-Tung, The Cave of the Thousand Buddhas, while tracing ancient caravan routes between China and the West. Stein became a British subject in 1904, and was knighted in 1912.

Albert W. Stevens

American balloonist who, in November 1935, ascended to a then-record altitude of 72,395 feet (19,150 m) in the balloon, *Explorer II.* During a previous ascent in July, 1935, Stevens's balloon ruptured at an altitude of 60,000 feet (15,870 m), nearly taking Stevens and his crew to their deaths before they bailed out at less than 6,000 feet (1,590 m). During their November ascent, however, no such problems occurred, and the crew was able to make a number of important scientific observations.

Matthew Williams Stirling
1896-1975

American archaeologist and ethnologist who parlayed a childhood interest in collecting arrowheads into a pioneering career in archaeology and, ultimately, the discovery of the lost Olmec civilization, one of the earliest cultures in Mesoamerica (far more ancient than the earliest Mayans). From 1928 to 1958 Stirling was Chief of the Bureau of American Ethnology. During this tenure he undertook a variety of investigations including his long-term field project (1939-46), supported by the Smithsonian and the National Geographic Society, studying the Olmec culture.

George Henry Hamilton Tate
1894-1953

English-born mammalogist who from 1921 until his death was affiliated with the American Museum of Natural History in New York. He made a number of specimen-collecting field trips to South America, Africa, and the Southwest Pacific and published extensively on various mammal groups. He also collected insects and plants and had research interests in geography and ecology. Tate's half-dozen books included accounts of South American mouse opossums and the mammals of the Pacific and eastern Asia. During World War II he was exploration chief for the American Rubber Development Corporation in Brazil, but the rapid development of synthetic rubber terminated his mission there. His career was cut short by leukemia and complications from several diseases contracted in Africa.

Miriam O'Brien Underhill
1898-1976

American mountaineer who pioneered all-female climbing expeditions. Considered the best American female mountain climber of her time, Underhill often ascended peaks without guides. After preliminary climbs of the Alps and North American Rockies, Underhill organized all-female ascents of the Aiguille de Grépon (in 1929) and Matterhorn (in 1932) with skilled climbers belonging to the Groupe de Haute Montagne. In addition to accompanying her husband, Robert Underhill, on climbs, she wrote *Give Me the Hills* (1956), contributed to *National Geographic,* and edited *Appalachia* magazine.

Bradford Washburn
1910-

American mountain climber, cartographer, and explorer whose adventurers have taken him to mountain peaks around the world, often as the first man to reach and map them. He made climbs at such sites as Mount Rainer, the French Alps, and Mount Kennedy. He is also an accomplished aerial photographer, and has held the position of director of the Boston Museum of Science for several decades. In 1999, he was part of a research team that set the height of Mount Everest at 29,035 feet (8,850 m), a figure that the National Geographic Society has adopted as the official elevation.

Alfred Lothar Wegener
1880-1930

German geophysicist whose theory of continental drift precluded the current understanding of plate tectonics. Born in Berlin, the son of an orphanage director, Wegener earned a doctorate in astronomy, and went on several harrowing expeditions to Greenland to study polar air circulation. While recovering from a World War II injury, Wegener studied the strange coincidences of identical plant and animal life existing on continents separated by oceans. He theorized the Earth had once consisted of a single land mass, which he called Pangaea (Greek for "all earth"), that broke into pieces some 200 million years ago. His "continental drift" theory was widely challenged at the time, but proved basically true after geophysicists established the idea of plate tectonics, several decades later.

Charles Leonard Woolley
1880-1960

British archaeologist known for his discoveries and excavations at Ur, a city in ancient Sumer (now a part of Iraq). Between 1922-34 he directed the work at Ur, revealing a previously unknown people, the Sumerians. Here he also found geologic evidence for a great flood, possibly the flood described in the Bible. His work greatly advanced knowledge of ancient Mesopotamian civilizations by revealing much about everyday life, art, architecture, literature, government, and religion.

Fanny Bullock Workman
1859-1925

American mountaineer who set altitude records. Supported by her wealthy Massachusetts family, Workman and her husband traveled abroad. They wrote eight books about their foreign adventures. The couple bicycled 20,000 miles through Asia. The Himalayas inspired Workman to pursue mountaineering. Her climb to the top of Pinnacle Peak (22,815 feet or 6,954 m) in 1906 was the highest any woman had reached.

Workman disputed rivals' claims, and her record was not broken until 1934. The Royal Geographic Society sponsored Workman's final 1912 expedition to Srachen Glazier.

Olga Yamshchikova

Soviet aviator who scored 17 kills during air combat in World War II. Yamshchikova was a flying instructor who volunteered for combat duty. Due in part to manpower shortages (the Soviet Union suffered millions of casualties, mostly among men) and partly because of the egalitarian ideals of Communism, over 1,000 women flew air-combat missions during the war. Yamshchikova shot down nearly as many planes as all the other pilots in her all-female 586th Fighter Regiment combined.

Sir Francis Edward Younghusband 1863-1942

British Army officer who explored western China, Afghanistan, and the mountains of northern India. He was the first European to explore a 1,200-mile (1,931-km) section of the Gobi Desert and in 1904 was the first European to enter Lhasa, Tibet, in more than 50 years. There he met with the Dalai Lama and negotiated a treaty that guaranteed British political and trade interests. He was the youngest person to be elected to the Royal Geographical Society in Britain.

Bibliography of Primary Sources

Books

Akeley, Delia. *Jungle Portraits* (1939). A volume of photographs taken by Akeley during her travels in Africa.

Bates, Daisy. *The Passing of the Aborigines* (1938). In this best-selling volume, a collection of reprinted newspaper articles, Bates recounts her life among the aboriginal tribes of Australia and describes their imperiled culture, offering new awareness for anthropologists.

Beebe, Charles. *Half Mile Down* (1943). In this work, Beebe recounts his record-setting 1934 deep-sea descent with Otis Barton aboard their bathysphere.

Bingham, Hiram. *Lost City of the Incas* (1948). In this work, Bingham discusses his expeditions to Machu Picchu and his studies of Inca civilization in Peru.

Evans, Arthur. *The Palace of Minos* (4 volumes, 1921-1936). In this work Evans described his work to uncover the ruins of the ancient city of Knossos in Crete, and with it a sophisticated Bronze Age civilization he named "Minoan."

Hedin, Sven. *Southern Tibet: Discoveries in Former Times Compared with My Own Researches, 1906-1908* (1917-22). A nine-volume survey of the exploration and mapping of Tibet up until the early twentieth century, including Hedin's landmark 1906-08 expedition. Hedin's most significant contribution to geography, the volumes contain impeccable scholarship, lavish illustrations, numerous maps, and much historical, ethnographic, meteorological, astronomical, geological, and botanical information about the entire central Asia region.

Hedin, Sven. *Trans-Himalaya* (1909-13). In this work, written for a nonscientific readership, Hedin recounts his influential 1906-08 expedition through Tibet. The three volumes (in the English edition) total nearly 1,300 pages and are illustrated with photographs and sketches made by Hedin.

Heyerdahl, Thor. *Kon-Tiki* (1950). In this work, Heyerdahl recounts his famous 1948 voyage from South America to Polynesia aboard a balsa raft.

Kidder, Alfred. *An Introduction to the Study of Southwestern Archaeology* (1924). This work by Kidder, an American archaeologist and anthropologist who pioneered work in the southwestern United States, represents the first comprehensive regional synthesis of North American archaeology.

Lawrence, T. E. *The Seven Pillars of Wisdom* (1926). This book made Lawrence—known ever after as "Lawrence of Arabia"—a legendary figure in the popular imagination. The book describes Lawrence's military exploits in Arabia during World War I.

Maillart, Ella. *Forbidden Journey* (1937). Maillart recounts her overland journey with journalist Peter Fleming through northern China and Tibet to Kashmir, a region largely unvisited by Western travelers.

Maillart, Ella. *The Land of the Sherpas* (1955). Maillart recounts her travels through Nepal, whose borders were opened in 1949.

Markham, Beryl. *West With the Night* (1942). In this memoir, Markham recounts her record-setting nonstop, solo transatlantic flight from England to the American continent, as well as her childhood and life in east Africa during the early twentieth century.

Peck, Annie Smith. *A Search for the Apex of America* (1911). In this memoir, Peck describes her experiences during her 1908 ascent of the Peruvian mountain Hauscarán.

Post, Wiley. *Around the World in Eight Days* (1931). Post recounts his record-setting flight around the world in 1931.

Rasmussen, Knud. *Across Arctic America* (1925). In this two-volume work, translated into English in 1927, Rasmussen recounts his extended expedition through Greenland, the Canadian Arctic, and Alaska, during which he studied the Eskimo people and culture, mapped their migration routes, and attempted to demonstrate his theory that the Eskimos as well as North American Indians had common ancestors in tribes that migrated from Asia across the Bering Strait.

Roosevelt, Theodore. *African Game Trails: An Account of the African Wanderings of an American Hunter-Naturalist* (1910). Recounts Roosevelt's 1909 hunting and scientific expedition with his son in Nairobi, Uganda, Sudan, and Khartoum.

Roosevelt, Theodore. *Through the Brazilian Wilderness* (1914). A chronicle of Roosevelt's three-month expedition with Cândido Mariano da Silva Rondon to the River of Doubt in Brazil, during which they produced the first map of the river, along with numerous animal specimens.

Stark, Freya. *Valley of the Assassins* (1934). Recounts Stark's journey through the distant region of the Elburz, a mountain range in Iran, where she sought information about an ancient Muslim sect known as the Assassins. This classic account was awarded the prestigious Gold Medal from the Royal Geographic Society.

Underhill, Miriam O'Brien. *Give Me the Hills* (1956). A memoir by the American mountaineer who pioneered all-female climbing expeditions.

Wilkins, George Hubert. *Flying the Arctic* (1928). Recounts Wilkins's record-setting early Arctic and Antarctic airplane flights.

JOSH LAUER

Life Sciences

Chronology

1901 Spanish histologist Ramón y Cajal revolutionizes understanding of the nervous system by showing that nerves are individual cells that do not touch; rather, signals jump a gap that comes to be known as a synapse.

1902 English physiologists Ernest H. Starling and William M. Bayliss isolate secretin in the duodenum, and Starling suggests a name for such substances: *hormones,* from a Greek term meaning "rouse to activity."

1903 American Walter S. Sutton and German Theodore Boveri, working independent of one another, present the theory that hereditary factors are passed down through chromosomes.

1905 German chemist Richard Willstätter discovers the structure of chlorophyll.

1909 Russian-American chemist Phoebus Levene discovers the chemical distinction between DNA (deoxyribonucleic acid) and RNA (ribonucleic acid.)

1912 Casimir Funk, a Polish-American biochemist, notes the existence of dietary substances in the amine group, which he dubs "life-amines" or "vitamines"; in the following year, Ernest V. McCollum discovers the first vitamin, A.

1926 James Batcheller Sumner, an American biochemist, conducts the first crystallization of an enzyme and proves that enzymes are also proteins.

1928 Frederick Reece Griffith, Jr., an American physiologist, shows that genetic information is transmitted chemically.

1928 Scottish bacteriologist Alexander Fleming discovers penicillin, the first antibacterial "wonder drug."

1935 Wendell Meredith Stanley, an American biochemist, first crystallizes a virus.

1935-38 Konrad Lorenz establishes the concept of filial imprinting, in which newly hatched geese will accept as their mother the first animal they see—even if that animal is a member of another species.

1944 Canadian scientists Oswald Theodore Avery, Maclyn McCarty, and Colin Munro Macleod discover that DNA carries a genetic "blueprint."

Overview: Life Sciences 1900-1949

Previous Period

Three giants of nineteenth-century life sciences set the stage for many of the biological investigations undertaken in the first half of the twentieth century. Charles Darwin's (1809-1882) *On the Origin of Species* (1859) established the theory of evolution by natural selection and focused attention on questions of how organisms are related to each other and how life on earth has changed over time. Claude Bernard's (1813-1878) research in physiology, particularly on the chemical functioning of animal tissue, and his writings advocating the experimental method helped make biology less of a purely observational science in which most emphasis was placed on classification and on describing the anatomy of organisms. Because of Bernard's influence, biology became more experimental in the twentieth century. Louis Pasteur's (1822-1895) research on infectious disease and on the chemistry of microbes made careful microscopic and biochemical studies more crucial to the biological sciences.

Genetics and Evolution

Gregor Mendel (1822-1884) was another important biologist of the nineteenth century, but his work remained unknown for years, having been published in an obscure journal in 1869. His research on heredity in pea plants was rediscovered in 1900, and when related to the work of contemporary geneticists, led to the development of modern genetics, the study of how traits are passed on from one generation to the next. Thomas Hunt Morgan (1866-1945) and his coworkers used fruit flies with particular mutations (hereditary defects) to discover the position of genes on chromosomes, the structures in cells that had been identified as the probable seat of genetic material. Also working with fruit flies, Hermann Muller (1890-1967) discovered in 1927 that mutations can be caused by X-rays, demonstrating that such mutations are essentially chemical changes and indicating the dangers of X-rays.

More theoretical work in genetics, particularly the development of statistical techniques by Ronald Fisher (1890-1962) and others on how rapidly a mutation could become common in a population, provided crucial information for relating genetics to the theory of evolution. This was an important contribution to what became known as the Modern Evolutionary Synthesis, which developed in the 1930s and 1940s. This synthesis was based on work not only in genetics, but also in ecology, paleontology (the study of fossils), and taxonomy (classification). All contributed to the idea that natural selection acting on random genetic variations could explain evolutionary change, and thus provided substantial support for Darwin's theory of evolution.

Developmental and Cellular Biology

Embryology, the study of the development of multi-cellular organisms, was one area of biology where a great deal of experimentation was done in the first half of the century. For example, Hans Spemann (1869-1941) found that tissue he called the "organizer," when transplanted from one embryo into another, could induce the recipient embryo to develop a whole new nervous system. There was also a strong focus on the composition and function of chromosomes, the structures that biologists linked with heredity since they seemed to reproduce and be passed on from one cellular generation to the next. Biologists also linked sex determination to specific chromosomes, which they designated the sex chromosomes. In an important discovery that set the stage for the development of genetics in the second half of the twentieth century, in 1944 Oswald Avery (1877-1955) and his coworkers identified DNA as the genetic material in bacterial chromosomes. This led to further research establishing DNA as the basis of heredity in all organisms.

Many other biochemical studies also contributed to the understanding of cellular processes. Fritz Lipmann (1899-1986) found that adenosine triphosphate (ATP) was an important energy molecule in the cell, and Hans Krebs (1900-1981) worked out the details of the citric acid or Krebs cycle, explaining how the chemical energy in sugar is transferred to ATP. Plant chemistry was also being explored, with the discovery of plant hormones that controlled many developmental processes and the identification of chlorophyll as the molecule that absorbs the Sun's energy in photosynthesis.

In terms of structural biology, Wendell Stanley (1904-1971) was the first to define the nature and organization of a virus in his research on the tobacco mosaic virus. This followed earlier work identifying bacteriophages as viruses that infect bacteria. Bacteriophages were later used in studies on genetics because they were so small and simple. For the much larger animal cells, James Gray (1891-1975) helped to establish the field of cytology, the study of cell structures, functions, and pathology. With the development of the electron microscope in the 1930s and 1940s, it finally became possible to look at viruses as well as to see much more clearly the intricate structures of cells.

Behavioral Biology

At the very beginning of the century, Santiago Ramón y Cajal (1852-1934) discovered that the nervous system was made up of a network of individual cells that do not touch, but instead send signals across microscopic gaps (synapses). Twenty years later, Otto Loewi (1873-1961) identified acetylcholine as one of the chemical signals that cross synapses. In other work on the nervous system, Charles Sherrington (1857-1952) discovered reflex arcs, in which nerve cells carry impulses into the spinal cord where other impulses are triggered that go directly out to the source of the stimulus, bypassing the brain. The other great control system of the body, the endocrine system, which releases hormones, also received a great deal of attention. William Bayliss (1860-1924) and Ernest Starling (1866-1927) discovered the first hormone, secretin, in 1902 and showed that it was involved in stimulating the secretion of digestive juices. Walter Cannon (1871-1945), one of the great physiologists of the twentieth century, found the connection between several hormones and the emotions.

It was during the first half of the twentieth century that ethology (the study of animal behavior) developed as a significant area of research in biology. Karl von Frisch (1886-1982) discovered that bees communicate through specific movements or "dances." Konrad Lorenz (1903-1989) studied imprinting in geese: the first thing they see when they hatch, usually the mother, is imprinted on their brains as the object to rigidly follow. Research was also done in behavioral ecology, the study of social insects and animal societies. The investigation of human societies flourished with anthropologists going all over the world to examine the customs of isolated indigenous populations. Franz Boas (1858-1942) studied the Indians of the Pacific Northwest and also the peoples of northern Asia, to whom these Indians are closely related. His student, Margaret Mead (1901-1978), became famous for her work on the Pacific island of Samoa.

Biology and Society

Research on human behavior showed the relevance of biology to societal issues, but during the first half of the twentieth century there were also several cases where the interpretation of biological concepts led to controversy outside of biology. One of the most famous cases was the trial of John Scopes (1900-1970), a biology teacher tried for breaking a Tennessee law that forbade teaching Darwin's theory of evolution. The public attention this trial received showed the extent of anti-evolution sentiment in the United States.

Eugenics, the concept that only those who have "good" genes should be allowed to reproduce, became a popular idea in both the United States and Europe, but by mid-century it had been discredited because of the way the idea was abused in Nazi Germany. In the Soviet Union, the director of the Institute of Genetics, Trofim Lysenko (1898-1976) distorted genetics research for his own purposes, condemning many noted Soviet geneticists and doing long-term harm to this science in his country. But there were also positive steps taken in establishing the relationship of science to society at large, especially in the United States where corporate and government support of science increased and major foundations became important patrons of scientific and medical research. There was also a great deal of agricultural research done by state universities and agro-industries that led to the development of hybrid corn and other enhanced plant strains, improved animal breeds, and the increased use of fertilizers.

The Future

The second half of the twentieth century saw the continuing growth of many areas of biology that had become important early in the century. Genetics blossomed still further after the structure of DNA was discovered in 1953; this led to huge growth in the field of molecular biology, which investigates the molecular basis of genetics. Ecology, which had grown as a science during the early part of the century with the development of such concepts as food webs and organismal communities, became an even more significant part of biology as the environmental movement emerged in the 1960s and 1970s.

ROBERT HENDRICK

The Rediscovery of Mendel's Laws of Heredity

Overview

In the 1860s, in an Augustinian monastery garden, Gregor Mendel (1822-1884) carried out a systematic experimental analysis of plant hybridization and inheritance patterns. Although Mendel published an account of his work and attempted to communicate with leading naturalists of his day, his work was essentially ignored for over 30 years. At the beginning of the twentieth century, however, Mendel and his laws were "rediscovered" by Hugo Marie de Vries, Karl Franz Joseph Correns, and Erich Tschermak von Seysenegg, firmly attaching Mendel's name to the basic laws of genetics. William Bateson, who came close to rediscovering Mendel's laws through his own experiments, became one of the leading advocates of Mendelian genetics.

Background

After exploring various animal and plant systems, Mendel conducted studies of 34 different strains of peas and selected 22 kinds for further experiments. He chose to study traits that were distinct and discontinuous and exhibited clear patterns of dominance and recessiveness. The "law of segregation," also known as Mendel's first law, refers to Mendel's proof that recessive traits reappear in predictable patterns. Crosses of peas that differed in one trait produced the now-famous 3:1 ratios. Complex studies that followed the variations of two or three traits led to the patterns of recombination now known as Mendel's second law, or the "law of independent assortment."

Mendel discussed his results at a meeting of the Brno Society for Natural History in March 1865 and published his paper "Research on Plant Hybrids" in the 1866 issue of the Society's *Proceedings*. He also sent reprints of his article to prominent scientists but received little attention and virtually no understanding.

Contemporaries tended to dismiss Mendel's "numbers and ratios" as merely empirical and devoid of a respectable theoretical framework. Sir Ronald A. Fisher (1890-1962), however, argued that the experimental design reported in Mendel's classic paper was so elegant that the experiments had to have been a confirmation, or demonstration, of a theory Mendel had previously formulated. Furthermore, Fisher claimed that Mendel's ratios are closer to the theoretical expectation than sampling theory would predict and he insisted that such results could not be obtained without an "absolute miracle of chance." Although Mendel was discouraged by the lack of response from the scientific community, he remained convinced of the fundamental value and universality of his work.

Impact

Although Gregor Mendel's name is now attached to the fundamental laws of genetics, his work was essentially ignored and misunderstood for over thirty years. Classical genetics, therefore, began not with the publication of Mendel's papers in the 1860s but at the beginning of the twentieth century with the independent rediscovery of the laws of inheritance by three botanists—Hugo de Vries (1848-1935), Carl Correns (1864-1935), and Erik Tschermak (1871-1962). Between the 1860s and 1900 developments in the study of cell division, fertilization, and the behavior of subcellular structures had established a new framework capable of accommodating Mendel's "ratios and numbers." A revival of interest in discontinuous, or "saltative," evolution was probably a significant factor in the approach the rediscoverers brought to the study of heredity; if the sudden appearance of new character traits led to the formation of new species, studies of the transmission of these traits using Mendelian breeding experiments might furnish the key to a new science of heredity and evolution.

Historians of science suggest that de Vries, Correns, and Tschermak attempted to emphasize their own creativity by claiming to have discovered the laws of inheritance before finding Mendel's paper. Citing Mendel's earlier work also helped them avoid a priority battle. During the 1890s de Vries had observed the 3:1 ratio from his own F2 hybrids, the reappearance of recessive traits, and independent assortment. By 1900 he had demonstrated his law of segregation in hybrids of 15 different species. Later, de Vries complained that it was unfair for Mendel to be honored as the founder of genetics.

Correns and Tschermak, however, were more generous towards Mendel than de Vries. They even suggested the use of the terms "Mendelism" and "Mendel's laws." Their work, however, had been anticipated by de Vries and they did not have as much at stake as he in any potential priority war. Correns admitted that the task of discovering Mendel's laws in 1900 was much simpler than it had been in the 1860s. Growing hybrids of maize and peas for several generations and analyzing new developments in cytology apparently had led Correns to think about the transmission of paired characters. A literature search led him to Mendel's paper and the realization that Mendel had anticipated his work by 35 years. Correns soon found out that Hugo de Vries was about to claim priority. While Correns praised Mendel, he was very critical of de Vries and often implied that de Vries was trying to suppress references to their predecessor. On the other hand, Tschermak often complained that his contributions to genetics had been slighted. Tschermak's interest in the question of hybrid vigor led to a series of experiments on the effects of foreign pollen. When he analyzed the results of these experiments, Tschermak observed the 3:1 ratio; after conducting a literature search, Tschermak also discovered Mendel's paper. Tschermak's work, however, was not as extensive as Mendel's. In March 1900 Tschermak received a reprint of de Vries's paper, "On the Law of Segregation of Hybrids," followed by a copy of Corren's paper, "Gregor Mendel's Law." Tschermak quickly prepared an abstract of his own work and sent copies of his article on artificial hybridization to de Vries and Correns to establish himself as a participant in the rediscovery of Mendel's laws.

The apparently simultaneous discovery of Mendel's laws by de Vries, Correns, and Tschermak suggests that by 1900 rediscovery had become inevitable. Indeed, others such as William Bateson (1861-1926) were also working along similar lines. In the 1890s Bateson had begun to analyze the inheritance patterns of discontinuous variations in the offspring of experimental hybrids; therefore, when Bateson read Mendel's paper, he was already thinking about the inheritance of discrete units. A reprint of de Vries's paper led Bateson to the 1865 Brno *Proceedings*. Bateson soon became a dedicated defender of Mendelian genetics. He coined and popularized many of the terms now used by geneticists, including the words "genetics" (from the Greek word for descent), "allelomorph" (allele), "zygote," "homozygote," and "heterozygote." The Danish botanist Wilhelm L. Johannsen (1857-1927) introduced the term "gene" to replace such older terms as "factor," "trait," and "character" and coined the terms "phenotype" and

Hugo de Vries examines flowers. *(AP/Wide World Photos. Reproduced with permission.)*

"genotype." In 1902 Bateson published *Mendel's Principles of Heredity.* The text included a translation of Mendel's paper and Bateson's assertion that Mendel's laws would prove to be universally valid. Indeed, further studies of the patterns of inheritance proved that Mendel's laws were applicable to animals as well as plants.

LOIS N. MAGNER

Further Reading

Bateson, W. *Mendel's Principles of Heredity—A Defense.* Cambridge, England: Cambridge University Press, 1902.

Bennett, J. H., ed. *Experiments in Plant Hybridization: Gregor Mendel.* Edinburgh: Oliver and Boyd, 1965.

Brink, A. P., and E. D. Styles, eds. *Heritage from Mendel: Proceedings of the Mendel Centennial Symposium.* Madison: University of Wisconsin Press, 1967.

Corcos, A. F., and F. V. Monaghan. *Gregor Mendel's Experiments on Plant Hybrids: A Guided Study.* New Brunswick, NJ: Rutgers University Press, 1993.

Darden, L. *Theory Change in Science: Strategies from Mendelian Genetics.* New York: Oxford University Press, 1991.

Dunn, L. C., ed. *Genetics in the Twentieth Century.* New York: Macmillan, 1951.

Mayr, E. *The Growth of Biological Thought.* Cambridge, MA: Harvard University Press, 1982.

Olby, R. *Origins of Mendelism.* 2nd ed. Chicago: University of Chicago Press, 1985.

Orel, V., and A. Matalová, eds. *Gregor Mendel and the Foundation of Genetics*. Brno, Czechoslovakia: The Mendelianum of the Moravian Museum, 1983.

Stern, C., and E. Sherwood, eds. *The Origin of Genetics: A Mendel Source Book*. San Francisco, CA: W. H. Freeman, 1966.

The Fruit Fly Group Contributes Key Discoveries to Genetics

Overview

A successful approach to proving that genes are located on the chromosomes in a specific linear sequence evolved in the laboratory of Thomas Hunt Morgan (1866-1945). Using the fruit fly, *Drosophila melanogaster,* as their model system, Morgan and his research associates in the famous "fly room" at Columbia University—Alfred H. Sturtevant (1891-1971), Calvin B. Bridges (1889-1938), Hermann J. Muller (1890-1968), Curt Stern (1902-1981), and others—exploited correlations between breeding data and cytological observations to define and map the genes on the chromosomes. Morgan's fly room became a distinguished center of genetic research and a magnet for both ambitious young scientists and well-established senior scientists.

Background

By 1910 experiments carried out by Walter S. Sutton (1877-1916), Theodor Boveri (1862-1915), Nettie M. Stevens (1861-1912), Edmund B. Wilson (1856-1939), and other scientists had provided suggestive evidence for the hypothesis that the inherited Mendelian factors might actually be components of the chromosomes. Advocates of the "chromosome theory" predicted that each chromosome would carry many genes because the number of traits that had been subjected to genetic analysis was greater than the number of chromosomes in any cell nucleus. Therefore, the factors carried by any particular chromosome might be inherited together, rather than independently. At first, Mendelian geneticists found little evidence of linkage and many biologists remained skeptical of the chromosome theory until work conducted by Thomas Hunt Morgan and his associates between 1910 and 1915 established the foundations of modern genetics.

Morgan was awarded the 1933 Nobel Prize in physiology or medicine for his contributions to the chromosome theory of heredity, but he was a zoologist with broad interests, including experimental embryology, cytology, and evolutionary theory. After earning his Ph.D. from Johns Hopkins University, Morgan replaced Edmund B. Wilson as head of the biology department at Bryn Mawr. In 1904 Morgan was appointed head of experimental zoology at Columbia. He moved to the California Institute of Technology (Caltech) in 1928.

Morgan's first research interest was experimental embryology, but, after visiting Hugo de Vries's garden, he became fascinated by mutation theory. At the beginning, Morgan was skeptical of Darwin's theory of natural selection and his emphasis on continuous variations. After attempts to study mutations in various animals, including mice, rats, pigeons, and lice, Morgan decided that the fruit fly, *Drosophila melanogaster,* provided the ideal system for studying the complex relationships among genes, traits, and chromosomes. Other scientists, including W. E. Castle at Harvard, W. J. Moenkhaus at Indiana University, and Nettie Stevens at Bryn Mawr, had been breeding *Drosophila* in the laboratory for some time before 1908, when Morgan established his famous "fly room." The fruit fly has a life cycle of about two weeks, produces hundreds of offspring, and is easy to maintain in the laboratory because thousands can be raised in a few bottles on inexpensive banana mash. Cytological studies of the fruit fly are simplified by the fact that it has only four pairs of chromosomes per nucleus. Natural populations of the fly provided many easily recognizable inherited traits.

Although Morgan had a long and distinguished career, most of his important work was carried out in the Columbia fly room between 1910 and 1915. At first, Morgan was quite skeptical about the universality of Mendel's work. Indeed, in 1910 Morgan sent to the journal *American Naturalist* a paper in which he argued that the chromosomes could not be the carriers of Mendelian factors because, if they were, characters on the same chromosome would be linked together. By the time this paper was published, however, Morgan's experiments on fruit flies had convinced him of the general validity of Mendel's laws. Morgan also realized that he could explain apparent deviations from Mendel's law of independent assortment in terms of the linkage between genes that were situated on the same chromosome.

Impact

The first definitive demonstration of linkage in *Drosophila* involved the sex-linked traits "white eye," "yellow body," and "rudimentary wings." (Female flies have two X chromosomes—XX—and males have only one X chromosome—XY.) White eye, yellow body, and rudimentary wing mutations were found almost exclusively in male flies; these genes are found on the X chromosome. Evidence for recombination of genes was obtained from experiments on double mutants. Recombination was explained in terms of exchanges between chromosomes. Franciscus Alphonsius Janssens (1863-1924), who proposed the "chiasmatype hypothesis" in 1909, had reported cytological evidence of such "crossing over." Chiasma seemed to be sites of chromosome breakage and repair where physical exchanges of chromosomal material occurred. Studies of recombinants would, therefore, link cytology and Mendelian genetics.

Proof that the behavior of chromosomes accounts for the physical basis of Mendelian genetics was established in studies of X-linked mutations. The degree of linkage established by breeding tests should reflect the distance between genes on a chromosome. In testing this hypothesis, Alfred H. Sturtevant constructed the first chromosome map. His report was published as "The linear arrangement of six sex-linked factors in *Drosophila,* as shown by their mode of association" in the *Journal of Experimental Zoology* in 1913. Within two years, Morgan and his associates described four groups of linked factors that corresponded to the four pairs of chromosomes. Sturtevant later described these years in the fly room as a time of great friendship and cooperation.

In 1926 Morgan published *The Theory of the Gene* as a summation of the developments in genetics since the rediscovery of Mendel's laws. The chromosome theory replaced the "bean bag" image of Mendelian factors with a model of genes as beads on a string. Morgan assigned five principles to genes: segregation, independent assortment, crossing over, linear order, and linkage groups. Morgan's "theory of the gene" was not

Hermann Muller experimented with bottled fruit flies and atomic rays. *(Library of Congress. Reproduced with permission.)*

immediately accepted by all geneticists, but by about 1930 the chromosome theory was essentially synonymous with classical genetics. The apparent exceptions to Mendel's laws could now be attributed to linkage groups, crossing over, and multiple alleles. Morgan and his associates in the "fly group" exerted a profound influence on the development of genetics and cytology.

The work of the fly group proved that mutants were valuable for genetic analysis, but the rate of natural mutation was too slow to allow direct studies of the process of mutation. To overcome this obstacle, Hermann Joseph Muller attempted to find a way to increase the rate of mutation in *Drosophila.* Muller had become interested in evolution and genetics as a student at Columbia. He learned about the chromosome theory from Edmund B. Wilson and performed his doctoral research in Morgan's laboratory at Columbia. Muller carried out research on crossing over, chromosome behavior, and genetic mapping. He was awarded a Ph.D. in 1915. When Muller began his study of mutations, the term "mutation" was applied to many phenomena. Muller insisted on redefining the term as a change within an individual gene. By bombarding flies with x rays, Muller was able to produce several hundred mutants in a short time. In classical breeding tests, most of these induced mutations behaved like typical Mendelian genes. In 1927, at the Fifth International Congress of Genetics in Berlin, Muller gave his report on x ray-induced mutations in fruit flies.

While working with Nikolai Timofeeff-Ressovsky (1900-1981) in 1932, Muller attempted to use induced mutation as a way of understanding the physical nature of genes. The results of these experiments were ambiguous, but they eventually inspired a collaborative effort between Timofeeff-Ressovsky and Max Delbruck, who proposed the "target theory" of mutation in 1935. After leaving Berlin, Muller worked with the Russian geneticist Nikolai Ivanovitch Vavilov (1887-1943) at the Soviet Academy of Sciences Institute of Genetics. From 1937 to 1940, at the University of Edinburgh, Muller studied the effect of radiation on embryological development. He returned to the United States in 1945 and became a professor of zoology at Indiana University. Muller was awarded the Nobel Prize in physiology or medicine in 1946 for his research into the effects of x rays on mutation rates.

Another member of the group of scientists associated with Morgan was the Russian population geneticist Theodosius Dobzhansky (1900-1975), who joined Morgan's group in 1927 in order to learn about fruit fly genetics and cytology. Dobzhansky was particularly interested in linking the genetic analysis of natural populations over time with evolutionary theory. Using the methods developed by Morgan and his associates, Dobzhansky was able to combine studies of natural populations of fruit flies collected from California to Texas with cytogenetic analyses of *Drosophila* in the laboratory. His work on population evolution was valuable in subjecting the theory of evolution by natural selection to experimental analysis. His research and his many books provided insights into the relationship between Darwin's theory of evolution and Mendel's laws of inheritance.

Morgan was also associated with George Wells Beadle (1903-1989), who shared the Nobel Prize in physiology or medicine with Edward L. Tatum (1909-1975) for work establishing the "one gene one enzyme" theory. Beadle had become interested in the cytology and genetics of maize as a graduate student at Cornell University. From 1931 to 1933 Beadle conducted postdoctoral work on fruit flies in Morgan's laboratory at Caltech. After Morgan's death in 1945, Beadle replaced him as head of the biology division at Caltech. With classical genetics well established, Beadle wanted to understand the biochemical basis for gene activity. After attempting to use tissue culture and tissue transplantation to study gene expression during the development and differentiation of *Drosophila,* Beadle eventually decided to work with *Neurospora,* the common bread mold, instead. In collaboration with Edward L. Tatum, Beadle irradiated *Neurospora* and collected mutants that could no longer synthesize the amino acids and vitamins needed for growth. Beadle and Tatum then identified the steps in the metabolic pathways that had been affected by the mutations, establishing the relationship between mutant genes and defective enzymes. They shared the Nobel Prize in physiology or medicine in 1958.

Morgan and his fly group would be pleased to know that at the end of the twentieth century, after geneticists had essentially abandoned the fruit fly, *Drosophila* was again playing a leading role as a model system for studies in such fields as molecular genetics, developmental biology, and neurobiology.

LOIS N. MAGNER

Further Reading

Allen, G. *Thomas Hunt Morgan: The Man and His Science.* Princeton, NJ: Princeton University Press, 1978.

Carlson, E. A. *The Gene: A Critical History.* Philadelphia: Saunders, 1966.

Darden, L. *Theory Change in Science: Strategies from Mendelian Genetics.* New York: Oxford University Press, 1991.

Ludmerer, K. *Genetics and American Society: A Historical Appraisal.* Baltimore, MD: Johns Hopkins University Press, 1972.

Mayr, E. *The Growth of Biological Thought.* Cambridge, MA: Harvard University Press, 1982.

Morgan, T. H. *The Theory of the Gene.* New Haven, CT: Yale University Press, 1926.

Morgan, T. H., A. H. Sturtevant, H. J. Muller, and C. B. Bridges. *The Mechanism of Mendelian Heredity.* New York: Henry Holt and Company, 1915.

Shine, I., and S. Wrobel. *Thomas Hunt Morgan: Pioneer of Genetics.* Lexington, KY: The University Press of Kentucky, 1976.

Sturtevant, A. H. *A History of Genetics.* New York: Harper and Row, 1965.

Wilson, Edmund B. *The Cell in Development and Inheritance.* 3rd ed. New York: Macmillan, 1925.

Hermann J. Muller and the Induction of Genetic Mutations

Overview

Hermann J. Muller (1890-1967) made many contributions to the understanding of genetic mutation, the gene, and radiation genetics, but he is primarily remembered for his demonstration that mutations could be artificially induced by x rays. Muller began his research career in the famous "fly room" of Thomas Hunt Morgan (1866-1945), who established the chromosome theory of heredity. When Muller began his research, the term "mutation" was applied indiscriminately to many different phenomena. Muller helped narrow the definition to its present meaning: an inheritable change in a specific gene. In 1946 Muller won the Nobel Prize for physiology or medicine for the induction of genetic mutations with x rays. He also established the laws of radiation genetics, demonstrating the relationship between the radiation dose and the frequency of gene mutations and chromosomal rearrangements.

Background

Thomas Hunt Morgan and his research associates in the fly room at Columbia University proved that genes are located on the chromosomes in a specific linear sequence. During this period of exciting discoveries, Morgan's students included Hermann J. Muller, Alfred H. Sturtevant, Calvin B. Bridges, and Curt Stern. Morgan won the Nobel Prize in 1933 for his contributions to the chromosome theory of heredity. Sturtevant remembered their days in the fly room as a time of friendship and cooperation, but others reported signs of interpersonal differences and tensions. Muller in particular seemed to feel that he did not receive proper credit for his work at Columbia and he accused Morgan of sabotaging his professional advancement.

Morgan and his students used the fruit fly *Drosophila melanogaster* for both breeding experiments and cytogenetics, because it has only four pairs of chromosomes per nucleus and scores of easily recognizable inheritable traits. Although *Drosophila* seemed a promising model for saltative evolution (the sudden appearance of new character traits leading the formation of new species), the "fly group" did not find mutations that established new species. In 1926 Morgan published *The Theory of the Gene* to summarize developments in genetics since the rediscovery of Mendel's laws. Based on statistical studies of inheritance in *Drosophila,* Morgan assigned five principles to the gene: segregation, independent assortment, crossing over, linear order, and linkage groups.

Critics of Mendelian genetics often pointed to the kinds of mutants studied by the fly group as evidence that mutations were basically pathological and could not, therefore, play an important role in evolution. Morgan and his colleagues admitted that the kinds of traits studied in the laboratory were generally deleterious, but they believed that there might be many subtle, undetected mutations that were physiologically advantageous. The mutations investigated by Morgan's fly group did not support Hugo de Vries's (1848-1935) hope that new species could be created in one step. When de Vries gave a lecture in the United States in 1904, he suggested that the recently discovered "curie rays" produced by radium might be used to induce mutations in plants and animals. Shortly afterwards, Morgan made some attempts to induce "Devriesian mutations" in various animals, including *Drosophila,* by subjecting them to radium, acids, alkalis, salts, sugars, and proteins. These experiments did not produce either significant macromutations or new species and Morgan decided that his results were not worth publishing.

Impact

Work in Morgan's laboratory had demonstrated the value of mutants in genetic analysis, but the natural rate of mutation was too slow for direct studies of the process. To overcome this obstacle, Hermann Joseph Muller attempted to find methods that would increase the mutation rate. At that time, the term "mutation" was indiscriminately used to describe any sudden appearance of a new genetic type. Muller discovered, however, that some mutations were caused by Mendelian recombination, some were abnormalities in chromosome distribution, and others were changes in individual genes. He argued that in the interests of scientific clarity the definition of the term mutation should be limited to inheritable changes in specific genes. Muller tested various agents in his attempts to increase the frequency of mutations and proved that x rays could induce mutations in *Drosophila*.

Muller's discovery of the mutagenic effect of x rays established the new field of radiation genetics, and the presentation of his report at the Fifth International Congress of Genetics in Berlin in 1927 received international attention. When Muller began his research, *Drosophila* geneticists had discovered about 100 spontaneous mutations. Using x rays, Muller produced several hundred mutants in a short time. Most of these induced mutations were stable over many generations and behaved like typical Mendelian factors when subjected to breeding tests.

Muller had been interested in the sciences, particularly evolution, while he was still a high school student. Majoring in genetics at Columbia University, he came under the influence of Morgan and Edmund B. Wilson (1856-1939). He earned his B.A. from Columbia in 1910 and enrolled in Cornell Medical School. Because of his interest in the possibility of consciously guided human evolution, Muller decided to join Morgan's *Drosophila* group so that he could select research topics that would provide a better understanding of the processes of heredity and variation. He earned his Ph.D. in 1915 for research on the mechanism of crossing over by following the interrelationships of many linked genes. At the same time, he undertook a sophisticated analysis of variable, multiple-factor characters by using "marker genes." This work provided further confirmation of chromosomal inheritance and the stability of the gene and led to Muller's theory of balanced lethals. Known for his ingenuity in experimental design, Muller clarified obscure aspects of chromosome behavior and genetic mapping and established the principle of the linear linkage of genes in heredity.

After three years at the Rice Institute in Houston, Texas, and a brief return to Columbia as an instructor, Muller became a member of the faculty of the University of Texas at Austin, where he remained until 1932. Although Muller felt isolated in Texas, the years that he spent in Austin were remarkably productive and culminated in the induction of genetic mutations through the use of x rays in 1926. Muller exposed fruit flies to x rays and proved that ionizing radiation produces mutations both within individual genes and as larger chromosomal aberrations. Later studies showed that other forms of radiation and various chemicals can also cause chromosomal aberrations and mutations in individual genes. Such mutagenic agents have been associated with aging, cancer, and genetic diseases.

Although he was elected to the National Academy of Sciences in 1931, he continued to feel isolated and depressed by tension related, at least in part, to his relationship with Morgan. In 1932 he suffered a nervous breakdown and took an overdose of sleeping pills. His personal and professional problems as well as pressure caused by his outspoken defense of radical ideas and socialist causes expedited his decision to leave Texas.

Working with Nikolai Timofeeff-Ressovsky (1900-1981) at the Kaiser Wilhelm (now Max Planck) Institute in Berlin, Muller tried to gain insights into the physical nature of the genetic material by bombarding it with radiation. These experiments led Timofeeff-Ressovsky and Max Delbrück to propose the "target theory" of mutation. Because of the growing threat of Nazism, Muller left Germany in 1933 and accepted Nikolai Ivanovitch Vavilov's (1887-1943) invitation to work as senior geneticist at the Institute of Genetics of the Academy of Sciences of the USSR. Muller and Vavilov, an eminent Russian geneticist, fought against the growth of Lysenkoism, a pseudoscientific form of "Marxist biology" established by Trofim Denisovich Lysenko (1898-1976). Thoroughly disillusioned by the losing battle against Lysenkoism, Muller left Russia. From 1937 to 1940, at the Institute of Animal Genetics in Edinburgh, Muller studied the chromosomal basis of embryonic death from radiation damage. World War II forced Muller to return to the United States. In 1945, after completing a study of the relationship between aging and spontaneous mutations, he became professor of zoology at Indiana University, where he continued his research on radiation-induced mutations, genetic analysis, and the mechanism by which radiation induces its biological effects. He remained at Indiana University until his death.

The Nobel Prize in physiology or medicine was awarded to Muller in 1946 for his research into the effects of x rays on mutation rates, but his interest in genetics went far beyond laboratory studies of the fruit fly. Muller used the publicity generated by his Nobel Prize to campaign against medical, industrial, and military abuse of radiation. After World War II, he was a leader of the radiation protection movement and he applied his expertise to the problem of radiation sickness. To raise awareness of the dangers of radiation, he published estimates of spontaneous and induced mutation rates in humans. He formulated the modern concept of spontaneous gene mutation, noting that most mutations are detrimental and recessive and involve accidental physical-chemical effects on the genetic material; he, therefore, concluded that the gene was the basis of evolution. Muller published more than 300 scientific articles. His books include *The Mechanism of Mendelian Heredity, Out of the Night—a Biologist's View of the Future,* and *Genetics, Medicine and Man.* He was a founder of the American Society of Human Genetics.

Although he remained a socialist, he was a vocal critic of Lysenko and resigned from the Soviet Academy of Science in 1947 as a protest. He was also interested in biology education, especially genetics and evolution, in secondary schools. Even as a college student, Muller had demonstrated a deep interest in evolution and human genetics, including a concern for the preservation and improvement of the human gene pool. Most mutations, he noted, were stable, deleterious, and recessive. Knowledge of mutation, therefore, had profound implications for eugenics and human reproduction.

Muller often condemned the mainline eugenics movement in the United States, calling it a pseudoscientific dogma based on racial and class prejudices that served the interests of fascists and reactionaries. Eugenic theory generally claimed that since civilized societies no longer permitted the natural elimination of the unfit by starvation and disease, such societies had to limit the reproduction of the unfit and encourage the fit to procreate. This approach assumed that inherited biological factors caused people to be unfit. Muller's commitment to eugenics was very strong, but his ideas were more subtle and complex than those who promoted the mainline creed. He called for detailed studies of twins in order to assess the possibility that nurture as well as nature affected human development. Muller believed that a true science of eugenics should guide human biological evolution in a positive direction. Indeed, he saw eugenics as a special branch of evolutionary science. As indicated by the "Geneticists' Manifesto," signed by Muller and about 20 other scientists, geneticists should encourage a rational change in attitudes towards sex and procreation in order to improve the human gene pool for the sake of future generations.

Because civilized societies had blunted the force of natural selection, Muller feared that undesirable genes would accumulate in the human gene pool until the pool became excessively burdened by defective genes. Muller expressed his concern in a 1949 presidential address on "Our Load of Mutations" to the American Society of Human Genetics. He argued that modern medicine and technology allowed defective genes to accumulate and, therefore, reduced the evolutionary fitness of advanced nations. Geneticists should guide civilized nations towards voluntary eugenic reproductive controls. People with defective genes should refrain from further burdening the gene pool. Those with a good genetic endowment should be encouraged to participate in positive eugenic programs, including "germinal choice"; suitable women should be artificially inseminated with sperm donated by great men. In order to be sure that the men involved were truly worthy, Muller suggested collecting and storing semen from promising candidates and using it only after history had passed its judgment on them.

Deliberately and consciously, Muller used his scientific successes to act as a gadfly to the scientific community and the eugenics movement. He raised serious questions about the dangers of ionizing radiation as well as the wisdom and propriety of attempting to apply eugenic measures to human beings. Muller's discovery of x-ray-induced mutations has had a profound impact on research in essentially every branch of genetics as well as on evolutionary theory and medicine.

LOIS N. MAGNER

Further Reading

Adams, M., ed. *New Perspectives in the History of Eugenics.* New York: Oxford University Press, 1988.

Bowler, P. J. *The Mendelian Revolution: The Emergence of Hereditarian Concepts in Modern Science and Society.* Baltimore, MD: Johns Hopkins University Press, 1989.

Carlson, E. A. *Genes, Radiation and Society: The Life and Work of H. J. Muller.* Ithaca, NY: Cornell University Press, 1981.

Kevles, D. J. *In the Name of Eugenics: Genetics and the Uses of Human Heredity.* Berkeley, CA: University of California Press, 1986.

Morgan, T. H., A. H. Sturtevant, H. J. Muller, and C. B. Bridges. *The Mechanism of Mendelian Heredity.* New York: Henry Holt and Company, 1915.

Muller, H. J. *Studies in Genetics: The Selected Papers of H. J. Muller.* Bloomington: Indiana University Press, 1962.

Muller, H. J. *Man's Future Birthright: Social Essays of H. J. Muller.* E. A. Carlson, ed. Albany, NY: SUNY Press, 1973.

The Genetic Foundation of Natural Selection

Overview

By 1900 scientists had been trying to explain how natural selection worked for 40 years. The idea proposed by Charles Darwin (1809-1882) was accepted, but the mechanism was unclear. When they turned to microscopic investigation, identifying structures and processes within cells, the mechanism of natural selection became clear. Scientists studied specific traits, followed populations in the wild, and applied chemical and statistical techniques and principles to the problem. By 1950 natural selection was supported by the new field of genetics, and DNA was about to be deciphered. The focus of genetics then turned to the structural makeup of the gene. By mid-century this would revolutionize medicine as well as genetics. It would lead to manipulation of genes, gene therapies, identification of individuals by their DNA, replacement of defective genes, and identification and location of specific genes.

Background

Darwin's theory of evolution of living species was published in 1859, suggesting that living species survive and change by natural selection. The offspring of organisms that possess favorable characteristics in a changing habitat will survive, leave the most progeny, and pass favorable traits to their descendants. Scientists greeted this idea with fascination and questions, though many in the religious community believed that God controlled creation and change in all species. Darwin specified no mechanism for the change that made evolution and natural selection work. Biologists, botanists and other scientists set to work trying to find it. For 40 years scientists tried to locate and define a mechanism to explain the phenomenon.

A number of theories were formulated, including the idea that mutations arise spontaneously and are passed on to offspring. Another was orthogenesis, or the idea that life evolved naturally from small to large organisms. A third proposal was saltation, or changes made in great sudden leaps, and a fourth was pangenesis, or the idea that small particles mingle to create new characteristics. Some scientists tried to confirm the reality of natural selection by doing field studies, while others used comparative anatomy or embryology, the study of the development of the embryo of an organism.

During this time, they were unaware that a large body of work on the subject of genes and the passing on of traits had already been done by Gregor Mendel (1822-1884) and had been published in an obscure journal in 1866. In 1900 this work was discovered by three men working on the problem independently. Mendel had experimented with sweet peas and formulated laws that governed the passing of traits to offspring. He recognized and named genes as the agent that passed the traits and attributed the results to the dominant or recessive character of each trait.

The two threads came together in 1900. Evolutionary change by way of natural selection was accepted by scientists, although they were still unable to explain the mechanism. Cells and their internal structures had been described by two German scientists in the mid-1850s. Chromosomes, thread-like substances on which genes are carried, were believed to pass traits to offspring as they split during cell division. This was confirmed in the 1930s. Researchers turned to microscopic, internal structures of the reproductive system and began to focus on the inheritance of specific traits. By 1910 it was known that species changed over time, that Mendel's laws were generally correct, and that more work was needed to understand how the mechanism worked.

In the first decades of the twentieth century a number of studies were undertaken in the United States, Great Britain, Germany, France, and Russia that led to the new field of biology called genetics. The study of the passing of traits of living organisms from one generation to the next was a new field of study. Originally called Mendelism, it was created by the rediscovery and republication of Mendel's work on transmission and inheritance of characteristics from parent to offspring. The field earned its modern name by the 1920s.

Many scientists had set the framework for this infant science. Early researchers approached it from several different angles. According to Darwin, natural selection is the conversion of variations among individuals into variations between groups. Some geneticists tried to reconcile the contrasting realities that organisms resemble their parents and differ from their parents at the same time. One early researcher, Thomas Hunt Morgan (1866-1945), used a small vinegar or fruit fly (*Drosophila melanogaster*) in a series of lab experiments. Since this insect is found everywhere, it was used in many labs. Low in maintenance, it needs little space, has a short life, and a measurable generation time, all advantageous characteristics in the study of genetics. Evolutionary change occurs in mammals over periods too long to study in a lab.

In 1908 Morgan began to raise *Drosophila* and to introduce specific mutations in succeeding generations. For this work, Morgan took on many assistants. Working in a lab at Columbia that came to be known as "the fly room," they produced the first genetic map and combined Mendelian and cytological, or cell, theories. This group developed their findings into a chromosome theory of heredity—that is, that genes are carried on the chromosomes within generative cells and a parent passes his genes to his offspring in predictable ratios. The idea spread to other researchers, and by 1930 a revolution in genetics was underway. It recognized genes as the mechanism of natural selection.

Many scientists were instrumental in creating the mature science of genetics. J. B. S. Haldane (1892-1964), working at Oxford University in England in the 1920s, developed a mathematical procedure for dealing with evolution on a larger scale (i.e. in whole populations). Such a study, he said, could follow dominant or recessive genes in a population and explain its changes. Haldane was a chemist interested in respiration, thermodynamics, and enzymes. He discussed the extent of linkage of characteristics to observed bands or intervals on a chromosome. He formulated laws in 1922 that explained that cross-breeding of animal species produces sterile offspring, he mapped the x chromosome, and he showed a genetic linkage between hemophilia and color blindness. His 1932 book *Causes of Evolution*, reinforced the conservative Darwinian theory that natural selection rather than mutation was the driving force of evolution. Haldane brought together ideas from many fields. His work was largely theoretical and remained unproven for years, but his ideas pointed in new directions and were a milestone in understanding the mechanism of evolution and using mathematics to study biology.

Two other researchers were inspired by the new genetics and, working along similar lines, introduced new ideas into the process. Ronald A. Fisher (1890-1962) had graduated from Cambridge with degrees in math and physics. Soon he became interested in biometrics, or the application of statistics to variations in populations. Combining theory and practice, Fisher used mathematics to show that Mendel's laws do actually lead to the conclusions he drew from them and that Mendel and Darwin are compatible. His statistical techniques led to major advances in the design of experiments and the use of samples. His methods were adopted and used wherever statistical analysis was applicable.

American geneticist Sewall Wright (1889-1988) studied inbreeding and cross breeding of animals. He developed a mathematical scheme to describe evolutionary development. He also formulated the idea of genetic drift—that genes could be lost and new species appear without natural selection. This is meaningful in small populations. Also in the 1930s, Barbara McClintock (1902-1992) demonstrated that a physical exchange of material occurs between chromosomes.

In the 1930s and 1940s Wright and Fisher, along with American Ernst Mayr (1904-), created what is called the synthetic theory of evolution. It is called "synthetic" because it is a combination, or synthesis, of ideas of natural selection and principles of genetics and related sciences on both observable and microscopic levels. Between 1900 and 1930 Russians were also making headway in genetics. Sergei Chetverikov (1880-1959) made the first systematic studies of genetics in wild populations. Along with Fisher, Haldane, and Wright, he is considered one of the founders of the study of population genetics and a pioneer in modern evolution theory. Chetverikov used statistical, biometric techniques in biological studies and set the agenda for a new synthesis of ideas.

Theodosius Dobzhansky (1900-1975), another Russian, had emigrated to the United states in 1927 in order to join Morgan's group at Columbia University. He later wrote a book that brought genetics and mathematics together and created a single theory of the processes of evolution. He said natural mutation, aided by variation, can lead to change when acted upon by natural selection. One example is the ability of an insect to become resistant to pesticides. He also believed new species could not arise from single mutations and must be isolated from others of its species by time, geography, habitat, or breeding season.

Impact

By the mid-twentieth century genetics was a mature science with three major areas of concentration. Molecular genetics focuses on the regulatory processes that cause genes to work. Transmission genetics, often called classical genetics, analyzes the patterns of inheritance and how traits are transmitted over generations. Population genetics investigates how mutations and other processes of natural selection work among a population over time. Modern genetics is the union of several systems of knowledge and techniques of investigation. The triumph of the field is that genetics can explain the constancy of inheritance from the visible individual to the component molecules. In the 1940s DNA was identified as the basic substance within the gene. Its structure was deduced and modeled in the 1950s, and its role in evolutionary change became clear. This led to an explosion of new ideas that would affect all life on Earth. Medicine, experimental biology, law enforcement, ethics, and government control would all be affected. Gene mapping, locating each gene on a chromosome and identifying each trait, took ten years of work by many scientists. DNA identification is routinely used in legal cases. Gene therapy, gene manipulation, replacing defective genes with non-defective ones, cloning, and enhancing traits by gene replacement are ethical questions still being debated. While biological and medical research go on, these questions are the focus of a great deal of controversy.

LYNDALL B. LANDAUER

Further Reading

Allen, Garland. *Thomas Hunt Morgan, the Man and his Science*. Princeton, NJ: Princeton University Press, 1978.

Bowler, Peter. *The Mendelian Revolution*. Baltimore: Johns Hopkins University Press, 1989.

Clark, Ronald. *The Life and Work of J. B. S. Haldane*. New York: Coward-McCann Inc., 1969.

Dobzhansky, Theodosius. *Genetics and the Origin of the Species*. New York: Columbia University Press, 1951.

Lewontin, R. C. *The Genetic Basis of Evolutionary Change*. New York: Columbia University Press, 1974.

Smith, John Maynard. *The Theory of Evolution*. New York: Cambridge University Press, 1993.

The Modern Synthesis of Evolutionary Theory

Overview

During the 1930s and 1940s a group of biologists and scientists in a variety of related fields assembled a new picture of biological change, mutation, and variation, merging genetics with Charles Darwin's (1809-1882) vision of natural selection, and refining and altering understanding of both. Contributions to the new understanding of evolutionary processes came from population geneticists, paleontologists, ornithologists, mathematical geneticists, and naturalists. Because it drew upon so many fields, the view of evolution that emerged was called the Modern Synthesis or New Synthesis.

Background

In some cases, the synthesis was also known as neo-Darwinism, or new Darwinism, because it helped resolve many of the problems some scientists had found in the mechanics of Darwin's view of evolution. (In its most proper usage, neo-Darwinism was the movement that flourished between the death of Darwin and the arrival of the Synthesis; this movement attributed all changes in species to the effects of natural selection.)

More than half a century after Darwin presented his theory of the evolution of life by natural selection, the actual processes by which that evolution took place were still not understood. While few scientists questioned evolution as a process, natural selection itself had fallen into some disrepute as the engine of biological change over time.

Put simply, natural selection argues that plants and animals develop and pass on to succeeding generations those characteristics that best adapt the organisms for survival in their environment. The dilemma for scientists seeking to explain natural selection lay in determining how those characteristics that best suited an organism to a particular environment were passed from one generation to the next, as well as how new characteristics could be acquired and, in turn, passed on.

Some scientists, notably Jean Lamarck (1744-1829), had wrestled with the question of inherited characteristics even before Darwin prepared his theory of evolution. Lamarck argued that survival characteristics could be acquired or developed by a living organism, and once acquired, passed on to succeeding generations. That is, a characteristic such as stretching to reach high leaves, could be developed by an animal in response to its environment and, once the characteristic proved effective, it would be passed on to that animal's offspring, which would possess longer necks.

Lamarckism, as the theory of acquired characteristics came to be known, enjoyed much popularity early in the nineteenth century, but by the time Darwin published *On the Origin Of Species by Means of Natural Selection* in 1859, the theory had begun to be doubted. By the twentieth century, Lamarckism had become almost completely discounted.

But no clear view of inheritance and change emerged to take its place. Scientists and philosophers found themselves confronted with fundamental questions, including uncertainty as to how species arrive in the first place. Ornithologist Ernst Mayr (1904-), who would be one of the major figures in the Modern Synthesis, pointed out that despite the title of his great book, Darwin did not actually address how species were originated. Scientists also questioned how existing species change, how those changes are transmitted to offspring, and, above all, what the rate of change was over time.

Several evolutionary theories perceived purpose and direction in the march of changes throughout the history of life. This interpretation of evolution, known as orthogenesis, implied that changes in species were directed toward the development of ever-higher species, culminating in human beings. The process of evolution, however it worked, guided life from the simpler to the more advanced, the humblest to the most sophisticated. Orthogenesis found, perhaps not surprisingly, its most devoted followers among scientists with a religious or philosophical bent.

Others were not so sure. Some suspected that species emerged gradually, as a result of small changes that compounded with subsequent generations. Again and again, the focus of scientific inquiry returned to the search for the exact mechanism of change within a species. Darwin himself had postulated a specialized type of cellular structure, which he called a "gemmule," that was responsible for inherited characteristics. Unfortunately, Darwin's "gemmule" also allowed for the inheritance of acquired characteristics. Darwin died still pondering the process of inheritance.

Ironically, though, the very mechanisms that would go far to explain the workings of evolution and inheritance, had been observed by a Silesian (now in the Czech Republic) scientist named Gregor Mendel (1822-1884), whose paper on inherited characteristics was published only six years after Darwin's *Origin of Species.*

Whereas Darwin had achieved worldwide fame immediately upon the publication of his *Origin of Species* (the book sold more than 1,000 copies the day it was printed,) Mendel's work went largely unnoticed. Having experimented with the breeding and cross-pollination of peas, Mendel showed that the species characteristics, encoded in genes, of a set of parents do not simply blend in offspring, as Darwin theorized, but are passed on independently of each other. That independence allows for many different ways for the genes to be recombined in offspring.

Around the turn of the twentieth century, Mendel's insights into the operation of genetics began to receive increasing attention as a mechanism that explained not only how traits and characteristics could be inherited, but also how changes in the genetic makeup of an organism could be passed on to successive generations, ultimately becoming a part of the entire species' genetic pattern.

The introduction of those changes into the genes became the focus of much scientific research and experimentation. Around the turn of the twentieth century, Dutch botanist Hugo de Vries (1848-1935), in addition to helping restore Mendel's insights to prominence, began to formulate a theory of the role played by mutation in the process of evolution.

According to de Vries, sudden evolutionary jumps, the appearance of new species, were the results of alterations, or mutations, in the parents' genetic material. The transmission of these alterations, and the ways in which they in turn combined with existing genetic material, could cause the immediate appearance of a whole new species.

In short, de Vries postulated that evolution moved rapidly, if fitfully, in great leaps, called mutationism. De Vries viewed evolutionary change as a process of large and dramatic changes in genetic material, macromutations, which resulted in new species. A similar theory, saltationism, also proposed large jumps as powering the engine of evolution, and attracted many adherents in the years after the publication of Darwin's *Origin of Species.*

Already, though, some scientists were beginning to question these interpretations of both the fossil record and the ways genes worked. Particularly important was the work of Sewall Wright (1889-1988), who in the 1920s developed a mathematical model for the effects of genes and genetic change in populations. His studies, along with those of J.B.S. Haldane (1892-1964) and Ronald A. Fisher (1890-1962), among others, laid the groundwork for an entire and important field of study—population genetics—as well as demonstrating the vital contributions mathematics could make to biological investigation.

Despite the increasing importance of genetics to evolutionary studies—some scientists felt that the rediscovery of Mendel had rendered Darwin's insights all but useless—a great debate raged on. Part of the problem derived from a split in the biological sciences between naturalists and experimentalists. Naturalists, such as Darwin, based their theories on careful observation of the natural world. Experimentalists derived their findings from laboratory work, often isolating the object of their investigations from its natural environment. The two branches rarely shared information.

Impact

By the 1930s, it was becoming clear that the various theories not only did not fit well with observations of nature, they were also becoming increasingly problematic in the lab. The stage was set for a new approach to gathering information and insight from diffuse sources, and assembling the whole into a more unified vision of the fundamental operations of inheritance, change, and evolution.

In the opinion of many, the great catalyst for this unification flowed from the Columbia University laboratory of Russian immigrant Theodosius Dobzhansky (1900-1975). Working with *Drosophila,* a common species of fruit fly whose short 10-day reproductive cycle permitted experiments with evolution in action, Dobzhansky studied the role of genetics in whole populations of a species, as well as the ways in which populations could become isolated from one another. As a result of his experiments and his observations of species in the natural world, Dobzhansky was able to show that small genetic changes in a population could result in that population's inability to interbreed with other populations of the same organism. In short, small changes, along with the effects of natural selection, could result in the emergence of a new species.

Dobzhansky's 1937 book *Genetics and the Origin of Species* attracted excited attention throughout the scientific community, and helped shift efforts toward unifying various fields of evolutionary inquiry into higher gear.

At roughly the same time, paleontologist George Gaylord Simpson (1902-1984) was using his studies of fossils as a basis for recreating the breeding patterns of prehistoric species. Specifically focusing upon the horse, Simpson showed, from the fossil record, how the species spread, how its population changed as its environments changed or as it entered new environments, how geographic separations resulted in the emergence of new species, the role natural selection played in adaptation to environments, and so on.

Ernst Mayr, working at Harvard during the same period, produced sharp insights into the effects of geographic and other factors on species populations. He was able to show that members of species, isolated from others of the same species, could as a result of mutation and natural selection eventually become an entirely new species, unable to interbreed with the original population from which it had become separated.

In England, Julian Huxley (1887-1975), the grandson of Darwin's great defender Thomas Henry Huxley (1825-1895), built upon his own substantial biological work, combining it with the insights of Dobzhansky and others to produce the 1942 book *Evolution: The Modern Synthesis,* which gathered together and presented the unified (although not completely in agreement) views of the architects of the revolution in evolutionary biology.

With the publication of Huxley's book, the great revolutionary period of the Synthesis gave way to the refinements, disagreements, reinterpretations, and ongoing scientific inquiries that have marked the field ever since.

What had been shown by all of the scientists involved in the development of the Synthesis was that evolution was not an either/or proposition. Both genetics and natural selection, as demonstrated by both Darwin and Mendel (and hundreds of thousands of subsequent scientists and researchers), were required.

As a result of the Modern Synthesis, natural selection returned once more to the heart of evolutionary studies, although this time informed by a more accurate understanding of how those processes worked, and the role of genetics and population in the emergence of new species.

But the Synthetic theory did more than just revolutionize the ways in which scientists thought about evolution. The development of Synthetic theory is one of the great examples of cooperative study by scientists in various disciplines. Communication was revealed to be as important as research.

Nor did the development of the synthesis end the debates over the workings of evolution. Rather, it provided a more level field on which scientists from all disciplines could present their theories and interpretations, which is still going on, with occasional vehemence, today.

The Modern Synthesis proved that evolution itself continues to evolve.

KEITH FERRELL

Further Reading

Eldredge, Niles. *Fossiles: The Evolution and Extinction of Species.* New York: Harry N. Abrams, 1991.

Mayr, Ernst. *One Long Argument: Charles Darwin and the Genesis of Modern Evolutionary Thought.* Cambridge: Harvard University Press, 1991.

Milner, Richard. *The Encyclopedia of Evolution: Humanity's Search for Its Origins.* New York: Facts On File, 1990.

Tattersoll, Ian. *Becoming Human: Evolution and Human Uniqueness.* New York: Harcourt Brace & Company, 1998.

The Scopes Trial Highlights the Battle over Evolution

Overview

The Scopes "Monkey Trial" is the best-known example of the conflict over the teaching of evolution in the United States. Even though most scientists in the 1920s were sure that biology could not be taught without reference to evolution, Christian fundamentalists saw evolutionary theory as a rejection of religious belief. In 1925 Tennessee passed the Butler Act, which made the teaching of evolutionary theory illegal within the state. High school teacher John Thomas Scopes (1900-1970) was tried and convicted in Dayton, Tennessee, for teaching "the theory of the simian descent of man." The Butler Act was not repealed until 1967.

Background

When Charles Darwin's (1809-1882) *On the Origin of Species* was published in 1859, both theologians and scientists were divided about Darwin's theory of evolution, especially his arguments about human origins. Even some supporters of evolutionary theory believed that the intellectual and spiritual traits of human beings constituted an unbridgeable chasm separating human beings from animals. Some Protestant theologians argued that since God operates through intermediate causes, such as the law of gravity, God could also use evolution as a means of bringing living beings into existence. After studying the relationship between church doctrine and the theory of evolution, Roman Catholic theologians decided that biological evolution was compatible with the Christian faith and that the Bible should not be read as if it were a scientific treatise.

In the United States, however, many Protestant fundamentalists held to a literal interpretation of the Bible and rejected any alternative readings. A belief in the infallibility of the Bible was also a major principle of the American Bible League, founded in 1902. The World's Christian Fundamentals Association, established in 1919, called for a campaign against modernism and the teaching of the theory of evolution, which was denounced as a foreign doctrine that tended to undermine Christianity. Fundamentalists asserted that a literal interpretation of the story of the creation in the Book of Genesis was incompatible with the gradual evolution of humans and other organisms. Moreover, they asserted that Christian beliefs about the immortality of the soul and the creation of man "in God's image" were incompatible with the theory that humans evolved from animals. Anti-evolution crusaders demanded that the states pass laws to ban the teaching of evolution in public schools.

During the 1920s more than 20 state legislatures attempted to pass laws prohibiting the teaching of evolution in their public schools. Arkansas, Mississippi, Oklahoma, and Tennessee succeeded in passing such laws.

Fundamentalist anti-evolution literature, such as T. T. Martin's *Hell and the High Schools: Christ or Evolution—Which?* (1923), equated Darwinism with atheism and fueled the drive to ban the teaching of evolution. (Pastor Martin showed up at the Scopes trial to promote his writings and expound on his philosophy.) In Kentucky attempts to pass a law prohibiting the teaching of "Darwinism, Atheism, Agnosticism, or the theory of Evolution as it pertains to man" were narrowly defeated, thanks to active opposition.

In contrast to the situation in Kentucky, opposition to the fundamentalists in Tennessee was negligible. A 1925 Tennessee law, written by state representative John Washington Butler and passed by a wide margin, made it illegal to teach "any theory that denies the story of the Divine Creation of man as taught in the Bible, and to teach instead that man has descended from a lower order of animals." The penalty for a teacher convicted of violating the Butler Act was a fine of not less than $100 nor more than $500. When Governor Austin Peay signed the bill on March 21, 1925, he said that in all probability the law would never be actively applied or enforced, but the *St. Louis Post-Dispatch* warned that the consequences were likely to be embarrassing.

Impact

John Thomas Scopes, who held a coaching and teaching position at Central High School in Dayton, Tennessee, a town with a population of about 1,800, volunteered to test the law. The biology textbook adopted in 1919, George William Hunter's *Civic Biology,* contained an evolutionary chart and a discussion of evolution. The American Civil Liberties Union (ACLU) had offered to provide legal assistance to a teacher who was willing to test the constitutionality of the Butler law. (The ACLU had been established in 1920 to protect constitutional rights and freedoms. In addition to working on cases already brought to court, the ACLU often initiated test cases such as the Scopes trial.) Scopes was asked by several people in Dayton to volunteer as a candidate for the ACLU test case. He agreed.

A small article about the case appeared in the Chattanooga paper. It was picked up by the Associated Press and became a national story. John Randolph Neal, a constitutional expert who ran a private law school in Knoxville, offered to serve as Scopes's lawyer. Neal had been a law professor and dean of the law school at the University of Tennessee, but had left the school in a feud over a textbook controversy with fundamentalists. He wanted to argue that the issue was not whether evolution was true or not but whether the Tennessee legislature had the power to prevent students from learning about the ideas of the world's greatest scientists and thus limiting freedom of thought.

William Jennings Bryan, a nationally known charismatic orator, skillful debater, and popular public lecturer who had been defeated in a bid for the U.S. Senate and in three attempts at the U.S. presidency, served as the state's special prosecutor. A fervent believer in the literal interpretation of the Bible, Bryan had promised fellow fundamentalists that they would drive Darwinism from the schools. (The Scopes trial was to be Bryan's last battle; five days after the trial he suffered a stroke and died.)

Shortly after Bryan joined the case, Neal received a telegram from Clarence Darrow and Dudley Field Malone expressing their desire to assist the defense. Darrow was America's best-known defense attorney, and he offered his services to Scopes at no charge—the only time in his career he ever did so. Even though Neal was listed as the defense's chief counsel (four attorneys represented Scopes), Darrow took the lion's share of the publicity and courtroom work. With two such well known figures as Darrow and Bryan about to do battle over the legitimacy of evolution, publicity about the case reached the national level, especially after H. L. Mencken of the *Baltimore Sun* (which helped pay for Scopes's defense) labeled the story the "Monkey Trial." Hordes of people descended on Dayton, and the trial began to take on a circus atmosphere.

The case was heard by Judge John T. Raulston during a special term of the Eighteenth Judicial Circuit. The trial was held in Dayton from July 10–21, 1925. A radio microphone was set up and a movie camera recorded the scene for the newsreels. Experts on both science and religion came to Dayton to testify against the Butler Act, but the prosecution moved to exclude expert testimony. Judge Raulston ruled for the Fundamentalists and denied the use of expert testimony in Scopes's defense. Denied the testimony of their expert witnesses, the defense lawyers called Bryan himself to the stand as an expert witness on the Bible.

Clarence Darrow (with fist raised) defends John Scopes, on trial for teaching evolution in a public school (1925). *(Bettmann/Corbis. Reproduced with permission.)*

Convinced that the case would be lost in Dayton, the defense team planned to appeal the case to the United States Supreme Court, where Charles Evans Hughes would represent the defense. Judge Raulston, however, ruled out any test of the law's constitutionality or of the validity of Darwin's theory. This ruling limited the trial to the single question of whether or not Scopes had taught evolution. Scopes was convicted and fined $100. Bryan magnanimously argued against a monetary fine, and even offered to pay Scopes's fee himself. On appeal in 1927, the state's supreme court upheld the constitutionality of the Butler Act and denied motions for a new hearing. The judgment at Dayton, however, was reversed on a technicality: the judge had imposed the fine when it should have been decided by the jury. In spite of the trial, Scopes was asked to stay in Dayton as both a coach and teacher. Instead, he used a scholarship fund donated by the defense to earn his Ph.D. in geology from the University of Chicago.

While the Scopes trial created a national sensation, the outcome was far from clear. Despite the publicity, subsequent anti-evolution laws were passed in other states, and teachers in Tennessee during the 1960s were still required to sign a pledge that they would not teach evolution. The teaching of evolutionary science at the high school level declined after the Scopes trial and most biology textbooks omitted any mention of Darwin or evolution. Evolution did not regain a significant place in biology textbooks until the Soviet Union's launch of *Sputnik* in 1957 led to a new interest in science education in the United States.

Thirty years after the Scopes trial, playwrights Jerome Lawrence and Robert E. Lee fictionalized the proceedings in *Inherit the Wind,* a play that has since become a classic of American theater. When the movie version premiered in Dayton on the thirty-fifth anniversary of the trial, John Scopes was given the key to the city. He was convinced, however, that little had changed—indeed, Tennessee educators still had to promise not to teach evolution. The Butler law was finally repealed in 1967, just one year before the U.S. Supreme Court overturned all state laws that banned the teaching of evolution. The Court ruled that it was unconstitutional to ban the teaching of evolution "for the sole reason that it is deemed in conflict with a particular religious doctrine."

The battles over the teaching of evolution have raised many questions about the separation of church and state, the teaching of controversial subjects in public schools, and the ability of scientists to communicate with the public. Many scientists find it impossible to believe that the theory of evolution, which lies at the core of modern biology, could be considered controversial. Long after the Scopes trial, however, questions of evolution and divine Creation continue to influence both public opinion and biology education.

LOIS N. MAGNER

Further Reading

Awbrey, Frank, and William Thwaites, eds. *Evolutionists Confront Creationists.* San Francisco, CA: American Association for the Advancement of Science, 1984.

Ginger, Ray. *Six Days or Forever? Tennessee v. John Thomas Scopes.* Boston, MA: Beacon Press, 1958.

Kitcher, P. *Abusing Science: The Case Against Creationism.* Cambridge, MA: Massachusetts Institute of Technology, 1983.

Larson, Edward J. *Trial and Error: The American Controversy over Creation and Evolution.* New York: Oxford University Press, 1989.

Moore, J. R. *The Post-Darwinian Controversies: A Study of the Protestant Struggle to Come to Terms with Darwin in Great Britain and America, 1870-1900.* New York: Cambridge University Press, 1979.

Nelkin, Dorothy. *The Creation Controversy: Science or Scripture in the Schools.* New York: W. W. Norton & Company, 1982.

Numbers, Ronald L., ed. *Creation-Evolution Debates.* New York: Garland, 1995.

Scopes, John T., and James Presley. *Center of the Storm: Memoirs of John T. Scopes.* New York: Holt, Rinehart and Winston, 1967.

Webb, George E. *The Evolution Controversy in America.* Lexington: The University Press of Kentucky, 1994.

Other

Clarence Darrow Questions William Jennings Bryan at the Scopes Trial (Monday, July 20, 1925). Transcript. http://socrates.berkeley.edu/~jonmarks/Darrow.html

Ianone, Carol. "The Truth About *Inherit the Wind.*" http://www.firstthings.com/ftissues/ft9702/iannone.html

Mencken, H. L. Writings on the Scopes Trial. http://www.aracnet.com/~atheism/hist/menck01.htm

Mencken, H. L. Writings on the Scopes Trial. http://www.aracnet.com/~atheism/hist/menck02.htm

Mencken, H. L. Writings on the Scopes Trial. http://www.aracnet.com/~atheism/hist/menck03.htm

University of Missouri at Kansas City Law School. Famous Trials in American History. Tennesse vs. John Scopes: "The Monkey Trial," 1925. "Biographies of Trial Participants." http://www.law.umkc.edu/faculty/projects/ftrials/scopes/biotp.htm

The Emergence of Endocrinology as a Medical Science

Overview

As scientific knowledge increased, our concepts of the human body have gone through a transformation. The nineteenth-century "Concept of Internal Secretions" gave way to a more thorough understanding of bodily secretions. When this information was combined with seemingly unrelated ideas and discoveries it evolved into the branch of medical science called endocrinology. Although there was significant work done in this area prior to the twentieth century, and some of this information dates to antiquity, endocrinology really did not blossom as a science until the turn of the century.

Background

Endocrinology is a branch of medicine that studies the role of hormones and other biochemical mediators in regulating bodily functions. It further provides treatment methods when hormone levels become unbalanced. The properly functioning endocrine system includes tissues and glands that secrete chemical mediators directly into the bloodstream, which then produce effects at distant target cells that have the proper receptors to bind the chemical mediator. Thus, imbalances in the system can result from too much or too little hormone secretion or from the inability of the body to utilize the hormone effectively.

The study of the endocrine system and its functions owes some degree of its development to the gifted French physiologist Claude Bernard (1813-1878). He reported that organisms go to great lengths to maintain the consistency of what he called the "milieu interieur" or internal environment. Walter Cannon (1871-1945) later coined the term homeostasis to describe this phenomenon. The endocrine system in conjunction with the nervous system helps to preserve homeostasis. While certain aspects of endocrinology such as the endocrine disorder diabetes mellitus have been known since antiquity, the emergence of it as an independent science is a fairly recent occurrence.

For thousands of years Chinese medicine has been effectively treating certain disorders that were endocrine in nature. While they did not understand the mechanism of action, they realized the effectiveness of the treatment. As an example, seaweed, a compound that is high in iodine, was prescribed for an enlarged thyroid gland (goiter). There is evidence from early Egyptian writings that they recognized that some people could have sugar in the urine. It is now thought that they were describing symptoms of the endocrine disorder diabetes mellitus. Castration of men was another example of practice that had a direct endocrine effect. It was used to safeguard the chastity of women living in harems and to preserve singing voices in young males. The obvious conclusion was that something released from the testicle was important for male characteristics.

The first recorded recognition of the endocrine system came from Friedrich Henle (1809-1885) in 1841. Henle reported that unlike glands that released their products into a duct, he had found a class of glands that released its product directly into the bloodstream. The famous physiologist Claude Bernard was able to differentiate the products released from ductless glands from other chemicals and termed these "internal secretions." This idea was the forerunner to our modern concept of a hormone. Charles Brown-Seguard used extracts from animal testes in an attempt to treat male aging. After injecting extracts from animal testes into his own body, Brown-Segued concluded that they contained some sort of vital substance. This first effort of endocrine therapy in 1889 was ineffective, but it provided the impetus for further research that was essential to the discovery of cortisone and thyroid hormone.

Impact

The first hormone to be purified was secretin in 1902. This event marked the birth of the science of endocrinology. The English physiologists William Bayliss (1860-1924) and Ernest Starling (1866-1927) discovered secretin. They injected dilute acid into the denervated intestine of a dog. The inner lining of the bowel was scraped, boiled, and filtered. The resulting compound was then injected into the blood of the dog. After the injection, there was a large increase in the release of pancreatic juice. There was something in the material that they injected that had caused this increase in pancreatic secretions. They gave this unidentified substance the name "secretin" after its mechanism of action. This experiment is significant in that it gave concrete evidence that chemicals can act at distant sites to regulate bodily functions. In 1905 Walter Cannon coined the term hormone (Greek for "to set in motion") to describe specific chemicals, such as secretin, that travel through the blood to stimulate a distant target cell.

Diabetes mellitus is an endocrine disorder caused by either the inability to release the hormone insulin or the inability of cell to properly respond to the presence of insulin. Diabetes mellitus is the most common endocrine disorder and causes deleterious effects on millions of people throughout the world. It took over 30 years of intensive effort to find the cause of this disease. In 1889 the German physicians Joseph von Mering and Oskar Minkowski removed the pancreas in dogs, which resulted in the disease. American pathologist Eugene L. Opie described degenerative changes within clumps of cells in the pancreas in 1901. Later Edward Sharpey-Schafer concluded that these cells in the pancreas secrete a substance that controls the metabolism of carbohydrate. The discovery of insulin in 1921 by Canadian surgeon Frederick Banting (1891-1941), with the assistance of Charles Best (1899-1978) and J.J.R. Macleod (1876-1935), was one of the most dramatic events in modern medicine. First, it provided needed therapy for those afflicted with diabetes mellitus. Patients now had the prospect of leading a long and healthy life, instead of the early demise that resulted from untreated diabetes. In addition, it helped to provide important clues regarding the endocrine function of the pancreas.

Another important discovery was that of cortisone and its use as an anti-inflammatory agent. Cortisone was isolated and purified by Edwin Kendall (1886-1972) in 1935. Philip Hench (1896-1965) and colleagues determined its therapeutic properties in 1949 at the Mayo Clinic. They discovered that a substance isolated from the adrenal gland had alleviated many of the symptoms associated with rheumatoid arthritis. Cortisone is a steroid hormone that has potent anti-inflammatory properties. It is useful in controlling many acute diseases and is a valuable compound in research.

An important realization regarding hormone action was purposed by Fuller Albright and colleagues in 1942. They reasoned that ineffective hormonal action on the target cell could produce symptoms similar to hormone deficiency. They made this assertion after studying a patient who exhibited all of the characteristics of a hormone deficiency, yet did not respond to treatment. This led Albright to conclude that the disturbance was actually the inability to respond to the hormone, not the lack of it. This idea paved the way for later important work.

The endocrine system has been shown to determine emotions. The James-Lange theory was an accepted, early influential hypothesis regarding the link between emotions and the body. This theory held that an emotion is the perception of phenomena from within the body. As an example, people are happy because they smile. Walter Cannon saw flaws with this approach and proposed an alternative theory with his colleague, Philip Bard, known as Cannon-Bard theory. According to this approach, the experience of an event leads to the simultaneous determination of emotion and changes in the body. The brain, upon receiving information from the senses, interprets an event as emotional while at the same time preparing the body to deal with the new situation by stimulating the nerves and releasing hormones. Thus, emotional responses and changes in the body are preparations for dealing with a potentially dangerous situation. Cannon demonstrated this was true through a series of experiments where he showed that emotions cause the excitation of nerves, which, in turn, lead to specific, but widespread, changes such as increases in heart rate and blood pressure. Many of these changes are mediated through the release of hormones.

Substantial progress was made in the first half of the twentieth century with all of the major classes of hormones. Many of these, like the discovery of insulin, led to treatment therapies for the afflicted and substantially increased our fundamental knowledge of physiology.

JAMES J. HOFFMANN

Further Reading

Books

Bliss, Michael. *The Discovery of Insulin.* Chicago: University of Chicago Press, 1984.

McCann, S.M., ed. *Endocrinology: People and Ideas.* American Physiological Society Book. New York: Oxford University Press, 1988.

Medvei, Victor Cornelius. *The History of Clinical Endocrinology: A Comprehensive Account of Endocrinology from Earliest Times to the Present Day.* New York: The Parthenon Publishing Group, 1993.

Periodical Articles

Grossmann, M.I. "A Short History of Digestive Endocrinology." *Advances in Experimental Medicine and Biology* 106 (1978): 5-10.

Hughes, A.F. "A History of Endocrinology." *Journal of the History of Medical and Allied Sciences* 32, no. 3 (July 1978): 292-313.

The Boveri-Sutton Theory Links Chromosomes to Heredity

Overview

Cytologists had established links between cell theory and theories of evolution and heredity by the beginning of the twentieth century. The work of Walter S. Sutton (1877-1916) and Theodor Boveri (1862-1915) provided sound cytological evidence for the individuality of the chromosomes and suggested that they might play a role as the carriers of hereditary factors. Studies of sex determination in insects by Nettie M. Stevens (1861-1912) and Edmund B. Wilson (1856-1939) also supported the idea that chromosomes served as the material basis of biological inheritance.

Background

By the beginning of the twentieth century, cytologists had established links between cell theory and theories of evolution and heredity. Painstaking investigations of the fine structure of the cell convinced scientists that the cell nucleus was very different in form and function from the cytoplasm. Indeed, in terms of theory as well as empirical observations, cytology played a major role in establishing the structural basis of Mendelian genetics. The availability of new methods for preparing and staining biological preparations made it possible to see the nucleus and the bodies known as chromosomes during cell division.

These observations indicated that the nucleus must have a fundamental role in heredity as well as the development and growth of the organism.

A valuable theoretical framework that guided studies of cytology and heredity had been proposed by the great theoretician of biology August Weismann (1834-1914) in the 1890s. Analyzing the key questions of evolutionary theory, variation, and inheritance, Weismann concluded that the stability of inheritance between generations was the most significant feature of heredity. Heredity had to be studied at both the level of the cell and the individual before it could be understood in terms of evolving populations. Weismann proposed the theory of the continuity of the germplasm. He also predicted and explained the necessity for a form of reduction division of the chromosomes during the formation of germ cells. Although Weismann knew nothing about the chemical nature of the germplasm, his theories provided a valuable framework for studies of the components of the cell, particularly the division of the nucleus and the chromosomes. When somatic cells divide, he theorized, the number of chromosomes must remain constant. When germ cells are produced, he continued, some mechanism is needed to divide the number of chromosomes by half; when fertilization occurs and the egg and sperm nuclei fuse, the normal chromosome number is restored.

By the turn of the century, microscopic observations of cell division seemed to vindicate Weismann's predictions. Moreover, both cytologists and geneticists were ready to accept the idea that there was a relationship between the factors studied in breeding experiments and the behavior of the chromosomes during cell division (mitosis and meiosis). Although cytologists knew that chromosomes occurred in pairs that seemed to join and separate during cell division, the nature of chromosomes was not understood. Some biologists thought that each chromosome carried all the hereditary units that were needed to form a complete individual, but it seemed possible that the maternal and paternal chromosomes might remain as distinct groups. By 1910 some biologists were sure that paired hereditary factors might actually be located on the chromosomes pairs contributed by the egg and the sperm. Cytological evidence for this conclusion was provided by the work of Walter S. Sutton and Theodor Boveri. Studies of sex determination in insects by Nettie M. Stevens and Edmund B. Wilson also supported the idea that the chromosomes provided the material basis of heredity.

Impact

Theodor Boveri used doubly fertilized sea urchin eggs to investigate the function of the chromosomes in germ cell formation and embryological development. By studying the fate of embryos with abnormal numbers of chromosomes, Boveri was able to test the idea that each chromosome carried all character traits. Boveri found correlations between the abnormal combinations of chromosomes and the peculiar development of these embryos. Although Boveri could not determine the role of specific chromosomes, his observations indicated that the chromosomes were not identical to each other. His early experiments were described in his classic paper, "Multipolar Mitosis as a Means of Analysis of the Cell Nucleus" (1902).

Walter Sutton was able to use studies of the behavior of the sex chromosomes of certain insects as a way to extend the observations that Boveri had made with random combinations of chromosomes. As a student of Clarence E. McClung (1870-1946), a pioneer in the study of the sex chromosomes, Sutton knew that he could focus on specific morphological differences among the chromosomes of grasshoppers. As early as 1902 McClung had suggested that a special chromosome, known as the "accessory chromosome," determined whether an individual was male or female.

Some cytologists thought that the apparent individuality of the chromosomes might be lost between cell divisions when the chromosomes seemed to disappear and become tangled masses of chromatin threads. Although there was no empirical evidence to prove that the same chromosomes appeared at the next cell division, Sutton believed that the regularity in the size and shape of the chromosomes at each cell division indicated that the chromosomes maintained their individuality. In 1903, while he was Edmund B. Wilson's graduate student, Sutton published his landmark paper, "The Chromosomes in Heredity." He asserted that the behavior of pairs of chromosomes indicated that they did indeed represent the physical basis for the Mendelian factors, or genes. Sutton also suggested that the random assortment of different pairs of chromosomes might explain the independent segregation of pairs of genes. Surprisingly, Sutton decided to become a practicing physician rather than pursue a career in research. Other researchers, however, were also involved in cytological studies of the nature of the sex chromosomes.

Within the first decade of the twentieth century, cytologists provided key insights into the relationship between specific chromosomes and sexual differentiation. Microscopic studies of specific "unpaired" chromosomes suggested that the presence or absence of these "accessory," or "X," chromosomes might determine sex in various insects. Nettie Stevens and Edmund Wilson confirmed this hypothesis in 1905. Nettie Stevens had earned her B.A. (1899) and M.A. (1900) at Stanford University. She performed her doctoral research at Bryn Mawr and was awarded her Ph.D. in 1903. Bryn Mawr was a small woman's college, but two of America's leading biologists, Edmund B. Wilson and Thomas Hunt Morgan (1866-1945), were faculty members. Morgan taught at Bryn Mawr until 1904 and kept in touch with Wilson, who had moved to Columbia University in 1891. After earning her Ph.D., Stevens won a fellowship that allowed her to study abroad and work with Theodor Boveri at the University of Würzburg. When she returned to Bryn Mawr, she received a grant from the Carnegie Institution of Washington that made it possible for her to devote all her time to her research on the sex chromosomes of the common mealworm. (Because Stevens used her initials when she published her research results, many scientists were surprised to discover that N. M. Stevens was a woman.) The objective of her research was to clarify the relationship between the behavior of the chromosomes and the laws of Mendelian genetics.

Stevens demonstrated that the somatic cells of female mealworms contained twenty large chromosomes, while the cells of the male mealworm contained nineteen large chromosomes and one small one. She reasoned that sex determination might be determined by some unique property of the unusual chromosome pair. Half of the spermatozoa produced by these male insects contained ten large chromosomes while the other half contained nine large chromosomes and one small one. The eggs that were fertilized by the sperm containing ten large chromosomes produced females; eggs fertilized by sperm carrying the small chromosome developed into males. (The large sex chromosome is conventionally referred to as "X" and the corresponding small chromosome is "Y." Females are XX and males are XY.) Stevens concluded that there must be a fundamental difference between the X and Y chromosomes and that this difference follows Mendelian rules of inheritance. Shortly after Stevens's paper on the chromosomal nature of sex determination was published, Wilson confirmed her conclusion through his research on a species in which the male had one fewer chromosome than the female, rather than a small Y chromosome.

The work of Boveri, Sutton, Stevens, and Wilson suggested that the chromosomes were distinctly individual in terms of their size, shape, and functional properties. Of course, cytological studies alone could not explain the precise nature of the chromosomes or the relationship between their chemical constituents and their role in carrying and transmitting genetic information. Nevertheless, these cytologists called attention to the way in which facts from cytology and Mendelian breeding studies could be integrated, forming the hypothesis that the hereditary Mendelian factors were physical components of the chromosomes. Moreover, the observations of these researchers suggested that each chromosome contained many Mendelian factors and that factors on the same chromosome would presumably be linked and thus always inherited together. This idea seemed to contradict the Mendelian law of independent assortment, but good evidence of linkage was discovered when more traits were subjected to breeding tests.

The Boveri-Sutton hypothesis explained or predicted many significant aspects of cytology and heredity, but many scientists remained skeptical of the implications of the chromosome theory of heredity. Wilson's experimental confirmation of the work of Sutton and Stevens was important in disseminating their conclusions to the scientific community. As author of the popular text *The Cell in Development and Heredity* and the friend and mentor of many prominent scientists, Wilson persuaded many skeptical scientists, including Thomas Hunt Morgan, to investigate the chromosome theory of heredity.

LOIS N. MAGNER

Further Reading

Books

Baltzer, F. *Theodor Boveri: Life and Work of a Great Biologist, 1862-1915.* Translated by D. Rudnick. Berkeley, CA: University of California Press, 1967.

Carlson, E. A. *The Gene: A Critical History.* Philadelphia: Saunders, 1966.

Darden, L. *Theory Change in Science: Strategies from Mendelian Genetics.* New York: Oxford University Press, 1991.

Dunn, L. C., ed. *Genetics in the Twentieth Century.* New York: Macmillan, 1951.

Mayr, E. *The Growth of Biological Thought.* Cambridge, MA: Harvard University Press, 1982.

Weismann, August. *The Germ-Plasm: A Theory of Heredity.* Translated by W. N. Parker and H. Ronnefeldt. New York: Scribner, 1893.

Wilson, Edmund B. *The Cell in Development and Inheritance.* 3rd ed. New York: Macmillan, 1925.

Periodical Articles

Brush, S. "Nettie M. Stevens and the Discovery of Sex Determination by Chromosomes." *ISIS 69* (1989): 163-172.

Elucidating the Structure and Workings of the Nervous System

Overview

The detailed study of the nervous system did not begin in earnest until the middle of the nineteenth century. As with other areas of science, it was not until innovative technology was able to keep up with the new ideas that the structure and workings of the nervous system began to be discovered. Prior to the important discoveries of Camillo Golgi (1843-1926) in 1873 and Santiago Ramón y Cajal (1852-1934) in 1889, very little was known about the nervous system. This was especially true with regard to the microscopic structure of how each nerve cell was connected to the next one. This lack of understanding was due, in part, to the lack of appropriate staining techniques to help visualize a nerve cell (neuron) in the microscope.

In 1873 Camillo Golgi invented a technique for staining neurons that made it possible to observe and distinguish the three principle parts of the cell—the cell body, dendrites, and axon—with a microscope. This technique was termed the "black reaction" (*reazione nera*), based on a process that first hardened the nervous tissue and later impregnated it with a metallic stain. The revolutionary process that Golgi developed, which for the first time allowed a clear visualization of an entire neuron, is still used with minor modifications today. This advancement allowed other researchers to begin to study the intricacies of the nervous system in much more detail than was previously possible. This important contribution helped to set the stage for acceleration in the advancement of our knowledge in the area of the nervous system.

Golgi's research did not provide the answer to every question, however, and controversy ensued over the components of the nervous system. Although the theory that tissues were made up of individual cells had been put forth decades earlier, there was much debate regarding whether this theory should be applied to the nervous system. Golgi did not believe that the nervous system was made up of individual cells. Based on his own observations, Golgi believed in the "reticular theory" for the nervous system. This theory postulated that the nervous system was an intricate network of interconnected fibers and that the nerve impulses propagated along this diffuse network. At the same time, Wilhelm Waldeyer put forth the "neuron theory" for the nervous system. He believed that the nervous system was actually made up of individual cells that were functionally and anatomically different. The main supporter of this theory was Santiago Ramón y Cajal, who used the staining technique of Golgi to visualize the components of the nervous system. It is interesting to note that despite these fundamental philosophical issues, Golgi and Ramón y Cajal were both awarded the Nobel Prize for physiology or medicine in 1906 for their work elucidating the structure of the nervous system.

The most important investigation that helped to clarify the issue regarding the components of the nervous system was an observation made in 1889 by Ramón y Cajal. He believed that his slides of the nervous system clearly demonstrated that it was composed of individual units (neurons) that were structurally independent of one another and whose internal contents did not come into direct contact (synapses). According to this hypothesis, which now serves as the central idea in modern nervous system theory, each nerve cell communicates with others through contiguity rather than continuity. That is, communication between adjacent but individual cells must take place across the space that separates them. The idea that communication in the nervous system is largely between independent nerve cells served to be the guiding principle for further study in the first half of the next century.

Background

While the exact mechanisms for nerve transmission were not worked until after the mid-twentieth century, substantial progress to that goal was made in the first half of the century. There were many important discoveries in that time period. In 1932 Sir Charles Scott Sherrington (1857-1952) and Lord Edgar Douglas Adrian (1889-1977) were awarded the Nobel Prize for physiology or medicine for their research into the functions of neurons. However, much of this work was performed as much as three decades earlier. They made major contributions to our knowledge of the function of neurons regarding reflexes, and they pioneered a technique called *electrophysiology* that directly recorded the electrical activity of neurons.

Sherrington studied reflexes in experimental mammals and concluded that a reflex must be regarded as an integrated activity of the total organism. His initial line of evidence supporting what came to be known as "total integration" was his demonstration of the concept of reciprocal innervation of muscle. This stated that when one set of muscles is stimulated, the muscles opposing that action are simultaneously inhibited. He expanded further on reflexes in 1906 when he distinguished between three types of sensory receptors. One type he called *proprioceptors*, which he found were important structures in maintaining the postural reflex, a reflex that allowed animals to stand upright against gravity. He noted that this reflex occurred even if part of the brain was removed and tactile nerves were severed, so it must be governed by receptors found in muscles and joints.

Another important area of research during this time was the chemical transmission of nerve impulses, which culminated in the Nobel Prize for physiology or medicine for Sir Henry Dale (1875-1968) and Otto Loewi (1873-1961) in 1936. They provided the seminal work that demonstrated that transmission of nerve impulses involved the release of chemicals (neurotransmitters) from the nerve endings. The majority of their research was performed on the autonomic nervous system, but later these principles were extended to the entire nervous system.

Loewi provided the initial evidence that chemicals were involved in the transmission of impulses from one nerve cell to another or to an end organ. Their classical experiment involved stimulating the nerves, which slowed the heart of a frog, and then perfusing a second frog heart with fluids from the first frog heart. When the second heart also slowed, it indicated the presence of an active substance, which was produced by the initial nerve stimulation of the first heart. This substance was shown to be the neurotransmitter acetylcholine by Dale in 1929.

A significant development regarding the functioning of the nervous system came in 1929 when German psychiatrist Hans Berger (1873-1941) revealed his technique for recording electric currents generated in the brain by placing electrodes on the exterior of the skull, a technique called the *electroencephalogram* (EEG). Further studies demonstrated that this activity changed according to the functional status of the brain, such as in sleep, and in certain nervous diseases, such as in epilepsy. Hans Berger had succeeded in founding an entirely new branch of medical science, called clinical neurophysiology, which would have tremendous impact on the diagnosis and treatment of neurological diseases.

The individual properties of nerve fibers were also an area of intense interest. In 1944 Joseph Erlanger (1874-1965) and Herbert Spencer Gasser (1883-1963) received the Nobel Prize in physiology or medicine "for their discoveries relating to the highly differentiated functions of single nerve fibers." Their exhaustive experiments demonstrated that peripheral nerves vary according to their physical and conductive properties, as well as their function. They categorized nerves into three main groups (A, B, and C) according to their conduction velocities, degree of myelination, and diameters. Increased diameter and the greater degree of myelination resulted in faster conduction of nerve impulses. In addition, it was further discovered that the type of information conveyed along the nerve fiber was dependent upon the size. For instance, fibers relaying pain signals to the brain tended to be smaller, slow-conducting fibers, while fibers going to muscles tended to be larger, fast-conducting fibers.

Impact

The work done in the area of the structure and workings of the nervous system during the first half of the twentieth century had an important impact on society from both a scientific and medical view. Important discoveries took place that helped to advance our knowledge in this field of study, and many had important clinical applications. Most disorders are at least partially mediated by the nervous system, so it is absolutely essential that we have an excellent understanding of its inner workings. Besides the important scientific discoveries previously discussed, there were numerous other studies that helped to lay the foundation for future researchers to be successful.

The histological work done by Golgi and Ramón y Cajal helped to provide vivid representations of the nervous system on the cellular level. This research was important in that it gave scientists a picture of the intricacies of the nervous system. It also provided a tool that served to be useful in elucidating structures at the cellular level for other systems as well. Sherrington and his colleagues contributed to the knowledge base in nearly every aspect of nervous function and directly influenced the development of brain surgery and the treatment of such nervous disorders as paralysis and atrophy. In addition, Sherrington came up with much of the terminology regarding the nervous system that is still used today. For example, he denoted the terms *neuron* for a nerve cell and *synapse* for the point where an axon of a neuron meets either another neuron or some other end organ. Lastly, Sherrington, along with Adrian, pioneered the branch of science called electrophysiology, which directly records the electrical activity of neurons. This development was crucial in the eventual elucidation of how nerves transmit impulses.

For their part, Loewi and Dale provided extensive information regarding the nature of neurotransmitters. This information has subsequently been used to help understand the nature of certain toxins and to develop drugs to enhance or block the activity of neurotransmitters. And Hans Berger succeeded in founding an entirely new branch of medical science with the use of his EEG, called clinical neurophysiology. This instrument had tremendous impact on the diagnosis and treatment of many neurological diseases and is an important diagnostic tool in modern medicine. Finally, the information provided on the conduction velocities and properties of nerve axons by Erlanger and Gasser proved to be invaluable in understanding the nervous system. It had direct application to many disease processes. It was through these and other discoveries that modern medicine has advanced as quickly as it has.

JAMES J. HOFFMANN

Further Reading

Books

DeFelipe, J. and E.G. Jones. *Cajal's Degeneration and Regeneration of the Nervous System*. New York: Oxford University Press, 1991.

Smith, R. *Inhibition: History and Meaning in the Sciences of Mind and Brain*. Berkeley: University of California Press, 1992.

Periodical Articles

Jones, E.G. "The Neuron Doctrine." *Journal of History of Neuroscience* 3 (1994): 3-20.

Raju, T.N. "The Nobel Chronicles. 1932: Charles Scott Sherrington (1857-1952), Edgar Douglas Adrian (1889-1977)." *Lancet* 353 (January 2, 1999): 76.

Developments in Embryology

Overview

At the beginning of the twentieth century, embryologists, following the principles established by Wilhelm Roux (1850-1924) and Hans Adolf Eduard Driesch (1867-1941), were actively investigating the question of how factors intrinsic or extrinsic to an egg could govern the development of the embryo. Hans Spemann (1869-1941) refined the techniques of experimental embryology and carried out systematic studies of embryonic development. The "organizer experiment," performed by Spemann's doctoral student, Hilde Mangold, demonstrated that, when the embryonic region known as the dorsal lip was grafted onto a host embryo, it induced the formation of a new embryo. Moreover, the secondary embryo was composed of a mosaic of cells from the host and the donor. The dorsal lip region was, therefore, called the "organizer region."

Background

Wilhelm Roux, who saw himself as the founder of a new discipline that he called "developmental mechanics," argued that embryologists must adopt experimental methods as the tools that would make possible the analysis of the immediate causes of development. The primary question Roux posed was whether development proceeded by means of self-differentiation or correlative dependent differentiation. Self-differentiation was defined as the capacity of the egg or of any part of the embryo to undergo further differentiation independently of extraneous factors or of neighboring parts in the embryo. In turn, correlative dependent differentiation was defined as being dependent on extraneous stimuli or on other parts of the embryo. These were operational definitions that could be tested by experiment, that is, by transplantation and isolation. Based on his most famous—but seriously flawed—experiment, Roux believed that self-differentiation served as the mechanism of development. In other words, Roux tended to visualize the fertilized egg as similar to a complex machine and development as a process that involved the distribution of parts of the machine to the appropriate daughter cells. In his experiment, Roux destroyed one of the cells of a frog embryo at the two-cell stage. The undamaged cell then developed into a half-embryo. This result supported Roux's belief that each cell normally develops independently of its neighbors and that total development is the sum of the separate differentiations of each part.

When other scientists repeated Roux's experiment with embryos of different species, they obtained quite different results. Nevertheless, as late as the 1930s embryologists still cited Roux as the scientist who had formulated the core questions of embryology. Hans Driesch, however, argued that his results provided definitive evidence against Roux's model of development. Driesch used sea urchin eggs instead of frog embryos and was able to separate early embryonic cells by shaking. The separated cells developed into embryos that were normal in configuration but half the size of their normal counterparts. Driesch's experiments suggested that at some fundamental level all the parts of the embryo were uniform; therefore, the fate of any given cell was a function of its relative position to the whole.

Roux and Driesch's followers were able to plan and perform increasingly subtle experiments on living embryos by means of isolation, transplantation, and tissue culture. Hans Spemann in Germany and Ross G. Harrison (1870-1959) in the United States were especially important for their role in refining the techniques of experimental embryology. Spemann carried out systematic studies of early organ determination, whereas Harrison established the experimental foundations of tissue culture and neurogenesis.

Impact

Hans Spemann was the son of Johann Wilhelm Spemann, a prosperous publisher and bookseller in Stuttgart. Spemann left school when he was nineteen and joined his father's business. After a year of military service, Spemann became fascinated by the writings of Goethe and Ernst Haeckel (1834-1919) and decided to resume his education. In 1891 he was admitted to the University of Heidelberg as a medical student. His mentors eventually convinced him to give up medicine and study general biology, comparative anatomy, and embryology at the Zoological Institute of the University of Würzburg with the cytologist Theodor Boveri (1962-1915), who introduced him to the use of the amphibian embryo as a model system. In 1895 Spemann received his degrees in botany, physics, and zoology. While enduring a rest cure for tuberculosis, Spemann read August Weismann's book *The Germ Plasm: A Theory of Heredity* (1892). He began to formulate an experimental approach, based on Roux's work, to the question of the relationship between the reproductive cells and embryological development and differentiation. Spemann, however, selected the salamander as his model system and developed his own methods of separating and rearranging the cells of the early embryo. His studies led to a series of papers called "Developmental Physiological Studies on the Triton Egg" (1901-1903) that introduced the technique of manipulating and constricting the egg with a loop of fine baby hair. Spemann had found that, if he constricted fertilized salamander eggs without completely separating the cells, he could produce animals with one trunk and tail but two heads. Passages in his autobiography reveal his fascination with watching the behavior of such "twin embryos." The mystery of this result and the pleasure derived from such experiments led to his total commitment to embryological research.

Spemann became co-director of the Division of Developmental Mechanics at the Kaiser Wilhelm Institute for Biology in Dahlem (a suburb of Berlin) in 1914. Although the institute provided good research facilities and stimulating colleagues, World War I delayed construction of Spemann's own laboratory and malnutrition related to the war had a severe impact on his health. In 1919 Spemann became the director of the Zoological Institute of the University of Freiburg.

Spemann called his work the study of the "physiology of development." The goal of his research program was to discover the precise moment when a particular embryonic structure became irrevocably determined in its path towards differentiation. The excitement generated by his results made the idea of determination a dominant theme in development research for many years. His colleagues and students considered him a master of the art of microdissection. Spemann guided his associates through a series of experiments in which selected parts of one embryo were transferred to a specific region of another. By using embryos of different species or varieties, Spemann could rely on color differences to follow the fate of the transplanted area during development.

Hilde Proescholdt, who had joined Spemann's laboratory in 1920, was assigned the task of transplanting a region known as the dorsal lip. In 1921 Hilde Proescholdt married Otto Mangold, who was also one of Spemann's students. Hilde Mangold appears as the second author of the landmark paper "Induction of Embryonic Primordia by Implantation of Organizers from a Different Species" (1924). Although Spemann's other students had been allowed to publish their thesis work as sole authors, Spemann insisted on adding his name to Hilde Mangold's thesis publication; he also insisted on being first author. After the discovery of the organizer region, Hilde Mangold abruptly disappears from accounts of the history of embryology. She died in 1924, at the age of 26, from severe burns caused by the explosion of a gasoline heater.

The "organizer experiment," performed by Hilde Mangold as her Ph.D. thesis, has been called the culmination of Spemann's achievements. When the region known as the dorsal lip was grafted onto a host embryo, it seemed to lead to the formation of a new embryo. Mangold's experiment involved grafting the dorsal lip from an unpigmented species onto the flank of a pigmented host embryo. Three days after the graft, an almost complete secondary embryo appeared on the host embryo.

Mangold proved that the secondary embryo was composed of a mosaic of host and donor cells. Therefore, the transplanted dorsal lip and the host embryo both participated in the formation of the secondary embryo. Because the tiny amount of tissue from the donor embryo was powerful enough to cause the formation of a new embryo, the dorsal lip was called the "organizer region." Keeping such embryos alive was very difficult, but Mangold described five experimental cases in detail and briefly noted several others. Some years later, using antibiotics and special growth media, Johannes Holtfreter and his coworkers were able to keep similar embryos alive into even more advanced stages. Remarkably, the secondary embryos in Holtfreter's experiments were as complete as the primary embryos of the hosts.

Through various refinements of the organizer experiment, Spemann discovered other organizer regions. He expected that this approach would eventually lead to a complete understanding of embryonic development. Spemann's discovery of the organizer and his concept of induction can be seen as the culmination of the approach advanced by Roux and Driesch. In 1935, the year in which he retired from the University of Freiburg-im-Breisgau, Spemann was awarded the Nobel Prize in physiology or medicine for his discovery of the "organizer effect" in the induction of embryonic development. One year later, Spemann published his final account of his ideas and experiments in *Embryonic Development and Induction.*

In the 1930s other embryologists demonstrated that the organizer region could induce a secondary embryo even after it had been killed by heating, freezing, or alcohol treatment. Spemann had previously discovered that the organizer was active after its cells had been killed by crushing but had not pursued an explanation for such observations. The idea that dead embryonic tissues could induce differentiation had a profound impact on experimental embryology, suggesting new biochemical approaches to development and differentiation. Investigators found that various bits of animal tissues—alive or dead—could induce complex structures, including heads, internal organs, and tails. In *Embryonic Development and Induction,* Spemann complained that a "dead organizer" was a contradiction in terms. Other scientists, however, hoped to find some "magic molecules" released by the inducing tissues and to discover the basis for their organizing and inducing powers. The failure to find such molecules eventually led to a loss of interest in organizer regions, but organizer theory had helped to stimulate a chemistry-based approach to embryology. The explosion of discoveries that followed the identification of DNA as the genetic material would make possible an entirely new approach to solving the fundamental questions of embryology.

LOIS N. MAGNER

Further Reading

Hamburger, V. *Heritage of Experimental Embryology.* New York: Oxford University Press, 1988.

Haraway, Donna Jeanne. *Crystals, Fabrics and Fields: Metaphors of Organicism in Twentieth-Century Develop-*

mental Biology. New Haven, CT: Yale University Press, 1976.

Horder, T. J., J. A. Witkowski, and C. C. Wylie, eds. *A History of Embryology.* New York: Cambridge University Press, 1986.

Nakamura, O., and S. Toivonen, eds. *Organizer: A Milestone of a Half Century From Spemann.* New York: Elsevier, 1978.

Needham, Joseph. *The Rise of Embryology.* Cambridge, England: Cambridge University Press, 1959.

Oppenheimer, J. *Essays in the History of Embryology and Biology.* Cambridge, MA: MIT Press, 1967.

Spemann, H. *Embryonic Development and Induction.* New Haven, CT: Yale University Press, 1938.

Willier, B. H., and J. M. Oppenheimer, eds. *Foundations of Experimental Embryology.* 2nd edition. New York: Hafner, 1974.

The Eugenics Movement: Good Intentions Lead to Horrific Consequences

Overview

For thousands of years, people have tried to rank each other according to perceived superiority or inferiority. This tendency reached its peak in the first half of the twentieth century in the eugenics movement, in which adherents called upon genetics and natural selection to lend scientific credibility to an age-old argument. Although eugenics began as an honest scientific attempt to understand the differences between individuals, nations, and races, it was hijacked by politics and used to justify all manner of objectionable acts. At its best, eugenics was seen as a way to improve humanity and the human condition. At its worst, it was a way for political ideologues and tyrants to justify oppressing those held to be inferior.

Background

One of the earliest mentions of inherent differences between groups of people is found in ancient Greece, when Socrates (c. 470-399 B.C.) suggested that some are born to lead, some to follow, and others to work. Later, others divided humanity along roughly the same lines—as rulers, aristocracy, and peasants. This trend was followed throughout Europe, in Asia, among many Native American tribes, and elsewhere. In all cases, those who led were perceived to possess some innate advantage over their followers, who, in turn, were deemed superior to those who toiled in the fields. Other cultures segregated people by race or religion, but the bottom line was always the same: some were born better than others.

None of this supposed superiority was demonstrable, however. One could say, for example, that kings and emperors were appointed by God and were, therefore, superior. But this could never be proven, and the list of kings and emperors overthrown by their subjects is long and distinguished. At the same time, there was no denying that talent, intelligence, and charisma (along with their negative counterparts) were distributed unequally among people. Some traits such as musical and mathematical talent, intelligence, criminality, and poverty seemed to run in families (especially if one overlooked certain social factors). And it was no secret that some cultures had advanced much further than others. Scientists wondered why these patterns existed; demagogues wanted to exploit them for personal or political purposes.

During the nineteenth century and the first half of the twentieth century, scientists began trying to quantify the differences that seemed so apparent between races and classes of people. By measuring skull volume, categorizing skull shapes, and performing other measurements, they believed that some "objective" criteria could determine which races were superior. Unfortunately, whether consciously or not, many of these measurements were biased in such a way as to give expected results. For example, it was obvious to all that Europe had achieved technology superior to that of most of Africa and among Native Americans. Since most of the scientists performing these experiments were European or of European extraction, their tendency was to assume that Europeans were superior to other groups, so more European skulls were deemed superior. Conducting further measurements only solidified this preconception.

In 1859 Charles Darwin (1809-1882) published his landmark book, *On the Origin of*

Species, in which he first laid out his theory of evolution by survival of the fittest. This was followed in 1870 by his book on human evolution, *The Descent of Man*. These volumes and the concepts contained in them were quickly seized upon as a scientific explanation for the "superiority" of some races and the apparent transmission of traits, both good and bad, among families or related groups of people. For example, if a tendency towards criminal behavior is hereditary, then one would expect "lower" classes to exhibit more criminal behavior because of the rarity of intermarriage between the very poor and the wealthy and because most prisoners were from the poor.

These lines of speculation were further fueled by papers and books published in the early years of the twentieth century in which the hereditary aspects of intelligence were touted. In some cases, virtually every negative aspect of human behavior was attributed to inherited lack of intelligence. With the advent of intelligence testing in the 1920s, scientists had another tool with which to study the effects of intelligence on human behavior. Unfortunately, this tool was neither as objective nor as accurate as promised.

All of this culminated in the eugenics movement. The term was first coined by British scientist Francis Galton (1822-1911) in 1883, when he suggested regulating family size and even marriage partners based on the genetic advantages enjoyed by the parents. Later, in the early 1900s, further testing was carried out in both Europe and America, culminating with suggestions that "inferior" people be sterilized to prevent them from passing on their genes to future generations. In this manner, it was felt, undesirable traits could be bred out of humanity, improving the human race and the human condition.

At least at the outset, eugenics was not used to justify discrimination, oppression, or other evil deeds. Most early scientists in this field were well-intentioned researchers who were honestly trying to understand what led people and races to differ so widely. If these differences could be tied to scientific fact, they believed, then people could consciously work to remove unfavorable traits from humanity. This, in turn, would lead to a better human, free of inherited physical and mental deficiencies that caused people to commit crimes or indulge in behaviors that did not benefit society. While these scientists' intentions were good and their field of study well respected, the eugenics movement was ultimately found to be based on fallacious assumptions,

Francis Galton, the founder of eugenics. *(Library of Congress. Reproduced with permission.)*

rendering findings and conclusions that were equally fallacious. Eugenics might have faded into obscurity like other incorrect scientific assumptions, but its tenets were used in several highly destructive ways during the early decades of the twentieth century. The impact of these destructive campaigns in still being felt today.

Impact

As mentioned above, eugenics was seen as a scientific way to remove unwanted traits from humanity, just as selective breeding is seen as a way to remove unwanted traits in livestock. Since so many of society's ills were thought to be inherited, the thinking went, the solution was simply to prevent the unintelligent, the criminal, and others with these traits from having children. In the United States, 24 states passed laws between 1911 and 1930 that restricted the right of the "unfit" to have children, either by restricting marriage or by out-and-out sterilization. This trend was followed to a lesser extent in England and to a greater extent in Germany, especially after the Nazi party's rise to power in 1933. The problem with this campaign was that there was no way to prove scientifically who was "unfit." In the end, the campaign was used in purely political fashion and for entirely political aims—to dominate and oppress those whom the ruling class deemed inferior.

Eugenics began to lose its appeal in the United States in the mid-1920s, when it was shown to be ill-advised and based on unsound scientific principles. Its popularity continued in Europe, however, especially in Nazi Germany, where scientists striving for German "racial purity" advocated and then put into effect sterilization, forced abortion, and other measures designed to limit the reproduction of Jews, Gypsies, Slavs, and Africans. Later, the Nazi regime took this campaign to a more extreme and horrific level: the mass execution of many of these groups in the Holocaust. Jews were the focus of this effort, and nearly 6 million of them are estimated to have been murdered in the 1930s and 1940s.

In more recent years, reexamination of many of the original studies upon which eugenics was based have shown them to be either blatantly falsified, or suffering from errors in judgment, faulty scientific understanding, erroneous assumptions, or some combination of the above. In any event, it is now accepted that the goals of the eugenics movement—bettering humanity through conscious breeding—is not only scientifically flawed but (more importantly) morally and ethically wrong.

In spite of this, there is no shortage of people today who still advance the same arguments that were discredited decades ago. In some cases, inaccurate or false information is purposely used by people because it serves their aims. In other cases, people with limited knowledge of science or genetics assume that humans can and should be bred the same way livestock are to improve the human species. Still others see genetics and Darwinism as justification for their preconceptions, hatreds, or prejudices. For whatever reason, many are willing to overlook the vast body of scientific, philosophical, moral, and ethical evidence against eugenics; others remain unaware that this evidence exists. Thus, the legacy of an originally well intentioned study of human differences became, and remains, a way to continue to foster hatred, prejudice, violence, and discrimination.

In addition to the use of eugenics to justify racial, religious, or national hatreds, there is currently some concern that similar concepts will be used for less sinister purposes today. With the availability of sophisticated genetics testing, we can now test children in utero (i.e. before they are born) for any number of genetic traits. As an increasing number of genes are linked to specific diseases or tendencies towards those diseases, the temptation grows to use this information in a number of ways. Parents, for example, routinely decide to terminate pregnancies because of the presence of serious conditions, but should they be given the option of terminating a pregnancy because of the "wrong" eye color or stature or the potential to have, say, diabetes? Can insurance companies refuse to extend medical or life insurance to unborn babies carrying the gene for Tay-Sachs disease, Gauchier's syndrome, or who are predisposed to certain types of cancer? And should people with these genetic traits be steered away from having children at all? These and similar questions remain with us, and are likely to remain for decades to come.

P. ANDREW KARAM

Further Reading

Books

Caplan, Arthur. *When Medicine Went Mad: Bioethics and the Holocaust.* Humana Press, 1992.

Gould, Stephen Jay. *The Mismeasure of Man.* New York: W.W. Norton & Company, 1981.

Reily, Philip. *The Surgical Solution: A History of Involuntary Sterilization in the United States.* Baltimore, MD: Johns Hopkins University Press, 1991.

Internet Sites

The University of Pennsylvania's Center for Bioethics. http://www.med.upenn.edu/~bioethic/library/papers.art/EugenicsNotreDame.html

Advances in the Understanding of Energy Metabolism

Overview

Investigation of the chemistry of life had begun in the nineteenth century with the discovery that organic molecules, the chemicals found in living things, were larger and more complex than inorganic molecules, those in the non-living world. By the end of the century, the major classes of organic molecules—the carbohydrates, lipids, proteins, and nucleic acids—had all been identified, but little was known about the cellular processes involved in making and breaking them down. Metabolism is a general term for the sum of all the chemical reactions that make up these processes. During the first half of the twentieth century, biochemists and biologists began to make significant progress in working out the steps in these chemical reactions, which turned out to be much more complex than had been originally suspected. The study of the reactions involved in providing energy for organisms resulted in some of the most important advances.

Background

In 1897 bacteriologist Hans Buchner (1850-1902) discovered that fermentation, the process by which yeast cells break down sugar, could occur even when the yeast cells had been ruptured; it could in fact occur in a soluble extract from the cells. Until that time, there had been a lively debate over whether or not intact cells were needed for the chemical processes related to life. The great bacteriologist Louis Pasteur (1822-1895) had argued that intact cells were necessary; but Buchner's finding proved him wrong, making it clear that biochemists could study the reactions of life in test tubes, thus calling into question the concept of vitalism, the idea that there was some life force necessary for chemical reactions in organisms.

Shortly after Buchner's discovery, Franz Hofmeister (1850-1922) presented the enzyme theory of life, arguing that all cellular reactions are controlled by specific chemicals called enzymes. This replaced the older protoplasmic theory, which held that the whole cell is needed for such processes. As the century progressed the idea of the enzyme became more clearly delineated, as it was found that enzymes were large protein molecules. It was in 1930 that the first enzyme was crystallized—that is, fully purified from all other cell components. Enzymes are catalysts, which means that they speed chemical reactions in living systems, but are the same chemically at the end of the reaction as they were at the beginning. So a single enzyme molecule can perform the same reaction many times, and thus in many cases not much of a particular enzyme is needed in living tissue. This posed a particular problem in early biochemical research because it meant that enzymes were often difficult to find with the crude detection methods then available, and often large quantities of tissue yielded only very small amounts of an enzyme after purification.

The disruption of cells and the isolation of specific enzymes were important tools in discovering how cells break down sugar, most commonly glucose, and release chemical energy from the bonds between the atoms in the glucose molecule. It became evident quite early in these investigations that this process in energy metabolism was a complex one, involving not a single reaction, but a series of reactions. It took several decades to work out all the steps in these reactions, and German biochemists made the most significant contributions to this research. Beginning in 1908, Otto Warburg (1883-1970) studied the use of oxygen in energy-releasing processes in cells, developing a number of techniques to carefully measure the uptake of this gas. His introduction of quantitative methods brought a level of precision to biochemistry that was essential to working with small amounts of material and to detecting subtle chemical changes. It was also Warburg who began experiments on tissue slices, a technique that was later used in studies of energy use in liver and muscle cells.

Working in the same Berlin research institute as Warburg, Gustav Embden (1874-1932) and Otto Meyerhof (1884-1951) discovered the steps by which the six-carbon glucose molecule is broken down into two three-carbon molecules of pyruvic acid. This series of reactions is anaerobic (does not require oxygen), and in yeast, the pyruvic acid is converted to alcohol, ethanol, and carbon dioxide in a process called alcoholic fermentation. In muscle cells, however, where an almost identical set of reactions occurs, if oxygen is insufficient, the pyruvic acid is converted to lactic acid, the buildup of which tires muscle. If oxygen is present, the complete breakdown of pyruvic acid can occur. Hans Krebs (1900-1981), who began his career working with Warburg, discovered the sequence of reactions involved in this process.

Impact

Krebs found that in the presence of oxygen, a pyruvic acid molecule entered a cycle of reactions, the first of which was the reaction of pyruvic acid with oxalacetic acid to form citric acid. This series of reactions thus became known as the citric acid cycle. It is also called the Krebs cycle in honor of his work, which was a significant achievement because it involved putting together findings from other researchers and then doing painstaking experiments to discover how the steps in the process fit together.

As Krebs later recalled, he built upon the work of another biochemist, Albert Szent-Györgyi (1893-1986), who had developed a way to separate intact mitochondria, the cell organelles involved in energy production, from the flight muscle of pigeons. Szent-Györgyi used the line of reasoning that if the addition of a substance to such a preparation greatly increased the chemical activity as measured by the use of oxygen, then this indicated that the substance was an intermediate in the breakdown of carbohydrate for energy. Of the many substances tested, only a few were active, but these substances, all four-carbon acids, did not bear any resemblance to the sugars found in food.

It was Krebs who discovered that these acids acted as carriers for pyruvic acid. Then, through a series of reactions, pyruvic acid was broken down and the energy in the molecule released. The four-carbon acids returned to their original form by the end of the reaction series, thus allowing for their recycling, and this is why the whole series of reactions is called a cycle. There are a number of such cycles found in metabolism. In fact, Krebs himself had discovered another cycle a few years earlier, one that creates urea, the form of nitrogen that is excreted from the body in urine. Krebs received the 1953 Nobel Prize for Physiology or Medicine for his research. Warburg received the same honor in 1931, indicating the importance of research on energy metabolism.

While some biochemists were discovering the intermediate steps in the breakdown of glucose, others were investigating the chemicals that served as the repositories for the energy stored in the chemical bonds in glucose. There is a great deal of energy in these bonds, and what happens in the breakdown of glucose is that this energy is distributed to other molecules, in packets of energy of more useable size. It was Fritz Lipmann (1899-1986) who identified adenosine triphosphate (ATP) as the high-energy product of the breakdown of glucose and other food stuffs, such as proteins and fats. As its name implies, there are three phosphate, or phosphorous-containing, groups in ATP, and the last phosphate is attached with a high-energy bond created using the energy from the Krebs cycle and from other energy-releasing processes. When glucose is completely broken down to carbon dioxide and water, 38 ATP molecules are produced, and it is ATP that provides the energy for muscle contraction, nerve impulse conduction, and most other energy-requiring processes in cells and in organisms.

The breakdown of lipids and proteins also results in the production of ATP by processes with final stages identical to those in the breakdown of sugar. It was again Lipmann who discovered the common denominator, a substance called acetyl CoA. Like Krebs, Lipmann was interested in how pyruvic acid was broken down in animal tissue. He discovered that a derivative of the B vitamin, pantothenic acid, was necessary for the process. It turned out to be, like other B vitamins, a coenzyme, that is, a small molecule that must be present in order for a specific enzyme to function. Lipmann called it coenzyme A, or CoA, which reacts with pyruvic acid to form acetyl CoA. This molecule is the form of pyruvic acid that enters the Krebs cycle by reacting with oxalacetic acid to form citric acid. Lipmann found that acetyl-CoA is formed not only in the breakdown of sugars, but in the breakdown of proteins and lipids as well. This means that, ultimately, these nutrients as well as carbohydrates can be broken down completely through the Krebs cycle and the chemical energy in these molecules converted to the high-energy bond in ATP.

Because acetyl-CoA proved so crucial in energy metabolism, Lipmann was awarded the Nobel Prize for Physiology or Medicine in 1953, the same year that Krebs received the award. The work of these men, along with that of many others, some of whom are mentioned above, accomplished a great deal during the first half of the twentieth century. In 1900 the concept of the enzyme as a highly specific protein molecule had yet to be developed, and the complexity of energy metabolism was unknown. By 1950 the basic steps in the breakdown of all the energy-rich nutrients—carbohydrates, lipids, and proteins—had been worked out by biochemists, the function of a number vitamins in these reactions was understood, and the crucial role of ATP appreciated. Biochemists had not only identified many chemicals involved in energy metabolism, but had also discovered how these chemicals related to each other, how one was converted into another, and in many cases had identified the enzymes responsible for these conversions.

MAURA C. FLANNERY

Further Reading

Bartley, W., H. Kornberg, and J. R. Quayle, eds. *Essay in Cell Metabolism*. New York: John Wiley & Sons, 1970.

Burk, Dean. "Otto Heinrich Warburg." In *Dictionary of Scientific Biography.* Vol. 14, edited by Charles Gillispie. New York: Scribner's, 1976.

Dressler, David, and Huntington Potter. *Discovering Enzymes.* New York: Scientific American Press, 1991.

Fruton, Joseph. Molecules and Life: Historical *Essays on the Interplay of Chemistry and Biology.* New York: John Wiley & Sons, 1972.

Holmes, Frederic. "Hans Adolf Krebs." In *Dictionary of Scientific Biography.* Vol. 17, supplement 2, edited by Frederic Holmes. New York: Scribner's, 1991.

Kleinhof, Horst, Hans von Döhren, and Lothar Jaenicke, eds. *The Roots of Modern Biochemistry: Fritz Lipmann's Squiggle and Its Consequences.* New York: Walter de Gruyter, 1988.

Krebs, Hans. *Otto Warburg: Cell Physiologist, Biochemist, and Eccentric.* Oxford, Clarendon Press, 1981.

Krebs, Hans. *Reminiscences and Reflections.* Oxford: Clarendon Press, 1981.

Lipmann, Fritz. *Wanderings of a Biochemist.* New York: John Wiley & Sons, 1971.

Advances in Understanding Viruses

Overview

At the beginning of the twentieth century, the term "virus" commonly referred to infectious agents that could not be seen under the microscope, trapped by filters, or grown in laboratory cultures. Today the term refers to a minute entity composed of an inner core of nucleic acid and an envelope of protein. The fundamental difference between a virus and other microbes is that a virus can only reproduce by entering a living host cell and taking over its metabolic apparatus. Studies of tobacco mosaic disease by Adolf Mayer, Martinus Beijerinck, and Dimitri Ivanovski led to the discovery of the tobacco mosaic virus, which was crystallized by Wendell M. Stanley. The discovery of bacterial viruses, or bacteriophages, by Frederick Twort and Félix-Hubert d'Hérelle provided one of the most important experimental tools in the development of molecular biology.

Background

The modern definition of a virus refers to a minute entity composed of an inner core of nucleic acid and an envelope of protein; a virus particle can only reproduce by entering a living host cell and taking over its metabolic apparatus. The meaning of the word "virus" has undergone many changes since its original usage in Latin for "slime," an unpleasant substance, but not necessarily dangerous. Eventually, "virus" assumed the connotation of a dangerous poison or venom or of a mysterious, unknown infectious agent. Eighteenth-century medical writers applied the term to the contagion that transmitted an infectious disease, for example, the "cowpox virus" or the "variolous virus." After the establishment of germ theory in the late nineteenth century, "virus" was generally used in reference to obscure entities with infectious properties. The infectious agents of some diseases displayed complex life cycles and could not be cultured by conventional techniques. Obviously, the failure of some infectious agents to grow in the laboratory might simply mean that they required special growth conditions.

Bacteriologists discovered that the causative agents of many diseases could be identified under the microscope, grown in the laboratory, and removed from growth media by using special filters, such as the filtration device developed by Charles Chamberland (1851-1908). The Chamberland filter was used to separate visible microorganisms from their culture medium and to prepare bacteria-free liquids. Chamberland was also instrumental in the development of the autoclave, a device for sterilizing materials that uses steam heat under pressure. Rigorous use of these techniques made possible the establishment of a new category of infectious agents, which were generally called "invisible-filterable viruses."

"Invisible viruses" were thought to cause many important diseases of humans and animals, but their nature remained obscure. Therefore, in the early twentieth century such infectious agents were defined operationally as "filterable" and "invisible" microbes. In other words, the viruses were defined operationally in terms of their ability to pass through filters that trapped bacteria and their ability to remain invisible under a light microscope. Criteria based on specific techniques, however, provided little insight into the biochemical nature of viruses. Moreover, further studies of the invisible-filterable viruses indicated that these two operational criteria were not necessarily linked. Eventually, some microbiologists began to consider the possibility that viruses might be a special category of microparasites that could only reproduce in the cells of other organisms.

Investigations of plant viruses, especially the tobacco mosaic virus, provided the basis for the new science of virology. Adolf Eduard Mayer (1843-1942), Martinus Beijerinck (1851-1931), and Dimitri Ivanovski (1864-1920) are generally regarded as the founders of virology. Plant virology can be traced back to 1886, when Mayer discovered that tobacco mosaic disease could be transmitted to healthy plants by inoculating them with extracts of sap from the leaves of diseased plants. Mayer could not culture the causative agent on artificial media, but he was able to prove that filtered sap from infected plants could transmit the disease.

In 1892 Ivanovski demonstrated that the infectious agent for tobacco mosaic disease could pass through the finest filters available. Using Chamberland's filter method, Ivanovski found that a filtered extract of infected tobacco leaves produced the same disease in healthy plants as an unfiltered extract. He read a paper on his discovery to the Academy of Sciences of St. Petersburg that same year. Ivanovski thought his observations might be explained by the presence of a toxin in the filtered sap, but he did not discount the possibility that unusual bacteria had passed through the pores of the filter. Unable to isolate a toxin or culture the "tobacco microbe," Ivanovski turned to research on alcoholic fermentations. Nevertheless, Russian historians consider Ivanovski the founder of virology.

Apparently unaware of Ivanovski's work, Beijerinck began research on tobacco mosaic disease as a graduate student under Mayer at the University of Leiden. In 1895 Beijerinck became professor of microbiology at the Polytechnical School in Delft. His research interests included botany, microbiology, chemistry, and genetics as well as tobacco mosaic disease. Like his predecessors, Beijerinck found that the sap of plants with tobacco mosaic disease remained infectious after passage through a filter. He was particularly intrigued by the ability of filtered sap to transmit the disease after passing through a series of a large number of plants; this fact indicated that the causative agent was an entity that could reproduce and multiply itself and was not a poison or toxin. Based on reports in the botanical literature, Beijerinck thought that many other plant diseases were caused by agents like the tobacco mosaic disease virus.

The fundamental difference between bacteria and virus was probably not size or filterability. In 1900 Beijerinck published an extensive review of his work and his theory. Unlike bacteria, the virus might be an obligate intracellular parasite, an agent that can only reproduce within living cells. In thinking about the way in which filtered plant sap could be used to transmit tobacco mosaic disease to a large series of plants, Beijerinck concluded that the disease must be caused by an agent that needed the living tissues of the plant in order to reproduce itself. In other words, the virus could be thought of as some kind of "contagium vivum fluidum," or contagious living fluid. The crucial difference between a virus and other microbes might not be simply size. Some microbes might be obligate parasites of living organisms that could not be cultured in vitro on any cell-free culture medium. Beijerinck, therefore, concluded that tobacco mosaic disease must be caused by an agent that had to incorporate itself into a living cell in order to reproduce. Beijerinck's theory of the "contagium vivum fluidum" was a first step towards a new way of thinking about the special nature of the virus.

One of the first detailed studies of an animal disease caused by a filterable virus was conducted by Friedrich Loeffler (1852-1915) and Paul Frosch (1860-1928) in 1898. Foot-and-mouth disease was an important disease of cattle, but all efforts to culture bacteria from fluid taken from lesions in the mouths and udders of sick animals were unsuccessful. Loeffler and Frosch were unable to isolate the causative agent, but they did prove that it could be transmitted to cattle and pigs with minute amounts of filtered, apparently bacteria-free, fluids from the vesicles of infected animals. These experimentally infected animals could transmit the disease to other animals. Using mixtures of blood and lymph from the vesicles of animals with the disease, Loeffler and Frosch were able to immunize healthy animals. These experiments suggested that a living agent, rather than a toxin, must have been in the filtrate. They concluded that the disease might be caused by a living agent small enough to pass through the pores of their filters. Their experiments suggested that only a living agent, one capable of reproducing itself, could continue to cause the disease after passage through a series of animals. Loeffler and Frosch suggested that other infectious diseases, such as smallpox, cowpox, and cattle plague, might be caused by

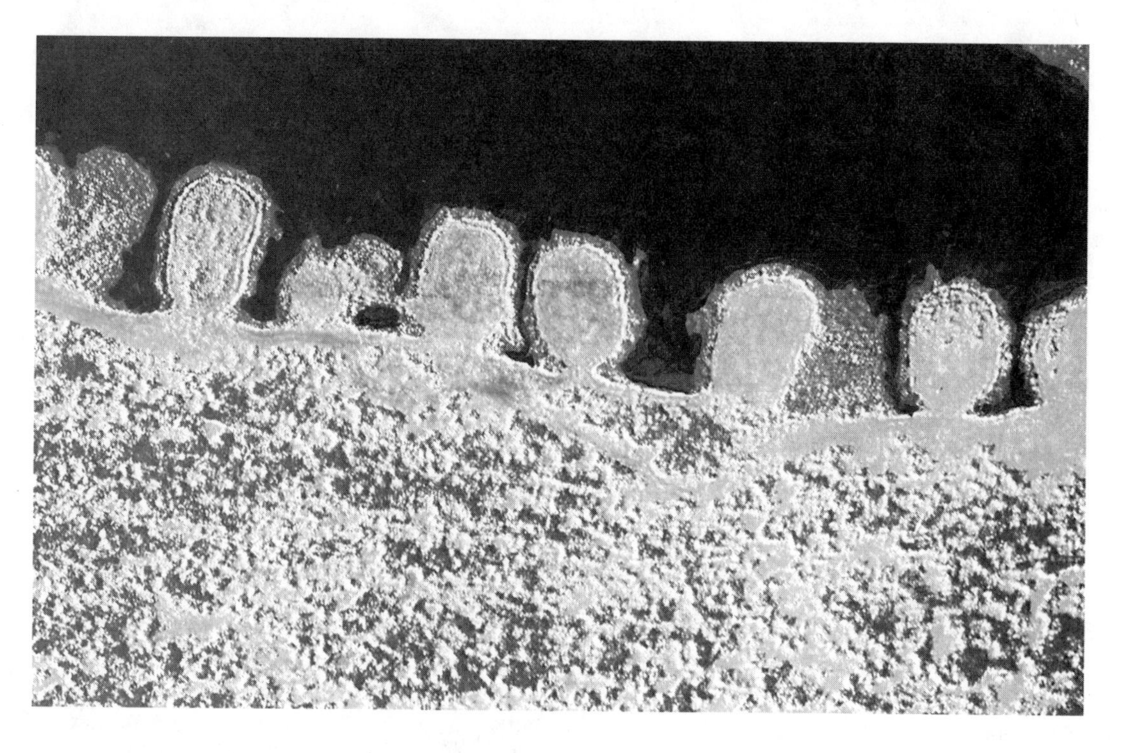

Transmission electron micrograph of influenza viruses budding from the surface of an infected cell. *(CNRI/Science Photo Library, National Audubon Society Collection/Photo Researchers, Inc. Reproduced by permission.)*

similar organisms. Nevertheless, they continued to think of the infectious agent as a very small and unusual microbe that had not yet been cultured in vitro, rather than a novel entity that could only multiply within the cells of its host.

Within a few years of these studies, filterable viruses were suspected of being the cause of various plant, animal, and human diseases. These predictions were confirmed in 1911, when Francis Peyton Rous (1879-1970) injected chicken with a cell-free filtrate from tumor cells and discovered a virus that could cause cancer. A farmer had noticed that one of his Plymouth Rock hens had developed a breast tumor and brought it to the Rockefeller Institute in New York. After performing an autopsy, Rous prepared an extract of the malignant tumor cells and injected it into healthy chickens. Even when the extract was filtered to remove whole cells, hens of the same purebred stock developed tumors. The virus became known as the Rous sarcoma virus. Rous was awarded a Nobel Prize in physiology or medicine in 1966 for his demonstration that a malignant tumor could be transmitted by a virus.

Impact

Demonstrations of the existence of plant and animal viruses were followed by the discovery of viruses acting as parasites of bacteria. In 1915 Frederick William Twort (1877-1950) discovered that even bacteria could be attacked by invisible viruses. While trying to grow viruses on artificial medium in agar plates, Twort noted that certain bacterial colonies contaminating the plates became glassy and transparent. If pure colonies of this micrococcus were touched by a tiny portion of material from the glassy colonies, they too became transparent. Twort later became obsessed with speculative work on the possibility that bacteria evolved from viruses that had developed from even more primitive forms. These viruses became known as "Twort particles. " Twort demonstrated that these microparasites were filterable, like the infectious agent of many mysterious plant and animal diseases. His landmark paper on the nature of ultra-microscopic viruses was published in the British medical journal *Lancet* in 1915. Unfortunately, Twort's research was interrupted by World War I and his paper had little immediate impact on microbiology.

In 1917, while working on the dysentery bacillus at the Pasteur Institute, Félix d'Hérelle (1873-1949) discovered what he called "an invisible microbe that is antagonistic to the dysentery bacillus." d'Hérelle obtained the invisible microbe from the stools of patients recovering from bacillary dysentery. When an active filtrate was added to a culture of Shiga bacilli, bacterial growth soon ceased and bacterial death and lysis (dissolution) followed. A trace of the lysate produced the same effect on a fresh Shiga culture. More than 50 such transfers gave the same results, indicating that a living agent was responsible for bacterial lysis. The invisible microbe could not grow on laboratory media or heat-killed bacilli but grew well in a suspension of washed bacteria in a simple salt solution. d'Hérelle concluded that the invisible anti-dysentery microbe must be an obligate parasite of the Shiga bacillus; therefore, he called it a "bacteriophage," that is, an entity that eats bacteria. Speculating on the implications of the phenomenon he had discovered, d'Hérelle predicted that bacteriophages for other bacteria would be found. He hoped that bacteriophages might be modified in the laboratory and used to destroy pathogenic bacteria. Unfortunately, d'Hérelle's hope that the bacteriophage would become the "microbe of immunity" did not come true. In honor of the scientists who had discovered them, bacterial viruses were sometimes called "Twort-d'Hérelle particles." Bacterial viruses remained mere laboratory curiosities until molecular biologists adopted bacteriophages as their primary experimental system, at which point the bacteriophage became known as the Rosetta stone of molecular genetics.

During the 1930s and 1940s researchers began to examine the biochemistry of viruses. By the 1940s improved biochemical techniques were making it possible to understand the nature of complex biological macromolecules. Biochemical studies supported the belief that the virus lay on the borderline between cells and molecules. At about the same time, x-ray crystallography and the electron microscope provided a new picture of the previously invisible viruses.

In 1935 Wendell M. Stanley (1904-1971) reported that he had isolated and crystallized a protein having the infectious properties of the tobacco mosaic virus. The title of his landmark paper in the journal *Science* was "Isolation of a Crystalline Protein Possessing the Properties of Tobacco Mosaic Virus." Stanley had seen that investigating the chemical nature of viruses was closely related to the problem of purifying an enzyme. Isolating a virus, however, was more difficult than purifying an enzyme because the chemical reaction catalyzed by an enzyme could be measured more easily than the biological activity of a virus. Nevertheless, Stanley attempted to purify the virus by methods similar to those that James Sumner (1887-1955) and John Northrop (1891-1987) had employed to isolate enzymes. Starting with literally a ton of infected tobacco leaves, Stanley obtained a spoonful of crystalline material that essentially consisted of protein but was still capable of producing tobacco mosaic disease. In 1945 he produced small amounts of crystals that were extremely active as infectious agents. He referred to the purified protein as a "molecular virus." Stanley found that even after several recrystallizations, tobacco mosaic virus retained its physical, chemical, and biological properties. He concluded that tobacco mosaic virus was an autocatalytic protein that could only reproduce and multiply in the presence of living cells. Further work showed that the purified virus contained nucleic acid as well as protein, but until the 1950s it was unclear whether protein, nucleic acid, or nucleoprotein was the infectious component. Stanley's work made possible the modern definition of viruses as particles composed of an inner core of nucleic acid enclosed in a protein overcoat. Stanley won the Nobel Prize in chemistry in 1946 for his demonstration that an apparently pure chemical substance could reproduce itself and multiply as if it were a living organism. Appropriately, he shared the prize with Northrop and Sumner. Northrop was awarded the prize for developing methods for the crystallization of proteins and Sumner was honored for providing the first convincing proof that enzymes are proteins.

Only a few years after Stanley characterized it, scientists obtained the first portrait of the tobacco mosaic virus by electron microscopy. A review of microbiology that set forth the fundamental differences between bacteria and viruses was published by Thomas Rivers (1888-1962) in 1927. Rivers helped to establish virology as a distinct field of study. During the 1930s the electron microscope made it possible to see viruses as well as bacteria. In 1934 Ladislaus Laszlo Marton published the first electron micrographs of biological materials. Three years later, he published the first electron micrographs of bacteria. In 1940 Helmuth Ruska prepared the first electron micrographs of a virus. In 1949 John Franklin Enders (1897-1985), Thomas H. Weller (1915-), and Frederick Chapman Robbins (1916-) developed a technique that made it possible to grow the polio virus in cultures of human tissue. Their methods provided a valuable tool for the isolation and study of viruses, which have been described as "living molecules." By the end of the twentieth century, however, scientists had discovered that some diseases were caused by other, previously unknown submicroscopic creatures, such as "slow" viruses, viroids, and prions.

LOIS N. MAGNER

Further Reading

Books

Fenner, F., and A. Gibbs, eds. *A History of Virology.* Basel, Switzerland: Karger, 1988.

Grafe, A. *A History of Experimental Virology.* Translated by E. Reckendof. Berlin: Springer-Verlag, 1991.

Hughs, S. S. *The Virus: A History of the Concept.* New York: Science History Publications, 1977.

Iterson, G. van, L. E. Den Dooren de Jong, and A. J. Kluyver. *Martinus Willem Beijerinck: His Life and His Work.* Madison, WI: Science Tech Publishers, 1983.

Scott, Andrew. *Pirates of the Cell: The Story of Viruses from Molecule to Microbe.* New York: Basil Blackwell, 1985.

Scholthof, Karen-Beth, John G. Shaw, and Milton Zaitlin, eds. *Tobacco Mosaic Virus: One Hundred Years of Contributions to Virology.* St. Paul, MN: The American Phytopathological Society, 1999.

Waterson, A. P. and L. Wilkinson. *An Introduction to the History of Virology.* New York: Cambridge University Press, 1978.

Periodical Articles

Kay, Lily E. "W. M. Stanley's Crystallization of the Tobacco Mosaic Virus, 1930-1940." *ISIS* 77 (1986): 450-72.

Lechavalier, H. "Dimitri Iosifovich Ivanovski (1864-1920)." *Bacteriological Reviews* 36 (1972): 135-146.

Stanley, W. M. "Isolation of a Crystalline Protein Possessing the Properties of Tobacco Mosaic Virus." *Science* 81 (1935): 644-645.

Twort, Frederick W. "An Investigation on the Nature of the Ultramicroscopic Viruses." *Lancet* 2 (1915): 1241-43.

Advances in Understanding Brain Behavior in Animals

Overview

Understanding the behavior of animals is essential to the work of scientists today. Animal models are used in many areas to study anatomy as well as behavior. Knowing the workings of animal brains led to the founding of psychology and neurology as distinct disciplines. Several scientists devoted their efforts to the study of brain behavior in animals. Russian physiologist Ivan Petrovich Pavlov (1839-1936) is known for his development of the concept of conditioned reflex. German-born American biologist Jacques Loeb (1859-1924) studied animal responses that he coined "tropisms." And American psychologist B. F. Skinner (1904-1990) experimented with scientific measurement of animal behavior that led to the development of a major branch called behaviorism. Behaviorism is not only a philosophy of the past but of the future. For example, computer-assisted instruction is based on the work of the behaviorists.

Background

The scientific study of brain behavior in animals has philosophical roots in the observations of French philosopher René Descartes (1594-1650). As a soldier in Amsterdam, Descartes lived next to a butcher's quarters and became fascinated with animal dissection. His dissections led him to develop an early mechanistic account of the nervous system that included a model of reflex concepts. Later, the study of animal behavior as a branch of psychology developed from the merging of two fields: 1) ethology, the study of characteristic behavior patterns of animals; and 2) comparative psychology, a field of comparing responses. Influenced by Charles Darwin (1809-1882), ethologists—mostly in Europe—studied the behavior of insects, fishes, and birds and became fascinated with the evolution of instinct and genetics. The comparative psychologists—mostly in America—focused on behavior in laboratory animals such as guinea pigs, mice, rats, and monkeys. They were much more interested in environmental aspects than genetics. These two fields merged in the 1950s with the realization that both environment and genetics are of fundamental importance.

Scientists who worked with animals also were not satisfied with just studying the behavior of animals but sought to relate this to human behavior. Ivan Pavlov is a famous name in psychology books for his development of the conditioned reflex. Pavlov, the son of a parish priest of the Russian Orthodox Church, was born in Ryazan, Russia, and received an M.D. at the Imperial Academy of Saint Petersburg. His research and dissertation on how the nervous system regulated circulation earned him a two-year travel scholarship to Germany. There, he was influenced by German physiologists who used animal models for their various studies.

Back in Russia, he met I. M. Sechenov (1829-1905), who had published his extensive work in his book *Reflexes of the Brain.* Pavlov had observed how many individual body functions are reflexive in nature, coordinated by the brain and nervous system. A reflex is an automatic response. For example, blinking and jerking away from a hot object are reflexes. Pavlov developed a "theory of nervism" that guided him through many experiments relating to digestion and behavior. A skilled surgeon, he developed procedures to study digestive processes in unanesthetized animals. He noted that when he would simply approach the laboratory dogs with a piece of meat, they began to produce saliva or salivate at the sight of food or even with the sounds of footsteps.

From about 1898 to 1930 Pavlov measured the salivary secretion of his dogs to study their brain function. For example, in one experiment, he rang a bell at the same time the food was presented; the dog would salivate. Then he would ring the bell without the food. The dog would still salivate. He called this a "conditioned reflex," in contrast to the unconditioned reflex that occurred when the dog had a natural flow of saliva. He proposed that these conditionings constituted the animal basis of behavior. Pavlov also found that the dogs would respond to bells of different pitch—one that would signal food and the other that would not be related to food. If the sounds of the bells were too much alike, the animal would receive mixed messages, become confused, and behave strangely. He called this erratic behavior "neurosis." Increased exposure to the confusing stimuli would cause great irritability in the nerves of the cerebral cortex. If this confusion continued, it affected the entire nervous system, and the behavior of the animal became abnormal.

During this time Pavlov came into contact with the work of British scientist Charles Sherrington (1857-1952). Sherrington had extensively studied spinal reflexes, which are any reflexes relating to the spinal cord. Sherrington proposed that the spinal reflex is made of integrated actions of several parts. Pavlov added to Sherrington's work the relation of the stimulus to reception in the cortex, the outer area of the brain, as well as the subcortex on the inner part of the brain. He proposed that sleep had a great effect and that neurotic disturbances were caused from the conflict or collision between being excited and then being inhibited at the same time.

By the 1920s Pavlov began to extend his theory of animal behavior to human psychology. He believed that all learned behavior related to the construction and integration of interconnected chains of conditioned reflexes. For example, language reflects these long chains of conditioning involving words. Excessive nerve excitation and inadequate inhibition of the cerebral cortex causes neurosis or psychosis. The person with the psychosis develops a protective mechanism—shutting out the external world that would include all the extreme stimuli he or she had received. From this Pavlov determined that a chief treatment for people with mental illness should be quiet rest, without the exciting stimuli. He supported institutional treatment for mental illness.

Jacques Loeb, German born and educated at the University of Strasbourg, moved to the United States in 1891. He was a mechanist who believed that many complicated life happenings could be explained in simple biochemical terms. Loeb applied the term "tropism" to mean a movement or response to a stimulus. He believed that all behavior of organisms is composed of tropisms or involuntary relationships. For example, some organisms naturally will go to the light while others will turn from the light. Other tropisms include darkness, heat, cold, or chemicals. Loeb was particularly interested in problems of instinct and free will. The unlearned behavior reaction to an environment is called a "stereotyped response." Instinct is one of these.

As the term *tropism* began to be used to relate to plants, later researchers used the term *taxes*—singular *taxis*—to refer to animal responses rather than those of plants. Burrhus Frederic (B. F.) Skinner, an American psychologist, became interested in the work of John B. Watson (1878-1958) as a student at Harvard. He studied intensely the work of Ivan Pavlov on conditioned reflexes and was fascinated by the essays of philosopher Bertrand Russell (1872-1970) on behaviorism. Watson, the founder of a school called behaviorism, became his mentor.

During World War II Skinner concentrated research on training pigeons and developed a device called the "Skinner box." This box enabled him to scientifically study the behavior of animals using controls and accurate measurements. He introduced the ideas of stimulus-response, or S-R connection. This proposed that the learning process is a basic matter of receiving a stimulus and being rewarded with a reinforced response. Skinner then applied his animal theories to humans. As a professor at Indiana University, Bloomington, he received public attention when he introduced his Air-Crib, sort of a Skinner box for human beings. The Air-Crib was a large, soundproof, germ-free, air-conditioned box that was a baby keeper for the first two years. It was supposed to provide the best environment for the growth of a child. In 1948 he published a controversial book called *Walden Two,* in which he showed a community modeled after these principles of social engineering. He went back to Harvard in 1948 and became one of the greatest influences on a generation of psychologists. He continued his work with animals and trained them to perform complex and even amazing feats—all using his S-R theories. For example, he trained a group of his pigeons to play ping-pong.

Impact

The work of Pavlov, Loeb, and Skinner had a profound impact on modern theories of animal behavior and brain development. Although Pavlov received the Nobel Prize for medicine in 1904 for his work on digestive secretions, he is most remembered for his pioneering studies in human behavior. He set the stage for other psychologists to study behavior using objective methods and describing in physiological terms. He also provided psychology with a basic unit that explains complex behavior. American psychologist John B. Watson used Pavlov's work as the foundation of behaviorist psychology, which is still a viable part of educational psychology. As for Skinner, his famous box is now used in pharmaceutical research to show how different drugs affect animal behavior.

The systematic theory of the behaviorists led Skinner to develop a system called programmed learning, using teaching machines. In programmed learning, a person is given the stimulus—a question. After the person answers, he/she is then given immediate reinforcement—positive (moving on to a new question) if the answer was correct or negative (not moving on) if the answer was incorrect. The student progresses through a systematic series of questions/answers at his or her own rate of speed. Central are the ideas of immediate reinforcement or reward and learning in limited, incremental steps.

Modern-day computerized-assisted instruction (CAI) is based on this same theory—using computers instead of the old teaching machines. The term now used to describe the behavior is *mastery learning* or *learning by objective.* During the first half of the twentieth century the experiments of Pavlov, Loeb, and Skinner advanced the study of animal behavior and brain development. The ideas of the behaviorists are prominent today in a school called neo-behaviorism, led by Albert Bandura.

EVELYN B. KELLY

Further Reading

Gray, Jeffrey A. *Ivan Pavlov.* New York: Penguin, 1981.

Skinner, B. F. *Science and Human Behavior.* New York: Macmillan, 1953.

Skinner, B. F. *The Shaping of a Behaviorist.* New York: Knopf, 1979.

Behavioral Studies Develop through Animal Observation and Experimentation

Overview

Early in the twentieth century, scientists became interested not only in discovering new organisms, but in understanding more about the behavior of known organisms. A group of biologists took special note of social behavior, and learned about the amazingly complex organizations of a wide range of animals from protozoa to bees, and ants to birds. One of the best-known behaviorists of the period was Karl von Frisch (1886-1982), who established how honey bees communicate with one another about the location of a food source. That study has become a staple in biology and behavior textbooks. Other major researchers who made important contributions to the study of animal behavior during this time included Warder Allee (1885-1955), Herbert S. Jennings (1868-1947), Nikolaas Tinbergen (1907-1988), Konrad Lorenz (1903-1989), and William Morton Wheeler (1865-1937).

Background

One of the most famous examples of the research on social behavior during this or any period came from Karl von Frisch, a zoologist from Vienna, and his study of honey bees. Originally, he was interested in how bees detected color and odors, but he soon turned his attention to the method by which a few scout bees were able to inform the hive of a newly found flower patch or other food source, and then give accurate directions to it. Frisch set out in the early 1920s to discover how this communication system worked. At first, he believed that the scout bees were informing the other foragers about the availability of a food source by carrying the food scent back to the hive. The others, then, simply sought out the source of the scent. It was only through additional experiments he conducted nearly two decades later that he discovered a more elaborate communication system involving so-called dances that conferred precise directions to the food.

At the turn of the century, American naturalist Herbert Jennings was studying protozoa, a group of single-celled microorganisms. In some of his most well-known behavioral work, he used heat, light, and other negative stimuli to demonstrate how protozoans respond to environmental changes. In his often-cited 1904 publication, "Contributions to the Study of the Behavior of Lower Organisms," he described how different protozoans respond to stimuli. In this work, he refuted the commonly held "theory of tropisms," which stated that the stimuli affected specific motor organs within an individual organism, and that the responses of those motor organs caused the organism to move away from the negative stimuli. Instead, he reported, "The responses to stimuli are usually reactions of the organisms as wholes, brought about by some physiological change produced by the stimulus; they cannot, on account of the way in which they take place, be interpreted as due to the direct effect of stimuli on the motor organs acting more or less independently. The organism reacts as a unit, not as the sum of a number of independently reacting organs."

Nikolaas Tinbergen's work with the fish called three-spined sticklebacks is another standard in animal-behavior textbooks. This researcher from the Netherlands found a stimulus-response pattern in which females give males cues, and vice versa, and ultimately generate a progressive series of courtship activities. He also found that male-to-male attacks were triggered by the color red. Using artificial models, he showed that the males would attack models as long as they had a red underside. He verified the importance of the color red by offering males different fish models, some of which closely resembled fish and others that looked little like fish.. The males preferentially attacked even the crudely made red-bellied models over much more accurate models that lacked the red bellies. The red, he concluded, was the stimulus—or the "releaser"—that triggers the attack behavior.

William Morton Wheeler, an American entomologist, did his behavioral research primarily on ants and other insects. In his studies, he detailed the complex social structure and division of labor within ants and within bees. Noting how well these groups of invertebrates worked together as a unit, he described the ant or bee colony as a superorganism made up of individuals playing their roles for the benefit of the overall colony.

Warder Allee and T. Schjelderup-Ebbe were the primary researchers in the area of dominance hierarchies, which include the so-called pecking order in animal societies. In such hierarchies, several patterns can emerge. The animals can arrange themselves in a linear fashion so that one member of the group is the most dominant, followed by the second-most dominant, the third-most dominant, and so on, with all those ranking lower in subordinate positions. More complex hierarchies also develop, with patterns where individual A is dominant over B, B over C, yet C over A. These hierarchies affect behavior in a number of ways. Once the hierarchies are formed, for instance, they are relatively firm, eliminating the need for constant infighting to determine dominance.

Allee's contributions to animal behavior extended well beyond his work on dominance, however. He also studied how loose associations between animals were important to their survival. In studies of planarian flatworms and goldfish, for example, he showed that these animals prefer to form aggregations in adverse environmental conditions and actually gain protective benefits from the associations. In perhaps his most important work, Allee summarized much of the known research on societal behavior among animals in his 1951 book *Cooperation among Animals*.

Impact

The experiments that led Frisch to discover the communication system of honey bees have often been cited as excellent examples of solid behavioral research. In these experiments, he set out a tray of scented food for the scout bees. When the scout found the food and returned to the hive, Frisch quickly set out another tray of the same food, but placed it nearer the hive. If the bees communicated by scent, he reasoned, the other foragers should leave the hive, pick up the stronger food scent, and go to the nearest food tray. Instead, he found that they went to the tray that the scout visited.

Next, he went to the hive to see what communication was occurring there. Using a glass-sided hive, he was able to watch the scout when it returned from the planted food to the hive. Frisch observed the scout engaging in either a "round dance" or a "waggle dance." Through a series of experiments, the researcher was able to determine that the round dance, in which the scout bee traces an approximate circle on a wall of the honey comb, told the other bees to search for food within about 164 feet (50 m) of the hive. As Frisch moved the food sources farther from the hive, the scout performed a more elaborate "waggle" dance. In this dance, the scout walked a figure-eight pattern on the vertical wall of the honeycomb, waggling its body when it reached the middle of the pattern. Frisch deduced that the waggle portion of the dance gave other bees the direction of the food source: a vertical dance meant the food source was located directly toward the sun, 20 degrees to the right of vertical meant the food source was 20 degrees to the right of the sun, and so on. He also learned that a faster dance related that the food was farther away. He described his bee studies in his 1967 book *The Language and Orientation of Bees*.

When Frisch's findings first became known, they were met with incredulity. Skeptics felt the lowly bee certainly could not have the capacity to engage in such a complicated communication system: Symbolic language was a human characteristic that surely was not shared by insects. Scientists also challenged his experimental design and his results. Despite the initial furor, numerous scientific studies have since confirmed Frisch's findings and conclusions, and opened the door to the realization that complex social behavior, including communication, extends well beyond humans.

Jennings and Wheeler's investigations of protozoans and insects, respectively, provided added evidence that a wide variety of organisms exhibit sometimes elaborate behaviors. Wheeler's work particularly expanded the current understanding of complex social structures. Tinbergen's work on releasers, and his collaborations with Austrian zoologist Konrad Lorenz, helped to develop a theoretical framework for the newly emerging field of ethology, or the systematic study of the function and evolution of behavior. In 1973, Tinbergen, Lorenz, and Frisch shared a Nobel Prize for their contributions to the study of animal behavior.

Karl von Frisch. *(AP/Wide World Photos. Reproduced with permission.)*

Through Allee's research, and particularly his book *Cooperation among Animals*, this researcher had a profound impact on the study of societal behavior. He investigated dominance and competition, as well as altruism and cooperation, and delineated social behaviors into several categories. These categories included the colonial behavior of some invertebrates, such as coelenterates; aggregations, like those he demonstrated with flatworms and goldfish; orientation to stimuli, as seen in the work by Jennings with protozoans; and the more complex social behaviors of some animals, like honey bees and chickens.

These researchers were only a few of the many scientists who helped promote the study of animal social behavior during the first half of the century, and helped to define animal behavior or ethology as a separate field of research.

LESLIE A. MERTZ

Further Reading

Books

Drickamer, Lee and Stephen Vessey. *Animal Behavior: Concepts, Processes and Methods.* 2nd ed. Belmont, CA: Wadsworth Publishing Co., 1986.

Jennings, Herbert S. *Contributions to the Study of the Behavior of Lower Organisms.* Carnegie Institution of Washington, 1904.

Manning, Aubrey and Marian S. Dawkins. *An Introduction to Animal Behaviour.* 4th ed. Cambridge: Cambridge University Press, 1992.

Mawler, Peter and William J. Hamilton. *Mechanisms of Animal Behavior.* New York: John Wiley & Sons, Inc. 1966.

McFarland, D., ed. *The Oxford Companion to Animal Behavior.* Oxford: Oxford University Press, 1987.

Muir, Hazel, ed. *Dictionary of Scientists.* New York: Larousse, 1994.

Imprinting and Establishment of Ethology

Overview

Although the term "ethology" dates back to 1859, it was only in the first half of the twentieth century that ethology—the systematic study of the function and evolution of behavior—expanded to become a recognized field of research. A key part of the development of this field was the widely recognized work of a number of animal behaviorists, notably zoologists Konrad Lorenz (1903-1989) of Austria, and Nikolaas Tinbergen (1907-1988) of the Netherlands. Lorenz gained prominence for his concept of filial imprinting, in which newly hatched geese will accept as their mother the first animal they see—even if that animal is a member of another species. Tinbergen is known for his research into the control and function of various animal behaviors. The two men, along with Austrian bee-communication researcher Karl von Frisch (1886-1982), shared the Nobel Prize in 1973 for their behavioral studies.

Background

At the turn of the century, the study of animal behavior was in its infancy. Animal behavior was not a cohesive and strong field on its own, but rather an outgrowth of numerous other disciplines, such as entomology or ornithology. As stated by Tinbergen in the foreword to *The Oxford Companion to Animal Behavior* (1987): "Biologists of my generation still remember the time when the study of animal behavior was not yet really accepted as a part of modern science, and was practiced by no more than a handful of individuals, who indulged in it either as a hobby or as a sideline to their 'real' work."

Lorenz, Tinbergen, Frisch, and others changed that viewpoint with their well-designed and provocative studies of animal behavior. One of these important research projects was a study of greylag geese by Lorenz. From 1935-38, he conducted field experiments of these birds and developed his concept of imprinting. For this work, he hatched goslings and presented himself to the goslings as they emerged from their shells. With the biological parents gone, the goslings fully accepted Lorenz as their mother, following him and calling him just as they would with their real mother. He later discovered that the goslings would accept any moving object as their foster mother, provided it was the first object they saw upon hatching. He called this behavior "imprinting." (Research in later years has indicated that some birds will even imprint on unmoving objects, provided they are of a color contrasting that of the environment.) Lorenz also identified this behavior in mallard ducks. In order to get the ducks to imprint on him, however, he also had to quack and squat so he was closer to their height. Through his work, he was able to demonstrate that imprinting occurred only during a very short time span, called a sensitive period.

In addition, Lorenz found that imprinting affected the goslings' and ducklings' selection of mates later in life. His research indicated that when certain animals are reared by foster parents of second species, the animals will approach members of the second species as mates when they mature. Even when these sexually imprinted animals were placed with members of their own species and eventually mated with them, they retained their imprinted preferences and, when given the choice, still approached the second species over their own. Sexual imprinting does not occur in all species, however. Cowbirds and cuckoos, for instance, typically lay their eggs in the nests of other birds, which then rear the young as their own. In the cases of these and other brood parasites, the young birds do not sexually imprint on the foster-parent species.

Tinbergen's studies of animal behavior extended from birds to insects. For his doctoral research, for example, he observed and dissected the behavior of digger wasps, and attempted to determine the causes of distinct activities. While conducting this work, he noted that each female wasp was able to pick out her own individual nest from the numerous nests located nearby. He then looked at the stimulus that keyed the female to the correct nest, and through various experiments, showed that the female located her nest via the tell-tale landmarks surrounding it. In addition, he found that the wasps identified their prey, which are honey bees, by first approaching an appropriately sized and moving object, then verifying that the target had the correct honey-bee odor, and finally gaining affirmation of its identity through either visual or tactile stimuli. Through this work, Tinbergen demonstrated the importance of distinct stimuli in triggering specific responses.

Influenced by other animal behaviorists, and in particular Oskar Heinroth (1871-1945), Lorenz and Tinbergen collaborated in the mid-1930s to flesh out the foundations of the emerging field of ethology, and promoted the field through their later writings.

Impact

The research of various animal behaviorists set the stage for the separate discipline of ethology. Oskar Heinroth made early contributions with his 1910-11 *Ethology of the Anatidae*, which discussed similarities and differences in the behaviors of different duck and goose species. Like Lorenz, he also observed imprinting in greylag geese.

Lorenz's work on imprinting identified irreversible patterns of behavior and illustrated the importance of learning in animal development. It also opened the doors to a wide array of studies on many animals, including humans, concerning the importance of exposure to various stimuli during the early stages of life. These topics, including the evolutionary processes that lead to the presence or absence of imprinting, are still very active areas of research today.

Konrad Lorenz. *(Corbis Corporation. Reproduced with permission.)*

Lorenz expanded his ideas with a 1935 paper in which he described how the action of one individual can spark a separate action in another individual, and so on, so that a series of behaviors occur among the individuals in a group. The triggering action is called a releaser, and the responding action is called an innate releasing mechanism. Tinbergen also conducted research to expand this hypothesis. Tinbergen and Lorenz collaborated on a variety of studies, with Lorenz typically coming up with the ideas and Tinbergen designing experiments to test them. One of these collaborations led to their notion of fixed action patterns, in which animals follow a specific set of behaviors in response to environmental cues. For example, they found that the red spot on a mother gull's bill prompts the young to peck at the red spot. This, in turn, triggers the mother to regurgitate food to the waiting young.

Combined, the work of Lorenz and Tinbergen promoted discussion by scientists of many disciplines about the instinct-mediated learning and genetically driven motor skills present in many species of animals. Their descriptions and discussions of this complexity in behavior was a catalyst for other researchers to consider the effects of releasers, innate releasing mechanisms, and fixed action patterns specifically, as well as genetics and evolution on a broader level.

Tinbergen's 1951 book, *The Study of Instinct*, became the first comprehensive introduction to and review of ethology to be printed in English. The book is often credited with launching the new field into the prominence it enjoys today. Another book of note was Lorenz's *On Aggression*. In this popular though controversial 1966 publication, Lorenz began to apply to humans what he had learned about behavior from other animals. He considered whether humans share the causes for their aggressive behaviors with other animals. He offered the supposition that human aggression was an instinctive means of survival, and one that had become particularly dangerous, because weapons had developed faster than humans could adapt behaviorally to counteract their potential consequences.

Other scientists, too, saw connections between animal behavior and human psychology, and began to consider whether some of the ethologists' findings were applicable to humans. For example, scientists today are still debating whether human infants have a "sensitive period" when they are highly susceptible to stimuli from the environment, and if so, whether stimuli in infancy can have an effect on how smart, how creative, or how musically inclined a child will become later in life.

Because of the work of Lorenz, Tinbergen, Heinroth, and many others early in the twentieth century, ethology became a discipline of its own by the 1940s. Several universities began creating separate faculty positions in animal behavior, and many new students enrolled in animal-behavior courses and programs. By the 1950s, ethology had become a dynamic and almost multidisciplinary field. As Tinbergen noted in *The Oxford Companion to Animal Behavior*, students in this discipline read the work of and collaborated with psychologists, physiologists, ecologists, evolutionists, and even human sociologists in attempting to gain a broader understanding of animal behavior. He added, "As a consequence, the study of animal (and human) behavior has now become a vast cooperative effort, in which both specialization and collaboration with related sciences have their parts to play."

In 1973, Lorenz, Tinbergen, and Karl von Frisch, who made groundbreaking findings about the communication system in honey bees, shared the Nobel Prize in physiology or medicine for "their discoveries concerning organization of individual and social behavior patterns."

LESLIE A. MERTZ

Further Reading

Books

Drickamer, Lee and Stephen Vessey. *Animal Behavior: Concepts, Processes and Methods.* 2nd ed. Belmont, CA: Wadsworth Publishing Co., 1986.

Manning, Aubrey and Marian S. Dawkins. *An Introduction to Animal Behaviour.* 4th ed. Cambridge: Cambridge University Press, 1992.

Mawler, Peter and William J. Hamilton. *Mechanisms of Animal Behavior.* New York: John Wiley & Sons, Inc. 1966.

McFarland, D., ed. *The Oxford Companion to Animal Behavior.* Oxford: Oxford University Press, 1987.

Muir, Hazel, ed. *Dictionary of Scientists.* New York: Larousse, 1994.

Advances in Botany

Overview

During the first half of the twentieth century, several areas of plant science or botany were particularly active fields of research. There was an increase in work on plant physiology, on how plant tissues function, and on the biochemistry of plants. There was also a great deal of research on photosynthesis, on how plants use the energy of the sun to make sugar; several plant hormones that controlled growth and development were identified; and photoperiodism, the control of plant processes by day length, was discovered.

As ecology, the study of relationships among species and between organisms and the environment, emerged as a new discipline, botanists investigated how plant communities change over time and how climate affects plant characteristics and community structure. Specialized surveys of the plants in particular geographic areas were also undertaken. In addition, older botanical fields still flourished with work continuing on the identification and classification of new species, and these taxonomic investigations were enriched by evolutionary theories concerning relationships between plants.

Background

Biochemistry, the study of the chemistry of living things, had developed in the nineteenth century, but the techniques for such investigations greatly improved during the first half of the twentieth century. In 1903 Russian botanist Mikhail Tswett (1892-1919) demonstrated that pigments from plant leaves could be separated by passing an extract of leaf material through powdered chalk. This led ultimately to the development of the technique called chromatography, and to the discovery by Richard Willstätter (1872-1942) of four plant pigments: chlorophylls a and b, carotene, and xanthophyll. Willstätter and his coworkers also determined the structure of chlorophyll, a complex, ring-shaped molecule that contains at its center a magnesium atom that is essential for chlorophyll's function in absorbing light energy in photosynthesis. He then moved on to the study of anthocyanin pigments that color many flower petals.

Richard Willstätter. *(Library of Congress. Reproduced with permission.)*

Other biochemists worked on the photosynthetic steps that followed light absorption. In 1930 Cornelis van Niel (1897-1985) found that photosynthesis involved two separate sets of reactions—"light" reactions that used light energy to split water and to provide the products used in "dark" reactions, consisting of the chemical conversion of carbon dioxide and water into sugar. In the 1920s light was found to have another role in plant physiology: the relative length of day and night controlled flowering in tobacco plants. Subsequent research showed that other plant processes, including leaf fall in the autumn and the activation of seeds from dormancy, were also controlled by relative length of day and night.

While some botanists were exploring the physiology of plants, others were investigating morphology, the study of structure, and some were tying the two fields together in their work. Morphology was an old branch of botany, but it changed considerably in the twentieth century as botanists focused more on microscopic studies of the cellular basis of plant structures. They worked out the organization of the root tip and of xylem and phloem, the vascular tissues in plants that respectively carry water and minerals up from the roots, and sugars and other nutrients down from the leaves.

Plant morphogenesis, the coordinated growth and differentiation of cells during development, was another important area of study at this time, and there were investigations of plant cell structure, including the discovery of the layered structure of the cell wall and localization of chlorophyll in cell organelles called chloroplasts. Studies on plant-cell chromosomes, the structures made of protein and DNA that determine an organism's traits, led to the discovery that some plants have more than the two sets of chromosomes, a condition called polyploidy, which turned out to be much more common in plants than in animals.

Impact

Generating polyploidy, that is, producing plants that have three or four sets of chromosomes (instead of the normal two sets) and sometimes as many as six sets, became a way to produce new plant strains, sometimes with interesting traits; in some cases economically important crop plants were created in this way. Another area of botanical study that ultimately had economic significance was research on plant hormones, chemicals produced by plants that are involved in growth and development as well as in responses to the environment. In 1926 Dutch plant physiologist Frits Went (1903-?) found that a chemical extracted from young seedlings could induce the bending of a growing shoot. He identified this substance as auxin, indoleacetic acid, and found that it causes bending by stimulating the elongation of cells along one side of the shoot, ordinarily the side that faces away from the light, thus causing the shoot to bend toward light. In the

1930s another class of plant hormones called gibberellins was discovered by Japanese botanists. These hormones work with auxins in controlling normal plant growth and development, including the development of fruit. Gibberellins are also involved in seed germination. In the 1940s Johannes van Overbeek found he could stimulate the growth of plant embryos with coconut milk, and this work led to the identification of still another class of plant hormones, the cytokinins.

While some botanists were investigating the internal environment of plants, others were studying the external environment. With work on vascular tissue came increased interest in the nutrients needed for plant growth that were dissolved in soil water and absorbed through the roots. Nitrate, sodium, phosphate, and potassium salts were all found to be necessary for normal growth; deficiencies in any of these could limit growth. Ammonia could replace nitrate as a nitrogen source, but could be toxic in high concentrations. Also needed were a number of minerals, such as boron, copper, and manganese, that had to be present in smaller or "trace" amounts. Botanists discovered that plant nutrition was a complex subject, with the nutrient requirements varying with the age of the plant, and that in plants that were becoming senescent or dormant at the end of a growing season, there was actually a movement of minerals out of the plant and back into the soil. Also, light was necessary for nutrient transport, indicating that this process required energy. Related to work on plant nutrition were investigations on nitrogen fixation, on how bacteria living in nodules on the roots of plants called legumes were able to "fix" nitrogen, that is, convert gaseous nitrogen, which cannot be used as a nitrogen source by plants, into useable nitrate.

Interest in how soils supported plant growth was tied to other questions about environmental factors related to plant survival and thus to ecology. Though its roots extend into the nineteenth century, much of the significant work in this relatively young science was done in the twentieth, with some key contributions made by botanists. Henry Chandler Cowles (1869-1939) was the first to work out complete successional series, that is, to describe how plant communities changed over time as the plants themselves altered conditions. For example, there are some plant species that can survive on barren sand dunes and thus are the first species to populate such terrain. As they put down roots, grow, and then die, these plants add organic material to the sand, enriching it, so that other species can then survive there. Often the latter species will crowd out the earlier, pioneer species, and will, in turn, eventually be replaced by others until a stable habitat develops.

Frederick Clements (1874-1945) created the idea of "climax" formations of species to describe such mature communities where a species equilibrium is reached that is dictated by the climate. He argued that each formation type had to develop in a particular way and saw the climax formation, of which he identified 14 different types in North America, as an organic entity, something like a super-organism. Clements's ideas were called into question by Henry Gleason (1882-1975), another American botanist, who contended that there was nothing inevitable about climax communities of plants and that it was not possible to categorize formations since the plant assemblages varied too much from one area to another. Gleason first presented his views at the International Botanical Congress in 1926, and thereafter Clements's ideas were slowly discredited as more habitats were studied in greater depth and their diversity became more apparent.

After publishing his criticism of Clements, Gleason turned away from ecological work and devoted himself to systematics, or taxonomy, the study and classification species. Beginning in 1919 he spent the next 20 years investigating the plants of South America, making many collecting trips there. At the time, little was known about the plants of this region, so every trip yielded many new species. Gleason described more than a thousand, and he was just one of many taxonomists exploring the plant world in the first half of the twentieth century. Though he specialized in flowering plants, others focused on mosses, on ferns, and on algae or aquatic plants. While taxonomy was considered a rather old-fashioned field compared to the more experimental field of physiology, which gained acceptance among botanists after 1900, taxonomists still made major contributions to botany, identifying species and documenting where they occurred naturally. This information became more important by the end of the century when so many habitats had been destroyed worldwide and so many species had become extinct.

Botanists interested in taxonomy were also investigating plant evolution. Dukenfield Henry

Scott (1854-1934) studied fossil plants, especially those of the Carboniferous era and found that many fern-like plants of this time were seed bearers and thus comprised a new order of fossil species, the Pteridosperms or seed-ferns. He also discovered the most ancient vascular land plants yet known, the Psilophytales. Frederick Bower (1855-1948), another student of plant evolution, focused on how living species were related to each other.

Still other botanists worked on genetics following the rediscovery at the turn of the century of Gregor Mendel's (1822-1884) research on heredity in pea plants. They investigated how specific traits were transmitted from one generation to the next, and made important contributions not only to genetics but also to cell biology. Barbara McClintock's (1902-1992) work with corn plants led to the discovery of how chromosomes exchange material with each other during meiosis, the process by which sex cells are produced. Genetic research by botanists also contributed to what came to be called the Modern Synthesis, in which genetics and evolutionary biology were closely tied together.

MAURA C. FLANNERY

Further Reading

Dawes, Ben. *A Hundred Years of Botany.* London: Duckworth, 1952.

Fruton, Joseph. "Richard Willstätter." *In Dictionary of Scientific Biography.* Vol. 14, edited by Charles Gillispie. New York: Scribner's, 1976.

Gabriel, Mordecai, and Seymour Fogel, eds. *Great Experiments in Biology.* Englewood Cliffs, NJ: Prentice-Hall, 1955.

Iseley, Duane. *One Hundred and One Botanists.* Ames, IA: Iowa State University Press, 1994.

Magner, Lois. *A History of the Life Sciences.* 2d ed. New York: Marcel Dekker, 1994.

Morton, A. G. *History of Botanical Science.* New York: Academic Press, 1981.

Reed, Howard. *A Short History of the Plant Sciences.* New York: Ronald Press, 1942.

Steere, William, ed. *Fifty Years of Botany.* New York: McGraw-Hill, 1958.

Developments in Ecology, 1900-1949

Overview

Through ancient descriptions of the relationship between organisms and the environment, the science of "ecology" can be traced to the philosophical interests of the Greeks during the time of Aristotle (384-322 B.C.). The Greek philosophical interest in nature paved the way centuries later for "naturalists" who—at home and in newly discovered lands—explored nature and classified its types. A true science of ecology did not emerge until the early twentieth century when scholars began looking at the natural environment as a place where there were "relationships." As ecology matured, ecologists began looking at animals and plants as communities and became concerned with population growth and its limits, cooperation and competition in nature, and, most recently, making evaluations of energy use and transfer in nature.

Background

"The science of nature" took a giant leap in the nineteenth century when naturalist Charles Darwin (1809-1882) published his views on evolution and natural selection, processes by which species change and forms succeed one another through adaptation to the environment to attain a better "fit." The direction of species change was called "succession" by evolutionists. The effects of environment and succession became fundamental concepts to ecology.

Late in the 1800s, a shift from exploration and classification toward trying to understand how nature works became a focus for scientists. With the new focus came an interest in studying plants and animals and their relationship with the environment. Scientists began to understand animals and plant populations as "communities." By the early 1900s, botanists and zoologists, both in England and the United States, studied plant and animal "communities" independently until a new science, "ecology," emerged that could embrace both fields.

While the new science had some foundation in Darwin's concepts of evolution and natural selection, most historians of ecology cite an indirect rather than direct role for Darwinism in the growth of the science of ecology. They suggest that internal changes in the fields of biology, botany, and zoology created ecology.

The independent science of ecology began in the 1890s. Many cite the German zoologist Ernst Haeckel (1834-1919) as the most likely "father" of ecology, based on his holistic approach to biology. Haeckel's ideas about energy flow in a closed system formed the foundation of modern ecology half a century after his work.

Impact

In the early 1900s, the new science of ecology struggled to establish its independence from biology and zoology. Its emergence was tied to changes in the other sciences that did not acknowledge the organism-environment relationship. Rather than just the study of organisms as individuals, the focus of ecology became the study of relationships in nature and how the relationships created a kind of "community."

In the United Kingdom, university scholars and new institutions, such as the British Ecological Society, founded in 1913, took the lead in ecology. World War I interrupted the development of ecology as many of the young scholars died in the war. Following the war, a general philosophical pessimism about the human role in nature and progress intervened. By the 1920s, however, scientists in a variety of biological, zoological, geological, mineralogical, and entomological studies began following a similar theoretical thread—that organisms and their environments were systematically, interdependently, and harmoniously balanced. How that balance worked became a subject of study for early ecologists.

In the first two decades of the twentieth century, the University of Chicago became a center for the study of ecology in the United States. For members of "The Chicago School," ecological interdependence in nature meant that nature evolved in such a way that animal and plant forms succeeded one another to benefit the community. Pioneering animal ecology at the University of Chicago at this time was Victor Shelford, who became the first president of the Ecological Society of America (1916). Shelford studied correlations between changes in animal populations and environmental changes.

New terms had to be created to discuss environmental relationships the ecologists described. Russian scientist Georgii Gause, for example, revived the term "niche," nearly forgotten from Darwinian studies. In 1921, Russian mineralogist Vladimir Ivanovich coined the term "biosphere" to offer a holistic view of nature, one in which all the parts were related. Ironically, plant and animal ecologists formed two distinct disciplines at this time.

Those interested in human populations were yet another group of scientists eventually found under the ecology umbrella. As far back as the eighteenth century, British clergyman Thomas Malthus (1766-1834) drew attention to the problem of rapid human population growth and a decreasing ability to feed more people. In his footsteps, early population ecologists, such as Raymond Pearl, Alfred J. Lotka, and V. Volterra, began developing mathematical formulas to study population growth and its limits.

Lotka published *Elements of Physical Biology* in 1925, inspiring a new aspect of ecology, called "population ecology," a science focused on not only population growth but competition between species occupying the same environment.

Early studies on the genetic control of growth—a field that would become known as ecological genetics—were carried out by Julian Huxley (1887-1975) and Edmund Ford in Great Britain from 1923-1926. Working with fresh water crustaceans, they found that genes controlled the time and occurrence of physiological processes and that animal population growth and change could be calculated.

In 1927, in his book *Animal Ecology*, British biologist Charles Elton popularized the concept of an animal's "niche" in local ecology, portraying niche as a kind of "job" an animal has in nature. Elton, a zoologist, rebelled against the emphasis on comparative anatomy, popular at the time, and focused on animals' natural habitat.

Studying in the Canadian Arctic with the Hudson Bay Company, Elton discovered simplifications in environmental relationships that may have been overlooked in more complicated ecosystems. He was interested in population size and its constraints, leading him to comparatively study populations of furbearing mammals and the records of Arctic fur trappers. In his next book, *Animal Ecology and Evolution* (1930), Elton stressed the idea that there was no "balance" in nature, that in times of environmental stress animals can change their habitat, a process just the opposite of natural selection by which the environment selects positive traits in animals. In 1932, Elton established the Bureau of Animal Population at Oxford.

In 1931, at the University of Chicago, Warder C. Allee, a professor of zoology, published "Animal Aggregation: A Study in General

Sociology." Allee, a founding figure in American ecology, maintained that animals cooperated for their group benefit. Allee challenged the animal behavorialsts who established the concept of "pecking orders" in nature. Allee said that pecking orders were not universal and a dangerous model on which to make comments on human societies. He stressed that success for populations—both in nature and in human societies—was to be found in cooperation. Population growth and its limits, said Allee, were controlled by an "optimum clumping factor," by which overcrowding limits population.

American ecologist Frederick E. Clements stressed in the 1930s that ecological relationships between species occupying a particular territory could form a coherent system able to be studied as a unit. Clements worked with Victor Shelford, who was the first president of the Ecological Society of America. Together they published *Bio-ecology*, which developed the concept of the "biome"—the community of plants and animals in a zone. While theories proposed by Allee and Clements stressed community cooperation in nature, their work conflicted with that of Pearl and Lotka, who stressed individuals and their intra-population competition.

At this same time, Canadian William Thompson argued that natural relationships were unpredictable and competitive, while Gause was trying to show that competing species could not occupy the same niche. Gause postulated what he called the "competitive exclusionary principle" by which populations became limited through competition for food.

Visiting the Galapagos Islands in the late 1930s, British young ornithologist David Lack restudied the finches that helped Darwin form his theory of natural selection. Lack's examination of their food and the environment refocused attention on Darwinian theory and helped further the popularity of Darwin's famous voyage on H.M.S. *Beagle* and reinforced Darwin's thesis that interaction between a species and the environment could mean extinction if the species could not adapt. Lack's book, *Darwin's Finches* (1947), addressed both the evolution and the principle of competitive exclusion and ecological effects.

During the 1940s and after World War II until mid-century, the science of ecology and the environmentalist movement began to converge as both ecologists and environmentalists felt that nature was something that humans must proactively protect. In the 1940s, the British government created a wildlife conservation committee and, in 1949, created from that committee the Nature Conservancy.

Just prior to 1950, there was a shift in the focus of ecology away from a descriptive examination of environmental niche toward analyses of energy use and distribution. In 1942, American ecologist Raymond Lindeman delivered a classic paper in which he demonstrated that the transfer of the Sun's energy in part of Cedar Lake Bog (Minnesota) was transferred to another part

THE GROWTH OF ENVIRONMENTALISM, 1900-1949

Ecology, the scientific study of relationships in the natural environment, is often confused with "environmentalism," a concern with conserving the natural environment by protecting it from the adverse affects of human use. Nineteenth-century American writers such as Henry David Thoreau expressed concern that development was spoiling natural areas. As the nineteenth century ended, public, state, and national parks developed in the United States to conserve areas of natural beauty. Before World War I President Theodore Roosevelt, an outdoorsman, established the national parks system to preserve wilderness in the early 1900s.

In the 1930s the American experience with the Dust Bowl of the Midwest showed that the destruction of the soil by reckless farming could permanently affect the land. By this time, the harmful effects of industrialization and urbanization also became apparent. Public concern about shrinking natural resources translated into some government action during the 1930s New Deal program, yet there was growing tension between those who wanted unrestrained exploitation of nature and those seeking to conserve it. The tension was also felt in Great Britain and in Europe, where conservationists often became associated with left-wing and socialist movements. Environmentalists' ideas likely grew out of what some historians call "the Haeckelian religion of nature," a philosophy based on the emotional aspects of German biologist Ernest Haeckel's work. Great Britain and Europe saw the rise of "green" political parties concerned with conservation vigorously opposed by right-winged industrialists, whom the environmentalists saw as destroying nature. Exploitation versus conservation became—and remains—an emotional and political issue.

RANDOLPH FILLMORE

of the lake, but with a loss of energy. By the end of the 1940s and using Lindeman's foundation, two American brothers, Eugene and Howard Odem, introduced the concept of "systems ecology." The work of Lindeman and the Odems had its roots in the concept of "ecosystem," a unit consisting of interacting organisms and all of the aspects of the environment in which they interact. The Odems worked with the idea that ecosystems could be represented and analyzed in terms of "energy flow." They and others tried to create mathematical models that would explain ecosystem energy and its flow.

Although the growth of the science of ecology and concern for the environment converged toward the middle of the twentieth century, ecology should not be confused with "environmentalism," although environmentalists and ecologists today accept the idea that a species can be wiped out by the destruction of their fragile ecosystem. The ideas ecologists and environmentalists share are a combination of Darwinism and introspective examinations of plants and animals and how and where they live. These concepts form both the foundation for the modern science of ecology as well as the emotional and political aspects of concern for the welfare of the Earth and its plant and animal inhabitants.

RANDOLPH FILLMORE

Further Reading

Books

Allee, Warder Clyde. *Cooperation Among Animals.* New York: Henry Schuman, 1951.

Boughey, Arthur. *Ecology of Populations.* New York: MacMillan, 1968.

Bowler, Peter J. *The Norton History of Environmental Science.* New York: W.W. Norton, 1992.

Bramwell, Anna. *Ecology in the 20th Century: A History.* New Haven: Yale University Press, 1989.

Kingland, Sharon E. *Modeling Nature.* Chicago: University of Chicago Press, 1985.

Developments in Anthropology, 1900-1949

Overview

Before 1900 anthropology was in one way a racist science. The nineteenth-century idea of evolution gave some European peoples (including those in North America) a reason to believe they had a superior culture because they had "evolved" more than other races. In the early twentieth century, the German anthropologist Franz Boas (1858-1942) challenged the conception that non-European cultures were inferior. He inspired a number of anthropologists in America and Europe to study all cultures with the belief that every culture is unique and should be studied on its own terms, instead of within the all-encompassing and judgmental evolutionary scheme. The variety of approaches that anthropologists developed in the period 1900-1949 reveals that they did as Boas suggested. Thus, by 1949, the notion that Europeans were superior to other cultures and races had been significantly challenged by anthropologists.

Background

Anthropology appeared as a distinct human science in the second half of the nineteenth century. During that period the separate social sciences as we know them today were emerging: economics, psychology, sociology, political science, and anthropology. These new sciences had as their goal the study of the growth and development of social relations among humans—hence the term social sciences. In its earliest stages, anthropology was to be the study of the peoples that nineteenth-century Europeans considered "primitive." These "primitive peoples" included many vanished, prehistoric cultures and existing tribes of natives in places like Africa, Australia, and South America. It was thought that these "primitive peoples" represented simple or undeveloped cultures—in contrast to the advanced cultures of Europe. In the nineteenth century, therefore, anthropology was based upon a European presumption of superiority.

This presumption of superiority derived in part from the doctrine of evolution. Ideas of evolutionary development had been around before Charles Darwin (1809-1882), but when Darwin published his *Origin of Species* in 1859, evolution was given scientific credibility. Also during the nineteenth century, Europeans, especially the British, had colonized much of the rest of the

Margaret Mead with artifacts from the Pacific region. *(Corbis Corporation. Reproduced with permission.)*

world. The non-European peoples they encountered seemed "primitive" by their own standards. This reconfirmed European beliefs that the process of evolution had made them a superior race. Thus nineteenth-century anthropology was a racist science. Ethnocentrism, the belief that one's own culture is superior, was given scientific credibility by evolution.

At the turn of the twentieth century, Boas challenged this ethnocentric view. In 1883 he spent time in northern Canada and became interested in the Eskimo peoples. In 1886 he returned to North America to study the natives of the pacific northwest. He then spent time in Chicago before settling in New York. He lectured and wrote on the results of his anthropological studies, and did museum work as well. Boas's primary interest was in *cultural anthropology*—the active study of individual cultures and the belief that each one had a unique and significant development. This view went against the evolutionary anthropologists, who studied cultures as being either advanced (usually meaning "European") or primitive (usually meaning "non-European").

Boas emphasized fieldwork and direct observation. He encouraged his students to go among the peoples whom they studied to learn their customs, record observations, and obtain artifacts. This was the beginning of a new method for anthropologists. Understanding a culture meant understanding individuals within that culture, according to Boas. He also encouraged his students to record the life histories of individuals in a culture, so that they could make connections between the larger culture and the individuals within it. Boas taught that each member of a culture, as an active, thinking being, helped shape the culture. Cultures were, in part, mental creations.

Boas's method of cultural anthropology had a component called *functionalism*. Functionalism meant that anthropologists should consider each culture to be like machines or organisms—complex objects with many interrelated parts. The goal of the functionalist anthropologist was to understand how the many parts of a culture (family life, food preparation, religious beliefs, etc.) functioned together as a working whole.

Impact

Boas's teachings would change the practice of anthropology in America, where he lived and taught. His first two doctoral students were Robert Lowie (1883-1957) and Alfred Louis Kroeber (1876-1960). Lowie, like Boas, rejected the evolutionary approach to the study of culture. Instead, he helped to advance Boas's notion that every culture is unique and has to be studied on its own terms. Kroeber, however, disagreed with Boas's view that individuals within a culture are as significant as the culture as a whole. Thus, for much of his career Kroeber was interested in the general patterns found in a culture instead of the actions of individual members.

One of Boas's most important students was Margaret Mead (1901-1978). She focused on Boas's teaching that culture is a mental creation. She incorporated psychology into her study of cultures to explore the interaction between individual personalities and their cultures. Boas encouraged Mead to focus on a small group of individuals within one culture in order to keep her work simple. Mead chose to study adolescent girls in the Samoa islands of the South Pacific. Her study, *Coming of Age in Samoa* (1928), became an important work in the theory of anthropology. Mead showed that the transition from childhood to adulthood for Samoan girls was relatively uncomplicated compared to the similar transition experienced by American girls. Mead argued that this was because Samoan girls were permitted considerable freedom in their interpersonal relationships, especially regarding sex. Her book shocked American readers, as the idea of adolescent promiscuity went against their cultural values. The book's most significant contribution to anthropology, however, was its support of Boas's view that each culture is unique, so each culture develops its own distinct set of morals. The idea that every culture could develop its own moral practices was called *cultural relativism.*

Another important anthropologist, Ruth Benedict (1887-1948), continued this argument for cultural relativism. Benedict entered the field of anthropology after she was 30. Like Mead, she had also met Boas and applied psychology to the study of culture. Benedict's most significant work was in comparing different cultures and noting their differences. She studied cultures in the American pacific northwest, the American southwest, and in the South Pacific. She then argued that different cultures developed different behavioral patterns depending on a variety of factors unique to each culture. What one culture considered "normal" behavior could be very different from the "normal" behavior of another culture. Where Mead was interested in the psychology of individuals within a culture, Benedict was interested in the psychology of a culture as a whole—that is, general cultural traits, or patterns. Thus, the title of her 1931 book in which she described the different patterns found among cultures was *Patterns of Culture.*

European anthropologists developed a variety of methods for studying cultures during the period 1900-1949. Some anthropologists looked to the psychological theories of Sigmund Freud (1856-1939). Freud suggested that much human action results from unconscious desires that have been repressed in the subconscious because of the fear of disapproval. The problem with Freud for anthropologists was that his theories were based on European psychology, and were not useful for studying non-European cultures. One anthropologist who modified Freud's theories was Abram Kardiner (1891-1981). In his book *The Psychological Frontiers of Society* (1945), Kardiner tried to show how Freud's theories, when properly altered, could help us understand how different types of personalities would develop out of different cultures.

In France, Emile Durkheim (1858-1917) brought a different view to anthropology. He argued that knowledge of culture could not come from individuals, and so psychology was a useless tool for anthropologists. Instead, cultures can only be understood on the basis of their social unity. If you want to know what a culture is, he contended, look for its common social bond—its "shared awareness" or "common understanding." In a slightly different way than that of Boas, Durkheim, too, viewed cultures as functioning organisms. Claude Lévi-Strauss (1908-) developed *structural anthropology.* Strauss studied cultures on the basis of their social structures—the relationships between men, women, families, and groups—and the way these social structures enabled cultures to function.

British anthropologists also wanted to understand cultures as wholes, like living organisms. Bronislaw Malinowski (1884-1942) looked closely at the biological and psychological needs of individuals within a culture—nutrition, shelter, and safety, for example. He studied how the individual search for those basic needs affected the way the entire culture functioned. A. R. Radcliffe-Brown (1881-1955) was not concerned with the needs of individuals, but rather with the needs of the culture as a whole. He studied the social relationships by which cultures function, in the same way that a biologist studies the relationships between parts of an organism to understand how that organism lives.

In the period from 1900-1949, anthropology became much more conscious of the variety and diversity of cultures in the world. This period in anthropology saw the initial decline of the belief that evolution had made European peoples superior to other peoples. Instead, many of these anthropologists argued for cultural relativism—the belief that each culture has its own valid and unique set of moral, religious, and social practices. They believed that it was unfair to characterize or judge other cultures by European standards. Though not all of the anthropologists mentioned above were followers of Boas, they all shared his view that each culture was unique. Anthropology in the first half of the twentieth century represents a significant change in the way that scientists have come to view the historical and geographical variety of cultures around the world.

STEVE RUSKIN

Further Reading

Books

Bohannan, Paul, and Mark Glazer, eds. *High Points in Anthropology,* 2nd ed. New York: McGraw-Hill, 1988.

Erickson, Paul A., with Liam D. Murphy. *A History of Anthropological Theory.* Peterborough, Ontario: Broadview Press, 1998.

Periodical Articles

Bennett, John W. "Classic Anthropology." *American Anthropologist* 100 (1998): 951-956.

Lewis, Herbert S. "The Misrepresentation of Anthropology and Its Consequences." *American Anthropologist* 100 (1998): 716-731.

The Disastrous Effects of Lysenkoism on Soviet Agriculture

Overview

The disastrous effects of Lysenkoism, a term used to describe the impact of Trofim Denisovich Lysenko's (1898-1976) influence upon science and agriculture in the Soviet Union during the first half of the 20th century, darkly illustrates the disastrous intrusion of politics and ideology into the affairs of science. Beyond a mere rejection of nearly a century of advancements in genetics, Lysenkoism—at a minimum—made worse the famine and deprivations facing Soviet citizens. Moreover, Lysenkoism brought repression and persecution of scientists who dared oppose Lysenko's pseudoscientific doctrines.

Background

Ten years after the 1917 Revolution in Russia, a plant-breeder in the struggling Soviet Union named Trofim Denisovich Lysenko observed that pea seeds germinated faster when maintained at low temperatures. Instead of concluding that the plant's ability to respond flexibly to temperature variations was a natural characteristic, Lysenko erroneously concluded that the low temperature forced its seeds to alter their species.

Lysenko's conclusions were based upon, and profoundly influenced by, the teachings of Russian horticulturist I.V. Michurin (1855-1935), who was a holdover proponent of the discredited Larmarckian theory of evolution by acquired characteristics (i.e., that organisms evolved through the acquisition of traits that they needed and used). In the nineteenth century, French anatomist Jean-Baptiste Lamarck (1744-1829) attempted to explain such things as why giraffe had long necks. Lamarck had reasoned that a giraffe, by stretching its neck to get leaves, actually made its neck lengthen and that this longer neck was then somehow passed on to offspring. According to Lamarck, the long-necked giraffe resulted from generation after generation of giraffe stretching their necks to reach higher into trees for food.

Unfortunately for Lysenko—and even more so for Soviet science—Lamarck's theory of evolution by acquired characteristics was incorrect. Scientists now understand that individual traits are, for the most part, determined by an inherited code contained in the DNA of each cell and are not influenced in any meaningful way by use or disuse. Further, Darwinian natural selection more accurately explains the long necks of the giraffe as a physical adaptation that allowed exploitation of a readily available food supply that, in turn, resulted in enhanced reproductive success for "long necks."

Despite the fact that Lamarck's theory of evolution by acquired characteristics had been widely discarded as a scientific hypothesis, a remarkable set of circumstances allowed Lysenko the opportunity to sweep aside more than 100 years of scientific investigation to advocate a "politically correct" way to enhance agricultural production. When Lysenko promised greater crop yields, a Soviet Central Committee—desperate after the famine in the early 1930s—listened with an attentive ear. The very spirit of Marxist theory, Lysenko claimed, called for a theory of species formation which would entail "revolutionary leaps." Lysenko attacked Mendelian genetics and Darwinian evolution as a theory of "gradualism."

Lysenko constructed an elaborate hypothesis that came to be known as the theory of phasic development. One of Lysenko's more damaging ideas was to "toughen" seeds by treating them with heat and high humidity in an attempt to increase their ability to germinate under harsh conditions. The desire to plant winter instead of spring forms of wheat was brought about by the need to expand Russian wheat production into areas that were climatically colder than traditional growing areas. In particular, the Nazi invasion during the Second World War made critical the planting of previously fallow and colder eastern regions. For centuries, winter wheat was the principal Russian crop. Deprived of its Ukrainian breadbasket, the Soviet Union struggled to survive following Hitler's onslaught. Faced with famine—and subsequently a massive movement of farming away from the onslaught of an advancing Nazi army—Soviet agricultural planners and farmers became unconcerned with the high politics of agricultural science or long-term scientific studies.

Lysenko ruled virtually supreme in Soviet science, and his influence extended beyond agriculture to all areas of science. In 1940, Stalin appointed Lysenko Director of the Soviet Academy of Science's Institute of Genetics. In 1948 the Praesidium of the USSR Academy of Science passed a resolution virtually outlawing any biological work that was not based on Lysenko's ideas.

Although thousands of experiments carried out by geneticists all over the world had failed to provide evidence for—and actually produced mounds of evidence against—Lysenko's notions of transmutation of species, Lysenko's followers went on to make increasingly grandiose claims regarding yields and the transformation of species. Not until 1953, following the death of Stalin, did the government publicity acknowledge that Soviet agriculture had failed to meet economic plan goals and thereby provide the food needed by the Soviet State.

Impact

Despite the near medieval conditions in which the majority of the population of Czarist Russia lived, the achievements of pre-Revolutionary Russia in science rivaled those of Europe and America. In fact, achievement in science had been one of the few avenues to the aristocracy open to the non-nobility. The Revolution had sought to maintain this tradition, and win over the leaders of Russian science. From the earliest days Lenin and Trotsky, in particular, fought, even in the midst of famine and civil war, to make available considerable resources to scientific research.

In the political storms that ravaged the Soviet Union following the rise of Stalin, Lysenko's idea that all organisms, given the proper conditions, have the capacity to be or do anything had certain attractive parallels with the social philosophies of Karl Marx (and the twentieth-century French philosopher Henri Bergson) that promoted the idea that man was largely a product of his own will. Enamored with the political correctness and with the "scientific merit" of Lysenko's ideas, Stalin took matters one step further by personally attacking modern genetics as counter-revolutionary or bourgeois science. While the rest of the scientific world could not conceive of understanding evolution without genetics, Stalin's Soviet Union used its political power to suppress rational scientific inquiry. Under Stalin, science was made to serve political ideology.

The victory of the Stalin faction within the ruling party changed the previously nurturing relationship between the Soviet State and science. Important developments in science (including what we would term today as social sciences) were terminated by state terror. During the 1930s and 1940s, scientists were routinely executed, imprisoned, or exiled. Soviet science was largely carried forward in specially built labor camps, where scientists denounced as "saboteurs" continued their work in total isolation from the outside world.

Under Stalin and Lysenko, scientific truth became both incompatible and inappropriate to political truth. Information on genetics was eliminated from Soviet biology textbooks, and Lysenko attempted to reduce his conflict with classical geneticists to political contradictions. Lysenko stated that there existed two class-based biologies: "bourgeois" vs. "socialist, dialectical materialist." The entire agricultural research infrastructure of the Soviet Union—a country with millions whose lives teetered on starvation—was devoted to a disproved scientific hypothesis, and inventive methods were used to falsely "prove" that there was no famine and that crop yields were actually on the rise.

Trofim Lysenko measures the growth of wheat at a collective farm in the Ukraine. *(Hulton-Deutsch Collection/Corbis. Reproduced with permission.)*

Soviet Central Committee support of Lysenko was critical. It was known that Stalin clearly expressed his positive attitude toward the idea of the inheritance of acquired characters and his overall support of Lamarckism. In such an atmosphere, some of Lysenko's supporters even denied the existence of chromosomes. Genes were denounced as "bourgeois constructs." Under Lysenko, Mendelian genetics was branded "decadent" and scientists who rejected Lamarckism in favor of natural selection became "enemies of the Soviet people."

Despite the stifling atmosphere created by Stalin's totalitarianism, some scientists resisted. Soviet geneticist Nikolay Ivanovich Vavilov (1887-1943) attempted to expose the pseudoscientific concepts of Lysenko. As a result, Vavilov was arrested in August 1940 and he died in a prison camp. Throughout Lysenko's reign there were widespread arrests of geneticists who were uniformly denounced as "agents of international fascism." In literal fear of their lives, many Soviet scientists cowered. Some presented fraudulent data to support Lysenko, others destroyed evidence that showed he was utterly wrong. It was not uncommon to read the public letters of scientists who had once advanced Mendelian genetics in which they confessed the errors of their ways and extolled the wisdom of the Party.

Based on his misunderstanding of genetics, Lysenko developed methods that falsely predicted greater crop yields through a hardening of seeds and a new system of crop rotation. Lysenko's system of crop rotation eventually led to soil depletion that required years of replenishment with mineral fertilizers. Under Lysenko's direction, hybrid corn programs based on successful U.S. models were stopped—and the research facilities destroyed—because Lysenko philosophically opposed "inbreeding."

When Nikita Khrushchev (1894-1971) assumed the post of Soviet Premier following the death of Stalin in 1953, opposition to Lysenko began to grow. Khrushchev eventually stated that under Lysenko, "Soviet agricultural research spent over 30 years in darkness." In 1964, Lysenko's doctrines were discredited, and intensive efforts made toward the reestablishing of Mendelian genetics and bringing Soviet agricultural, biological, and genetics research into conformity with Western nations.

Ultimately, the rejection of Lysenkoism was a victory for empirical evidence. Science also triumphed over Lysenkoism because there was a manifest rejection of incorporating Marxist rhetoric into the language and theoretical underpinning of genetics.

K. LEE LERNER

Further Reading

Books

Graham, L. *Science in Russia and the Soviet Union.* Cambridge University Press, 1993.

Joravsky, D. *The Lysenko Affair.* Harvard University Press, 1970.

Lysenko, T.D. *Soviet Biology: A Report to the Lenin Academy of Agricultural Sciences.* Birch Books, NY: International Publishers, 1948.

Lysenko, T.D. *Agrobiology.* Moscow: Foreign Language Press, 1954.

Soyfer, V. *Lysenko and the Tragedy of Soviet Science.* Rutgers University Press, 1994.

Periodical Articles

Darlington, C.D. "T.D. Lysenko." (Obituary) *Nature* 266 (1977): 287-88.

Advances and Trends in the Agricultural Sciences

Overview

After their emergence as a distinct realm of professional scientific research in the nineteenth century, the agricultural sciences continued their impact in the first half of the twentieth century. Led by developments in genetics, animal nutrition, bacteriology, and agricultural chemistry, farmers were able to produce more food from existing lands, and also to extend their production into lands that had previously been beyond the realm of cultivation. As a consequence, successful agricultural production became increasingly dependent upon access to the information and capital associated with the agricultural sciences.

Background

Although most developments in nineteenth-century agricultural sciences originated in western Europe, three important pieces of legislation shifted the stage to the United States by the early twentieth century. The Hatch Act of 1887 provided funds for the establishment of agricultural experiment stations in each American state and territory, the Adams Act of 1906 doubled these stations' funding and allowed for increased emphasis on basic research in the agricultural sciences, and the Smith-Lever Act of 1914 supported agricultural extension programs and thereby lessened scientists' duties to work directly for farmers. The United States Department of Agriculture thus became one of the world's centers for scientific research, using a network of extension agents, publications, and, by the 1930s, weekly radio shows to disseminate scientific knowledge among practicing farmers and to the public at large. In the United Kingdom the Development Act of 1910 and similar bills also authorized increased public funding for basic research in the agricultural sciences at the university level, most notably at Cambridge. In Germany, meanwhile, the relative position of the agricultural sciences declined in the face of economic hardships, political turmoil, and diminishing funding for universities and research institutions.

The rediscovery of Gregor Mendel's (1822-1884) theories on genetics was perhaps the most important development in the agricultural sciences at the turn of the century. The American plant breeder Luther Burbank (1849-1926) developed hundreds of new plant varieties without aid of the Mendelian theory, but other scientists dismissed his work as unscientific or unimportant. Supporters of the Mendelian theory turned instead to the systematic inbreeding of desirable varieties as a more predictable method. In 1918 Donald F. Jones (1890-1963) of the Connecticut Agricultural Experiment Station discovered that by crossing four pure-bred lines over two generations, new corn varieties could be produced that accentuated desirable traits like disease resistance, climatic adaptation, and higher yield. By 1950 hybrid seeds were commonly used for other crops as well.

Plant breeding took a different turn in the Soviet Union, where T. D. Lysenko's (1898-1976) explicitly anti-Mendelian interpretation of genetics was intended to fit Marxist ideology and Stalinist land policies. His theory of "vernalization" involved attempts to adapt Russian crops to local growing conditions, but it resulted in the purge of Mendelian geneticists and an increasing the gap between western and Soviet food production.

A basket of hybrid corn. *(Bettman/Corbis. Reproduced with permission.)*

In animal breeding, researchers found increasingly efficient strategies for selecting desirable traits such as physiological response to work, efficiency of feed utilization, and resistance to disease. The American Breeders' Association also gained national attention for its work popularizing the controversial field of eugenics. Artificial insemination techniques, common in the United States by the 1940s, permitted the rapid emergence of new livestock breeds tailored for the consumer markets.

Discoveries in biochemistry and animal nutrition changed ideas on animal feeds. Elmer Vernor McCollum (1879-1967) of the Wisconsin Agricultural Experiment Station discovered that trace minerals and organic substances, now known as vitamins, are essential in animal nutrition. By mid-century researchers had discovered that the addition of synthetic hormones to animal feeds could improve feed efficiency, regulate sexual activity, and stimulate lactation in dairy animals. In the process, poultry science, dairy science, and veterinary medicine dramatically altered the traditional work of animal husbandry.

Beginning with Louis Pasteur's (1822-1895) mid-nineteenth century research on silkworm parasites, anthrax vaccines, and microbes in foods and dairy products, the science of bacteriology also emerged from work on agricultural topics. Control of bacteria in dairy and other products virtually eradicated one of the primary causes of human disease and improved the safety of commercial foodstuffs. Soil bacteriology was another important branch of the field, led by Selman Waksman's (1888-1973) discovery of streptomycin and other antibiotics derived from soil microbes. The Dutch scientist Martinus Willem Beijerinck (1851-1931) found applications for soil microbes in the industrial production of yeast, alcohol, and various chemicals.

Though agricultural chemists were not as dominant in the agricultural sciences as in the nineteenth century, the chemicalization of agriculture continued unabated. Well aware that the supply of Chilean nitrates, the major source of nitrogenous fertilizers, was in decline, many scientists sought a solution to the western world's "nitrogen question." By 1913 German chemist Fritz Haber (1868-1934) and his colleague Carl Bosch (1874-1940) developed a process that used very high temperatures and pressures to synthesize ammonia from atmospheric nitrogen and hydrogen. American consumption of synthetic fertilizers had increased more than tenfold between 1900 and 1945, and the market for natural organic fertilizers virtually vanished.

Farmers also turned to powerful chemicals to control pests, such as calcium cyanides used against rodents and the arsenates used to fight the cotton boll weevil. During World War II scientists devoted great energy into finding chemicals that could kill the insects that threatened soldiers with typhus and malaria. The resultant research on DDT (dichloro-diphenyl-trichloroethane) had promising implications for pest control on the farm. By the late 1940s farmers, governments, and private industries all promoted DDT and other agricultural chemicals as panaceas that brought both higher yields and pest control with no apparent risk to human health. The broadleaf herbicide 2,4-D (2,4-dichlorophenoxyacetic acid), developed in 1941, was also significant—it has properties that mimic plant hormones and thus artificially speed up the metabolism of weed plants to force them to "starve themselves" by using up all of their stored food.

Impact

Developments in the agricultural sciences continued to increase farm productivity, reduce threats of disease, and permit the cultivation of previously untilled lands, and thereby contributed to the rapid growth of the world's population. At the same time, the percentage of the world's population engaged in agricultural pursuits declined steadily.

These circumstances suggest that the agricultural sciences helped accelerate the ascendancy of the farmers and businesses with better access to the knowledge and capital required to use hybrid seeds, scientific feeds, pesticides, herbicides, and similar goods. Farmers' skill and experience became a less important factor in their success, and declining rural populations also reduced farmers' influence in the political and cultural arenas.

These trends also signaled a relative decline in the professional significance of the agricultural sciences. At the beginning of the century, agricultural scientists at government agencies like the USDA and institutions like the land grant universities and agricultural experiment stations held important positions within the scientific community and were often influential in public policy debates. Major discoveries in genetics, bacteriology, public health, and other disciplines emerged from laboratories of the agricultural scientists. Scholars and public officials alike expressed great enthusiasm for the agricultural sciences, confident that safe and abundant food supplies would soon be available to all.

By the middle of the century, however, a small number of critics expressed concern about the increasing interconnections between farming and its costly inputs. Yet agricultural issues also often fell out of the public dialogue, as much of scientific research shifted to the large universities associated with "Big Science." In the Atomic Age research in plants and animals lacked the same status in the public imagination as physics, chemistry, and aeronautics.

The agricultural sciences also had growing significance for farmers of non-Western nations in the early twentieth century. Rural elites and colonial administrators used the rhetoric and techniques of science in order to establish experiment stations, to develop hybrids, and to adapt local farming techniques in ways that contributed to the commodification of tropical nature. As elsewhere, the expansion of agricultural science in the European colonies increased production to such a degree that those who were well connected and well capitalized could maintain profitable access to the marketplace.

The expansion of the hybrid corn and agricultural chemicals illustrate some of these concerns. Hybrid corn became commercially viable in the United States in the 1930s, and it expanded so rapidly that, by 1943, hybrid seeds were used on 90% of the American corn acreage. Yet inbred hybrids lose their desirable traits in just one generation, so farmers have little choice but to purchase their seed from increasingly large and powerful seed conglomerates. Though they might have developed alternative seed technologies, government-funded institutions like experiment stations gradually abandoned research in crop genetics and left it to the control of private sector seed companies. These companies have developed strains that stand more upright, allowing more seeds to be planted per acre, and hybrid plants that have enough uniformity and durability to make them suitable for farm machines like harvesters and combines. Similarly, farmers have become ever more dependent on artificial fertilizers and chemical pesticides in order to keep their lands under virtually continuous cultivation. Though farmers today face smaller risks of crop failure, their increasing specialization in fewer and fewer profitable commodities has led to an ominous reduction in the diversity of the world's germplasm.

MARK R. FINLAY

Further Reading

Books

Borth, Christy. *Pioneers of Progress: The Story of Chemurgy.* Indianapolis: Bobbs-Merrill, 1939.

Dunlap, Thomas R. *DDT: Scientists, Citizens, and Public Policy.* Princeton: Princeton University Press, 1981.

Fitzgerald, Deborah. *The Business of Breeding: Hybrid Corn in Illinois, 1890-1940.* Ithaca: Cornell University Press, 1990.

Kloppenburg, Jack Ralph. *First the Seed: The Political Economy of Plant Biotechnology.* Cambridge: Cambridge University Press, 1988.

Russell, E. John. *A History of Agricultural Science in Great Britain, 1620-1954.* London: Allen and Unwin, 1966.

Storey, William Kelleher. *Science and Power in Colonial Mauritius.* Rochester: Rochester University Press, 1997.

Wharton, James C. *Before Silent Spring: Pesticides and Public Health in Pre-DDT America.* Princeton: Princeton University Press, 1975.

Periodical Articles

Kimmelman, Barbara A. "The American Breeders' Association: Genetics and Eugenics in an Agricultural Context." *Social Studies of Science* 13 (1983): 163-204.

Palladino, Paolo. "Wizards and Devotees: On the Mendelian Theory of Inheritance and the Professionalization of Agricultural Science in Great Britain and the United States, 1880-1930." *History of Science* 32 (1994): 409-444.

Key Scientific Research Institutions Are Founded

Overview

As late as the nineteenth century, many scientists pursued their field of work as a side occupation to a more financially reliable profession. During this time universities were the main centers of scientific research, with scientific "societies" fueling the exchange of knowledge across academic and geographic borders. By the late 1800s, however, as scientific knowledge grew at an incredible pace, scientists were turning more to governments and private philanthropists to fund their progress and support their dedication to a specialized field of study. Science and scientists entered the twentieth century divided into specialties, influenced by fast-paced innovation, and driven by an intense focus on research. In many cases, the kind of institution—private industrial, militaristic, or medical—that supported a field of research dictated or restricted the path of that specialty. Today's scientific research community grew from the successes and mistakes made in the first half of the twentieth century, and continues to change as research modifies science as we know it.

Background

In the early 1800s, the full-time scientist of the day typically worked in a university or a medical hospital. Research, or what many call "pure science," was difficult to support without an applicable purpose. And, in turn, many applied sciences were accidental: inventions such as the sewing machine, the steam engine, and agricultural machinery were produced by non-scientists who were simply improving upon existing technology.

Many governments were investing more in the "scientific" expeditions of explorers and their pursuit to discover, map, and reveal the unknowns of mysterious lands than the pursuit of laboratory science. Some of these missions involved trained scientists, but most did not. When the United States funded an expedition to explore the western U.S., for example, they hired Meriwether Lewis (1774-1809), an Army officer with no formal scientific training. He was given a crash course in natural sciences before he left.

By the middle of the nineteenth century, there was a sharp division in science, with the "philosophical" scientists on one side, who were frequently part of scientific societies, exchanging high-minded theories and arguments, and "professional" scientists on the other, who made improvements in their lives or worked for the government on some aspect of applicable science. It wasn't until the German university system emerged that the idea of combining the two in a research forum, free of government expectation but encouraged through the structure of academia, became acceptable. German universities began building immense research structures, focusing as much on the grandeur of the physical surroundings as the prestige of its scientists. Germany also expanded into state-sponsored pure research institutes, most notably the Pasteur Institute, founded in 1888, and the Koch Institute, founded in 1891.

State-sponsored and privately funded institutes modeled after the German research university sprung up across Europe and the U.S. Johns Hopkins University became the first research-focused university in America, and Cold Spring Harbor, one of the few research institutes in the United States, funded by the Brooklyn Institute of Arts and Sciences, was built in 1890. Its proximity to marine life and salt and freshwater made it the ideal setting to study one of the more popular subjects of "pure science" at the time: Darwin's principles of natural selection.

Soon the wealthiest philanthropists in America fixed on the burgeoning interest in research at the turn of the century, with John D. Rockefeller (1839-1937) and Andrew Carnegie (1835-1919) providing generous endowments to research. By the end of the nineteenth century, science became the focus of several channels of funding, particularly industry and government.

Impact

By the beginning of the twentieth century, financial support for scientific research seemed abundant, particularly in the industry-booming United States. By 1902, the Rockefeller Board had contributed more than $1 million for scientific research, forming what became (and still is) the Rockefeller Institute for Medical Research. At the same time, Andrew Carnegie established the Carnegie Institute of Washington, contributing $10 million by 1911. American industry saw that it could benefit from funding its own bank of scientific experts, with General Electric and American Telephone and Telegraph (today's AT&T) bankrolling industrial research centers.

In Europe, the Kaiser Wilhelm Institut, founded in 1911, was becoming world-renowned for its focus on "pure science," something the American scientists could not claim. In the United States, it was estimated that less than 4,000 scientists were practicing pure research. America was still focused on science for utility, with the private endowments of Rockefeller and Carnegie mainly going to medical research, and government funding directed toward technical work. The Kaiser Wilhelm Institut, on the other hand, was formed on the stated purpose of "promoting the sciences."

But the influence of government on scientific research had, and still has, its disadvantages, even within the "pure" sciences. The most extreme example of this is in the Kaiser Wilhelm Institut, where the turn toward studying psychiatry became a grotesque example of the inhumanity of the Nazi regime. In 1924, the Institut began to concentrate on brain research and mental illness, focusing on heredity and genetics. By the onset of WWII, the bank of prisoners that Germany held became research subjects. Children, in particular, were routinely euthanised, their brains dissected and studied.

Likewise, although not as deliberately sinister, Cold Spring Harbor in the United States took a dark turn, as well. Under the guidance of Charles Davenport (1866-1944), what started as a study of "experimental evolution" became a center for studying human genetics. Davenport set out to prove that genetics explained racial stereotypes, and he promoted a need for sterilization campaigns to keep "bad" genes out of the population. It was the middle of a worldwide scientific focus on eugenics, or the effort to improve the human species through controlling hereditary characteristics. The work of these sensational projects momentarily overshadowed some of the groundbreaking genetics work at the time. At Cold Spring Harbor, for example, scientists funded by the Carnegie Institute developed the first hybrid corn, isolated the "pregnancy hormone" prolactin, and pioneered work on the genetic basis for certain cancers.

In Russia, where Communism had taken total control of scientific research, a focus toward efficiency and industrial technology all but eliminated the pure sciences. Similarly, the war and a push toward improving technology are what drove most government-funded institutions around the world. At the end of World War II, Russia and the U.S. began working on developing sophisticated atomic weapons. This resulted in government-sponsored research focused solely on militaristic gains, with the formation in the U.S. of the Atomic Energy Commission, which led to the establishment of laboratories dedicated to this purpose, such as the Los Alamos National Laboratory in New Mexico.

Many of the once-famous institutions of Europe—especially those of Germany—were desiccated during the war, but the American scientific community was strengthened by the forced application of science and technology. What followed were the moral and ethical complications of scientists developing work for killing. J. Robert Oppenheimer (1904-1967), seen as a sort of hero following his leadership at Los Alamos during the war, argued with legislators about scientific creative freedom, and the dangers of working solely for the military needs of the government. This was difficult for politicians to accept, accustomed to accounting for every dollar of government money spent, but in due time, Oppenheimer's request would be recognized.

The role of government in fundamental research beyond its pet projects—space, military technology, and atomic energy—has grown considerably since WWII. The National Science Foundation, formed in 1950, is one of the great achievements of the U.S. government that represents this change of focus. The major institutes formed, and reformed, during the early twentieth century continue to play a major role in scientific research, and remain a reminder of the importance of ethical considerations in research. The Rockefeller Foundation, as well as the Carnegie Institute, remain two of the most important foundations in the world, donating enormous funds and support into scientific work worldwide. While scientists still struggle to conduct "pure" research in a world where applicable science can promise enormous financial reward, the structure established after the turn of the century guarantees that fundamental scientific research will carry on.

LOLLY MERRELL

Further Reading

Books

Dupree, Hunter S. *Science in the Federal Government.* Massachusetts: Harvard University Press, 1957.

Greenburg, Daniel S. *The Politics of Pure Science.* New York: The New American Library, 1967.

Miller, Howard S. *Dollars for Research.* Seattle: University of Washington Press, 1970.

Internet Sites

The Carnegie Institute. http://www.ciw.edu

Cold Spring Harbor. http://www.cshl.org/public/history.htm

The Rockefeller Foundation. http://www.rockfound.org

Biographical Sketches

Oswald Theodore Avery
1877-1955
Canadian-American Bacteriologist and Physician

DNA's role in genetics and heredity was one of the focal points of biological inquiry in the second half of the twentieth century, but through most of the first half of the century there was little interest in this molecule. The person who changed all that was Oswald T. Avery. In 1944, he and his coworkers published evidence that DNA carried genetic information in bacteria, and this research first brought attention to this paper launched a renewed interest and research into DNA.

Avery was born in 1877 in Halifax, in the Canadian Province of Nova Scotia, and several years later, his family emigrated to the United States. Avery graduated from Colgate University in 1900 and then went to medical school at Columbia University College of Physicians and Surgeons in New York, graduating in 1904. After a brief period of medical practice and research at the Hoagland Laboratory in Brooklyn, Avery moved to the Rockefeller Institute for Medical Research in 1913. He spent the rest of his career there, retiring in 1948 to Tennessee, where he died in 1955.

At Rockefeller, almost all of Avery's research was on *pneumococcus*, the bacterium that causes pneumonia, which at the turn of the century was a leading cause of death. By the time Avery was doing his work, several different types of *pneumococcus* had been identified. In 1917 Avery and Alphonse Dochez (1882-1964) found that these differences were due to substances on the surface of the bacteria, which also appeared in the blood of patients. Six years later, working with Michael Heidelberger (1888-1991), Avery identified these substances as polysaccharides, sugar molecules linked together. The different *pneumococcus* types produced polysaccharides with different combinations of sugars. A patient's immune system reacted against the bacteria's polysaccharides, which explained why immunity was specific for each *pneumococcus* type. This discovery was the first indication that the immune system could respond to polysaccharides; until this time it was assumed that only proteins could stimulate such a response.

In Britain, Fred Griffith (1877-1941) researched another difference among *pneumococci* strains: those with the polysaccharide coat formed large smooth colonies (the S form) when grown on agar gel, while those that lacked the coat formed rough-looking colonies and were incapable of causing infection (the R form). Griffith discovered that when live type II *pneumococci* of the R form were mixed with dead type I of the S form, and this mixture was injected into mice, the mice died of pneumonia, but of the type I kind. In other words, something—some chemical information—from the dead S-form bacteria had been transferred to the live R-form bacteria, allowing them to make the type I polysaccharide.

While Avery was skeptical of this result, biologists in his lab were able to reproduce it, and so he then set out to identify the substance responsible for the transformation of one type of *pneumococcus* bacterium into another. This effort began in 1928 and eventually involved the assistance of two collaborators, Colin MacLeod (1909-1972) and Maclyn McCarty (1911-). The results weren't published until 1944 because it proved difficult to achieve reproducible results. But finally they were able to show convincingly that what Avery called the "transforming factor" was DNA. Biologists long knew that DNA, along with protein, made up chromosomes, the cellular structures that appeared to carry genetic information. But the assumption had been that protein, with its 20 different building blocks, was the likely genetic material, and that DNA, with only 4 different building blocks, simply served as a structural foundation. Avery's work called this assumption into question, and while it took several years for his findings to be widely accepted, his research did spur others to at last look more closely at DNA. Among these researchers were James Watson (1928-) and Francis Crick (1916-), who worked out the structure of DNA in 1953. This discovery led to the tremendous level of interest in the molecular basis of genetics that characterized biology in the second half of the twentieth century.

Oswald T. Avery. *(Library of Congress. Reproduced with permission.)*

MAURA C. FLANNERY

William Bateson
1861-1926
English Biologist

William Bateson was one of the scientists who, at the turn of the twentieth century, expanded the views of evolution and helped to describe the force of heredity and variation. He promoted the ideas contained in the newly rediscovered paper by Gregor Mendel (1822-1884) about inheritance of characters, and applied those ideas to Darwinian evolution. Bateson also coined the term "genetics," and conducted a wide range of experiments that broadened understanding in this area.

Bateson was born on August 8, 1861, in Whitby, Yorkshire, England. Throughout his childhood, he maintained a limited interest in the natural world, considerably different from his father, who was a classics scholar at St. John's College in Cambridge. The young Bateson enrolled at St. John's in 1879. His childhood interests came to the forefront three years later when he won an honors examination in the natural sciences. While studying for that test, he became intrigued by the acorn worm, an elongate marine animal, and spent the next two summers in America studying whether it was actually a primitive chordate. (Current classification places it under the phylum Hemichordata, which has some of the characteristics of chordates but lacks a notochord.) After publishing his research, he earned a position as a fellow at St. John's College in 1885.

Bateson's investigations of the acorn worm led him to challenge the then-current biological thought, going so far as to dispute the accepted phylogeny of organisms, which he felt lacked sufficient scientific evidence to back it up. These controversial views stalled his career a bit, but by 1908, he had accepted the position of professor of biology at Cambridge University. In 1910 he became director of the John Innes Horticultural Institution at Merton Park in Surrey, England, a position he held for the rest of his life. In 1912, he also took a two-year appointment as Fullerian professor of physiology at the Royal Institution.

Through the years, Bateson continued to investigate variation, or the differences between animals, and the role of heredity in that variation. When Mendel's 1866 paper on inheritance in peas resurfaced in 1900, Bateson began to consider whether the inheritance characters Mendel described were at work in animal variation. He felt that these characters, or genes, could explain his view that evolution is discontinuous, or proceeds by sudden jumps, rather than gradual and minute changes as Darwinian evolution proposed. He also felt that Mendel's paper helped explain how some characters—those traits associated with recessive genes—might disappear for a generation or two, then reappear. In all, he believed Mendel's paper, which Bateson ultimately translated into English, filled what he perceived as gaps in the current ideas about evolution.

He continued his studies by conducting a series of experiments in which he discovered that some genes were inherited together. That phenomenon is now known as "linkage," and occurs when certain genes are linearly arranged on the same chromosome. During his work on variation and heredity, Bateson in 1909 introduced the term "genetics" for this dynamic and growing research area. The word quickly caught on.

He continued his studies of variation, frequently challenging widely accepted concepts along the way. He was honored with the Darwin Medal in 1904, and the Royal Medal in 1920. Bateson died on February 8, 1926.

LESLIE A. MERTZ

Franz Boas
1858-1942
German-American Anthropologist

Franz Boas is primarily remembered for his pioneering work as an anthropologist and ethnologist. Boas was the founder of the culture-centered (but still scientifically based) approach to anthropology. He subjected the premises of physical anthropology to rigorous and critical analysis. According to Boas, all cultures must be studied in their totality, including their language, religion, art, history, physical characteristics, environmental conditions, diseases, nutrition, child-raising customs, migrations, and interactions with other cultures. Based on his research of many different cultures, he concluded that no truly pure races exist and that no so-called "race" is innately superior to any other.

Boas was born in Minden, Germany, where his father was a merchant. As a child, he was interested in books and the natural sciences. While he was a student at the Minden Gymnasium, he became interested in cultural history. After studying at the universities of Heidelberg and Bonn, he earned his baccalaureate degree from the University of Heidelberg and his Ph.D. in physics and geography from the University of Kiel. After participating in a scientific exploration of the Baffin Island region of the Arctic from 1883 to 1884, he found positions in an ethnological museum in Berlin and on the faculty of geography at the University of Berlin. In 1886, after a study of the Kwakiutl and other tribes of British Columbia, he immigrated to the United States and obtained a job as an editor of the magazine *Science*. While teaching anthropology at Clark University in Worcester, Massachusetts, he worked on the preparation of the anthropological exhibitions for the 1893 Columbian Exposition. In 1899 he became the first professor of anthropology at Columbia University, where, as a specialist in North American Indian cultures and languages, he was an influential teacher and a leader in his profession. His work contributed to the fields of statistical physical anthropology, descriptive and theoretical linguistics, and American Indian ethnology. Many of his students, including Ruth Benedict (1887-1948), Margaret Mead (1901-1978), Melville Herskovits (1895-1963), Edward Sapir (1884-1939), and Alfred L. Kroeber (1876-1960), became famous anthropologists in their own right. From 1896 to 1905 Boas was curator of ethnology at the American Museum of Natural History in New York, where he directed studies of the relationships between the aboriginal peoples of Siberia and North America. He established the *International Journal of American Linguistics* and the American Anthropological Association and served as president of the American Association for the Advancement of Science.

Boas can also be considered a pioneer in visual anthropology. He was one of the first anthropologists to use the motion picture camera to collect data in the field in order to study gesture, body movement, and dances. He had been using still photography in the field since 1894. In 1930, when he was 70 years old, he took a motion picture camera and a wax cylinder sound-recording apparatus on another field trip to the Kwakiutl.

Boas made many trips to study the Native American tribes of British Columbia, but he is best known for his work with the Kwakiutl Indians from northern Vancouver and the adjacent mainland of British Columbia. While studying the Kwakiutl, he established a new concept of culture and race, emphasizing the importance of observing all aspects of a culture. Boasian anthropological theory encompassed cultural relativism, which argues that the observed differences among populations are the results of unique historical, social, and geographic conditions. Each culture is, therefore, complete and equally developed in its own framework and is not a stage in a universal scheme of cultural evolution. Boas and his followers opposed the evolutionary view of culture advocated by such ethnologists as Lewis Henry Morgan (1818-1881) and Edward Tylor (1832-1917), who believed that each race, or culture, went through certain "evolutionary" stages as it developed towards "higher" stages of culture according to certain universal laws. Boas also opposed race-based explanations and argued that culture, rather than race, was the fundamental factor in understanding the uniqueness of human societies. For personal as well as theoretical reasons, he was a passionate opponent of anti-Semitism and racial discrimination, especially of Nazi propaganda claiming to provide "scientific" explanations for the racial inferiority of non-

Franz Boas. *(Library of Congress. Reproduced with permission.)*

Aryans. He argued that no race was innately superior to any other. In the 1930s the Nazis burned his book *The Mind of Primitive Man* (compiled from a series of lectures on race and culture and published in 1911) and rescinded his Ph.D. from Kiel University.

LOIS N. MAGNER

George Washington Carver
1861?-1943
American Chemist and Agronomist

George Washington Carver is credited with the development of innovative crop-rotation methods that preserved soils and allowed sustainable levels of increased agricultural productivity. During a long and productive scientific and teaching career, Carver discovered hundreds of uses for crops and his work revitalized a Southern economy left barren and depressed by war and land mismanagement.

Carver was born a slave in Missouri. A frail and sickly child, he was orphaned during the Civil War, and nursed back to health by his former owners, Moses and Susan Carver. Young George lived with the Carvers until he was 10 or 12, when he left to pursue an education at a segregated school nearby. He lived with a black family, doing chores for room and board. Carver was well into his twenties before he was able to move from a one-room schoolhouse to become an outstanding student at Minneapolis High School in Kansas. Though subsequently denied admission to the school of his choice because of this race, Carver eventually undertook his college studies at Simpson College in Indianola, Iowa.

Carver set out to be a scientist and a few years into his studies he transferred to Iowa Agricultural College (now Iowa State University), where he earned his baccalaureate in 1894 and his master's degree in 1897. Shortly thereafter, Booker T. Washington, founder of what was then known as the Tuskegee Normal and Industrial Institute for Negroes (now Tuskegee University) located in Tuskegee, Alabama, asked Carver to accept a post as the school's director of agricultural studies.

Carver accepted and moved to Tuskegee, where he taught and spent research time developing improved crop-rotation methods. Carver's training in chemistry allowed him to spot a glaring deficiency in traditional uses of soils. He measured dramatic depletions of critical minerals and nutrients in fields planted with the same crops season after season. Carver designed a plan whereby nitrate producing crops, such as such as peanuts and peas, would be alternated annually with cotton. Although cotton was the most important cash crop for Southern farmers, repeated planting left fields depleted and unable to sustain or produce profitable crops. When farmers followed Carver's advice—alternating annual crops of cotton and peanuts—the soil nutrient balances remained within acceptable levels. In addition, healthy soils reduced erosion and water contamination by runoff. In a South sorely pressed by the boll weevil, Carver's planting techniques actually proved to be an effective means of pest control.

Unfortunately, farmers were initially unable to find profitable uses for their off-year crops and many had to return to planting cotton, despite the long-term risks. In response, Carver rescued his soil rotation plan by find more than 300 economic uses for the peanut and more than 100 uses for the sweet potato. The resulting availability of two cash crops offered new hope for Southern farmers who had suffered one deprivation after another since the Civil War.

During his lifetime, Carver garnered great honor and became an internationally respected scientist. Carver maintained a simple lifestyle and remained dedicated to his work at Tuskegee. He repeatedly turned away offers and opportunities to enrich himself. Thomas A. Edison

George Washington Carver in his laboratory. *(Library of Congress. Reproduced with permission.)*

(1847-1931) reportedly offered Carver a salary greater than that earned by the president if Carver would come to work with Edison.

Carver was awarded an honorary doctorate and was made a member of the British Royal Society of Arts. In 1923 he received the Spingarn Medal, awarded annually by the National Association for the Advancement of Colored People (NAACP) to the person who has made the greatest contribution to the advancement of his race.

Despite his simple outlook on life, Carver was sought out by some of the most influential leaders in the world. At various times Carver's views were sought by such disparate figures as Franklin D. Roosevelt, Joseph Stalin, and Mohandas Gandhi. As per his wishes, Carver's life savings, following his death, went toward the establishment of a research institute for agriculture at Tuskegee. Carver was buried on the Tuskegee campus beside Booker T. Washington.

BRENDA WILMOTH LERNER

Erwin Chargaff

1905-

American Biochemist

Erwin Chargaff has been a pioneer in the study of the chemical nature of nucleic acids. Chargaff's work proved that, contrary to the prevailing views about nucleic acids in the 1940s, DNA was actually a complex and highly variable molecule that could serve as the genetic material. He also established the relationships, now known as "Chargaff's rules," among the four bases found in DNA.

Chargaff was born in 1905 in Czernowitz, which at the time was a provincial capital of the Austrian monarchy. His parents, Hermann Chargaff and Rosa Silberstein Chargaff, had been moderately well off, but they were financially ruined during the Great Inflation after World War I. As a child, Chargaff was a voracious and eclectic reader. In 1914, at the outbreak of World War I, his family was forced to move to Vienna to avoid the Russian occupation of Czernowitz. Chargaff received most of his education at the Maximiliansgymnasium in Vienna. He enjoyed school and excelled in literature and the classical languages. Chargaff, who is well known for his literary style, remembers having a special relationship to language ever since his childhood.

At the University of Vienna, Chargaff decided to major in chemistry. His dissertation dealt with organic silver complexes and the action of iodine on azides. Eventually, he was able to utilize some aspects of his thesis research in identifying the presence of sulfur-containing amino acids on paper chromatographs. While at the university, he met his future wife, Vera Broido. After Chargaff received his doctoral degree in 1928, he obtained a research fellowship at Yale University, where he worked with Rudolph J. Anderson on the lipids of tubercle bacilli. Chargaff and Vera Broido were married in New York in 1929 and returned to Europe in 1930. Their son was born in 1938.

Between 1930 and 1933 Chargaff carried out further research on tubercle bacilli at the University of Berlin's Institute of Hygiene. Chargaff was involved in the investigation of the "Lübeck scandal," which involved the prosecution of several physicians who had caused the deaths of several babies by using cultures of virulent tubercle bacilli instead of BCG vaccine. Chargaff's work helped to prove that BCG itself had not been responsible for the disaster. In 1933 Albert Calmette (1863-1933) invited Chargaff to work at the Pasteur Institute. One year later, Chargaff returned to the United States to work at the Mount Sinai Hospital in New York, before accepting a research position in the department of biochemistry at Columbia University, where he carried out research on the biochemistry of blood coagulation. He also studied lipids, lipoproteins, phospholipids, inositols, hydroxy amino acids, and conducted the first synthesis of a radioactive organic compound. Chargaff remained at Columbia for the rest of his professional career. He became an assistant professor at 33, an associate professor at 41, and a professor at 47. He retired in 1974 and is currently Professor Emeritus and Special Lecturer in Biochemistry and Molecular Biophysics.

In 1944 Chargaff read Oswald T. Avery's (1877-1955) famous paper on the transforming principle and became convinced that DNA was the "hereditary code-script." He had recently read Erwin Schrödinger's (1887-1961) *What Is Life?* and was already thinking about the biochemical nature of genes. The combined impact of Avery's paper and Schrödinger's book convinced Chargaff to study nucleic acids. Work previously carried out by Phoebus Aaron Levene (1869-1940) suggested that DNA was a repetitious polymer, but Chargaff disproved that assumption by conducting precise analyses of nucleic acids from many sources, such as beef thymus, spleen, liver, human sperm, yeast, and tubercle bacilli. Using methods that first became available after World War II—paper chromatography, UV spectrophotometry, and ion-exchange chromatography—he proved that the four bases were not present in equal amounts, contrary to Levene's "tetranucleotide hypothesis." Chargaff called attention to the noteworthy fact that in all DNA molecules the molar ratios of total purines to total pyrimidines, and, more specifically, the ratios of adenine to thymine and of guanine to cytosine, were about one. The meaning of this "regularity" could not be determined at the time, but it was later explained in terms of the complementarity rules of the DNA double helix.

Chargaff was elected to the National Academy of Science in 1965. His voluminous writings include a two-volume work called *The Nucleic Acids* (1955, 1960), *Essays on Nucleic Acids* (1963), *Voices in the Labyrinth* (1977), and the autobiographical *Heraclitean Fire* (1978). Although the Nobel Committee has failed to acknowledge his crucial work on the biochemistry of DNA, he was awarded the National Medal of Science in 1975. At the end of the twentieth century, Chargaff was still actively thinking about science and the meaning of life, as indicated in his essay "In Dispraise of Reductionism," which was published in the journal *Bioscience* in 1997. Here, he analyzed issues related to the term "reductionism" and their implications for science.

LOIS N. MAGNER

Carl Franz Joseph Erich Correns
1864-1933
German Geneticist and Botanist

Carl Correns was one of three scientists who simultaneously rediscovered the work of Gregor Mendel (1822-1884). Correns investigated and confirmed the validity of many Mendelian laws and showed a deep understanding of Mendelian genetics. Correns's research was the first to correlate Mendelian segregation with the reduction division of chromosomes, and he illustrated several important examples of deviations in Mendelian inheritance, such as natural variations in the dominant-recessive heredity pattern.

Correns was born in Munich to a Swiss mother and German father who was a member of the Bavarian Academy of Art. An only child who was orphaned at age 17, Correns contracted a case of tuberculosis that delayed the degree he finally received from the University of Munich in 1889. He married Karl von Nageli's (1817-1891) niece Elizabeth Widmer in 1892. He served as a lecturer, then assistant professor, at the University of Leipzig, became full professor at Munster, and finally in 1913 was appointed the first director of the Kaiser-Wilhelm Institute of Biology in Berlin. He served at this post until his death.

Working with Nageli during his early career, Correns was at least partially restricted to investigations that sought to extend the work of his mentor Nageli. His research focused on heterostylism, in an attempt to counter and critique Charles Darwin's (1809-1882) views on adaptations to outbreeding. Correns's research denied the idea that pollen grain size variations were adaptations to flower styles, and he further challenged Darwin regarding his conclusions on the effect of increasing temperature on the sensitivity of leaves. Correns's research also included a study of plant cell wall growth, as well as floral morphology, the physiology of climbing plants and plant sensitivity, leaf primordia, and moss and liverwort vegetative reproduction.

In the 1890s Correns examined the hereditary effects of pollen on developing embryos by cross breeding varieties of peas and maize. He discovered that his pea stock progeny had produced color traits that exhibited very simple ratios, while his maize research produced a much more complicated set of results. As Correns was attempting to reconcile the meaning of his conflicted pea and maize breeding results, he read Mendel's previously overlooked and forgotten work, which verified his own results. In 1900 Hugo de Vries (1848-1935) published his own work on hybridization and his rediscovery of Mendel's research, and Correns was forced to quickly write up and publish his already completed work.

Correns dedicated his remaining career to investigating the validity of Mendelian laws. He was able to correlate the reduction division of chromosomes with Mendel's law of allelic segregation. Correns subsequently discovered an example of coupling of hereditary traits in a strain of maize that conjugated self-sterility with blue coloration. This data was evidence of variation of Mendel's law of independent assortment. Correns also developed a theory of chromosome inheritance that allowed for the occurrence of crossing over, or gene exchange between chromosomes. Correns also predicted and experimentally proved the Mendelian inheritance of sex.

Correns concluded his research career studying the inheritance of variegation in leaves, and he documented the occurrence of cytoplasmic inheritance in the form of plastid DNA. His experiments illustrated that plastids were extranuclear units of inheritance that could produce variegated leaves, and that variegation could also be inherited through nuclear genes or even be a non-inheritable form. Correns was able to prove that the plastid source was predominantly the egg and not the sperm. Plastids have since been shown to carry important genes such as those related to disease and drug resistance.

Though considered an average lecturer, Correns was hard working and dedicated to science. He had an approach to Mendelian genetics that was thorough, well developed, and considerably advanced for his time. Correns's place in history consists of both his rediscovery of Mendel's work and his eloquence, skill, and sophistication as a Mendelian geneticist.

KENNETH E. BARBER

Hugo Marie de Vries
1848-1935
Dutch Botonist

Hugo Marie de Vries was a Dutch botanist whose work with plant species led him to rediscover Gregor Mendel's theories of heredity, and brought his concept of mutation into evolutionary theory.

De Vries was born in Haarlem, the Netherlands, on February 16, 1848. He became interested in botany, the study of plants, at an early age. In 1870, he received his Ph.D. from the University of Leiden, then continued his studies at the University of Heidelberg, where he was fortunate enough to work with noted German plant physiologist Julius von Sachs (1832-1897). He continued his research on the physiology of plant cells while serving as professor at the University of Amsterdam in 1878, a position he held until 1918.

By the late 1880s, controversy was mounting over the application of evolutionary theory to plant heredity. In 1886, while conducting experiments involving the hybridization, or cross-breeding, of plants, de Vries discovered wild varieties of the evening primrose (*Oenothera lamarckiana*) that differed from the cultivated species. This led him to a new method of investigating evolution—by experimentation, rather than observation alone.

De Vries found that unique variations appeared randomly within the cultivated primrose specimens, a phenomenon he labeled mutations. de Vries found that these mutations, or variations, could produce a new species within a single generation. Mutations which were favorable to the survival of the species would persist until replaced by other, even more beneficial variations. While our modern understanding of heredity and mutation have since diverted from De Vries's theories, his concepts of species variation was well received at the time, as it provided the first alternative to Darwin's theory of natural selection. Darwin's widely accepted supposition held that new species developed slowly, through tiny variations, whereas De Vries proposed that a new species could emerge more quickly.

De Vries re-discovered the research papers of Austrian monk Gregor Mendel (1822-1884), which outlined Mendel's evolutionary theory. De Vries brought Mendel's theories back into the scientific mainstream, giving his predecessor full credit. Two other botanists, Karl Correns (1864-1933) and Erich Tschermak von Seysenegg (1871-1962), simultaneously made the same discovery, but only de Vries held to his own theory of heredity. De Vries's theory strayed from that of Mendel when he proposed that independent units called pangenes carried hereditary traits. He summarized his research in his *Die Mutationstheorie* (The Mutation Theory) in 1901-1903.

De Vries's other work included a study of the role of osmosis in plant physiology. In 1877, he demonstrated the relationship between osmotic pressure and the molecular weight of substances in plant cells. His other writings included *Intracellular Pangenesis* (1889) and *Plant Breeding* (1907).

De Vries retired in 1918 from the University of Amsterdam, but continued his studies for several years. He died in Amsterdam in May 1935.

STEPHANIE WATSON

Theodosius Grigor'evich Dobzhansky
1900-1975

Russian-American Population Geneticist and Entomologist

Theodosius Dobzhansky (originally Feodosy Grigorevich Dobrzhanskii), one of the most influential biologists of the twentieth century, was born in the small city of Nemirov, Ukraine, on January 25, 1900. He was educated at home until 1910, when his parents moved to the outskirts of Kiev. He began to collect butterflies during his second year of Gymnasium (formal school) and by age 12 had decided to become a biologist. He attended his first amateur scientific excursion and was impressed by reading Charles Darwin's (1809-1882) *Origin of Species* at the age of 15.

At the Gymnasium Dobzhansky met Victor Luchnik (1892-1936), who invited Dobzhansky to join the local entomological society. On Luchnik's advice Dobzhansky decided to specialize in ladybird beetles (*Coccinellidae*) and began to collect and dissect them. His participation in this entomological circle influenced his later scientific ideas, leading him to study the morphology of sex organs as the defining trait for species.

In 1917 Dobzhansky graduated from the Gymnasium and began to work with Sergei Kushakevich (1878-1920), a professor at the University of Kiev. In the fall of that year Dobzhansky entered the Physico-Mathematical department of the university, where he received a broad education in natural science. He played an important role in assisting the geologist Vladimir Vernadsky (1863-1945). In 1921 Dobzhansky completed all of his courses at the university, although he never received a diploma. In 1918 he lost his father and two years later his mother.

Partially influenced by the cytologist Grigorii Levitskii (1878-1942) and the genetics publications of Yurii Filipchenko (1882-1929), Dobzhansky began investigating the genetics of *Drosophila* (fruit flies). His discovery of the correlation between the shape of genitalia and a geographic locality won him an invitation from Filipchenko to St. Petersburg University. Dobzhansky married Natalia Sivertsov, also a student of genetics, in 1925. On Filipchenko's nomination Dobzhansky won a Rockefeller Foundation fellowship to study with Thomas H. Morgan (1866-1945) at Columbia University. Dobzhansky departed for New York on December 6, 1927—never again to return to his homeland.

At the Morgan laboratory Dobzhansky worked with Alfred Sturtevant (1891-1970) and Calvin Bridges (1889-1938) in mapping *Drosophila* chromosomes. In 1928 the Morgan laboratory moved to the California Institute of Technology in Pasadena. During the 1930s Dobzhansky began to investigate the genetics of natural *Drosophila* populations. Based on his research he defined the formation of a new species in biological terms—as the change that occurs when two groups of a particular species (that were able to reproduce with each other) eventually became two distinct species (that could no longer produce fertile offspring).

In 1935 Dobzhansky delivered the Jessup Lectures at Columbia University, publishing them as *Genetics and the Origin of Species* (1937). This book became the foundation of the evolutionary "modern synthesis," which combined Mendelian genetics and the Darwinian theory of natural selection. Dobzhansky applied the theoretical work of Sewall Wright (1889-1988) to the study of natural populations, allowing him to address evolutionary problems in a novel way.

Dobzhansky was a synthesizer of genetic thought. In his nearly 600 publications, especially his series *Genetics of Natural Populations* (1938-1975), he drew on and extended the work of many researchers, including Nikolai Timofeev-Ressovsky (1900-1981) and Sergei Chetverikov (1880-1959). He also published several books on the relationship between biology, society, and religion. In *Mankind Evolving* (1962), for example, he discussed the moral concerns of natural selection when it is applied to human beings.

Between 1940 and 1962 Dobzhansky moved from Caltech to Columbia University in New York City, where he became professor of zoology. In 1962 he became professor emeritus at the Rockefeller Institute, a position he held until 1970, when he accepted his final position—adjunct professor at the University of California, Davis. Dobzhansky outlived his wife by six years, dying from heart failure in 1975.

LLOYD T. ACKERT JR.

Theodosius Dobzhansky. *(Library of Congress. Reproduced with permission.)*

Sir Ronald Aylmer Fisher

1890-1962

British Geneticist

Sir Ronald Aylmer Fisher was a British geneticist and statistician. To create genetic experiments that yielded greater results with less effort, he pioneered the use of statistics in experimentation, and came up with the now widely used concepts of variance and randomization.

Fisher graduated from the University of Cambridge with a B.A. in astronomy in 1912. While there, he gained an interest in the theory of errors in astronomical observations, which eventually led him to a career in statistical research.

In 1919, he accepted a position as statistician for the Rothamsted Agricultural Experimental Station near Harpenden, Hertfordshire. His work on plant-breeding experiments combined biology and statistics. At Rothamsted, he developed a new technique by which scientists could vary different elements in an experiment to determine the probability that those variations would yield different results. He published his findings in the book *Statistical Methods for Research Workers* (1925).

While at Rothamsted, Fisher also introduced new theories about randomization and variance, included in his work *The Genetical Theory of Natural Selection* (1930), which are now widely used in the field of genetics. His goal was to design plant-breeding experiments to yield the maximum results while using the least amount of time, effort, and money. One problem he discovered was biased selection of materials, which could lead to inaccurate results. To avoid this, Fisher introduced the concept of randomization, which provided that experiments must be conducted among a random sample of the entire population, and must be repeated on a number of control subjects to ensure validity.

Fisher also introduced his concept of variance. At the time, scientists were only able to vary one factor at a time in experiments, allowing for only one potential result. He proposed instead a statistical procedure by which experiments would be designed to answer several questions at once. This was accomplished by dividing each experiment into a series of sub-experiments, each of which differed enough to provide several unique outcomes. Fisher summed up his statistical work in his definitive work, *Statistical Methods and Scientific Inference* (1956).

In 1933 Fisher became Galton professor of eugenics (the use of selective breeding to improve the heredity of the human race) at the University of London. From 1943 to 1957 he served as Balfour professor of genetics at his alma mater, the University of Cambridge.

For his achievements, Fisher was elected a Fellow of the Royal Society in 1929, and was awarded the Royal Medial of the Society in 1938, as well as the Darwin Medal of the Society in 1948 "in recognition of his distinguished contributions to the theory of natural selection, the concept of its gene complex and the evolution of dominance."

Fisher was knighted in 1952. His last years were spent conducting research in Australia, where he died on July 29, 1962.

STEPHANIE WATSON

John Burdon Sanderson Haldane

1892-1964

British Geneticist

John Burdon Sanderson Haldane was born in Oxford, England, but spent much of his childhood in Scotland. He was the son of an eminent physiologist, John Scott Haldane. His sister became the well-known novelist Naomi Mitchison. He went to Eton and then Oxford University, where he began to study classics, but then switched to the study of genetics. He was particularly interested in the mathematical study of biological questions and eventually became Professor of Biometry at University College, London, for 20 years from 1937. His father had worked on poison gases and the efforts to devise an effective gas mask during the First World War. The young Haldane and his sister often served as experimental subjects for these studies. As a young boy he went into mines with his father to test devices for detecting dangerous gas buildups, and took a 39 foot (12 m) dive off the Scottish coast to study the process of decompression as he surfaced. He was even called back from the Black Watch regiment on the Western Front during the First World War to serve as an experimental subject in his father's chlorine gas experiments.

After the First World War, Haldane argued that poison gas weapons were a relatively humane means of waging war. He based this on a mathematical analysis of the casualties caused by the chemical weapons such as mustard gas, calculating that only 4,000 or 1 in 40 of the 150,000 British soldiers who had been gassed had died, whereas conventional weapons such as bayonets, shells, and incendiary bombs killed one out of three men they hit. So he concluded that these weapons were disabling rather than killing mechanisms and therefore preferable. He did not consider the evidence that many of the disabilities were very long term and deeply affected the quality of life of the survivors. He dismissed the possibility that these gases might cause cancers in the victims. His critics suggested that experimentation in his father's laboratory with chlorine gases was rather different than the experience of fighting men in the trenches facing mustard gas and that while he was free to offer himself for experiments conducted by his father, the soldiers were in rather different circumstances.

In the inter-war years Haldane became interested in Marxism as a political philosophy, which suggested that human experience could be understood analytically as something that was regulated by its own forces, just as he felt the natural world was. He joined the British Communist Party in the 1930s and was chairman of the board of its newspaper, the *Daily Worker*, throughout the 1940s. But during the Second World War he returned to government service at the Admiralty and concentrated on understanding the effects of submarine service on the blood and breathing of navy personnel. He continued

to experiment on himself to determine dangerous levels of carbon dioxide buildup.

Haldane left the Communist Party in 1956 in disagreement with its promotion of the dubious science of the Russian plant geneticist Trofim Denisovich Lysenko (1898-1976). The next year he left Britain in protest against the Anglo-French invasion of Suez and went to India, where he became Director of the Genetics and Biometry Laboratory in Orissa. He became an Indian citizen in 1961 and died of cancer in that country in 1964. He had become a very well-known writer and broadcaster on scientific questions and had made major contributions to both genetics and physiology.

SUE RABBITT ROFF

Wilhelm Ludvig Johannsen

1857-1927

Danish Geneticist and Botanist

Wilhelm Johannsen is considered one of the founders of genetics. His research provided evidence supporting the mutation theory of Hugo de Vries (1848-1935), which holds that there are sudden, spontaneous appearances of new characters or traits in existing species. Johannsen's research appeared to counter Charles Darwin's (1809-1882) theory that changes in species occur slowly through the process of natural selection alone. Johannsen came to believe that both natural selection and mutations in the hereditary units of germ cells act as influences on subsequent progeny, thereby changing species through the evolutionary process.

Born the son of a Danish army officer and German mother, Johannsen's schooling in Copenhagen was stopped short of the University level because his father did not have the financial means to provide for his tuition. Johannsen served as a pharmacy apprentice in Denmark and Germany, and passed the pharmacist's examination in 1879. Johannsen's keen mind allowed him to become self-taught and trained in the fields of chemistry and botany while he worked as a pharmacist. Johannsen never attained a formal university degree, but was awarded several honorary degrees, became a professor of botany and plant physiology at the University of Copenhagen, served as rector of the university, and became a member of the Royal Danish Academy of Sciences. Johannsen was considered to be a well-read person of the humanities, aesthetics, philosophy, and several languages.

In 1881 he began working at the Carlsberg Laboratory researching plant physiology problems, particularly the metabolism of plants, buds, tubers, and seeds during the processes of germination, dormancy, and ripening. This research included his discovery of a method for breaking the dormancy period of winter buds (1893). At this point Johannsen shifted the focus of his research from plant physiology to plant heredity.

Greatly influenced by both Darwin and Francis Galton (1822-1911), Johannsen would soon become a leading authority on heredity, studying self-fertilizing plants and creating "pure lines" of plants whose hereditary make-up would all be identical. Johannsen focused his analysis on hereditary characters that expressed continuous variation, such as fertility, size, and responses to environmental stimuli—traits that were most useful to plant and animal breeders. Any variation that Johannsen observed in the "pure line" progeny that was not heritable would be due to environmental factors, while physical and physiological differences that were heritable would be the result of a mutation in the germ cells. Johannsen concluded that the process of mutation could produce a new character in a species that natural selection would then act upon, either maintaining the new character or removing it from the population. Johannsen recognized the importance of the genetic structure of populations to evolutionary biology, and the concepts of heredity that he developed at this time both preceded the discovery of DNA and survived all subsequent technical advances made since then.

Johannsen published his text *The Elements of Heredity* in 1905, in which he introduced the terms "gene," to refer to a unit of heredity, "genotype" as the totality of an individual's genes, and "phenotype" as the overall appearance and processes of an individual. The phenotype represents the summation of the genotype and the environmental conditions encountered by the individual. It was the first and most influential textbook of genetics in Europe, and contained many mathematical and statistical methodologies useful for analyzing quantitative data from genetics experiments. Johannsen's text and his subsequent role as a science historian and critic served to both establish the modern sciences of genetics and evolutionary biology and to remove outdated concepts and ideas that did not stand up to the rigors of objective scientific inquiry. Johannsen used his position to act as a purging filter for any nineteenth-century myths of biology that were based on mysticism, superstition, and teleology, and thus performed an overdue, lasting service for biology.

KENNETH E. BARBER

Hans Adolf Krebs

1900-1981

German-born British Biochemist

Hans Adolf Krebs won the 1953 Nobel Prize for physiology or medicine, which he shared with American biochemist Fritz Albert Lipmann (1899-1986), for his studies of intermediary metabolism, especially his discovery of the metabolic pathway known as the tricarboxylic acid (TCA) cycle, or Krebs cycle, the major source of energy in living organisms. In the presence of oxygen, the reactions of the TCA cycle result in the conversion of the metabolic products of sugars, fats, and proteins into carbon dioxide, water, and energy-rich compounds. Krebs was also involved in the discovery of the metabolic pathway known as the urea cycle. The urea cycle allows ammonia to be converted into urea, a less toxic substance that is excreted in the urine of most mammals. This cycle also serves as a source for the amino acid arginine.

Krebs was born in Hildesheim in Hanover, Germany. His father, Georg Krebs, was an ear, nose, and throat surgeon. After completing his studies at the Hildesheim Gymnasium, Krebs decided to study medicine and join his father's successful practice. He was drafted into the army in 1918, but was discharged shortly afterward when World War I ended. From 1918 to 1923 he studied medicine at the universities of Göttingen, Freiburg-im-Breisgau, and Berlin. During this period, he became interested in the new field of intermediary metabolism. Krebs obtained his M.D. from the University of Hamburg in 1925 and spent one year of required medical service at the Third Medical Clinic of the University of Berlin. After taking a special chemistry course for medical students in Berlin, he became assistant to Otto Warburg (1883-1970) at the Kaiser Wilhelm Institute for Biology at Berlin-Dahlem. Here, Krebs learned techniques, involving manometry and tissue slices, that Warburg had developed to study glycolysis and cellular respiration. Krebs remained at the Institute until he returned to hospital work in 1930. During the next three years, he combined research on the biosynthesis of urea with clinical work at the Municipal Hospital at Altona and the Medical Clinic of the University of Freiburg-im-Breisgau.

In June 1933 the National Socialist government terminated his appointment and he left Nazi Germany for England. He was invited by Sir Frederick Gowland Hopkins (1861-1947) to join the School of Biochemistry at the University of Cambridge. Krebs held a Rockefeller Studentship until 1934, when he became a demonstrator of biochemistry at Cambridge. In 1935 he was appointed a lecturer in pharmacology at the University of Sheffield; three years later, he was put in charge of the department of biochemistry and married Margaret Cicely Fieldhouse, a teacher of domestic science, with whom he had three children. In 1945 Krebs was promoted to professor and became director of a Medical Research Council unit established in his department. In 1954 he was appointed Whitley Professor of Biochemistry at the University of Oxford, and the Medical Research Council's Unit for Research in Cell Metabolism was transferred there. During World War II, Krebs carried out research on special diets that might be useful as supplements in times of scarcity. From 1954 to 1967 Krebs served on the faculty at Oxford.

During his long and distinguished research career, Krebs investigated many problems related to intermediary metabolism, including the synthesis of urea in the mammalian liver, the synthesis of uric acid and purine bases in birds,

Hans Adolf Krebs in his laboratory. *(Library of Congress. Reproduced with permission.)*

the intermediary stages of the oxidation of foodstuffs, the mechanism of the active transport of electrolytes, and the relationship between cell respiration and the generation of adenosine polyphosphates. One of the great landmarks in biochemistry was his demonstration of the existence of a cycle of chemical reactions involved in the breakdown of sugars and resulting in the formation of citric acid; the TCA cycle liberates carbon dioxide and electrons, which are used in the creation of high-energy phosphate bonds in the form of adenosine triphosphate (ATP), the chemical-energy reservoir of the cell.

Krebs earned many other awards and honors in addition to the Nobel Prize. He was elected a fellow of the Royal Society of London in 1947 and knighted in 1958. He was awarded the Copley Medal; the Gold Medal of the Netherlands Society for Physics, Medical Science, and Surgery; and honorary degrees from many universities.

LOIS N. MAGNER

Fritz Albert Lipmann
1899-1986
German-American Biochemist

Fritz Albert Lipmann was a German-born, American biochemist who helped discover the biochemical processes by which organisms produce and use energy. In particular, he discovered an essential substance called coenzyme A, which is a crucial intermediary in metabolism. It is necessary for the conversion of carbohydrates, fats, and amino acids into usable energy. For the work cited above, Lipmann was awarded the 1953 Nobel Prize in Physiology or Medicine, which he shared with German biochemist Sir Hans Adolph Krebs (1900-1981). Lipmann was also the first to propose that adenosine triphosphate (ATP) was the common form of energy used in many cellular reactions, a concept now thoroughly accepted and documented.

Lipmann was born in Königsberg, East Prussia (now Kaliningrad, Russia) in 1899. While his early education was unremarkable, he went on and studied medicine at the Universities of Königsberg, Berlin, and Munich from 1917 to 1922. Lipmann was enamored with the subject of biochemistry. He continued his studies at the University in Berlin and received his medical degree in 1924. His doctorate in chemistry was completed there in 1927. He then joined famous physiologist Otto Meyerhof (1884-1951) in Heidelberg, to further research the biochemical reactions that occur in muscle.

In 1930 Lipmann went back to the Kaiser Wilhelm Institute in Berlin to work as a research assistant in the laboratory of Albert Fischer, who was interested in applying biochemical methods to tissue culture. Lipmann also briefly traveled to New York where he worked extensively with a class of chemicals known as phosphates. Lipmann accompanied Fischer to the Institute in Copenhagen in 1932. While at Copenhagen, Lipmann studied the metabolism of fibroblasts, which led to important publications regarding the role of glycolysis in the metabolism of the cells of embryos.

Fearing the political environment in Germany, Lipmann emigrated to the United States in 1939 and became Research Associate in the Department of Biochemistry, Cornell Medical School, New York. In 1941 he joined the research staff of the Massachusetts General Hospital in Boston. Lipmann became a United States citizen in 1944. He received an appointment as Professor of Biological Chemistry at Harvard Medical School in 1947. In 1957, he was appointed a Member and Professor of the Rockefeller Institute in New York.

Lipmann spent most of his career investigating cellular metabolism in some form or another. He studied how food is converted into energy for the cells. While he is most famous for his discovery of coenzyme A, which netted him a Nobel Prize, he has made a significant amount of other contributions to the field. In 1941, he was the first scientist to propose that a cellular compound known as adenosine triphosphate (ATP) is the universal energy source for cellular metabolism. Lipmann also spent a significant amount of time examining other metabolic cycles, the structure of cancer cells, and the activity of the hormone secreted by the thyroid gland. His most important discovery occurred in 1945 when Lipmann and his colleagues identified a substance in pigeon liver extracts that needed to be present for specific metabolic reactions to occur. He isolated and characterized the substance that now has become known as coenzyme A. This molecule is one of the most important in the body and is necessary to help generate cellular energy from ingested foods.

In 1931, Lipmann married Elfreda M. Hall, and they had one son, Steven. Fritz Albert Lipmann passed away in 1986 after a lifetime of scientific achievement.

JAMES J. HOFFMANN

Konrad Zacharias Lorenz
1903-1989
Austrian Zoologist and Ethologist

Konrad Lorenz is most well-known as a founder of the field known as ethology, and for his research into animal behavior. As the study of animal behavior became more extensive after the turn of the century, Lorenz and other scientists helped to establish ethology as the systematic study of the function and evolution of behavior. His behavioral research focused mainly on social structure and instinctive behaviors, including imprinting, in which newly hatched birds become attached to the first moving thing they see.

Lorenz was born in Vienna, Austria, on November 7, 1903, as the second of two sons to a well-known orthopedic surgeon and his wife. As a youngster, Lorenz studied at a private elementary school and at one of the city's finest secondary schools. His interests in animal behavior began and flourished at his family's summer home in Altenburg, Austria, where he kept numerous pets. His interests expanded when, at the age of 10, he read about Charles Darwin's (1809-1882) theory of evolution. Following his father's wishes, Lorenz went to New York's Columbia University for premedical training, and transferred to the University of Vienna to continue his medical studies. There, he also took up zoology, and began to conduct informal animal-behavior studies, such as observing his pet bird and keeping a comprehensive diary of its activities. The diary was published in an ornithological journal in 1927. A year later, he earned his medical degree from Vienna University and became a professor's assistant at the university's anatomical institute. In his 1982 book *The Foundations of Ethology*, Lorenz noted that the methods used in evolutionary comparison and comparative morphology "were just as applicable to the behavior of the many species of fish and birds I knew so thoroughly, thanks to the early onset of my love for animals."

After receiving his medical degree, he continued his studies in the zoology program at Vienna University, earned his Ph.D. in the field in 1933, and turned his full attention to animal behavior studies. From 1935-38, he spent what he called his "goose summers" at the Altenburg family home. There, he conducted detailed observations of greylag geese, and demonstrated that newly hatched geese would readily accept a foster mother—even Lorenz himself. He termed this behavior "imprinting." He found that another bird species, mallard ducks, would imprint on him if he also quacked and squatted so that he appeared shorter.

In 1937, he accepted a position as lecturer of comparative anatomy and animal psychology at the University of Vienna, and three years later moved to the comparative psychology department at the University of Königsberg. His career was interrupted in 1943 when he was drafted into the German army and taken prisoner by the Russians from 1944-48. Upon his release, he went to the Max Planck Institute for Marine Biology as head of the research station, studying the physiology of behavior.

During the ensuing years of research into animal behavior, Lorenz and fellow scientist Nikolaas Tinbergen (1907-1988) investigated the triggers that led to a series of instinctive behaviors called fixed-action patterns. They called the triggering actions "releasers," and the responding action an "innate releasing mechanism." In 1973, Lorenz, Tinbergen, and Karl von Frisch (1886-1982), who described an elaborate communication system in honey bees, shared a Nobel Prize for their work in animal behavior.

One of a series of scientists who had a role in establishing the field of ethology, Lorenz is often credited as the person who synthesized the range of behavioral studies and who, with Tinbergen, helped to develop modern ethology, also known as comparative ethology, which encompasses observational field studies and behavioral experiments. Lorenz died on February 27, 1989, of kidney failure at his home in Altenburg.

LESLIE A. MERTZ

Trofim Denisovich Lysenko
1898-1976
Russian Horticulturist

Trofim Denisovich Lysenko was born in Karlovka near Poltav, Russia. Despite good grades in school, Lysenko, a peasant's son, struggled to receive an education in pre-Revolutionary Czarist Russia. He became a gardener and, after the 1917 Revolution, undertook studies at the Uman School of Horticulture. After graduating in 1921, Lysenko received an appointment to work at the Belaya Tserkov Agricultural Selection Station. In 1925, Lysenko received his doctorate from the Kiev Agricultural Institute. Following his graduation, Lysenko worked at the Gyandzha Experimental Station and, subsequently, at the Ukrainian All-Union Institute of Selection and Genetics in Odessa. Eventually, with Soviet dictator Joseph Stalin's (1879-1953) patronage, Lysenko became the director of the Odessa All-Union Selection and Genetics Institute.

Late in the 1920s, Lysenko began to advocate an outmoded Lamarckian view of genetics to harden Soviet crops against the brutal Russian winters. Lysenko's conclusions were based upon, and profoundly influenced by, the teachings of Russian horticulturist I.V. Michurin (1855-1935), and the discredited Lamarckian theory of evolution by acquired characteristics (i.e., that organisms evolved through the acquisition of traits that they needed and used) put forth by nineteenth-century French anatomist Jean Baptiste Lamarck (1744-1829).

Based on his misunderstanding of genetics, Lysenko developed methods that falsely predicted greater crop yields through a hardening of seeds and a new system of crop rotation. Lysenko's theories drew the attention of the Soviet Central Committee desperate to avoid elongation of a string of politically disastrous famines. Moreover, Lysenko's Lamarckian-based teachings were thought by Stalin to be harmonious with the goals of a worldwide revolution of workers.

Championed by Stalin, Lysenko steadily ascended the rungs of power within Soviet science. In 1938 Lysenko took the reins as president of the powerful and influential Lenin Academy of Agricultural Sciences. As president of the Academy, Lysenko began to persecute scientists who did not agree with his theories. Backed by the Soviet Central Committee Lysenko was responsible for the exile, torture, and death of many talented Soviet scientists. Under Lysenko, Mendelian genetics was branded "decadent," and scientists who rejected Lamarckism in favor of natural selection were denounced as "enemies of the Soviet people." Lysenko's supporters even denied the existence of chromosomes, and genes were denounced as "bourgeois constructs."

During his career, Lysenko published articles in which he stressed the role of, what he termed, "sudden revolutionary leaps" in the origin of new species. He asserted that many species of cultivated plants could spontaneously transform, even under natural conditions, into other, quite different species (e.g., wheat into rye). The very spirit of Marxist theory, Lysenko claimed, called for a theory of species formation that would entail "revolutionary leaps." Lysenko attacked Mendelian genetics and Darwinian evolution as "a theory of all round gradualism." In 1940, Stalin appointed Lysenko Director of the Soviet Academy of Science's Institute of Genetics. Although Stalin died in 1953, Lysenko continued to act as director of the academy until 1964.

For a generation of scientists trained under Lysenkoism, Michurin "science," an exploration of man's power over nature, became the focus of theoretical development. Lysenko constructed an elaborate hypothesis that came to be known as the theory of phasic development. One of Lysenko's failed programs involved the treatment of seeds with heat and high humidity in an attempt to increase germination under harsh conditions. Although crop yields declined, Lysenko's programs were considered politically correct, and Lysenko was awarded the Order of Lenin and two Stalin prizes, and was nominated Vice Chairman of the Supreme Soviet.

Lysenko ruled virtually supreme in Soviet science, and his influence extended beyond agriculture to basic science and medicine. In 1948 the Praesidium of the USSR Academy of Science passed a resolution virtually outlawing any biological work that was not based on the teachings of Michurin and Lysenko. It was not until Nikita Khrushchev's (1894-1971) premiership following the death of Stalin in 1953 that opposition to Lysenko was tolerated. Under Khrushchev, Lysenko lost control of the Lenin Agricultural Academy and, following Khrushchev's political demise in 1964, Lysenko's doctrines were discredited, and intensive efforts were made toward the reestablishment of orthodox genetics and science in the U.S.S.R.

K. LEE LERNER

Ernst Mayr
1904
German-American Evolutionary Biologist

Ernst Mayr was one of the scientists who helped develop the so-called "modern synthesis" of evolutionary theory, combining evolution, genetics, and speciation. He also developed and promoted the widely accepted biological concept of species, which also provided an explanation of how new species evolve.

Mayr was born in Kempten, Germany, on July 5, 1904. The son of a judge, he had a comprehensive early education and took a particular interest in birds. In college, he first studied medicine at the University of Greifswald in Berlin from 1923-26, but became more and more intrigued by zoology, particularly after he met noted ornithologist Erwin Stresemann. Mayr's meeting with Stresemann was prompted by Mayr's 1923 sighting of a bird that had not been seen in Europe for decades. With encouragement from Stresemann, Mayr transferred to the zoology program at the University of Berlin, and earned his doctorate in the field *summa cum laude* in 1926.

Upon graduation, Mayr remained at the University of Berlin as assistant curator in its zoological museum until 1932. While in that position, Mayr headed three scientific expeditions to New Guinea and the Solomon Islands from 1928-30, and spent much of the remainder of his time identifying and classifying the birds he saw and collected on those trips. His reputation as a skilled ornithologist grew, and he accepted a position as associate curator in 1932, and a promotion to curator in 1934, at the American Museum of Natural History in New York, where he stayed until 1953. During this time, he also became a U.S. citizen. From the museum, Mayr moved on to Harvard University in 1953 as Alexander Agassiz Professor of Zoology and in 1961 also took on the title of director of the Museum of Comparative Zoology. He retired from Harvard in 1975, and now holds the title of professor emeritus from that university.

During his long and accomplished career, Mayr intertwined evolutionary concepts with his studies of animal variation. One of the first major problems he addressed was the definition of a species: When was one organism sufficiently different from another so that they could be termed separate species? With an eye toward the ideas behind evolution, and particularly the role of genetics, he promoted a biological concept of species. Under this definition, two different species cannot interbreed. This reproductive isolation is necessary to prevent gene flow between them. Different species may be reproductively isolated for a number of reasons. For example, the two species may live in different habitats, they may breed at different times of the year, or the males of one species may not exhibit the courtship behaviors that attract females of the other species. Before Mayr's definition, species were delineated in a more subjective manner, such as whether they looked fairly similar.

Following this definition, he also explained how new species could evolve. For example, he explained how a small population of one species could be separated from the main population. Because the small population had a limited gene pool, its successive generations could develop different characteristics. If the offshoot group's characteristics became sufficiently different from the original group so that individuals of the two groups could not reproduce, the offshoot population would, by definition, be a separate species.

Mayr's 1942 book, *Systematics and the Origin of Species*, further coalesced evolution, genetics, and speciation, and paved the way for the "modern synthesis" of evolutionary theory.

The author of 20 books, Mayr has received a number of prestigious honors, including the Balzan Prize in 1983, the International Prize for Biology in 1994, and the Crafoord Prize for Biology in 1999.

LESLIE A. MERTZ

Margaret Mead
1901-1978
American Anthropologist

Margaret Mead was the first major anthropologist to study the fine details of child-rearing practices and learning theory within social groups. Based on her observations, Mead proposed that children learned through "imprinting." Much of Mead's work was interdisciplinary in nature, borrowing freely from psychology, anthropology, sociology, and other fields.

Mead was born in Pennsylvania, the oldest of four children. Her parents were teachers who taught Mead progressive values (by early twentieth-century standards) regarding the role of women in society. Mead's mother was a suffragette (a lobbyist for women's right to vote) who encouraged Mead to mix with children of all racial, ethnic, and economic backgrounds.

Mead undertook her college studies at Barnard College and became enthralled with anthropology. After graduating in 1923, she married and started graduate school at Columbia University. While still a graduate student, Mead made her first serious field observations on an extended, yearlong visit to the Samoan islands. After repeated trips, Mead became deeply aware of the fact that Samoan cultural and sexuality patterns differed greatly from those in the West. Mead concluded that human values depend on time and environment, not inherited traits. This phraseology and disposition were to become Mead trademarks. Her conclusions profoundly influenced the subsequent course of sociological research and thought.

Mead's 1928 publication of *Coming of Age in Samoa,* a study of adolescent behavior in a Polynesian society, signaled major changes in approaches to the formulation of social theory. Mead became an instant celebrity over the book's frank and honest descriptions of Samoan female sexual behavior.

Mead is considered one of the founders of the "culture and personality" school among anthropologists. Throughout her career she maintained her belief in cultural determinism.

Mead's celebrity allowed her to become a sought after, though controversial, authority on American social culture. Mead's observations on sex and family structure made her simultaneously loved and loathed by millions.

Mead went through several marriages and had a daughter by her third husband in 1939. During World War II, Mead worked to foster British and American relations. In 1942 she published a critical book about American culture titled, *And Keep Your Powder Dry: An Anthropologist Looks at America.*

Mead's unique perspectives on child rearing prompted her to allow her daughter to become a test subject for Dr. Benjamin Spock's new theories of child rearing. Spock advocated overturning what he described as overly rigid child-rearing practices. (For example, he contended that children should be fed on demand rather than on schedule.) Mead's daughter was thus among the first of millions of "Spock babies" to follow. In 1969 *Time* magazine named Mead "Mother of the Year."

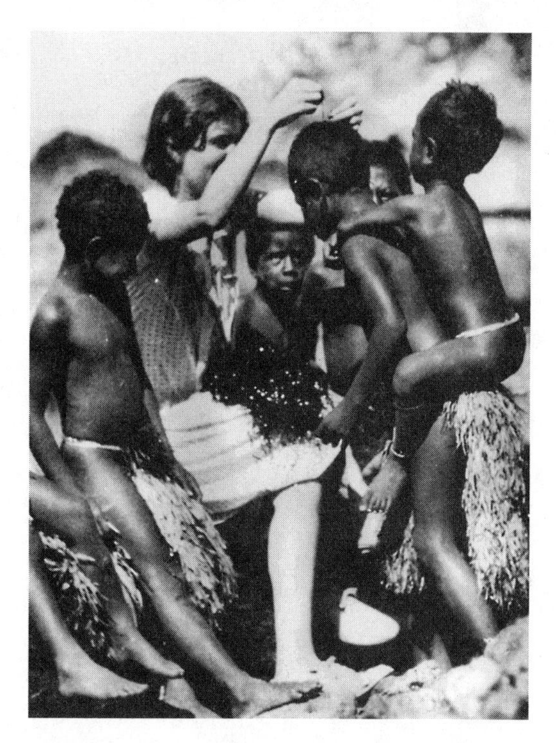

Margaret Mead. *(Corbis Corporation. Reproduced with permission.)*

Mead was anything but a remote and lofty scholar who could be easily dismissed as out of touch with the American public. For nearly two decades, through the tumultuous 1960s and 1970s, Mead co-authored an advice column for *Redbook* magazine. In her column Mead often offered commentaries about changing gender roles and the decline of the extended family. Mead also tackled various contemporary issues, such as the 1960s "generation gap" and environmentalism, with her usual frankness. In particular, Mead became concerned with the ill-effects of overpopulation and became an outspoken advocate for birth control.

Mead served in various capacities with the American Museum of Natural History in New York, including curator. At age 72, Mead was elected and served as president of the American Academy for the Advancement of Science. In 1979 she was posthumously awarded the United States' highest civilian honor, the Presidential Medal of Freedom. Mead died of cancer in 1978.

BRENDA WILMOTH LERNER

Thomas Hunt Morgan

1866-1945

American Geneticist and Zoologist

Thomas Hunt Morgan was an American geneticist who was awarded the Nobel Prize for Physiology or Medicine in 1933 for his work spanning a 17-year period at Columbia University with mutations in the fruit fly *Drosophila*. His research specifically established the chromosome theory of heredity. He demonstrated that genes are linked in a series on chromosomes and are responsible for identifiable, hereditary traits. Morgan's work played a key role in establishing the field of genetics.

Thomas Hunt Morgan. *(Library of Congress. Reproduced with permission.)*

Thomas Hunt Morgan was born in 1866 in Lexington, Kentucky. Morgan's father, Charlton Hunt Morgan, was a U.S. consul and his uncle, John Hunt Morgan, had been a Confederate army general. As a child Morgan had shown an immense interest in natural history and by the age of 10, he was an avid collector of birds, eggs, and fossils. He was educated at the University of Kentucky, where he obtained his B.S. degree in 1886. During the years 1888-1889, he was engaged in research for the United States Fish Commission at Woods Hole, a laboratory with which he always maintained a close association, making expeditions to Jamaica and the Bahamas during that time. Morgan did his postgraduate work at Johns Hopkins University, where he was awarded his doctoral degree in biology in 1890. Most of his graduate research was concentrated in the area of embryology. In that same year, he was awarded the Adam Bruce Fellowship and visited Europe, working at the Marine Zoological Laboratory at Naples. While there, he met Hans Driesch (1867-1941) and Curt Herbst. Their influence, especially that of Driesch with whom he later collaborated, no doubt helped him to focus on the field of experimental embryology. When he returned in 1891, Morgan accepted a teaching post at Bryn Mawr College.

In 1904, Morgan left Bryn Mawr College and accepted a professorship at Columbia University. He stayed there until he moved to the California Institute of Technology (Caltech) in 1928. Morgan began his revolutionary genetic investigations of the fruit fly *Drosophila* in 1908 after reading through the recently rediscovered research of the father of genetics, Gregor Mendel (1822-1884). Morgan initially suspected that Mendel's research was inaccurate, but once he performed rigorous experiments that demonstrated genes were indeed discrete chromosomal units of heredity, he changed his mind and vigorously pursued the field of genetics. In 1910 he discovered sex-linked inheritance in *Drosophila*, and postulated that a connection might exist between eye color in fruit flies and the human trait of color blindness. Morgan's research laboratory was often referred to as the "Fly Room," and his group mapped the relative positions of genes on *Drosophila* chromosomes. They published an important book titled *The Mechanisms of Mendelian Heredity* in 1915. In 1928 Morgan and many of his colleagues from Columbia moved to Caltech to continue *Drosophila* research. Morgan received the Nobel Prize for Physiology or Medicine in 1933 for his cumulative research contributions to the field of genetics. He remained at Caltech until his death, performing administrative duties and pursuing investigations of inheritance in *Drosophila*, mammals, birds, and amphibians.

In addition to this work in genetics, Morgan also made contributions to experimental embryology and regeneration. He also left a tremendous legacy of research at Columbia and Caltech. Numerous colleagues who worked with Morgan also made significant contributions in their respective fields, and a number of them even went on to receive Noble Prizes themselves.

Morgan married a former student of Bryn Mawr College, who often assisted him in research, named Lilian Vaughan Sampson. They had one son and three daughters. While still active at Caltech, Professor Morgan passed away in Pasadena in 1945.

JAMES J. HOFFMANN

Hermann Joseph Muller

1890-1967

American Geneticist

Hermann Joseph Muller was a geneticist whose experiments with fruit flies demonstrated that x rays increase genetic mutations by altering gene structure. Upon discovering this connection, he began publicizing the danger of x rays and radiation, promoting radiation-safety measures, and fighting against nuclear bomb tests. His research on x rays was rewarded with the 1946 Nobel Prize for physiology or medicine.

Muller was born in New York City on December 21, 1890. He attended Morris High School in the Bronx, and then went on to Columbia University, where he received his bachelor's in 1910, his master's in 1911, and, after transferring briefly to Cornell University, his doctorate in 1916. During his years at Cornell, Muller was a member of Thomas Hunt Morgan's (1866-1945) research team. In 1910, the team verified that fruit flies had a fourth chromosome.

Before he earned his doctorate from Columbia, Muller became a biology instructor at Rice Institute and continued to hold a position there from 1915-1918. Next, he took a zoology instructor position at Columbia and then went on in 1920 to become an associate professor and then a professor of zoology at the University of Texas, where he remained for more than a decade.

During his changing faculty positions, Muller continued his genetics research. He was particularly interested in genetic mutations and set about trying to obtain a greater number of mutations than would occur naturally in the fruit flies he was studying. After finding in 1919 that heat heightened the mutation rate, he began to consider the reasons behind that increase and how he might enhance the effect. He deduced that x rays might exaggerate heat-related mutations, and over the next several years confirmed through experimentation that an x-ray application could increase the mutation rate by 150-fold. His finding, which he reported during an international scientific meeting, propelled him to scientific stardom. Using his technique, geneticists could now generate and study mutations in their research.

Hermann Joseph Muller. *(Library of Congress. Reproduced with permission.)*

Despite the boon for genetics investigations, Muller saw the harmful side of this discovery as well, and began a crusade to publicize the dangerous nature of x rays, and to promote the safe use of radiation, particularly in health-care settings. When nuclear war loomed in the mid-1950s, he and many other scientists campaigned against the use of such weapons, with Muller concentrating on the potential the weapons posed for genetic mutations.

In addition, controversy swirled around him following the publication of his 1935 book *Out of the Night*, in which he described how the human race could be improved via artificial insemination using the sperm of particularly accomplished men.

On the professional side, Muller spent several years overseas, first as a researcher in Leningrad, and then in Moscow. When Soviet politics began to affect genetics research, Muller left Moscow and briefly served in a medical unit during the Spanish Civil War before moving on to spend two years at the Institute of Animal Genetics at the University of Edinburgh. Afterward, he returned to the United States, where he spent five years at Amherst College and then accepted a position as professor of zoology in 1945 at Indiana University, where he remained for the rest of his career. Over the years, he received many honors, including the Kimber Genetics Award in

1955, the Darwin Medal in 1959, and the Alexander Hamilton Award in 1960, in addition to the Nobel Prize in 1946. Muller died on April 5, 1967.

LESLIE A. MERTZ

Ivan Petrovich Pavlov

1849-1936

Russian Physiologist

Ivan Petrovich Pavlov was born in Ryazan, Russia. After attending a local theological seminary, Pavlov traveled to the University of St. Petersburg, where he undertook the study of chemistry and physiology. Pavlov earned his medical degree from the Imperial Medical Academy in 1879. A year later, Pavlov assumed a professorship at the academy, a post he was to maintain throughout the tumultuous 1917 Russian Revolution.

Pavlov began his postgraduate work with studies on the mechanistic physiology of the circulatory system. Subsequently, Pavlov's work centered on nervous control of the heart. Eventually, Pavlov moved on to studies of the digestive system, and his efforts resulted in the publication of a well-received 1897 work titled *Lectures on the Work of the Digestive Glands*.

Although Pavlov's work on digestive secretions earned a 1904 Nobel Prize for physiology or medicine, Pavlov is best known for his work with conditioned reflexes. Pavlov's classic experiment involved training a hungry dog to salivate at the sounding of a bell, rather than at the food itself. Pavlovian conditioning involves associating one event with another in such a manner that the response to one event is associated, or transferred, to the other event.

Pavlov's work laid the foundation for the scientific analysis of behavior. Pavlov's work was seminal in the formulation of modern psychology and many psychiatric treatment programs. Classical conditioning is often referred to as Pavlovian conditioning, and much of the terminology associated with those concepts derives from Pavlov's famous experiments using quantitative measurement of salivary secretion in dogs as a measure of reflex or response to food stimulus. More specifically, Pavlovian refers to the conditioning of responses to previously neutral stimuli.

According to Pavlov, an unconditioned stimulus (meat) was a stimulus that elicited an unconditioned response (salivation). A conditioned stimulus (Pavlov's bell) was a previously neutral stimulus that did not normally elicit a response until after being paired or otherwise associated with an unconditioned stimulus (meat). Ultimately Pavlov was able to observe a conditioned response (salivation to a bell).

Ivan Pavlov. *(Library of Congress. Reproduced with permission.)*

Important psychological principles were ultimately derived from Pavlov's work, including concepts related to habituation (the loss of potency by a stimulus caused by overuse of that stimulus so that greater or more intense presentation of the stimulus is required to elicit response), generalization (responses to stimuli other than the conditioned stimulus in a manner similar to the conditioned stimulus), discrimination (Pavlov demonstrated that dogs could discriminate between high and low pitch tones as conditioning stimuli), and extinction (the disappearance of a conditioned response when a conditioning stimulus is repeatedly presented without the unconditioned stimulus). Other important Pavlovian-derived psychological models involve spontaneous recovery, higher order conditioning, sensory preconditioning, and counter conditioning.

Pavlov was often at odds with the post-revolutionary communist government. At great peril to himself, he often argued passionately against the triumphant Bolsheviks. Vladimir Lenin (1870-1924) personally refused to allow Pavlov to move his work abroad. In response, Pavlov resisted the Bolshevik hierarchy by refusing to accept the privileges accorded other scientists of his ability and achievement. For instance, Pavlov refused extra food rations not available to his co-workers.

In 1924, Pavlov resigned his professorship in protest of expulsions of students whose families refused to quit the clergy. Even under the iron dictatorial hand of Lenin's successor, Joseph Stalin (1879-1953), Pavlov continued to be a thorn to the communist government. By 1927 Pavlov found himself isolated in the Soviet Academy of Sciences, where he railed against increasing political intrusions into matters of science. At a time when many were put to death for far less, Pavlov bravely defied Stalin and his commissars (Pavlov actually began to deny them actual physical access to his laboratory). By the mid-1930s, however, Pavlov's criticisms of the Soviet government gradually diminished as Pavlov balanced the dangers of Nazi fascism looming on the Western border and the growing agitation with an Imperial Japan to the East with his personal dislike for the communist system.

BRENDA WILMOTH LERNER

Santiago Ramón y Cajal
1852-1934
Spanish Histologist

Santiago Ramón y Cajal. *(Library of Congress. Reproduced with permission.)*

Putting to use an interest in art that he had displayed earlier in life, Santiago Ramón y Cajal developed a method for staining individual nerve cells that improved on that developed by Italian scientist Camillo Golgi (1843-1926). For his advances in histology, a field of anatomy concerned with tissue structures and processes, Ramón y Cajal received the 1906 Nobel Prize in physiology or medicine.

Born on May 1, 1852, in a remote country village in Spain, Ramón y Cajal was the son of Justo Ramón y Casasús, a barber-surgeon, and Antonia Cajal. Despite the father's lack of money or training, he was an individual of extraordinarily strong will who managed to rise above his circumstances, obtain a medical degree, and become a professor of anatomy.

The young Ramón y Cajal was equally strong-willed, and determined to become an artist rather than a doctor, as his father intended him to be. Ultimately the father got his way, enrolling 16-year-old Ramón y Cajal at the University of Zaragoza. Five years later, in 1873, the young man earned an undergraduate degree in medicine. Some time thereafter, he joined the army and served a year as an infantry surgeon in Cuba, where he contracted malaria. This led to his discharge, and while he was still recovering in 1879, he passed the examinations for his doctorate in medicine. During the following year, he married Silveria Garcia, with whom he would have three daughters and three sons.

In the end, Ramón y Cajal found something between his father's ambition for him, and his own plan of becoming an artist: a career in anatomical research, which perhaps called to mind the anatomical studies made by Leonardo da Vinci (1452-1519) and other artists. Working initially with a discarded microscope at the University of Zaragoza, Ramón y Cajal gradually gained recognition and prestige, rising through a series of positions. In 1892 he assumed the chair of histology and pathologic anatomy at the University of Madrid, a post he would retain for three decades.

During this period, Ramón y Cajal conducted his Nobel-winning research, refining Golgi's method of staining tissue samples and thereby isolating the neuron as the nervous system's basic component. He also distinguished the neuron from other types of cells in the body, and made discoveries that supported the neuron theory. At that time, most scientists studying the nervous system were "reticularists" who maintained—incorrectly, as it turned out—that the nervous system was continuous and interconnected.

Ramón y Cajal, by contrast, asserted that the boundless networks of the nervous system ultimately terminate in "buttons" that never actually touch the nervous cells. This proposition, which in fact was correct, led to a fierce debate with Golgi. When the Nobel Committee awarded its 1906 Prize in physiology or medicine, Ramón y Cajal and Golgi were co-recipients, despite the fact that their methods and conclusions diverged widely.

Ramón y Cajal also studied tissues of the ear, eye, and even brain, and was interested in the regeneration of nerve tissues and fibers. Furthermore, he developed his own photographic process for the reproduction of his intricate histological drawings. In 1920 Spain's King Alfonso XIII commissioned the building of the Insituto Cajal, a leading histological research center in Barcelona. Ramón y Cajal worked there from 1922 until his death in Madrid on October 18, 1934.

JUDSON KNIGHT

George Gaylord Simpson
1902-1984
American Paleontologist

A pioneer in the application of statistical methods to paleontology, George Simpson added immensely to scientific knowledge concerning prehistoric life. In the course of a long career that took him to varied destinations around the globe, he analyzed fossil remains, and from these derived information about migratory patterns, evolutionary histories, and other facts of the distant past. He was also a prolific writer who produced several important texts.

The youngest of three children, Simpson was born in Chicago on June 16, 1902, to Joseph and Helen Kinney Simpson. When Simpson was a baby, his father, a lawyer, took a job as a railroad claims adjuster in Denver, Colorado; later he became a land speculator. Simpson credited his father for taking him on many hikes and camping trips, which engendered in him a love of the outdoors that would aid him throughout his career.

A brilliant student, Simpson finished high school several years early, and in 1918 entered the University of Colorado. Later, a professor convinced him to make the transition to Yale, from which he graduated with a bachelor's degree in 1923. Also in 1923, he married Lydia Pedroja, with whom he had four daughters.

George Gaylord Simpson. *(Library of Congress. Reproduced with permission.)*

In 1926 Simpson earned his doctorate at Yale. His dissertation concerned fossils in the Peabody Museum collection dating from the Mesozoic era, a period that marked the first appearance of mammals. Later, he received a fellowship to study Mesozoic mammals at the British Museum in London. By 1927 he was back in the United States, where he took a position as assistant curator at the American Museum of Natural History in New York.

Simpson would remain at the museum for more than three decades, during which time he conducted his most important research. He eventually became curator of fossil mammals and birds, as well as chairman of the department of geology and paleontology, and, from 1945 to 1959, he taught at Columbia University.

In April 1938 Simpson and Lydia divorced, and a month later he remarried to Anne Roe, a psychologist with whom he had been friends since childhood. With Roe's knowledge of statistics and Simpson's expertise in paleontology and zoology, the two collaborated on a number of projects, including the textbook *Quantitative Zoology,* published in 1939. Simpson followed this in 1944 with *Tempo and Mode in Evolution,* his most important work, in which he demonstrated that fossil findings could be quantified. Furthermore, he showed that the fossil record could be shown to align with emerging knowledge at the nexus of population genetics and natural history.

From 1942 to 1944 Simpson served in World War II as an army officer, with tours of duty in North Africa and Italy. After he returned to the United States, he conducted extensive fieldwork in New Mexico and Colorado, searching for mammal fossils from the Eocene (54 million years ago) and Paleocene (65 million years ago) eras. Most prominent among his finds were 15 inch (38.1 cm) high creatures he named Dawn Horses. In 1949 he published *The Meaning of Evolution,* a text that presented the complexities of evolutionary theory in easy-to-understand language.

Misfortune struck Simpson while conducting fieldwork in Brazil in 1956. A tree felled by an assistant clearing a campsite fell on him, leaving him with a concussion and such severe injuries that he could not walk for two years. He was forced to resign his position at the American Museum, but in the quarter-century that followed he traveled and wrote extensively. Among his most significant expeditions was a 1961 trip to Kenya with Louis Leakey (1903-1972), during which Leakey discovered a highly significant skull fragment. The fragment was subsequently linked with *Ramapithecus,* believed to have been a human ancestor from 14 million years ago.

In 1967 Simpson and his wife move to Tucson, Arizona, where he took a position with the University of Arizona. The couple also established the Simroe Foundation, a nonprofit agency intended to disseminate the knowledge they had gathered. He received a number of awards and belonged to several professional associations. In 1982 Simpson retired, and, on October 6, 1984, he died of pneumonia at a hospital in Tucson.

JUDSON KNIGHT

Burrhus Frederic Skinner
1904-1990
American Psychologist

During the mid-twentieth century, B. F. Skinner became the most widely known proponent of behaviorism in psychology, the theory that human behavior is primarily a matter of conditioned responses to stimuli. In fact, Skinner represented the much narrower, and more radical school of thought known as operant behaviorism, or operationism. Among his many writings and ideas, perhaps the one most frequently associated with this highly public intellectual was the "Skinner box," a creation of the 1950s.

B. F. Skinner. *(The Granger Collection Ltd. Reproduced with permission.)*

Skinner was born on March 20, 1904, in Susquehanna, Ohio. He went to Hamilton College, then enrolled at Harvard, from which he received his master's degree in 1930. During the following year, he earned his doctorate in experimental psychology, and in 1936 began teaching at the University of Minnesota. Also in 1936, he married Yvonne Blue, with whom he had two daughters.

Skinner published *Behavior of Organisms* (1938), before his ideas on operationism had reached their mature form, as they would appear in *Verbal Behavior* (1957). In fact, Skinner began writing the latter volume in 1941, thanks to a Guggenheim fellowship that permitted him the freedom to devote considerable attention to it, and was largely finished with the book by 1945. In 1948 he published *Walden Two,* one of his more widely read books, in which he presented his ideas concerning a utopian community.

Skinner became chairman of the psychology department at Indiana University in 1954, and four years later began publishing the *Journal of the Experimental Analysis of Behavior.* During the 1950s, he presented many of his most well-known ideas and constructs, including the air crib, which permitted babies much greater freedom of movement by not keeping them swaddled in heavy blankets.

The Skinner box also made its debut during that decade. This experiment involved placing a rat inside a box, where it eventually learned that by pressing a bar it would receive a pellet of food. Gradually, the "knowledge" that it would receive food for pressing the bar became ingrained in the rat, and it no longer needed a stimulus in order to perform the operation. Thus its actions went from respondent behavior to operant behavior, and this, in Skinner's view, was the model for learning both in rats and in human beings.

In 1957 Skinner published two of his most important works, *Verbal Behavior* and *Schedules of Reinforcement.* The former examined questions of speech and learning, and the latter showed that the frequency of reinforcement (i.e., reward, punishment, or indifference) influences learning and behavior.

Skinner gained many admirers and critics. To critics, he had oversimplified the complexities of human behavior, reducing it to a level that failed to admit even the existence of consciousness. To admirers, he had made the process of learning comprehensible. During the course of his career, Skinner received a wide array of honors, most notably the National Medal of Science in 1968. He died on August 18, 1990, in Cambridge, Massachusetts.

JUDSON KNIGHT

Nettie Maria Stevens

1861-1912

American Zoologist and Geneticist

Nettie Stevens provided evidence that sex is determined through the inheritance of specific chromosomes in germ cells, and she later discovered that chromosomes exist as pairs in cells of the body. Stevens worked with various types of insects, and in her studies of their germ cells she was able to illustrate two different systems of chromosomal inheritance that controlled the sex of offspring during reproduction.

Nettie Stevens worked as a school teacher from 1883-92 and a public librarian from 1893-95, when she returned to school at age 31 to earn a B.A. from Normal School in Westfield, Massachusetts. Stevens went on to complete both her B.A. in 1899 and her M.A. at Stanford University in 1900. After receiving her Ph.D. from Bryn Mawr College in 1903, she began working there as a research fellow in 1903, then became a reader in experimental morphology until 1905 and finally an associate in experimental morphology from 1905-12. She also spent time at the Naples Zoological Station in 1901 and 1905, where she received an award for study, and the University of Würzburg in 1901 and 1908-09. Stevens was a pioneer among women biologists and was recognized during her lifetime as a major contributor to biology through her important research and discoveries.

While working at Bryn Mawr, Stevens initially studied the regenerative processes in planarian flatworms and hydroids. In 1905 she began a working with insect germ cells and discovered sperm cells that had either 6 or 7 chromosomes, while the egg cells in that species always had 7 chromosomes. The sperm with 6 chromosomes always produced male offspring and those sperm with 7 always produced females. Thus, having the extra chromosome gives the offspring 14 total chromosomes and makes them develop as female, and having only 13 chromosomes makes the individual develop as a male. The female-inducing chromosome became known as the X chromosome, and the inheritance pattern was termed the XO mechanism of sex determination.

Before this discovery, most theories focused on external factors such as nutrition and temperature as the sex determinants. This work was of great significance to genetics. In 1906 Stevens and E.B. Wilson (1856-1939) reported that they had independently discovered a second chromosomal inheritance pattern in other insects in which sperm had equal numbers of chromosomes, but the sperm that produced male offspring had a newly discovered Y chromosome. The smaller Y chromosome confers the male sex to offspring, and this second system is called the XY mechanism of sex inheritance. The XY mechanism is now known to be the inheritance pattern characteristic of most of the higher animals.

Stevens continued her cytology research and subsequently discovered that the chromosomes of somatic cells exist as paired structures. Further examination of insect cells led her to discover supernumery chromosomes in the cells of certain insects. Although Stevens was fully credited and honored for her landmark discoveries during her lifetime, later texts have focused credit more onto E.B. Wilson and his colleagues, undeservedly reducing her stature, or worse yet failing to mention Stevens at all. Her short research career includes more that 40 publications and several critical discoveries that linked cytology to heredity and advanced the science of genetics.

KENNETH E. BARBER

James Batcheller Sumner

1887-1955

American Biochemist

On his way to winning the 1946 Nobel Prize for Chemistry, James B. Sumner overcame not only a physical handicap, but prevailing scientific opinion. At that time the received wisdom held that, first of all, it was impossible to isolate an enzyme, and furthermore, that enzymes were not proteins. Sumner proved both assertions wrong.

Born in Canton, Massachusetts, on November 19, 1887, Sumner was the son of Charles and Elizabeth Kelly Sumner. They were a wealthy family whose New England lineage went back all the way to 1636, and they lived on a large estate. There the young Sumner enjoyed hunting and shooting, a hobby that resulted in tragedy when he lost his left forearm and elbow in a shooting accident. This was a particularly great misfortune due to the fact that he was left-handed, but he learned to work with his right.

Sumner originally enrolled at Harvard University to study electrical engineering, but soon discovered that he preferred chemistry, and in 1910 graduated with a degree in that discipline. After working briefly in a family business, followed by a stint as a chemistry teacher, he went on to his doctoral studies at Harvard in 1912. In 1914 he received his Ph.D. with a thesis on "The Importance of the Liver in Urea Formation from Amino Acids," published in the *Journal of Biological Chemistry.*

In 1915 Sumner married Bertha Louise Ricketts, with whom he would have five children. By then he had taken a position as assistant professor in the department of biochemistry at the Ithaca Division of Cornell University Medical College. Sumner would spend his entire career at Cornell.

Due to a heavy teaching load, Sumner lacked the time, as well as the research funds, to conduct lengthy or cost-intensive studies. Therefore, he reasoned that enzyme research would be a good undertaking, since he could fit the project into his time and money constraints. Furthermore, popular doubts about this enterprise meant that the rewards of success in such a long-shot effort would be great. Therefore, in 1917 he began his efforts to isolate an enzyme.

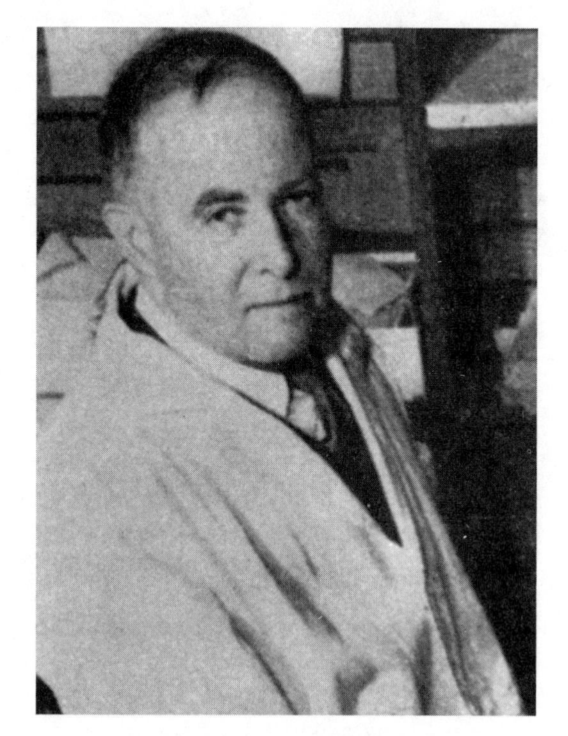

James Batcheller Sumner.*(Library of Congress.)*

It took Sumner nine years to crystallize urease, which catalyzed the breakdown of urea into ammonia and carbon dioxide. Because it contained relatively large quantities of the enzyme, he chose to work with the jack bean. In 1926 he published his findings in the *Journal of Biological Chemistry,* reporting that he had isolated what he believed to be urease—and that he had proven it to be a protein.

Sumner's report earned him plenty of detractors, most notably German chemist Richard Willstätter (1872-1942), winner of the 1915 Nobel Prize for his work on chlorophyll. Willstätter, who himself had tried unsuccessfully to isolate an enzyme, contended that Sumner had merely crystallized the carrier of the enzyme rather than the enzyme itself. Yet, when John Howard Northrop (1891-1987) of the Rockefeller Institute crystallized pepsin in 1930, this bolstered Sumner's argument.

In 1946 Sumner and Northrop—along with Wendell Meredith Stanley (1904-1971), whose work concerned virus proteins—jointly received the Nobel Prize for Chemistry. In the meantime, Sumner remained busy with laboratory work, researching other enzymes and publishing his results in more than 125 papers and books. He also had a number of personal changes during the 1930s and 1940s. In 1930 he and Bertha divorced, and Sumner remarried to a Swedish woman named Agnes Lundquist the following year. Sumner and Agnes were later divorced, and in 1943 he married Mary Morrison Beyer, with whom had two sons, one of whom died in childhood.

In addition to his Nobel, Sumner received numerous honors from Sweden, Poland, and the United States. He belonged to a number of associations, among them the National Academy of Sciences and the American Association for the Advancement of Sciences. Sumner was preparing to retire from Cornell and to organize an enzyme research program at a medical school in Brazil when he was diagnosed with cancer. He died on August 12, 1955, at a hospital in Buffalo, New York.

JUDSON KNIGHT

Sir D'Arcy Wentworth Thompson
1860-1948
Scottish Zoologist

In his studies of natural history, D'Arcy Wentworth Thompson combined on the one hand a penchant for the application of mathematics, and on the other hand an enthusiastic interest in the classics. His most significant work was *On Growth and Form* (1917), in which he departed from the prevailing wisdom of his time by showing mathematical properties in the form of various natural structures.

Born in Edinburgh, Scotland, on May 2, 1860, Thompson was the son of a classics professor also named D'Arcy Wentworth Thompson. His mother, Fanny, died giving birth to him. From an early age, Thompson was influenced by the mother's family, the Gamgees, which included many scientists and doctors. In addition, his father's encouragement ensured a lifelong interest in Greek and Latin, both languages in which Thompson became highly adept.

While a 19-year-old student of medicine at the University of Edinburgh, Thompson published his first papers. In these he discussed hydroid taxonomy, the classification of invertebrate animals, as well as a particular fossil seal from the Pleistocene era. Later, he went on to Trinity College, and during this time published a translation of a book by German biologist Hermann Muller. The volume is notable for its preface by Charles Darwin (1809-1882), one of Darwin's last published writings.

Thompson took a position as professor of biology at University College in Dundee, where in 1884 he established a teaching museum of zoology. University College merged with the University of St. Andrews in 1897, and Thompson accepted its chair of natural history, a position he would hold for the remainder of his life. In 1901 Thompson married Maureen Drury, with whom he had three daughters.

From 1885 onward Thompson began writing on a wide array of zoological subjects, discussing everything from the recently discovered ear of the sunfish to a fossil mammal thought to be a close relative of whales, but which Thompson showed to be more closely related to seals. In 1896 the British government sent Thompson to Alaska to help settle a dispute with the United States over fur-seal fisheries, and in 1902 he was sent as the British representative to the newly established International Council for the Study of the Sea.

Drawing on his interest in the classics, Thompson published *A Glossary of Greek Birds* (1895), which he followed a half-century later with *A Glossary of Greek Fishes* (1947). He also published an annotated translation of Aristotle's *Historia Animalium*. Around the same time, he began performing his most important work, applying a new mathematical approach to morphology, or the biological study of structure and form. This led to the publication of *On Growth and Form,* in which he showed that the structures in a wide array of natural phenomena, such as honeycombs or the flight patterns of moths, followed mathematical principles.

Elected to the Royal Society in 1916, Thompson was knighted in 1937, and received a number of other awards, among them the Darwin Medal in 1946. In that year, he went to India as a member of a Royal Society delegation to the India Science Congress, but while there he contracted pneumonia and never recovered. Thompson died on June 21, 1948.

JUDSON KNIGHT

Edmund Beecher Wilson
1856-1939
American Biologist

Though he is best known for his discovery of the XX and XY sex chromosomes, Edmund Beecher Wilson also deserves credit for helping transform biology into a true scientific discipline. In the late nineteenth century, the field was characterized primarily by passive description of natural phenomena on the one hand, and by fanciful speculation and wild theories on the other. Wilson was a leader in the generation that helped tame such speculation within the rigors of a careful, yet highly active and curious, theoretical framework.

Born on October 19, 1856, Wilson was the second of Isaac and Caroline Clark Wilson's four surviving children. His father became a circuit court judge in Chicago when Wilson was two, and the family moved. However, the boy's childless aunt and uncle, Mr. and Mrs. Charles Patten, were so fond of him that Wilson's mother left him with them. Thus Wilson had the good fortune to grow up with two homes and two sets of parents.

Wilson was educated in a one-room schoolhouse in Geneva, and due to a shortage of teachers, he taught for a year at age 16. A year later, he entered Antioch College in Ohio to study biology, but in the fall of 1874 began taking courses at the University of Chicago. His plan was to go on to Yale, which had been recommended to him by his cousin and close friend Samuel Clarke. In 1875 he enrolled at Yale, where he received his bachelor's degree three years later.

In 1881 Wilson earned his Ph.D. at Johns Hopkins University, and spent the next year traveling in Europe, followed by a year at the Zoological Station in Naples, Italy. During 1883-84, he taught at Williams College in Massachusetts. This was the result of an agreement with Clarke, whereby one would spend a year at Williams while the other worked in Naples, then they would switch.

Wilson worked for another year at the Massachusetts Institute of Technology, where he collaborated with William T. Sedgwick on a textbook entitled *General Biology* (1886). Between 1885 and 1891, Wilson taught at Bryn Mawr College, a women's school in Pennsylvania. There he became involved in research regarding cell differentiation, or the means by which the fertilized egg produces the wide variety of cells contained in a fully formed organism. This question remained the focal point of Wilson's work, from his time at Yale to the end of his life.

Wilson spent a year working in Munich and Naples before becoming an adjunct professor of zoology at Columbia University in 1892. Shortly thereafter, he gave a series of lectures on cell structure that became the basis for another textbook, *The Cell in Development and Inheritance*. As with *General Biology*, the book was destined to become highly influential.

Woods Hole, Massachusetts, had a marine biological laboratory where Wilson spent his summers for a half a century. The town was also home of Anne Maynard Kidder, whom he married in 1904. They later had a daughter, Nancy, who became a professional cellist. Wilson himself was an enthusiastic amateur flautist and cellist.

Soon after his marriage, in 1905, Wilson presented the most important findings of his career: the discovery, which he and Nettie Maria Stevens (1861-1912) of Bryn Mawr had made simultaneously but independently, that X and Y chromosomes in the sperm determined the gender of the offspring.

Wilson continued to study cell structure for the remainder of his career, and in 1925 wrote a third edition of *The Cell in Development and Inheritance*. This edition ran to over 1,200 pages and was essentially an entirely different book from the first edition, thus reflecting the transformation of biology that had taken place during his career. Wilson's health was failing by that point. When he retired from Columbia in 1928, he was 72 years old. He died on March 3, 1939, in New York.

JUDSON KNIGHT

Sewall Wright
1889-1988
American Geneticist

As author of the "shifting-balance" theory—concerned with how certain gene combinations spread throughout a population—Sewall Wright transformed scientific views concerning evolution. Much of his research was done on guinea pigs, not previously viewed as useful for scientific research; Wright was the first to make use of the creature whose name became virtually synonymous with "laboratory animal." Wright also contributed to genetics by encouraging the use of statistical analysis.

The oldest of three children, Wright was born on December 21, 1889, to Philip, a college professor, and Elizabeth Sewall Wright. A child prodigy who wrote a small book called "The Wonders of Nature" at age seven, Wright was an exceptional student. Intending to study languages, he began his higher education at Lombard College, where his father taught, but a teacher interested him in biology instead. In the summer after his graduation from Lombard in 1911, he worked at Columbia University's Cold Spring Harbor laboratory on Long Island, New York. Wright earned his M.S. in zoology at the University of Illinois in 1912, and his Sc.D. in that discipline from Harvard in 1915.

At Harvard's Bussey Institution, a research facility, Wright had begun working with guinea pigs, which he considered useful for research despite their relatively long reproductive cycles. He discovered gene sequences that produced various effects on coat and eye color in the guinea pigs, and wrote his dissertation on that subject.

Wright's studies of guinea pigs continued after he became senior animal husbandman at the U.S. Department of Agriculture (USDA) in Washington, D.C., in 1915. There he analyzed the results of a massive guinea pig inbreeding study that the USDA had begun in 1906, developing a mathematical theory of inbreeding that he published in 1921 as *Correlation and Causation.*

Back at Cold Spring Harbor in 1920, Wright met Louisa Williams, an instructor at Smith College in Massachusetts. The two married in 1921, and later had two sons, Richard and Robert. In 1925 Wright left the USDA to become a professor of zoology at the University of Chicago. The late 1920s saw Wright engaged in debate with geneticist Ronald A. Fisher (1890-1962), who maintained that the success of natural selection was a function of population size—i.e., the larger the sample, the faster and more effective the development of mutations.

By 1931 this debate had sparked Wright to present his own theory concerning evolution. This, the "shifting balance" theory, held that a random gene-frequency drift within subpopulations leads to an increase in the preferred combination of genes, and in turn brings about the dispersal of the preferred gene throughout the larger population. Later in the 1930s, Wright worked with geneticist Theodosius Dobzhansky (1900-1975), who adapted Wright's mathematical methods in his highly influential book *Genetics and the Origin of Species* (1937).

Wright continued to conduct research and publish his findings in the 1940s. In 1954 he left the University of Chicago for the University of Wisconsin, where he devoted most of his efforts to the four-volume *Evolution and the Genetics of Populations* (1968-78). He received the National Medal of Science (1967), the Darwin Medal (1980), nine honorary doctorates, and numerous other awards.

In 1975 Louisa died of pneumonia, and five years later Wright began to lose his vision. He remained physically active, however, and died in 1988 at the age of 99. His wrote and published his last paper in the year of his death. In 1991, three years after his death, *Science* published the results of an experiment by a pair of geneticists that confirmed Wright's shifting-balance theory.

JUDSON KNIGHT

Biographical Mentions

Edgar Douglas Adrian
1889-1977

English physiologist who won the 1932 Nobel Prize for Physiology or Medicine, along with Charles Sherrington, for his studies of nerve impulses. During World War I, Adrian worked primarily with patients suffering from nerve injuries and nervous disorders, and in 1919 began the work for which he would become famous. He was able to amplify nerve impulses by electronic means to achieve a more sensitive measure of these impulses, enabling him to isolate impulses from a single sensory nerve fiber and to study the electrical impulses that cause pain. Adrian also studied the electrical activity of the brain, and investigated abnormalities as revealed by an encephalogram. His work led to progress in the study of epilepsy and the location of other types of brain lesions.

Warder Clyde Allee
1885-1955

American biologist and zoologist whose work spanned an interest in freshwater crustacean populations to analyses of how animals cooperate, rather than compete, for the betterment of their communities. Allee felt that animal cooperation was a natural function and pointed out the implications for human societies. He was professor of zoology at the University of Chicago and ended his career as a biologist at the University of Florida.

Willard Myron Allen
1904-1993

American physician who discovered progesterone. Earning a medical degree from the University of Rochester, Allen specialized in obstetrics and gynecology and taught at the Washington University School of Medicine and the

University of Maryland Hospital. His endocrinological research focused on the biochemistry of sex hormones. With George Washington Corner, he isolated progesterone, a hormone created by the ovaries, and verified that the hormone was crucial to prepare and sustain the uterus during pregnancy. Allen and Oskar Wintersteiner produced crystalline progesterone.

Roy Chapman Andrews
1884-1960

American naturalist and explorer who led a series of expeditions to Mongolia in the 1920s and discovered the first known dinosaur eggs. With the backing of the American Museum of Natural History and several New York businessmen (including J.P. Morgan), Andrews mounted the Central Asiatic Expedition, with the goal of proving that Central Asia—Mongolia in particular—was the cradle of mammalian and thus human evolution. The expedition lasted from 1921 until 1930. They discovered fossil mammals and dinosaurs, including the first known dinosaur eggs, but were unable to find any human fossils. He retired in the 1940s and died at home in California in 1960. Andrews may have been the inspiration for the fictional character of Indiana Jones.

Florence Bailey
1863-1948

American ornithologist who promoted wildlife conservation. At Smith College, Bailey established an early Audubon Society chapter. Shocked by the slaughter of birds to obtain feathers for hats, she promoted public education about the preservation of threatened species and habitats. Unlike her colleagues, Bailey did not kill specimens to examine. She taught amateur birdwatchers and wrote books and articles describing birds she observed during field trips with her scientist husband, Vernon Bailey. Florence Bailey inspired other women to pursue ornithology professionally.

Erwin Baur
1875-1933

German botanist who was one of the earliest advocates of Mendelian genetics in Germany. He organized some major genetics research institutes, including the Kaiser Wilhelm Institute for Plant Breeding and Genetic Research, where he did important experiments on genetic mutations. In the field of agriculture, he produced disease-resistant breeds of plants. He was an avid proponent of eugenics, the attempt to improve human heredity by controlling reproduction.

William Maddock Bayliss
1860-1924

English physiologist who, in collaboration with his colleague Ernest Henry Starling, discovered secretin, a hormone produced by the intestinal mucosa. Bayliss and Starling demonstrated nervous control of the peristaltic wave (the muscle action that causes the movement of food through the intestine). In 1902 they isolated secretin, a substance that stimulates the secretion of pancreatic digestive juice into the intestine. In 1904 Bayliss and Starling coined the term "hormone." Bayliss published the widely used textbook *Principles of General Physiology* in 1915.

George Wells Beadle
1903-1989

American biologist who shared the Nobel Prize in physiology or medicine with Edward Tatum for the research that established the "one gene one enzyme" theory. Beadle and Tatum irradiated the common bread mold, *Neurospora,* and collected mutants that could no longer synthesize the amino acids and vitamins needed for growth. They then identified the steps in the metabolic pathways that had been affected by the mutations, establishing the relationship between mutant genes and defective enzymes.

Hugh Hammond Bennett
1881-1960

American soil conservationist who fought passionately against careless use of the soil. Bennett began his career with the United States Department of Agriculture's Bureau of Soils, coordinating soil surveys and studying connections between soil types and crop production in the United States and in the tropics. As soil erosion worsened and the dust storms of the 1930s dramatized the problem, Bennett became the first chief of the Soil Conservation Service and led its efforts to bring contour tillage and other soil conservation practices to American agriculture.

Johannes Hans Berger
1873-1941

German neurologist who is known as the "Father of Electroencephalography" because he was the first person to record human brain electrical activity in 1924 and later invented the electroencephalograph (EEG) to measure that activity. The measurement of brain wave patterns revolutionized the diagnosis of neurological disorders and helped to establish normal standards. He is also considered to be a founder of psychophysiology, a branch of science that helped lead to our modern understanding of brain dysfunction.

John Desmond Bernal
1901-1971

British physicist who made major contributions to the determination of molecular structures using X-ray crystallographic methods. Bernal is regarded as one of the founders of X-ray crystallography. He was a professor at the University of London (1938-63), where he developed new methods and influenced a generation of crystallographers. He also made important contributions to the study of molecular biology and theories of the origin of life and the structure of the surface layers of the earth.

Davidson Black
1884-1934

Canadian anatomist and paleoanthropologist credited with the discovery of the hominid fossil known as Peking Man in the 1920s. Black received medical training in Canada and then studied with one of England's leading anatomists, Grafton Eliot-Smith, in 1914. Under Eliot-Smith's tutelage Black became interested in human origins and finding the earliest humans. He, along with many other scientists, believed that the first humans had appeared in Central Asia. In 1919 Black was offered a position at the Peking (Beijing) Union Medical School in China. In 1926 several fossil human teeth were discovered outside Peking at Zhoukoudian. Black examined the site briefly, but no more material was discovered. Convinced more fossils waited to be found, he received a grant to continue work there in 1927. By 1929 another tooth and skull fragments were unearthed. When enough material was recovered, Black determined that it was an unknown type of hominid and labeled it *Sinanthropus pekinensis* (Peking Man). It was later determined that Peking Man was a member of the group *Homo erectus.*

Calvin Blackman Bridges
1889-1938

American geneticist who worked in Thomas Hunt Morgan's famous "fly room" at Columbia University. Using mutants of the fruit fly, *Drosophila melanogaster,* Bridges helped establish the chromosome theory of genetics. Bridges exploited the giant salivary gland chromosomes of the fruit fly to demonstrate a relationship between specific bands on the chromosomes and the linear sequence of genes on linkage maps. His recognition of chromosome banding was one of the key events in the development of modern cytogenetics.

Elizabeth Britton
1858-1934

American botanist who promoted plant preservation. Britton grew up on a Cuban sugar plantation, where she became interested in plants. After graduating from New York's Normal College, she accompanied her husband, Nathaniel Lord Britton, on botanical research trips in the West Indies. An expert on mosses, she published articles on the topic. She helped to establish the New York Botanical Garden and the Wild Flower Preservation Society of America to protect indigenous plants. She organized boycotts against using native holly for Christmas decorations.

Paul Ernst Christof Buchner

German zoologist who did important work in the area of symbiosis. Symbiosis is the process by which two species live together in a mutually beneficial arrangement. Buchner helped elucidate the manner in which this situation can occur and described many symbiotic arrangements that are found in nature. In particular, Buchner described the transmission of symbiotic yeasts from one generation of scale insect to the next.

Luther Burbank
1849-1926

American plant breeder known as the "Wizard of Santa Rosa." Burbank had experimental gardens in Santa Rosa and Sebastopol, California. He used artificial selection and hybridization on a massive scale to produce new variations of fruits, vegetables, and flowers, including the Shasta daisy and the spineless cactus. Evolutionary scientists such as Hugo DeVries, David Starr Jordan, and Vernon Kellogg often visited his gardens to collect data that would help them better understand evolution.

Adolf Friedrich Johann Butenandt
1903-1995

German chemist whose isolation of sex hormones earned him a share of the 1939 Nobel Prize for Chemistry. Butenandt received his Ph.D. from the University of Göttingen in 1927, and spent several years teaching at Göttingen and at the Institute of Technology in Danzig (now Gdansk, Poland). In 1929 Butenandt isolated estrone, a hormone responsible for sexual development in females. In 1931 he isolated its male counterpart, androsterone, and in 1934, progesterone, which plays an important role in female reproduction. He later discovered a way to synthesize both progesterone and the male hormone testosterone, which paved the way for the development of steroids and birth control pills.

Walter Bradford Cannon
1871-1945

American physiologist who coined the word "homeostasis" to describe the conditions that maintain the constancy of the interior environment of the body. Cannon explained that homeostasis does not refer to a fixed and unchanging state but to a complex, relatively constant condition. As a physiologist, Cannon investigated the mechanisms that allow the body to maintain this state. Cannon helped to popularize ideas about physiological regulatory mechanisms through his voluminous writings, including his best-known book, *The Wisdom of the Body* (1932).

Britton Chance
1913-

A multifaceted American chemist and physiologist who is renowned for his work in the field of enzymatics. His primary work in this area consisted of demonstrating how reactants and enzymes interact to form a complex and the study of enzyme kinetics. Chance is also well known for his work on how mitochondria function to make cellular energy, as well as for his efforts in developing automated instruments and circuits for radar and navigation during World War II.

Sergei Sergeevich Chetverikov
1880-1959

Russian geneticist who helped initiate the modern evolutionary synthesis and founded population genetics. He connected fluctuating population size to Darwinian evolution in *Waves of Life* (1905). His interest in *Drosophila* (fruit fly) genetics was partially stimulated by T. H. Morgan's research and by Herman J. Muller's 1922 visit to Moscow. In "On Certain Aspects of the Evolutionary Process from the Point of View of Evolutionary Genetics" (1926) he introduced Mendelian genetics into the study of the evolutionary process.

George Washington Corner
1889-1981

American physician, anatomist, endocrinologist, and embryologist whose research into the female endocrine system proved the hormonal basis of menstruation. After receiving his M.D. from Johns Hopkins University in 1913, he taught anatomy at the University California at Berkeley, Johns Hopkins, and the University of Rochester before becoming Director of Embryology at the Carnegie Institution of Washington, D.C., in 1940. He and Willard Myron Allen reported their joint discovery of progesterone in "Physiology of the Corpus Luteum" (1929).

John Merle Coulter
1851-1928

American botanist who established botany academically in the United States. Receiving a University of Indiana doctorate, Coulter became director of the University of Chicago's botany department in 1896. He was a mentor to hundreds of graduate students, who formed the first group of American-educated botanists. Coulter led reform efforts to transform botany from a descriptive process to a system based on laboratory experimentation. He wrote numerous books about morphology and plant evolution. He also created and edited the *Botanical Gazette*.

Henry Hallett Dale
1875-1968

British physiologist who shared a Nobel Prize in physiology or medicine in 1936 with Otto Loewi for his work on the role of endogenous chemicals in the regulation of normal and abnormal functioning of the body. His work contributed to the development of experimental pharmacology and chemotherapeutics. His research into the effect of ergot on the activity of epinephrine was significant in the development of the beta-blocking drugs as well as a better understanding of the physiological effects of epinephrine. Dale identified chemicals, now known as "sympathomimetic," that are involved in the process of the transmission of impulses along the nerves.

Cyril Dean Darlington
1903-1981

English biologist who performed research showing that chromosomes are responsible for carrying hereditary information. Before Darlington's work, the role of chromosomes was largely unknown, with some speculating that they helped provide structural support for the nucleus. Darlington was able to demonstrate their role in heredity and evolution and went on to show how these two roles are interconnected. This achievement was very important in helping give evolution theory more scientific rigor.

Raymond Arthur Dart
1893-1988

Australian paleontologist and anthropologist whose discoveries of early hominid fossils led to developments in evolutionary theory. In 1924, while working as a professor of anatomy at the University of Witwatersrand in Johannesburg,

South Africa, Dart heard of a fossilized baboon skull found at a nearby limestone quarry, and delved into an exhumation. The young primate, which he named *Australopithecus africanus*, exhibited human-like features, and Dart believed it to be an intermediary between apes and humans. His discovery first met with skepticism by the scientific community, but finally gained acceptance in the 1940s, when more australopithecine remains were found in South Africa, which helped to establish the species as a precursor to man.

Charles Davenport
1866-1944

American geneticist who was an early promoter of Mendelian genetics. He advocated using statistical methods (biometry) in biology. He taught zoology at Harvard and the University of Chicago, but then convinced the Carnegie Institution to endow a research institute to promote eugenics. With these funds he established a genetics research institute in 1904 and a few years later the Eugenics Record Office at Cold Spring Harbor, New York, where he collected data on hereditary illnesses, as well as other hereditary physical and mental traits.

Max Delbruck
1906-1981

German-born American physicist who won the Nobel Prize for physiology or medicine in 1969 for his discoveries concerning the mechanism of viral replication and the genetic structure of bacterial viruses. His early attempts to link quantum physics and the structure of the gene inspired Erwin Schrödinger's book *What is Life?* (1944), a landmark in the history of molecular biology. Delbruck was the leader of the "phage school" of molecular biologists, who considered the bacteriophage the ideal experimental system for solving the riddle of the gene.

Félix Hubert D'Hérelle
1873-1949

Canadian-French physician who, independently of Frederic William Twort, discovered the existence of bacterial viruses. He published his observations on "an invisible microbe that is antagonistic to the dysentery bacillus" in 1917. (He obtained the invisible microbe from the stools of patients recovering from bacillary dysentery.) D'Hérelle concluded that the antidysentery microbe was an "obligate bacteriophage," that is, an "eater of bacteria," and he predicted that viruses would be found for other pathogenic bacteria.

Edward Charles Dodds
1899-1973

British chemist who developed DES (diethylstilbestrol) and synthesized some of the sex hormones. DES, given to mothers for 30 years to help them with some of the complications of pregnancy, was later found to cause some very specific birth defects in the children of these women. For that reason, it was banned in most nations, although it has lately been prescribed for some conditions in non-pregnant women. In addition to DES, Dodds's work in synthetic sex hormones was important in the development of the birth control pill.

Edward Adelbert Doisy
1893-1986

American biochemist known for his isolation and synthesis of vitamin K, as well as his research on sex hormones. In 1939 Doisy isolated pure Vitamin K, which expedites blood clotting, from natural sources and figured out a way to produce K-1 synthetically. He also isolated the sex hormones estrone, estriol, and estradiol. In 1943 he was awarded the Nobel Prize for Physiology or Medicine, which he shared with Henrik Dam.

Benjamin Minge Duggar
1872-1956

American botanist who developed antibiotics from fungi. Duggar earned a Cornell University doctorate, studying plants without seeds or flowers, such as mosses and ferns. He conducted laboratory work in Germany regarding spore germination and developed a graduate program at the University of Wisconsin, encouraging interdisciplinary botanical research and unique applications of plant science. He focused on how radiation affects plants. After his forced retirement at age seventy, Duggar worked as a consultant for corporations, assisting in the derivation of penicillin substitutes from plants.

John Sydney Edkins
1863-1940

British physician who first identified the role of gastric enzymes in food digestion. Unfortunately, his ideas fell into disfavor and he left research for teaching. However, his research was vindicated shortly after his death and launched an entire field of inquiry into the hormonal regulation of cells that secrete such enzymes. Sadly, by the time this took place, Edkins had been dead for over two decades.

Paul Ehrlich
1854-1915

German bacteriologist who was awarded a Nobel Prize in physiology or medicine in 1908 for his contributions to chemotherapy and immunology. Ehrlich discovered a synthetic drug that acted as a "magic bullet" against syphilis. (Magic bullets were drugs that destroyed pathogens without damaging the host.) Ehrlich's studies of serum-mediated immunity led to his discovery of antibodies, nature's own magic bullets. He developed the "side-chain" theory of immunity to explain the induction of antibodies and the specificity of antibody-antigen interaction.

Herbert McClean Evans
1882-1971

American anatomist known for his research on female reproduction and the pituitary gland. In 1915 Evans was appointed chairman to the University of California's Department of Anatomy. He described the estrous cycle of the female rat, which led to research on the influence of vitamins on fertility, confirmed the pituitary gland's role in growth, and isolated growth hormone from the anterior pituitary secretion. He is also noted for incorrectly identifying 48 chromosomes in human cells in 1918.

David Fairchild
1869-1954

American botanist and agricultural explorer who brought more than 20,000 plant varieties into the United States. Trained in plant pathology and mycology, Fairchild directed the United States Department of Agriculture's plant introduction efforts, and completed many journeys around the world, particularly in the tropics, in a search for plants of potential economic importance. His important introductions include varieties of soybean, bamboo, mango, date, nectarine, avocado, and the flowering cherry trees of Washington, D.C.

Hans Fischer
1881-1945

German chemist whose investigation of the properties of blood and bile led to the synthesis of bilirubin, a compound produced by the breakdown of hemoglobin from red blood cells. Fischer worked at the Second Medical Clinic in Munich, and the First Berlin Chemical Institute, before becoming a lecturer on internal medicine and later physiology in Munich. His studies of blood pigments, bile, and leaves, specifically haem and haemoglobin, earned him the 1930 Nobel Prize for Chemistry.

E. B. Ford

American biologist and geneticist who helped confirm some important aspects of population genetics. Early researchers in population genetics, Godfrey Hardy and Wilhelm Weinberg had developed relationships describing the distribution of genetic traits in populations. In a classic experiment, Ford conducted the first definitive study (using wing spotting in scarlet tiger moths) to demonstrate the accuracy of the Hardy-Weinberg Law.

Karl von Frisch
1886-1982

Austrian zoologist who won a Nobel Prize in physiology or medicine in 1973 for his studies of sensory discrimination and communication in bees. Frisch became interested in bees while a graduate student in biology. He trained bees to respond to specific stimuli, such as colors, and analyzed the ability of bees to distinguish between scents. Frisch described the "dances" that bees use to communicate information about food sources. Frisch was a founder of the science of ethology (animal behavior).

Archibald Edward Garrod
1857-1936

British physician who introduced the concept of "inborn errors of metabolism" with his report on alcaptonuria in 1902. The urine of babies with alcaptonuria darkens when exposed to air because it contains homogentisic acid. Individuals with alcaptonuria lack the enzyme homogentisic acid oxidase, which is needed for the complete breakdown of tyrosine. Garrod's analysis of the family trees of affected individuals suggested that alcaptonuria was inherited as a Mendelian recessive gene. Similar patterns were found for other human "chemical abnormalities," including albinism, cystinuria, and pentosuria.

Richard Benedict Goldschmidt
1878-1958

A German-born American zoologist and geneticist credited with many years of important research in the field of genetics. He is best known for being a pioneer in population genetics and putting forth ideas that challenged thinking at that time, such as the belief that chromosomes determine inheritance to a greater degree than do individual genes. He worked extensively with the genetics of moths, fruit flies, and worms to show that geographical variations within animal species are caused by genetic factors.

Camillo Golgi
1844-1926

Italian physician and cytologist who shared the 1906 Nobel Prize for physiology or medicine with Santiago Ramón y Cajal for his investigations of the fine structure of the nervous system. Golgi introduced the silver nitrate method for staining nerve tissue and demonstrated the existence of a nerve cell now known as the Golgi cell. He also discovered the entity known as the Golgi tendon organ and the subcellular network of small fibers, vesicles, and granules known as the Golgi complex or Golgi apparatus. The Golgi complex plays an important role in the modification and intracellular transport of proteins and the export of secretory proteins and glycoproteins.

Sir James Gray
1891-1975

British zoologist who analyzed the mechanisms of cell and animal locomotion. His research on cellular movements are contained in *Ciliary Movements* (1928) and his classic text *Experimental Cytology* (1931), which helped unify the field of cytology. Gray applied mechanical engineering principles to solve biological problems of animal locomotion, and his publications *How Animals Move* (1953) and *Animal Locomotion* (1968) led to greater understanding of biological locomotion. Gray was honored with the royal medal of the Royal Society in 1948 and was knighted in 1954.

Fred Griffith
1877-1941

English bacteriologist whose studies of pneumococci, the bacteria that cause pneumonia, suggested that bacteria could undergo some kind of genetic transformation. The virulence of pneumococci depends on the presence of a polysaccharide capsule. Virulent strains with a polysaccharide coat were known as type S; bacteria without the capsule were type R. When mice were inoculated with a mixture of living R and heat-killed S bacteria, they became infected. Griffith was able to isolate living S bacteria from these animals. Later researchers proved that the transforming factor was DNA.

Allvar Gullstrand
1862-1930

Swedish ophthalmologist who received the 1911 Nobel Prize for Physiology or Medicine for his research on the mechanics of the human eye, especially the process of refraction. Gullstrand completed his medical studies in Stockholm, and held various positions as a doctor and lecturer before being appointed first professor of ophthalmology at Uppsala University in 1894. He employed mathematical calculations to determine how the muscles and ligaments of the eye are used to focus on objects near and far. He also invented the slit lamp, used for focal illumination, and improved the ophthalmoscope.

Sir Arthur Harden
1865-1940

English biochemist who shared the 1929 Nobel Prize for Chemistry with Hans von Euler-Chelpin for research on the process of fermentation and its relationship to metabolism. Harden studied at the University of Manchester's Owens College before traveling to Germany to complete his Ph.D. He served for many years as a lecturer and professor, but was best known for his research on the fermentation of sugar by yeast juice, which significantly advanced the study of metabolism. He was knighted by the British crown in 1926.

Godfrey Harold Hardy
1877-1947

English mathematician who left his mark in the field of population genetics in addition to mathematics. He is known for his development of the Hardy-Weinberg law, which dictates the transmission of dominant and recessive genetic traits in large populations. The law bears his name with that of Wilhelm Weinberg because they developed their ideas concurrently, yet independent of each other. Hardy is best known as a prominent figure in the field of mathematical analysis.

Ross Granville Harrison
1870-1959

American zoologist who improved tissue-culture techniques. Receiving a doctorate from Johns Hopkins University, Harrison focused on embryology experiments. He was particularly interested in the development of the nervous system. Harrison observed the growth of frog nerve fibers, proving that cultured tissues could grow independently of the body. He encouraged medical applications of this discovery. Harrison also was interested in the symmetry of developing embryos. He experimented with grafting cells to host embryos to determine how this transfer affected growth.

Charles Holmes Herty
1867-1938

American chemist who was a crusader for the American chemical industry. After several years in academia, Herty served as president of the American Chemical Society on the eve of World War I. As a consequence, Herty was involved in American research on chemical warfare, and promoted the American chemical industry in an era of its ascendancy. Herty later lobbied for the establishment of the forerunner of the National Institutes of Health, and he helped develop the technology that allowed the production of pulp and paper from southern pine trees.

Richard Hesse
1868-1944

German zoologist whose textbook *Ecological Animal Geography* (1924) set the standard for biogeography study, covering the distribution of animal life in terrestrial, freshwater, and marine ecosystems. Hesse recognized that the distribution patterns of animal life are correlated to the complex relationships that exist between all species of an ecosystem and all the nonliving factors of each ecosystem. Hesse emphasized the need for careful observation and experimentation to foster the young science of ecological animal geography.

Sir Archibald Vivian Hill
1886-1977

English physiologist who received the 1922 Nobel Prize for Physiology or Medicine, along with German biochemist Otto Fritz Meyerhof, for research on the physiology of muscles. Hill studied mathematics at Trinity College, Cambridge, but later shifted his interest to the study of physiology. In 1909 he began investigating the physiology of muscles, specifically the changes in heat associated with muscle function.

Johannes Friedrich Karl Holtfreter
1901-1992

A German-born American biologist who was considered the foremost embryologist in the world for over 40 years. He is best remembered for his theories regarding the formation of specialized cells, tissues, and organs from the undifferentiated early embryo. Holtfreter demonstrated that embryo development was due to directed cell movement. His research dealt primarily with the development and differentiation of amphibian embryos, and he contributed many lines of research that are still active today.

Frederick Gowland Hopkins
1861-1947

English biochemist who shared the 1929 Nobel Prize for physiology or medicine with Christiaan Eijkman for discovering the "accessory factors" that are needed for growth and health. These essential dietary factors are now known as vitamins. Hopkins proved that, contrary to prevailing opinion, animals could not live on a diet that only contained pure protein, carbohydrates, and fats, even if mineral salts were added. Hopkins also isolated the essential amino acid known as tryptophan and the tripeptide called glutathione.

Julian Huxley
1887-1975

British biologist who specialized in ornithology, but is best known for his contribution to the neo-Darwinian synthesis. He was the grandson of the famous Darwinist T.H. Huxley, and became a professor at King's College in London. In 1942 he published *Evolution, the Modern Synthesis*, which argued that Darwinian natural selection and Mendelian genetics could be combined to explain biological evolution. Huxley was involved in founding the United Nations Educational, Scientific, and Cultural Organization, becoming its first director in 1946. He also wrote many works trying to integrate his vision of atheistic Darwinian evolution into a total world view.

Herbert Spencer Jennings
1868-1947

American botanist who was trained at Illinois Normal School and the University of Michigan, then spent a year studying protozoans with Max Verworn at the zoological station in Naples. In the early twentieth century, Jennings began studying inheritance and evolution in protozoans and introduced new experimental methods for laboratory study. By 1920 he had left the laboratory to popularize genetics and harmonize the relationship between biology, religion, and the humanities.

Donald F. Jones
1890-1963

American agronomist who developed commercial hybrid corn. Earning a Harvard University doctorate, Jones continued his plant genetics research at the Connecticut Agricultural Experiment Station. He selected corn plants with desirable dominant traits to create hardier hybrids that produced higher yields. Jones genetically sterilized some strains to eliminate the need for detasseling. He identified genes to restore fertility so plants could grow viable seed. Jones's patent for this process was the first patent to include genetics. His research enabled "global agriculture" to expand.

David Starr Jordan
1851-1931

American botanist and administrator. After attending Cornell University, Jordan secured a permanent position at Indiana University, where he quickly rose to president. Jordan instituted major reforms that substantially raised the quality of the institution's faculty and students. In 1892 he was chosen as the first president of Stanford University, a position that earned him a national reputation as a progressive reformer and administrator. Trained as a botanist, Jordan made his scientific reputation as an ichthyologist by studying and cataloging fish in the upper Midwest and the Pacific Coast. He was also a strong advocate for Darwinian evolution, and he argued for the significance of isolation as a factor in evolution as well as the importance of natural history in studying evolution.

Paul Kammerer
1880-1926

Austrian biologist who claimed that he had demonstrated the inheritance of acquired characteristics. Kammerer reported that he had induced the inheritance of pigmented thumb pads in the male midwife toad, which normally lacks such pads. His preserved specimens were examined in 1926 and the pads were found to have been artificially colored with India ink. Kammerer denied having altered his specimens, but he committed suicide shortly after the story became public. The controversy was examined by Arthur Koestler in *The Case of the Midwife Toad* (1971).

David Keilin
1887-1963

Russian-English biochemist and parasitologist best known for his work on the life-cycle of the parasitic and free-living Diptera. Born of Polish parents, Keilin studied in Moscow, Warsaw, Liege, and Paris. Through his biochemical research he discovered cytochrome, an intracellular respiratory pigment, and hemocuperein, a copper protein in red corpuscles. He researched and published on a wide range of topics—from the suspended animation of living things to the history of biology—and aided the careers of many younger scientists.

Vernon Lyman Kellogg
1867-1937

American entomologist educated at the University of Kansas, Cornell University, and the University of Leipzig in entomology and evolution. In 1894 David Starr Jordan hired Kellogg as professor of entomology at the newly established Stanford University. There he experimented on silkworms to establish the causes of variation and test Mendelian inheritance. In 1915 Kellogg resigned from Stanford to join his former student, Herbert Hoover, on the Commission for Relief in Belgium and then the United States Food Administration. After the war he ran the National Research Council until poor health forced his retirement in 1931.

Edwin Calvin Kendall
1886-1972

American biochemist who was awarded the Nobel Prize for Physiology or Medicine in 1950 with Philip Hench and Tadeus Reichstein for their work regarding the structure and function of adrenal cortex hormones. He is best remembered for his isolation of the steroid hormone, cortisone, from the adrenal gland, which was successfully used to treat rheumatoid arthritis. Considered to be one of the foremost endocrinologists of his time, Kendall made several other important contributions to his field.

Shack August Krogh
1874-1949

Danish physiologist who was awarded the Nobel Prize in physiology or medicine in 1920 for his research into the regulation of circulation in the capillaries. Although the prevailing belief was that blood flowed continuously through all the capillaries, Krogh thought that it would be more efficient for some capillaries to close while a muscle or organ was at rest. His work on circulation, blood volume, and oxygen supply led to a better understanding of total body metabolism.

Peter Kropotkin
1842-1921

Russian geographer best known for his anarchist political theory, but also as a student of biology. Because of his political views, he spent much of his adult life in exile, mostly in England. He was an advocate of biological evolution, but opposed Darwin's theory of natural selection. He believed that Darwin's theory of natural selection ignored many cooperative aspects of organic life, so he wrote *Mutual Aid* in 1902 to advance his own theory of evolution by cooperation rather than competition.

Joshua Lederberg
1925-

American geneticist who shared the Nobel Prize for Physiology or Medicine in 1958 with George Beadle and Edward Tatum for his work with bacteria and genetic recombination. He pioneered the use of bacteria in genetics research by discovering the process of transduction, where genetic material could be transferred from one bacterium to another. While his bacterial research set the stage for genetic engineering, he is also remembered for many other contributions to science.

(Rand) Aldo Leopold
1887-1948

American wildlife ecologist and educator who was, with brief interruptions, associated with the United States Forest Service in various capacities from 1909, the year he completed his Master of Forestry degree at Yale, until 1928. Leopold directed a game survey for the Sporting Arms and Ammunition Manufacturer's Institute between 1928 and 1932, publishing a *Report on a Game Survey of the North-Central States* (1931). This was a pioneering regional study of the status of game animals and what was being done to foster their conservation. In 1933 he organized the first department of game (later wildlife) management in the United States at the University of Wisconsin and for fifteen years held the first professorship in the subject there. His textbook *Game Management* (1933) was the first to examine population dynamics within the context of ecological relationships. His posthumously published *Sand County Almanac and Sketches Here and There* (1949) expounded his "Land Ethic" and broadly defined many issues that have since characterized the modern environmental movement.

Phoebus Aaron Theodor Levene
1869-1940

A Russian-born American chemist who is best known for his pioneering studies in the area of nucleic acids. While he worked with a variety of compounds, his most important findings were the isolation of the sugars, d-ribose from the ribonucleic acid (RNA) molecule and 2-deoxyribose from the deoxyribonucleic acid (DNA) molecule. Also noteworthy were his studies of DNA structure that helped lead to the deduction of its complete configuration by Watson and Crick.

Lewis, Margaret
1881-1970

American cytologist who improved tissue-culture methodology. Lewis was Thomas H. Morgan's research assistant before she accepted a position at the Carnegie Institution. Scientists adopted Lewis's tissue-culture techniques. With her husband, Warren H. Lewis, she developed what was called the Lewis-Locke Solution, a clear, nonsalty fluid. Lewis observed cells' physiological activity and demonstrated that monocytes and macrophages represent varying physiological stages instead of separate cell types. She also experimented with dye for use in chemotherapy.

Choh Hal Li

Chinese-American chemist and endocrinologist who was the first to synthesize a human hormone. Li was also the first to show that hormones, in this case, ACTH (adrenocorticotrophic hormone), was comprised of a total of 39 amino acids in a specific order, or sequence. In later years, Li was the first to isolate human growth hormone and, in 1970, to synthesize it. This was one of the first hormones used in the treatment of dwarfism.

Jacques Loeb
1859-1924

German-born American biologist who was the leading spokesperson for the mechanistic philosophy of biology in the early twentieth century. He was primarily interested in replacing explanations of behavior that were based on instincts, which he considered vitalistic, with explanations based on chemical tropisms. Extremely well known to the public as a materialist, mechanist, and socialist, Loeb became a popular figure for other American dissenters, such as H. L. Mencken and Thorstein Veblen.

Otto Loewi
1873-1961

German-born American physiologist and pharmacologist who received the Noble Prize for Physiology or Medicine in 1936 with Henry Dale for their work regarding the chemical nature of nerve signals. Loewi primarily studied the physiological action of chemicals and is best known for his extensive work regarding the autonomic nervous system. One important discovery was that the neurotransmitter acetylcholine was released directly on to the heart by nerve endings where it exerted a physiological effect.

Alfred James Lotka
1880-1949

British demographer and statistician who developed statistical methods for analyzing the dynamics of biological populations. Lotka formulated a growth law for two competing populations using differential equations that produced accurate modeling. His major book, *Elements of Physical Biology,* treated all biology from a mathematical-physical viewpoint. He was an early advocate of ecological concerns. He left science in

1924 and joined the statistical branch of the Metropolitan Life Insurance Company in New York, where he did research on life expectancy.

Keith Lucas
1879-1916

British physiologist who established the fundamental "all or nothing" law for skeletal muscle. His university career was interrupted by the death of a close friend in the Boer War. After a long break, Lucas pursued physiological research using a homemade camera setup for recording muscle contraction. During World War I, he used his camera techniques to analyze airplane maneuvers, designed an improved bombsite, and constructed a new magnetic compass. He was tragically killed in a mid-air collision.

Colin Munro Macleod
1909-1972

Canadian bacteriologist whose research with Oswald Avery determined that DNA was the molecular agent responsible for transforming harmless avirulent bacteria into disease-causing virulent bacteria. Macleod and his colleagues isolated DNA from heat killed, virulent *Streptococcus pneumoniae* and exposed it to living, avirulent cells. The DNA from the dead, virulent cells was able to transform the live, harmless strains into encapsulated virulent ones. The ability to absorb and incorporate DNA from dead bacteria occurs in several important bacterial genera and involves traits such as encapsulation, pathogenicity, and drug resistance.

Rudolf Magnus
1873-1927

German physiologist and pharmacologist who studied the regulation of balance in the inner ear and the many automatic reflex actions by which a human body maintains posture. While a medical student, he published a method for measuring blood pressure in an exposed artery. Specializing in pharmacology, Magnus studied the role of arsenic in the gut, and water-balance in tissues. In 1904 he devised the standard method for studying the responses of isolated muscle. At the time of his sudden death at age 53, he was under consideration for the Nobel Prize.

Clarence Erwin McClung
1870-1946

American biologist who studied the role of chromosomes in heredity beginning in the 1890s and provided the basis for the study of heredity during the twentieth century. McClung examined the role of chromosomes in sex determination and, based on his belief in the significance of development, argued that the environment played a significant role in determining an individual's sex. He was also active as an administrator and held positions in several prominent scientific organizations, including the National Research Council and the Marine Biological Laboratories at Woods Hole.

Elmer Verner McCollum
1879-1967

American biochemist who pioneered the study of vitamins. McCollum's research began at the Wisconsin Agricultural Experiment Station, where he developed innovative methods of studying animal nutrition with colonies of white rats. His findings exposed shortcomings in the existing understanding of the nutritive value of proteins and fats. This led him to conclude that trace organic substances, now known as vitamins, as well as a number of trace minerals, are essential in animal nutrition. McCollum was also an outspoken advocate for greater public understanding of nutrition issues.

Lemuel Clyde McGee
1902-

American biochemist who performed important research into the chemistry of various digestive enzymes. McGee's research into gastrointestinal function and the chemistry of jejunal juices helped lay the foundation for all subsequent research into the functioning of the digestive tract. In other work, McGee examined the role of toxins in the work place, helping establish the fields of toxicology and industrial medicine.

Clinton Hart Merriam
1855-1942

American zoologist and a founder of the National Geographic Society. Merriam was noted for bringing major attention to the study of North American land vertebrates and to the biogeographical settings in which they lived. Born in upstate New York, he was trained as a physician at Columbia University. Beginning in 1885 he directed the newly formed Division of Entomology (later the Bureau of Biological Survey) in the federal Department of Agriculture. He left in 1910 to study tribes of disappearing native Americans in California. His thorough field methods, and the many regional biological surveys and major revisions of mammalian genera published under the aegis of the Biological Survey, were very influential well into the twentieth century. His conclusion that temperature was the major determinant in controlling the geographical distribution of animals later underwent considerable revision at other hands but is still considered useful in the western United States.

Elie Metchnikoff
1845-1916

Russian embryologist and bacteriologist who studied phagocytosis and successfully promoted the theory of cellular immunity. He applied the term "phagocytes" to the cells that removed unnecessary tissues during larval invertebrate metamorphosis. Metchnikoff recognized that through phagocytosis of bacterial invaders, white blood cells are capable of defending the body and provide immunity against disease. His years of research led to a shared Nobel Prize in medicine with Paul Ehrlich in 1908. Metchnikoff's *Immunity in Infectious Disease* (1905) is a classic text of comparative immunology and the theory of phagocytosis.

Gerrit Smith Miller, Jr.
1869-1956

American mammalogist who for more than half a century was associated with the United States National Museum, serving as Curator of Mammals there from 1908 until 1940. He was an authority on small mammals, particularly bats and shrews, and is best known for several important classified lists of European and American mammals. His *Catalog of the Land Mammals of Western Europe* appeared in 1912. Several earlier keys and historical summaries of American mammals culminated in *List of North American Land Mammals* (first published in 1912, with revisions in 1923 and 1954, the latter with Remington Kellogg). These were valued references on which zoologists relied for many decades. In later years his interests widened to include larger mammals, along with man. He also authored many papers on botanical, anthropological, and musical subjects.

Joseph Needham
1900-1995

English biochemist who documented the history of science. Educated at Cambridge University, Needham specialized in embryology. He conducted research at the Cambridge Dunn Institute of Biochemistry. Needham wrote several books, including *A History of Embryology* (1934). Traveling through China on a scientific mission, Needham became fascinated with Chinese medicine and technology. He compiled the comprehensive, seven-volume *Science and Civilisation in China* (1954). Needham also served as visiting professor at several universities and as director of Cambridge's Needham Research Institute.

Arnold Joseph Nicholson
1912-

American biologist and zoologist whose ecological studies of the wood mouse and the fruit bat helped launch the scientific study of ecosystems. Nicholson and other researchers helped establish the scientific basis for ecological studies by elevating them beyond a simple description of observed events. In addition to his early ecological studies, Nicholson also worked at restoring populations of game animals that were depleted by overhunting.

Alexandr Oparin
1894-1980

Russian biochemist who was appointed head of the plant biochemistry department at Moscow State University and who achieved worldwide recognition for his studies which advanced the earlier work of Charles Darwin. In addition to his purely academic research, he contributed much to the Soviet economy with biochemical solutions to problems and the subsequent improvement of tea, sugar, tobacco, wine, and sugar production methods.

Henry Fairfield Osborn
1859-1935

American naturalist, evolutionary biologist, and long-time head of the American Museum of Natural History. Osborn studied natural history and biology with some of the premier scientists of the nineteenth century, including Arnold Guyot, T.H. Huxley, and Edward Drinker Cope. In 1891 he was offered simultaneous positions teaching biology at Columbia University and running the vertebrate paleontology department at the American Museum of Natural History in New York. He helped transform the museum from a Victorian curiosity cabinet to one of the world's leading centers of scientific research. He developed a theory of human evolution that stressed an interaction of what he called an internal guiding principle, or "race plasm," and the individual's conscious effort to overcome hardship in the environment. He saw different groups of people having different levels of this race plasm, making some better equipped to survive and prosper than others. As a result of this thinking Osborn became involved with the racist Eugenics movement, and immigration restriction legislation.

Thomas Park
1908-

American zoologist whose experimental work was important in establishing the field of population biology. Park's work actually spanned fields, and his work in population biology carried over into studies of ecosystems and the ecology of animal groups. Park's work in these areas helped define these fields for both contemporary and present-day scientists studying the interactions of populations of animals under changing conditions.

Raymond Pearl
1879-1940

American statistician who took his A.B. at Dartmouth College, his Ph.D. at the University of Michigan, then spent two years in London studying with statistician Karl Pearson. He worked at the University of Pennsylvania and then the Maine Agricultural Experiment Station, where he used statistical methods to study the heredity and reproduction of poultry and cattle. Pearl did his most significant work at Johns Hopkins University, which included pioneer work in human population statistics and life expectancy.

Wilder Graves Penfield
1891-1976

American-born Canadian neurosurgeon who is best known for his contributions to the area of neurology. He founded the world-renowned Montreal Neurological Institute to research and further the knowledge of the human brain. Penfield was the first researcher to systematically map the brain by observing conscious patients' responses to electrical stimulation in specific areas of the brain. Through his research, he developed surgical procedures for treating debilitating neurological diseases such as epilepsy.

Ludwig Plate
1862-1937

German zoologist who avidly promoted Darwin's theory of natural selection. He succeeded Ernst Haeckel as professor of zoology and curator of the Phyletische Museum in Jena in 1909, after teaching and directing a biological museum in Berlin. He conducted experiments in the newly emerging field of Mendelian genetics. As a founding member of Society for Race Hygiene and co-editor of its journal, he was an influential eugenics advocate.

Reginald Crundall Punnett
1875-1967

British geneticist who extended the understanding of Mendelian genetics and used sex-linked plumage color genes to bio-engineer the first "autosexing" chicks. This application of genetic recombination saved critical resources for the British government during World War I, because female chicks could be immediately identified. Punnett identified examples of autosomal linkage and confirmed classical Mendelian principles through his research and instruction at Cambridge University, where he was honored with the first Arthur Balfour Chair of Genetics, a Royal Society Fellowship, and a Darwin Award.

Emil G. Racovita
1868-1947

A Romanian biologist who is considered to be the father of biospeleology (cave biology) for his pioneering work with the subterranean domain. Racovita was an eminent biologist who excelled in many areas, but it was his dedication and systematic approach to biospeleology that is considered his most noteworthy accomplishment. Under his direction, over 50,000 previously unidentified cave dwelling species were catalogued. Also noteworthy, he taught the first college general biology course in Romania (1920).

Bernard Rensch

German biologist whose work on aspects of animal intelligence was of great importance. Rensch was able to show that brain size confers some degree of evolutionary advantage if all other factors are equal. He then suggested this finding to be a reason why species tend to increase in size over time, although not all were convinced. Rensch's work with elephants, showing them to have phenomenal memories, was of particular public interest.

Francis Peyton Rous
1879-1970

American pathologist who discovered that a virus could induce tumors and be mechanically transmitted among chickens. Initially derided, then belatedly honored with the Nobel Price in 1966, Rous successfully extracted a submicroscopic agent from tumors, and induced the sarcoma by injecting it into chickens. Rous also postulated that carcinogenesis consists of initiation and promotion and could be induced by chemicals, radiation, and viruses. He also developed methods for cellular and viral culturing, as well as blood preservation.

Richard Semon
1859-1918

German zoologist who developed a Lamarckian view of heredity that was influential in the early twentieth century, but was later discredited. He was heavily influenced by Ernst Haeckel, under whom he began his biological studies in Jena in 1879. Semon never received a full professorship, so after various biological expeditions and teaching positions, he became a private scholar. His most important book was *Mneme* (1904), a book arguing that an organism's heredity can be influenced by acquired characteristics.

Ernest (Evan) Thompson Seton
1860-1946

English-born American artist and naturalist who was a notable author of animal and Indian stories, mainly for children and young adults, from the 1880s until his death. His forty books—one of which, *Wild Animals I Have Known*, has been continuously in print for over a century—included two multivolume studies of North American mammals. All his publications were based on his own field observations and were accompanied by his unique drawings and paintings. He pioneered in developing methods of identifying animals in the field, later elaborated on by other naturalists. In the early 1900s he organized the Woodcraft Indians, on which the Boy Scouts of America was largely based. He was Chief Scout Executive of the BSA from 1910 to 1915. In his later years he was a student of American Indian culture, establishing the Seton Institute in New Mexico.

Erich Tschermak von Seysenegg
1871-1962

Austrian botanist who was one of three scientists to re-introduce Gregor Mendel's laws of heredity. Erich Tschermak von Seysenegg completed his doctorate at the University of Halle, then spent several years working at various seed-breeding institutions. He joined Vienna's Academy of Agriculture in 1901, where he would remain as a professor for the majority of his career. In 1898, Tschermak was conducting breeding experiments on the garden pea at the Botanical Garden of Ghent, when he discovered that his findings were similar to those of geneticist Gregor Mendel. Mendel's work with the hybridization of pea plants had lain virtually ignored for more than 30 years. Upon reviewing Mendel's papers, Tschermak realized that his findings indeed duplicated those of his predecessor. Around the same time, botanists Hugo de Vries and Carl Erich Correns also reported on their individual re-discoveries of Mendel's work. Tschermak applied Mendel's theories to the development of new plant hybrids, such as wheat-rye and a disease-resistant oat, as well as new plants, including Hanna-Kargyn barley.

George Harrison Shull
1874-1954

American geneticist who showed an early interest in botany and agriculture. Home schooling and scant formal education qualified him to teach at Ohio public schools and Antioch College, where he also received a B.S. in 1901. Three years later he earned a Ph.D. from the University of Chicago. Shull was interested in statistical studies of variation, which brought him to the attention of Charles Davenport and earned him a position at the Station for Experimental Evolution. After spending several years studying the products and procedures of Luther Burbank, he spent the rest of his life teaching and researching genetics at Princeton University. His primary contribution was his work with corn and his development of hybrid corn, one of the most significant agricultural advances of the twentieth century.

Hans Spemann
1869-1941

German zoologist and embryologist who received the 1935 Nobel Prize for Physiology or Medicine for his discovery of embryonic induction. Fascinated with newts, his early experiments were on the evolution of the eye lens of amphibians. This delicate work led him to develop small, precise surgical instruments and techniques—the foundation of microsurgery. His experiments with newt's eggs showed that at early stages cells were not specialized, and he discovered the first known example of a causal mechanism controlling the development of an embryo.

Wendell Meredith Stanley
1904-1971

American biochemist who shared the Nobel Prize for Chemistry in 1946 with John Northrup and James Sumner for their work demonstrating the molecular structure of viruses. Specifically, Stanley crystallized the tobacco mosaic virus, a disease-causing agent in plants, to demonstrate its basic structure. Other scientists used this information to ascertain its precise molecular structure. A recognized authority on viruses, he made many important contributions to the field, including the development of an influenza vaccination.

Ernest Henry Starling
1866-1927

English physiologist who was considered by many to be the foremost physiologist of his time. Starling made significant contributions to numerous areas involving body function. Most importantly, he formulated Starling's hypothesis of the capillaries, which modeled the forces dictating fluid movement through capillary walls. He devised his "law of the heart" which stated that the force of cardiac contraction was directly related to amount of blood within the chamber, and coined the term hormone.

George Ledyard Stebbins
1906-2000

American plant geneticist who created an artificial grass species through induced polyploidy, the presence of two or more sets of chromosomes in an organism. Stebbins studied angiosperm evolution under the context of natural selection, reproductive isolation, and the modern synthetic theory of evolution. This theory focuses on the basic properties of gene action, mutation, and recombination, as well as changes in chromosome structure and number, all occurring over geological time periods. Stebbins proposed that morphological changes in specific plant structures result from mutations in genes that control mitosis.

Alfred Henry Sturtevant
1891-1970

American geneticist who is best remembered for his research involving the fruit fly *Drosophila*. Among his contributions, he determined that genes are arranged in linear order on chromosomes, demonstrated the role of chromosomal crossing over (genes switching chromosomes) in mutations, and the role genetics plays in sexual selection. He was also one of the first scientists to be concerned with the possible problems radiation from nuclear testing would have on human genetics and evolution.

Walter S. Sutton
1877-1916

American molecular biologist who laid the foundation for the chromosomal theory of heredity—a science that has altered the lives of humans, animals, and plants. He worked under Clarence McClung, who discovered the sex-determining chromosome, while obtaining his master's degree at the University of Kansas. His most important work, *The Chromosomes in Heredity*, was published in 1903.

Theodor Svedberg
1884-1971

Swedish physical chemist whose researches on colloids, biological molecules, and radiochemistry significantly affected both theoretical and applied chemistry. In order to investigate the behavior of very small particles, Svedberg invented the ultracentrifuge, which he also used to study proteins and carbohydrates. His researches on Brownian motion helped to establish the existence of molecules. Svedberg was one of the first to recognize the existence of isotopes. He received the 1926 Nobel Prize for his work on colloidal systems.

Jokichi Takamine
1854-1922

Japanese-American chemist who isolated the chemical adrenaline from the suprarenal gland. Jokichi Takamine graduated from the College of Science and Engineering of the Imperial University of Tokyo in 1879. The university then sent him to Glasgow, Scotland, to complete his postgraduate study. Upon his return to Japan, Takamine took a position with the Ministry of Agriculture and Commerce and eventually rose to the head of its chemistry division. In 1887, he left the public sector to establish his own factory, the Tokyo Artificial Fertilizer Company, which manufactured superphosphate fertilizers. In his laboratory, Takamine developed a digestive agent, known as diastase, which caught the interest of American brewing companies, who invited him to develop his enzyme in the United States. Takamine found a new home in America, where he patented his new enzyme and began work on isolating adrenaline (now called epinephrine), a hormone used as a cardiovascular stimulant.

Edward Lawrie Tatum
1909-1975

American geneticist who shared the Nobel Prize for Physiology or Medicine in 1958 with George Beadle and Joshua Lederberg for their work showing genetic transmission in bread molds. Using normal and mutated chromosomes, they demonstrated that specific genes are responsible for particular enzymes. This "one-gene-one-enzyme hypothesis" helped confirm his suspicion that all biochemical processes are genetically regulated. Because of this and other contributions to the field, Tatum is considered a founder of molecular genetics.

Nikolaas Tinbergen
1907-1988

Zoologist from the Netherlands who conducted a wide range of animal-behavior studies, and is credited as a founder of the field of ethology, the systematic study of the function and evolution of behavior. Tinbergen is widely recognized for his studies of behavioral patterns and the individual environmental triggers, or "releasers," that cause specific actions in organisms. He, Konrad Lorenz, and Karl von Frisch jointly accepted the Nobel Prize in 1973 for their work in animal behavior.

Mikhail Semyonovich Tswett (also Tsvet)
1872-1919

Russian botanist who developed the chromatography method for extracting plant pigments from leaves. Tswett received his doctorate in Geneva, Switzerland, in 1896, as well as a degree from the University of Kazan, Russia, in 1901. He was hired as a laboratory assistant at the University of Warsaw, and in 1908 began teaching botany and microbiology at the Warsaw Technical University. In 1917, he became director of the botanical garden at Yuryev (later Tartu) University in Estonia. Tswett is known as "the father of chromatography," the technique he developed for separating plant pigments from leaves by passing a solution of ether and alcohol through a chalk column. He published two papers describing his method in 1906. Tswett discovered several new forms of the green plant pigment chlorophyll, as well as carotenoids, the pigments that give many fruits and flowers their color.

Frederick William Twort
1877-1950

English microbiologist and bacteriologist who discovered a virus that attacked bacteria, later known as bacteriophages. While the superintendent of a veterinary dispensary in London, Twort isolated himself and engaged in solitary research for 35 years, interrupted only by World War I. Little recognized in his life, he discovered bacteriophages while studying cultures of *Staphylococcus aureus* (a bacterium responsible for boils), and in a 1907 paper established the idea of mutation and adaptation of bacteria. He also did pioneering work on the nutritional needs of bacteria. In 1944 his laboratory was destroyed by wartime bombing.

C.H. Waddington
1905-1975

British embryologist and geneticist who made great strides in the study of embryonic development. Waddington was born in Evesham, Worcestershire, England, and studied geology at the University of Cambridge. He later turned his attention to the study of biology, teaching zoology at Strangeways Research Laboratory, Cambridge, and animal genetics at the University of Edinburgh. In the 1930s, he studied the embryonic development of birds and mammals, and investigated the role played by genes in regulating tissue and organ development. After World War II, Waddington established the Unit of Animal Genetics for the British Agricultural Research Council, where he studied livestock breeding. His many published works include *Principles of Embryology* (1956) and *The Ethical Animal* (1960).

August Weismann
1834-1914

German biologist who was an early adherent of Darwin's theory of evolution and became famous for his studies on heredity. He denied that organisms could inherit acquired characteristics, touching off an important debate between his followers and the opposing neo-Lamarckians, who believed organisms could inherit acquired characteristics. Weismann's views gradually won the day in the early twentieth century, especially through the advent of Mendelian genetics.

Walter Frank Raphael Weldon
1860-1906

English biologist and statistician recognized as one of the founders of biometrics. Welson used statistical methods to test Darwinian ideas, and published a number of classic papers on the statistical variation of various characteristics in natural populations. His study of the death rates in crabs (1894) argued that natural selection could take place for small variations, and started a debate between those who saw natural selection as occurring in big, discontinuous steps (Mendelians), and supporters of the new statistical methods showing small continuous changes (Biometrics).

(Frank) Alexander Wetmore
1886-1978

American ornithologist who was a biologist with the United States Bureau of Biological Survey from 1910 to 1924 and later became Superintendent of the National Zoological Park in Washington, D.C. (1924-1925), assistant secretary of the Smithsonian Institution (1925-1944), and its Secretary from 1945 to 1952. An authority on the living and fossil birds of North and South America, he also proposed and published several editions of *Systematic Classification for the Birds of the World* between 1934 and 1960. His major work was *Birds of the Republic of Panama* (3 volumes, 1965-1972), with a fourth and final volume completed by colleagues in 1984. He described 189 previously unknown species and subspecies of birds and was active in many national and international scientific organizations. Wetmore's published output of over 700 books and monographs has been described as "staggering."

William Morton Wheeler
1865-1937

American zoologist and a foremost authority on the ecology, behavior, and classification of ants and social insects. Wheeler's insect embryology work is considered classic, and he developed a comparative analysis approach for studying the sociology and psychology of insects. Well liked and respected as a learned biologist, researcher, theorist, and philosopher, Wheeler spent most of his career at Harvard University. His career there included positions as professor of entomology, Dean of the graduate school, and associate curator of insects at the Museum of Comparative Zoology.

Heinrich Otto Wieland
1877-1957

German chemist who received the 1927 Nobel Prize for his research on the constitution of bile acids. Weiland's determination of the molecular structure of these acids later led to the discovery of steroids related to cholesterol. Among his important contributions to structural organic chemistry was the discovery that different forms of nitrogen in organic compounds could be detected and distinguished from one another.

Richard Willstätter
1872-1942

German organic chemist who did foundational research on chlorophyll and its role in photosynthesis, for which he received the 1915 Nobel Prize for Chemistry. Willstätter also analyzed the chemical structure of cocaine and other alkaloids, and discovered new quinones and quinone imines. After World War I, he concentrated on enzyme research. He later resigned his post as professor of chemistry at Munich in protest over the treatment of a fellow Jewish scholar, and fled to Switzerland in 1939 to escape the Nazis.

Joseph Henry Woodger
1894-1941

English biologist whose work helped to establish the field of theoretical biology. His 1929 book, *Biological Principles*, was the first significant analysis of theoretical biology and helped to establish the field. This was followed in 1937 by *The Axiomatic Method in Biology*, which helped to introduce a heightened degree of logical rigor to the field of theoretical biology.

Almroth Edward Wright
1861-1947

British pathologist, bacteriologist, and immunologist who developed a vaccine against typhoid fever. Wright studied literature and medicine, specializing in pathological anatomy. In 1892 he joined the Army Medical School and began a decade of research in blood coagulation and bacteriology. His typhoid vaccine was tested on British troops in India, and later proved effective when used by British soldiers during the Boer War in South Africa. During World War I, Wright scientifically justified the early closure of wounds to reduce infection. Blunt and unconventional, he made many enemies, but his students dominated British immunology for decades.

Bibliography of Primary Sources

Books

Allee, Warder. *Cooperation Among Animals* (1951). Allee's most important work, in which he summarized much of the known research on societal behavior among animals.

Bateson, William. *Mendel's Principles of Heredity—A Defense* (1902). Includes a translation of Gregor Mendel's groundbreaking paper on heredity and Bateson's assertion that Mendel's laws would prove to be universally valid. Indeed, further studies of the patterns of inheritance proved that Mendel's laws were applicable to animals as well as plants.

Boas, Franz. *The Mind of Primitive Man* (1911). An important anthropological study of race and culture, compiled from a series of Boas's lectures on the subject. This book was denounced and burned by the Nazis during the 1930s.

Chetverikov, Sergei. *Waves of Life* (1905). In this work, Chetverikov connected fluctuating population size to Darwinian evolution.

De Vries, Hugo Marie. *Die Mutationstheorie* (The Mutation Theory) (1901-03). Contains a summary of De Vries's research on heredity, whose laws he rediscovered decades after Gregor Mendel; de Vries's work aided the rediscovery of Mendel's overlooked earlier work. De Vries studied the role of mutations in plant evolution and proposed that independent units called pangenes carried hereditary traits, a concept that differed from Mendel's theory.

Dobzhansky, Theodosius. *Genetics and the Origin of Species* (1937). The published version of Dobzhansky's "Jessup Lectures," delivered at Columbia University in 1935, this book became the foundation of the evolutionary "modern synthesis," which combined Mendelian genetics and the Darwinian theory of natural selection.

Fisher, Ronald Aylmer. *Statistical Methods for Research Workers* (1925). In this work, Fisher described his development of a new technique by which scientists could vary different elements in an experiment to determine the probability that those variations would yield different results. His work on plant-breeding experiments combined biology and statistics.

Fisher, Ronald Aylmer. *The Genetical Theory of Natural Selection* (1930). In this work, Fisher introduced new theories about randomization and variance, which are now widely used in the field of genetics.

Fisher, Ronald Aylmer. *Statistical Methods and Scientific Inference* (1956). Fisher's definitive work on statistical research methods, which included the concept of variance. Prior to Fisher's work, scientists were only able to vary one factor at a time in experiments, allowing for only one potential result. Fisher instead proposed a statistical procedure by which experiments would be designed to answer several questions at once. This was accomplished by dividing each experiment into a series of sub-experiments, each of which differed enough to provide several unique outcomes.

Huxley, Julian. *Evolution: The Modern Synthesis* (1942). In this book, Huxley merged his own substantial biological work with that of Theodosius Dobzhansky and others to present a unified (although not completely in agreement) view of evolutionary biology known as the "modern synthesis."

Martin, T. T. *Hell and the High Schools: Christ or Evolution—Which?* (1923). In this book, Martin equated Darwinism with atheism and fueled the drive to ban the teaching of evolution.

Mayr, Ernst. *Systematics and the Origin of Species* (1942). In this work, Mayr further coalesced evolution, genetics, and speciation, and paved the way for the "modern synthesis" of evolutionary theory.

Mead, Margaret. *Coming of Age in Samoa* (1928). An anthropological study of adolescent behavior in a Polynesian society, which signaled major changes in approaches to the formulation of social theory. Mead became an instant celebrity over the book's frank and honest descriptions of Samoan female sexual behavior.

Morgan, Thomas Hunt. *The Theory of the Gene* (1926). A summation of developments in genetics since the rediscovery of Mendel's laws. Based on statistical studies of inheritance in *Drosophila,* Morgan assigned five principles to the gene—segregation, independent assortment, crossing over, linear order, and linkage groups.

Needham, Joseph. *A History of Embryology* (1934). One of several books by Needham, an acclaimed historian of science associated with Cambridge University for most of his career.

Needham, Joseph. *Science and Civilisation in China* (1954). A comprehensive, seven-volume work on Chinese medicine and technology, compiled by Needham after his travels through China on a scientific mission.

Papanicolaou, George, and Herbert F. Traut. *Diagnosis of Uterine Cancer by the Vaginal Smear* (1939). In this paper, Papanicolaou and Traut argued that cancerous cervical lesions could be detected by observable and measurable cellular changes while the cells were still in a preinvasive phase. Accordingly, Papanicolaou's diagnostic technique made it possible to diagnose asymptomatic patients.

Sabin, Florence. *An Atlas of the Medulla and Mid-Brain* (1901). A classic text on brain anatomy. While still in medical school, Sabin studied anatomy and helped transform the field of anatomy from a purely descriptive science to an academic discipline concerned with the relationships between form and function.

Skinner, B. F. *Schedules of Reinforcement* (1957). In this work, Skinner showed that the frequency of reinforcement (i.e., reward, punishment, or indifference) influences learning and behavior.

Skinner, B. F. *Verbal Behavior* (1957). Contains Skinner's examination of speech and learning.

Skinner, B. F. *Walden Two* (1948). One of Skinner's more widely read books, in which he presented his ideas concerning a utopian community based on scientific principles.

Tinbergen, Nikolaas. *The Study of Instinct* (1951). As the first comprehensive introduction to ethology to be printed in English, this book is often credited with launching the then-new field of ethology into prominence.

Periodical Articles

Boveri, Theodor. "Multipolar Mitosis as a Means of Analysis of the Cell Nucleus" (1902). In this classic paper, Boveri described his experiments on embryos with abnormal numbers of chromosomes, during which he tested the idea that each chromosome carries all character traits. Boveri found correlations between the abnormal combinations of chromosomes and the peculiar development of these embryos. Though he could not determine the role of specific chromosomes, his observations indicated that the chromosomes were not identical to each other.

Chetverikov, Sergei. "On Certain Aspects of the Evolutionary Process from the Point of View of Evolutionary Genetics" (1926). In this work, Chetverikov introduced Mendelian genetics into the study of the evolutionary process.

Corner, George Washington, and Willard Allen. "Physiology of the Corpus Luteum" (1929). In this text, Corner and Willard reported their joint discovery of progesterone. Corner's research into the female endocrine system proved the hormonal basis of menstruation.

Spemann, Hans. "Developmental Physiological Studies on the Triton Egg" (1901-03). A series of papers in which Spemann introduced the technique of manipulating and constricting an egg with a loop of fine baby hair. Spemann found that, if he constricted fertilized salamander eggs without completely separating the cells, he could produce animals with one trunk and tail but two heads.

Spemann, Hans. "Induction of Embryonic Primordia by Implantation of Organizers from a Different Species" (1924). A landmark paper that describes the growth of new embryos and specific organs out of embryo parts grafted onto other embryos of the same species as well as different species. This experiment, performed by Spemann's student Hilde Mangold as her doctoral thesis, demonstrated the phenomena known as embryonic induction.

Stanley, W. M. "Isolation of a Crystalline Protein Possessing the Properties of Tobacco Mosaic Virus" (1935). In this work, Stanley reported that he had isolated and crystallized a protein having the infectious properties of the tobacco mosaic virus. Stanley's work made possible the modern definition of viruses as particles composed of an inner core of nucleic acid enclosed in a protein overcoat.

Sturtevant, Alfred. "The Linear Arrangement of Six Sex-Linked Factors in *Drosophila,* as Shown By Their Mode of Association." (1913). This report, first published in the *Journal of Experimental Zoology,* contains the first chromosome map. Sturtevant identified the position of genes in mutated fly chromosomes by establishing the degree of linkage among recombined genes.

Twort, Frederick W. "An Investigation on the Nature of the Ultramicroscopic Viruses" (1915). In this landmark paper, first published in the British medical journal *Lancet,* Twort, the discoverer of bacteriophages, demonstrated that ultramicroscopic viruses are filterable, like the infectious agent of many mysterious plant and animal diseases.

JOSH LAUER

Mathematics

Chronology

1900 At the International Congress of Mathematicians in Paris, David Hilbert stimulates mathematical research for decades when he presents 23 unsolved problems.

1905 In his *Sur Quelques Points du Calcul Fonctionnel,* French mathematician Maurice-René Fréchet establishes the principles of functional calculus.

1906 William Henry Young and Grace Chisholm Young, a British husband-wife team, publish *The Theory of Set Points,* the first comprehensive textbook on set theory.

1908 Dutch mathematician Luitzen Egbertus Jan Brouwer founds the intuitionist school, which treats mathematics as a mental construct governed by self-evident laws.

1910 Alfred North Whitehead and Bertrand Russell publish the three-volume *Principia Mathematica,* in which they attempt unsuccessfully to construct an entirely self-contained mathematical system from first principles.

1928 Austrian-American mathematician Richard von Mises advances probability studies with the philosophical approach contained in *Probability, Statistics, and Truth.*

1928 In writing *The Theory of Games and Economic Behavior* (published 1944), John von Neumann and Oskar Morgenstern establish the principles of game theory, which will come to prominence in the latter half of the twentieth century.

1931 Austrian mathematician Kurt Gödel presents his incompleteness theorem, which states that within any rigidly logical mathematical system there are propositions that cannot be proved or disproved by the axioms within that system.

1932 The International Mathematical Congress establishes the Fields Medal, which becomes the equivalent of the Nobel Prize in mathematics.

1936 American mathematician Alonzo Church offers Church's theorem, which states that there is no single method for determining whether a mathematical statement is provable or even true.

1937 English mathematician Alan Mathison Turing puts forth the idea of the imaginary "Turing machine," which can solve all computable problems, and uses this to prove the existence of undecidable mathematical statements.

1939 A group of French mathematicians, working under the collective pseudonym of "Nicholas Bourbaki," begins writing what it considers to be a definitive synthesis and survey of all mathematics.

Overview: Mathematics 1900-1949

Abstraction and generalization have often been useful elements of the mathematician's toolkit. They gained in prominence in the nineteenth century and became perhaps the most distinguishing features of mathematics during the period under consideration. For example, Galois's theory of polynomial equations with real-number coefficients was developed by Ernst Steinitz in 1910 over arbitrary fields. The definition of distance between points in Euclidean space was extended by Maurice Fréchet in 1906 to arbitrary sets, giving rise to the important concept of a metric space. This also enabled mathematicians to extend the notion of continuity of a function to such a space. About a decade later Felix Hausdorff generalized these ideas further by noting that distance is not needed to define continuity: "neighborhoods" would do. This was done in the context of his definition of a topological space. The basic notion of function also underwent a grand generalization, embodied in the theory of distributions (or generalized functions) of Sergei L'vovich Sobolev and Laurent Schwartz, introduced in the 1930s and 1940s, respectively. These, of course, were not generalizations for generalizations' sake. They shed light on what is essential, thus extending the range of applicability of concepts and results.

The driving force behind this movement toward high levels of abstraction and generalization was the axiomatic method. While in the past it was used primarily to clarify and codify specific mathematical systems (for example, Euclidean or projective geometry, and the natural numbers), it now became a unifying and abstracting device. Thus, mathematical entities defined by systems of axioms admitting different (nonisomorphic) interpretations (models), such as groups, topological spaces, and Banach algebras, now became fundamental objects of study. The greatest triumph of the axiomatic method has perhaps been in algebra (groups, rings, fields, vector spaces), but the method has also been essential in such diverse areas as analysis (Banach and Hilbert spaces, normed rings), topology (topological, metric, and Hausdorff spaces), set theory (Zermelo-Fraenkel and von Neumann axioms), and even probability (axiomatized by Andrei Nikolaevich Kolmogorov in 1933). Of course, what is abstract to one generation may be concrete to another. In 1945 Samuel Eilenberg and Saunders MacLane introduced category theory, which is an abstraction of an abstraction: instead of studying a group or a topological space, one now studied the class of all groups or of all topological spaces. This idea turned out to be very fruitful in the second half of the twentieth century.

Problems and theory are two opposing pillars of the mathematical enterprise. But they are very much interconnected: attempts to solve problems often give rise to important theories (in fact, this is how theories usually arise), and these, in turn, aid in the solution of problems. Because the supply of mathematical problems is immense, one must choose fruitful ones for investigation (of course, what is fruitful is usually known only in retrospect). The mathematics of the twentieth century began auspiciously with an address by David Hilbert entitled "Mathematical Problems," delivered in 1900 before the second International Congress of Mathematicians in Paris. Although Hilbert favored abstraction and axiomatics and developed important mathematical theories, he was very conscious of the vital role of problems in the subject, noting that "as long as a branch of science offers an abundance of problems, so long is it alive; a lack of problems foreshadows extinction or the cessation of independent development." He proposed 23 problems for the consideration of twentieth-century mathematicians, and he was confident that they would all be amenable to solution. As he put it: "There is the problem. Seek its solution. You can find it by pure reason, for in mathematics there is no *ignorabimus*." Indeed, many of the problems *were* solved during the century, leading to important concepts and theories—though some have resisted solution these past 100 years!

Another pair of opposing pillars of the mathematical enterprise is pure versus applied mathematics—mathematics for its own sake versus mathematics as a tool for understanding the world. There is no doubt that pure mathematics has often turned out to be useful. For example, the two foremost theories in physics of the first half of the twentieth century, general relativity and quantum mechanics, used ideas from abstract mathematics, namely Riemannian geometry and Hilbert-space theory, respectively. But some prominent mathematicians (notably Courant) lamented the fact that mathematics was becoming too abstract (witness the Bourbaki phenomenon, discussed later in this chapter) and thus sterile, not—as in past centuries—maintaining its intimate connection with the physical world, from which it often drew inspiration. As John von Neumann put it: "At a great distance from its empirical source, or after much 'abstract' inbreeding, a mathematical subject is in danger of degeneration." Although there *were* examples of mathematical developments prompted by nonmathematical considerations, such as the founding of game theory, Minkowskian geometry, linear programming, and aspects of differential geometry, there was undoubtedly some validity to Courant's and others' concerns. The balance, however, was restored with great vigor in the last decades of the twentieth century.

The nineteenth century saw the creation of new fields; these were expanded and extended, and some were axiomatized, in the first half of the twentieth century. Among the numerous new advances were abstract algebra, the Lebesgue integral, ergodic theory, metamathematics, class field theory, Fourier analysis on groups, and the theory of complex functions of several variables. A most important field—topology—was (essentially) newly founded during this period. It joined algebra and analysis as one of three pillars underpinning much of modern mathematics. A central feature of the period was the cross-fertilization of distinct areas, resulting in new departures—new fields, for the most part. Important examples are algebraic topology, algebraic geometry, algebraic logic, topological groups, Banach algebras, functional analysis (analysis and algebra), Lie groups (algebra, analysis, and topology), and differential geometry. Note the predominance of algebra in this list. Some have spoken of the "algebraization of mathematics" as a major trend of this period.

Yet another new development was the emergence of the foundations of mathematics as an important field of *mathematical* (rather than philosophical) study. The impetus came from the paradoxes of set theory, the use of the Axiom of Choice, and the emergence in the nineteenth century of various geometries and algebras, resulting in the loosening of the close connection between mathematics and the "real world." The questions that were asked had to do with the consistency and completeness of major branches of mathematics defined axiomatically, and, more broadly, with the nature of mathematical objects and the methods of mathematical reasoning. The results were the founding of three "philosophies" of mathematics—logicism, formalism, and intuitionism—and Kurt Gödel's Incompleteness Theorems. The latter showed that large parts of mathematics (including elementary arithmetic and Euclidean geometry) are incomplete, and their consistency can not be established (within the given system). These were most profound results, placing (at least *theoretical*) limits on the axiomatic method, indispensable in much of modern mathematics. Another great achievement in mathematical logic, soon to be of profound *practical* impact, was the work of Alan Turing, Emil Post, and Alonzo Church in the 1930s, giving (independently) precise expression to the vague notions of algorithm and computability, thus providing the necessary mathematical underpinnings for the advent of the computer.

Finally, a comment on professionalization in mathematics. The subject grew enormously in the first half of the twentieth century, so much so that its output during that period was (likely) larger than that of all previous centuries combined! The age of the mathematical universalist, such as David Hilbert or Jules Henri Poincaré, who could master substantially the whole subject, came to an end at mid-century. Strong new mathematical centers began to flourish (e.g., in the United States, the former Soviet Union, Japan, China, India, Poland), with the center of gravity of mathematical activity gradually moving away from Western Europe. Many universities were established in various countries, doing both teaching and research in mathematics. The research was channeled into formal institutions: hundreds of mathematical periodicals were founded (there were only a handful in the nineteenth century), three of which were "reviewing" journals, acquainting mathematicians with research done worldwide. Numerous mathematical societies came into being in various countries, including the International Mathematical Union. National meetings were held regularly, and the International Congress of Mathematicians, which began in 1897, took place every four years. At each Congress "Fields Medals" (mathematical counterparts of the Nobel Prizes) were awarded to outstanding mathematicians. Mathematics was thriving!

ISRAEL KLEINER

The Foundations of Mathematics: Hilbert's Formalism vs. Brouwer's Intuitionism

Overview

Different philosophical views of the nature of mathematics and its foundations came to a head in the early twentieth century. Among the different schools of thought were the logicism of Gottlob Frege (1848-1925), the formalism of David Hilbert (1862-1943), and the intuitionism of Luitzen Egbertus Jan Brouwer (1881-1966).

Background

Frege, who founded modern mathematical logic in the 1870s, claimed that mathematics is reducible to logic. That is, if all logic were understood perfectly, then all mathematics could be derived from it, or considered part of logic. This view, logicism, was always controversial, but it started important lines of inquiry in philosophy, logic, and mathematics. Bertrand Russell (1872-1970) and Alfred North Whitehead (1861-1947) adopted a weaker version of logicism than Frege's. Ludwig Wittgenstein (1889-1951) defended logicism in *Tractatus Logico-Philosophicus* (1921).

In 1889 Giuseppe Peano (1858-1932) introduced five axioms for the arithmetic of the natural numbers. These "Peano postulates" had extensive influence on investigation into the foundations of mathematics. The formal system of Russell and Whitehead in *Principia Mathematica* (1910-1913) offered rigorous, detailed derivations of Peano's arithmetic and other mathematical theories in terms of propositional logic, artificial language, and "well-formed formulas" (known as *wffs*).

Peano was a formalist. Formalism attempts to reduce mathematical problems to formal statements and then prove that the resulting formal systems are complete and consistent. A mathematical system is complete when every valid or true statement in that system is provable in that system. A theory is consistent if it contains no contradictions, that is, if some statement p does not imply both q and not-q.

Platonism in mathematics is the belief that mathematical objects exist as ideals, independent of worldly experience, and can be discovered by thinking. Constructivism opposes Platonism, claiming that mathematical objects exist only if they can be constructed, that is, only if proofs or axioms can be invented for them. For constructivism, a mathematical proof is acceptable only if it arises from the data of experience.

At the Second International Congress of Mathematics in Paris in 1900, Hilbert challenged his colleagues with 23 problems. This "Hilbert program," with modifications through the 1920s, became the agenda of formalism.

Georg Cantor (1845-1918) developed set theory in the 1870s. Hilbert was bothered by the many paradoxes and unanswered questions that attended Cantor's concept of infinite sets. Hilbert's essay, "On the Infinite" (1925), attempted to resolve some of these issues, and the Hilbert program encouraged other mathematicians to think about them, too.

Among Cantor's results was the theorem that the power set of any set S (the set of all subsets of S) has a greater cardinality (more members) than S. This theorem led to difficulties with the notion of infinity. The set of natural numbers is infinite. Call its cardinality $\aleph_0$. Obviously some sets, such as the set of real numbers, the set of points on a line, the power set of the natural numbers, etc., have greater cardinalities than $\aleph_0$, that is, greater than infinity. This is the "continuum problem."

Hilbert's version of formalism accepted Platonism, but some formalisms are closer to constructivism, holding that mathematical statements are just sequences of "chicken scratches" and do not represent any actual objects.

Finitism recognizes the existence of only those mathematical objects that can be demonstrated in a finite number of steps or proved in a finite number of wffs. Hilbert and many of his followers were finitists.

Intuitionism claims, against logicism, that logic is part of mathematics; against Platonism, that the only real mathematical objects are those that can be experienced; against formalism, that mathematical proofs are assertions of the reality of mathematical objects, not just series of wffs; and against finitism, that proofs are necessary, not just sufficient, to assert the reality of mathematical objects. Intuitionism frequently makes common cause with constructivism.

Reacting against Platonism and formalism, the intuitionism and constructivism of Brouwer rejected the law of the excluded middle as established by Augustus De Morgan (1806-1871) and questioned Cantor's concept of infinity because of its unintelligibility. As Brouwer devised it in the first two decades of the twentieth century, intuitionism demanded the actual mental conception of mathematical objects, rather than just the possibility that they could be proved.

Hilbert and Brouwer were each influenced by the philosophy of mathematics of Immanuel Kant (1724-1804), who held that there are two, and only two, forms of intuition: space and time. For Brouwer, since the infinite continuum cannot be understood in space and time, it must be just an idea of pure reason, or what Kant would call a "regulative ideal." Brouwer emphasized Kant's intuition of time, while Hilbert used a Kantian epistemology, or theory of knowledge, to support his finitism.

Impact

Many important results in mathematical and symbolic logic stemmed from the formalism vs. intuitionism debate.

Emil Leon Post (1897-1954) proved the consistency of the Russell/Whitehead propositional logic in 1920. Wilhelm Ackermann (1896-1962) and Hilbert did the same for first-order predicate logic in 1928. Consistency proofs for other logics were offered by Jacques Herbrand (1908-1931) in 1930 and Gerhard Gentzen (1909-1945) in 1936. In the 1930s John von Neumann (1903-1957) solved the "compact group problem," the fifth of the 23 challenges that Hilbert had given to the worldwide mathematical community. Jan Lukasiewicz (1878-1956) and Post proved decidability for propositional logic in 1921. A system or theory is decidable if there is a way (algorithm) to determine whether a given sentence is a theorem of that system or theory. In various degrees, these results all tended to favor formalism.

Results favoring Brouwer and intuitionism also appeared. In 1930 Arend Heyting (1898-1980) formulated a propositional logic in accordance with intuitionistic/constructivistic principles. In 1939 John Charles Chenoweth McKinsey (1908-1953) demonstrated that Heyting's axioms for intuitionistic propositional logic are self-sufficient.

The two incompleteness proofs of Kurt Gödel, a Platonist, dealt formalism a telling blow in 1931.

Beyond the negative result of Gödel's incompleteness proofs, Alonzo Church (1903-1995) and Alan Turing (1912-1954) each proved independently in 1936 that first-order logic is undecidable. This result, called Church's theorem, is not to be confused with what is known as Church's thesis or the Church-Turing thesis, propounded the same year, which has to do with possible algorithms for the computability of functions.

Before World War II, most of the investigators of the relation between logic and the foundations of mathematics were mathematicians, and the German-speaking world field dominated the field. After Hilbert's death, and especially since the 1960s, mathematicians generally lost interest in these foundational issues. Mathematical logic and the logic of mathematics became mostly the province of logicians and philosophers, with the English-speaking world dominating.

ERIC V.D. LUFT

Further Reading

Books

Allenby, R.B.J.T. *Rings, Fields and Groups: An Introduction to Abstract Algebra*. London: Edward Arnold, 1983.

Benacerraf, Paul, and Hilary Putnam, eds. *Philosophy of Mathematics*. Cambridge: Cambridge University Press, 1983.

Beth, E.W. *The Foundations of Mathematics*. Amsterdam: North-Holland, 1959.

Brouwer, L.E.J. *Philosophy and the Foundations of Mathematics*. Amsterdam: Elsevier, 1975.

Chaitin, Gregory J. *The Unknowable*. New York: Springer, 1999.

Epstein, Richard L., and Walter A. Carnielli. *Computability: Computable Functions, Logic, and the Foundations of Mathematics*. Pacific Grove, CA: Wadsworth & Brooks, 1989.

Sowa, John F. *Conceptual Structures: Information Processing in Mind and Machine*. Reading, MA: Addison-Wesley, 1983.

Sowa, John F. *Knowledge Representation: Logical, Philosophical, and Computational Foundations*. Pacific Grove, CA: Brooks Cole, 1999.

van Heijenoort, Jean, ed. *From Frege to Gödel*. Cambridge, MA: Harvard University Press, 1967.

Periodical Articles

Tarski, Alfred. "The Semantic Conception of Truth and the Foundations of Semantics." *Philosophy and Phenomenological Research* 4, no. 3 (March, 1944): 341-76.

Bertrand Russell and the Paradoxes of Set Theory

Overview

Bertrand Russell's discovery and proposed solution of the paradox that bears his name at the beginning of the twentieth century had important effects on both set theory and mathematical logic.

Background

At about the same time in the 1870s, Georg Cantor (1845-1918) developed set theory and Gottlob Frege (1848-1925) developed mathematical logic. These two strains of theory soon became closely intertwined. Cantor recognized from the beginning that set theory was replete with paradoxes. He, along with other mathematicians and logicians or philosophers of mathematics such as Bertrand Russell (1872-1970), Alfred North Whitehead (1861-1947), and Edmund Husserl (1859-1938), tried to resolve these difficulties.

Cantor himself discovered one of the earliest paradoxes of set theory in 1896. The power set of any set S is the set of all subsets of S. The power set of S cannot be a member of S. But consider the set U of all sets. The power set of U would, by the definition of U, be a member of U. This contradiction is Cantor's paradox. One possible solution is to stipulate that the set of all sets cannot itself be a set, but must be treated as something else, a "class."

Burali-Forti's paradox, or the ordinal paradox, discovered in 1897 by Cesare Burali-Forti (1861-1931), is akin to Cantor's. It says that the greatest ordinal is greater than any ordinal and therefore cannot be an ordinal. In set theory ordinal numbers refer to the relationship among the members of a well-ordered set, that is, any set whose non-empty subsets each have a "least" or "lowest" member.

In 1901 Russell discovered the paradox that the set of all sets that are not members of themselves cannot exist. Such a set would be a member of itself if and only if it were not a member of itself. This paradox is based on the fact that some sets are members of themselves and some are not.

Russell's paradox is related to the classic paradox of the liar, attributed to either the Cretan philosopher Epimenides (7th century B.C.) or the Greek philosopher Eubulides (4th century B.C.). Is someone who says, "I am lying," telling the truth or lying? In other words, is that person a member of the set of liars or the set of truth-tellers?

Russell's paradox is effectively illustrated by the barber paradox. Divide all the men in a certain town into two non-intersecting sets: the set X of those who shave themselves and the set Y of those who are shaved by a barber. Such a barber (unless she is a woman) cannot exist. If he exists, then he both shaves and does not shave himself; that is, he is a member of both set X and set Y. He is part of the definition of set Y, but at the same time he is expected to be a member of either set X or set Y.

These paradoxes are self-referential because they conflate the definition with what is being defined. They confound the properties that define a set with the members of the set. They result from their implicit confusion between the "object language," which talks about the individual members of sets, and the "metalanguage," which talks about the sets themselves. Russell's discovery led immediately to much research in set theory and logic to define the nature of sets, classes, and membership more accurately.

Russell's theory of types may solve these paradoxes by clarifying the distinction between object language and metalanguage. It says that if all statements are classified in a hierarchy, or "orders," according to the level of their subject matter, then any talk of the set of all sets that are not members of themselves can be avoided. Thus first-order logic quantifies over individuals; second-order, over sets of individuals; third-order, over sets of sets of individuals; and so on. To confuse orders is to make a "category mistake."

There are many variants of Russell's paradox and many associated paradoxes. Richard's paradox, announced in 1905 by Jules Antoine Richard (1862-1956), deals with problems of defining sets. Berry's paradox, a simplified version of Richard's, was introduced by Russell in 1906 but attributed to George Berry, a librarian at Oxford University. The Grelling-Nelson paradox, sometimes called the heterological paradox, was stated in 1908 by Kurt Grelling (1886-1942) and Leonard Nelson (1882-1927). It says that some adjectives describe themselves and some do not. Adjectives that do not describe themselves are heterological. Is the adjective "heterological" itself heterological? If so, then it is not; and if not, then it is.

Russell first mentioned his theory of types in a 1902 letter to Frege. He published it in 1903 and revised it in 1908. The earlier version is called the simple theory of types, and the later version, specifically directed at the liar and Richard's paradox, is called the ramified theory of types.

Impact

Logical paradoxes are generally of two kinds: set theoretic and semantic. Set theoretic paradoxes such as Cantor's, Burali-Forti's, Russell's, and the barber, expose contradictions or complications in set theory. Semantic paradoxes, such as the liar, Richard's, Berry's, and the Grelling-Nelson, raise questions of truth, definability, and language. The demarcation between these two kinds of paradoxes is not clear. Frank Plumpton Ramsey (1903-1930) believed that all logical paradoxes were set theoretic paradoxes and occurred only in the object language, while semantic paradoxes occurred only in the metalanguage and involved only meanings, not logic.

In the wake of and partially in reaction to Russell's theory of types, mathematicians made several attempts to refine set theory. In 1908 Ernst Zermelo (1871-1953), building upon the work of Cantor and Richard Dedekind (1831-1916), formulated the axioms that became the basis of modern set theory. These axioms, modified by Abraham Fraenkel (1891-1965), are known as "ZF." With the addition of the axiom of choice, which states that for any two or more non-empty sets there exists another non-empty set containing exactly one member from each, they are called "ZFC."

In 1925 John von Neumann (1903-1957) offered an alternative set theory with an axiom disallowing any set containing itself as a member. This theory, modified by Paul Isaac Bernays (1888-1977) and Kurt Gödel (1906-1978) is called "NBG."

The logician Alfred Tarski (1901-1983) refined the distinction between object language and metalanguage and thus was able to resolve semantic paradoxes without relying upon Russell's theory of types. Yet his solution of the liar paradox resembles Russell's and can be regarded as a variant of it.

ERIC V.D. LUFT

Further Reading

Clark, Ronald W. *The Life of Bertrand Russell*. New York: Alfred A. Knopf, 1976.

Halmos, Paul R. *Naive Set Theory*. New York: Springer, 1987.

Kasner, Edward, and James R. Newman. "Paradox Lost and Paradox Regained." In James R. Newman, ed., *The World of Mathematics*. New York: Simon and Schuster, 1956: 1936-56.

Russell, Bertrand. *Principles of Mathematics*. Cambridge: Cambridge University Press, 1903.

Russell, Bertrand. *Introduction to Mathematical Philosophy*. London: George Allen and Unwin, 1919.

Russell, Bertrand. *Logic and Knowledge*. London: George Allen and Unwin, 1956.

Russell, Bertrand, and Alfred North Whitehead. *Principia Mathematica*. 3 vols. Cambridge: Cambridge University Press, 1910-1913.

van Heijenoort, Jean, ed. *From Frege to Gödel*. Cambridge, MA: Harvard University Press, 1967.

Mathematical Logic: Proofs of Completeness and Incompleteness

Overview

Philosophers, logicians, and mathematicians in the first half of the twentieth century made significant progress toward understanding the connections between mathematics and logic. Major results of these investigations include the Löwenheim-Skolem theorem (1920), Gödel's first and second incompleteness theorems for axiomatic systems (1931), and Henkin's completeness theorem for first-order logics (1949).

Background

Logicians use the terms "formal system," "formal theory," "formal language," and "logic" almost interchangeably to refer to any set K whose members are in practice subject to rules of inference. Each formal system has three components: grammar, deductive system, and semantics, which together determine the logic of K. The grammar consists of the members of K, such as symbols, logical operators, constants, variables, etc., and the rules governing the formation of terms. A "well-formed formula" (known as a *wff*) is any statement made according to these formation rules. The deductive system is the process by which proofs are generated. A proof in a formal system is a series of wffs ($p_1 \ldots p_n$) where p_n=q and where q is the statement which was to be proved. The semantics is the meaning of statements in K. The same grammar and semantics can have different deductive systems; similarly, the same grammar and deductive system can have different semantics.

These three components of actual practice are designated respectively by "G,D,S"; the components of the model or formal structure of practice are designated by the corresponding Greek letters: "Γ,Δ,∑." A model is a planned structure that exhibits in a formal system a precise way. Both mathematics and logic contain formal systems that can be modeled.

The logic of mathematics is metamathematics, and generally has two aspects: proof theory, which studies Γ and Δ, and model theory, which studies Γ and ∑. The major founders of proof theory and model theory were, respectively, David Hilbert (1862-1943) and Alfred Tarski (1901-1983).

Among the concerns of metamathematics are the completeness and soundness of formal systems.

In Δ, the desired characteristic of sentences is provability or deducibility; in ∑, truth or logical consequence. A proved sentence in Δ is a theorem of the system. A valid sentence in ∑ is a logical truth or a logical consequence of the system. Logicians designate the provability of q by $\vdash$ q and the logical truth of q by $\vdash$ q.

A formal system is complete if every valid or true statement in that system is provable within the system. Completeness can be either strong or weak. Strong completeness says that when q is a logical consequence of premises p in the semantics of K (p $\vdash$q in $\sum_K$), then q is provable from p in the deductive system of K (p $\vdash$ q in Δ_K), then q is a logical consequence of p in the semantics of K (p $\vdash$ q in $\sum_K$). Weak soundness says that when there is a proof of q in K from no assumptions ($\vdash$ q in Δ_K), then q is logically true in K ($\vdash$ q in $\sum_K$).

One of the most important results of model theory is the Löwenheim-Skolem theorem, produced in 1915 by Leopold Löwenheim (1878-1957) and improved in 1920 by Thoralf Albert Skolem (1887-1963). It says that any mathematical theory that has a model has a countable model. That is, the set of all the members of the model can be put into one-to-one correspondence with either the set of positive integers or some subset of it. The Skolem paradox (1922) of this theorem is that if an uncountably infinite set, such as the set of real numbers, has a model, then in that model the real numbers would be countable.

There are many completeness proofs. Paul Isaac Bernays (1888-1977) proved completeness for propositional logic in 1918. Kurt Gödel (1906-1978) proved it for first-order predicate logic in 1930. Garrett Birkhoff (1911-1996) offered a proof of strong completeness for equational logics. Probably best known is the proof of Leon Henkin (1921-). Henkin did not use the "weak/strong" terminology at that time, but in 1996 he claimed that in his 1947 doctoral dissertation at Princeton University he had proved strong completeness for the first-order calculus. Weak completeness follows trivially as a corollary to Henkin's proof.

In 1931 Gödel proved incompleteness for formal systems. His first incompleteness theorem states that for any consistent formal system adequate for number theory, there exists at least one true statement that is not provable or decidable within the system. His second incompleteness theorem, a corollary to the first, states that no such system can serve as its own metalanguage, i.e., it cannot be employed to show its own consistency.

Impact

Gödel's incompleteness proofs had wide-ranging and serious impact. They disproved Hilbert and Bernays' belief that all mathematics could be formulated as a single internally consistent and coherent theory. They proved that the internal consistency of several basic mathematical theories, including the arithmetic of Giuseppe Peano (1858-1932) and the set theory of Ernst Zermelo (1871-1953) and Abraham Fraenkel (1891-1965) cannot be proved within their own languages, but only in stronger metalanguages. This negative result does not mean that these theories are inconsistent, but only that their consistency must be established by theories other than themselves.

Completeness and soundness are important in computer science, especially with regard to software design. Results in mathematical logic are applicable to the theories of monoids, strings, groups, semi-groups, rings, fields, Boolean algebras, and lattices. Logic illuminates the mathematics of very large or inaccessible cardinals, the connection between ordinality and cardinality, countably infinite sets, number theory, and recursion theory.

ERIC V.D. LUFT

Further Reading

Books

Detlefsen, Michael, David Charles McCarthy, and John B. Bacon. *Logic from A to Z*. London: Routledge, 1999.

Hofstadter, Douglas R. *Gödel, Escher, Bach: An Eternal Golden Braid*. New York: Vintage, 1989.

Other

Birkhoff, Garrett. "On the Structure of Abstract Algebras." *Proceedings of the Cambridge Philosophical Society* 31 (1935): 433-54.

Corcoran, John. "Three Logical Theories." *Philosophy of Science* 36, no. 2 (June, 1969): 153-77.

Henkin, Leon. "The Completeness of the First-Order Functional Calculus." *Journal of Symbolic Logic* 14, no. 3 (September, 1949): 159-66.

Henkin, Leon. "The Discovery of My Completeness Proofs." *Bulletin of Symbolic Logic* 2, no. 2 (June, 1996): 127-58.

Nagel, Ernest, and James R. Newman. "Goedel's Proof." In James R. Newman, ed., *The World of Mathematics*. New York: Simon and Schuster, 1956: 1668-95.

Zach, Richard. "Completeness before Post: Bernays, Hilbert, and the Development of Propositional Logic." *Bulletin of Symbolic Logic* 5, no. 3 (September, 1999): 331–66.

David Hilbert Sets an Agenda for Twentieth-Century Mathematics

Overview

In 1900 David Hilbert (1862-1943), one of the acknowledged leaders of pure mathematics at the turn of the century, identified what he considered to be the most important problems facing contemporary mathematicians at an address to the Second International Congress of Mathematicians. While some of the questions concerned purely technical issues, a number addressed the foundations of mathematics. These led to major changes in the philosophy of mathematics and ultimately to the development of digital computers and artificial intelligence.

Background

The nineteenth century was a period of restructuring in mathematics. Once considered a study of self-evident truths about quantity and space, the paradoxes of set theory and the discovery of non-Euclidean geometries and generalized kinds of numbers motivated mathematicians to reexamine the most fundamental ideas in each area of mathematics. Many mathematicians thought that the best way to avoid any possible problems was to adopt the axiomatic method, allowing each field to have very few undefined concepts and unproved statements about it, called axioms, which were taken to be true. All other statements could only be considered true if they could be proved in a strictly logical way from the axioms.

One of the masters of this new, more formal, mathematics was David Hilbert, Professor of Mathematics at the University of Göttingen. Hilbert had established his reputation with a treatise on algebraic number theory in 1897. In 1899 he published a book, *The Foundations of Geometry*, which placed the subject on a far more rigorous footing than the classic *Elements* of Euclid (c. 330-260 B.C.), which had dominated geometric thinking since antiquity. In doing so he emphasized the importance of considering the simplest geometric concepts—point, line and plane—to be undefined terms.

In 1900 Hilbert was invited to present the inaugural address at the Second International Congress of Mathematics, to be held in Paris, France. Hilbert used the occasion to list the most important issues he considered to be facing mathematicians at the beginning of the twentieth century. Hilbert's list of 23 problems, published in a German mathematical journal and translated into English and republished by the American Mathematical Society, would serve as a stimulus to mathematical research for decades to

David Hilbert. *(Corbis Corporation (New York). Reproduced with permission.)*

come. A number of these problems are described below.

Hilbert's first problem came from the theory of sets with an infinite number of elements. Georg Cantor (1845-1918) had shown that the set of all real numbers could not be put into a one-to-one correspondence with the counting numbers, 1, 2, 3.... In contrast, the set of positive rational numbers (fractions) could be placed in countable order by expressing each fraction in lowest terms, and then listing them in the order of the total of numerator and denominator, $^0/_1$, $^1/_1$, $^1/_2$, $^2/_1$, $^1/_3$, $^3/_1$, and so on. Cantor guessed that there were only two types of infinite sets of numbers—those that could be put into correspondence with the counting numbers and those that could be put into correspondence with the real numbers, which includes rational numbers and irrational numbers as well. Hilbert called for a proof or disproof of Cantor's conjecture.

The second problem posed was a challenge to mathematicians to prove that the axioms of arithmetic could never lead to a contradiction, or, more precisely, that there existed a set of axioms for arithmetic that would yield the familiar properties of the whole numbers without leading to a contradiction. As mathematicians had worked on axiomatic formulations of calculus and geometry in the nineteenth century, it became clear that these fields would be free of contradiction provided that arithmetic could be put on such an axiomatic basis. Hilbert had been particularly interested in the attempt of Italian mathematician Giuseppe Peano (1858-1932) to provide an axiomatic basis for arithmetic, and he was fairly confident that this approach would eventually eliminate any possible contradiction.

Hilbert's sixth question concerned the possibility of providing a fundamental set of rules, or axioms, for the science of physics. As established by Sir Isaac Newton (1642-1727), the laws of motion provided a powerful means of analyzing what was happening in any system of bodies, but as in the case of arithmetic it was not clear whether one might come up with contradictory predictions about the same physical situation. The question had become more important as new laws of physics had been discovered to describe electric, magnetic, and thermal phenomena.

The eighth question posed by Hilbert involved the distribution of prime numbers. A prime number is one that can only be divided by the number one and itself. It had been proved in ancient times that there was no largest prime number, but no one had a definite way of calculating the number of prime numbers between any two numbers, short of testing each number by dividing it by all smaller prime numbers.

Hilbert's tenth question asked if it was possible to determine whether a given set of Diophantine equations could be solved in whole numbers. Diophantine equations, named after the Greek mathematician Diophantus (fl. c. 250 A.D.), are sets of equations involving whole number multiples of several unknown quantities and their squares or cubes or higher powers. Mathematicians thought that it should be possible by manipulating the numbers appearing in the set of equations to determine if a solution in integers could be found, but no one had discovered such a method.

The twenty-third and last challenge posed by Hilbert was to further develop the calculus of variations, that branch of mathematics concerned with finding the one function of space and time for which a specified overall property takes on a minimum or maximum value. Problems of this type include that of the catenary, the shape taken by a chain suspended at the ends, and the geodesic, the shortest curve between two points on a given surface. Nineteenth-century mathematicians had demonstrated that Newton's laws of mechanics could be cast in the form of a minimum principle, further stimulating the development of this mathematical field.

Impact

The status of a scientific discipline can often be gauged by the social organization of the field: the number of professional organizations, the character of professional meetings, the number of journals, and the number of universities offering advanced degrees in the field. Hilbert's list of 23 questions appeared at a time when the field of mathematics had moved from a matter of correspondence by letter to periodic meetings of leading mathematicians. The First International Congress of Mathematicians had been held in Zurich, Switzerland, in 1897. Subsequent meetings have been held every four years with the exception of the periods during the two world wars (1914-1918 and 1939-1945).

Possibly the question of the greatest fundamental importance posed by Hilbert was the second, about the possibility of contradictions arising from the axioms of arithmetic. Austrian (later American) mathematician Kurt Gödel (1906-1978) in 1931 ruled out the possibility that a set of axioms for arithmetic could be found that would never lead to a meaningful statement being both true and not true. He did this by proving that any set of axioms that led to the accepted properties of integers would necessarily allow for statements that could neither be proved nor disproved. Gödel's work caused many mathematicians and philosophers to reexamine their own concepts of mathematical truth.

Gödel also provided half of the answer to Hilbert's first question in 1938 by proving that the continuum hypothesis could not be disproved. The other half was provided in 1963 by American mathematician Paul J. Cohen (1934-), a former student of Gödel who showed that the hypothesis could not be proved, either.

The question that may have had the greatest impact on twentieth-century technology may have been Hilbert's tenth, about being able to decide whether a set of Diophantine equations had an integer solution. Hilbert generalized this at the Eighth International Congress, held in Bologna, Italy, to a general question of decidability—that is, whether there might exist a generally effective procedure that would allow a mathematician to determine whether any mathematical statement was true by manipulating the symbols appearing in it. To answer this question English mathematician Alan Matheson Turing (1912-1954) developed the idea of a machine that could implement any effective procedure for manipulating symbols. This machine, which could read the symbols on a tape one at a time and, following an internal set of instructions, compute a new symbol to be written on a tape and move the tape forward or backward, would become the model for the digital computer. With this type of machine carefully described, Turing converted Hilbert's question about a problem being decidable to a question about whether such a machine would halt or continue in an endless loop, given any problem translated into a suitable code. Turing was then able to prove that the desired general effective procedure could not possibly exist.

Hilbert's sixth question would turn out to be of great significance for twentieth-century physics. As physicists were forced to abandon familiar notions of location and trajectory in describing the behavior of subatomic particles, many came to accept the necessity of an axiomatic approach. Hilbert himself would play an important role in developing a sound mathematical formulation for the quantum theory. In this theory, each state of a physical system corresponds to a direction in an abstract geometrical space, called Hilbert space in his honor, and all observable quantities are given geometrical interpretation in this space.

DONALD R. FRANCESCHETTI

Further Reading

Boyer, Carl B. *A History of Mathematics*. New York: Wiley, 1968.

Fang, Joong. *Hilbert: Towards a Philosophy of Modern Mathematics II*. Hauppauge, NY: Paideia Press, 1970.

Kline, Morris. *Mathematics: The Loss of Certainty*. New York: Oxford University Press, 1980.

Penrose, Roger. *The Emperor's New Mind*. New York: Oxford University Press, 1989.

Reid, Constance. *Hilbert*. New York: Springer-Verlag, 1970.

The Bourbaki School of Mathematics

Overview

Nicholas Bourbaki is the pen name of a group of mathematicians, most of them French, who have undertaken the writing of a definitive treatise of modern mathematics. The Bourbaki volumes emphasize the highest degree of mathematical rigor and the structures common to different areas of mathematics. The group perpetuates itself by continually electing new members and requiring that current members must leave the group at age 50.

Background

Nicolas Bourbaki is the invention of two French mathematicians, Claude Chevalley (1909-1984) and André Weil (1906-1998), who decided to write a more modern calculus text for French-speaking students than the ones that were typically used. The choice of the Bourbaki name may have had its origin in a student prank. At some time in the 1930s, students at the Ecole Normale Supérieure in Paris were invited to a lecture by a famous mathematician named Nicholas Bourbaki. The lecturer was really a student in disguise, and the lecture given was pure double-talk. There was, however, an important Bourbaki in French history, General Charles Denis Sauter Bourbaki, a French military officer of Greek descent who experienced a decisive defeat in the Franco-Prussian war of 1870. It may be that the choice of name honors the Greek roots of mathematics, but it may also simply reflect the memory of an unusual name, memorialized in a statue of General Bourbaki at Nancy, France, where some of the Bourbaki group had taught.

A new calculus book was needed because of the many developments that had occurred in mathematics over the preceding half century. In particular, important insights had been gained into the system of real numbers and the process of taking limits that are fundamental to the calculus. Calculus deals with derivatives, which are the limits of the ratios of small quantities as the quantities become infinitesimally small, and integrals, which are the limits of sums of quantities where the number of terms in the sum becomes infinitely large while the terms themselves become infinitely small.

Chevalley and Weil invited a number of other prominent French mathematicians to join in their project. The first Bourbaki group thus came to include Henri Cartan (1904-), Jean Dieudonné (1906-1992), Szolem Mandelbrojt, and Rene de Possel. Although the group intended to finish its writing project in two years, they soon realized that to provide the detailed and rigorous treatment that they felt the subject required would take many years. The group thus made provisions for new members to be elected and agreed upon a rule that members would retire from the group at the age of 50. This rule reflected a consensus that mathematicians are at their most creative in their twenties and thirties, a belief widely held even today, despite some important examples to the contrary. Membership has primarily been drawn from French mathematicians, and the group has numbered between 10 and 20 members at a time. A very few American mathematicians have been members, however. Initially the group met monthly in Paris. Soon they changed to a slightly less demanding schedule, with Bourbaki congresses held three times a year.

The main product of the Bourbaki group is a set of volumes, the *Eléments de mathématique*, published beginning in 1939 and now amounting to more than 30 volumes, many of them revised more than once. The title is undoubtedly meant as a reference to Euclid's *Elements*, the first attempt to summarize all mathematical knowledge. The fiction of a single author is maintained throughout. There was no public acknowledgment of the multiple authorship, though the fact became public knowledge in 1954, after an article on Bourbaki by the American mathematician Paul R. Halmos (1916-) appeared in the magazine *Scientific American.*

Most of the published volumes of the *Elements* come from the first part of the series, *The Fundamental Structures of Analysis,* which will consist of six "books" of from four to ten chapters each. A number of chapters of an unnamed second part have also appeared. While the contents of the first part are arranged from most fundamental to more specialized, the chapters are not being published in order. The first published chapters, appearing in 1940, are the first two in Book Three, dealing with general topology. The first two sequential chapters, on formal mathematics and the theory of sets, did not appear until 1954. This is not an illogical procedure, however, as the writers needed to decide how they would present the more advanced topics and then arrange the fundamentals accordingly.

For the Bourbaki group the foundation of mathematics is set theory. The original formulation of set theory, that of German mathematician Georg Cantor (1845-1918), was plagued by paradoxes resulting from the existence of infinite sets and the possibility of sets belonging to themselves. To avoid these problems, the Bourbaki group adopted the axiomatic form of set theory devised by Ernst Zermelo (1871-1953) and Adolf Abraham Fraenkel (1891-1965). Among these axioms is a rule that no set can belong to itself as well as the axiom of choice, which allows one to construct a set containing exactly one member from any collection of nonempty sets. Within the axiomatic approach, the earlier troubling paradoxes can be avoided. The Bourbaki writers do not claim, however to have settled the question of the foundations of mathematics permanently; rather, they believe it will be possible to make adjustments as the need arises, without having to discard 2,500 years of mathematical thinking.

ACTUALITÉS SCIENTIFIQUES ET INDUSTRIELLES

846

ÉLÉMENTS DE MATHÉMATIQUE

PAR

N. BOURBAKI

I

PREMIÈRE PARTIE

LES STRUCTURES FONDAMENTALES DE L'ANALYSE

LIVRE I

THÉORIE DES ENSEMBLES

(FASCICULE DE RÉSULTATS)

PARIS

HERMANN & Cie, ÉDITEURS

6, Rue de la Sorbonne, 6

1939

Title page of the first edition of the Bourbaki group's *Eléments de mathématique.*

Impact

Like the *Elements* of Euclid (c. 335-270 B.C.), the *Elements* of Bourbaki is not so much one of new creation as much as it is one of synthesis and systematic exposition. The Bourbaki writers are involved in solidifying the mathematical insights of their times, and thus the conclusions obtained in the *Elements* are often those of other mathematicians. The Bourbaki approach is thus not limited to the group. When features found in the Bourbaki approach are reflected in the work of other mathematical schools or in mathematics education, it is difficult to tell whether Bourbaki has originated the trend or is just following it.

There has been a persistent controversy over the proper place of rigor in mathematics education throughout the twentieth century. The Bourbaki members do not claim that their writings should be the basis for elementary mathematics instruction or even the university education of scientists and engineers who will use mathematics as a tool without attempting to add to mathematical knowledge. Since the 1950s there have been a number of efforts by mathematicians in the United States to base general mathematical education on notions from set theory and on the laws obeyed by the common arithmetic operations of addition and multiplication. These attempts, called the "new math" in the elementary grades, have been generally disappointing and have largely been abandoned.

The Bourbaki group's practice in accepting communal credit for their work is quite unusual in modern times. For most mathematicians and scientists, the assignment of credit for new discoveries is a serious matter. This was not always so. In late classical times a writer might have attached the name of an earlier philosopher to his work to give it more credibility. In the Middle Ages authors might have preferred to see their works remain anonymous as an act of humility. With the invention of printing and the publication of the first scientific journals, priority became important. When English physicist Robert Hooke (1635-1703) discovered the basic law governing the motion of elastic bodies, he first made it known as an anagram, a letter puzzle. Thus Hooke was able to prove his own priority and yet keep the discovery to himself for a while. The great English physicist and mathematician Isaac Newton (1642-1727) became engaged in a bitter dispute with German mathematician and philosopher Gottfried Wilhelm Leibniz (1646-1716) over credit for the invention of the calculus. In contrast to this overall trend, the willingness of the Bourbaki members to share or forego credit for a substantial amount of work is truly remarkable. The Bourbaki practice may, however, have anticipated some of the issues facing "big science" towards the end of the twentieth century, in which the number of creative contributors to a short research paper may number in the hundreds.

Scholars make a distinction between the primary literature of a field and the secondary literature. Since the appearance of the first scientific journals in the seventeenth century, the primary literature has consisted primarily of research papers, written immediately after each new discovery or observation. The secondary literature includes monographs, encyclopedias, and textbooks, in which current understanding of a field is systematized for the benefits of workers in the field. Most works in the secondary literature merely summarize the current status of a field, often with many authors each writing a specific chapter or section. The Bourbaki attempt to synthesize, rather than summarize, such an extensive body of knowledge is unique.

DONALD R. FRANCESCHETTI

Further Reading

Boyer, Carl B. *A History of Mathematics*. New York: Wiley, 1968.

Fang, Joong. *Bourbaki: Towards a Philosophy of Modern Mathematics I*. Hauppauge, NY: Paideia Press, 1970.

Kline, Morris. *Mathematics: The Loss of Certainty*. New York: Oxford University Press, 1980.

Lebesgue's Development of the Theories of Measure and Integration

Overview

Henri Lebesgue (1875-1941) revived the troubled field of integration. His generalization of integration, and the complex theory of measure he introduced to accomplish this, countered the criticisms and challenges to the field that threatened it at the end of the nineteenth century.

Background

Integration can be thought of in two ways. First, as the opposite of differentiation, so an integral is an anti-derivative. However, this is a very abstract concept. Second, integration between two points can be seen as the method of calculating the area of a shape where at least one side is not straight, but varies according to some function. While the calculation of the area of a square or triangle is straightforward, the area of, for example, a "D"- or "B"-shaped area is much more challenging mathematically. Often, these problems are thought of in terms of finding the area under a curve on a plotted graph, which is how they are generally presented in textbooks.

Many Greek mathematicians were concerned with such problems. Methods for calculating the areas of squares, triangles, and other regular shapes were understood, but once an object had a curve it seemed impossible to compute. Various methods were considered by thinkers such as Archimedes (287-212 B.C.). One popular method used the idea of upper and lower limits. If you take the shape you wish to measure and draw a regular polygon that would fit just inside the irregular shape, this gives a lower limit to the area of the irregular shape. Then draw a regular polygon that only just completely surrounds the irregular shape. This gives an upper limit to the shape, so the actual area must lie between the two areas drawn. If one regular shape was not close enough, then several packed together like a jigsaw could be used. While this method could get close approximations, it was in a sense a self-defeating process, as you could never get the correct answer.

Centuries later European mathematicians such as Johannes Kepler (1571-1630) introduced the use of infinitesimals to their calculations of areas. Infinitesimals were thought of as quantities smaller than any actual finite quantity, but not quite zero. While this is a somewhat strange idea, it was very influential. The first textbook on integration methods was published in 1635 by Bonaventura Cavalieri (1598-1647), but in a form quite unrecognizable today. Further modifications were made by various thinkers, each using their own special method for approximating the area under a curve. Many methods worked well for one kind of curve but poorly for others. The modern form of integration began to take shape when both Isaac Newton (1643-1727) and Gottfried von Leibniz (1646-1716) worked independently on calculus and developed a more general method for finding the area under a curve. A bitter argument ensued over who had discovered the idea first. Newton won in the short term; his anonymously written attacks on Leibniz were numerous, insulting, and vicious. However, in the long term it was the simpler style and notation of Leibniz that had a larger impact on mathematics.

Integration was further refined by mathematicians such as Augustin Cauchy (1789-1857) and Karl Weierstrass (1815-1897). The idea of infinitestimals was replaced with the concept of limits, in a sense going back to Archimedes' ideas. Cauchy gave elementary calculus the form it still holds today. It was from his concept of the integral as a limit of a sum, rather than simply an anti-derivative, that many modern concepts of integration have come.

In the middle of the nineteenth century Bernhard Riemann (1826-1866) defined integration more precisely, expanding the functions for which the method was useful. His work was in turn tinkered with over the next few decades by a number of theorists, so that it applied a little more generally.

However, towards the end of the nineteenth century many exceptions to Riemann integration became obvious. In integration and many other fields of mathematics, younger mathematicians took great delight in finding cases that broke the rules, cases that were termed "pathological examples." For Riemann integration these were often functions that had many discontinuous points, that is where a curve was not smooth and flowing but had gaps or points of extreme, instantaneous change. Another problem was the number of cases in which the inverse relationship between integration and differentiation did not appear to hold. Since one way of treating integration was as the opposite of differentiation, finding cases where this was not true suggested that something was wrong with the whole field. A number of established mathematicians feared mathematics was crumbling around them, as the whole basis of the discipline seemed under attack. Some critics called for the abandonment of integration altogether.

Henri Lebesgue was an unlikely savior of integration. He had a fine but not outstanding, university career. After completing his studies at the Ecole Normale Supérieure in Paris, France, he worked in its library for two years, during which time he read the thesis of another recent graduate, René Baire (1874-1932). Baire's doctorate, written in 1899, contained work on discontinuous functions, just the type of pathological cases that could not be solved using Riemann integration. Baire's poor health limited further work, but Lebesgue saw the potential of the ideas.

Lebesgue was also influenced by the work of a number of other mathematicians. Camille Jordan (1838-1922) had extended the Riemann integral by using the notion of measure. Emile Borel (1871-1956) radically revised what the notions of measure and measurability were in regards to sets. The mathematical idea of measure is not easily explained in words, as it is mathematically complex. It is an extension of the familiar idea of measuring areas (length, area, volume) generalized to the wider system of sets, and so deals with abstract spaces.

Borel had used radical new ideas in the field of set theory in his work, specifically the controversial work of Georg Cantor (1845-1918). Cantor's set theory was strongly opposed by some mathematicians, as it dealt with infinities in ways that contradicted some philosophical and religious doctrines. Mathematicians, and the wider intellectual community, had long been uncomfortable with the notion of infinity, and for some Cantor's argument that some infinite quantities were bigger than others seemed to go too far.

By putting all these divergent elements together Lebesgue revolutionized integration. In 1901 he introduced his theory of measure. This extended the ideas of Jordan and Borel and greatly increased the generality of the concept. The definition he presented is still called the Lebesgue measure. More importantly, however, it also had implications for integration.

Impact

In 1902 Lebesgue used his concept of measure to broaden the scope of integration far beyond the definitions of Riemann. He realized that the Riemann integral only applied in exceptional cases and that he needed to make integration into a more robust and general concept if it was to include the ideas he had read in Baire's thesis. At a stroke Lebesgue answered most of the questions that had been asked of integration, and his work paved the way for further refinement. He showed how his new definition of the integral, called the Lebesgue integral, could now handle the pathological cases that had been used to undermine the field. The Lebesgue integral had far fewer cases where integration was not the inverse of differentiation, and it gave hope that more work would reduce these further.

Lebesgue's work helped to reassert the fundamental importance of integration and resolved a century-long discussion of the differentiability properties of continuous functions. His work on measure and integration greatly expanded the scope of Fourier analysis and is one of the foundations of modern mathematical analysis.

Lebesgue's radical ideas were presented in his thesis, submitted at the University of Nancy in 1902. At first his work attracted little attention. However, in 1904 and 1906 he published two books that helped popularize his arguments. Strong criticism of his work emerged, particularly as it relied on notions taken from Cantor's controversial set theory. Slowly, however, more and more mathematicians began to use the Lebesgue measure and the Lebesgue integral in their work. As the Lebesgue integral became more accepted, it helped boost the validity of Cantor's set theory.

By 1910 Lebesgue's work was well received enough that he was offered a position at the Sorbonne in Paris. However, Lebesgue did not concentrate on the field he had revived. Rather he moved on to other areas of mathematics, such as topology, potential theory, and Fourier analysis. He thought his work on the integral had been a striking generalization, and he was fearful of what generalizations were doing to mathematics. He said, "Reduced to general theories, mathematics would be a beautiful form without content. It would quickly die." Yet the field he created flourished in the hands of others.

Arnard Denjoy (1884-1974) extended the definition of the Lebesgue integral (the Denjoy integral), and Alfred Haar (1885-1933) did further work (the Haar integral). Lebesgue's ideas were combined with that of his contemporary T. J. Stieltjes (1856-1894) by a third mathematician to give the Lebesgue-Stieltjes integral—yet another extension of integration, allowing even more functions to be included. The work done by Lebesgue and those that followed him led some to suggest that while the concept of integration was at least as old as Archimedes, the *theory* of integration was a creation of the twentieth century.

Lebesgue's ideas are complex and abstract. Indeed, his ideas of measure and integration are generally only taught at post-graduate level in most universities. For the majority of cases the Riemann integral is still used, and the vast majority of undergraduate calculus courses worldwide still use Riemann's definition to explain integration. Most elementary calculus courses do not even go as far as Riemann's ideas, and the format of Cauchy's calculus is still taught today. However, while Lebesgue's ideas may be abstract and difficult to comprehend, the continued success and mathematical viability of the field of integration—at all levels—owes a great deal to his work in reviving the discipline.

DAVID TULLOCH

Further Reading

Halmos, Paul R. *Measure Theory*. New York: Van Nostrand, 1950.

Hawkins, Thomas. *Lebesgue's Theory of Integration: Its Origins and Development*. Madison, WI: University of Wisconsin Press, 1970.

Lebesgue, Henri. *Measure and the Integral*. K. O. May, ed. San Francisco: Holden-Day, 1966.

Development of the Fundamental Notions of Functional Analysis

Overview

In the early twentieth century, mathematicians learned to give a geometrical interpretation to sets of functions that met certain overall conditions. This interpretation allowed mathematicians to assign a norm or "length" to each function in the set and to provide a measure of how much two functions differed from each other. One set of conditions on the functions in a set defined a Hilbert space, which could be treated as a vector space of infinite dimensions. Somewhat more general conditions were allowed in a Banach space. Both types of function spaces are of great importance in modern applied mathematics, and the ideas of Hilbert space play a particularly important role in quantum physics.

Background

At the close of the nineteenth century many mathematicians believed that the theory of sets, introduced by the Russian-born German mathematician Georg Cantor (1845-1918), would provide a reliable and unifying basis for all mathematics. While not all mathematicians shared this view, it was embraced and popularized by the great German mathematician, David Hilbert (1862-1943), who in 1926 wrote, "No one shall expel us from the paradise which Cantor has created for us." By that time a small army of mathematicians had been at work, establishing a rigorous basis for set theory itself and providing set theoretic interpretations for algebra, geometry, the calculus, and the theory of numbers.

The most powerful notion used in applying set theory to other areas of mathematics is that of a mapping between sets. If A and B are sets, then a mapping from A to B associates with each member of A, or a subset of A, a particular member of B. A function, like $y(x) = 7x + 2$ can be thought of as a mapping from the set of real numbers into itself, since for every real number assigned to x a real number y is produced. The function $y(x) = \sin(x)$ is then a mapping from the set of real numbers to the set of real numbers between -1 and 1.

Another important notion is that of ordered pair—that is, a pair of elements, with the first drawn from a set A and the second from a set B. A function can then be thought of as a set of ordered pairs. The set of ordered pairs of real numbers can be taken as a description of the plane assumed in Euclidean geometry. The set of pairs of ordered pairs of real numbers, that is, the set of objects of the form ((a,b), (c,d)), where a, b , c, and d are each real numbers, then describes the set of directed line segments that can be drawn in the plane. Also, the length of each directed line segment, which we could write as length ((a,b), (c,d)), as determined by the Pythagorean theorem, could be thought of as a mapping from this set to the set of nonnegative real numbers. Likewise, the cosine of angle between the two line segments, which we can write as cos ((a,b), (c,d)), could be thought of as a mapping between the set of pairs of directed line segments and the real numbers between 1 and -1.

A number of mathematicians, including the great German mathematician Bernhard Riemann (1826-1866), had, as early as 1851, speculated that a geometrical interpretation could be given to sets of functions. The first systematic effort in this direction was provided in 1906 by French mathematician Maurice Frechet (1878-1973) in his doctoral thesis at the University of Paris. Frechet had first to concern himself with the idea of the limit of a sequence of points, so that he could speak of sets of functions as being complete if they contained the limit of every sequence of functions. Such a complete set of functions, then, would be called a *metric space* if there existed a mapping, which could be called the called the *metric*, between the set of all ordered pairs of functions and the set of nonnegative real numbers that had three important characteristics of length, namely:

metric (f,g) = metric (g,f) and is always nonnegative.

metric (f,g) = 0 if and only if f = g.

metric (f,g) + metric (g,h) is always greater or equal to metric (f, h).

The third requirement is called the *triangle inequality* because the sum of the lengths of any two sides of a triangle is always greater than the length of the remaining side.

David Hilbert became interested in sequences of functions as he began to consider the problem of expressing a given function as the sum of an infinite set of other functions. Hilbert's interest was ultimately a further development of the discovery by French engineer Jean-Baptiste Joseph Fourier (1968-1830) that any periodic function could be expressed as the sum of sine and cosine functions of the same period and whole number multiples of that period. Mathematicians had generalized Fourier's ideas to encompass other sets of functions, now called *orthogonal functions*, that could be used to express an arbitrary function defined over an interval of real numbers. Thus, given a set of orthogonal functions, any other function could be described over the same region by an infinite series of numbers, each of which described the contribution of the corresponding orthogonal function.

The fact that functions could be described by infinite series of numbers suggested that the set of all functions over a given interval could be treated as a vector space with an infinite number of dimensions. The implications of this important discovery were worked out primarily by German mathematician Erhardt Schmidt (1876-1959). Schmidt introduced the notion of an inner product space of functions. Such a space is characterized by a mapping, called the *inner product*, from the set of all pairs of functions to the set of real numbers. The inner product of a function with itself was always positive and could be taken as the square of a metric for the space. The inner product of two different functions could be given a geometrical interpretation as well, revealing how similar or "parallel" the two functions are. Despite the important role played by Schmidt, inner product spaces are usually called *Hilbert spaces* today.

Polish mathematician Stefan Banach (1892-1945) is generally credited with taking the most general approach to function spaces. Banach introduced the notion of a normed linear space, where the term *norm* denotes a generalized notion of length, and the space need not have an inner product. Such spaces, which include all Hilbert spaces as well as many others, are now called *Banach spaces*.

Impact

The field of mathematics introduced by Frechet, Hilbert, and Banach is termed *functional analysis* today. The name is slightly inaccurate, as it reflects an earlier focus on functionals, which are mappings between sets of functions and the set of real numbers. The evaluation of a function for one value is a functional, as are the operations in calculus of taking the definite integral of a function over an interval and evaluating a derivative at a point. The theory of function spaces includes a number of results important in dealing with functionals, but it is now valued more as a general theory of the transformations that are possible among functions belonging to a Banach space.

Functional analysis is an area of mathematics that has been strongly influenced by the physical sciences, going back to the work of Fourier, whose studies on the expression of functions as sums of other functions appear in a work on the flow of heat, a subject of great importance to early nineteenth-century engineers. Another principle that had drawn the attention of mathematicians was the principle of least action in mechanics, stating that between any two points in space and time, a particle would follow that path that minimized a linear functional, the action integral, of the path.

The greatest interaction of functional analysis with physical sciences, however, occurred in the development of the quantum theory of matter in the 1920s. Beginning in 1900 physicists had discovered that many aspects of the behavior of matter could only be explained by assuming that certain mechanical properties could only take on certain values. Thus, the angular momentum of any subatomic particle as measured about a given axis could only be a whole number multiple of a fundamental value. Further, it was discovered that material particles, like the electron, had wavelike properties, and that their behavior in atoms could better be understood by treating the electrons as standing waves. Finally, it was discovered that the act of measurement of a given property would generally change the state of the system.

As was determined primarily by Hungarian-born mathematician John von Neumann (1903-1957), the allowable states of a quantum mechanical system can be viewed as represented by functions constituting a Hilbert space. The physically observable quantities such as energy and angular momentum are then each represented by transformations, called *linear operators*, that map the Hilbert space into itself. The effect of a measurement is to convert the function representing the system before the measurement into one of a selected number of states for which the representative function is simply multiplied by a real number under the transformation. Such functions are called *eigenfunctions of the transformation*. The number is the value of the physical quantity that is observed measured, and the probability of each possible value is given by the inner product of the original state function and the corresponding eigenfunction. Thus, Hilbert space theory provides the needed mathematical basis for the study of quantum phenomena.

DONALD R. FRANCESCHETTI

Further Reading

Bell, Eric Temple. *Development of Mathematics*. New York: McGraw-Hill, 1945.

Boyer, Carl B. *A History of Mathematics*. New York: Wiley, 1968.

Kline, Morris. *Mathematical Thought from Ancient to Modern Times*. New York: Oxford University Press, 1972.

The Decimation of Mathematics in Hitler's Germany

Overview

When the Nazi regime came to power in Germany in the 1930s, Jews and those who tried to protect them were systematically persecuted, exiled, or killed. As the Nazis attempted to root out "Jewish influence" on German life, university science and mathematics departments, where Jews were particularly numerous, became favorite targets. In mathematics, Germany had been the center of the international research community in the first decades of the twentieth century. With the rise of Adolf Hitler, some mathematicians were murdered, and many others fled. The focus of the mathematical world shifted to the United States and Canada.

Background

The Enlightenment of eighteenth-century Europe brought new ways of thinking to Western Europe. Political freedom as well as the freedom to ask questions about the way the world worked became at least ideals to strive for, even if they could not be immediately attained. New opportunities became available to ethnic and religious minorities, although restrictions were still common.

The blossoming of science and mathematics opened many new doors. At the universities, traditional fields such as the classics were seen as the most prestigious and were the most reluctant to admit those outside the cultural mainstream. The scientific fields, on the other hand, provided a new domain with fewer established hierarchies.

This came at a good time for the Jews in Germany. Influenced by the Enlightenment, many had begun to break away from strict religious observance and social isolation. They became more involved in German civilization, widely viewed at that time as among the most advanced in the world. At the same time, they retained a tendency toward scholarship, an important value in Jewish culture. Judaism encourages inquiry and disputation, and so contemplating shifts in scientific thinking about the universe did not particularly threaten the Jewish worldview. Bringing the long Talmudic tradition of abstract thought to the new methodology of scientific reasoning and research, many Jews were attracted to fields such as mathematics and theoretical physics.

For most of the nineteenth century, the anti-Semitism still found in many sectors of German life was primarily religious rather than racial. Among those Jews who had become completely secular and assimilated into German society, there were some who chose baptism in an attempt to solve this problem, reasoning perhaps that they could ignore one religion as easily as another. However, the economic depression that hit Germany in the 1870s led to resentment of successful people of Jewish extraction, regardless of their religious observance or lack thereof. The new racial anti-Semitism waxed and waned with the times. When economic hardships again arose in the 1920s following Germany's defeat in World War I, Jews provided a convenient scapegoat against which Adolf Hitler could marshal his armies of hate.

Many intellectuals in Germany did not believe it possible that Nazism was a serious threat. From the vantage point of the university, the Germany they saw was a center of culture and learning. Who would take a barbarian like Hitler seriously? They had an illusion of security provided by the company of others like themselves, but they were tragically mistaken, and some did not realize it until it was too late.

In 1933, a few months after Hitler became *Reichskanzler* (prime minister), the Nazi influence began to appear in the Berlin classrooms of what was then called the Friedrich Wilhelm University (now Humboldt University). Among the scholars there was a mathematics professor named Ludwig Bieberbach (1886-1982), who would go on to denounce "alien mathematics" in a series of racist speeches. Students also began showing up for class in the brown shirts of the storm troopers, and anti-Semitic signs were posted on campus.

Issai Schur (1875-1941) was a famous algebraist at the university. He was among those Jewish academicians who idealized the cultural and intellectual life of Germany. He was sure that his compatriots would soon come to their senses. He declined many invitations from colleagues in the United States and Britain who would have helped him leave the country, enduring years of Nazi persecution and daily humiliations. Finally, he escaped to Palestine in 1939. Broken in body and spirit, he died there two years later.

On April 7, 1933, the "Law for the Restoration of the Civil Service" was passed in Germany, mandating that anyone with one or more Jewish grandparents and/or leftist political leanings be "purged" from state employment. Since German universities are state institutions, professors and lecturers were covered by this decree. At first, Jews who had fought for Germany in World War I were exempted, although many resigned when ordered to fire others. In any case the veterans' exemption was eliminated after the Nuremberg Laws were implemented in September 1935. After the takeover of Austria on March 13, 1938, the law was extended to that country as well.

"Aryans" were made to swear oaths of loyalty to Hitler in order to retain their positions. Non-Jewish faculty seldom protested the treatment of their expelled colleagues. There were a number of reasons for this. Some were themselves anti-Semitic, and some were simply frightened. Many appreciated the opportunity to fill the professorial seats that had been suddenly vacated.

Impact

The mass migration of mathematicians to the United States and Canada resulted in a dramatic shift of high-level mathematics research from Europe to North America. Almost 200 mathematicians were expelled from Germany and Austria alone, including eleven students of Issai Schur. Hundreds more fled the Nazis' march through Europe, beginning with Czechoslovakia, where dismissals took place after the German occupation in March 1939. Most found it difficult to gain admittance to other countries because of the worldwide economic depression. Nations were reluctant to admit refugees when there was widespread unemployment and poverty among their own people. Prominent professors were forced to live as illegal aliens, accept support from charities, or take menial jobs. Many lost what should have been the most productive years of their professional lives. Yet those were the lucky ones, as they were able to escape.

About 75 German mathematicians, many of them internationally famous, emigrated to the United States. Many other European mathematicians joined them. Once established in universities, they trained a new generation of American mathematicians. They played the major part in making North America the new center of the world's mathematical community.

For example, before the rise of Hitler, the mathematical journal of record was the *Jahrbuch über die Fortschritte der Mathematik.* In 1931 it was joined by the *Zentralblatt für Mathematik und ihre Grenzgebiete,* intended to address complaints that the *Jahrbuch* publication process was too slow. Germans dominated mathematical peer-reviewing, and the primary language of mathematical discourse was German. Attempts in the 1920s to establish comparable American, British, or international journals, research-oriented and covering the whole of abstract mathematics, were unsuccessful without German contributions.

This all changed after 1933. The *Jahrbuch* was a semi-governmental institution, published by the Berlin Academy of Sciences. The Nazi Bieberbach was among its editors, and he dismissed all the Jewish reviewers. Mathematical physics, considered a Jewish discipline, was expunged from the journal's pages in 1936. At the *Zentralblatt,* the managing editor, Otto Neugebauer (1899-1990), fled to Copenhagen and later accepted a position at Brown University in Providence, Rhode Island. Tullio Levi-Civita (1873-1941), the co-editor and an Italian Jew, was dismissed in 1938. At last the publisher, Springer, adopted a policy prohibiting Jews from reviewing papers written by Germans. With the majority of the German journals' reviewers by then living in North America, the American Mathematical Society established the *Mathematical Reviews* in 1940, with Neugebauer at the helm. American mathematics began to dominate the field, and remains its center today.

In August 1998 the International Congress of Mathematicians met in Germany for the first time since 1904. The organizer of the meeting, the Deutsche Mathematiker Vereinigung (DMV), marked the significance of the event by presenting an exhibition entitled "Terror and Exile." The meeting took place in Berlin, where more than 50 mathematicians were expelled and three, the schoolteacher Margarete Kahn (b. 1880), the algebraist Robert Remak (1888-1940), and the logician Kurt Grelling (1886-1942), were murdered. After a solemn opening ceremony, the few surviving mathematicians who had been victims of the Nazi era were invited to speak as honored guests. The DMV representatives spoke of their hope that someday the world of German mathematics centered at Göttingen, Munich, and Berlin could be restored to the glory it enjoyed before it, like so much else of value, was swept away by Nazism.

SHERRI CHASIN CALVO

Further Reading

Periodical Articles

Fosdick, R.B. "Hitler and Mathematics." *Scripta Mathematica* 9 (1943): 120-22.

Reingold, N. "Refugee Mathematicians in the United States of America, 1933-1941." *Annals of Science* 38 (1981): 313-338.

Richards, P.S. "The Movement of Scientific Knowledge to and from Germany under National Socialism." *Minerva* 28 (1990): 401-425.

Siegmund-Schultze, Reinhard. "Scientific Control in Mathematical Reviewing and German-U.S.-American Relations between the Two World Wars." *Historia Mathematica* 21 (1994): 306-329.

Siegmund-Schultze, Reinhard. "The Emancipation of Mathematical Research Publishing in the United States from German Dominance (1878-1945)." *Historia Mathematica* 24 (1997): 135-166.

Other

Fleming, D. and Bailyn, B. *The Intellectual Migration: Europe and America, 1930-1960.* Cambridge, MA: Harvard University Press, 1969. (Book.)

Siegmund-Schultze, Reinhard et al. *Terror and Exile: The Persecution and Expulsion of Mathematicians from Berlin between 1933 and 1945.* Berlin: Deutsche Mathematiker-Vereinigung, 1998. (Pamphlet written for an exhibit at the 1998 meeting of the International Congress of Mathematicians in Berlin.)

Operations Research

Overview

Operations research is, by its very nature, a multidisciplinary field in which a group of experts attempts to manage and optimize complex undertakings. These undertakings can be military operations, industrial systems management, governmental operations, or commercial ventures. The heart of operations research is the mathematical modeling of part or all of these ventures, which can provide more insight into the interactions of certain variables of interest. The technique of linear programming was developed, in part, to help solve some of these problems. Operations research proved invaluable in solving some military problems in World War II and in subsequent wars and has had a significant impact on the way complex operations are planned and managed.

Background

With the Industrial Revolution in the nineteenth century, production systems became more automated and mechanized, many activities became more complex, and people began to supervise and manage machines that produced goods. This led to the field of industrial engineering, in which designers tried to make machines and production lines as efficient as possible to maximize overall efficiency and corporate profitability.

During World War II, problems arose that required new methods to reach a satisfactory solution. One of these problems was finding German submarines in the open ocean; another was maximizing the utility of Britain's early radar warning networks; yet another was planning complex operations such as the invasion of Normandy. Each of these was exceedingly complex and required the integration of several specialties. In addition, it turned out that each of these problems could be solved, in part, using mathematical analysis to arrive at an optimum solution. Put more succinctly, operations research involves the scientific and mathematical examination of systems and processes to make the most effective use possible of personnel and machines.

Please note that, in this context, "optimum" does not necessarily mean "ideal"; it simply means the best solution possible given the constraints under which the problem is solved. For example, when all is taken into account, the optimum amount of time in which to move three divisions of soldiers and all of their equipment from the United States to the Middle East might be three weeks or to move *only* the soldiers in two days. If the soldiers *and* their equipment are required in three days, the "optimum" solutions to this problem are hardly what the planners would desire, but they are the best ones possible, given certain factors such as the number of transport planes available, the number of soldiers and equipment that can fit into a single transport plane, the number and length of runways for both takeoff and landing, the loading and unloading capacity at all airports involved in the mobilization, the amount of fuel available on short notice, the airspeed of the aircraft, the availability of aircrews to load and fly the planes, and many other factors.

It was quickly noted that many operations research problems are very complex. In theory, a large number of these problems cannot be solved easily because the number of possibilities increases exponentially with the number of factors to consider. In actuality, however, many possibilities can be disregarded because, although mathematically valid, they are not considered in real life. For example, in trying to find the optimum routing for supplies from a variety of production locations to a central shipping location, it is mathematically valid to consider shipping supplies from one location to that same location, although in real life this would not be done. Similarly, it is not reasonable to assume that an attacking airplane will originate its flight from over friendly territory, so attack vectors that originate behind friendly lines may be disregarded in planning an air defense system. These and other similar possibilities may be disregarded when setting up such a problem, simplifying it to the point of being solvable.

The inaugural use of operations research came in the late 1930s when future Nobel laureate P.M.S. Blackett (1897-1974) was asked to develop a way to make best use of new radar gunsights for anti-aircraft guns being introduced to the British air defense system. Blackett assembled a team that included physiologists, physicists, mathematicians, an astronomer, an army officer, and a surveyor. Each person brought unique insights and job skills to the task, collectively reaching a solution in a relatively short period of time. This success was noticed quickly by other branches of the British military and by the Americans, all of whom adopted operations research by the early 1940s. Other problems addressed in this manner included hunting submarines in the Atlantic Ocean, designing the mining blockade in Japan's Inland Sea, and helping to plan many major amphibious operations during the war.

After the end of the war, operations research was also adopted by many American businesses, correctly seeing this as a powerful tool that could be adapted to nonmilitary use. In the business world, operations research helped design efficient flow paths for parts being incorporated into a final product, designed transportation systems, and optimized electrical distribution systems. At this time, too, the first mainframe computers were being built, giving researchers new capabilities. More sophisticated models were made possible by high-speed computers and the use of linear programming (a mathematical technique in which operations are described in equations in which no variables are raised to exponential powers or are multiplied together). Provided a system could be described in mathematically simple terms without losing touch with reality, linear programming made it possible to solve problems exactly, rather than by guesswork and experimentation.

Two milestones in the field of linear programming were the development of the technique by Soviet mathematician Leonid Kantorovich and the discovery of "polynomial-time" algorithms for many problems by another Soviet mathematician, Leonid Khachian. Both of these breakthroughs made the solution to many previously intractable problems possible. Linear programming is used to construct an equation that describes the interactions of various aspects of a problem, such as the military mobilization problem mentioned above. For example, if the number of men (n_m) that can be transported in a single plane is 200 and five planes (n_p) are available, then no more than 1,000 men can be transported at a single time. If the time (t_l) to load a plane, fly to the destination, unload it, refuel, and return to the point of origin is 18 hours, then in three days (t_t), the number of men that could be moved is equal to:

$$N = n_m \times n_p \times t_t / t_l = 200 \text{ men} \times 5 \text{ aircraft} \times 72 \text{ hours} / 18 \text{ hours} = 3000 \text{ men}$$

In addition, certain inequalities will be set up as constraints on the problem. For example, a plane cannot hold more than 200 people, so we can say that $n_m = 200$. By programming a computer to solve all of these equations simultaneously while recognizing the inequalities, such problems can be "solved" by coming up with approaches that make the best possible use of available resources to reach a desired goal. Similar, albeit more complex analyses can be used to determine the best trade-off in this example, suggesting how to get the highest number of fully armed soldiers on the ground in the shortest period of time.

Impact

Operations research had an immdediate effect on the military success of Allied forces in the Second World War. While the Allies are likely to have won the war in any case, these techniques probably helped shorten the war and allowed it to be won on more favorable terms with fewer casualties. The same is true for subsequent wars, including the Persian Gulf War. In fact, operations research is increasingly important to the U.S. military in an era of declining enlistment, decreasing numbers of planes, ships, and tanks, and more complex military problems in a wider variety of scenarios. This combination requires the most careful possible planning to meet both national priorities and military obligations around the world.

Operations research has also influenced business and industry, giving companies the ability to analyze their operations for the most efficient use of personnel, transportation systems, and production systems. This, in turn, has increased productivity, reduced the use of resources, and streamlined operations, adding to profitability and corporate efficiency. Thes techniques have also been used by government organizations to increase efficiency in many areas.

Finally, in the area of science, operations research was perhaps the first truly interdisciplinary field of inquiry. As such, it not only pioneered many of mathematical techniques now used in other fields, but also helped pave the way for the current increase in interdisciplinary research. Assembling scientists from different specialties offers many different views of a problem and provides a wider variety of intellectual tools that can be brought to a solution. In a way, it is the equivalent of parallax: looking at an object from different vantage points to fix its position in space. This "intellectual parallax" not only gives scientists a better overall view of a problem, but allows the team a much better chance of reaching a solution because they can more quickly and more accurately outline the problem's boundaries.

P. ANDREW KARAM

Further Reading

Books

Devlin, Keith. *Mathematics: The New Golden Age*. Columbia University Press, 1999.

The Development of Computational Mathematics

Overview

In the post-World War II years, computers became increasingly important for solving very difficult problems in mathematics, physics, chemistry, and other fields of science. By making it possible to find numerical solutions to equations that could not be solved analytically, computers helped to revolutionize many areas of scientific inquiry and engineering design. This trend continues as supercomputers are used to model weather systems, nuclear explosions, and airflow around new airplane designs. Since this trend has not yet leveled off, it is still too soon to say what the final result of these computational methods will be, but they have been revolutionary thus far and seem likely to become even more important in the future.

Background

The first attempts to invent devices to help with mathematical calculations date back at least 2,000 years, to the ancestor of the abacus. "Rod numerals," used in India and China, were not only used for counting, but also for simple calculations, too. These gave way to the abacus, which was used throughout Asia and beyond for several hundred years. However, in spite of the speed and accuracy with which addition, subtraction, multiplication, and division could be done on these devices, they were much less useful for the more complex mathematics that were being invented.

The next step towards a mechanical calculator was taken in 1642 by Blaise Pascal (1623-1662), who developed a machine that could add numbers. Thirty years later, German mathematician Gottfried Leibniz (1646-1716) invented a machine that could perform all four basic arithmetic functions to help him avoid the tedium of manually calculating astronomical tables. In developing this machine, Leibniz stated, "it is unworthy of excellent men to lose hours like slaves in the labor of calculation which could safely be relegated to anyone else if machines were used." Unfortunately, mechanical inaccuracies in Leibniz's machine—unavoidable, given the technology of the times—rendered it undependable for any but the simplest calculations.

In 1822 English inventor Charles Babbage (1792-1871) developed a mechanical calculator called the "Difference Engine." Unlike previous machines, Babbage's invention could also solve some equations in addition to performing arithmetic operations. In later years, Babbage attempted to construct a more generalized machine, called an Analytical Engine, that could be programmed to do any mathematical operations. However, he failed to build it because of the technological limitations under which he worked. Others tried to carry Babbage's vision through to fruition, also without success because they, like he, were limited to constructing mechanical devices that had unavoidable inaccuracies built into them due to the technology of the day.

With the development of electronics in the 1900s, the potential finally existed to construct an electronic machine to perform calculations. In the 1930s, electrical engineers were able to show that electromechanical circuits could be built that would add, subtract, multiply, and divide, finally bringing machines up to the level of the abacus. Pushed by the necessities of World War II, the Americans developed massive computers, the Mark I and ENIAC, to help solve ballistics problems for artillery shells, while the British, with their computer, Colossus, worked to break German codes. Meanwhile, English mathematician Alan Turing (1912-1954) was busy thinking about the next phase of computing, in which computers could be made to treat symbols the same as numbers and could be made to do virtually anything.

Turing and his colleagues used their computers to help break German codes, helping to turn the tide of the Second World War in favor of the Allies. In the United States, simpler machines were used to help with the calculations under way in Los Alamos, where the first atomic bomb was under development. Meanwhile, in Boston and Aberdeen, Maryland, larger computers were working out ballistic problems. All of these efforts were of enormous importance toward the Allied victories over Germany and Japan, and proved the utility of the electronic computer to any doubters.

The first purely scientific problem taken on by the electronic computers was that of solving Schroedinger's "wave equation," one of the central equations used in the then-new field of quantum mechanics. Although this equation, properly used, could provide exact solutions to many vexing problems in physics, it was so complex as to defy manual solution. Part of the reason for this involved the nature of the equation itself. For a simple atom, the number of calculations necessary to precisely show the locations and interactions of a single electron with its neighbors could be up to one million. For several electrons in the same atom, the number of calculations increased dramatically, pushing such problems beyond the range of even most of today's computers. Attacking the wave equation was one of the first tasks of the "newer" computers of the 1950s and 1960s, although it was not until the 1990s that supercomputers were available that could actually do a credible job of examining complex atoms or molecules.

Through the 1960s and 1970s scientific computers became steadily more powerful, giving mathematicians, scientists, and engineers ever-better computational tools with which to ply their trades. However, these were invariably mainframe and "mini" computers because the personal computer and workstation had not yet been invented. This began to change in the 1980s with the introduction of the first affordable and (for that time) powerful small computers. With the advent of powerful personal computers, Sun and Silicon Graphics workstations, and other machines that could fit on a person's desk, scientific computing reached yet another level, especially in the late 1990s when desktop computers became nearly as powerful as the supercomputers of the 1970s and 1980s. At the same time, supercomputers continued to evolve, putting incredible amounts of computational power at the fingertips of researchers. Both of these trends continue to this day with no signs of abating.

Impact

The impact of computational methods of mathematics, science, and engineering has been nothing short of staggering. In particular, computers have made it possible to numerically solve important problems in mathematics, physics, and engineering that were hitherto unsolvable.

One of the ways to solve a mathematical problem is to do so analytically. To solve a problem analytically, the mathematician will attempt, using only mathematical symbols and accepted mathematical operations, to come up with some answer that is a solution to the problem. For example, the problem $0 = x^2 - 4$ has an exact solution that can be arrived at analytically. By rewriting this problem as $x^2 = 4$ (which is the result of adding 4 to each side of the equality), we can then determine that x is the square root of 4, which means that $x = 2$. This is an analytical so-

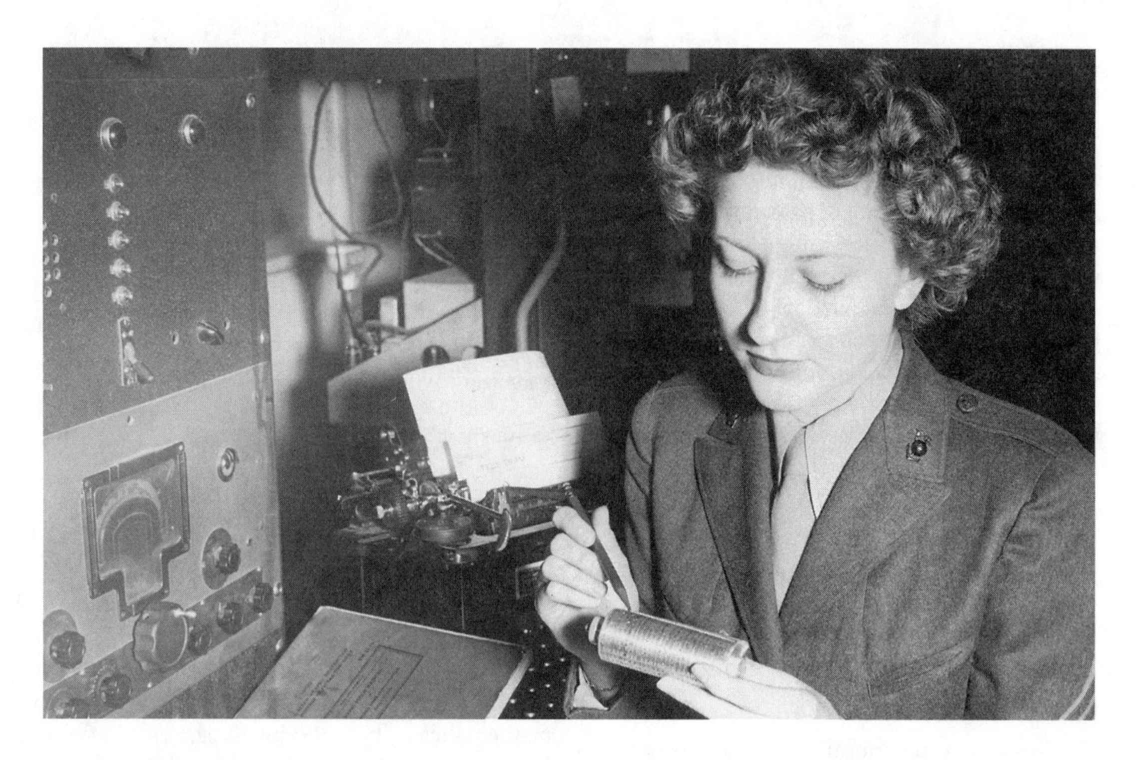

Cryptography, which became important during World War II, was spurred by developments in computational mathematics. Here a woman works on a cryptographic assignment during World War II. *(Bettmann/Corbis. Reproduced with permission.)*

lution because it was arrived at by simple manipulations of the original equation using standard algebraic rules.

On the other hand, more complex equations are not nearly so amenable to analytical solution. Schroedinger's wave equation, for example, cannot be solved in this manner because of its extreme complexity. Equations describing the flow of turbulent air past an airplane wing are similarly intractable, as are other problems in mathematics. However, these problems can be solved numerically, using computers.

The simplest and least elegant way to solve a problem numerically is simply to program the computer to take a guess at a solution and, depending on whether the answer is too high or too low, to guess again with a larger or smaller number. This process repeats until the answer is found. In the above example, for instance, a computer would start at zero and would notice that (0)2 - 4 = -4. This answer is too small, so the computer would guess again. A second guess of 1 would give an answer of -3, still too small. Guessing 2 would make the equation work, ending the problem. Similarly, computers can be programmed to take this brute force approach with virtually any problem, returning numerical answers for nearly any equation that can be written.

In other cases, for example, in calculating the flow of fluids, a computer will be programmed with the equations showing how a fluid behaves under certain conditions or at certain locations. It then systematically calculates the different parameters (for example, pressure, speed, and temperature) at hundreds or thousands of locations. Since each of these values will affect those around it (for example, a single hot point will tend to cool off as it warms neighboring points), the computer is also programmed to go back and recalculate all of these values, based on its first calculations. It repeats this process over and over until satisfied that the calculations are as accurate as they can be.

Consider, for example, the problem of trying to calculate the temperatures all across a circuit board. If the temperature of any single point is the average of the four points adjacent to it, the computer will simply take those four points, average their temperatures, and give that value to the point in the middle. However, when this happens, the calculated temperature of all the surrounding points will change because now the central point has a different temperature. When they change, they in turn affect the central point again, and this cycle of calculations continues until the change in successive iterations is too small to matter much. This is called "finite difference" computation, and it is a powerful tool in the hands of engineers and scientists.

The bottom line is that computer methods in the sciences have had an enormous impact on mathematics, the sciences, engineering, and our world. By freeing skilled scientists from the drudgery of endless calculations, they have freed these people to make more and more important discoveries in their fields. And by making some complex problems solvable for the first time, they have helped us to design better machines, to better understand our world and universe, and to make advances that would have otherwise been impossible.

P. ANDREW KARAM

Further Reading

Banchoff, Thomas. *Beyond the Third Dimension.* New York: Scientific American Library, 1990.

Kaufmann, William, and Larry Smarr. *Supercomputing and the Transformation of Science.* New York: Scientific American Library, 1993.

The Origins of Set-Theoretic Topology

Overview

Topology was a generalization of geometry that began to assume a separate identity in the nineteenth century. The study of set theory was first made a serious part of mathematics at the end of the nineteenth century. The two were combined in the twentieth century to the advantage of both, as set theory (which until then had seemed to have little to do with ordinary mathematics) found an application in topology and topology (which was capable of wandering off in many directions from geometry) could have a place of its own with the help of set theory. Many of the traditional areas of mathematics could be restated in general terms with the help of set-theoretic topology, which gave a new range of application to terms like "function."

Background

Geometry was one of the earliest branches of mathematics to assume identity as a separate discipline, largely thanks to the efforts of Euclid (fl. c. 300 B.C.) in his work the *Elements*. Euclid remained the source for all matters geometric well into the nineteenth century, although some by that time had come to worry about the specific form that the axioms (the assumptions from which Euclid started) took. In general, the material of the *Elements* covered topics like the areas of plane figures and the volumes of solids as well as the more abstract domains of geometry. Those in other branches of science could use the results of geometry even if they did not need to follow the proofs that Euclid supplied.

In the early nineteenth century several mathematicians began to investigate what would happen if some of the axioms of Euclidean geometry were dropped or altered. This non-Euclidean geometry did not seem to have any immediate application to the external world, but it could be connected with a variant of ordinary geometry called "projective" geometry. This study had attracted the attention of painters who were seeking to represent on a flat surface a three-dimensional image. Girard Desargues (1591-1661) had written an entire volume on projective geometry, which still seemed remote from the geometry of Euclid ordinarily taught in the classroom In the nineteenth century mathematicians like Arthur Cayley (1821-1895) in England recognized the possibility of connecting projective geometry with the rest of mathematics by treating it as an interpretation of non-Euclidean mathematics.

One of the central notions of Euclidean geometry was that of congruence of figures, which had to be the same size and shape. In projective geometry the idea of size dropped out of consideration and the notion of shape became central. For example, the Sun is known to be much bigger than a coin, but if the coin is held at an appropriate distance from the eye it will block out the sun exactly to the viewer. What is important for projective geometry is that the two objects could be made to appear the same size, a notion that was formalized in the work of Cayley and others.

German mathematician Felix Klein (1849-1925) used the work on projective and non-Euclidean geometries to produce a grand scheme for classifying all the various sorts of geometry that had begun to appear. In Euclidean geometry two figures could be considered the same if they had the same size and shape. In projective geometry it was enough if they had the same shape, even if they were not the same size. Still more general approaches to geometry could be introduced into this scheme, including topology, which took two figures to be the same if one could be turned into the other by bending or stretching but without tearing. (In technical terms this kind of transformation was known as a *continuous deformation*.)

The other ingredient that contributed to the creation of set-theoretic topology was the discipline of set theory, introduced into mathematics by Georg Cantor (1845-1918). Ideas about the infinite had been in existence in philosophical and popular scientific literature for many years, but there seemed to be paradoxes involved that would prevent their being used in mathematical applications. Cantor was able to define the central features of infinite sets in a way that would make sense of the paradoxes. Cantor had invented his set theory for use with the representation of functions, although there were other mathematicians at the time who felt quite uneasy about the use of the infinite in mathematics itself.

Impact

In Euclidean geometry straight lines and more complicated figures were made up of points, although the connection between the points and the line they made up was obscure. After all, the line was made up of an infinite number of points, and if a single point were omitted from the line, that did not seem to change the line itself. What was needed was precisely the idea of infinite set with which Cantor dealt. This provided the basis for the generalization of the terminology from ordinary geometry to the newer kinds of relationships and figures studied in topology. The word *topology* came from the Greek word *topos,* which referred to a place, and the word *topography* had been in existence for the making of maps. The objects with which topology had to deal were much more abstract, however, than the kind of maps put together in an atlas. For example, points in the new discipline of topology frequently did not represent just locations in a plane. The points could be used to represent functions and other mathematical objects.

Perhaps the most important contributor to the emergence of set-theoretic topology as a discipline was René Maurice Fréchet (1878-1973). Although his best-known book on abstract spaces did not appear until 1951, his ideas played a role in topology throughout the first half of the twentieth century. He recognized how to extend the ideas usually applied to the real numbers to the points that were generated from more complicated objects. Of these, perhaps the most crucial was that of distance, since the distance between two real numbers was easily defined as the absolute value of their difference. In more than one dimension, the Pythagorean theorem could be used to calculate the distance between points as the square root of the sum of the squares of the distances in the individual dimensions.

Fréchet observed that there were certain features about the distance function that were crucial to proceed to the rest of mathematics—for example, the triangle inequality held in ordinary geometry, which claimed that the sum of the lengths of two sides of any triangle was always greater than the length of the third side. Fréchet laid down that inequality as a requirement for the distance function when applied to points in the general spaces he had created. When this was combined with a couple of other requirements, a more general kind of geometry was possible, thanks to the additional abstractness of considering functions as points.

The work of Cantor allowed for even more general spaces than those for which a distance function was designed. German mathematician Felix Hausdorff (1868-1942) wrote a textbook on topology in 1914, perhaps the first to deal with the new subject. Cantor had defined a subset of a set to be a collection of objects contained in the set, so that (as an example) the even numbers are a subset of the whole numbers. In ordinary geometry one defined the objects near another object in terms of distance. For Hausdorff a neighborhood of a set could be defined in terms of subsets without even having a distance function with which to start. The basis for the choice of subsets that would define the neighborhood was the geometry of the real numbers, but the new definition could be given in an entirely abstract setting.

Hausdorff, Fréchet, and other contemporaries went to work trying to translate as much of the mathematics of standard geometry into a form applicable for the more general spaces of topology. In the language of the real numbers, an interval was the set of numbers between two given points. For topology this notion had to be replaced by that of a connected set, one that could not be separated into two sets by use of the neighborhoods that had been defined. One of the most important notions in calculus as performed on functions on the real numbers was that of continuity, which was often expressed in terms of being able to draw the graph of a function without having any gaps. In topology the notion of continuity was expressed in terms of open sets, a generalization of the open intervals from the ordinary real numbers. All the theorems of ordinary calculus had to remain true under the new definitions of topology, which encouraged even those taking more traditional mathematics as their domain to give the definitions a form more easily translated into the general setting.

Not all centers for mathematical research were equally quick to pick up set-theoretic topology as a subject worth investigation. After all, there had been plenty who objected to Cantor's introduction of set theory in the first place, and topology was not so thoroughly accepted as to carry much confidence. It is all the more striking that a relatively new mathematical culture like that of Poland (which had only become independent again after many years of foreign government in 1918) made a conscious effort to specialize in the areas of set-theory and topology. With the support of mathematicians elsewhere who were interested in the subject and of Polish mathematicians who had gone abroad, research into functional analysis—the application of set-theoretic topology to generalizations of problems from calculus—boomed. The periodical *Fundamenta Mathematica* may have been Polish in place of publication but its contributors were from all over the mathematical world.

Many members of the Polish school were forced either to go abroad during the Second World War or fell victim to the German invasion, but their results were disseminated through the centers for research elsewhere as well. Waclaw Sierpiński (1882-1969) was for many years editor of *Fundamenta Mathematica* and spent much of that time on set-theoretic topology. Of even greater importance was the work of Kazimierz Kuratowski (1896-1980), which proved to be even more important, as he used the ideas of Boolean algebra (the ideas of basic set theory familiar from before Cantor) to produce even more general spaces. In ordinary geometry one thinks of a boundary of a set as the points that define the shape of the figure. Then the closure of a set is the original set together with its boundary. Kuratowski took the idea of a closure operator more generally and defined the topology of an abstract space entirely independent of the notion of points. This proved to be more fertile than Fréchet's approach.

The area of set-theoretic topology blossomed in the hands of mathematicians from many countries. In addition to those mentioned above, Russia, England, and Hungary all were home to mathematicians who made contributions to set-theoretic topology. It remains an interesting historical example of a discipline that was appropriated by the mathematicians of one country (Poland) and made their specialty. One kind of abstract topological space is even called a Polish space in recognition of the central role played by the centers of research in Poland in promoting the subject.

The early part of the twentieth century saw the generalization of many branches of mathematics, much to the consternation of those who did not see the need to go beyond the traditional divisions. One of the lessons of the century has been that even problems stated in traditional terms may have a solution most easily found using weapons from a more abstract arsenal. Set-theoretic topology has supplied many of the tools involved in both stating new mathematical challenges and in solving old ones.

THOMAS DRUCKER

Further Reading

Manheim, Jerome. *The Genesis of Point Set Topology*. New York: Macmillan, 1964.

Moore, Gregory H. *Zermelo's Axiom of Choice: Its Origins, Development, and Influence*. New York: Springer-Verlag, 1982.

Richards, Joan. *Mathematical Visions: The Pursuit of Geometry in Victorian England*. San Diego, California: Academic Press, 1988.

Temple, George. *100 Years of Mathematics*. London: Gerald Duckworth and Company, 1981.

Hermann Minkowski Pioneers the Concept of a Four-Dimensional Space-Time Continuum

Overview

There is no doubt that Albert Einstein's (1879-1955) relativity theory changed the way we view the universe. Less well known is the extent to which Einstein's thinking was influenced by his former professor, Hermann Minkowski (1864-1909). Minkowski was the first to propose the concept of a four-dimensional space-time continuum, now a popular phrase in science fiction. Minkowski later became an influential proponent of Einstein's theories, helping them to gain acceptance despite their radical view of physics and the universe. Although Einstein consolidated work from many physicists and mathematicians in constructing his theory, Minkowski's contributions are noteworthy because of his influence over the young Einstein and physicists of his day.

Background

For millennia, mathematicians recognized that space could be divided into three dimensions—length, width, and height. These form the basis of Euclid's (330?-260? B.C.) geometry and virtually all subsequent geometry. In fact, it was not until 1826 that Russian mathematician Nicolai Lobachevsky (1793-1856) developed the first non-Euclidean geometry; that is, the first geometry not based on Euclid's postulates. For example, Euclid theorized that straight lines intersect only at a single point. However, in some non-Euclidean geometries, lines may intersect each other multiple times. Consider, for example, the surface of a sphere, on which all nonparallel lines must intersect each other twice.

Another long-established notion was that time was separate from any other phenomenon in the universe. For centuries, time was more of a philosophical concept than a physical one, something for philosophers to ponder rather than scientists to compute. With the ascendance of the physical sciences in the seventeenth and eighteenth centuries, time began to acquire its current meaning, but it was still considered something apart from the rest of the physical universe, something not well understood.

In the latter part of the nineteenth century, this concept of time began to evolve further, coming even closer to our current concept. At the turn of the century, Minkowski first proposed the interwoven nature of space and time being, stating: "Henceforth space by itself, and time by itself, are doomed to fade away into mere shadows, and only a kind of union of the two will preserve an independent reality." This amazing concept influenced Einstein, who would carry it to heights not considered by Minkowski or any others of that time.

Explaining this conception of space-time is actually not difficult, when we view it with the advantage of a century's hindsight. Consider, for example, taking a trip. If you travel from New York to Chicago you go not only through physical space, but through time as well, because it takes a finite amount of time to make this journey. The trip cannot be viewed simply as movement through space, but must be seen as movement through time, as well. Similarly, if you traveled at the speed of light from the Earth to the Moon you would traverse a quarter million miles of physical space and a second and a half of temporal space. Even if you didn't move at all you'd travel through the time dimension.

Minkowski's conceptualization of space-time was a set of four axes, the familiar x, y, and z axes of high-school geometry class and a fourth, t axis, upon which time is marked. In this system, then, your trip from New York to Chicago would take you along all four axes; standing perfectly still would still take you along the t axis, moving into the future without moving physically through space. It should also be noted that movement along the t axis is not time travel, at least not as long as the rate of movement is exactly the same as the rate that time normally flows. Although parts of relativity theory address ways to change the rate at which time flows along this axis, these are not yet significant from the standpoint of human experience.

Impact

The impact of Minkowski's conception of space-time is hard to describe briefly because of its influence on Einstein's thinking and the subsequent impact of relativity theory. However, it is safe to say that his work influenced philosophy, physics, and popular culture during the mid- to late-twentieth century.

From the philosophical point of view, Minkowski's work led to some interesting and, in some cases, unsettling results. As mentioned above, for millennia time had been the province of philosophers as much as physicists, and theories abounded as to the reasons time existed, its nature, and why we perceive it the way we do. To see time treated like any of the physical dimensions was shocking to many who had spent time speculating about the nature of time and who had treated it as something special for so long.

Also shocking was that, unlike the physical dimensions, time had special properties. Chief among them was that it could only be traveled in one direction, and at only a given rate of speed. True, as later relativity theory was to show, the rate at which time passed was related to the observer's frame of reference, but the key issue was that, to the observer, time always passed at the same rate and it was the rest of the universe that seemed to experience time at a faster or slower rate, depending on the observer's velocity relative to the rest of the universe. These points, as much physical as philosophical, were very disturbing to many, and it took years until they were widely accepted. In fact, only after the widespread acceptance of Einstein's work was Minkowski's notion of space-time taken to be an accurate description of our universe.

In the world of physics, Minkowski's work, and its effect on Einstein's, was even more revolutionary. By showing time to be an inseparable part of space-time, Minkowski not only inspired parts of Einstein's work, but also helped set the stage for further levels of abstraction such as string theory in physics. In this theory, all elementary particles are viewed as vibrating "loops" that occupy no fewer than 11 physical dimensions, most of which are "compactified," or shrunken to the point of being unnoticeable. Virtually every aspect of string theory depends on looking at the universe in a vastly different way than in previous centuries—a way made possible in part by Minkowski's uniting the visible dimensions of space with the invisible dimension of time.

Minkowski not only changed modern physics, he greatly influenced the theory of relativity, which describes the rate at which time passes, and how this rate changes. At relatively low speeds, such as those we experience in our daily lives, this change is not noticeable. However, at high speeds (approaching the speed of light), these changes are very evident. This is because the speed of light appears to be the same to an observer, regardless of the observer's speed. So, for example, if you were in a rocket traveling at nearly the speed of light and you shined a light in the direction the rocket is traveling, you would not see the beam crawl towards the front of the rocket. Instead, you would see the beam move at the same speed as if you were standing still. That same beam of light seen by a stationary observer would also seem to move at the speed of light. In short, two observers, looking at the same beam of light will see it move at exactly the same rate, regardless of their speed relative to each other or the beam of light. The only way that this can happen is if time for the rapidly moving observer slows down so that, with respect to him or her, the beam seems to be moving at its "normal" speed. This prediction has been proven with a very high degree of accuracy in experiments performed in space and on earth and is held to be generally true throughout the universe.

Finally, Minkowski's conceptualization of space and time as inseparable has become part of the popular culture. In fact, the term "space-time continuum" has become a staple of science fiction and, in this guise, has become part of the vocabulary for many people who otherwise have no knowledge of physics. As a punch line in jokes, a plot gimmick in science fiction movies and books, or a phrase used to impress people at parties, it has entered the lexicon and is familiar to virtually everyone who reads or keeps up with the media. This widespread usage does not seem to have improved the general public's understanding or appreciation of theoretical physics, but then, a large number of people can also quote Einstein's famous equation, $E = mc^2$ without understanding it or its importance, either. However, simply knowing the term and understanding that it has something to do with physics and the universe is more than anyone in previous centuries knew, which is a significant step forward in the public's understanding and appreciation of physics. From this standpoint, Minkowski's work still influential.

P. ANDREW KARAM

Further Reading

Naber, Gregory L. *The Geometry of Minkowski Space-Time: An Introduction to the Mathematics of the Special Theory of Relativity.* New York: Springer-Verlag, 1992.

Schutz, John W. *Foundations of Special Relativity: Kinematic Axioms for Minkowski Space-Time.* New York: Springer-Verlag, 1973.

Thompson, Anthony C. *Minkowski Geometry.* Cambridge: Cambridge University Press, 1996.

Weaver, Jefferson Hane, ed. *The World of Physics: A Small Library of the Literature of Physics from Antiquity to the Present.* Vol. 2. New York: Simon and Schuster, 1987.

Seeking the Geometry of the Universe

Overview

The German mathematician Georg Friedrich Riemann died shortly before his fortieth birthday and long before the importance of his work was truly recognized. He left the equivalent of only one volume of writings. Yet these provided the tools with which, 50 years after Riemann's death, Albert Einstein formulated his general theory of relativity.

Background

In traditional Euclidean geometry, codified in about 300 B.C., a straight line is the shortest distance between two points. Euclidean geometry is easily visualized on a two-dimensional flat surface, or plane (like a piece of paper). Spaces of any number of dimensions in which Euclidean geometry holds are called *Euclidean* or *flat spaces*.

On a curved surface, however, the shortest distance between two points is a curve called a *geodesic*. If the surface is spherical, the geodesic falls on an arc of a great circle, the intersection of the surface of the sphere with a plane passing through its center. The usefulness of curved-surface geometry for such important pursuits as navigation seems obvious. However, progress was delayed for centuries by religious proscriptions against admitting that Earth was round, even in the face of observations by astronomers and mariners. The classic work on curved-surface geometry, *General Investigations of Curved Surfaces*, was published by Carl Friedrich Gauss (1777-1855) in 1827. Georg Friedrich Riemann (1826-1866) was among his students.

Riemann generalized geometry so that it applied to spaces of any curvature and in any number of dimensions. He began with the concept of coordinates, invented by the French mathematician and philosopher René Descartes (1596-1650). Coordinates are ordered sets of numbers that describe the position of a point. On a plane, which has two dimensions, the point is described by an ordered pair (x_1, x_2). In three-dimensional space, an ordered set of three numbers (x_1, x_2, x_3) is required.

This was as far as Descartes deemed it necessary to go, since three dimensions defined his view of the physical world. Riemann extended Descartes' invention to accommodate any number of dimensions *n*, with *n-tuples*, coordinates defined as $(x_1, x_2, \ldots x_n)$. Two *n*-tuples $(x_1, x_2, \ldots x_n)$ and $(y_1, y_2, \ldots y_n)$ are equal if all their elements are equal; that is, $x_1 = y_1, x_2 = y_2, \ldots x_n = y_n$.

If we let all the elements in an *n*-tuple vary over some range of real numbers, the set of all the distinct *n*-tuples we obtain is called a *manifold of n dimensions*. If the elements can only take discrete values, such as integers, we have a *discrete manifold*; otherwise, it is a *continuous manifold*. The intermediate possibility, in which some elements are continuous and some discrete, can always be handled as the sum of a continuous and a discrete manifold, so it need not be considered separately.

Just as the Earth has mountains and valleys, vast featureless plains and the Grand Canyon, the geometry of space must allow for the possibility that the curvature is not a constant. Distance must therefore be computed piecewise, as differentials, and added up. Such techniques of the calculus as applied to manifolds are called *differential geometry*.

For any Riemannian manifold, distances are computed in quadratic form. This means that the differential terms are squared, just like the terms in the familiar Pythagorean theorem for computing distances in flat space. The coefficients of the differentials define the metric, or theory of measurement, for that manifold. These coefficients may be functions of any type, so an infinite number of Riemannian geometries are mathematically possible.

Riemann developed formulas describing the curvature of manifolds in terms of the coefficients of the differentials. These were extended by his fellow German Hermann von Helmholtz (1821-1894) and Englishman William Clifford (1845-1879). There are three basic possibilities for the curvature of a space or a region of space. In positive or spherical curvature, straight lines eventually return back upon themselves. Such a space would have no end but would still not be infinitely large. Negatively curved or flat (zero curvature) space both have straight lines running out to infinity, but they differ in the behavior of parallel lines.

The mathematics of curved space were to prove useful to Albert Einstein (1879-1955) decades later. Riemann had perhaps been ahead of his time, but finally a physicist came along with the genius to apply his mathematics to the universe.

Impact

Relativity theory is based on four-dimensional space-time, consisting of the familiar three spatial dimensions with time as the fourth. The "distance" between two events is thus a space-time interval, and Einstein hypothesized that this interval is invariant for continuous transformations of the four coordinates, referred to as a change in *frame of reference*. Any observer, in other words, should come up with the same answer. Einstein's first theory of relativity was "special" as opposed to "general"; it was restricted to flat space-times and observers moving with constant velocity with respect to each other.

An accelerated reference frame presented Einstein with a more difficult problem, because acceleration implies a force, and classical mechanics demands that a force have an effect. How could the space-time interval between events thus remain invariant? Fortunately, Riemann had conveniently provided Einstein with another tool, the tensor calculus. He had developed it for a problem on heat conduction he addressed for a French Academy of Sciences competition. Later it was elaborated by Matteo Ricci (1835-1925) and Tullio Levi-Civita (1873-1941).

A tensor is a general way of expressing a physical or mathematical quantity and its transformation properties. A scalar, or constant, which is invariant under all transformations, can be thought of as a tensor of "rank 0." A vector quantity, which is invariant under coordinate transformations, is a tensor of first rank. The quantities commonly called "tensors" are those with rank of two or greater. Tensor analysis helps in identifying differential equations that are invariant for all transformations of their variables, which in the case of relativity are the 4-tuple coordinates of a point in space-time. If the system of equations results in a tensor that goes to zero, the system is invariant under transformations of its variables.

Einstein postulated that the square of the space-time interval is invariant, and zero in the case of the path of light. The path of a free particle was assumed to be a geodesic in space-time. Next came Einstein's famous thought-experiment of the man in the elevator. His accelerating reference frame produces effects that are to him indistinguishable from changes in the force of gravity. Similarly, Einstein reasoned, gravity is equivalent to a transformation of coordinates in a local region. The curvature of space-time in a local area, which is perceived as gravity, is a result of a mass in that area. This theory of general relativity has been supported by evidence of light's path curving in the vicinity of a large mass such as a star, and other observations.

While the Riemannian geometries were more general than anything that had been developed before, they did not cover all possibilities. The advent of general relativity, however, concentrated work in this area. Still, it is conceivable to envision geometries in which the differentials are not quadratic, or the transformations work differently. Hermann Weyl (1885-1955) was an important contributor to the discipline of non-Riemannian geometry.

SHERRI CHASIN CALVO

Further Reading

Bell, Eric Temple. *Mathematics: Queen and Servant of Science*. New York: McGraw-Hill, 1951.

Bochner, Solomon. *The Role of Mathematics in the Rise of Science*. Princeton: Princeton University Press, 1966.

Coolidge, Julian Lowell. *A History of Geometrical Methods*. New York: Dover, 1963.

Katz, Victor J. *A History of Mathematics*. New York: HarperCollins, 1993.

The Organization of the Mathematics Community

Overview

Along with the establishment of international fellowships, institutes and prizes such as the Fields Medal, both the formal and the informal formation of an international community of scholars of mathematics proved a powerful shaping force to the course of twentieth-century mathematical thought and research. In essence, the rise of an international mathematics community in the early part of the twentieth century was, with particular regard to scholarly communication, analogous to the Internet revolution at the end of the century. In both cases new communications media revolutionized mathematics teaching, publishing, and institutions.

Background

During the nineteenth century mathematical methods and applications became increasingly useful to science, engineering, and economics. Mathematicians also developed mathematical logic and abstract definitions, complex relations, and theorems of pure mathematics. Although there were subtle divisions of mathematics at the beginning of the nineteenth century, by the end of the century there were full and formal divisions of pure and applied mathematics. Moreover, within these divisions, mathematicians took on increasingly specialized roles that resulted in a rapid compartmentalization and specialization of mathematics. This often left mathematicians disconnected from their scholarly colleagues.

During the early decades of the twentieth century, the specialization in mathematics accelerated. Mathematicians also increasingly worked in the shadows of tremendous advancements in the characterization of physical law. What was once pure mathematical theory found new expression and emphasis in the advent of relativity and quantum theories. In stark contrast to the insulated and isolated traditions of prior centuries, in such an intellectually tumultuous age, increased communication among scholars was a prerequisite for serious work.

Compared with the nineteenth century, the mathematics community of the early twentieth century seemed relatively cohesive. This cohesion resulted from nineteenth-century efforts to establish an international community of mathematicians and to increase communication between mathematical scholars. There were, for example, efforts to organizing an international meeting of mathematicians during the 1893 Chicago's World Fair. Finding the exchange of information and ideas invigorating, scholars moved to establish the International Congress of Mathematicians.

Beginning with an 1897 meeting in Switzerland, the International Congress of Mathematicians resolved to meet every four years, one of their prime objectives being to prevent the complete isolation of mathematics' diverging branches. Another goal of the early international congresses was to facilitate international communication and cooperation, although they often devolved to national competitiveness regarding the display of mathematical prowess.

The process for selecting the site of the quadrennial congress eventually came under the control of the International Mathematical Union (IMU). The site-selection process eventually became similar in many regards to the selection process for the Olympic Games. IMU members evaluate bids by potential cites to host the congress, and many political and economic factors can influence the ultimate decision regarding site selection. These matters are more than logistical details: decisions concerning where to host a conference can profoundly influence the content and tone of the proceedings. Because the general intent of the congress is to facilitate communication and exchange between mathematical scholars, the location must be accessible both physically and politically for potential participants.

Even more controversial are decisions regarding the content of congressional proceedings. Selected members of the IMU gather to screen potential participants and issue invitations to speakers for the lectures to be delivered at the congress. This also can profoundly influence the mathematics community. Nowhere was this more important than in the proceedings of the 1900 International Congress of Mathematicians in Paris. At the Paris Congress German mathematician David Hilbert (1862-1943) presented 23 famous problems that spurred mathematics research for scholars around the world. Hilbert's articulation of the major problems facing mathematicians profoundly influenced the course of twentieth-century mathematics. Indeed, because of the broad scope of many of Hilbert's problems, some problems may continue to provides challenges to mathematicians for centuries to come. During the later half of the twentieth century it became traditional to award the Fields medals and the Nevanlinna Prize (for information science) on the first day of each Congress.

Organizing the International Congress of Mathematicians was only one of many steps in the formation of an international mathematics community. Prior to the outbreak of World War II, various local and national organizations helped provide an international forum for scholars. The publication of the journal *Acta Mathematica,* for example, also marked a major step in linking international scholars.

Despite the success and popularity of the International Congress of Mathematics, the organization of the mathematics community was not instantaneous. Progress was made in fitful spurts and often, especially during times of political turbulence and discord, various mathematicians were called upon to renew efforts to establish permanent international ties. During the first half of the twentieth century, however, continuing and bitter feuds between France and Germany not only resulted in two World Wars but also spilled over into the academic world, carving deep divides between these two countries, both of which had outstanding mathematical traditions.

War played an important part in shaping the character of the international mathematics community. Unlike the emergent political institutions painted on the post-World War I political map, most of the international institutions dedicated to the advancement of mathematics were already in place and functioning—at least to a limited degree—early in the twentieth century. The goal of these societies, especially the International Mathematical Congress, was often to simply weather the storms of war.

War-heightened nationalism crept into what otherwise should have been purely academic decisions and affected both the submissions and published content of both mathematical societies and journals. Following World War I, the International Mathematical Union formally bound only the former Allied powers and most of the neutral nations. The vanquished countries, including Germany and Austria, were excluded from membership. As a result of these exclusions research work done in Germany was often ignored or shunned. To the dismay of many scholars, the International Congress also excluded scholars from former other Central Power nations as well.

Most scholars bristled against such exclusions, however, and the isolations were short-lived. By 1926 scholars of all nations were once again welcomed into the IMU. Nationalism was not, however, exclusive to the victorious powers, and many German mathematicians declined the invitation. In an effort to heal wounds, the 1928 conference, held in Italy, adopted a policy that attendance at the International Congress must be free of political and ethnic restrictions. A dispute between the organizers of the Italian conference and the governing powers of the IMU led to a loss of control by the IMU over the organization of the congress. In 1932 the IMU essentially ceased to function, and attempts to revive the union were thwarted by the gathering clouds of the Second World War.

In the aftermath of World War II, scholars found themselves caught in the crossfire of a very different Cold War. Early Cold War tensions prevented a complete reformation of the IMU (along less political lines) until mid-century.

Impact

One benefit of the growth of an international community of mathematicians during the twentieth century was the proliferation of peer-reviewed journals both of a general and specialized nature. In addition, beginning in the later portion of the nineteenth century there were increasing numbers of foreign-based contributions to institutional publications.

A growing international community of mathematical scholars also provided additional outlets for the dissemination of work that, prior to that time, was critiqued only at the more provincial level of the university or academy. The rise of the international community also tended to shunt the dissemination of work among selected scholars—often along nationalistic lines.

After the two World Wars, the world entered an era of superpower military industrialization, in which exhausted nations such as France could no longer maintain a dominant international position. Accordingly, these nations increasingly relied on scientific, mathematical, and cultural advancement as a source of pride. The Société Mathématique de France and the Association Française played in important role in providing a forum for furthering contacts between the rich French mathematical culture and the international mathematical community.

The repressive policies of Nazi Germany had two immediate effects on the international mathematics community. The first was a practical reisolation of Germany and German mathematicians. Few Germans participated in the 1936 Oslo International Congress. The second was a mass exodus of mathematicians from Germany in the 1930s and 1940s, an exodus and dissemination of talent to all parts of the Western world—especially to the United States and Canada. This migration provided the basis for improving postwar international communication.

Progress in formulating a truly international community of mathematicians was also not limited to the West. During the twentieth century the IMU and the International Congress forged links between Western and Eastern nations and, following World War II, Japan and China sent increasing numbers of mathematicians for training in the West.

Especially in a modern world where international dialogue is taken for granted as a byproduct of Internet development, it is important to note the very different conditions that existed at the start of the twentieth century, when mathematicians labored in relative isolation or—at best—with other members of their academic institutions. The rise of the international community provided new forums for communication that further revolutionized mathematics teaching, publishing, and research.

Work to establish an award for mathematical achievement culminated with the institution of the International Medal for Outstanding Discoveries in Mathematics, more commonly known as the Fields Medal. More than just an international research award, the Fields Medal was also designed to promote promising academic talent. For mathematicians, the medals eventually became the equivalent of a Nobel Prize. The awarding of the medals, and the spotlight of publicity brought to the honored mathematicians and topics, profoundly influenced the course of mathematics research during the twentieth century.

That such a general trend toward international cooperation in mathematics could occur in spite of strident nationalistic fervor and the distractions of war was testimony both to the higher ideals of scholarship and to the practical benefits of scholarly discourse.

K. LEE LERNER

Further Reading

Boyer, C. B. *A History of Mathematics*. Princeton, NJ: Princeton University Press, 1985.

Dauben, J. W., ed. *The History of Mathematics from Antiquity to the Present*. New York: Garland Press, 1985.

Kline, M. *Mathematical Thought from Ancient to Modern Times*. New York: Oxford University Press, 1972.

The Establishment of the Fields Medal in Mathematics

Overview

In the period between the two World Wars, despite frequently divisive international bitterness and rancor, Canadian mathematician John Charles Fields (1863-1932) worked to establish an award for mathematical achievement designed to underscore the international character of mathematics and promote promising academic talent. For mathematicians, the International Medal for Outstanding Discoveries in Mathematics, more generally known as the Fields Medal in mathematics, eventually became an equivalent of a Nobel Prize.

Background

Fields undertook his studies in mathematics in Canada, France, and Germany before spending the majority of his academic career at the University of Toronto. This international experience made him sensitive to political intrusion into academic life and aware of the need to unify the mathematics community. In 1924 Fields became president of the International Congress of Mathematics in Toronto. Far from being a truly international conference, however, Fields found that the scars of World War I had not yet sufficiently healed to permit the inclusion of German mathematicians. Disturbed at this, Fields proposed establishing an award that would renew international ties within the mathematics community by recognizing exceptional work and promise in mathematics. (Traditionally, Fields medals have gone only to younger mathematicians under the age of 40.) Following Fields's death in 1932, his will endowed such an award. The 1932 International Congress of Mathematicians in Zurich, Switzerland, adopted Fields's plan for an international award in mathematics, and the first Fields medals were awarded at the 1936 congress in Oslo, Norway.

It was Fields's intent, as expressed in a letter that preceded his will, that at least two gold medals be awarded at each meeting of the International Congress of Mathematics (which are held every four years). Fields stressed that the awards should be open to mathematicians from all countries and that the final decision on the awarding of the medals must be left to an international committee. In 1966 the ICM increased the possible number of medals awarded to four.

Although Fields intended the international committee awarding the medals to have wide discretion in their choices, he also made it clear that he wished the medals awarded not simply for academic achievement (i.e., work and research already performed) but also as an "encouragement for further achievement on the part of the recipients and a stimulus to renewed effort on the part of others." In addition to bestowing academic honor, the monetary award accompanying the Fields Medal (established from funds left over from the fractionated 1924 Toronto Congress) offers medal recipients a sum to facilitate their further work and research in mathematics.

Because Fields so adamantly wanted the medals to be a unifying incentive and influence for the mathematics community, he specified that no country, institution, or individual should be linked to the medals. Ironically, of course, the International Congress award eventually became known—contrary to Fields's own wishes—as the Fields Medal. Officially, the medal is titled the International Medal for Outstanding Discoveries in Mathematics. Even more bitterly ironic for an award designed to promote peaceful discourse and recognition, the Fields Medals awarded in 1936 were the last given until 1950, when international tensions following World War II subsided enough to allow the resumption of awards.

With his characteristic attention to detail, Fields also specified the nature of the medals, asserting that they should be of significant value (i.e., worth more than $200 in gold—a considerable sum during the Depression). Although Fields did not specify any special inscription, the medals bear the likeness of Archimedes (c. 287-212 B.C.) and carried a Greek inscription translated as a charge to "transcend one's spirit and to take hold of the world."

Impact

The International Congress of Mathematics committee that selects Fields Medal recipients eventually wielded a profound power to shape mathematics. This result was not unforeseen by Fields, who intended from the outset that the medals be used to stimulate new research and promote contributions to mathematics in areas selected by the committee. In this regard the international congress, working through the Fields Medal selection committee, was able to significantly influence the evolution of and emphasis on certain areas of mathematics. Subsequent to the establishment of the Fields Medal, various branches and subdisciplines of mathematics, for example topological analysis, received a substantial boost from the awarding of Fields medals.

Beyond directly influencing the course of work in mathematics though monetary awards, the Fields Medal also exerts a more subtle—yet substantial—influence. Serving essentially as a Nobel Prize for mathematics, the Fields Medal also garners public attention and stimulates a broader dissemination of information. Many academic mathematicians receive their only brush with public recognition and exposure through the receipt of a coveted Fields Medal. In addition, the committee's shifting recognition of work in certain fields attracts talented young mathematicians and helps focus attention on rapidly developing areas. The spotlight of attention cast by the Fields Medal has also illuminated work done at a few leading institutions, especially Princeton University's Institute for Advanced Study, where a substantial number of medal winners have held appointments.

The awarding of Fields medals is not without controversy. The selection committees are, of course, comprised primarily of mathematicians able to evaluate the difficult and often abstract mathematical concepts involved in nominated work. Although the committee mitigates individual academic biases and prejudices, some academicians charge that areas of work such as logic are too often ignored. In the later half of the twentieth century, especially during the height of the Cold War tensions, other mathematicians expressed a growing concern regarding bias for, or prejudice against, certain mathematicians or schools of mathematics based on nationalism or parochial scholarly interest.

As a result of attention to various fields and subdisciplines, the Fields Medal both reflects and stimulates trends in the highest levels of mathematics research. Because the nominating and selecting committees have wide latitude, they are often able to reward and stimulate work in areas that are far removed from practical application. As a direct result of the awarding the Fields Medal—especially as its influence and prestige grew throughout the course of twentieth century—diverse and abstract fields such as algebraic topology received a considerable boost. Especially for theorists, the challenge of mathematics that involved solutions to German mathematician David Hilbert's (1862-1943) famous list of 23 problems (posted at the International Congress held in Paris in 1900) found renewed appreciation.

As the twentieth century progressed, advances in science, especially in physics and cosmology, became increasingly dependent upon advances and application of mathematics. Accordingly, stimulation of some areas of mathematics work, particularly in topological geometry, was needed to keep up with growing demands of science. In 1936 the medals were awarded for work dealing with Riemann surfaces and the plateau problem. In the 1950s work in functional analysis, number theory, algebraic geometry, and topology drew favor from the committee. In the 1960s research in partial differential equations, topology, set theory, and algebraic geometry found favor. In the 1970s, in addition to continued stimulation of topology, new and important work in number theory, group theory, Lie groups, and algebraic K-theory found recognition. In the 1980s research in topology continued to garner recognition as did work with the Mordell and Poincaré conjectures. Links to modern physics were strengthened with the 1990s recognition of work with superstring theory and mathematical physics.

By the end of the twentieth century the Fields Medal would recognize work in chaos theory, a fusion of scientific and mathematical efforts to seek order in complex and seemingly unpredictable systems and phenomena including population growth, the spread of disease (epidemiology), explosion dynamics, meteorology, and highly complex and intricate complex chemical reactions.

K. LEE LERNER

Further Reading

Monastyrsky, M. *Modern Mathematics in the Light of the Fields Medal.* Wellesley, MA: A.K. Peters, 1997.

Tropp, Henry S. "The Origins and History of the Fields Medal." *Historia Mathematica* 3 (1976): 167-181.

Advances in Game Theory

Overview

Game theory provides ways to understand and predict behavior in situations where participants are driven by goals. Developed by John von Neumann (1903-1959) and Oskar Morgenstern (1902-1977), game theory reveals patterns of advantage for participants and has been applied broadly in areas such as jury selection, investment decisions, military strategy, and medical analysis. Some game theory models, like the prisoners' dilemma and zero-sum games, have become part of common discourse.

Background

Von Neumann was actively involved in a wide range of mathematical activities. In the 1920s he became aware of a series of papers by Emile Borel (1871-1956) that gave mathematical form to mixed strategies (combining rational selection and chance). Von Neumann also discovered a limited minimax (from minimum/maximum) solution for certain games. In 1928, von Neumann provided a proof for the minimax theorem, establishing one of the pillars of game theory.

The minimax theorem applies to zero-sum games (one in which the total payoff of the game is fixed). For two players, everything that one person gains is lost by the other. Minimax states that using a mixed strategy, the average gain or loss over time for a given game can be calculated. It also concludes that, for a large class of two-person zero-sum games, there is no point in playing. The long-term outcome is determined entirely by the rules, rather than any clever play.

The proof of minimax was a neat piece of mathematics, incorporating both topology and functional calculus, but it was up to von Neumann to expand the work and apply it to the real world.

The idea that games like poker, chess, or betting on a coin toss could provide deep insights into complex economic and social behavior is not obvious, and von Neumann's intuitive grasp of this may be his greatest contribution. The impact comes from three factors: First, games provide a ready reference point since game playing is an almost universal human endeavor. Second, games provide a model that incorporates strategy. Third, games can be varied and tested with real human players.

One example of an insight that can come from game theory is the concept of dominance. You and an opponent may seem to have a dozen choices each in a game. However, if all the outcomes are put into a matrix, you will probably find some choices that have a worse outcome in every single case. You will eliminate these choices from consideration and, looking at your opponent's point of view, eliminate his or her consistently unfavorable choices. Choices that are consistently worse are said to be dominated by other choices. They can be excluded as possibilities, immediately simplifying your analysis.

Von Neumann slowly built upon game theory, turning to applications in economics in 1939 after meeting economist Oskar Morgenstern. The two coauthored *Theory of Games and Economic Behavior* in 1944. The book was pivotal in establishing the field, laying the foundations for all subsequent work in game theory. It not only brought zero-sum ideas to a practical application (e.g., economics), it also looked at cooperative games and brought the notion of utility (quantification of preference) to game theory.

Impact

Economists immediately recognized the significance of game theory, but initially found it difficult to apply. Its first important, practical adoption was in military strategy. During the Cold War, game theory was used both for analysis of specific operations (e.g. dogfights) and overall strategy, such as the doctrine of mutual assured destruction that came to dominate nuclear politics for decades. The latter inspired some opposition to game theory as a dehumanizing tool. Looking at battles only in terms of victories or defeats without regard to human suffering was criticized as a dangerous simplification.

Some of the most important contributions toward realizing the potential of game theory in economics (and, by extension, other practical endeavors) came from John Forbes Nash (1928-), who, through his thesis in 1950 and an influential article the next year, encouraged its adoption. In particular, Nash developed bargaining theory and was the first to distinguish cooperative games, where binding agreements are feasible, from noncooperative games, where such agreements are not feasible. His solution for noncooperative games provided a way that all players' expectations could be fulfilled and their strategies optimized. This solution is now known as the Nash equilibrium.

One of the most fruitful games used for predicting the outcomes of strategic interactions has been A. W. Tucker's prisoners' dilemma. In this scenario, two men have been captured for burglary and are being questioned separately. If both keep quiet, they will get one year in jail each on weapons charges. If both confess and implicate the other, they will get 10 years each. If one agrees to testify against the other, who is keeping quiet, the cooperative one goes free and the other one goes to jail for 20 years. The "rational" solution is to confess and avoid the 20-year sentence, with a chance of immediate freedom. But it's hard to ignore the fact that if both stay quiet, they only get a year each, the best outcome possible for the team. The situation is paradoxical, but it reflects real-world circumstances, such as OPEC nations trying to limit oil production, budgeting for defense during an arms race, and compliance with anti-pollution laws. In each case, cheating can lead to big payoffs for individuals, but everyone is worse off if cheating is widespread.

Game theory continues to move closer to real-world situations. Further developments include the study of static games (where players play simultaneously), dynamic games (where players play over time), and games of incomplete information. In 1967, there was a serious challenge to von Neumann and Morgenstern's work in cooperative games. Since publication, most experts had assumed that, universally, each cooperative game had at least one solution. Neither von Neumann nor Morgenstern had been able to find a counterexample. However, William Lucas was able to prove that for one specific, complicated 10-person game, no solution existed. While this exception points to a fundamental incompleteness in the understanding of cooperative games, it does not invalidate the work of von Neumann and Morgenstern work, which continues to be useful both for theoretical studies and practical applications.

Game theory has become a dominant tool in economics. It has been used to regulate industry, providing, for instance, a basis for antitrust legislation and action. It has also been used to decide where best to site a new factory and to calculate the best price for new products. Game theory provides business strategists with guidance on how best to organize conglomerates and how to work cooperatively within their industries. Outside economics, game theory has been used to explain evolutionary biology and to formulate strategies to curb epidemics, encourage immunization, and test new medicines. In government and politics, game theory has been helpful in creating policies and strategies for voting, forming coalitions, and moderating the deleterious effects of majority rule. Attorneys have applied game theory to decide when to use their right to dismiss prospective jurors, and laws related to the fair distribution of inheritances have been formulated using game theory analyses. Game theory was seriously applied to philosophy as early as 1955.

Experimental economics, a growing field important to understanding commerce on the Internet, got its first foothold thanks to game theory. Von Neumann and Morgenstern's expected utility theory inspired serious testing as early as 1951. These experiments led to new theories in economics and extended the range and techniques of empirical testing in economics.

In 1994, the Nobel Prize for economics went to John Nash, John Harsanyi (1920-), and Reinhard Selten (1930-) for their contributions to game theory. Nash was honored for his investigations into noncooperative games and the results of seeing into an opponent's strategy. Harsanyi made original contributions to understanding dynamic strategic interactions; Selten demonstrated how games with incomplete information, where players don't know each other's objectives, could be analyzed; and both have provided mathematical proofs for game theories.

PETER J. ANDREWS

Further Reading

Books

Davis, Morton D. *Game Theory.* New York: Basic Books, Inc., 1983.

Gibbons, Robert. *Game Theory for Applied Economists.* Princeton, NJ: Princeton University Press, 1992.

McMillan, John. *Games, Strategies and Managers.* Oxford: Oxford University Press, 1996.

Other

McCain, Roger. "Strategy and Conflict: An Introductory Sketch of Game Theory." http://www.lebow.drexel.edu/economics/mccain/game/game.html

Advances in the Field of Statistics

Overview

Statistics provides a theoretical framework for analyzing numerical data and for drawing inferences from such data. Statistics uses the concept of probability—the likelihood that something will happen—to analyze data that are produced by probabilistic processes or to express the uncertainty in the inferences it draws.

In the first part of the twentieth century statisticians elaborated a number of theoretical frameworks within which statistical methods can be evaluated and compared. This led to a number of schools of statistical inference that disagree on basic principles and about the reliability of certain methods. Users of statistics are often happily unaware of these disagreements, and are taught a hybrid mixture of various approaches that may serve them satisfactorily in their practices.

Background

In the course of the nineteenth century huge masses of data were gathered by state agencies as well as private organizations and individuals on social phenomena like poverty and suicide, on physical phenomena like the heights of mountain tops, daily rainfall, and agricultural yields, and on repeated measurements, like the speed of light in a vacuum or the intensity of the gravitational field. But the means to analyze these data were found to be wanting, and where they did exist they lacked an overarching theoretical framework within which they could be compared and justified.

Statistics was often called the science of mass phenomena, and the ideal statistical investigation involved a complete enumeration of the mass that was studied: all the suicides, all the poor, the yields of all the fields in a certain region. Sampling, or the representative method, came into use towards the end of the nineteenth century, and then only in the form of purposive selection, where the sample is carefully constructed to mimic the whole population, rather than the more modern form of random sampling that gives a statistical control of error bounds.

Within agricultural experimentation comparison of means was common, but researchers were unclear how variability of yields affected the firmness of the conclusions they drew. While studying anthropological data comparing the height of children with the average height of their parents, Englishman Francis Galton (1822-1911) went beyond a simple comparison of averages and introduced the concepts of regression (1885) and later of correlation. Galton was motivated in this work by his interest in eugenics, the idea that a country's population could be improved by having people with desirable characteristics have more children, and people with undesirable characteristics, such as mental illness, have fewer, or no, children. Later British statisticians, such as Karl Pearson (1857-1936), who founded the influential journal *Biometrika* and the Biometric Laboratory at University College, London, and statistician and geneticist Ronald A. Fisher (1890-1962), were similarly motivated.

In the 1890s Karl Pearson introduced a system of frequency curves that differed in shape from the normal distribution in symmetry and peakedness. He also introduced a statistical test, the chi-square test, to determine the most fitting curve for a body of data. Earlier statisticians had developed a barrage of tests for the identification and rejection of "outliers"—data that were suspected to have errors of such a size that an analysis would be better if they were first thrown away. But the rules lacked a convincing justification and were often incompatible.

By 1950 the situation had changed dramatically. A number of different theoretical frameworks for statistical analysis had been elaborated: by Ronald A. Fisher, who stressed the design and analysis of comparative experiments using randomization, significance testing, analysis of variance, and the likelihood method; by Jerzy Neyman (1894-1981) and Egon Pearson, Karl Pearson's son, who stressed the basic idea that statistical inference is trustworthy when it derives, with high frequency, true conclusions from true premises; and by Harold Jeffreys (1891-1989) and Bruno de Finetti, who elaborated an analysis of statistical inference in which uncertain conclusions are expressed as probabilities, and inference is approached through looking at conditional probabilities—the probability of a hypothesis that is conditional on the collected data. In the meantime, statistical analysis had become essential to all the empirical sciences, from psychology to physics, and moreover had become indispensable for control processes in industry and management as stressed by the American W. Edwards Deming (1900-1993).

Ronald A. Fisher is the central figure in the founding of modern statistics. Educated as a mathematician and geneticist in Cambridge, England, in the early 1910s, Fisher was hired as a statistician by the venerable Rothamsted Experimental Station, an agricultural research institute, to analyze the backlog of data the institute had collected. Soon, he realized that it is not possible to derive reliable estimates from experiments that are not well designed, nor is it possible to calculate a measure of the reliability of the estimates. He laid down three fundamental principles—randomization, replication, and local control—to be followed in designing experiments. Randomization means that treatments are randomly allocated within the group of comparable experimental subjects. Replication is the repetition of the same treatment, and local control refers to the insight that only subjects that agree on covariates should be directly compared. These three principles made a calculation of valid error bounds possible and minimized the variance of the estimates.

The design of experiments is possibly Fisher's most outstanding contribution to statistics, and it has been incorporated into the theory of the various schools of statistical inference. He was the first to draw a clear distinction between a population and a sample drawn from it, and he introduced the classification of statistics problems into model specification, estimation, and distribution.

In his later work Fisher stressed that there are various forms of quantitative statistical inference and that a monolithic structure of statistical inference as was developed by the two rival schools, the frequentist school of Neyman and Pearson and the Bayesian school of Jeffreys and de Finetti, is not possible. The nature of the problem dictates the assumptions one can objectively make. When one can make few assumptions, a test of significance is appropriate. Here, one calculates the improbability of a deviation as large as observed assuming the truth of a so-called *null hypothesis*. Alternately, when one may assume a full parametric model, more powerful means of analysis come into play. The method of mathematical likelihood gives an ordering of rational belief, in which case the maximum likelihood estimate gives the most likely value of the parameter. In rare instances one may use the method of fiducial probability. This controversial derivation depends on an inversion of a pivotal quantity, a function of parameters and data that has a known distribution that is independent of the data. Fisher believed that future generations of statisticians may come up with further methods of inference, and he saw the future of statistics as necessarily open-ended.

Fisher occupies a position in between the two rival schools of probabilistic inference, the frequentist school of Neyman and Pearson and the Bayesian school of Jeffreys and de Finetti. The frequentist school rejects the idea that there is such a thing as statistical inference altogether, in the sense of data giving partial support to a hypothesis. The Bayesian school, named after English mathematician Thomas Bayes (1702-1761), relates all forms of statistical inference to the transition from a prior probability to a posterior probability given the data.

The Pole Jerzy Neyman was also trained as a mathematician and also worked, although briefly, for an agricultural research station. But Neyman had more of an affinity with a rigorous approach to mathematics. When Neyman came to England in the 1920s, he was disappointed by the low level of mathematical research in Karl Pearson's Biometric Laboratory. Being intrigued by Ronald Fisher's conceptual framework, he set out with his research partner, Egon Pearson, to provide a rigorous underpinning to Fisher's ideas, thus infuriating Fisher, who felt that the subtle points in his thinking were disregarded.

In a defense of his theory of statistical inference, Neyman argued that it would be better to speak of a theory of "inductive" behavior, since "inference" wrongly suggested that there is logical relation of partial support between data and hypothesis. Neyman rejected significance testing and replaced it by hypothesis testing. In significance testing the exact level of significance is a measure of discordance between the data obtained and the null hypothesis. A low level of significance will tend to make an experimenter reject the null hypothesis, but a significance test can not lead to the acceptance of the null hypothesis, since there are many other hypotheses under which the data may fail to be significant. In Neyman-Pearson hypothesis testing, one needs to have at least two hypotheses, and the goal of the statistician is to devise a data-dependent rule for the rejection of one of these hypotheses and acceptance of the other. Accepting means behaving as if a hypothesis is true without necessarily having a belief about the hypothesis, one way or another. The statistician will try to identify rules with good "operating characteristics," such as a high frequency of getting it right and low frequency of making an error. Fisher believed that this theory may be appropriate for testing batches of lightbulbs in industrial quality control programs but not for scientific inference, where a scientist should weigh what data means for the various hypotheses he entertains.

A framework for statistical inference in which Bayes's theorem is central was developed in England by the astrophysicist Harold Jeffreys and in Italy by Bruno de Finetti. Jeffreys tried to work out a version of objective Bayesianism in which the prior probability over the unknown parameters has an objective status dictated by the structure of the problem assuming further ignorance. Bruno de Finetti's version of Bayesian inference asserted famously that "probability does not exist." By this he meant that a statistician does not have to assume that there are stable frequencies or chances out there in nature in order to use the calculus of probability as a measure of personal uncertainty. We can measure this uncertainty by considering the various odds we are willing to accept on any event that we can observe in nature. Thus, when we repeatedly toss a coin, we are uncertain as to whether it will come up heads or tails, and we can express that uncertainty by a probability distribution. But talking of the chance of heads as an objective propensity inherent in the coin introduces a nonobservable property about which no bets can be settled. In his famous representation theorem, de Finetti showed that we can artificially introduce a probability distribution over such an unobservable parameter if we judge the sequences of heads and tails to be exchangeable—that is, that the probability is independent of the order of heads and tails. Especially after the 1950s the Bayesian school of scientific inference has come to great fruition and has become both conceptually and technically very sophisticated.

Impact

Users of statistical methods tend to be unaware of the great disputes that have occurred between the various schools of statistical inference. The popular statistical textbooks one studies in college tend to present a uniform hybrid theory of statistics in which hypothesis testing is interpreted as significance testing, but with the twist of the possibility of accepting a null hypothesis.

Notwithstanding these foundational disputes, the empire of probability greatly expanded in the first half of the twentieth century. Descriptive statistics became common fare in every newspaper, and statistical inference became indispensable to public health and medical research, to marketing and quality control in business, to accounting, to economic and meteorological forecasting, to polling and surveys, to sports, to weapon research and development, and to insurance. Indeed, for practitioners in many areas of the biological, social, and applied sciences, standardized procedures from inferential statistics virtually define what it means to use "the scientific method."

ZENO G. SWIJTINK

Further Reading

Box, Joan Fisher. *R. A. Fisher: The Life of a Scientist*. New York: Wiley, 1978.

Gigerenzer, Gerd, Zeno Swijtink, et al. *The Empire of Chance: How Probability Changed Science and Everyday Life*. Cambridge and New York: Cambridge University Press, 1989.

Reid, Constance. *Neyman—From Life*. New York: Springer-Verlag, 1982.

Emergence of Women at the Highest Levels of Mathematics

Overview

Throughout history women have made important contributions to the field of mathematics. Ada Byron Lovelace (1815-1852) wrote the first computer program in 1844, and Florence Nightingale (1820-1910) invented the pie chart. Despite their accomplishments, however, women mathematicians have faced almost insurmountable odds; some were persecuted, and one was even martyred. Hypatia (c. 370-415), the first mathematician to formulate the idea of conic sections, was brutally killed in 415 A.D. because she was a powerful intellectual. Sophie Germain (1776-1831) has a theorem named after her, but was barred from classes at the Paris Polytechnique in eighteenth-century France because she was a woman. French mathematician Emilie de Breteuil (1706-1749) received her excellent education only because her family thought her too tall and ugly to get married. Sonya Kovalevskaya (1850-1891) also has a theorem named after her, but as a woman she could neither enroll in classes in Berlin nor later obtain a university position in Germany or her native Russia. In nineteenth-century England, Mary Somerville (1780-1872) needed her husband's approval just to write a book about mathematics; her outstanding research about violet light magnetizing a steel needle had to be presented to the Royal Society by her husband because women were not accepted as members. For centuries, women mathematicians faced incredible oppression. Nevertheless, their genius, determination, and courage paved the way for their successors in the twentieth century.

Background

The main obstacle facing female mathematicians at the turn of the century was that few universities accepted women as doctoral candidates. One exception to this, however, was Göttingen Mathematical Institute in Germany. An important mathematical center, Göttingen was a magnet for both male and female mathematicians. Grace Chisholm Young (1868-1944) chose to study there because graduate schools in her native England did not yet admit women. In 1896, Young became the first woman to receive an official doctorate in Germany. (Sonya Kovalevskaya's 1874 doctorate was not official because she was not enrolled in classes at the time.) Young published her own book on geometry in 1905; it included patterns for geometric figures that are still used in math classes. The next year she and her husband, mathematician William Young (1863-1942), published the first book to provide comprehensive applications of problems in mathematical analysis and set theory.

Emmy Amalie Noether (1882-1935), a brilliant abstract algebraist, also attended lectures at Göttingen. When German law finally allowed women to be regular university students, Noether transferred to Erlangen where her father, a famous mathematics professor, taught. There she received her Ph.D. in 1907, *summa cum laude*, her thesis listing systems of more than 300 covariant forms. After her degree, Emmy Noether faced the second major obstacle that confronted women mathematicians in the first half of the twentieth century: few universities accepted them as research professors. Noether was not allowed to teach at the university in Erlangen because, professors argued, young German soldiers returning from World War I would be shocked to find a woman lecturing to them at their university classes. David Hilbert (1862-1943) convinced Noether to return to Göttingen in 1915. There she and Hilbert collaborated on the general theory of relativity, with Noether providing the mathematical formulations that became the basis of several concepts of Albert Einstein's general theory of relativity. A paper on differential equations that she coauthored in 1920 firmly established her reputation as a mathematical genius; the paper explained the axiomatic approach and made an invaluable contribution to the field of mathematics, specifically abstract algebra. By 1930, Noether was a vital part of the Göttingen mathematical team, a stimulating lecturer, and a prolific publisher.

The German political climate, however, forced Noether's dismissal from the university because she was Jewish. Colleagues found her a position in 1933 at Bryn Mawr, a women's college in Pennsylvania. Unfortunately Noether died there only two years later. Albert Einstein wrote in the *New York Times* that Emmy Noether "...was the most significant creative mathematical genius thus far produced since the higher education of women began." Today mathematicians study structures called Noetherian rings, named to honor Emmy Noether, who had developed them.

American Anna Pell Wheeler (1883-1966) was given a fellowship to study mathematics at Göttingen. She completed all the coursework but left before receiving her doctorate because of a dispute with Hilbert. Wheeler was nonetheless awarded her Ph.D. from the University of Chicago in 1909, using the thesis she had begun in Germany. She, too, taught at Bryn Mawr where she later became full professor and head of mathematics. The focus of her work was integral equations and infinite dimensional linear spaces. Wheeler was the first woman to give colloquium lectures at meetings of the American Mathematical Society.

In 1915, American mathematician Olive Clio Hazlett (1890-1974) also received her Ph.D. from the University of Chicago where she was her thesis advisor's second female student. She spent her career at the University of Illinois and was considered one of only two important women in America in mathematics. Hazlett wrote 17 research papers on nilpotent algebras, division algebras, modular invariants, and the arithmetic of algebras, more than any other pre-1940 female American mathematician. She was authored an article about quaternions for the *Encyclopedia Britannica* and was the editor of the *Transactions of the American Mathematical Society* from 1923-1935.

Mary Lucy Cartwright (b. 1900) attended Oxford from 1923 to 1930, after Oxford agreed in 1921 to allow women to take final degrees. Cartwright received her Ph.D. in mathematics from Oxford in 1930 and then accepted a fellowship at Girton College, the women's college at Cambridge University. Cartwright discovered the phenomena subsequently known as chaos. She received innumerable honors and was the first female mathematician elected as Fellow of the Royal Society in 1947. She was also elected President of the London Mathematical Society in 1951. Cartwright was awarded several medals honoring her mathematical achievements and became Dame Cartwright in 1969.

A major force of modern computer technology was Grace Murray Hopper (1906-1992). After earning her 1934 Ph.D. in mathematics from Yale University, Hopper taught at Vassar until World War II, when she joined the U.S. Navy Reserve. She pioneered computer software and coined the term "computer bug" after she discovered that an insect had shorted out two tubes. Rear Admiral Dr. Grace Hopper also invented the modern subroutine, the computer language APT, and played a significant role in verifying the computer language COBOL.

Olga Taussky-Todd (1906-1995) has the distinction of being the first woman to hold a formal appointment at the California Institute of Technology; in 1971 she became the first female full professor there and 10 years later, professor emeritus. Her doctorate in mathematics was from the University of Vienna. Taussky-Todd worked with Emmy Noether at Göttingen and then taught in England. Before moving to California, Taussky-Todd worked again with Noether at Bryn Mawr; the two often traveled to Princeton to lecture, where they frequently met Albert Einstein (1879-1955). Taussky-Todd's research focus was matrix theory and its application to emerging computer technology. She authored more than 200 papers and received innumerable honors, including the Ford Prize, the Austrian Cross of Honour for Science and Art (Austria's highest scientific honor), an honorary D.Sc. from the University of Southern California, and Fellow of the American Association for the Advancement of Science; she also served as Vice-President of the American Mathematical Society.

Nazi politics also forced mathematician Hanna Neumann (1914-1971) to flee Germany. Neumann was a doctoral student at Göttingen in 1938 when she realized that she would have to leave the country to marry her Jewish fiancé Bernhard Neumann, a fellow mathematician who had immigrated to Britain. Hanna Neumann continued her studies at Oxford University, where Olga Taussky-Todd supervised her research. Neumann's thesis studied the problem of determining the subgroup structure of free products of groups with an amalgamated subgroup. After her degree from Oxford in 1944, she taught and continued her research on Hopf's problem or, as it is now known, Hopf's property, questioning whether the free product of finitely many Hopf groups is again a Hopf group; she also worked on near-rings. Neumann and her husband accepted offers from the Australian National University, where in 1963 she became the chair of Pure Mathematics, a new position created for her. In recognition of her great contribution to the field of mathematics, Hanna Neumann was elected in March 1969 to the Fellowship of the Australian Academy of Science.

Impact

These brilliant, pioneering women forged the way for female mathematicians in the latter half of the twentieth century. Some examples: Evelyn Boyd Granville (1924-) obtained her doctorate from Yale in 1949, one of the first two African-American women to receive a Ph.D. in mathematics. Louise Hay (1935-1989), after her doctorate from Cornell in 1965, was named head of the department of mathematics at the University of Illinois at Chicago; at that time she was the only woman to head a major research-oriented university mathematics department in America. In 1971 Hay, along with Alice T. Schafer (1915-), were co-founders of the Association of Women in Mathematics.

The first award for Distinguished Service to Mathematics from the Mathematical Association of America went to Mina Rees (1902-1997) in 1962; eight years later she was elected the first woman president of the American Association for the Advancement of Science. Twenty years later, Berkeley mathematics professor Julia Robinson (1919-1985) became the first woman president of the American Mathematical Society.

By the close of the twentieth century, women had made a definite and permanent contribution to the field of mathematics. They had followed their predecessors and refused to allow prejudice or bias to thwart their ambitions. Many female mathematicians successfully combined career and marriage, raising children who often grew up themselves to be scientists and mathematicians. These extraordinary women were an inspiration to their sons and daughters, to their students, and to their colleagues.

ELLEN ELGHOBASHI

Further Reading

Books

McGrayne, Sharon Bertsch. *Nobel Prize Women in Science.* New York: Carol Publishing Group, 1993.

Morrow, Charlene and Teri Perl, eds. *Notable Women in Mathematics.* Westport, CN: Greenwood Press, 1998.

Osen, Lynn. M. *Women in Mathematics.* Cambrige: MIT Press, 1974.

Internet Sites

Agnes Scott College. Biographies of Women Mathematicians. "Other Resources on Women Mathematicians and Scientists." http://www.agnesscott.edu/Lriddle/women/resource.htm

Association for Women in Mathematics (AWM). "Emmy Noether Lectures. http://www.math.neu.edu/awm/noetherbrochure/Introduction.html

Four Thousand Years of Women in Science. http://crux.astr.ua.edu/4000WS/4000WS.html

History of Mathematics. http://aleph0.clarku.edu/~djoyce/mathhist/mathhist.html

University of St. Andrews Scotland. School of Mathematics and Statistics. "Emmy Amalie Noether." http://www-groups.dcs.stand.ac.uk/~history/Mathematicians/Noether_Emmy.html

University of St. Andrews Scotland. School of Mathematics and Statistics. The MacTutor History of Mathematics Archive. http://www-groups.dcs.st-and.ac.uk/history/index.html

University of St. Andrews Scotland. School of Mathematics and Statistics. "An Overview of the History of Mathematics." http://www-history.mcs.st-and.ac.uk/history/HistTopics/History_overview.html

The Emergence of African-Americans in Mathematics

Overview

Today, fewer than 1% of professional mathematicians are African-American. Compared to their overall proportion of the general population (slightly more than 10%), this is a very small number. There are a number of social and economic factors that explain this relative underrepresentation, one of which is that until the twentieth century few African-Americans were allowed to earn graduate degrees in any technical or scientific field. In spite of this, or perhaps because of it, African-American mathematicians have had an influence that extends beyond their research and teaching. By their example, these mathematicians have also demonstrated that Africans can succeed in complex, abstract fields, in spite of many whites who felt (or still feel) otherwise.

Background

No African-American was awarded a doctorate in mathematics until 1925, when Elbert Cox (1895-1969) became the first black in the world to earn this distinction. In that year only 28 mathematics doctorates were awarded in the entire United States. For a black American to achieve a place among this very select group just 60 years after the end of the Civil War at a time when even many advantaged students did not complete an undergraduate degree was impressive in the extreme. After Cox came an increasing stream of African-American mathematicians, augmented at times by very talented blacks from outside the United States. Within the United States, David Blackwell (1919-), Earnest Wilkins (1923-), Marjorie Lee Browne (1914-1979), and Evelyn Boyd Granville (1924-) each won distinction in their pursuit of mathematics, and deserve special mention.

David Blackwell has been called the greatest black mathematician. He earned his Ph.D. at the University of Illinois in 1941, only the seventh African-American to do so. During his career he published over 90 scientific papers and mentored over 50 graduate students in mathematics. He was the first African-American to be a faculty member at Princeton University, the first to become a fellow of the National Academy of Sciences, and the first African-American president of the American Statistical Society. His accomplishments are even more significant in light of the pre-Civil Rights era in which he achieved them.

The first woman to earn a Ph.D. in mathematics was Evelyn Boyd Granville, who graduated from Yale in 1949. Following on her heels was Marjorie Lee Browne, whose Ph.D. was completed at the same time, but not awarded until 1950. Other notable African-American women in mathematics include Etta Falconer and Fern Hunt, both of whom have made significant contributions in a number of areas. As black women they had to overcome double hurdles in their pursuit of a graduate education in mathematics. Not only were blacks often considered incapable of succeeding in mathematics, but the field was also traditionally thought to be the province of men only. The fact that these women not only achieved their goals, but did so in the face of this dual prejudice is impressive and speaks volumes of their intelligence and determination.

Through much of the 1800s and even into the twentieth century Africans were thought incapable of high intellectual achievement. In the United States, this is partly a result of slavery, which created a negative image of blacks in the public mind, a view that was reinforced by pseudoscientific findings that "proved" the inferiority of Africans. While there are many other reasons for these erroneous views, the truth is that blacks are as capable as members of any other race, a fact that is amply demonstrated by the success of the African-American mathematicians noted above, their colleagues, and African-Americans in other fields ranging from astronomy to engineering to the astronaut corps.

Impact

The influence of these pioneering African-American mathematicians cannot be understated. First, they made significant contributions to the field of mathematics through their research, teaching, and mentoring. These achievements alone refuted the argument that Africans lack the mental ability to comprehend advanced mathematics; by succeeding in this field they demonstrated their intellectual equality instead of merely arguing it. Finally, by their example, they showed other younger blacks that it is possible for them to succeed, too, in fields offering intellectual challenge. This also undercuts those who stereotype any racial group as inferior.

In addition, the scientific accomplishments of African-American mathematicians cannot not be underemphasized. David Blackwell's 90 papers in various branches of mathematics helped him become a member of the National Academy of Sciences, the first African-American to be so honored. This is not a "token" honor bestowed merely to balance racial disparity among recipients of such elite honors. Rather, it reflects his status as a great mathematician, regardless of skin color. Similarly, Nigerian mathematician George Okikiolu, who has published three books and nearly 200 scientific papers, has made significant and lasting contributions to the discipline of mathematics; his daughter Katherine promises more of the same.

As important as these scientific contributions are, it is likely that the example set by these men and women is even more important to society. As mentioned above, the centuries-long stereotype that blacks are not capable of performing at a high intellectual level has given innumerable bigots an excuse to discriminate against blacks in the job market, to place them in less skilled jobs, to keep them out of skilled military specialties, and more. When the first black earned a Ph.D. in mathematics, a field of extreme difficulty and abstraction, he showed that at least one black was capable of working at this rarified level of thought. As other African-Americans followed his example, they showed the world that he was not a fluke, but that blacks were as intellectually capable as any other race in the world. Fortunately, today increasing numbers of blacks occupy technical and scientific positions, which suggests that this stereotype may finally be eroding.

In addition to showing American society that African-Americans can succeed in mathematics, Blackwell, Hunt, Browne, and others have shown fellow blacks that they are equally capable. This is not only a point of pride for the black community, but a clear demonstration to all blacks that they, too, can do this level of work. In many ways, it is more discouraging to tell yourself, "I can't," than to hear someone else say, "You can't." Telling yourself that a goal—whether intellectual, career, financial, or otherwise—is impossible is to decide to not attempt it at all. Every African-American scientist who has succeeded has shown convincingly other blacks can attain similar success, limited only by their education, desires, determination, and talents.

It has been too short a time to determine how far and how rapidly the influence of these pioneering mathematicians will spread. We can hope that, as they continue to mentor students of all races and to produce high-quality research, they will continue to encourage young African-Americans to enter mathematics, engineering, and other technical fields. Everyone who can contribute to society should also have the opportunity to do so, and this opportunity, if seized upon, will benefit all.

P. ANDREW KARAM

Further Reading

Books

Dean, Nathaniel, ed. *African-Americans in Mathematics: DIMACS Workshop, June 26-28, 1996.* Providence, RI: American Mathematical Society, 1997.

Gould, Stephen Jay. *The Mismeasure of Man.*New York: Norton, 1996.

Newell, Virginia K., ed. *Black Mathematicians and Their Works.* Ardmore, PA: Dorrance, 1980.

Other

Williams, Scott W. University of Buffalo. "Mathematicians of the African Diaspora." http://www.math.buffalo.edu/mad/mad0.html.

Modern Probability As Part of Mathematics

Overview

Probability theory developed into a branch of abstract mathematics during the first 30 years of the twentieth century. Until the late nineteenth century probabilities were treated mostly in context, be it as the probability of testimony or arguments, of survival or death, of making errors in measurement, or in statistical mechanics. This is the era of classical probability. It was rife with paradoxes and had a low mathematical status. In the early twentieth century various efforts were made to develop a probability theory that was independent of applications and possessed a provably consistent structure. The theory that found near universal acceptance tied probability theory to measure theory. The Russian mathematician Andrei Kolmogorov (1903-1987) gave in 1933 a definitive axiomatic formulation of measure theoretic probability. Probability is now defined as a measure over an algebra of subsets of an abstract space, the space of elementary events.

Background

Probability theory as mathematics goes back to the 1650s when Blaise Pascal (1623-1662) solved a problem about the fair division of an interrupted game of chance. At that time the expression "calculus of chances" was used. Soon afterwards a connection was made with the art of conjecture, an analysis of "probable arguments," in which the conclusion is not conclusively, but only partially, established by the premises. Towards the end of the eighteenth century the term "theory" was sometimes used, but "calculus of chances" or "calculus of probabilities" remained the dominant expression until the end of the nineteenth century.

When much of mathematics was made more abstract and rigorous in the course of the nineteenth century, the calculus of probabilities was left behind as a part of so-called mixed mathematics. Besides games of chance, various other applications were tried, like trustworthiness of witnesses, different forms of insurance, and the statistics of suicides, but the most prominent applications were in the theory of making errors in measurement and in statistical mechanics.

In the theory of errors probabilistic assumptions were made about the frequency distribution of errors to justify using, in the most simple case, the arithmetical mean as the best estimate. Probability also entered in the calculation of the probable error of an estimate.

Later in the nineteenth century probabilities were used in statistical mechanics. In this theory the temperature of a gas is identified with the average velocity of the gas molecules. This allows the application of classical mechanics. However, the laws of mechanics are temporally symmetric: for every mechanical process the reversed process is equally possible. But observation tells that two gases easily mix but do not spontaneously separate. Probabilities were introduced to bridge the temporally symmetric treatment of gas phenomena within classical mechanics with the obvious time-directedness of mixing. Unmixing is highly improbable but not impossible. That is why it is never observed.

None of these applications implied that chance was anything real or that natural processes were indeterminate. In fact, due to the successes of science, determinism became more and more ingrained in the nineteenth century. Probabilities were only used because of ignorance of the true causes.

Ignorance was the hallmarks of the early or classical probability theory. With it came the principle of indifference that says that two possibilities are equally probable if one is equally ignorant about them. Many inconsistencies occurred in classical probability theory because of its connection with the principle of indifference. Different formulations of the same problem led to different pairs of possibilities about which one would be equally ignorant, and thus the principle of indifference gave different results dependent on how one had formulated the problem.

Another hallmark of classical probability theory was its concern with a finite number of alternative possibilities. Even the classical limit theorems—like Bernoulli's theorem, which says that, in coin tossing, for instance, the relative frequency of heads approaches, with probability going to 1, the probability of heads on a single toss when the number of tosses grows indefinitely—preserves the finite nature of classical probability, since it is formulated in terms of a limit of finitary probabilities rather than the probability of a limit. Later, in the twentieth century, after mathematicians had learned how to treat the probabilities of limits, results of this type concerning limits of finitary probabilities were called weak laws of large numbers.

It still comes as a surprise to learn that David Hilbert (1862-1943), in his call for an axiomatic treatment of probability theory during his famous lecture of 1900, listing important open problems in mathematics, still discussed probability theory as an applied field under the heading "Mathematical treatment of problems in physics." Hilbert mentioned in particular probability in the context of averages in statistical mechanics.

For Hilbert, a mathematical or physical theory constitutes a system of ideas possessing a certain structure. As the theory matures, certain key ideas emerge, serving as foundational principles from which the remaining results can be derived. But these are provisional. As the theory develops further results will require the reformulation of the axiomatic foundations. In a logical axiomatic treatment of any part of mathematics, the three prime considerations are: internal consistency, mutual independence, and completeness. The study of these properties for a mathematical theory is called metamathematics.

Internal consistency means that the various axioms do not contradict each other, and forms, for Hilbert, a proof of the existence of the mathematical concepts that are said to be implicitly defined by the axioms. Independence of axioms shows that none of the axioms are redundant or could have been derived as theorems. Completeness of a system of axioms means that the system is sufficiently strong to derive all results of the field as theorems.

At the heart of Hilbert's philosophical outlook stood his belief in the fundamental unity and harmony of mathematical ideas. One purpose of axiomatics is to show how the particular field is part of the whole of mathematics. The axiomatization of probability that was accepted 30 years after Hilbert's lecture does exactly that.

After Hilbert's call for a rigorous axiomatic treatment of probability theory, a number of his students worked on this problem. But the approach to the foundations of probability theory that attracted the most attention in the early twentieth century came from the German applied mathematician Richard von Mises (1883-1953). He developed an empirical frequency theory of probability, taking up earlier ideas of Wilhelm Lexis (1837-1914) and Heinrich Bruns (1848-1919). For von Mises, "the theory of probability is a science of the same order as geometry or theoretical mechanics. (...) just as the subject matter of geometry is the study of space phenomena, so probability theory deals with mass phenomena and repetitive events." To give a mathematical treatment of probability, von Mises considered an idealized situation, an infinite sequence of trials. Probability, according to his frequency theory, applies to the outcomes of an infinite sequence of trials if, first, the ratio of successes/trials has a limit, and, second, this limit is the same for all blindly chosen infinite subsequences. Von Mises calls such a sequence a *collective*.

The first condition corresponds to the idea that a probability is a stable frequency, although the stability may express itself only in the limit. The second condition is a randomness condition. A probability sequence should be highly irregular—no gambler who follows a gambling strategy, like betting heads every fifth time, or every time five tails have appeared, should be able to increase his odds of winning.

Various objections were raised against von Mises's frequency theory. Only later was it shown that the concept of blindly chosen subsequence can be defined in a consistent and satisfactory manner. The objection that a sequence which approaches its limit from above can be random was harder to answer. Such a sequence is not a typical probability sequence, since one expects a running average to fluctuate around the limit, not to hover constantly above it.

A more consequential problem with von Mises's approach may have been that it was very tedious to develop the known mathematics of probability theory within his framework. A probabilistic process in which the probability of an outcome is dependent on an earlier outcome has to be modeled, within von Mises's frequency theory, as a combination of dependent collectives. The measure theoretic approach of Andrei Kolmogorov would give a more elegant treatment of such dependencies.

The measure theoretic approach to probability derives from the measure theoretic study of asymptotic properties of sequences of natural numbers. Originally astronomers had studied these sequences in their efforts to prove that our solar system is a stable system in which planets could not suddenly run off into the depth of space. Around 1900 mathematicians started asking such questions as: How many rational numbers are there relative to all the real numbers? Or, put probabilistically, if one picks a real number at random what is the probability that it is rational? Or, formulated in measure theoretic terms, what is the measure of the set of rational numbers between 0 and 1, if the set of real numbers between 0 and 1 has measure 1?

In 1933 the Russian mathematician Andrei Kolmogorov published a book in German titled "Basic Concepts of the Calculus of Probability." This influential monograph transformed the character of the calculus of probabilities, moving it into mathematics from its previous state as a collections of calculations inspired by practical problems. Whereas von Mises's frequency theory had focused on the properties of a typical sequence obtained in sampling a sequence of independent trials, Kolmogorov axiomatized the structure of the underlying probabilistic process itself, and independence of successive trials is only a special condition on the probabilistic structure.

Kolmogorov axiomatics starts with a basic set, the event space of elementary events, Ω. Events are identified as subsets of the elementary event space. In tossing a dice the event space consists of six elements {1, 2, 3, 4, 5, 6}, corresponding to the various numbers of eyes one can obtain. Getting an even number of eyes is then the subset {2, 4, 6}. But typically the space of elementary events will be a product space, corresponding to various combinations of outcomes, as when a diced is tossed repeatedly. Kolmogorov requires that the space of all events, F, is an algebra. This means that if some subsets are events their union is also an event and the complement of an event is also an event. Moreover, the set of all elementary events is an event. A probability is defined as a measure function on the event space: each event should have a measure or probability, a number between 0 and 1. The largest event, the set of all elementary events, should have probability 1. The probability of the union of a (countable) number of disjunct events should be the sum of the probabilities of these disjunct events.

A consequence of this approach is that an event can have probability 0 without being impossible. In the measure theoretic approach the link between probability 0 and impossibility is broken, just as the set of rational numbers within the real interval [0, 1] has measure zero, but obviously is not empty. This led to a problem for Kolmogorov: how to define conditional probability for those cases where the conditioning event has probability 0. In classical probability theory conditional probability is defined as a ratio: P(A given B) = P(A and B)/P(B). But if the probability of the event B is 0 the definition is ill defined. In a startling innovation, Kolmogorov was able to define conditional probabilities as random variable, and prove that all the defining characteristics of probability could be satisfied.

Impact

Kolmogorov's measure theoretic axiomatization of probability theory opened up many new avenues of research, but also earlier work was expressed in it more precisely. Andrei Markov (1856-1922) had introduced in 1906 what are now called Markov chains with discrete time: sequences of trials on generally the same event space in which the probability of outcomes depends solely on the outcome of the previous trial. Kolmogorov's work made it possible to define the Markov property precisely. Problems from physics motivated the generalization to stochastic processes with continuous time. The general theory of stochastic processes became the central object of study of modern probability theory.

ZENO G. SWIJTINK

Further Reading

Books

Fine, Terrence L. *Theories of Probability: An Examination of Foundations.* New York : Academic Press, 1973.

Gigerenzer, Gerd, Zeno Swijtink, Theodore Porter, Lorraine Daston, John Beatty, and Lorenz Krueger. *The Empire of Chance: How Probability Changed Science and Everyday Life.* Cambridge (Cambridgeshire); New York: Cambridge University Press, 1989.

Von Plato, Jan. *Creating Modern Probability: Its Mathematics, Physics, and Philosophy in Historical Perspective.* Cambridge; New York: Cambridge University Press, 1994.

Investigations into the Irrationality and Transcendence of Various Specific Numbers

Overview

Number theory has often been thought of as one of the bastions of "pure" mathematics, unsullied by application to real-world problems and in which mathematics is pursued for the sheer beauty of the concepts involved. Starting in the latter part of the nineteenth century, however, some actual applications of number theory began to emerge. In the first half of the twentieth century, this trend continued. In addition, mathematicians began to prove some important points about specific numbers, such as π, *e* (the base for natural logarithms), the square root of two, and others. This work led to some very interesting and important advances in the way these numbers are viewed. In addition, since many of these irrational and transcendental numbers are widely used in physics, engineering, and other practical disciplines, this work has also helped to shed some light on interesting phenomena outside the rarefied realm of number theory.

Background

The area of a circle is calculated by multiplying the circle's diameter by the number π. The area of a square is obtained by simply multiplying the length of one side by itself (squaring the length of a side). From the time of the ancient Greeks, professional and amateur mathematicians struggled to find some way, using only an unmarked straight edge and a compass (a device for making circles or for marking distances), to create a square having *exactly* the same area as a circle. For two millennia, this quest was unsuccessful. In 1882, Ferdinand Lindemann (1852-1939), a German mathematician, was finally able to prove conclusively that the quest was impossible because π is a transcendental number, a class of numbers that cannot be calculated by algebraic means. While it may not be immediately obvious, using a straight edge and compass are algebraic methods of solving a problem, although the algebra is disguised as numbers describing the length of the straight edge or the radius of the compass opening. Since π is transcendental (can't be calculated exactly using algebra) and irrational (the decimal neither repeats nor terminates), it is mathematically impossible to "square the circle," in spite of all those who felt they proved otherwise for over 2,000 years.

Irrational numbers are numbers that cannot be expressed as a ratio of two whole numbers. The square root of two, for example, is an irrational number. Another way to look at it is that the decimals neither end nor repeat (⅓, for example, is written as 0.33333.... with the number "3" repeating endlessly while ¼ terminates as the decimal 0.25). *Transcendental numbers* are numbers that cannot form the roots, or solutions, of algebraic equations in which the exponents are real integers. For example, if we look for the solution to the equation $x^2 + x + 1 = 0$ (that is, for what values of x is this equation true?), the answer will be two *algebraic* numbers, because the numbers are solutions to an equation with whole-number exponents. On the other hand, π is not the solution to such an equation, so it is a transcendental number. This was shown conclusively by Lindemann. By so doing, he also proved that the ancient problem of squaring the circle was mathematically impossible.

All transcendental numbers are irrational, but not all irrational numbers are transcendental. In *Mathematics: The Science of Patterns*, Keith Devlin notes that the square root of two, while irrational, is not transcendental because it is the solution to the algebraic equation $x^2 - 2 = 0$. Joseph Liouville (1809-1882), the famous French mathematician, was the first to demonstrate the existence of transcendental numbers when, in 1844, he showed the limits of how precisely algebraic numbers could be approximated by rational numbers (i.e., by numbers that are the product of two integers and, as such, either terminate or repeat). By going through successively more accurate approximations, Liouville was able to demonstrate that algebraic numbers could not provide exact solutions to some mathematical problems. This proof, in turn, required that transcendental numbers must exist because these problems (such as determining the value of π, which is the ratio of a circle's diameter to its circumference) were known to have solutions.

Many efforts have been made to develop very precise approximations for the value of π. One of the more accurate was also one of the earliest; the ancient Greeks realized that a circle could be approximated by a polygon with an ever-increasing number of sides. Since the length of each side could be calculated, this gave an ever-closer approximation of the value of π. Other approximations include expressing π as a fraction (22/7 is the most common). However, regardless of the degree of sophistication, these are only approximations. As of 1999, the value of π had been calculated to over one billion decimal places with no end in sight. In fact, mathematically, an end to the sequence is not possible.

In an address at the Second International Congress for Mathematics, held in Paris in 1900, German mathematician David Hilbert (1862-1943) posed a series of problems to the assembled mathematicians, problems whose solutions would help to advance the field of mathematics. A total of 23 such problems were eventually presented, although not all at this congress. Hilbert's seventh problem dealt with the mathematics of transcendental numbers and has yet to be solved in its entirety. However, in 1934, Russian mathematician Aleksander Gelfond (1906-1968) proved a specific case to be true, which may someday lead to a general solution.

Hilbert's seventh problem asks: given an algebraic number α and an irrational number ß, will the number represented by $\alpha^{ß}$ always be a transcendental number? Gelfond was able to show was that, if ß is an irrational number (but not necessarily a transcendental number) and if

α is not equal to either 0 or 1, this number will always be transcendental, but the more general case has yet to be proved.

Impact

One possible reaction to all of this is, of course, "So what?" To some extent, this response is not inappropriate. However, irrational and transcendental numbers are extremely useful in a number of scientific disciplines, are used extensively in engineering (including electrical and electronic engineering), and it may behoove us to better understand them.

One minor problem, that of squaring the circle, could only be resolved by understanding transcendental numbers and their significance. While the solution to this problem will not feed the hungry or lead to world peace, it does help to show some of the limits inherent in one branch of mathematics—and that we will never know π with perfect precision.

Many computer graphics programs, including those used for computer-assisted design, will calculate the area of a circle by adding up the number of pixels the circle covers on the screen and then assigning a unit area to each pixel. However, this is an algebraic method of calculating area and, since π is not an algebraic number, all such areas are inherently inaccurate. While we may be able to refine this algorithm (and have) to the point at which the error is negligible, there will always be error in this process. In fact, it is impossible to reach an exact numerical solution for any calculation using irrational or transcendental number because the numbers do not terminate or repeat.

Other commonly used numbers that fall into one or both of these categories are the square root of two (irrational), *e* (transcendental and irrational), and many other square roots, cube roots, and higher roots (irrational, but not transcendental). Roots in particular, whether square, cube, or higher, may be irrational, but are always algebraic (not transcendental). This is easy to see because an equation such as the one given above ($x^2 - 2 = 0$) is the same form as any equation representing one of these roots. Since, by their nature, such equations are algebraic, their solutions, the roots, cannot be transcendental.

This, in turn, implies that other problems cannot be solved using only straight edge and compass. For example, it is not possible to create a cube with exactly twice the volume of a unit cube using only these tools because the cube root of 2 requires solving a cubic equation (an equation in which one variable is raised to the third power). This is beyond the abilities of such simple tools. However, such a problem *can* be solved if a *marked* straight edge is permitted. Similarly, dividing an angle into three equal angles can be solved with a marked straight edge, but not with the simpler tools. Both of these problems have obvious implications for computer-aided design as well. In effect, this work helps to show the limits of what we can achieve with simple tools.

Gelfond's work was important to mathematics, as has been virtually all of the work done on most of Hilbert's problems. In fact, that was the whole point of posing these problems—to try to spark mathematicians to new and greater heights at the start of a new century. In the case of Gelfond's work, by developing his partial solution to Hilbert's seventh problem, he was able to advance the field of number theory enormously. His techniques and methods were accepted by his colleagues and found use in other aspects of number theory, while his construction of new classes of transcendental numbers helped to advance that particular sub-specialty of mathematics as well.

P. ANDREW KARAM

Further Reading

Books

Beckmann, Peter. *History of Pi.* St. Martin's Press, 1976.

Devlin, Keith. *Mathematics: The Science of Patterns.* Scientific American Library, 1994.

Maor, Eli. *e: The Story of a Number.* Princeton University Press, 1998.

Other

Wolfram Research. "Eric Weisstein's World of Mathematics."http://mathworld.wolfram.com

The Thorough Axiomatization of Algebra

Overview

Late-nineteenth-century concerns about the meaning of number led, in the early twentieth century, to attempts to provide an axiomatic foundation for algebra and the theory of numbers. Building on the earlier work of Niels Abel and Evariste Galois on algebraic equations and using the framework provided by Cantor's theory of sets, a group of mathematicians led by Emmy Noether formalized definitions for a number of algebraic structures. Emphasis in mathematical research shifted from finding solutions to equations to the structures that such sets of solutions exhibit. The theory of groups, in particular, provided an important tool for theoretical physics.

Background

As mathematics has developed over the centuries, the concept of number has been broadened and generalized. Originally only the counting numbers, 1,2,3,..., or positive integers were recognized. The introduction of the zero and the negative integers turned arithmetic into a far more powerful tool for commerce, while the introduction of rational numbers or fractions allowed for a notion of proportion that was essential to the development of geometry and architecture.

Geometry, and especially the Pythagorean theorem, brought to light the existence of irrational numbers that could not be expressed as the ratio of two integers. Greek mathematicians knew that the square root of two is irrational but were disturbed by the knowledge. The disciples of Pythagoras (582-497 B.C.) treated this fact as a secret to be kept within the group.

With the development of algebra during the Renaissance, there appeared a need for numbers to represent the square roots of the negative numbers. A fully satisfactory treatment of complex numbers was not attained until the early nineteenth century, by which time other interesting generalizations of the number concept—quaternions and matrices, for example—were appearing.

A further classification of numbers arose from studies of algebraic equations involving rational coefficients and whole number powers of one unknown. Exact solutions had been found for the quadratic, cubic, and quartic equations by the sixteenth century. These solutions were expressed in terms of rational numbers and radicals—the square root, cube root, or fourth root of a positive or negative rational number. It came then as somewhat of a surprise when Norwegian mathematician Niels Henrik Abel (1802-1829) published a proof that the general fifth order equation did not have a solution of this form. Studies of the cases in which higher order equations were so solvable by the young mathematical prodigy Evariste Galois (1811-1832) involved studying the behavior of the solutions under a group of interchange operations, and gave rise to the group concept.

The group concept would be formalized further, making use of the set theory introduced by German mathematician Georg Cantor (1845-1918). A group is a set on which an operation is defined that matches each pair of elements in the set with one element of the set. Using the notation $\odot$ to indicate the operation, so that "$a \odot b = c$" means that element c is produced when a is allowed to operate on b, the set is said to form a group if three conditions are met:

(1) for all elements a, b, and c, in the set, $a \odot (b \odot c) = (a \odot b) \odot c$, where the operation in parenthesis is understood to be performed first

(2) there is an identity element i, such that $a \odot i = i \odot a = a$ for every element a in the set

(3) for every element a in the set there is an inverse element a', such that $a \odot a' = a' \odot a = i$

These properties may then be taken as the axioms defining a group. An example of a group is provided by the integers—positive, negative, and zero—with ordinary addition as the operation and zero as the identity. The first property is termed the associative property since it states that the result of two operations is independent of how the terms are grouped. Groups with an operation that obeys a commutative axiom are called Abelian groups after Abel. These are written thus:

(4) for all elements a, b in the group $a \odot b = b \odot a$

A ring is a set for which two operations have been defined, which we might denote as $\oplus$ and $\otimes$, with the set being an Abelian group under the $\oplus$ operation, while the operation $\otimes$ is associative, and there is an identity operation for it as well. Further, there is a distributive axiom

(5) a ⊗ (b ⊕ c) = (a ⊗ b) ⊕ (a ⊗ c)

The integers under ordinary addition and multiplication are a ring. The power of the ring concept was highlighted in 1907, when Joseph Wedderburn (1882-1948), a professor of mathematics at Princeton University, published a paper treating hypercomplex numbers, generalizations of the complex numbers and quaternions, from the field point of view.

A field is a ring for which the ⊗ operation is commutative and for which

(6) if c is not zero, and a ⊗ c = b ⊗ c, then a = c

(7) every element except zero has an inverse under ⊗

The real numbers, but not the integers alone, constitute a field.

An ideal is a subset of a ring that is itself a ring and has the property that when any of its elements is multiplied by any element of the parent ring, a member of the subset results.

As the real numbers are an example of a ring, so the subset of real numbers that can appear as solutions of a given type of algebraic equation generate an ideal. German mathematician Emmy Noether (1882-1935) adapted the theory of rings and ideals to the study of algebraic equations, in the process developing the abstract description of these structures. Austrian-born mathematician Emil Artin (1898-1962), who spent a year working with Noether at the University of Göttingen, continued to contribute to field theory throughout his long career. Dutch mathematician Bartel Leendert van der Waarden (1903-) was also an important contributor to the axiomatic school that developed around Noether.

Impact

The career of Emmy Noether reflects many of the difficulties facing female scholars at the beginning of the twentieth century, and also the rapid change in the status of women at about the time of the First World War. The daughter of a university mathematics professor, she was allowed to attend classes at the University of Erlangen but not permitted to enroll as a student because of her sex. By 1907 conditions had changed somewhat so that she could receive the doctorate degree from the university. From then until 1915 she wrote mathematical papers, supervised the research of advanced students, and occasionally took her father's place in the classroom, but without any official position or salary. In 1916 she joined the great German mathematician David Hilbert (1862-1943) at the University of Göttingen. Despite Hilbert's strenuous efforts, he was unable to obtain a salaried position for her until 1923. For the next decade she was one of the most influential mathematicians in Europe, until, like so many other intellectuals of Jewish descent, she fled the Nazi regime in 1933 for the United States. In America she accepted a teaching position at Bryn Mawr College, a select college for women, and became affiliated with the Institute for Advanced Study in Princeton, New Jersey, but died following surgery two years later.

Noether made important contributions to theoretical physics as well as to pure mathematics. Group theory provides a natural description of the symmetries of objects and of space itself. The set of possible rotations of a three-dimensional coordinate system, for example, form non-commuting groups. Noether was able to show that the independence of the laws of physics under rotations in space leads directly to one of the fundamental conservation laws in physics, that of the conservation of angular momentum. Further, the independence of the laws of physics under displacements in space or time, leads, respectively, to the laws of conservation of linear momentum and energy. Modern particle physics relies heavily on the connection between conservation laws and symmetry in developing theories of the behavior of elementary particles. Several new conservation laws have since been discovered, each connected to a basic symmetry of the underlying equations of motion. The importance of the fundamental relationship between symmetry and conservation is generally recognized, and the basic idea is included in many texts for physics undergraduates. Ironically, the individual responsible for this important theoretical insight is usually not identified.

Enthusiasm among mathematicians for group theory was perhaps excessive. The eminent French mathematician Jules Henri Poincaré (1854-1912) had said, "The theory of groups is, as it were, the whole of mathematics stripped of its subject matter and reduced to pure form." By 1930 it was clear that group theory, though important , would not continue to be seen as the essence of mathematics. At about the same time, group theory was embraced by the new field of quantum physics as an essential tool in understanding the quantum states of atoms, molecules, and nuclei. The more complex structures of rings, fields (in the mathematical sense), and ideals, have had some impact on mathematical physics, but not as much.

The axiomatization of algebra and its emphasis on the associative, commutative, and distributive laws also had an impact on elementary and high school mathematics teaching in the United States In the 1950s and 1960s a movement often described as the "New Math" spread through many school districts. Designed by university professors of mathematics and research mathematicians, the new curriculum emphasized understanding of the basic concepts of mathematics and the properties of numbers rather than the memorization of addition and multiplication facts. After some criticism from within the mathematics community, including the publication of a popular book, *Why Johnny Can't Add,* by New York University Professor Morris Kline, it became generally accepted that grade school students were not yet ready to acquire such advanced concepts. The effort was gradually replaced with an emphasis on the discovery of mathematical ideas by solving practical problems.

DONALD R. FRANCESCHETTI

Further Reading

Bell, Eric Temple. *Development of Mathematics.* New York: McGraw-Hill, 1945.

Boyer, Carl B. *A History of Mathematics.* New York: Wiley, 1968.

Crease, Robert P. and Charles C. Mann. *The Second Creation.* New York: Macmillan, 1986.

Kline, Morris. *Mathematical Thought from Ancient to Modern Times.* New York: Oxford University Press, 1972.

Kline, Morris. *Why Johnny Can't Add: The Failure of the New Math.* New York: St Martin's Press, 1973.

Advances in Number Theory between 1900 and 1949

Overview

The latter part of the nineteenth century and the first half of the twentieth saw major advances in many branches of mathematics, including the theory of numbers. Work on classic problems in the field led to important advances in our understanding of numbers and their relationships to each other. New work by both renowned and emerging mathematicians pointed the way to solutions of other classic problems, and also promised future advances. Much of this work focused on prime numbers, numbers that are only divisible by 1 and themselves. Although prime numbers have enchanted mathematicians for centuries, only in the last century or so did many of their properties begin to be better understood, including how to generate them and how to factor very large numbers into their prime components. Recent improvements in computer technology has made prime numbers central to many methods of encrypting information for personal, business, or governmental security.

Background

A number is an abstraction. It is difficult to explain to someone what, for example, the number 5 is without directly or indirectly referring to the number 5 itself. However, there are few arguments in daily life about the properties of the number 5. Once we learn to count, we all agree that 5 is greater than 4 and less than 6. After studying a little mathematics in school, we also agree that 5 is a prime number, one that is evenly divisible only by itself and the number 1. Further study may reveal other properties of this number, such as, when multiplied by even numbers, the product invariably ends in a 0 while, when multiplied by odd numbers, the product ends in a 5.

If we extend our inquiries a bit further, we may also notice that there are other prime numbers in the "vicinity" of 5; 2 and 3 as well as 7 and 11. We might also notice that prime numbers are not evenly grouped (2, 3, 5, 7, 11, 13, 17, 19, 23, 29, 31, 37, etc.), that (with the exception of 2) they are all odd numbers, and as numbers grow larger, primes generally become fewer and farther between. There are many other ways that numbers form groups, patterns, or have certain properties. The field of mathematical inquiry that studies these patterns and properties of numbers is called number theory.

Number theory is a relatively new subdiscipline of mathematics, which was first recognized in the nineteenth century as a formal field of study. Earlier mathematicians had, of course, done work in the field, including, most famously, Pierre de Fermat (1601-1665), Edward Waring (1734?-1798), Christian Goldbach (1690-1764), and others. In the first half of the twentieth century, equally brilliant mathematicians, including David Hilbert (1862-1943), Ivan Vinogradov (1891-1983), Srinivasa Ramanujan (1887-1920), Godfrey Hardy (1877-1947), and John Littlewood (1885-1977), made great advances in some areas of number theory, in some cases finding solutions to problems originally posed a century or more earlier.

In 1742, Christian Goldbach proposed that all even numbers larger than 4 were the sum of two odd prime numbers. Goldbach's conjecture, as this came to be called, has been proven numerically (that is, by computation) for a great many numbers, up to very large values, but the general case (i.e., for *all* numbers) has remained unproven to this day. A partial solution to Goldbach's conjecture, however, was proven in 1937 by the Russian Vinogradov, when he showed that every sufficiently large odd number is the sum of three odd prime numbers. This is, admittedly, a far cry from Goldbach's conjecture because it is less general. However, it was much closer to a solution than had been previously achieved and, perhaps, it will inspire mathematicians in the direction of a true general solution to this problem.

Another eighteenth-century mathematician, Edward Waring, proposed in 1770 that any number is the sum of no more than 9 cubes (a number multiplied by itself three times) or 19 fourth powers. Unfortunately, Waring proposed this without mathematical proof and he stopped at fourth powers, without extending his proposition any further. Waring's problem, as it came to be called, occupied mathematicians because it was something that seemed as though it should be true for all numbers, but that seemingly defied either proof or extension into higher powers.

In 1909 the great German mathematician David Hilbert was able to find a solution to Waring's problem, making a number of significant advances in number theory while doing so. Hilbert extended the boundaries of the problem by finding a general solution, valid for any number and for any power of equation.

Other important work in number theory was undertaken by English mathematician God-

Srinavasa Ramanujan. *(Granger Collection, LTD. Reproduced with permission.)*

frey Hardy, in collaboration with both John Littlewood and Srinivasa Ramanujan. With Littlewood, Hardy published a several important papers on number theory in which they examined the distribution of prime numbers (as discussed very briefly above) to see if any patterns could be discerned. For example, prime numbers become more scarce as numbers increase, but is there any way to determine how many primes exist between, say, 0 and 1,000 compared with the interval 9,000 to 10,000? Or, is there any way to predict the occurrence of "twin" primes; prime numbers pairs like 41 and 43 that are consecutive odd numbers? And, for that matter, are there any formulae that can be used to reliably generate prime numbers? For example, the formula $x^2 + x + 41$ generates 40 consecutive prime numbers, and then it stops being nearly as effective.

Harding's other major collaboration was with the Indian mathematical prodigy Srinivasa Ramanujan, a largely self-taught mathematician whose pursuit of mathematics at the expense of all his other studies led to his dismissal from the University of Madras in 1903. Following a year's correspondence and several important publications, Hardy arranged for Ramanujan to travel to England for joint work and tutoring. Their work, in conjunction with Ramanujan's own efforts, led to significant advances in a variety of areas, including number theory. In some areas, however, his lack of formal mathematics education showed, particularly in some aspects of prime number theory. However, these errors were not important compared to his mastery of other aspects of mathematics and, upon his death from tuberculosis at the age of 33, was generally regarded by his fellow mathematicians as one of those true mathematical geniuses who only grace us every century or so.

Impact

Many of the advances in number theory outlined above have little applicability to what most of us call "real life." Some, like the proof of Fermat's Last Theorem, arouse public interest for a short time, and then fade into obscurity. Others fail to create even a minor stir in the general public. From that perspective, advances that are of utmost importance and interest to mathematicians are often of no import to the nonmathematical world because nonmathematicians often fail to see any practical impact of these discoveries on their lives.

This attitude, while understandable, may result from the perception that mathematicians want to keep their field aloof from the rest of the world or from any practical application. While this undoubtedly has some justification, it is simply not correct. Even if there is no direct application, for example, of the Goldbach conjecture on daily life, it should be of at least passing interest because it tells us something about our world. We all use numbers daily, in counting change, picking flowers, numbering the pages of a book, and so on. We may not all sit and contemplate the fact that 2, 3, and 5 are the only prime numbers that close to one another, but knowing this and other facts about the numbers that crop up so frequently in our lives enriches us in at least a small way. In addition, there is at least one very important and very practical way that research into prime numbers that affects many people: the use of large prime numbers in encrypting data.

Prime number theory, although of interest to mathematicians, was of little practical utility until recently. In the 1990s, however, with the advent of very fast and inexpensive desktop computers, prime numbers became very important in the field of encryption and data security. In fact, the ability to find, multiply, and divide by very large prime numbers is the basis for generating secure military, commercial, and private encryption systems. Any algorithm that makes generating large prime numbers faster and easier aids encryption efforts while any algorithm that makes factoring very large numbers faster and easier aids those trying to break such codes. The most secure system currently available to the public (at the time this was written) uses very large prime numbers (larger than 3×10^{38}) to encrypt messages. The only way to decrypt the message is to factor this large number into its component primes and use them to restore the message to its original form. Obviously, any research in number theory that helps to generate or factor such large primes is relevant to cryptography, making data either more or less secure and making governments either more or less assured.

P. ANDREW KARAM

Further Reading

Books

Peterson, Ivars. *The Mathematical Tourist.* W.H. Freeman & Company, 1988.

Reid, Constance. *Hilbert.* New York: Springer-Verlag, 1970.

New Levels of Abstraction: Homological Algebra and Category Theory

Overview

Two of the more abstract branches of mathematics are homological algebra and category theory. Important progress was made in both during the first half of the twentieth century. Indeed, since the fields both arose in the latter part of the nineteenth century, virtually all work in them took place in the twentieth. While their practical effects may not be as great as their mathematical importance, research is still worth pursuing because the field of mathematics provides such an accurate description of the universe in which we live. This leads to the assumption that, even if these fields are seemingly of little import, the future may hold something more.

Background

According to the website Eric Weisstein's "World of Mathematics," category theory is "the branch of mathematics which formalizes a number of algebraic properties of collections of transformations between mathematical objects (such as binary relations, groups, sets, topological spaces, etc.) of the same type, subject to the constraint that the collections contain the identity mapping and are closed with respect to compositions of mappings. The objects studied in category theory are called categories." This definition includes a link to "category," about which it is said, "A category consists of two things: a collection of objects and, for each pair of objects, a collection of morphisms (sometimes called 'arrows') from one to another." While these definitions sound slightly abstruse, they can be understood without much effort.

To start, an object is simply any mathematical construct, be it a group, a space, a manifold, or anything else that can be defined mathematically. While we tend to think of an object as a tangible "thing," mathematical objects are no less real, even if they are less tangible. For example, a ball is a sphere. An equation can be written that, when plotted in three-dimensional space, will produce a sphere. Although the ball can be touched and the mathematical construct is intangible, is the graphed sphere any less real than the ball? And, by extension, aren't all surfaces, lines, curves, shapes, and other objects described by mathematical equations equally real?

If two objects can be related to one another through some sort of consistent mathematical relationship, this relationship is called a "morphism," or a map. Thinking about it, the term "map" is not unreasonable so much as it is unexpected. The typical image that comes to mind when hearing the word "map" is of a street map, not a mathematical relationship. However, a street map is precisely that—the "real world" constructed as (or related to) an image on a sheet of paper by means of a mathematical expression, in this case, simply a factor by which distances and sizes are divided. In a 1:250,000 map, each real world distance is divided by a factor of 250,000, giving a map distance. Of course, although we know by experience that the real world is mapped onto our street map, mathematically it makes as much sense to say that the map served as a template for the real world; in other words, we could speculate that our home city was created by simply scaling up the features shown on our map. The mathematical relationships used to map mathematical objects onto one another are much more complex than simple multiplication or division, but this added complexity does not make the result any less a map in the mathematical sense.

Category theory understands and explicitly states the rules that link these mathematical objects together by mapping one onto the other. By formalizing relationships in this manner, these collections of related mathematical objects can be categorized.

Homological algebra, a somewhat related topic, is defined by Weisstein as "[a]n abstract algebra concerned with results valid for many different kinds of spaces. Modules (another sort of mathematical object in which members can be added together in certain ways) are the basic tools used in homological algebra." It's also worth noting that algebra is a type of mathematics that studies systems of numbers and the mathematical operations that can be performed between the members of these systems. Incidentally, when mathematicians speak of an algebra, they are describing a set of mathematical rules and operations much more complex than what is taught in secondary schools (which is usually referred to as arithmetic); used mathematically, algebra means something much more complex.

The last term that needs defining is "homology," which was first used by Henri Poincaré (1854-1912) to define a relation between manifolds that was mapped onto a manifold of higher dimension in a particular way. Although this, too, sounds abstruse, the three-dimensional world is more complex than a two-dimensional piece of paper. So building a house, for example, from a blueprint would be one way of mapping a manifold (of sorts) onto a manifold of higher dimension in a particular way described in the instructions to the blueprint. In recent years, however, homology has come to mean what used to be called a homology group, which is a slightly looser definition of the term, extending it to a space rather than simply a manifold (which is a kind of surface).

Taking all of this together, we can see that homological algebra is a way to describe the mathematical rules by which objects can be related to (or mapped onto) each other, just as a blueprint is mapped into a house by following the general rules of construction.

Impact

One of the more interesting impacts of category theory lies in its application to one of the more philosophical problems facing mathematicians: whether mathematics and the physical world are composed of sets and set theory operations on these sets (i.e., unions of sets, intersections of sets, etc.). Under set theory and much of formal mathematics, this is the case. However, category theory is an alternative to this way of thinking. In category theory, there is no distinction between objects and the operations on them (i.e., transformations, morphisms, etc.). Whether category theory is a more complete or more accurate way of mathematically describing the world remains to be seen, but it is an alternate way that, along with other alternatives, has forced mathematicians to look more closely at their work and, by so doing, to understand it better.

Homological algebra is perhaps more abstract than category theory, although both disciplines are seemingly far removed from everyday life. Homological algebra, like category theory, primarily affects the manner in which mathematicians view their work and the effect their work has on the rest of mathematics. In this sense, both category theory and homological algebra, while not of earth-shattering importance, are of more than passing interest to the field of mathematics.

Much of the interest in homological algebra lies in the fact that multi-dimensional spaces are very important to both mathematics and the real world. We live in a multidimensional space—a space with three linear dimensions and a time dimension. Some physics theories require 10 dimensions to work correctly, positing that six of them are "compactified" in the same manner a box can be smashed into a flat sheet, seeming to lose a dimension in the process.

From this perspective, it is important to be able to map features from 10 dimensional space onto three or four, just as having the mathematics to map 10- dimensional space might be useful to physicists exploring this realm. This is one of the ways that seemingly abstract mathematics can connect with the real world: the world is described with mathematics and, no matter how complex the mathematics seem to be, the world is still more complicated.

In general, too, we must remind ourselves that the laws of the universe seem written in mathematics. There have been a great many instances in which arcane mathematics have been found to be precisely what was needed to permit a more appropriate mathematical description of real physical phenomena. In fact, while we think of Albert Einstein (1879-1955) as a great physicist, his contemporaries thought of him as a mathematician because his breakthroughs were made possible by the application of new mathematical techniques that allowed him to formalize and describe his insights in a manner that was convincing to the scientific world. From that perspective, it might just be that homological algebra and category theory, while currently without much direct impact outside the world of mathematics, may end up being as useful to future generations as were the mathematics that gave us relativity theory.

Even without a direct physical impact, however, it is still important to research even the most abstract and arcane mathematics. One of the things that sets us apart from other animals is our ability to consciously explore beyond our immediate surroundings, even when we are fed and comfortable. By pushing against the boundaries, be they physical, geographic, or mathematical, we are engaging in the activities that make us human. And we tend to learn lessons that help us in our everyday life, even if this process takes awhile.

Studies into homological algebra or category theory may never lead to world peace or feed the hungry. But, then, on the other hand, maybe they will. Thus far in our history as an intelligent species, we have had much success at turning seemingly abstract theory into practical reality. But, even if these fields never affect the average person, they are still worth pursuing because, someday, we or our descendants will be able to take some measure of pride in having understood our universe even a little better.

P. ANDREW KARAM

Further Reading

Books

Weibel, C.A. *An Introduction to Homological Algebra.* Cambridge University Press, 1994.

Other

Eric Weisstein's World of Mathematics. http://mathworld.wolfram.com

Biographical Sketches

Lars Valerian Ahlfors

1907-1996

Finnish-American Mathematician

Finnish-born mathematician Lars V. Ahlfors became one of the first two people to receive the Fields Medal in 1936. In awarding the medal, the International Congress of Mathematicians (ICM) cited Ahlfors's complex analysis work, and in particular his investigations of Riemann surfaces, schematic devices for mapping the relation between complex numbers according to an analytic function. Ahlfors's studies in this area led to developments in meromorphic functions—functions that are analytic everywhere except in a finite number of poles—which in turn spawned a new field of analysis.

Ahlfors's father, Axel, was a mechanical engineering professor at the polytechnical institute in Helsingfors, Finland, where Ahlfors was born on April 8, 1907. Ahlfors's mother, Sievä Helander Ahlfors, died giving birth to him, and the boy grew up close to his father. From an early age, the young Ahlfors took an interest in mathematics, teaching himself calculus from his father's engineering books.

In 1924, 17-year-old Ahlfors entered the mathematics program at the University of Helsingfors, where he came under the tutelage of two outstanding mathematicians, Ernst Lindelöf and Rolf Nevanlinna. Lindelöf, who taught all the significant Finnish mathematicians of Ahlfors's generation, took a paternal role with his students. Ahlfors's most intensive contact with Lindelöf was during his undergraduate years; after graduating in 1928, he followed Nevanlinna to the University of Zürich, where the professor was replacing the distinguished Hermann Weyl (1885-1955).

As a student of Nevanlinna, Ahlfors attracted notice when he proved Denjoy's conjecture on the number of asymptotic values of an entire function. He did this by using conformal mapping, a function involving the angle of intersection between two curves. This was a formative period intellectually for Ahlfors, who had never before been out of Finland or had any exposure to what he later called "live mathematics." Aside from Nevanlinna, number theorist George Pólya (1887-1985) had a strong influence on Ahlfors during this time.

After leaving Zürich, Ahlfors spent three months in Paris, where he developed a geometric interpretation of Nevanlinna's theory of meromorphic functions. He then took a job at Abo Akademi, a Swedish-language university in Finland, and began work on his thesis, which concerned conformal mapping and entire functions. After finishing his thesis in 1930, he earned his Ph.D. in 1932. Also in that year, he was named a fellow of the Rockefeller Institute, and as a result had an opportunity to study in Paris.

A much greater award was to follow, but in the meantime Ahlfors returned to his hometown of Helsingfors to take a job as an adjunct professor at the university there in 1933. Also in 1933, he married Austrian-born Emma Lehnert, and the couple eventually had three daughters. Ahlfors began a three-year assignment as an assistant professor at Harvard University in 1935, and in the following year he attended the International Congress of Mathematicians (ICM) at Oslo, Norway.

The ICM announced that it would begin awarding a new prize, the Fields Medal—equivalent to a Nobel in mathematics—to recognize mathematicians under the age of 40 who had produced outstanding work. Much to the surprise of the 29-year-old Ahlfors, the ICM presented him one of its first two Fields medals, citing his "research on covering surfaces related to Riemann surfaces of inverse functions of entire and meromorphic functions."

Ahlfors returned to Finland in 1938, taking a position as a professor at the University of Helsinki. At the latter part of World War II—which had forced the closing of the university, during which time Ahlfors continued his studies of meromorphic curves—Ahlfors accepted a position at the University of Zürich. This took him away from the fighting, and returned him to the place where his most fruitful work had begun; but after the war, Ahlfors returned to Harvard, where he began teaching in 1946.

He spent the remainder of his career at Harvard, during which time he published his influential *Complex Analysis* in 1953. Ahlfors retired in 1977, and five years later won the Steele Prize for his three editions of *Complex Analysis*. He died of pneumonia on October 11, 1996, near his home in Boston.

JUDSON KNIGHT

George David Birkhoff

1884-1944

American Mathematician

Founder of the modern theory of dynamical systems, which investigates the interrelation of separate motions in individual bodies and their impact on one another, George David Birkhoff is considered one of the most significant mathematicians of the twentieth century. Continuing the work of the distinguished French mathematician Jules Henri Poincaré (1854-1912) on celestial mechanics, he first attracted international attention when in 1913 he proved a geometrical theorem that Poincaré had proposed but not proved. He also made contributions in the study or relativity, quantum mechanics, and the four-color theorem, and developed a mathematical theory of aesthetics.

Birkhoff was born in Oversiel, Michigan, on March 21, 1884, the eldest of six children born to David Birkhoff, a physician, and his wife Jane Droppers Birkhoff. When he was two years old, his family moved to Chicago, where they lived during most of his childhood.

He studied at the Lewis Institute in Chicago, now the Illinois Institute of Technology, from 1896 to 1902, and went on to the University of Chicago. Birkhoff did not remain there long, however: in 1903 he transferred to Harvard University, where he earned his B.A. in 1905. In the following year, he received his M.A.

Birkhoff returned to the University of Chicago to begin work on his doctorate with a thesis on differential equations, which he wrote under the direction of Eliakim Hastings Moore. In 1907, he was awarded his Ph.D. degree with summa cum laude honors.

From 1907 to 1909, Birkhoff taught mathematics at the University of Wisconsin. During this time, he married Margaret Grafius of Chicago, with whom he had three children. In 1909 he took a job as assistant professor at Princeton University, and in 1912, he moved to Harvard, where he continued to teach for the remainder of his career. Among his students were some of the most notable American mathematicians of the later period, including Marston Morse and Marshall Stone.

The year 1912 also marked the death of Poincaré, whose work Birkhoff continued, first by proving Poincaré's geometrical theorem in 1913. The latter concerned the problem of three bodies in celestial mechanics, involving the trajectories and orbits of entities and the effect of their movement on one another. These investigations led him to his contributions in dynamical systems, a field of study for which he laid the foundations by defining and classifying possible types of dynamic motions.

George David Birkhoff. *(Library of Congress. Reproduced with permission.)*

Birkhoff wrote extensively during the 1920s and in following years, producing *Relativity and Modern Physics* (1923), in which he contributed to the growing study of relativity theory; *Dynamical Systems* (1928); *Aesthetic Measure* (1933), in which he applied Pythagorean notions in an attempt to reach a mathematical understanding of beauty; and the textbook *Basic Geometry* (1941). In 1931, he proved the ergodic theorem, concerning the behavior of large dynamical systems, an issue that had confounded physicists for half a century.

Birkhoff received the first Bôcher Memorial Prize from the American Mathematical Society in 1923. An active member of the Society, he served as its vice president in 1919, and its president from 1925 to 1926. He also edited its journal, *Transactions of the American Mathematical Society,* from 1921 to 1924. Birkhoff earned a number of other international awards. On November 12, 1944, Birkhoff died of a heart attack in Cambridge, Massachusetts.

JUDSON KNIGHT

Félix Edouard Justin Emile Borel

1871-1956

French Mathematician

Though he is noted for his work on complex numbers and functions, Emile Borel in fact wrote, researched, and taught on a variety of subjects. His *Space and Time* (1922), for instance, helped make Albert Einstein's theory of relativity comprehensible to non-technically educated readers, and his work extended far beyond the world of mathematics. As an influential figure in French politics, he helped direct that country's policy toward scientific and mathematical research and education.

Borel was born the son of Honoré, a pastor, and Emilie Teissié-Solier Borel, in the French town of Saint-Affrique on January 7, 1871. His was a world heavily influenced by the recent humiliation of the Franco-Prussian War, and on a personal level, Borel, as the younger brother of two sisters, had to fight to distinguish himself.

Distinction came early, however, with an invitation to study at several prestigious preparatory schools in Paris. Borel went on to the Ecole Normale Supériere, a preeminent school in science and mathematics with which he would remain connected for most of his life. After earning his doctorate in 1894, he returned to the Ecole Normale to teach. In 1901, when he was 30, he married Marguerite, the daughter of mathematician Paul Appell. The couple never had children, but adopted one of Borel's nephews, who was later killed during the First World War.

Borel in 1899 became the first mathematician to develop a systematic theory for a divergent series. His work with complex functions, or functions involving complex numbers such as the square roots of negative numbers, led him to prove Picard's theorem, which concerns the number of possible values for a complex function. He also provided an increased understanding of how to measure complicated two-dimensional surfaces, which are said to be "Borel-measurable." In his development of a general notion of area, Borel laid the foundations for what became known as measure theory.

In the realm of topology, Borel articulated a theorem of compactness, concerning the ways in which an infinite set mimics the simpler qualities of a finite set. His theorem became known by the rather inaccurate name of the Heine-Borel theorem—inaccurate because German mathematician Heinrich Heine (1812-1881) actually never formulated a theorem of compactness.

A prolific writer, Borel produced some 300 papers and books, among them *Lessons in the Theory of Functions* (1898), a classic in measure theory that he published when he was only 27 years old. In 1906, after winning the Petit Prix d'Ormoy, Borel used the prize money to start *La revue du mois* (Monthly Review), a journal of general interest which he edited with his wife.

During World War I, Borel conducted important investigations into probability theory that helped lay the groundwork for what later became game theory. He also earned the distinguished Croix de Guerre in 1918 for his work during the war. In the interwar period, he served in France's Chamber of Deputies (1924-36) as a representative of Saint-Affrique, and as minister of the navy in 1925. He also became instrumental in directing a number of efforts in scientific and mathematical research and education, for instance by helping establish the Institut Henri Poincaré in 1928 and serving as its president until his death.

During World War II, Borel was briefly imprisoned by the pro-Nazi Vichy regime, and afterward received the Resistance Medal for his efforts. In 1950, he earned the Grand Croix de la Légion d'honneur, among the most coveted prizes available to a French citizen. By then he had retired from the Sorbonne in Paris, where he had taken a position earlier, but he remained active in the world of mathematics up to the time of his death, on February 3, 1956.

JUDSON KNIGHT

Luitzen Egbertus Jan Brouwer

1881-1966

Dutch Mathematician

Dutch mathematician L. E. J. Brouwer made two principal contributions to the study of mathematics, though one—his development of a topological principle known as Brouwer's theorem—received far more attention. In the realm of logic, he approached concepts of concern to philosophers as well as to mathematicians. His intuitionist school, though it never became highly influential, challenged the two prevailing schools of thought regarding the nature of mathematical knowledge.

The son of Egbert and Henderika Poutsma Brouwer, the future mathematician was born in

Overschie, Holland, on February 27, 1881. He studied at the Haarlem Gymnasium, and in 1897 entered the University of Amsterdam, an institution with which he was to remain connected throughout his career. At the university, he first distinguished himself in 1904, when he was twenty-three years old, with his studies on the properties of four-dimensional space. Also in 1904, Brouwer married Reinharda de Holl. The couple had no children.

Showing the diversity of his interests, in 1905 Brouwer published *Leven, Kunst, en Mystiek,* in which he examined the role of individuals in society. He earned his doctorate in 1907 with a thesis "On the Foundations of Mathematics," in which he laid the groundwork for intuitionism. At that time, the philosophy of mathematics was dominated by two schools: logicism, which maintained that mathematical concepts had an existence independent of the human mind's conception of them; and formalism, which was concerned with the rules by which such concepts were to be interpreted. Brouwer's intuitionism offered a third viewpoint by maintaining that mathematical truths are a matter of common sense, apprehended *a priori* through intuition.

Though Brouwer remained committed to intuitionism, which he considered his principal achievement, the intuitionist school never gained many adherents, in his lifetime or thereafter. By contrast, his other significant contribution, in topology, was a highly influential one. Brouwer's fixed-point theorem, which he formulated in 1912, states that for any transformation affecting all points on a circle, at least one point must remain unchanged. He later extended the application of this theorem to three-dimensional objects.

Also in 1912, Brouwer became a professor of mathematics at the University of Amsterdam, after having served there for three years as an unsalaried tutor. He maintained that position until his retirement in 1951, and in the meantime earned a variety of awards and honors. The latter included election to the Royal Dutch Academy of Science (1912), the German Academy of Science (1919), the American Philosophical Society (1943), and the Royal Society of London (1948). He was also awarded a knighthood in the Order of the Dutch Lion in 1932. In 1959, Reinharda died, and Brouwer himself passed away in Blaricum, Holland, on December 2, 1966.

JUDSON KNIGHT

Elie Joseph Cartan

1869-1951

French Mathematician

The career of Elie Joseph Cartan brought together four disparate mathematical fields: differential geometry, classical geometry, topology, and Lie theory. The latter was the creation of Norwegian mathematician Marius Sophus Lie (1842-1899), and concerns the application of continuous groups and symmetries in group theory. A successful and highly admired teacher, Cartan influenced a number of younger mathematicians—including his son, Henri Paul Cartan.

Born on April 9, 1869, Cartan's background was that of a peasant: his father, Joseph, was the village blacksmith in Dolomieu Isére, a town in the French Alps. Joseph and his wife Anne Cottaz Cartan had four children, of which Elie was the second. A talented student, the young Cartan attracted the attention of an inspector of primary schools, Antonin Dubost, when the latter visited Cartan's school. Dubost assisted Cartan in earning a scholarship to a lycée, or secondary school, a rare opportunity for someone of his humble origins. Cartan's advancement inspired his youngest sister Anna, who also went on to become a teacher of mathematics and author of several geometry texts.

Cartan entered the Ecole Normale Supériere in 1888, and in 1894 earned his doctorate. During this period, he first became interested in Lie's theory of continuous groups, which at that time had not attracted much attention among mathematicians. Cartan's doctoral thesis concerned the classification of semi-simple algebras, an undertaking begun by Wilhelm Killing earlier, and applied this to Lie's algebras.

His further investigations of group theory would have to wait, however, because Cartan was drafted into the French army following his graduation. He served for a year, and was discharged as a sergeant. Afterward he accepted a position as lecturer, first at the University of Montepellier, and in 1896, at the University of Lyons. Cartan remained at Lyons until 1903, during which time he continued to explore Lie's theory. It was at this point that he first began bringing it together with differential geometry, classical geometry, and topology. Also during this phase, he helped establish the foundation for the calculus of exterior differential forms, which he later applied to a variety of geometric problems.

In 1903, Cartan married Marie-Louise Bianconi, and became a professor at the University of Nancy. He remained there until 1909, at which point he moved to the Sorbonne, or the University of Paris. There he became a full professor in 1912, and he would continue at the Sorbonne for the remainder of his career.

Cartan's work continued in 1913, with the discovery of spinors, complex vectors used for developing representations of three-dimensional rotations. These in turn gained application in the development of quantum mechanics. He also continued to explore his applications of Lie theory, and its intertwining with the three other disciplines to which he had applied it.

Cartan's marriage was a happy one, and it produced four children, of whom the most famous was Henri, born in 1904. His daughter Hélène also became a mathematician. Cartan's other two children, however, met with tragedy: Jean, a composer, died of tuberculosis when he was just 25 years old, and Louis, a physicist, was executed by the Nazis in 1943 for his activities with the French Resistance.

On his 70th birthday in 1939, just before the Nazi invasion brought an end to the peaceful world he had known, Cartan was honored with a celebratory symposium at the Sorbonne. He earned a number of other honors during his career, and though he retired from the Sorbonne in 1940, in retirement he taught math at the Ecole Normale Supériere for Girls. Cartan died on May 6, 1951, in Paris.

JUDSON KNIGHT

Alonzo Church
1903-1995
American Mathematician

Like his more famous pupil Alan Turing (1912-1954), Alonzo Church contributed significantly to the foundations of computer science. He is credited, along with Turing, with formulating a key principle concerning computer logic involving recursion, or the recurring repetition of a given operation. His other principal achievement was Church's theorem (1936), which maintains that there is no decision procedure—i.e., no fail-safe method for ensuring that one will always reach correct conclusions—in mathematics.

Born in Washington, D.C., on June 14, 1903, Church was the son of Samuel and Mildred Letterman Church. He studied at Princeton University, where in 1924 he received his B.A. in mathematics. On August 25, 1925, Church married Mary Julia Kuczinski, with whom he had three children: Alonzo, Mary Ann, and Mildred.

Church earned his doctorate at Princeton in 1927, after which he spent a year as a fellow at Harvard. He followed this with a year of study at the prestigious University of Göttingen in Germany, then returned to the United States to take a position as professor of mathematics at Princeton in 1929. Church would remain at Princeton for nearly four decades, during which time he taught a number of important students—including Turing, then studying for his Ph.D., who would later go on to a brilliant if tragic career. (Turing, whose hypothesized "Turing machine" provided an early theoretical model for the computer, committed suicide in 1954 after being arrested for homosexuality, then a punishable offense in Great Britain.)

During his early years at Princeton, Church approached the question of decidability, first notably raised by German mathematician David Hilbert (1862-1943). A problem in logic, decidability revolves around the question of whether there exists a method that could in principle be used to test any proposition and thereby ensure a correct answer as to that proposition's truth or falsehood. Hilbert had believed decidability to be possible, but Church showed in "An Unsolvable Problem of Elementary Number Theory," which he published in the *American Journal of Mathematics* in 1936, that no general method could be used for all possible assertions. In so doing, he extended the investigations of Kurt Gödel (1906-1978) regarding the incompleteness of mathematics, or indeed of any system of thought.

Also in the 1930s, Church undertook work that would prove significant in the development of a machine that did not yet exist: the computer. He discovered that in order for a problem to be calculable, it must admit to recursiveness. In other words, for the problem (100 - 50) to be solvable, it must be capable of being broken down into a recursive series, as indeed it is: 100 - 1, 99 - 1; 98 - 1, and so on. This 1 calculus, as it came to be known, would later provide the basis for the operation of computer problem-solving, which must be broken down into recursive rules and terms.

In 1967, Church made a big move, from Princeton on the East Coast to the University of California at Los Angeles on the West. At that point he was already 64, an age when most people prepare for retirement; however, he continued teaching at Los Angeles until 1990, when he was 87 years old. His wife died in 1976. A number of honors attended Church's career, among them membership in the American Academy of Arts and Sciences and the National Academy of Science. He died on August 11, 1995, in Hudson, Ohio.

JUDSON KNIGHT

George Bernard Dantzig

1914-

American Mathematician

George Dantzig is the founder of linear programming, a mathematical method that has had extensive applications in a variety of areas, from logistics management to computer programming. Simultaneous with his development of linear programming was his discovery of the simplex method, a problem-solving algorithm that has proven highly efficient in the linear programming of computers. His interest in the application of his ideas has taken him into a variety of pursuits, including the coauthorship of a seminal text in urban planning, *Compact City* (1973).

Dantzig's father, Tobias, participated in the abortive Revolution of 1905 in Russia. After spending nine months in a Russian prison, the elder Dantzig emigrated to Paris, where he studied mathematics at the Sorbonne. He married Anja Ourisson, moved to the United States in 1909, and settled in Portland, Oregon, where George was born on November 8, 1914. Tobias went on to write *Number: The Language of Science,* an influential book on the evolution of numbers in relation to that of the human mind.

George studied mathematics at the University of Maryland, receiving his B.A. in 1936. Also in 1936, he married Anne Shmumer, with whom he had three children, David, Jessica, and Paul. After his graduation, he was appointed Horace Rackham Scholar at the University of Michigan, and earned his M.A. there in 1938. Dantzig went to work for the U.S. Bureau of Labor Statistics for two years, then enrolled at the University of California, Berkeley, where he would later obtain his Ph.D.

World War II interrupted Dantzig's doctoral studies, however, and in 1941 he left the university to take a position as chief of the combat analysis branch at the U.S. Army Air Corps's statistical control headquarters. In 1944, he received the Exceptional Civilian Service Medal from the War Department. He returned to Berkeley to complete his doctoral studies under statistician Jerzy Neyman, and earned his Ph.D. in 1946.

Following the completion of his doctoral work, Dantzig returned to employment with the Air Corps, which in 1947 became the United States Air Force. The latter had initiated Project SCOOP, or Scientific Computation of Optimum Programs, which was intended to optimize the deployment of forces. While working on the project, Dantzig found that linear programming could be applied to all manner of planning problems. He also discovered the simplex method, an algorithm which, with its wide applicability to programming problems, further revealed the range of uses inherent in linear programming.

Dantzig's research was aided by the development of computers then taking place under the aegis of the Department of Defense, as the War Department had been renamed in 1947. Among the areas of interest to him were "best value" calculations, a means of finding the optimal value for a set of variables in a complex problem. "Best value" calculations would enable the Air Force, for instance, to deploy the most effective number of personnel in a manner that made optimal use of geography and available resources.

In 1952, Dantzig went to work with the RAND Corporation, but by 1960, the growing development of operations research as a field of academic studies led the University of California, Berkeley, to offer him a position as chairman of its Operational Research Center. He published *Linear Programming and Extensions,* a highly influential work on linear programming, whose origins he traced to the work of Fourier more than a century earlier.

Dantzig in 1966 became professor of operational research and computer science at Stanford, and served as acting chairman of the university's operational research department from 1969 to 1970. Along with mathematician Philip Wolfe, he developed the decomposition principle, a method for solving extremely large equations. Dantzig and Thomas L. Saaty conducted extensive research on urban planning with an aim toward providing more livable urban communities, and published their findings in *Compact City* (1973). In the following year, he was appointed to an endowed chair at Stanford.

During the 1970s, Dantzig received a number of honors, most notably the National Medal of Science, awarded in 1975 by President Gerald R.

Ford in recognition of his work in linear programming. He continued to travel and lecture widely.

JUDSON KNIGHT

Paul Erdös
1913-1996
Hungarian Mathematician

Not only was Paul Erdös one of the most extraordinary mathematical geniuses who ever lived, he was also a standout in a profession already noted for its eccentrics. Erdös, who said he had never buttered his own bread until he was 21 years of age, in some ways remained an eternal child, traveling the world and popping in on friends, whom he expected to host him and join him in grappling with mathematical complexities. His work in number theory, discrete mathematics, and combinatronics (a branch of mathematics involving arrangement of finite sets) won him many admirers; but to his many friends and acquaintances, the most amazing thing about Paul Erdös was Paul Erdös.

Born in Budapest, Hungary, on March 26, 1913, Erdös was an only child by default, his two older sisters having died of scarlet fever. Both Lajos and Anna Erdös were mathematics teachers, and in the face of her earlier loss, Anna became an extremely protective mother. The parents quickly realized that they had a genius on their hands—Erdös could multiply three-digit numbers in his head at age three, and discerned the existence of negative numbers when he was just four—and undertook his education at home with the aid of a governess.

Erdös graduated at age 21 after four years at the University of Budapest, not in itself an extraordinary feat—except that he left with a Ph.D., not a B.A. He went on to the University of Manchester in England, where he undertook a postdoctoral fellowship, and decided not to return to Hungary. By then it was 1934, and the Nazis were in power in Germany. Erdös, a Jew, recognized the threat, since his own nation had a quasi-fascist government that would join the Axis powers in World War II. Later, several of his relatives were murdered, and his father died of a heart attack in 1942.

Erdös went to the United States in 1938, but the democratic nations of the world were slow in giving asylum to Jews, so he decided to relocate to the part of Palestine that later became Israel. In fact Erdös never truly settled anywhere, but spent much of his life travelling around the world, appearing at mathematical conferences. It was an activity to which he was uniquely suited, since he was not susceptible to jet lag, and he once remarked that the perfect death would be to "fall over dead" while delivering a lecture.

A specialist in number theory, Erdös remained intrigued by problems that seemed simple, yet were far from it. His talent for raising perplexing questions tended to exceed his ability to solve them, and he was fond of exhorting younger mathematicians to work out problems he had not solved, establishing awards for those who succeeded. Despite his lack of concern for practical matters, he always seemed to have money, partly because the Hungarian Academy of Sciences provided him with a small stipend (he never renounced his citizenship), and partly from awards. In 1983, he won the Wolf Prize in Mathematics, which carried a $50,000 award, and he used most of this money to provide endowed scholarships in his parents' names.

Erdös, who never married, remained close to his mother, and was deeply affected by her death in 1971, when he was 58 years old. In his depression, he began taking amphetamines, and though his many friends and admirers pressed him to stop, he continued his habit. Nonetheless, he lived to the age of 83, dying on September 20, 1996, in circumstances much like the "perfect death" he had described: he passed away while attending a mathematics conference in Warsaw, Poland.

In its obituary, the *New York Times* discussed the "Erdös number," an inside joke among his many friends. The closer one was to Erdös, the lower one's Erdös number: thus someone who had worked with him on a paper (he wrote more than 1,500) had an Erdös number of 1, whereas a friend of a friend received a 2, and so on. According to the obituary, 458 persons had an Erdös number of 1, and another 4,500 could claim the honor of a 2 in a number system that went all the way up to 12.

JUDSON KNIGHT

Maurice René Fréchet
1878-1973
French Mathematician

A pioneer in the field of topology, one of the more intellectually challenging disciplines of mathematics, Maurice Fréchet removed the intricacies of topological studies even further from intuitive understanding by introducing an unprecedented degree of abstraction. To mathematicians of the early twentieth century, who were just beginning to accept topology in its more general, concrete form, this development hardly seemed like a step forward. Applications of Fréchet's ideas, however, revealed the value of his methods for solving concrete problems that had longed bedeviled mathematicians.

Fréchet was born on September 10, 1878, in the provincial town of Maligny. His parents were in the business of looking after other people: in Maligny, his father, Jacques, operated an orphanage; and in Paris, to which the family moved soon after Fréchet's birth, his mother, Zoé, ran a boardinghouse that appealed to foreigners. There, Fréchet gained a wide exposure to different cultures, endowing him with the cosmopolitan quality that characterized him—despite the fact that he spent much of his career teaching in the provinces rather than in Paris.

In his high school, or *lycée,* Fréchet had the good fortune to study under a distinguished mathematician, Jacques Hadamard (1865-1963), the author of an important proof regarding the prime-number theorem. Hadamard helped prepare him for the prestigious Ecole Normale Supérieure, which he entered in 1900. Following his graduation, Fréchet went to work with Emile Borel (1871-1956), who, though he was only seven years older than Fréchet, had gained so much experience that he took on the role of a mentor.

In 1908 Fréchet married Suzanne Carrive, with whom he would have four children. He had meanwhile undertaken his doctoral work with Hadamard as his advisor and in 1906 wrote his dissertation on the concept of metric space. This work was a groundbreaking one in topology, which he linked to the set theory of German mathematician Georg Cantor (1845-1918).

At that point, the French mathematical community was just warming up to the more concrete presentation of topology offered by Jules Henri Poincaré (1854-1912). Poincaré had examined the generalization of geometry within the framework of standard Cartesian coordinates, including the *x, y,* and *z* axes. Fréchet, however, took the study of topology into even more rarefied territory, applying it to situations in abstract space, where it was impossible to assign coordinates. In Cartesian space, one could use a distance function based on the Pythagorean theorem, but Fréchet's application required that distance functions be identified in far more generalized terms—what he called "metric spaces."

All of this was a bit too avant-garde for the mathematical establishment of Fréchet's era, though subsequent applications of his ideas proved exceedingly fruitful. Fréchet himself, however, remained at the fringes, teaching in the provinces until 1928, at which point he took a position at the Sorbonne. He remained there until 1949, during which time he saw his ideas become an inspiration to a new generation of mathematicians—only they were Polish, not French. Symbolic of his country's lack of regard for him was the fact that Poland elected him to its academy of sciences in 1929, but France did not give him the same honor until 1956. Awards did come, however, and he eventually received the highly prestigious Legion of Honor Medal. Fréchet, almost 95 years old, died in Paris on June 4, 1973.

JUDSON KNIGHT

Aleksandr Osipovich Gelfond

1906-1968

Russian Mathematician

In 1929 and 1934 Aleksandr Gelfond published papers on his inquiries into transcendental numbers, or numbers that are not the solution to an algebraic equation with rational coefficients. He was specifically concerned with Hilbert's seventh problem, which revolved around the assumption that a^b is transcendental if a is any algebraic number other than 0 or 1 and b is any irrational algebraic number. Using linear forms of exponential functions, he solved the problem and established what became known as Gelfond's theorem.

Gelfond was born in 1906 in St. Petersburg, Russia, which would later become Leningrad, U.S.S.R. His father, a physician with an interest in philosophy, was an acquaintance of Lenin, who in 1917 became leader of the new Communist state. In 1924, the year Lenin died, Gelfond entered Moscow University, where he completed his undergraduate studies in mathematics three years later. He performed his postgraduate studies between 1927 and 1930, at which point he took a position at Moscow Technological College. A year later, he became a professor of mathematics at Moscow University, where he remained until his death.

An ardent student of mathematical history, Gelfond was intrigued by a proposition made by Leonhard Euler (1707-1783) in 1748 stating that logarithms of rational numbers with rational bases are either rational or transcendental. David Hilbert (1862-1943), who in 1900 presented the mathematical community with 23 problems that would provoke debate for decades to come, had built on Euler's conjecture by proposing, in his seventh problem, that a^b is transcendental if a is any algebraic number other than 0 or 1 and b is any irrational algebraic number. In his 1929 paper, Gelfond drew connections between the arithmetic nature of a number's values and the properties of an analytic function. He followed this paper with a 1934 work on Hilbert's seventh problem, in which he showed that a^b is indeed a transcendental number—a statement that came to be known as Gelfond's theorem.

Concurrent with his teaching at Moscow University, Gelfond received a post at the Soviet Academy of Sciences Mathematical Institute in 1933, and in 1935 he earned his doctorate in mathematics and physics. Those times were frightening ones in the Soviet Union, with the atmosphere of ever-present terror created by Stalin's dictatorship, but Gelfond—a Communist Party member—kept a low profile.

During the 1930s, Gelfond was peripherally associated with the Luzitania, a coterie of students and admirers who gathered around the mathematician Nikolai Luzin (1883-1950). Though Luzin was later charged with ideological sabotage and nearly subjected to a trial before receiving a pardon, Gelfond himself escaped censure. He risked endangering himself, however, for Ilya Piatetski-Shapiro, a Jewish student denied admission to the Moscow University graduate school in the wave of Stalin-inspired anti-Semitic hysteria that followed World War II. Piatetski-Shapiro remained a devoted admirer, and was with Gelfond when he died on November 7, 1968, in a Moscow hospital.

JUDSON KNIGHT

Kurt Friedrich Gödel

1906-1978

Austrian-American Mathematician

The names of most mathematical innovators, while they may loom large within their own community, are hardly ever known to the outside world. Kurt Gödel's name and achievements, however, are a part of the framework necessary to an understanding of the postmodern worldview. His incompleteness theorem, which holds that within any axiomatic system there are propositions that cannot be proved or disproved using the axioms of that system, would prove as earth shattering as the better-known relativity theory of his friend Albert Einstein (1879-1955).

Kurt Gödel. *(AP/Wide World. Reproduced with permission.)*

The younger son of Rudolf (a textile worker) and Marianne Gödel was born on April 28, 1906. His hometown of Brünn—now Brno in the Czech Republic—was then part of Moravia, a province of the Austro-Hungarian Empire. Gödel's family were themselves German speakers and thus would remain more closely tied to the world of the Austrian capital than to that of their Moravian surroundings.

A devout Lutheran throughout his life, Gödel studied at a Lutheran school in Brünn and in 1916 entered a *gymnasium,* where he studied until 1924. Planning to study physics, he enrolled at the University of Vienna, but when a professor interested him in number theory he changed his major to mathematics. Gödel, however, retained a lifelong interest in physics.

In 1929 Gödel's father died, and his mother and older brother moved to Vienna. Meanwhile, Gödel was hard at work on his dissertation concerning the incompleteness theorem, for which he received his doctorate in 1930. Inspired by the self-contained system of Euclidean geometry, in which every precept could be derived from a few initial axioms, mathematicians had long been intrigued by the question of just how many axioms would be necessary to prove all true statements in geometry. In his dissertation, Gödel showed that the set of all true statements and the set of all provable statements were essentially the same—a reassuring answer to those who hoped to apply Euclidean methods to the whole of mathematics.

Among those hopefuls were two eminently qualified contemporaries, Bertrand Russell (1872-1970) and Alfred North Whitehead (1861-1947), who had earlier published their ponderous *Principia Mathematica.* The work, a three-volume tome whose publication costs Whitehead and Russell had to undertake personally, has been called "one of the most influential books never read." In it, the two distinguished mathematician-philosophers set out to derive all mathematical principles from a given set of axioms. Then, in September 1930, the 24-year-old Austrian mathematician who had already made a name for himself with his incompleteness theorem dropped a bomb—a paper called "On Formally Undecidable Propositions of *Principia Mathematica* and Related Systems."

In it, Gödel applied a new method, assigning numerical values to the symbols of logic, to prove that Russell and Whitehead's quest was doomed. At the heart of mathematics, Gödel showed, were statements that were clearly true but could not be proved by axioms, which, in turn, meant that the set of true statements and the set of provable statements were not identical. At some point, self-referential statements—logic loops on the order of "This statement does not say what it says"—would be inevitable.

The incompleteness theorem shook the foundations of mathematicians' assumptions about the absolute truths undergirding their discipline and forever changed the terms in which mathematics was discussed. Having turned the world upside-down, Gödel—an introverted young man now 27 years old—joined the Institute for Advanced Study at Princeton, where he was free to work without teaching responsibilities. He divided his time between Princeton and Austria and during this period devoted himself to the study of set theory. On September 20, 1938, he married a partner who was altogether his opposite yet with whom he proved compatible—Adele Nimbursky, a nightclub dancer. The two never had children.

In 1939 Gödel, alarmed by the changing situation in Austria, which had recently been absorbed by Nazi Germany, left his homeland for good. He remained at Princeton for the rest of his life and there he enjoyed the companionship of Einstein and mathematical economist Oskar Morgenstern (1902-1977), one of the founders of game theory. In his latter years, he was primarily concerned with questions of philosophy rather than mathematics.

Gödel received honorary degrees from Yale and Harvard as well as the Einstein Award in 1951 and the National Medal of Science in 1975. Increasingly eccentric in his latter years, he refused to eat for fear of poisoning and died of malnutrition on January 14, 1978.

JUDSON KNIGHT

David Hilbert

1862-1943

German Mathematician

Perhaps the most famous event from the long and fruitful career of David Hilbert was his 1900 address to the International Congress of Mathematicians (ICM) in Paris. A new century was dawning, and Hilbert, at 38 already well known, proposed a set of 23 problems. These problems, he suggested, would keep mathematicians occupied throughout the coming century and, indeed, some did. But long before the century had reached its midway point, Hilbert's brilliant career had peaked and entered a headlong dive—another casualty of the political nightmare unfolding in Hitler's Germany.

Hilbert was born on January 23, 1862, in Königsberg, Prussia (now Kaliningrad in Russia). His father, Otto, was a lawyer, and his parents were protective of their social standing. This fact would become clear years later, when they expressed their disapproval of Hilbert's friendship with fellow mathematician Hermann Minkowski (1864-1909), who not only came from the lower classes but was a Jew.

Hilbert entered the University of Königsberg in 1880 and earned his doctorate in 1885. Following his graduation, he went to work at the university, becoming a professor in 1893. Two years later, he took a chair at the University of Göttingen, where he would spend his entire career. In 1892 he married Kathe Jerosch and they had one child, a son named Franz, who appears to have had a mental disorder.

Hilbert's first important work was on the theory of invariants, involving the expression of an entity that remains the same throughout a variety of transformations. Classifying invariants through calculation had proven a herculean task, but Hilbert cut through the haze by showing that such calculations were unnecessary. These revelations proved so unsettling to the mathematical community that invariant theory itself went out of vogue for a time, but, when it was later reopened for study, mathematicians discovered that Hilbert's observations were accurate.

At this early stage in his career, Hilbert struck up a friendship with Minkowski, and together they labored over a summary of the state of number theory, commissioned in 1893 by the German Mathematical Association. Minkowski eventually abandoned the enterprise, but Hilbert's 1897 "Number Report" helped formulate the terms by which the theory of numbers would be discussed throughout the twentieth century.

Hilbert turned from this task to a reassessment of Euclidean principles, a project that had become necessary in the face of non-Euclidean geometry—not to mention the fact that mathematicians had become increasingly aware of the many untested assumptions at the heart of Euclidean theory. Hilbert proposed that in order to prevent untested assumptions from entering into geometry it was necessary to focus on the form of an axiom. This led him into debate with Gottlob Frege (1848-1925), one of the founders of mathematical logic, and the ensuing controversy only served to promote Hilbert's name further.

At the high point of his career, Hilbert addressed the ICM. Among the most enduring of the problems he presented was the first, involving the question of how many real numbers there were compared to the number of whole numbers, and the seventh, on the irrationality of certain real numbers. The first was not solved until 1963 and the seventh until 1934, when Aleksandr Gelfond (1906-1968) presented his theorem concerning transcendental numbers. All in all, Hilbert's problems created a veritable industry, with numerous books and conferences devoted to individual questions.

In the years that followed, Hilbert contributed to the study of mathematical analysis. During this period, he became something of a celebrity, noted for his Panama hat and his beard. In the increasingly tense environment of Germany during World War I and afterward, he repeatedly showed himself to be a mathematician and a human being first and a German second. (He defended Emmy Noether [1882-1935], a Jewish mathematician, against whom the German mathematical establishment discriminated.)

Hilbert later devoted himself to theoretical physics and formalism, neither of which proved to be pursuits of enduring impact. Though formalism at one time was an influential theory of mathematics, attacks from many sides—most notably from the incompleteness theorem of Kurt Gödel (1906-1978)—rendered it obsolete. Hilbert began to find himself increasingly isolated in the 1930s as more Jewish members of German academia fled the country. He refused to go at his age, and he died in the middle of World War II on February 14, 1943, in Göttingen. Only about a dozen people attended his funeral.

JUDSON KNIGHT

Andrei Nikolaevich Kolmogorov

1903-1987

Russian Mathematician

In discussing great mathematicians, the word "genius" is so often used that its effect sometimes seems blunted, yet in the case of Andrei Nikolaevich Kolmogorov the term is exceedingly apt. Though he is best known as the founder of probability theory, Kolmogorov contributed to virtually every area of mathematics as well as to fields ranging from physics to linguistics.

Kolmogorov's father, a Russian agriculturist named Nikolai Kataev, was killed in World War I, and his mother, Mariya Kolmogorova—who was not married to Nikolai—died giving birth to their son in the town of Tambov on April 25, 1903. Kolmogorov was raised by his mother's sister in the village of Tunoshna. He showed his brilliance at the age of five when he noted the pattern $1=1^2$, $1+3=2^2$, $1+3+5=3^2$, and so on.

At age 17, Kolmogorov enrolled at Moscow University and worked his way through college as a secondary school teacher. As an undergraduate, he published a number of important papers, including one wherein he formulated the first known example of an integrable function with a Fourier series that diverged almost everywhere. He was only 19, and he soon improved on this work by extending his result to a series that divulged everywhere—work that quickly brought him to the attention of the international mathematics community.

After receiving his doctoral degree in 1925, Kolmogorov went to work as a research assistant at Moscow University and at the age of 28 became a full professor. In 1933, when he was just 30, he became director of the institute of mathe-

Andrei Kolmogorov. *(UPI/Corbis-Bettmann. Reproduced with permission.)*

matics at the university. He married Anna Egorova in 1942.

Kolmogorov published one of his most important works, "General Theory of Measure and Probability Theory," while he was still a research assistant. The paper, called the "New Testament" of mathematics by one of his admirers and often compared to Euclid's work for its foundational character, provided the framework for probability theory. Kolmogorov also contributed to the understanding of stochastic processes, which involve random variables. These achievements were not purely academic pursuits; during World War II, the Soviet military made use of his stochastic theory in planning its placement of barrage balloons in order to thwart Nazi bombing raids. As for probability theory, it led to progress in a number of areas, including Kolmogorov's own development of reaction-diffusion theory, the analysis of how events such as cultural changes spread through a group.

During the decades that followed, Kolmogorov worked tirelessly in mathematical studies and their applications to a wide array of fields. In 1939 he became one of the youngest full members of the Soviet Academy of Sciences and in 1946 he was chosen to direct the Turbulence Laboratory of the Academy Institute of Theoretical Geophysics. During the years 1970 to 1972, he sailed around the world to study the properties of ocean turbulence. He also is credited with being codeveloper of the Kolmogorov-Arnold-Moser (KAM) theorem for analysis of stability in dynamic systems.

The list of Kolmogorov's achievements and discoveries is seemingly endless. Among the ones in which he took the most pride was his solution of Hilbert's thirteenth problem, which involved the representation of functions of many variables in terms of a combination of functions possessing fewer variables. Kolmogorov, however, remained modest to the end, and, in line with his belief that a mathematician could no longer conduct valuable research after the age of 60, he retired in 1963, spending 20 years teaching high school. Long concerned with the state of mathematical education, he chaired the Academy of Sciences Commission on Mathematical Education and helped the Soviet Union move to the forefront in mathematics during the 1960s. Kolmogorov died in Moscow on October 20, 1987.

JUDSON KNIGHT

Edmund Georg Hermann Landau
1877-1938
German Mathematician

Noted primarily for his work in number theory, Edmund Landau investigated the distribution of prime numbers and prime ideals. In all, he wrote some 250 papers and books, but his career was cut short by the Nazis' accession to power.

Landau was born in Berlin on February 14, 1877, the son of Leopold (a gynecologist) and Johanna Landau. His father was Jewish, but, like many German Jews of his era, he had become as fully assimilated as possible and embraced German patriotism. Leopold remained committed to Jewish causes, however, and in 1872 helped establish a Jewish academy in Berlin.

Young Edmund studied at Berlin's Französische Gymnasium, or "French School," graduating at the age of 16. He then entered Berlin University, where in 1899 he earned his Ph.D. with a dissertation on number theory, written under the direction of Georg Frobenius (1849-1917). In 1901 Landau began teaching at the university, where his enthusiasm for his subject made him a popular instructor.

In 1903 Landau made his first major contribution to mathematics with his simplification of the proof for Gauss's prime-number theorem, which had been demonstrated independently by Jacques Hadamard (1865-1963) and C. J. de la Vallée-Poussin (1866-1962) in 1896. Landau extended the application of the theorem to algebraic number fields, in particular the distribution of ideal primes within these fields.

Landau married Marianne Ehrlich, daughter of Nobel laureate Paul Ehrlich (1854-1915), in 1905. The couple had four children, two girls and a boy, of whom one died before the age of five. In 1909 Landau took a position as professor of mathematics at the University of Berlin, but in the following year he moved to the University of Göttingen, where he would remain for the rest of his career.

During these years, Landau published a number of books, providing the first systematic discussion of analytical number theory. He continued to be a dynamic lecturer, but his style became more demanding at Göttingen and in some cases elicited complaints from students.

Like his father, Landau remained connected to his Jewish heritage and in 1925 delivered a lecture in Hebrew at the Hebrew University in Jerusalem. He even considered staying in Jerusalem, but, after a sabbatical in 1927 and 1928, he returned to Göttingen. He continued to publish, bringing together various branches of number theory into one comprehensive form.

Within a few years, the Nazi takeover made Landau's situation at home increasingly uncomfortable. Late in 1933, less than a year after Hitler had assumed the chancellorship, he was forced to leave his post at Göttingen. Though he tried to resume teaching, SS troops were on hand to ensure that his students stayed out of the classroom. By then, the anti-Semitic hysteria had taken on a life of its own and his students sent him a letter explaining that they no longer wished to be indoctrinated in the ways of Jews.

Thus, Landau's career ended abruptly, at a point when he should have been making a smooth transition toward retirement. He did not attempt to leave Germany, although he traveled out of the country to lecture at Cambridge in 1935 and in Brussels in 1937. Unlike many of his fellow Jews at that time, Landau died of natural causes, in Berlin on February 19, 1938.

JUDSON KNIGHT

Henri Léon Lebesgue
1875-1941

French Mathematician

Like many another mathematical innovator, Henri Lebesgue was not immediately recognized for his principal achievement, in his case a new approach to integral calculus. As so often happens with groundbreaking ideas, his integration theory offended prevailing sensibilities. In time, however, Lebesgue would see his ideas accepted—not only in Poland and America but even in his homeland.

Born in the town of Beauvais on June 28, 1875, Lebesgue was the son of a typographical worker and an elementary school teacher. In 1894 he entered the Ecole Normale Supérieure, where he became distinguished both for his sharp mind and for his rather cavalier approach to his studies—an early sign of the unorthodox attitude that would characterize his work on integrals. After graduating in 1897, he worked for two years at the school's library before going on to a teaching position at the Lycée Central in Nancy.

During the period between his graduation and the end of his time at Nancy in 1902, Lebesgue conducted some of his most important work with regard to integral calculus. The integral, which relates to the limiting case of the sum of a quantity that varies at every one of an infinite set of points, is fundamental to the study of calculus. Typically, the integration of a function is represented as a curve; some kinds of functions, however, reveal a non-continuous curve, a curve with jumps and bumps along its trajectory. Half a century earlier, Bernhard Riemann (1826-1866) had attempted to extend the concept of integration to apply to these discontinuous curves, but the Riemann integral did not apply adequately to all functions.

It was Lebesgue's achievement to transcend the difficulties of the Riemann integral, and in this aim he was aided by the contributions of Emile Borel (1871-1956), with whom he became personally acquainted. Borel's theory of measure helped Lebesgue define the integral geometrically and analytically, thus making it possible to include more discontinuous functions in his theorem than Riemann had earlier. He presented his findings in his 1902 doctoral dissertation, a paper that has been highly praised. That praise, however, came later. In the early twentieth century, Lebesgue's work shocked many other scholars.

After receiving his doctorate at the Sorbonne, Lebesgue took a position at the University of Rennes and in 1906 went on to the University of Poitiers. As his reputation grew, he was asked to return to the Sorbonne, and later he became a professor at the Collège de France. The following year saw his recognition by the French academic community with his election to the Académie des Sciences. The adoption of his integration concepts in France, however, was slower. By 1914, however, students at Rice University in the United States were already studying his work, and Polish schools at Lvov and Warsaw adopted his ideas just after World War I.

Lebesgue published some 50 papers, along with several books, and in the last two decades of his life saw his integral become widely accepted as a standard tool for analysis. He died in Paris on July 26, 1941, leaving behind a wife, son, and daughter.

JUDSON KNIGHT

Hermann Minkowski
1864-1909
Russian-German Mathematician

Hermann Minkowski. *(Corbis-Bettmann. Reproduced with permission.)*

Hermann Minkowski established the framework for modern functional analysis, expanded the understanding of quadratic forms, developed the geometry of numbers, and even contributed to Albert Einstein's theory of relativity. His achievements are all the more remarkable in light of his relatively short career.

Born on June 22, 1864, in the town of Alexotas (then in the Russian Empire and now, under the name of Kaunas, a part of Lithuania), Minkowski was the son of a German-Jewish rag merchant. In 1872, when Minkowski was eight years old, the family returned to Germany, settling in the town of Königsberg. (Ironically, Königsberg, renamed Kaliningrad, is now part of Russia.)

Minkowski's brother Oskar (1858-1931) would later become famous as the physiologist who discovered the link between diabetes and the pancreas, and Minkowski himself showed a prodigious talent for mathematics as a student at the University of Königsberg. He was only 17 years old when he set about to win a prize offered by the Paris Académie Royale des Sciences for a proof describing the number of representations of an integer as a sum of five squares of integers. Both Minkowski and the British mathematician H. J. Smith (1826-1883), who had first approached the problem in 1867, won the prize, though many observers since then have maintained that Minkowski's proof was superior in its use of more natural and general definitions.

After receiving his doctorate from Königsberg, where he began a lifelong friendship with fellow mathematician David Hilbert (1862-1943), Minkowski took a position at the University of Bonn. He remained there until 1894, then returned to Königsberg to teach for two years before moving to the University of Zurich.

Among Minkowski's most significant achievements was the geometry of numbers, which he introduced in 1889. The geometry of numbers involved the application of geometrical concepts to volume, which made it possible to prove a variety of theorems without recourse to numerical calculations. Starting with quadratic forms, he extended the method's application to ellipsoids and a variety of convex shapes, such as cylinders. From this development emerged studies in packing efficiency, that is, the most efficient packing of a space given different shapes, a task which has applications in a number of scientific fields.

Hilbert had arranged for the University of Göttingen to create for Minkowski a new professorship, which he took in 1902. There, his most distinguished pupil was Albert Einstein (1879-1955), who took more courses with Minkowski than with any other professor. This fact, however, did not indicate a personal liking between the two men. Indeed, Minkowski, who took an interest in relativity theory himself, viewed his own approach not as an augmentation of Einstein's but as an entirely independent method that he considered superior because of his greater understanding of mathematics.

In 1905, the same year that Einstein produced his first important work in relativity theory, Minkowski undertook a study of electrodynamics in which he compared the subatomic theories of both Einstein and Hendrik Lorentz (1853-1928). It was Minkowski, not Einstein, who first realized that both men's theories required an understanding of four-dimensional, non-Euclidean space.

Einstein later adopted aspects of Minkowski's findings into his general theory of relativity. By the time he published it, however, Minkowski was long gone, having died from a ruptured appendix on January 12, 1909, in Göttingen. He was just 44 years old.

JUDSON KNIGHT

Emmy Amalie Noether

1882-1935

German-born American Mathematician

In the world of early twentieth-century German mathematics, Emmy Noether had two strikes against her: she was a woman, and she was Jewish. Through perseverance and outstanding contributions to abstract algebra and other disciplines, Noether overcame the barrier posed by the first fact. Ironically, just when it seemed she had gained acceptance, the Nazi takeover of Germany forced her to begin her career again in America.

Noether was born on March 23, 1882, in Erlangen, Germany, the daughter of Max and Ida Kaufmann Noether. Max was a well-known mathematics professor at the University of Erlangen, noted for his work on the theory of algebraic functions. Later, two of Noether's three brothers earned distinction in mathematics and science—Fritz as a mathematician, and Alfred as a chemist.

Such careers seemed blocked to Noether, who could not enroll in the gymnasium, or college preparatory school, simply because girls were not allowed to do so. Instead, she entered a teachers' college for women, and in 1900 passed the Bavarian state teaching examinations. Determined to study at the University of Erlangen, however, Noether received permission to audit classes, and in 1903 was able to take the matriculation exam, which she passed easily.

She even had the opportunity to attend the distinguished University of Göttingen, where she studied under a number of influential figures. Among these was David Hilbert (1862-1943), destined to become a lifelong friend. By 1904, Erlangen had opened its doors to women, so she undertook her doctoral studies under Paul Gordan, a friend of her family. In 1908, she earned her Ph.D. *summa cum laude* with the dissertation "On Complete Systems of Invariants for Ternary Biquadratic Forms."

On the invitation of Hilbert and Felix Klein (1849-1925), Noether joined them at the Mathematical Institute in Göttingen to work on problems relating to the general theory of relativity. Specifically, Noether was concerned with the problem of invariance under transformation, or laws that remain the same regardless of changes in space and time. This led her to the development of Noether's Theorem, which established the correspondence between an invariance property and a law of conservation. She also formulated a theorem concerning the relationship between invariance, equations of motion, and the existence of certain integrals.

Despite this groundbreaking work, Noether was unable to obtain a permanent, paying job in mathematics; fortunately, family money allowed her to work without pay at Göttingen. Hilbert also helped her immensely, arranging for her to lecture his classes in mathematical physics. After a long struggle in the late 1910s and early 1920s, Noether—a popular and enthusiastic instructor—obtained a position as adjunct teacher.

Because so much of her early career had been devoted to simply gaining the right to participate in mathematics, Noether defied the usual pattern for mathematicians, who typically produce their most important work in their thirties or even their twenties. Beginning with the publication of a paper in 1920, Noether began to delve into an abstract and generalized approach to the axiomatic development of algebra, a method dubbed "epoch-making" by renowned mathematician Hermann Weyl (1885-1955).

Weyl tried but failed to help her gain a better position at Göttingen, yet in 1932 Noether was honored as co-recipient of the Alfred Ackermann-Teubner Memorial Prize. Also in that year, the algebra faculty at Göttingen held a celebration in honor of her fiftieth birthday, and Helmut Hasse (1898-1979) presented a paper validating one of her concepts of non-communicative algebra.

It seemed that Noether had finally gained acceptance. Then, less than nine months after she turned 50, the Nazis took power. On April 7, 1933, she was officially informed that she could no longer continue teaching in Germany.

Through the help of the Emergency Committee to Aid Displaced German Scholars, as well as a grant from the Rockefeller Foundation, Noether obtained a professorship at Bryn Mawr, a women's college in Pennsylvania. She soon became involved with the Institute for Advanced Study at Princeton, where she delivered weekly lectures. Once again, it seemed that she had begun a new career, but on April 10, 1935, she developed an extremely high fever following surgery to remove a uterine tumor. Losing consciousness, she died four days later.

JUDSON KNIGHT

Karl Pearson

1857-1936

British Mathematician

A man of wide-ranging interests, Karl Pearson became the father of statistics by an indirect route that led him from mathematics to other disciplines and back again. Among the innovations associated with his establishment of the discipline was his formulation of the concept of the "standard deviation."

Born in London on March 27, 1857, to William, a lawyer, and Fanny Smith Pearson, the future mathematician was a sickly child. His health forced him to conduct part of his early education at home, but at age 17 he went to Cambridge, and followed this with studies in a wide array of subjects at King's College. These subjects included philosophy, religion, literature, and, mathematics, his major, in which he graduated with honors in 1879.

Over the coming years, Pearson traveled throughout Germany, became a socialist, and in 1884 took a position as Goldsmid Professor of Applied Mathematics and Mechanics at University College in London. In 1890 he married Maria Sharpe, with whom he had three children—Egon, Sigrid, and Helga.

Intrigued by *Natural Inheritance,* a study of heredity by Francis Galton (1822-1911), Pearson reasoned that there must be some mathematical means of testing the correlation between two phenomena. Numbers would show clearly whether the apparent correlation resulted from mere chance, or from a consistent factor. From the beginning, he understood this pursuit primarily in terms of applied rather than pure mathematics, instantly grasping the possible applications of statistical theory for a wide array of disciplines.

It so happened that during this period Pearson met W. F. R. Weldon, a professor of zoology who spurred him on to develop statistical methods for application to the study of heredity and evolution. In the course of these studies, Pearson established the idea of a standard deviation. Standard deviation is a measure of the variance within a sample or population that is based on the average distance from the mean score to any score within the set, and makes it possible to predict the average variance within a set.

In 1901 Pearson founded the journal *Biometrika* for the publication of studies on statistics. A decade later, he became Galton Professor of Eugenics (the study of genetic factors as a means of preventing disease and other physical impairments), as well as director of a new department of applied statistics at University College. His staff served the British war effort in World War I, providing valuable calculations, and in 1925 he established the journal *Annals of Eugenics*. He published some 300 works in the course of his career.

Pearson's first wife died in 1928, and he remarried to Margaret V. Child, who also taught at University College, the following year. In 1933, at age 77, he retired, and on April 27, 1936, he died.

JUDSON KNIGHT

Laurent Schwartz

1915-

French Mathematician

French mathematician Laurent Schwartz is known primarily for his work in functional analysis and the theory of distributions, an expansion of concepts relating to differential and integral calculus. He won the Fields Medal in 1950.

Born in Paris on March 5, 1915, Schwartz entered the Ecole Normale Supériure in 1934, and graduated in 1937 with the Agrégation de Mathématique. He then began his doctoral work at the University of Strasbourg, where he earned his Ph.D. in 1943. For a year between 1944 and 1945, Schwartz served as lecturer at the University of Grenoble, then took a professorship at the University of Nancy.

Schwartz's principal area of concern, and the one for which he earned the Fields Medal in 1950, was functional analysis. Earlier, British mathematicians Oliver Heaviside (1850-1925) and Paul Dirac (1902-1984) had generalized the differential and integral calculus for special applications. Among those who used these applications were physicists working with mass distributions. Prior to Schwartz's work, the most useful formula in this area was Dirac's Delta function, which had its limitations. In a 1948 paper, Schwartz presented a generalized function built on a stronger and more abstract foundation, one which expanded the range of applications to include areas such as potential theory and spectral theory.

In 1953, Schwartz became a professor at the Sorbonne, where he remained until 1959, then moved to the Ecole Polytechnique in Paris. In addition to the Fields Medal, he received prizes from the Paris Academy of Sciences in 1955, 1964, and 1972. Also in 1972, he was elected a member of the Academy. He has also been awarded honorary doctorates from universities around the world, among them Humbolt (1960), Tel Aviv (1981), and Athens (1993).

Schwartz left the Ecole Polytechnique in 1980 to teach at the Sorbonne for three years before retiring in 1983. Among the concerns of his later work is stochastic differential calculus.

JUDSON KNIGHT

Atle Selberg

1917-

Norwegian-American Mathematician

In 1950, Atle Selberg received the Fields Medal for his generalizations of Viggo Brun's sieve methods, and for a proof relating to the Riemann zeta function. He is also noted for a proof of the prime number theorem, on which he worked with Paul Erdös (1913-1996).

Born on June 14, 1917, in Langeslund, Norway, Selberg was the son of Ole and Anna Skeie Selberg. Inspired by reading about the Indian mathematical prodigy Ramanujan as a child, he decided on a career in mathematics, and was further inspired by a lecture he heard at the International Mathematical Conference in Oslo in 1936. He performed his doctoral studies at the University of Oslo, where he earned his degree in 1943. A year earlier he had been appointed a resident fellow of the university, and would remain in that position for five years.

In 1947, Selberg emigrated to the United States. Also in 1947, he married Hedvig Liebermann, with whom he later had two children, Ingrid and Lars. He spent his first year in America at the Institute for Advanced Study at Princeton, then taught for a year as associate professor of mathematics at Syracuse University before returning to the Institute as a permanent member in 1949.

During this period, Selberg did much of his most important work, a great deal of it in the area of generalizing the findings of others. Thus he generalized the number-sieve methods of Viggo Brunn, a study that was linked to his investigations in the Riemann zeta function. With regard to the latter, Selberg was able to prove that a positive proportion of the zeroes in the zeta function satisfy the Riemann hypothesis concerning them.

Also during this time, Selberg and Erdös both attacked the proof of the prime number theorem, a problem of long standing that Bernhard Riemann (1826-1866) himself had tried unsuccessfully to complete. The theorem had actually been proven, independently and simultaneously, in 1896 by Jacques Hadamard (1865-1963) and Charles de la Vallée-Poussin (1866-1962). However, they had used complex analysis, whereas Selberg and Erdös were able to provide an elementary proof that required no complex function theory. For reasons that are unclear, Selberg published his findings before Erdös, and received most of the credit for the proof.

After receiving the Fields Medal in 1950, Selberg became a professor at Princeton. Among his other awards were the Wolf Prize in 1986, and an honorary commission of Knight Commander with Star from the Norwegian Order of St. Olav in 1987. In later years, Selberg has traveled extensively, participated in commemorative events, and taken an interest in mathematics education among high-schoolers.

JUDSON KNIGHT

Olga Taussky-Todd

1906-1995

Austrian-American Mathematician

The two principal areas of interest to Olga Taussky-Todd, in a long and varied career, were number theory and matrix theory. The former is the study of integers and their relationships; the latter concerns the study of sets of elements in a rectangular array, which are subject to operations such as addition or multiplication according to specified rules. Like Emmy Noether (1882-1935), whom she knew, Taussky-Todd was a Jewish woman attempting to gain acceptance in the early twentieth-century world of German mathematics; yet her story was a much happier one than Noether's. Not only was Noether 24 years older than she, and thus practically of a different generation with fewer opportunities for women, but Taussky-Todd quickly left the world of Germany and Austria behind when the Nazis took power.

The second of Julius and Ida Pollach Taussky's three daughters was born on August 30, 1906, in Olmötz, Austro-Hungarian Empire (now Olomouc, Czech Republic). Her father was an industrial chemist and journalist who encouraged his daughters to use their minds, and later when he went to work as manager of a vinegar factory in Vienna, he used Olga's help in calculating water-to-vinegar ratios for various mixtures. During World War I, the family moved again, this time to Linz, and while Taussky-Todd was still in high school, her father died.

She worked for a time to help support the family, then entered the University of Vienna, where she received her Ph.D. in mathematics in 1930. Afterward she worked briefly at the University of Göttingen, but she quickly sensed that Germany was becoming an increasingly unpleasant place for someone of Jewish ancestry.

Therefore Taussky-Todd left Central Europe behind for good, taking a three-year science fellowship with Girton College at Cambridge University in England. The first year was spent at Bryn Mawr College in Pennsylvania, where she worked with Noether. After completing her two years at Girton, she took a teaching position at one of the women's colleges at the University of London. There she met British mathematician Jack Todd (1908-1994), whom she married in 1938.

During World War II, the couple moved frequently. In 1943, Taussky-Todd took a research position with the Ministry of Aircraft Production, studying problems of flutter and stability in combat aircraft. These investigations directly related to matrix theory, an area of primary interest to Taussky-Todd. After the war, both she and her husband took positions with the National Bureau of Standards in California.

Olga Taussky-Todd. *(Olga Taussky-Todd. Reproduced with permission.)*

The couple had already been living in California for a decade when in 1957 they made their last major career move, both accepting professorships at the California Institute of Technology. In the years that followed, Taussky-Todd received a number of awards: in 1964, for instance, she was named *Los Angeles Times* "Woman of the Year," and in 1970 she received the Ford Prize for an article on the sums of squares. She retired in 1977, and died on October 7, 1995, at her home in Pasadena, California.

JUDSON KNIGHT

Alan Turing
1912-1954
English Mathematician

The English mathematician Alan Turing is best known for his pioneering work in the area of computers and artificial intelligence. His concept of the Universal Turing Machine helped the allies break the German Enigma code in WWII, and his inquiries into artificial intelligence established the core theoretical and philosophical problems for AI research.

Turing was born into a middle-class English family, both parents being involved with the British colony in India. He was left in the care of another couple when his parents returned to India in 1913, and this, along with the realization of his own homosexuality as he grew older, caused Turing to develop an aloof nature and a disregard for social niceties. As a boy his intelligence was displayed through his interest in organic chemistry and cryptography, though he also showed an impatience for learning fundamentals when deeper problems piqued his interest.

In 1931 Turing won a scholarship in mathematics to King's College, Cambridge, where he became fascinated with problems of mathematical logic. While working on problems associated with Gödel's theorem, Turing first theorized the Turing Machine. Such a machine could be fed instructions on a paper tape, perform calculations, and present an answer. These calculations could be anything from the solution to a math problem, to how to play chess. The Universal Turing Machine could be programmed to imitate the behavior of any other type of calculating machine, just as a personal computer today can be programmed with software to act as a calculator, word processor, or game. The results of Turing's research were published in the paper "On Computable Numbers, with an application to the Entscheidungs problem," and were immediately recognized as major breakthroughs in the area of machine computing.

In WWII Turing was enlisted to help crack the Enigma code created by the Germans. To this end, he worked with a group of British mathematicians and engineers to create Colossus, arguably the first electronic Universal Turing Machine. Using Colossus, Turing and his team were able to break the code and provide the Allies with intelligence information that was critical in winning the war.

In 1948 Turing became Deputy Director of the computer laboratory at Manchester University, home of MADAM, the Manchester Automatic Digital Machine, the first functional digital machine that ran a stored program. Turing taught MADAM to play chess and write love letters, and in 1950 he published "Computing Machinery and Intelligence." Here he presented the "Turing test" as a means of demonstrating that machines could display the same intelligence as humans. In this test, a human being interrogates a machine using a keyboard and monitor. Turing thought that as long as the machine could imitate a human response to a question, it should be considered intelligent, because this was the standard by which humans judged other beings to have intelligence. Turing dedicated the rest of his career to fundamental questions on the nature of intelligence and its artificial creation.

Alan Turing. *(The Granger Collection Ltd. Reproduced with permission.)*

The complication in Turing's career was his homosexuality, which was still illegal and socially unacceptable. In 1952, Turing was arrested on charges of "gross indecency," put on probation, and forced to undergo hormone treatments. The treatment had disastrous physical consequences for Turing, and his career suffered as well. In 1954, following a deep depression, Turing committed suicide by eating a poisoned apple.

Turing's impact on the world of computing and artificial intelligence is incalculable. His Universal Turing Machine is the forerunner of today's personal computer, and his work in the area of artificial intelligence laid the groundwork for today's research. More important, however, are the philosophic questions that Turing raised about what it means to be a human being.

PHIL GOCHENOUR

John von Neumann
1903-1957
Hungarian-born American Mathematician

One of the most influential figures of the twentieth century, John von Neumann was also one of the most creative minds of any era. Much of what modern computer users take for granted—for instance, the use of a central processing unit, or CPU—has its roots in von Neumann's foundational computer science work. Of his many other achievements, the most noteworthy include his work on game theory, quantum physics, and the development of the atomic bomb.

Born Janos von Neumann in Budapest, Hungary, on December 28, 1903, the future mathematician adopted the name John when he emigrated to the United States. His father, Max, was a prosperous banker, and he and von Neumann's mother, Margaret, soon recognized that they had a genius in the family; therefore they arranged to have him tutored at home. When he attended the Lutheran Gymnasium for boys, von Neumann's teachers placed him with a tutor from the University of Budapest, mathematician Michael Fekete.

Von Neumann studied in Budapest and Zürich, and also spent a great deal of time in Berlin. In 1926 he earned his Ph.D. in mathematics at the University of Budapest with a dissertation on set theory. He went to work at the University of Berlin as the equivalent of an assistant professor, reportedly the youngest person to hold that position in the university's history. In 1926 von Neumann received a Rockefeller grant to conduct postdoctoral work with mathematician David Hilbert (1862-1943), who long remained a powerful influence, at the University of Göttingen. Already a rising star in the world of mathematics, von Neumann transferred to the University of Hamburg in 1929, the same year he married Mariette Kovesi. The two had a daughter, Marina, in 1935.

In his early work as Hilbert's student, von Neumann assisted his mentor in attempting to show the axiomatic consistency of arithmetic—a project doomed to failure by Gödel's incompleteness theorem. Hilbert was not destined to make significant contributions to quantum physics, but von Neumann's involvement in his teacher's attempts to apply the axiomatic approach to that discipline led him to an abstract unification of Schrödinger's wave theory and Heisenberg's particle theory. He was the first to affect such a union.

During the 1930s, von Neumann went to Princeton, where he became part of the newly formed Institute for Advanced Study. There he developed what came to be known as von Neumann algebras, published *The Mathematical Foundations of Quantum Mechanics* (1932), still considered essential reading on the subject, and investigated a number of other areas. In 1937 von Neumann, having become a naturalized citizen of the United

John von Neumann. *(AP/Wide World. Reproduced with permission.)*

States, began the first of many projects for the military, acting as a consultant in ballistics research for the army. During the same year, Mariette divorced him, but in 1938 he remarried to Klara Dan, who, like Mariette, was from Budapest.

During the early part of World War II, von Neumann worked on several defense-related projects. In 1943 he became involved in the development of an atomic bomb at Los Alamos, New Mexico. There he convinced J. Robert Oppenheimer (1904-1967) to investigate the use of an implosion technique in detonating the bomb. Simulation of this technique would require extensive calculations, which would be hopelessly time-consuming using old-fashioned methods. It was then that von Neumann began looking into the army's recently developed ENIAC (Electronic Numerical Integrator and Calculator)—the world's first computer.

Von Neumann and others improved on ENIAC with EDVAC (Electronic Discrete Variable Automatic Computer), which incorporated von Neumann's groundbreaking concept of the stored program. This became the foundation for computer design, and later, in developing a computer for scientific use at Princeton, von Neumann established a number of features now essential to all computers: random access-memory (RAM), the use of input and output devices operating in serial or parallel mode, and other elements.

In the midst of his other work, von Neumann became immersed in game theory, a concept that had first intrigued him in Germany. Working with mathematical economist Oskar Morgenstern (1902-1977), who shared von Neumann's conviction that the mathematics of the physical sciences was not adequate for the study of social sciences, von Neumann began applying the idea of an analogy between games and complex decision-making processes. Today, game theory is used in a variety of fields, from business organization to military strategy.

Von Neumann remained involved in defense technology after the war, and contributed to the development of the hydrogen bomb as well as other weapons. He held a number of key positions, including a seat on the Atomic Energy Commission, to which President Dwight D. Eisenhower appointed him in 1954. Eisenhower also awarded von Neumann the Medal of Freedom, one of many prestigious honors he received. Diagnosed with bone cancer in 1955, von Neumann was confined to a wheelchair, but continued to work feverishly. He died on February 8, 1957, in Washington, D.C., at the age of 53.

JUDSON KNIGHT

André Weil

1906-1998

Mathematician

A member of the Nicolas Bourbaki circle, André Weil advanced studies of mathematics in a variety of areas, including algebraic geometry, group theory, and number theory. Considered one of the twentieth century's preeminent mathematicians, Weil in 1980 received Columbia University's Barnard Medal, whose past recipients include Albert Einstein (1879-1955) and Niels Bohr (1885-1962).

Weil was born on May 6, 1906, in Paris, the son of Bernard, a physician, and Selma Reinherz Weil, Jewish parents who no longer observed Jewish traditions. Weil's sister Simone would later become famous as a member of the French Resistance in World War II. Weil himself proved a mathematical prodigy, reading geometry for fun at age eight and solving difficult problems by the age of nine. During World War I, Weil's father was drafted into the medical service, and the family followed him to various posts around France.

Weil entered the prestigious Ecole Normale Supériere at age 16, and earned his doctorate there at age 22 in 1928. From 1930 to 1932, he taught at Aligarh Muslim University in India, returning to France and a post as professor of mathematics at the University of Strasbourg. In 1937 he was married, and two years later, he was in Finland when World War II began.

Weil refused to return to military service in France, maintaining that he could best serve as a mathematician; but through a series of mishaps, the Finns—whose country was then under attack by Stalin's troops—mistook him for a Soviet spy. This very nearly led to his execution, but instead he was returned to France and a trial. The outcome was that he had to go fight, but in the confusion attending the Nazi invasion of his homeland, he and his wife Eveline fled the country.

After holding professorships at a number of universities in America and Brazil (during which time a daughter, Sylvie, was born), in 1947 Weil was invited to take a position at the University of Chicago. By then he had long since made a name for himself, first with his theory of "uniform space," which the *Science News Letter* dubbed one of the most important mathematical discoveries of 1939. Uniform space, a mathematical construct, bears little resemblance to three-dimensional space, and indeed is beyond the power of intuition to apprehend.

Another difficult concept with which Weil grappled was what came to be known as "Weil conjectures." These were formulas in algebraic geometry that provided the numbers of solutions to equations in a finite field. Decades later, Weil's advances in geometry and algebra would find a variety of applications in fields as diverse as Hollywood special-effects technology and computer modeling of black holes by astronomers. Weil was also involved with the Nicolas Bourbaki group, a circle of more than a dozen mostly French mathematicians, from the 1930s onward. Among the innovations associated with "Nicolas Bourbaki" (a collective pseudonym) was the development of what came to be called the "new math" when it was introduced in American schools during the late 1960s and thereafter.

Weil taught at Chicago until 1958, when he went to work at Princeton's Institute for Advanced Study. In 1976, he retired. He received a wide array of awards and honors, ranging from a fellowship at the Royal Society of London in 1966 to the Wolf Prize in 1979 to the Kyoto Prize in Japan in 1994. He died in Princeton on August 6, 1998.

JUDSON KNIGHT

Hermann Klaus Hugo Weyl

1885-1955

German-American Mathematician

Hermann Weyl served as a link between an older generation of mathematicians, dominated by David Hilbert (1862-1943) and others at the University of Göttingen, and the world of refugees from Nazism who fled to America and there helped establish the world of the future. His mathematical work concerned a variety of topics, ranging from the most purely philosophical pursuits in the foundations of the discipline to highly practical applications in physics. Many scholars consider Weyl among the immortals of twentieth-century mathematics.

Born on November 9, 1885, in the town of Elmshorn near Hamburg, Weyl was the son of a bank clerk, Ludwig, whose wife Anna Dieck had come from a wealthy family. While studying in the gymnasium at Altona, Weyl's abilities caught the attention of a headmaster who was related to Hilbert. He went on to study with the latter at Göttingen, where he earned his doctorate in 1908.

In 1913, the same year he married Helene Joseph, Weyl accepted a professorship at the National Technical University (ETH) in Zurich, where sons Fritz and Michael were raised. There he became acquainted with Albert Einstein (1879-1955), about whose relativity theory he would write in *Space, Time, Matter* (1918), a book comprehensible to non-technically educated readers. He also examined tensor calculus, involving functions on a number of vectors that take a number as their value, and his findings in this area helped clarify some of the still-untidy mathematical underpinnings of Einstein's theory.

During this extremely fruitful period, Weyl investigated the boundary conditions of second-order linear differential equations—that is, the behavior of functions within a given region in which the behavior at the boundary has been defined. He also became interested in Hilbert spaces, another realm in which he considered the behavior of different functions at different points. In addition, Weyl brought together classical methods of geometry and analysis with the relatively new discipline of topology, and in 1913 published a highly readable exposition on Riemann surfaces.

In the period leading up to World War I, during which he served briefly in the German army, Weyl examined the properties of irrational numbers—numbers, such as the square root of a negative, that cannot be expressed as a ratio of two whole numbers. He discovered that the fractional parts of an irrational number and its integral multiples appeared to be evenly distributed in the interval between 0 to 1.

Hermann Weyl. *(AP/Wide World, Inc. Reproduced with permission.)*

In the 1920s, Weyl became involved in a battle over the foundations of mathematics. In this he found himself torn between the formalism of his mentor Hilbert and the intuitionism of Luitzen Jan Brouwer (1881-1966), which he found appealing. The saga underlying this conflict, and the arguments for both sides, are intriguing; however, they are academic, because the incompleteness theorem of Kurt Gödel (1906-1978) in 1931 would render moot any attempt to provide an all-encompassing explanation of mathematics.

Weyl, whose wife was Jewish, left newly Nazified Germany for the United States in 1933, and became involved with Princeton's Institute for Advanced Study. He became an American citizen, and remained in America for the rest of his life, though he traveled widely and frequently. In 1948, his first wife died, and in 1950 he married Ellen Bär. During the last three decades of his life, Weyl continued to investigate a wide array of mathematical questions, and remained active up to the very end. He died of a heart attack on December 8, 1955, a month after his 70th birthday, in Zurich.

JUDSON KNIGHT

Grace Emily Chisholm Young
1868-1944
British Mathematician

Several years before Emmy Noether (1882-1935) entered the University of Göttingen, Grace Chisholm Young earned her doctorate there in 1895, becoming the first woman to officially receive a Ph.D. in *any* field from a German university. Later, she conducted research in derivatives of real functions, which yielded the Denjoy-Saks-Young theorem. Other important work appeared as a collaboration between Young and her husband, William Henry Young (1863-1942).

The youngest of Henry and Anna Bell Chisholm's three surviving children, Young was born on March 15, 1868, in Haslemere, Surrey, England. Her father was a civil servant who became director of Britain's bureau of weights and measures, and her brother Hugh later earned distinction as editor of the *Encyclopedia Britannica*'s eleventh edition. Initially she wanted to study medicine, something her mother forbade, so in 1889 Young entered Girton College, a school for women at Cambridge. She graduated in 1892 with first-class honors, and she later sat informally for the mathematics finals at Oxford. She placed first.

Thus Young was eminently qualified to become a pioneer for women's education, and in 1893 she transferred to what was then the leading European center for mathematical education: Göttingen. Working under the distinguished mathematician Felix Klein (1849-1925), in 1895 she earned her doctorate, magna cum laude, with a thesis on "The Algebraic Groups of Spherical Trigonometry."

Back in London, she married her former tutor at Girton, William Henry Young, and the couple soon moved to Göttingen. In time they would have six children, yet Grace managed to raise them while conducting mathematical work of her own, collaborating with her husband—*and* studying anatomy. The Youngs in 1905 coauthored a well-known textbook on set theory, and in the period between 1914 and 1916 she published papers on derivatives of real functions.

Grace Chisholm Young. *(Sylvia Wiegand. Reproduced with permission.)*

These contributed to the formulation of what became known as the Denjoy-Saks-Young theorem.

In 1908 the family moved to Geneva, Switzerland, which remained their base thereafter. Financial needs, however, forced Henry to take teaching posts at a variety of places ranging from Wales to Calcutta, and this in turn kept him separated from the family for long periods of time. Meanwhile their children grew up. Their son Frank, an aviator, was killed during World War I, but the others went on to successful careers in fields that included medicine, chemistry, and mathematics.

During the late 1920s, by which point she was entering her sixties, Young largely abandoned mathematics for other pursuits, including language studies, music, medicine, and literature. She was particularly fond of writing, and took an interest in becoming an author of children's fiction. In the spring of 1940, she visited England while her husband remained in Switzerland. This parting would turn out to be their last, because the Nazi invasion of France prevented either from traveling to see the other. William died in 1942, and Grace followed him by two years, succumbing to a heart attack at age 76.

JUDSON KNIGHT

Biographical Mentions

Abraham Adrian Albert
1905-1972

Albert was born in 1905 in Chicago, Illinois, the son of Russian immigrants. He earned his Ph.D. in 1928 from the University of Chicago, with a dissertation on algebras and their radicals. He was eventually elected President of the American Mathematical Society. Albert's book, *Structure of Algebras*, published in 1939, is considered a classic in the field. He died in 1972 in Chicago, Illinois.

James Waddell Alexander
1888-1971

Alexander was born in Sea Bright, New Jersey, in 1888. He was fortunate to study mathematics and physics under Professor Veblen at Princeton University. After earning a B.S. and an M.S. in mathematics, he was invited to become a faculty member of the mathematics department at Princeton. Alexander achieved recognition in his field when he generalized what was known as The Jordan Curve Theorem and, later, when he discovered the Alexander Polynomial, which is still in wide use in knot theory. The American Mathematical Society awarded Alexander its coveted Bocher Prize for his memoir, *Combinatorial analysis situs*. He remained at Princeton until retirement in 1951 and was a life member of the Institute for Advanced Studies at Princeton. He died in 1971.

Pavel Sergeevich Alexandrov
1896-1982

Born in Bogorodsk, Russia, in 1896, Alexandrov and his family moved to Smolensk State Hospital one year later, where his father became a well-known surgeon. Alexandrov's studies took him to the University of Moscow, where he became known as the leader of the Soviet topologists. During his amazing career, Alexandrov published some 300 scientific papers on topology and homology. Among the many honors awarded him were: president of the Moscow Mathematical Society; vice president of the International Congress of Mathematicians; memberships in the Academies of Science in eight different countries (including the American Philosophical Society); and numerous Soviet awards such as the Stalin Prize and Five Orders of Lenin. He died in Moscow in 1982.

Emil Artin
1898-1962

Born in Vienna, Austria, Artin served in his country's army during World War I before entering the University of Leipzig, receiving a doctorate in mathematics at this famous school. However, he was forced to leave Austria in 1937 because his wife was Jewish. It was America's good fortune that he emigrated there and taught at Notre Dame, Indiana University, and at the prestigious Princeton University. Artin was invited to return to Germany in 1958, where he was welcomed to the University of Hamburg faculty. His best known books are *Geometric Algebra* (1957) and *Class Field Theory* (1961). The American Mathematical Society awarded its coveted Cole Prize in number theory to Artin.

René-Louis Baire
1874-1932

Baire was born in Paris, France, in 1874 and is best known for his work on the theory of functions and the concept of a limit. When preparing himself for a doctorate in mathematics, he was fortunate to study under the famous Italian mathematician Vito Volterra, who was dismissed from his chair in Rome for refusing to pledge allegiance to the Fascist government. Volterra guided Baire during his postgraduate studies. Baire had a very short academic career because of poor health, but was appointed to the University of Montpellier and to Dijon for a few years. However, he was able to produce two important analysis books before retiring: *Théorie des nombres irrationels, des limites et de la continuité* (1905) and *Leçons sur les théories générales de l'analyse* (1907-1908). Baire died in Chambery, France, in 1932.

Stefan Banach
1892-1945

Banach was born in 1892 in Krakow, Austria-Hungary, which is now Poland. When he had completed his schooling in Krakow, he decided to abandon his study of mathematics for other subjects, but they did not hold much appeal, so he soon returned to lecture in mathematics at the Institute of Technology in Lvov and later became a professor at the University of Lvov. Banach is known as the founder of modern functional analysis, but his other specialties also brought him a measure of fame: the theory of topological vector space, and what is today known as Banach space as well as Banach algebras. Banach died in 1945 of lung cancer.

Nina Karlovna Bari
1901-1961

Born of Russian parents in Moscow in 1901, Bari showed an early aptitude in mathematics. This led to an invitation to join the Faculty of Mathematics and Physics at Moscow State University in 1918. While teaching, she worked for her doctorate on the theory of trigonometrical series and received her Ph.D. in 1926. She then became a research assistant at the Institute of Mathematics and Mechanics in Moscow. Following two years of study in Paris, she returned to a full professorship at Moscow State University. Among her numerous papers and publications were two textbooks: *Higher Algebra* (1932) and *The Theory of Series* (1936). She also translated Lebesgue's famous book on integration into Russian. Bari died in Moscow in 1961.

Paul Bernays
1888-1977

Swiss mathematician who contributed to mathematical logic and set theory, particularly in the framework of David Hilbert's systematic analysis of proofs. In his early work on propositional logic (the logic of sentential connectives such as "and," "or," "if-then," and negation), he showed that the system of Bertrand Russell and A. N. Whitehead was complete, in the sense that all valid formulas are provable. Later Bernays developed an axiomatization of set theory in first order logic with three fundamental notions—membership, set, and class—in order to avoid the contradictions of earlier systems.

Ludwig Georg Elias Moses Bieberbach
1886-1982

Bieberbach was born in Goddelau, Germany, in 1886 and was best known for a mathematical function known as the Bieberbach Conjecture (1916). His name is also synonymous with the study of polynomials, which now bears his name. Although he was reputed to be an inspiring teacher, he was also said to be rather disorganized in his methods. Perhaps the most damaging aspect of his personality was his outspoken anti-Semitism. This bigotry resulted in his publishing of a "German" style of mathematics as opposed to the abstract "Jewish" style of analysis. He went so far as to found a journal called *Deutsche Mathematik*. Bieberbach died in Germany in 1982.

Garrett Birkhoff
1911-1996

American algebraist and applied mathematician whose most important work is his theory of lattices, which influenced the development of quantum mechanics. Birkhoff believed strongly in applying mathematical concepts and had much success deriving observed fluid flows from hydrodynamic principles. An enthusiastic supporter of computer use for solving complex physical problems, he devoted considerable effort to developing methods for computing numerical fluid flows. During World War II he worked on radio-wave ranging of targets and antitank charges.

Harald August Bohr
1887-1951

Danish mathematician and the brother of famous Danish physicist Niels Bohr. Harald Bohr was a professor in Copenhagen, Denmark; his early research was on the theory of numbers, especially the Dirichlet series. Later, he collaborated with German mathematician Edmund Landau. Together they devised a theory called the Bohr-Landau theorem (1914), about the distribution of zeros.

Karol Borsuk
1905-1982

Born in Warsaw, Poland, in 1905, Borsuk earned his doctorate in geometry from the University of Warsaw in 1930. Other than several one-year appointments spent in the United States, Borsuk lectured in Warsaw for the balance of his career. He is best known for his work in founding the theory of retracts and the important concept of neighborhood retracts and notion of cohomotopy groups. The Nazi occupation of Poland during World War II was a difficult time for the entire academic community. Borsuk did his best to keep the University going and was imprisoned by the Nazis for his efforts. Fortunately, he escaped but remained in hiding through the rest of the occupation. He died in Warsaw in 1982.

Arthur Lyon Bowley
1869-1957

English statistician and mathematical economist known for his use of quantitative methods in economics. Bowley held many distinguished posts, including Professor of Mathematics, Economics and Statistics at the Universities of Reading and London, and Director of the University of Oxford Institute of Statistics. His most famous works were *Mathematical Groundwork* (1924), which resurrected the work of Irish economist and statistician Francis Edgeworth, *Wages and Income in the UK Since 1860* (1937), and *Three Studies on the National Income with Stamp* (1938).

Constantin Caratheodory
1873-1950

Caratheodory was born in Berlin, Germany, in 1873, where he received the major portion of his academic training. He received his doctorate in mathematics from the University of Gottingen in 1904 under the guidance of Hermann Minkowski, a famous Russian-born German mathematician. After lecturing in several German locales, Caratheodory was sought out by the Greek government to take up a chair at the University of Smyrna. He did so and was present when the Turks attacked Smyrna in 1922. Through his efforts, he was able to save the university library, which he moved to Athens. Caratheodory made significant contributions to the mathematics community by his work in the calculus of variations, and the theory of functions of a real variable. He also made worthy contributions in related fields such as in the special theory of relativity, thermodynamics, and geometrical optics. Caratheodory died in Munich, Germany, in 1950.

Henri Paul Cartan
1904-

Born in 1904 in Nancy, France, Cartan went on to teach at several prominent universities in his native land: Lycée Caen, Lille, Strasbourg, Paris, and Orsay. As the son of Elie Cartan (who modernized differential geometry), he devoted his life to mathematics. Among his accomplishments (under the name Bourbaki) were more than 30 volumes on *Eléments de mathématique*, illustrating the axiomatic structure of modern mathematics. He was elected to a Fellowship in the Royal Society of London (1971) and was given an honorary membership in the London Mathematical Society in 1959.

Mary Lucy Cartwright
1900-1998

Cartwright was born in Aynho, Northamptonshire, England, in 1900. At that time in history, it was unusual for women to receive recognition in academic subjects like mathematics and the sciences, but Cartwright graduated from Oxford in 1923 with a major in mathematics. She taught for four years before receiving her Ph.D. and went on to lecture in math at Cambridge University. Honored by many, Dame Cartwright was elected to the Royal Society, received the De Morgan Medal from the London Mathematical Society in 1968, as well as the Sylvester Medal from the Royal Society in 1964. She had a long, eventful life and died in England in 1998.

Guido Castelnuovo
1865-1952

Venice, Italy, was the birthplace of Guido Castelnuovo. He studied mathematics at the University of Padua and—following his graduation—spent the next year in Rome on a scholarship. His first academic appointment was at the University of Turin and, from there, he advanced to the University of Rome as the Chair of the Analytic and Projective Geometry Department. His most important publishing work was produced in 1903: *Geometria analitica e prolettiva*. During the following years, he produced a series of papers on algebraic surfaces. He also wrote on probability and the theory of relativity. Castelnuovo died in Rome, Italy, in 1952.

Eduard Cech
1893-1960

Czechoslovakian mathematician who made significant contributions in the fields of topology and differential geometry. Through his theories in topology and its application to several forms of mathematics, Cech became a worldwide lecturer and received numerous awards for his innovative work. The latter years of his life were devoted to differential geometry, and he published 17 papers on that topic before his death in Prague in 1960.

Shiing-shen Chern
1911-

Chern was born in 1911 in Jiaxing, China, where he eventually studied at Nankai University in Tientsin. He was offered a scholarship to study in America but instead chose the University of Hamburg because of his high regard for Prof. Blaschke, whom he had met when the latter visited China in 1932. After World War II when Chern returned to China to work, the civil war made life too difficult and he opted to accept an attractive offer from Princeton, where he had previously spent two years on staff. From there, he held the chair of geometry at the University of Chicago until 1960, when he moved to the University of California at Berkeley. Chern won many prizes and medals, including the National Medal of Science and the Wolf Prize. He was also a Fellow of the Royal Society of London.

Claude Chevalley
1909-1984

Chevalley was born in Johannesberg, Transvaal, South Africa, in 1909, the son of Abel and Marguerite Chevalley—authors of the *Oxford Concise French Dictionary*. Chevalley had an unusual education in that he graduated from the Ecole Normale Supérieuer in Paris, moved to Germany, then on to the United States to work at the Institute for Advanced Study at Princeton. He joined the faculty of the latter school, then moved on to Columbia University in New York, then back to the University of Paris, France. Chevalley worked in several mathematical areas: class field theory; algebraic geometry; and early studies of the theory of local rings. He won the Cole Prize of the American Mathematical Society in 1941 and was a life member of the London Mathematical Society.

Richard Courant
1888-1972

German-American mathematician who significantly influenced the development of physics, computer science, and other disciplines through applied work on variational calculus and numerical analysis. In *Methods of Mathematical Physics* (1924-7) he and David Hilbert developed techniques later important in quantum mechanics and nuclear physics. Courant was a principal coordinator and catalyst of twentieth-century mathematical research, as the impact of the Mathematics Institute in Göttingen and Courant Institute of Mathematical Sciences at New York University attest.

Elbert Frank Cox
1895-1969

American mathematician who became the first African-American to receive a Ph.D. in mathematics. Elbert Frank Cox was born in Evansville, Indiana, in December 1895. He earned his bachelor's degree in mathematics from the University of Indiana in 1917. After serving in the U.S. Army in France during World War I, he returned to the States to teach mathematics at public schools in Kentucky, and later at Shaw University in Raleigh, North Carolina. In 1925, Cox became the first African-American to receive a Ph.D. in Mathematics, from Cornell University. That same year, he accepted a teaching position at West Virginia State College, then in 1929, he transferred to Howard University. Cox remained at Howard until his retirement in 1965, serving as chairman of its mathematics department from 1957-1961.

Gertrude Mary Cox
1900-1978

Born in Dayton, Iowa, in 1900, Gertrude Mary Cox took an unusual path for women in those times: she earned a master's degree in statistics at Iowa State College. She continued her graduate work in California at Berkeley and returned to her alma mater, where she was appointed assistant professor of statistics. Among her many honors, Cox was appointed Director of Statistics at the Research Triangle Institute in Durham, North Carolina, where she remained until her retirement in 1964. Cox was the first woman elected into the International Statistical Institute. In 1956, she was elected President of the American Statistical Association and in 1975 was elected into the National Academy of Sciences.

W. Edwards Deming
1900-1993

American statistician and management consultant who developed an influential theory of management based on statistical quality control. Deming, who was trained as a physicist, became interested in management theory while working for the Department of Agriculture, the Census Bureau, and other United States government agencies. His theory of quality control, which was especially popular in the Japanese business world, stresses improving the interdependencies within a system and reducing variation on the basis of statistical sampling data.

Leonard Eugene Dickson
1874-1954

Dickson was born in Independence, Iowa, in 1874. After studying at the University of Texas, he earned his Ph.D. at the University of Chicago, where he remained for the major portion of his academic career. During his tenure there, he was considered to have made the university a major center for study and research in mathematics in America. Dickson is best known for his work in group theory and linear associative algebras. In 1901, he published his now-famous book on linear groups and, over the years, followed this with 17 more books on related aspects of the same subject. When the American Association for the Advancement of Science decided to give an annual prize for the best current work in the applicable fields, Dickson was the first winner of this coveted honor. He also received honorary degrees from Princeton, Harvard, and other institutions.

Jean Alexandre Eugène Dieudonné
1906-1992

Dieudonné was born in 1906 in Lille, France. He attended schools in Paris and earned his B.S. and Ph.D. at the Ecole Normale Supérieure and was considered one of the leading French mathematicians of his era. Much of his work was devoted to abstract analysis and algebraic geometry, but he began his career by concentrating on the analysis of polynomials. Dieudonné was a founder of the Bourbaki Group and published essays on algebraic geometry and topology, which are still held in high regard by his followers. Among his writings, the best known today are *La Géométrie des groupes classiques* (1955), *Foundations of Modern Analysis* (1960), and *Algèbre linéaire et géométrie élémentaire* (1964). Jean Dieudonné died in 1992.

Luther Pfahler Eisenhart
1876-1965

American mathematician who, through study of differential geometry, developed a theory of deformation of surfaces and systems of surfaces. He showed that the deformation of a surface defines the congruence of lines connecting a point and its image point. Eisenhart, who was home schooled by his mother and later studied mathematics mainly through guided independent reading, promoted the option of obtaining a college degree through independent study and the preparation of a thesis, as adopted by some American colleges.

Federigo Enriques
1871-1946

Italian mathematician and philosopher of science who made important contributions to the theory of algebraic geometry. He argued that topology, metric geometry, and projective geometry are linked to specific orders of sensation, those being the general tactile-muscular, the special sense of touch, and vision, respectively.

Agner Krarup Erlang
1878-1929

Born in 1878 in Lonborg, Jutland, Denmark, Erlang showed early promise in mathematics. At age 14, he passed the required tests in Copenhagen to teach. When he finally attended the University of Copenhagen in 1896, he was so poor that the school supplied him with food and lodgings while he studied there. His major was mathematics but he chose physics, astronomy, and chemistry as minor studies. While attending meetings of the Mathematical Association, he met the chief engineer of the Copenhagen Telephone Company, who persuaded Erlang to use his mathematical skills to benefit his employer. Erlang applied theories of probability to help solve problems in several areas of the company's operations and eventually provided a formula for loss and waiting time, which drew interest from many other countries in Europe, including the British Post Office. Erlang died in 1929 in Copenhagen.

Mary Celine Fasenmyer
1906-1996

American mathematician, teacher, and nun whose work is used in today's computer methods in proving hypergeometric identities. Fasenmyer graduated from St. Joseph's Academy in Pennsylvania, where she later taught math for 10 years. She earned her A.B. degree from Mercyhurst College. Her Catholic order encouraged her to earn a master's degree at Pitt, which she received in 1937. She later received a doctorate from the University of Michigan.

Kate Fenchel
1905-1983

German mathematician whose work focused on finite nonabelian groups. Although she attended the University of Berlin, Fenchel was unable to complete the advanced courses necessary for a Ph.D. She was able to teach when she came to the attention of a Danish mathematics professor who invited her to lecture at Aarhus University in Denmark.

John Charles Fields
1863-1932

Fields was born in 1863 in Hamilton, Ontario, Canada. After earning his B.A. in mathematics at the University of Toronto, he elected to work for his Ph.D. at Johns Hopkins University in Maryland. For the next three years, Fields traveled to Europe, where he was able to study with at least five of the best mathematicians on that continent. He then returned to Canada, where he was appointed lecturer at the University of Toronto—a position he held until his death in 1932. Before he died, Fields had been a strong proponent of an international medal for mathematical excellence and had provided financially for its adoption. It was formally announced at the International Congress of Mathematicians in Zurich in 1932 with the first medals being awarded in Norway at the Oslo Congress in 1936.

Ronald Aylmer Fisher
1890-1962

English statistician and geneticist considered one of the founders of modern statistics. Fisher introduced the analysis of variance and the concepts of randomization and confounding into the design of comparative experiments. Other contributions included significance testing and estimation with methods suitable for small samples using exact distributions, which he derived for many efficient sample statistics, and the important concept of likelihood, as distinct from probability. As a geneticist he contributed to understanding of the probabilistic evolution of dominance by natural selection. Politically, he was concerned about the differential fertility of the various social classes.

Irmgard Flugge-Lotz
1903-1974

When Flugge-Lotz was born in 1903 in Hameln, Germany, there were very few women being educated in mathematics and engineering. However, because her mother's family was active in construction engineering, she visited many sites during her early years. In 1923, she attended the Technical University of Hanover, majoring in math and engineering. She earned a doctorate in 1929 when she completed a dissertation on the mathematical theory of heat. After her marriage to Wilhelm Flugge (a civil engineer), they lived and worked in the German area which later became part of France. Both were offered positions in Paris at the National Office for Aeronautical Research. In 1948, both journeyed to America at the invitation of Stanford University, where Flugge-Lotz became Stanford's first woman Professor of Engineering in 1961. She died in California in 1974.

Maurice René Fréchet
1878-1973

French mathematician who introduced the concept of a metric space in topology, together with the properties of completeness, compactness, and separability, and connected topology with algebra into what is now known as functional analysis. Fréchet also had an interest in probability theory and statistics, as evident in his early work with sociologist Maurice Halbwachs and his cooperation with Emile Borel and Paul Levy in exploring analogies between the measure theory of real variables and strong laws of large numbers.

Eric Ivar Fredholm
1866-1927

Swedish mathematical physicist known for his work on integral equations and spectral theory, with his main contribution being the development of the theory of Fredholm integral equations. His few but carefully written papers quickly earned Fredholm great respect throughout Europe.

Hilda Geiringer
1893-1973

Austrian-American applied mathematician who discovered the fundamental Geiringer equations for plastic deformations of the plane in mathematical plasticity theory. Geiringer also studied the concept of the probability of a hypothesis and applied probability theory to population genetics, applying Mendelian principles. After the death of her husband, Richard von Mises, she edited, defended, and elaborated upon his frequentist theory of probability and completed his unfinished manuscript on a mathematical theory of compressible fluid flow.

William Sealy Gosset
1876-1937

British research chemist and statistician who in 1908 derived the first exact distribution of sample statistics—the probable error of a mean—suitable for statistical analysis of samples met in experimental work. Gosset, who worked for the Guinness brewery company in Dublin, published under the name of "Student." His ideas were subsequently incorporated into a full-fledged theory of experimental design and analysis by Ronald Fisher.

Jacques Hadamard
1865-1963

French mathematician who founded the branch of mathematics known as functional analysis, and developed a formula for determining the number of prime numbers. The ancient Greeks had already proved that for every prime number there is a larger one. Using his work on the Riemann zeta function, Hadamard was able to demonstrate that the number of prime numbers less than x is asymptotically equal to $x/\log x$. He made many other fundamental contributions to mathematics and published a book entitled *The Psychology of Invention in the Mathematical Field.*

Philip Hall
1904-1982

Hall was born in Hampstead, London, England, in 1904 and received the major portion of his education at King's College in Cambridge. In 1932 Hall achieved his greatest recognition for an extraordinary paper he published on the theory of groups of prime power order. It has withstood the test of time and is still one of the basic sources of modern group theory. Hall was honored several times in his native country—election to the Royal Society in 1942, then earning its Sylvester Medal in 1961. Also, because of his innovative work and researches in algebra, he was elected President of the London Mathematical Society for two years and was honored by receiving its De Morgan Medal in 1965.

Godfrey Harold Hardy
1877-1947

English mathematician known for his work on the Fourier series and the Riemann zeta function. He studied mathematics at Cambridge; his happiest years, however, were spent at Oxford, where he was Savilian Professor of Geometry. Hardy received numerous awards, including the prestigious Fellow of the Royal Society. He later returned to Cambridge, where he died in 1947.

Helmut Hasse
1898-1979

Born in 1898 in Kassel, Germany, Hasse attended various secondary schools until and through World War I. Upon leaving the German navy, he entered the University of Göttingen, where his superior work in mathematics became evident to his associates. While there, he developed a "local-global" principle which brought him international recognition and was the basis for his doctoral dissertation at the university. World War II was a difficult period for mathematicians at Göttingen. When the Nazi order came to expel all Jews from the faculty, the Mathematical Institute lost 18 prominent members. Later, in 1949, Hasse took a post as professor at the Humboldt University in East Berlin. In 1950, he moved to Hamburg, where he taught until he retired in 1966. Hasse died in Germany in 1979.

Felix Hausdorff
1869-1942

Hausdorff was born in Breslau, Germany (now Wroclaw, Poland) in the latter part of 1869. He attended local schools and graduated from the University of Leipzig in 1891 and was asked to remain there as a teacher. Within a year, he received an invitation to join the staff at the prestigious Göttingen University but elected to teach at Bonn instead. Hausdorff's work was mainly centered on topology and set theory, and his remarkable grasp of this area of mathematics was apparent to all who worked with him. Unfortunately for Hausdorff, his Jewish ancestry cost him his position at the university. Early on, he and his family managed to avoid capture but, in 1942, Hausdorff, his wife, and her sister were sent to one of the internment camps, where Hausdorff committed suicide in 1942.

Olive Clio Hazlett
1890-1974

American mathematician often referred to as "one of the two most noted American women in the field of mathematics." Hazlett spent the majority of her professional career as a mathematics instructor at the University of Chicago, where she taught calculus, plane trigonometry, and analytic geometry. She also taught graduate studies in modern algebra. The renowned mathematician authored 17 research papers, more than any other pre-1940 American women mathematician. Her research on invariants of nilpotent algebras appeared in the *American Journal of Mathematics* in 1916.

Jacques Herbrand
1908-1931

French mathematician remembered for his outstanding contributions to mathematical logic and for his theorem on testing for sentential validity. From an early age, Herbrand was an outstanding student in mathematics. He was admitted to the Ecole Normal Supérieure when he was only 17 years of age and received his Ph.D. when he was 21. He received a Rockefeller fellowship which allowed him the time and money to travel and study at Berlin and Göttingen universities, where the most famous mathematicians were gathered. Unfortunately, he died at age 23 in a vacation accident in the Alps, and his promising career came to an early end in 1931.

William Vallance Douglas Hodge
1903-1977

Hodge was born in Edinburgh, Scotland, in 1903. While attending the university at Edinburgh, he was advised by his teacher, Sir Edmund Taylor Whittaker, to continue his studies at Cambridge in England (Sir Edmund's alma mater). Hodge held a chair at Cambridge, where his work concentrated on the relationship between geometry, analysis, and topology. He was one of the founders of the Mathematical Colloquium, a yearly event in which mathematicians visit various universities in Great Britain. "[I]n recognition of his pioneering work in algebraic geometry, notably in his theory of harmonic integrals," the Royal Society of London awarded Hodge its Copley Medal in 1974, three years before his death, in Cambridge, in 1977.

Heinz Hopf
1894-1971

German mathematician known for his work in algebraic topology. Hopf became interested in mathematics while serving in World War I. Later, he studied at the University of Berlin and in 1925 received his Ph.D. He accepted a chair in Zurich, Switzerland, and is remembered for his formula about the integral curvature and for what came to be known as the "Hopf invariant."

Witold Hurewicz
1904-1956

Hurewicz was born in Lodz, Russian Poland, in 1904. He was educated in Vienna under the tutelage of Hans Hahn and Karl Manger and received his Ph.D. in 1926. After a stint in Amsterdam (eight years), he traveled to America and decided to remain there. He accepted an offer to work at a school in North Carolina and went on to teach at the Massachusetts Institute of Technology in Boston. Hurewicz's research was mainly concerned with set theory and topology, but he is primarily associated with his discovery of exact sequences. He is also credited with "setting the stage" for homological algebra. Hurewicz had a fairly short academic career. During a conference side trip in Uxmal, Mexico, in 1956, he fell off a pyramid and died.

Sof'ja Alexsandrovna Janovskaja
1896-1966

A few years after her birth in 1896 in Poland (now Kobrin, Belarus), Janovskaja and her family moved to Odessa and, after an early education in the classics, she entered the Higher School for Women in Odessa in 1915. She received her doctorate in mathematics from the Moscow State University, where she was already teaching. Janovskaja devoted her research to the philosophy of logic and mathematics, but her ongoing interest was the actual history of mathematics, a subject in which she excelled and published several important papers. In 1951, Janovskaja was awarded the Order of Lenin and eight years later, she headed the new Department of Mathematical Logic at Moscow State University. She died in Moscow in 1966.

Pelageya Yakovlevna Polubarinova Kochina
1899-

Russian mathematician who was renowned in the Soviet Union for her work in applied mathematics. The Russian Revolution in 1917 ended gender discrimination, which allowed Kochina to pursue her mathematics degree from Russia's Petrograd University (1921). She went on to complete her doctorate in Physical and Mathematical Sciences in 1940. Over the course of Kochina's career, she held several professional posts, including Director of the University of Moscow's division of hydromechanics, Director of the Department of Applied Hydrodynamics at the Hydrodynamics Institute, and head of the Theoretical Mechanics Department at the University of Novosibirsk. Kochina was a high-ranking member of the USSR Academy of Sciences, and was considered among the Stalinist scientific elite.

Cecilia Kreiger
1894-1974

Polish mathematician and physicist who was the first woman to earn a mathematics doctorate from a Canadian university. Kreiger's graduate studies included courses in modular elliptic functions, minimum principles of mechanics, theory of sets, theory of numbers, and theory of functions. Her thesis on the summability of trigonometric series with localized properties was published in two parts in the *Transactions of the Royal Society of Canada*. The Canadian Mathematical Society awards the CMS Krieger-Nelson Prize Lectureship for Distinguished Research by Women in Mathematics in honor of Krieger and her colleague, Evelyn Nelson.

Christine Ladd-Franklin
1847-1930

American mathematician who made important contributions to symbolic logic and vision color theory. Her early interest was physics, however, since women were not provided with graduate laboratory facilities, she switched to mathematics. After receiving her degree, she taught secondary school science and published mathematics research in the *British Educational Times*. When she later applied to John Hopkins, her name was recognized from her articles and she was allowed to attend despite current enrollment restrictions against women. She was a strong advocate of women's suffrage and educational reform.

Solomon Lefschetz
1884-1972

Russian mathematician best known for his work on the algebraic aspects of topology. He earned a degree in engineering at the Ecole Centrale in Paris and elected to travel to America when he was 21 years of age. While working at Westinghouse in Pittsburgh, he was in a laboratory accident that cost him both hands. Lefschetz turned to the study of mathematics, and, as a fellow of Clark University, he earned a doctorate in mathematics in 1911. He taught at Nebraska, Kansas, and Princeton universities. He edited the *Annals of Mathematics* for 30 years and was President of the American Mathematical Society for three years.

Jean Leray

French mathematician who made important contributions in many areas of mathematics. In his early years, Leray was active in algebraic topology, where he discovered spectral sequences. Later, Leray worked on the mathematical aspects of fluid mechanics, and many of his discoveries retained their impact for over 60 years. Later, his work on the theory of complex variables was of great significance in that branch of mathematics.

Stanislaw Leshniewski
1886-1939

Russian-born Polish mathematician who published numerous papers on his theories of logic and the foundations of mathematics. Leshniewski attended several universities before earning a doctorate in 1912 at the University of Lwów (now Lviv) in Poland. He held teaching posts in Moscow and Warsaw, where he headed the department of the philosophy of mathematics. After his death, one of his former students (Tarski) reconstructed his unpublished work, which was destroyed in World War II.

Tullio Levi-Civita
1873-1941

Italian mathematician best know for contributing to the development of relativity. He and Gregorio Ricci-Curbastro developed absolute differential calculus (1900). This was employed by Albert Einstein in formulating the general theory of relativity (1915). Levi-Civita's most significant contribution to mathematics was to extend the notion of parallel lines to curved spaces (1917). This and other work influenced the evolution of absolute differential calculus into tensor calculus, which enables unified treatments of electromagnetic and gravitational fields.

Paul Pierre Lévy
1886-1971

French mathematician who made contributions to probability, partial differential equations, and functional analysis. Lévy's work in geometry extended Laplace transforms to broader function classes. He published several books, notably *Leçons d'analyse fonctionelle* (1922), *Calcul des probabilitiés* (1925), and *Processus stochastiques et mouvement brownien* (1948), and taught at the Ecole Polytechnique in Paris from 1913 to 1959.

John Edensor Littlewood
1885-1977

English mathematician who, in collaboration with G. H. Hardy, produced an acclaimed body of work, including contributions to pure mathematics and the theory of series and functions. Littlewood became a Fellow of the Royal Society in 1915 and received the organization's Royal Medal (1929), Sylvester Medal (1943), and Copley Medal (1958).

Nikolai Nikolaevich Luzin
1883-1950

Russian mathematician who made significant contributions to the theory of functions, sets, and boundaries. Luzin earned a doctorate at Moscow University, where he was granted a professorship in the Department of Pure Mathematics just before the 1917 Revolution. In 1927 Luzin won membership in the Academy of Sciences of the U.S.S.R, and in 1935 became head of the Department of the Theory of Functions of Real Variables at the Steklov Institute.

L.A. Lyusternik

Russian mathematician whose work in geometry and topology was of great importance. Lyusternik's work in polyhedra (solid shapes with many sides) included shapes that inspired Buckminister Fuller to invent the geodesic dome, after whom the pure carbon molecule "buckminsterfullerne (C_{60}") was named.

Sheila Scott Macintyre
1910-1960

Scottish mathematician. Macintyre won several scholarships at the University of Edinburgh, where she earned an M.A. with first class honors in mathematics and natural philosophy. She later married A. J. Macintyre, who lectured in mathematics at Aberdeen University, where she earned a doctorate in mathematics. Sheila Macintyre had an active academic career while raising several children and, in 1958, moved to the United States, where she and her husband taught at the University of Cincinnati until her untimely death from cancer in 1960.

Saunders MacLane
1909-

American mathematician. After graduating from Yale in 1930, MacLane traveled to Germany to enroll at the prestigious Göttingen University—the center of European mathematics at that time. However, when the Nazis came to power in the 1930s, he quickly completed his doctorate with his dissertation, "Abbreviated Proofs in the Logical Calculus," and immediately returned to America. In 1947 he was appointed professor of mathematics at Chicago University and, five years later, was given the chair of that department, which he held until his retirement.

Andrei Andreevich Markov
1856-1922

Russian mathematician who was the first to provide a rigorous proof of the central limit theorem, and who discovered that this theorem and the law of large numbers also hold for certain dependent trials. A Markov process refers to a probabilistic process in which the probability of an outcome depends only on the outcome of the previous trial. This concept is used to model a wide range of physical phenomena, from gas diffusion to traffic problems.

Helen Merrill
1864-1949

American mathematics instructor who was appointed chairperson of the mathematics department at Wellesley University, a position she held until her retirement in 1932. Merrill authored or co-authored three mathematical textbooks. Her last, *Mathematical Excursions: Side Trips along Paths Not Generally Traveled in Elementary Courses in Mathematics*, was written for the general public. Merrill was also very active with the Mathematics Association of America from its inception. She also served as vice-president (1920-1921) of the Mathematical Association of America. In addition to her mathematical interests, Merrill was also a published poet, writing several poems about her field of study.

Winifred Edgerton Merrill
1862-1951

American mathematician who was the first American woman to receive a Ph.D. in mathematics. Although the Board of Trustees at Columbia University refused Merrill's first application for a doctorate, she prevailed with the support of the university president and by personally meeting with each of the trustees. Merrill made important contributions to the mathematics of astronomy and taught college mathematics for many years at a number of schools.

Richard Martin Edler von Mises
1883-1953

Austrian-American applied mathematician and philosopher of science who developed the first mathematically precise frequency theory of probability and contributed to the theory of powered airplanes, plasticity, elasticity, and turbulence. He was concerned with the application of abstract mathematical theories to observational data. Probability, according to his frequency theory, is an ideal notion that applies to the outcomes of an infinite sequence of trials if 1) the ratio of successes/trials has a limit, and 2) this limit is the same for all blindly chosen infinite subsequences.

Eliakim Hastings Moore
1862-1932

American mathematician often credited with establishing the University of Chicago as a world-class center of mathematical study, which it remains to this day. Moore earned his doctorate at Yale in 1885, then traveled to Europe to study, primarily at the University of Berlin, where mathematics research was at its height. After teaching at Northwestern and Yale, Moore was made head of the mathematics department at the University of Chicago when it first opened in 1892. An atypical genius, Moore was not only a progressive, innovative thinker, but also inspired others to do their best work. He was first vice-president, then president, of the American Mathematical Society, and editor of the *Transactions of the Society* from 1899 to 1907.

Robert Lee Moore
1882-1974

American mathematician who worked in the field of geometry. He received his doctorate for his work entitled "Sets of Metrical Hypotheses for Geometry" from the University of Chicago. He taught at various schools, including University of Tennessee, Princeton, Northwestern and Pennsylvania universities, and the University of Texas. He was president of the American Mathematical Society for two years and was chief editor of the *Colloquium Publications* for seven years.

Louis Joel Mordell
1888-1972

American mathematician who made contributions to number theory and discovered the finite basis theorem. Mordell was educated at Cambridge and subsequently taught at Manchester University for 23 years. In 1945 he took a post at Cambridge, where he worked in collaboration with Harold Davenport. Together, they made significant advances in the geometry of numbers. Mordell became a Fellow of the Royal Society in 1924 and was awarded the Sylvester Medal (1949). He was also president of the London Mathematical Society, from which he received the De Morgan Medal (1941).

Oskar Morgenstern
1902-1977

German-born American economist who applied John von Neumann's "Theory of Games" to business strategy. Morgenstern studied, then taught economics at the University of Vienna from 1929 to 1938. He also taught at Princeton University (1938-70) and New York University (1970-77). During the 1930s, he was a member of the so-called "Austrian circus," a group of Austrian economists who met regularly to discuss issues in the field. He and mathematician John von Neumann co-wrote the book *Theory of Games and Economic Behavior*, which used mathematics to analyze competitive business situations. Their Theory of Games suggested that often the outcome of a "game," or business situation, depends on several parties, or "players." Each player calculates what all of the other players will do before determining their own strategy. This theory was applied to several fields, including economics, politics, business, law, and science.

Ruth Moufang
1905-1977

German mathematician who made contributions to theoretical and applied geometry, and became the first German woman with a doctorate to be employed in industry. Moufang earned a Ph.D. in projective geometry and introduced certain planes and non-associative algebraic systems that bear her name. Despite her extraordinary skills, the Nazi minister of education prohibited Moufang from teaching anywhere because of her gender. Thus restricted, she turned to industrial mathematics, the only area in which she could apply her knowledge, and focused on elasticity theory. In 1946, after the Nazi defeat, she was granted a professorship in Frankfurt.

Rolf Herman Nevanlinna
1895-1980

Finnish mathematician who established the study of harmonic measure and formulated a theory of value distribution, known as Nevalinna theory, in 1925. Nevanlinna took an early interest in the classics and mathematics, and studied at Helsinki University, where he was greatly influenced by one of his father's cousins, Ernst Lindelöf. Nevanlinna completed his doctorate in 1919 and taught at Helsinki University. The Rolf Nevanlinna Prize, established in 1982, is an annual award presented by the International Congress of Mathematicians for advances in mathematical aspects of information science.

Giuseppe Peano
1858-1932

Italian mathematician and logician best known for introducing into mathematics the axioms that bear his name. Peano's axioms provided for a purely logical derivation of the arithmetic of natural numbers. He also submitted to rigorous criticism the foundations of projective geometry and presented the first statement of vector calculus. Peano was a mathematics professor at the University of Turin (1890-1932) and spent much of his time after 1900 developing an international language (Interlingua).

Rózsa Péter
1905-1977

Hungarian mathematician who was the first Hungarian female to become an Academic Doctor of Mathematics. Péter's work helped establish the modern field of recursive function theory as a separate area of mathematical research. Her book, *Recursive Functions in Computer Theory*, was considered indispensable to the theory of computers. It became the second Hungarian mathematical book to be published in the Soviet Union. Péter taught mathematics from 1945 until her retirement in 1975. In 1973 she was elected as the first female mathematician to the Hungarian Academy of Sciences.

Jules-Henri Poincaré
1854-1912

French mathematician who contributed to all fields of mathematics but may be best known for his work in celestial mechanics, which led to the development of the new field of topology and forms the basis for modern chaos theory. In 1879 Poincaré received an engineering degree and submitted his mathematics research as a doctoral thesis to the University of Paris. There he took a faculty position that he would hold for the rest of his life.

George Polya
1887-1985

Hungarian-born American mathematician and mathematics educator who made a number of contributions to the theory of probability, being the first to study the "random walk." Polya studied mathematics at the University of Budapest, receiving his doctorate in 1912. He taught in Switzerland before immigrating to the United States in 1940. A prolific writer on mathematical topics, he is best known for his small book *How to Solve It,* which has helped generations of students learn how to approach mathematical problems.

Lev Semenovich Pontryagin
1908-1988

Russian mathematician who contributed to the study of overlapping aspects of algebra and topology, and later, control theory. After Pontryagin was blinded in an explosion, his mother devoted herself to his education and he subsequently earned a degree from the University of Moscow in 1929. Pontryagin joined the Steklov Institute in 1934, and the next year was named head of its Department of Topology and Functional Analysis. He was elected to the Academy of Sciences in 1939, and further honored when the International Union elected him vice president in 1970. Pontryagin published several significant works, including *Topological Groups* (1938) and *The Mathematical Theory of Optimal Processes* (1961), for which he received the Lenin Prize.

Srinivasa Ramanujan
1887-1920

Indian mathematician notable for his brilliant mathematical intuition, which inspired him to create numerous new theorems and insights into analytical number theory. Born into a poor Brahmin family, Ramanujan was largely self-taught. His genius was recognized by the British mathematician G. H. Hardy, who arranged for Ramanujan to come to England and publish many of his results on arithmetic functions, infinite series, and highly composite numbers. Ramanujan also developed a fast method for computing the value of π.

Frank Plumpton Ramsey
1903-1930

English mathematician best known for his criticisms of Alfred North Whitehead and Bertrand Russell's *Principia Mathematica.* Ramsey showed how to eliminate the axiom of reducibility, which they had introduced to deal with paradoxes arising from their theory of types. Ramsey also published two studies in economics, the last of which John Maynard Keynes described as "one of the most remarkable contributions to mathematical economics ever made." Ramsey's contributions to philosophy were small but significant, including work on universals and scientific theories.

Mina Spiegel Rees
1902-1997

American mathematician who held high-level military research posts in the early 1950s and, in 1970, became the first woman to be elected president of the American Association for the Advancement of Science. Rees earned a doctorate in associative algebra at the University of Chicago in 1931. During World War II, she left a teaching post at Hunter College to work for the Office of Scientific Research and Development. After the war, Rees was named director of the mathematics division of the Office of Naval Research, and served as its deputy science director from 1952-53.

Georges de Rham
1903-1990

De Rham was born in 1903 in Roche, Canton Vaud, Switzerland. When he entered the University of Lausanne in 1921, he had every intention of majoring in chemistry, physics, and biology. However, numerous questions came up in physics that required more advanced studies in mathematics. He found this field so intriguing that he finally abandoned the sciences and devoted himself exclusively to mathematics. He studied for his doctorate in Paris and, upon earning the degree, he returned to his alma mater in Lausanne. He eventually was granted full professorships at both the University at Lausanne and the University of Geneva. In addition to many honors bestowed upon him, he received honorary degrees from the universities of Strasbourg, Lyon, Grenoble, and l'Ecole Polytechnique Fédérale Zurich. He was also president of the International Mathematical Union for two years.

Frigyes Riesz
1880-1956

Hungarian mathematician best known as a founder of functional analysis, integral equations, and subharmonic functions. Working with Bela Szokefalvi-Nagy, he published what would become a classic in the mathematics community: *Lessons of Functional Analysis* in 1953.

Marcel Riesz
1886-1969

Polish mathematician who was the younger brother of Frigyes Riesz. Educated in Budapest, Riesz was singled out for his innovative work when he won the Lorand Eotvos competition in 1904. He published only one paper with his brother, Frigyes, on boundary behavior of an analytic function but worked with others in joint research projects. In 1908, he was contacted by Mittag-Leffler in Sweden to teach and research there. He was appointed to a position in Stockholm and chaired a mathematics department at Lund. Riesz spent the rest of his life in Sweden and died in Lund in 1969.

Bertrand Arthur Russell
1872-1970

British mathematician and philosopher. The son of an earl, Russell was educated at Trinity College, Cambridge. In 1903 Russell published a book on the *Principles of Mathematics*. Over the years 1910-13 Russell and philosopher Alfred North Whitehead published *Principia Mathematica*, a three-volume work in which they attempted to establish an axiomatic basis for arithmetic without using the principle of induction. In later life Russell wrote numerous books on philosophy for the popular reader and actively supported world disarmament and social reform.

Charlotte Angas Scott
1858-1931

English mathematician who served as editor of the *American Journal of Mathematics* and published the noted textbook *An Introductory Account of Certain Modern Ideas and Methods in Plane Analytical Geometry* (1894). Scott won a scholarship to Girton College, Cambridge University, where she earned a doctorate in algebraic geometry in 1885. News of her accomplishments spread and she was summoned to the United States to become the first head of the mathematics department at the newly established Bryn Mawr College in Pennsylvania. Scott was also a member of the Council of the American Mathematical Society, serving in 1905 as its vice president.

Claude Elwood Shannon
1916-

American mathematician who founded the field of information theory with his 1948 paper "A Mathematical Theory of Communication," published in the *Bell System Technical Journal*. Shannon graduated from the University of Michigan and proceeded to MIT for further study. He was intrigued by the use of George Boole's algebra to analyze and optimize relay switching circuits, a topic he later wrote about. In 1941 he accepted an offer to work for Bell Telephone as a research mathematician; he remained with the company until 1972. In 1952 he proposed an experimental system that demonstrated the capabilities of phone relays. Shannon was awarded the National Medal of Science in 1966.

Carl Ludwig Siegel
1896-1981

German mathematician who won recognition for his work on the theory of numbers. Born in Berlin, Siegel held professorships at Frankfurt and Göttingen from 1922 to 1940. He subsequently traveled to the United States and worked at Princeton University until 1951, then returned to Göttingen. Siegel built upon the work of Pierre de Fermat, and much of his ensuing work on quadratic forms was based upon the earlier endeavors of Joseph Lagrange, Carl Gauss, Ferdinand Eisenstein, and Charles Hermite.

Waclaw Sierpinski
1882-1969

Polish mathematician who produced 724 papers and 50 books on the subject of mathematics. At a time when the Russians were aiming their efforts at keeping the Poles illiterate and generally uninformed, Sierpinski entered the Department of Mathematics and Physics at the University of Warsaw, where he proved to be an outstanding student, winning the department's gold medal for a dissertation on Voronoy's contribution to number theory. Sierpinski received dozens of medals, awards, and honors from the Royal Societies of most of the European nations, and founded several publications, among them the *Fundamenta Mathematicae*, which is important to this day.

Thoralf Albert Skolem
1887-1963

Norwegian mathematician best known for his contributions to mathematical logic. The Skolem-Löwenheim theorem established that any class of formulas simultaneously satisfiable is also satisfiable in a denumerably infinite domain (1920). Skolem realized this led to a paradox—the apparent conflict between a set's magnitude in axiomatic theory and its magnitude in the more limited domain it is modeled in (1922). Skolem resolved the paradox by treating certain set-theoretic concepts, such as non-denumerability, as relative.

Norman Steenrod
1910-1971

American mathematician whose work in topology led to many important discoveries. In particular, Steenrod was active in homology theory, including helping to define homology groups and elaborating on the concept of fiber bundles, an important concept in topology. Steenrod's most lasting contribution, however, may be the "Steenrod volumes," in which he compiled, classified, and cross-referenced many important papers in the field. This has proven an invaluable reference tool in this specialized field.

Ernst Steinitz
1871-1928

German mathematician best known for providing the first abstract definition of a field, published in *Algebraische Theorie der Korper* (1910). This work proved that every field has an algebraically closed extension field. Steinitz completed his doctoral dissertation in 1894, and received an assignment as Privatdozent at the Technische

Hochschule Berlin in Charlottenburg. Six years later, he accepted a professorship at the Technical College of Breslau, where he remained for 10 years. He later served as head of the mathematics department at the University of Kiel.

Marshall Harvey Stone
1903-1989

American mathematician who contributed to the study of linear differential equations and spectral theory. The son of a prominent lawyer, Stone attended Harvard, where he earned a doctorate in mathematics in 1926. During World War II, the government invited Stone to participate in undercover war work—a position he retained until 1946. Technically, he was still on staff at Harvard but elected to assume the department chair in mathematics at the University of Chicago. His reputation attracted world-class mathematicians to Chicago and the department flourished under his management. Following his retirement in 1968, he took up a lifelong hobby of travel; he died in Madras, India, while visiting in 1989.

Gabor Szego
1895-1985

Hungarian mathematician who published numerous articles and contributed to the study of orthogonal polynomials and Toeplitz matrices. After graduating from Budapest University, Szego studied at the German universities of Berlin and Göttingen and served in the Austro-Hungarian cavalry during World War I. In 1926 he was invited to join the staff at Königsberg, where he remained until the Nazis made academic life in Germany untenable for Jews. He left for the United States in 1934 and secured a teaching position at Washington University in St. Louis, Missouri. Four years later, he relocated to Stanford University, where he spent the rest of his academic life.

Teiji Takagi
1875-1960

Japanese mathematician who made contributions to class field theory, a subject on which he wrote two comprehensive papers in 1920 and 1922. After graduating from Tokyo University, Takagi studied for several years in Germany at the universities of Berlin and Göttingen, then returned to Tokyo to complete his doctorate in 1903. He received a professorship at Tokyo the next year and remained at that institution until his retirement in 1936. Takagi produced two important works—*Introduction to Analysis* (1938) and *Algebraic Number Theory* (1948).

Alfred Tarski
1901-1983

Polish-American mathematician-logician best known for developing the semantic method. Tarski's method made it possible to discuss the relationships between expressions and the extralinguistic objects they denote. This required making a distinction between the syntax of the language being studied and the metalanguage used to describe it. Tarski's researches also yielded a mathematical definition of truth in languages. He made other important contributions to mathematics, including the theory of inaccessible cardinals and the Banach-Tarski paradox.

Paul Julius Oswald Teichmuller
1913-1943

German mathematician who contributed to the field of geometric function theory. Teichmuller studied under Helmut Hasse at Göttingen, where he completed a thesis on linear operators on Hilbert spaces, then undertook further studies under Ludwig Bieberbach in Berlin. In a six-year period Teichmuller researched and produced some 34 papers and introduced quasi-conformal mappings and differential geometric methods into complex analysis. An ardent Nazi, Teichmuller joined the German army in 1939 and was killed in battle in the Dnieper region of the Soviet Union.

Stanislaw Ulam
1909-1984

Polish-born American mathematician. Ulam was educated at the Polytechnic Institute in Warsaw, where his mathematics teachers challenged him with special problems in set theory after realizing the extent of his mathematical talents. He immigrated to the United States in 1935, becoming a citizen in 1941, and taught at Harvard and the University of Wisconsin. He interrupted his academic career in 1944 to work on the hydrogen bomb, remaining active in weapons-related work for 23 years.

Pavel Samuilovich Urysohn
1898-1924

Russian mathematician who contributed to the study of integral equations and topology. Born in the Ukraine, Urysohn initially studied physics at the University of Moscow, where he attended lectures by Nikolai Luzin and Dimitri Egorov, both active in the mathematics department. Their work inspired Urysohn to bypass physics for math and, after his graduation in 1919, he became an assistant professor at the university.

His work came to the attention of the Mathematical Society of Göttingen, and in 1924 he traveled to Germany, Holland, and France to meet with other mathematicians. While on this trip, he accidentally drowned off the coast of Brittany; he was 26 years old.

Charles Jean Gustave Nicolas de la Vallée Poussin
1866-1962

Vallée Poussin was born in 1866 in Louvain, Belgium. His father (professor of geology at the University of Louvain) encouraged him to become a teaching Jesuit, but found he was not suited for the seminary. He became interested in engineering and although he graduated in that venue, he soon devoted his academic life to pure mathematics. His position as assistant at the University of Louvain allowed him to work with Louis Claude Gilbert (one of his former teachers) and when Gilbert died unexpectedly at the age of 26, Vallée Poussin was promoted to Gilbert's chair in 1893. Although one of Vallée Poussin's papers on differential equations was a prize winner, he was best known for proving the prime number theorem. He also worked on algebraic and trigonometric polynomials. Vallée Poussin chaired the mathematics department at the University of Louvain for 50 years before he died in 1962.

Oswalk Veblen
1880-1960

American mathematician who made important contributions to the study of topology and projective and differential geometry. After completing his undergraduate degree at the University of Iowa, Veblen spent a year at Harvard, then transferred to the University of Chicago, where he completed his doctorate in 1903. Veblen taught mathematics at Princeton from 1905 to 1932, then at the Institute for Advanced Study, which he helped establish in 1932. Veblen's research gave Princeton the honor of being one of the world's premier centers of topology research. Veblen served as both vice president and president of the American Mathematical Society and delivered the organization's Colloquium Lecture, on the topic of topology, in 1916.

Ivan Matveevich Vinogradov
1891-1983

Russian mathematician who made contributions to number theory and worked on quadratic residues. After completing a master's degree at St. Petersburg in 1915, Vinogradov taught at the State University of Perm and the St. Petersburg Polytechnic Institute and University. In 1934 he was appointed the first director of the Steklov Institute in Moscow, a position he retained for the rest of his life. Although he left the Soviet Union only once to attend a mathematical conference in Edinburgh, many of the world's foremost mathematicians traveled to Moscow to visit him and discuss his research. Even today, work being done on the same types of problems studied by Vinogradov confirms his brilliance. Vinogradov was not only a mathematical genius, but a strong advocate of physical fitness. He was proud of his own athleticism and remained active and in good health well into his nineties.

Bartel Leendert van der Waerden
1903-1996

Dutch-born mathematician best known for his classic *Modern Algebra* (1930), which helped disseminate the new algebra developed by David Hilbert, Emil Artin, Emmy Noether, and others. Waerden worked on abstract algebra, algebraic geometry, combinatorics, number theory, probability theory, statistics, topology, and group theory. Having become interested in physics, he wrote an influential book applying group theory to quantum mechanics (1932). Waerden also published many historical works on mathematics and science in the ancient world.

Joseph Henry MacLagen Wedderburn
1882-1948

Scottish mathematician who achieved international prominence with his 1907 paper on the classification of semisimple algebras, in which he showed that a simple algebra was a matrix algebra over a division ring. After earning a degree in mathematics from Edinburgh University in 1898, Wedderburn went on to study at both Leipzig and Berlin universities, followed by a year in the United States at the University of Chicago on a Carnegie scholarship. After serving in the British Army during World War I, he was invited to teach at Princeton, where he remained until his early retirement in 1945. Wedderburn was elected to a fellowship in the British Royal Society in 1933, and was made an honorary Fellow of the Edinburgh Maths Society in 1946.

Anna Johnson Pell Wheeler
1883-1966

American mathematician who contributed to the study of integral equations and infinite dimensional linear spaces. After completing an undergraduate degree in mathematics at the University of South Dakota, Wheeler earned a M.A. at the University of Iowa and a Ph.D. at the University of Chicago in 1909. She subsequently taught at Mount Holyoke College and Bryn Mawr, where she remained until her retirement in 1948. Wheeler received numerous honors and in 1927 became the first woman to deliver the prestigious Colloquium Lecture at the annual meeting of the American Mathematical Society.

Alfred North Whitehead
1861-1947

English philosopher and mathematician who made valuable contributions to the fields of pure and applied mathematics. Whitehead was educated at Cambridge and became a professor at the University of London and later at Harvard. He authored several important works, including *A Treatise of Universal Algebra* (1898), *The Principle of Relativity* (1922), which challenged Einstein's general theory of relativity, and *Process and Reality* (1929). He also co-authored the landmark book *Principia Mathematica* (1910-13) with philosopher Bertrand Russell.

Hassler Whitney
1907-1989

American mathematician who is best known for his contributions to topology, graph theory, and chromatic polynomials. Whitney received his first degree in mathematics at Yale, then went on to earn a doctorate at Harvard in 1932 with a dissertation entitled "The Coloring of Graphs." He taught at Harvard until 1952, when he accepted an appointment at the Institute for Advanced Study at Princeton, where he remained until his retirement in 1977. Whitney was awarded the National Medal of Science (1976), the Wolf Prize (1983), and the Steele Prize (1985).

Norbert Wiener
1894-1964

American mathematician best known for establishing the science of cybernetics, which is concerned with the mathematical analysis of, analogies between, and information flow within mechanical and biological systems. This work was influenced by his previous application of statistical methods to antiaircraft fire control and communications engineering. Wiener also derived a physical definition of information related to entropy. He developed a mathematical theory of Brownian motion and advanced the study of integrals, quantum mechanics, and harmonic analysis.

Ernest Julius Wilcynski
1876-1932

German-born American mathematician who contributed to mathematical astronomy and projective differential geometry. Wilczynski emigrated with his family to the United States and attended high school in Chicago. Upon graduation, he returned to Germany to study at the University of Berlin, where he earned a doctorate in 1897. Returning to America the next year, he eventually secured a teaching position at the Berkeley campus of the University of California, followed by posts at the University of Illinois and Chicago. Wilcynski served as vice president of the American Mathematical Society and, in 1919, was elected to the National Academy of Sciences.

William Henry Young
1863-1942

British mathematician whose research included studies in orthogonal series (including Fourier). All books on advanced calculus now in use retain his approach to functions of many complex variables. Young was president of the London Mathematical Society from 1922 to 1924, after having earned the society's De Morgan Medal in 1917. He was also a Fellow of the Royal Society and was awarded its coveted Sylvester Medal in 1928. During World War II, Young was working in Switzerland when France fell to the Nazis. He died there in 1942, separated from his wife (the famous mathematician Grace Chisholm) and his family.

Oscar Zariski
1899-1986

Russian-American mathematician known for his work in algebraic geometry. Born Ascher Zaritsky in Kobrin, Belarus, he changed his name to Oscar Zariski after moving to Rome, Italy, in the early 1920s. He in turn fled the fascist regime in Italy by immigrating to the United States in 1927. Once in America, he taught at Johns Hopkins University. In 1947 he moved to Harvard University, where he taught until his retirement in 1969. Among his many honors are the Cole Prize in Algebra from the American Mathematical Society (1944), the National Medal of Science (1965), and the Steele Prize (1981) for the body of his work. He died in Brookline, Massachusetts, in 1986.

Ernst Friedrich Ferdinand Zermelo
1871-1953

German mathematician best known for his work in set theory. He was the first to prove the well-ordering theorem (1904), a result important for many applications of set theory. Widely challenged, Zermelo responded with a second proof of the theorem in 1908. The same year he produced the first axiomatic formulation of set theory. Zermelo later placed certain restrictions on set formation to avoid paradoxes that otherwise would have rendered his axiomatic system useless.

Bibliography of Primary Sources

Books

Ahlfors, Lars. *Complex Analysis* (1953). A Fields Medal recipient for his work in complex analysis, Ahlfors later won the Steele Prize for his three editions of this influential book.

Birkhoff, George David. *Relativity and Modern Physics* (1923). In this work Birkhoff contributed to the growing study of relativity theory.

Birkhoff, George David. *Aesthetic Measure* (1933). Here the author applied Pythagorean notions in an attempt to reach a mathematical understanding of beauty.

Borel, Felix. *Space and Time* (1922). This book helped make Albert Einstein's theory of relativity comprehensible to non-technically educated readers.

Bourbaki, Nicolas. *Eléments de mathématique* (1939-). This series is the main product of the Bourbaki group (a group of mathematicians who write under the pseudonym Nicolas Bourbaki). It now amounts to more than 30 volumes, many of them revised more than once. The title is undoubtedly meant as a reference to Euclid's *Elements*, the first attempt to summarize all mathematical knowledge. The fiction of a single author is maintained throughout.

Courant, Richard, and David Hilbert. *Methods of Mathematical Physics* (1924-27). Here Courant and Hilbert developed techniques later important in quantum mechanics and nuclear physics.

Hilbert, David. *Grundlagen der Geometrie* (1899). This book (the title means "The Foundations of Geometry") placed geometry on a far more rigorous footing than the classic *Elements* of Euclid, which had dominated geometric thinking since antiquity. In doing so Hilbert emphasized the importance of considering the simplest geometric concepts—point, line and plane—to be undefined terms.

Lebesgue, Henri.*Leçons sur l'intégration et la recherche des fonctions primitives* (1904) and *Leçons sur les séries trigonmétriques* (1906). In these two works Lebesgue revolutionized the field of integration by introducing his theory of measure and offering his definition of an integral, called the Lebesgue integral.

Russell, Bertrand, and Alfred North Whitehead. *Principia Mathematica* (1910-1913). This classic book offered rigorous, detailed derivations of Peano's arithmetic and other mathematical theories in terms of propositional logic, artificial language, and "well-formed formulas" (known as *wffs*).

Waerden, Bartel van der. *Modern Algebra* (1930). Waerden's 1930 book helped disseminate the new algebra developed by David Hilbert, Emil Artin, Emmy Noether, and others.

Wittgenstein, Ludwig. *Tractatus Logico-Philosophicus* (1921). Wittgenstein here defended logicism—the view that all mathematics is reducible to logic.

Articles

Brouwer, L. E. J. "On the Foundations of Mathematics" (doctoral thesis, 1907). Brouwer's paper laid the groundwork for intuitionism, the mathematical philosophy maintaining that mathematical truths are a matter of common sense, apprehended *a priori* through intuition.

Church, Alonzo. "An Unsolvable Problem of Elementary Number Theory" (1936). In this journal article Church addressed the decidability question, showing that no general method could be used for all possible mathematics assertions.

Gödel, Kurt. "On Formally Undecidable Propositions of Principia Mathematica and Related Systems" (1930). In this article he 24-year-old Austrian mathematician shocked the mathematical establishment with his critique of the classic book by Bertrand Russell and Alfred North Whitehead. Gödel applied a new method, assigning numerical values to the symbols of logic, to prove that Russell and Whitehead's quest to derive all mathematical principles from a given set of axioms was doomed.

Hilbert, David. "On the Infinite" (1925). Hilbert's essay attempted to resolve issues in set theory, particularly as they related to Georg Cantor's concept of infinite sets.

NEIL SCHLAGER

Medicine

Chronology

1900 Austrian-American physician Karl Landsteiner discovers the four blood types (A, B, AB, and O), thus making possible safe and practical transfusions.

1910 German bacteriologist Paul Ehrlich discovers a drug then called Salvarsan as an effective treatment for syphilis, marking the beginnings of modern chemotherapy.

1921 Insulin is discovered by a primarily Canadian team composed of Frederick Banting, Charles Best, John J. R. Macleod, and James Collip; a year later it is administered for the first time to a patient with diabetes mellitus.

1922 U.S. Army doctor Fernando Rodriguez lays the foundation for modern preventive dentistry when he demonstrates the cycle of tooth decay and the role of bacteria.

1928 Greek-American physician George Nicholas Papanicolaou invents the Pap test, a simple and painless procedure for the early detection of cervical and uterine cancer.

1929 The iron lung, created by American industrial hygienist Philip Drinker, becomes one of the first inventions to keep alive people who are unable to breathe.

1935 An artificial heart is first used in surgery when John Heysham Gibbon, Jr., an American surgeon, demonstrates its use on a dog.

1936 Americans Perrin Hamilton Long and Eleanor Albert Bliss introduce sulfa drugs, which will have a wide range of applications.

1937 The first U.S. blood bank is established.

1937 For the first time, lung cancer is linked to smoking in a study by U.S. surgeons Alton Ochsner and Michael Ellis DeBakey.

1943 The U.S. Army in North Africa makes the first wide-scale use of penicillin, laying the groundwork for massive introduction of the drug into civilian medical practice after the war.

1944 Building on research conducted by Helen Brooke Taussig, American surgeon Alfred Blalock is the first to successfully correct the "blue baby syndrome"—congenital heart disease—in surgery.

1945 Willem J. Kolff invents the kidney dialysis machine.

Overview: Medicine 1900-1949

Overview

The twentieth century was a period of rapid scientific development and unprecedented progress in the biomedical sciences. During the first half of the twentieth century, advances in science—especially in microbiology, immunology, biochemistry, endocrinology, and nutrition—revolutionized medical theory and practice. Even though anesthesia and antiseptic techniques had transformed surgery in the nineteenth century, surgeons could not cope with blood loss, shock, or postsurgical infections until well into the twentieth century. Advances in biochemistry and physiology led to the development of more precise diagnostic tests and more effective therapies. The development of new instruments and laboratory techniques accelerated the growth of specialization within the medical profession and led to major changes in clinical medicine.

The identification and isolation of the microbial agents that caused many of the most important infectious diseases facilitated public health campaigns against epidemic diseases. Advances in preventive and therapeutic medicine correlated with remarkable changes in disease patterns and mortality rates. Striking increases in life expectancy also reflected broader social factors, such as improvements in nutrition, housing, sanitation, and health education. At the turn of the century, life expectancy in the United States was only about 50 years. By the 1950s, however, life expectancy was approaching 70 years. Similar changes occurred in other industrialized nations. In general, life expectancy increased for all groups, but significant differences were closely correlated with race and gender. In the United States, life expectancy was longer for women than men and longer for whites than nonwhites. American medical records, including reports of births, deaths, and specific diseases, however, are not very accurate for the period before the 1930s. In 1900 influenza and pneumonia, tuberculosis, and gastritis were the top killers, but by 1950 heart disease, cancer, and cerebrovascular disorders were the major causes of death.

Research on the nature of infectious diseases and their means of transmission led to the discovery and classification of many pathogenic organisms. In addition to bacteria, scientists identified the rickettsias, which cause such diseases as typhus, and pathogenic protozoans, such as those that cause malaria. Some infectious diseases were attributed to mysterious microbial agents that were invisible under the microscope and small enough to pass through filters that trapped bacteria. These entities were operationally defined as invisible, filterable viruses. Although viruses clearly were able to reproduce and multiply in plant and animal hosts, they could not be cultured in the laboratory. Early studies of viruses included Peyton Rous's demonstration in 1910 that a virus could transmit a malignant tumor in chickens. In 1916 Frederick Twort and Félix-Hubert d'Hérelle independently discovered bacteriophages, which are viruses that attack bacteria and multiply within them. Wendell M. Stanley showed that the tobacco mosaic virus could be purified and crystallized and still maintain its infectious qualities.

Advances in immunology and chemotherapy followed Paul Ehrlich's successful search for a "magic bullet," that is, a chemical that could destroy pathogenic microbes without seriously damaging the host. In 1910 Ehrlich and Sahachiro Hata discovered that arsphenamine, or Salvarsan, was effective in the treatment of syphilis. The next major advance in chemotherapeutics was a report by Gerhard Domagk in 1932 that the red dye Prontosil was effective against streptococcal infections in mice and humans. After other researchers found that the active antibacterial component in the dye was sulfanilamide, many derivatives were synthesized and tested for efficacy and safety.

Alexander Fleming had discovered penicillin in 1928, before the introduction of the sulfanilamides, but Fleming was unable to purify and test the antibiotic. About 10 years later, Howard Florey, Ernst Chain, and others isolated penicillin and tested its potency and toxicity. Production of penicillin began during World War II and by 1944 the drug was being used to treat soldiers with infected wounds and infectious diseases. Unfortunately, penicillin was not effective against certain bacteria, including *Mycobacterium tuberculosis*, the bacillus that causes tuberculosis, which was still a major public health problem in the 1940s. In 1944 Selman A. Waksman and his colleagues isolated streptomycin and demonstrated that it was active against *M. tuberculosis.* The widespread use of antibiotics soon led to the emergence of drug-resistant strains of bacteria, which pose a significant threat to victims of disease and infection. Research on the mechanism of immunity led to diagnostic tests for several diseases, including syphilis (the Wassermann test) and tuberculosis (the tuberculin test).

Within a few years of the discovery of x rays by Wilhelm Conrad Röntgen in 1895, numerous medical applications for x rays were established. X rays were used to analyze fractures and locate stones in the urinary bladder and gallbladder. Doctors introduced radio-opaque substances into the body to study the kidneys, spinal cord, gallbladder, ventricles of the brain, the chambers of the heart, and the coronary arteries.

Military medicine, especially during major armed conflicts, provided direct demonstrations of the value of advances in sanitation, nutrition, and medical techniques. For example, the effectiveness of a vaccine against typhoid fever, developed in the 1890s, was established during World War I. The use of tetanus antitoxin for all wounded men during World War I virtually eliminated the threat of battlefield tetanus.

During the 1930s the use of vaccines, or toxoids, led to the control of tetanus and diphtheria. Widespread immunization of children against diphtheria virtually eliminated the disease in the United States and other industrialized nations. A live, but weakened (attenuated) vaccine known as bacillus Calmette-Guérin (BCG), which was developed by Albert Calmette and Camille Guérin, has been used in Europe since the 1930s; since the infamous Lübeck disaster of 1930, however, its efficacy and safety have been questioned. In 1930, after 249 infants had been vaccinated with BCG vaccine in Lübeck, Germany, 73 died. After intensive investigations, scientists concluded that the vaccine was safe if properly prepared. The vaccine used in Lübeck had been contaminated by virulent bacteria.

Little progress was made in understanding viruses until the 1930s, when researchers learned to use tissue-culture techniques to grow viruses in the laboratory and the electron microscope to provide portraits of viruses. Scientists were then able to produce vaccines for yellow fever, influenza, and poliomyelitis. One of the most destructive epidemics in history was the influenza pandemic that killed more than 15 million people between 1918 and 1919. Because the influenza virus has the ability to transform itself from one epidemic year to another, outbreaks of influenza throughout the world must be carefully monitored so that appropriate vaccines can be developed.

The science of endocrinology evolved rapidly after 1905, the year that Ernest H. Starling introduced the term "hormone" for the internal secretions of the endocrine glands. During the first two decades of the twentieth century, various hormones, including epinephrine (adrenaline), were isolated and identified. The most notable event in this field was the discovery of insulin by Frederick Banting, Charles H. Best, and J. J. R. Macleod in 1921. In 1935 Edward C. Kendall isolated cortisone and in the 1940s Philip S. Hench and his colleagues demonstrated the beneficial effect of cortisone on rheumatoid arthritis. Cortisone and its derivatives proved to be potent anti-inflammatory agents that could be used in the treatment of rheumatoid arthritis, acute rheumatic fever, certain diseases of the skin and the kidneys, and some allergic conditions, including asthma.

Although interest in the relationship between foods and health is at least as ancient as Hippocratic medicine, the science of nutrition developed in the early twentieth century with the discovery of the "accessory factors," or vitamins. Frederick Gowland Hopkins proved unequivocally that a diet containing only proteins, carbohydrates, fats, and salt did not allow animals to grow and thrive. The classic experiments published by Hopkins in 1912 stimulated the isolation and characterization of the vitamins essential to health. Improvements in diet and the use of vitamin supplements made it possible to prevent specific vitamin-deficiency diseases, such as rickets (vitamin D), scurvy (vitamin C), beriberi (thiamine), and pellagra (vitamin B_3 or niacin). The work of George H. Whipple and George R. Minot in the 1920s showed that raw beef liver was useful in the treatment of pernicious anemia, a previously fatal disease. The active principle, vitamin B_{12}, or cobalamin, was isolated from liver in the 1940s.

Significant progress towards controlling major tropical diseases was made possible by the introduction of new drugs and vaccines, but mosquito control and other environmental interventions were also essential. After World War I, several synthetic antimalarial drugs, including quinacrine (Atabrine), joined quinine in the prevention and treatment of malaria. These new drugs were quite effective in reducing the burden of malaria among Allied troops during World War II. The hope that malaria might be eradicated by a direct attack on its mosquito vector was raised by the successful use of the insecticide dichlorodiphenyltrichloro-ethane (DDT) during World War II. Although DDT was initially quite effective in the battle against malaria and yellow fever, mosquitoes became resistant to DDT and ecologists found evidence of environmental damage caused by the insecticide.

Twentieth-century surgeons gradually adopted antiseptic and aseptic techniques, including the ritual of scrubbing up, the sterilization of all instruments and dressings, and the use of gloves and masks during surgery. Anesthesia was accepted, but its administration was often left in the hands of untrained assistants. Although chloroform was clearly more dangerous than ether, it was often used because it was easier to administer. Eventually, surgeons replaced chloroform with a combination of nitrous oxide, ether, and oxygen. In operations that required the relaxation of the abdominal muscles, deep anesthesia was used—despite the danger this posed to the patient. Improvements in anesthesia included intravenous anesthesia using the barbiturate thiopental sodium (Pentothal) and the injection of curare to induce muscular paralysis. The introduction of inhaled anesthesia administered under pressure was essential to the development of thoracic (chest) surgery. Eventually, anesthesiologists became skilled specialists, making complex operations possible. Harvey Williams Cushing's pioneering operations for brain tumors, epilepsy, trigeminal neuralgia, and pituitary disorders established neurosurgery as a model for other surgical specialists.

During World War I surgeons who had learned to practice asepsis had to treat contaminated wounds while working under primitive conditions. They were forced to revert to antisepsis and ancient wound treatments, such as debridement (removal of morbid tissue and foreign matter). To fight putrefaction and gangrene, Alexis Carrel and Henry Dakin introduced a method of antiseptic irrigation (the Carrel-Dakin treatment); antitoxin and antiserum were used to prevent tetanus and gangrene from developing after wound treatment. Military surgeons also realized the importance of the patient's rehabilitation. This discovery led to important developments in orthopedic and plastic surgery.

Even after surgeons learned to cope with pain and infection, shock remained a major problem both during and after surgery. When physiologists discovered that shock was essentially due to a decrease in the effective volume of blood circulation, doctors attempted to fight shock by transfusions. Safe blood transfusions were made possible by Karl Landsteiner's 1901 discovery of the ABO blood groups. During World War II blood banks were organized to support the increased use of blood transfusions.

A few isolated attempts at heart surgery occurred in the first decades of the century, but most such interventions were failures. The medical profession generally considered heart disease a problem requiring medical management, rather than surgical intervention. During World War II Dwight Harden demonstrated that it was possible to save the lives of wounded soldiers by removing shrapnel from the chambers of the heart. After the war, several surgeons performed operations to correct congenital defects of the heart. It was not until the 1950s, however, that surgeons were able to stop the heart while still supplying the body with oxygen in order to perform more complex operations.

As modern medicine succeeded in controlling many of the infectious, epidemic diseases, the chronic, degenerative diseases—heart disease, stroke, and cancer—emerged as the major causes of death in the wealthy, industrialized nations.

LOIS N. MAGNER

The Development of Schools of Psychology Leads to Greater Understanding of Human Behavior

Overview

Although speculation about the nature of human thought and behavior are very ancient, psychology did not entirely separate itself from philosophy and physiology until relatively recent times. By the end of the nineteenth century psychologists had established several different approaches to the study of the mind, including structuralism, associationism, and functionalism. During the early twentieth century several schools of psychology developed distinct and competing approaches to the study of mental processes. Through the work of Sigmund Freud, psychoanalysis replaced structuralism, but conflicts within the Freudian community gave rise to new schools of psychology, such as behaviorism and humanistic psychology. Proponents of humanistic psychology considered their work a revolt against the reductionist and deterministic approach of previous schools.

Background

Psychology emerged as a field distinct from philosophy with the establishment of the school of psychology known as structuralism by Wilhelm Wundt (1832-1920) and Edward B. Tichener (1867-1927). Using the technique of introspection, structural psychologists attempted to analyze the "anatomy of the mind" in terms of simple components and discover how these simple components combined to create complex forms. According to Tichener, conscious experience could be adequately described in terms of sensations and feelings. After Tichener's death, the movement was superceded by several rival schools of psychology.

In contrast to the reductionist approach of structuralism, psychoanalytic theory (sometimes called "depth psychology") emphasizes unconscious mental processes. The psychoanalytic movement was founded by the Austrian psychiatrist Sigmund Freud (1856-1939). In 1895 Freud and his colleague Josef Breuer (1842-1925) described the technique of "free association of ideas" in *Studies in Hysteria.* Breuer had been using hypnosis in the treatment of neurotic patients, but while treating a patient referred to as "Anna O." he discovered that when his patient was encouraged to talk about her symptoms she seemed to find some relief. Breuer and Freud argued that the "talking cure" allowed patients to discharge emotional blockages that caused hysterical symptoms. Freud realized that when his patients expressed random thoughts associated with specific queries, he could uncover hidden material from the unconscious mind. The degree of difficulty experienced during this process indicated the importance of the repressed material and the strength of the patient's defenses. In contrast to Breuer, Freud concluded that the most strongly repressed materials were sexual in nature. He argued that the struggle between sexual feelings and psychic defenses was the source of the patient's neurotic symptoms. Freudian theory provided explanations for neuroses, such as hypochondria, neurasthenia, and anxiety neurosis, and psychoneuroses, such as obsessive-compulsions, paranoia, and narcissism.

In 1896 Freud coined the term *psychoanalysis* to describe his theories and technique. After focusing on hysteria and female sexuality, Freud turned his attention to the male psyche and the development of a general theory of the mind. Much of his theory was based on generalization from his own self-analysis. Freud realized that the analysis of dreams was an important source of insights into the unconscious. All of Freud's disciples underwent a training analysis before they were considered competent to analyze their own patients.

In 1905 Freud published *Three Essays on the Theory of Sexuality,* the work that established him as a pioneer in the serious study of sex, mind, and behavior, both normal and pathological. Although Freud's theories were controversial from the beginning, he attracted many disciples. By 1902 the Psychological Wednesday Circle that met in Freud's office included Alfred Adler (1870-1937), Wilhelm Stekel, Sándor Ferenczi (1873-1933), Carl Gustav Jung (1875-1961), Otto Rank (1884-1939), Ernest Jones (1879-1958), and A. A. Brill. The group became the Vienna Psychoanalytic Society in 1908 and held its first international congress. In 1909 Freud, Jung, and Ferenczi gave a series of influential lectures at Clark University in Worcester, Massachusetts. Freud's lectures were published as *The Origin and Development of Psychoanalysis* (1910).

Because of Freud's assumption that male sexuality was the norm of development and his controversial assumption of penis envy, his theories evoked considerable opposition and feminist criticism. As anthropologists investigated other cultures, Freudian claims of universality were also criticized. Many of Freud's closest associates eventually rejected various aspects of his theories and methods. When the Nazis took over Germany and attacked "Jewish science," Freud's books were among the first to be burned. Although psychotherapy itself was not banned, many of the pioneers of psychoanalysis escaped to North America and England, where they developed new schools of thought.

Impact

Disagreements about psychoanalytic theory and techniques led some of Freud's associates to revise Freudian theories or to establish competing schools of psychology. These theorists included Carl Jung, Otto Rank, Alfred Adler, Erik Erikson (1902-1994), Karen Horney (1885-1952), Erich Fromm (1900-1980), and Harry Stack Sullivan (1892-1949). Initially, all psychoanalysts were physicians and held credentials in psychiatry, but eventually nonmedical therapists found a role in the evolving professional world of psychology

Alfred Adler is primarily remembered for developing the concept of the inferiority complex. Adler's work suggested means of compensation that could overcome the inferiority complex. Adler's *Practice and Theory of Individual Psychology* (1924) is the landmark treatise for the school of thought known as individual psychology.

Hermann Rorschach (1884-1922), the Swiss psychiatrist who devised the inkblot test that is widely used in diagnosing psychopathology, was one of the early advocates of psychoanalysis in Switzerland. When shown ambiguous stimuli, such as inkblots, human beings tend to project their own interpretations and feelings onto them. Instead of relying entirely on psychoanalysis and the talking cure, Rorschach argued that trained observers could use such clues to diagnose hidden personality traits and impulses in patients taking such tests. In 1921 Rorschach published his landmark treatise *Psychodiagnostics,* which summarized his work with 300 mental patients and 100 normal subjects. His method became widely used as a tool for psychological evaluation and diagnosis.

Carl Gustav Jung broke with Freud and founded a new school of thought known as analytic psychology. Jung is best known for his concepts of the extroverted and introverted personality types, archetypes, and the collective unconscious. His major treatise in this area was *Psychology of the Unconscious* (1916). Many of Jung's relatives were clergymen, and he initially intended to become a minister, but he decided to study medicine and become a psychiatrist instead. While working at a mental asylum in Zurich, Switzerland, Jung experimented with association tests. His results convinced him that the peculiar responses elicited by certain words were caused by clusters of unconscious and unpleasant associations. Jung used the term *complex* to describe these conditions.

Initially, Jung believed that his results confirmed many of Freud's ideas. From 1907 to 1912 Jung and Freud were close collaborators. Personal and professional conflicts eventually severed their relationship. Primarily, Jung rejected Freud's ideas about the sexual bases of neurosis. Through studies of his own dreams and fantasies, Jung developed his famous theory of the collective unconscious, which postulates that everyone's unconscious mind contains similar impulses and memories that arise from the inherited structure of the brain. (The collective unconscious is separate from the personal unconscious, which is based on each individual's experience.) Jung's interest in the psychology of religion led to his theory of archetypes, which he saw as universal instinctive patterns of behavior and images. The relationship between psychology and religion became Jung's major interest and led him to explore new psychotherapeutic methods. Jung believed that psychotherapy could help middle-aged and elderly people through a process he called individuation.

Max Wertheimer (1880-1943), along with his assistants Kurt Koffka (1886-1941) and Wolfgang Köhler (1887-1967), was the founder of Gestalt psychology. Wertheimer's interests included music, the philosophy of law, the psychology of courtroom testimony, the perception of complex and ambiguous structures, and experimental psychology. His doctoral research led to the development of a lie detector and a method of word association. Studies of the problem-solving abilities of feebleminded children led Wertheimer to the basic concepts of the Gestalt school of psychology.

Gestalt psychology examines psychological phenomena as structural wholes, instead of breaking them down into components. In 1921 Wertheimer and his associates founded the journal *Psychological Research,* which became the voice of the Gestalt movement. Critical of conventional educational approaches, Wertheimer called for an emphasis on problem-solving processes that required grouping and reorganization, so that students could deal with problems as structural wholes. Shortly before the Nazis came to power in Germany, Wertheimer fled to the United States and became a professor at the New School for Social Research in New York City. His last book, *Productive Thinking,* was published posthumously in 1945.

Followers of Gestalt psychology always considered the whole of anything to be greater that the sum of its parts. Gestalt theorists called for a holistic description of psychological experience and a humanistic approach to psychology that could encompass the qualities of form, meaning, and value. Although primarily interested in perception, Gestalt psychologists applied their approach to problem solving, learning, thinking, motivation, social psychology, personality, aesthetics, economic behavior, ethics, political behavior, and the nature of truth.

Twentieth-century behaviorism as a psychological theory evolved from much older philosophical and physiological sources. While some behavioral theorists, such as William James (1842-1910) and William McDougall (1871-1938), gave increased attention to the instinctive component of human behavior and paid less attention to the concept of will, others, such John B. Watson (1878-1958), emphasized learning and dismissed instinct and will. Watson's school of behaviorism regarded behavior as a response to changes in environmental stimuli and explained motivation in terms of drive. The psychological term *drive,* which is linked to the physiological concept of homeostasis, was first used by American psychologist Robert S. Woodworth (1869-1962) in 1918. Drive theory was developed by Clark Hull (1884-1952) in the 1940s, but by the end of the twentieth century it was no longer widely accepted. Until the 1960s behaviorism dominated research on motivation, although cognitive psychologists regarded this approach as overly mechanistic. Cognitive psychologists argued that behaviors could be purposive and that motivation was based on the expectation of future events and choices among alternatives.

Humanistic psychology, based on the principle that psychologists should recognize and treat human beings as unique individuals, developed in opposition to behaviorism and psychoanalysis. Humanistic psychologists rejected the laboratory models favored by behaviorists and the deterministic orientation of psychoanalysis, which postulates that adult behavior is determined by early experiences and drives. Humanists believe that the individual must be understood as a feeling, thinking being, capable of making major changes in attitude and behavior. The goals of humanist psychology are to facilitate maturation, to establish of a system of values, and to enhance the individual's capacity for love, fulfillment, self-worth, and autonomy.

Eventually, Freud's ideas about female sexuality, gender issues, and child psychology were extended, revised, or rejected by other psychoanalysts, including Helene Deutsch, Karen Horney, Clara Thompson (1893-1958), Melanie Klein (1882-1960), and Erik H. Erikson.

Anna Freud (1895-1982), the youngest daughter of Sigmund Freud and his wife Martha, became a distinguished psychoanalyst and one of the founders of child and adolescent psychology. Her book *The Ego and Mechanisms of Defense* (1937) was an important contribution to ego psychology. With her American associate, Dorothy Burlingham, Anna Freud published several studies of the psychological impact of war on children. Her major theories were summarized in *Normality and Pathology in Childhood* (1968).

PARAPSYCHOLOGY: SCIENTIFIC PARIAH

Parapsychology is the scientific study of unexplained phenomena. Parapsychological events include psychokenesis (motion via psychic powers), telepathy (communication through means other than the senses), clairvoyance (the power to perceive things out of the natural range of human senses), and post-death experiences. Extrasensory perception (ESP) includes telepathy and clairvoyance in either an historic (retrocognition) or prophetic (precognition) mode, but not psychokenesis (PK), which involves affecting the outcome of real events (e.g., games of chance) and is studied separately. Parapsychology is often associated with many sorts of unexplained phenomena (UFOs, astrology, etc.), but most researchers limit their research to ESP or PK phenomenon.

Scientific interest in parapsychology dates from 1882 with the establishment in London of the Society for Psychical Research, a group of scholars dedicated to examining "without prejudice or prepossession and in a scientific spirit" phenomena not readily explicable by contemporary knowledge. Generally hostile to parapsychology, academic psychologists began conducting experiments in telepathy in the United States as early as 1912 (J. E. Coover at Stanford University). Important work from the Parapsychology Laboratory at Duke University began in 1927. Today many dozens of respected scientists engage in psychic research.

Setting aside the possibility of outright fraud, two theories currently dominate thinking on the subject of parapsychology: the first demands a reconceptualization of space-time such that objects or information distant in a three- or four-dimensional system are accessible; the second proposes a set of laws that govern psychical events apart from those that govern the physical world. Experimental data for fantastic acts of ESP and PK exist, though few fields have had their claim to scientific credibility as routinely or fiercely challenged as psychical research and, despite its popular appeal, the scientific community overwhelmingly rejects current theories.

DAVID D. LEE

Austrian-born British psychoanalyst Melanie Klein (born Melanie Reizes) was primarily known for her work with young children. Klein argued that observations of the content and style of play provided insights into the child's unconscious fantasy life and psychological impulses. After undergoing psychoanalysis with Sándor Ferenczi, one of Freud's closest associates, Klein began her studies of young children. She published her first paper on child psychoanalysis in 1919 and joined the Berlin Psychoanalytic Institute, at the invitation of Karl Abraham (1877-1925). In 1926 she moved to London. Her theory of child analysis was published in 1932 as *The Psychoanalysis of Children*. Klein's insights and methods are now widely used to help troubled children. In the 1930s Klein extended her methodology to work with adult patients.

In contrast to orthodox Freudian theory, German-born American psychoanalyst Karen Horney (born Karen Danielsen) suggested an environmental and social basis for the development of personality disorders. After undergoing analysis with Karl Abraham, Horney joined the teaching staff of the Berlin Psychoanalytic Institute. In 1932 she took a position at the Chicago Institute for Psychoanalysis. Two years later she moved to New York, where she established a private practice and taught at the New York Psychoanalytic Institute and the New School for Social Research. Horney publicly criticized orthodox Freudianism and called for new studies of psychosexual issues specific to women. At the time, her theories about the importance of social and cultural factors in human development were considered heretical in Freudian circles. Rejecting Freud's ideas about female psychology, the libido, the death instinct, and the Oedipus complex, Horney argued that the male-dominated culture that produced Freudian theory was the major source of female psychiatric problems. In contrast to Freud's assumptions about penis envy, Horney proposed the concept of womb envy, that is, male envy of women's ability to produce new life. Although Horney's theories were considered controversial during her lifetime, she was subsequently recognized as a major figure in the history of psychoanalytic theory. Clara Mabel Thompson also left the orthodox Freudian community in the 1940s. Thompson was influential as a teacher, writer, and researcher with a special interest in female sexuality. Her major book was *Psychoanalysis: Evolution and Development* (1950).

Despite the differences between the various twentieth-century schools, clinicians, experimenters, and theoreticians were able to contribute to the growth of a shared body of psychological knowledge that has provided valuable insights into human thought and behavior. By the 1950s competing schools of psychology had essentially converged into a broad professional community sharing the insights of predecessors and contemporaries. The actual success of psychoanalytic therapy for the treatment of psychoneuroses is, however, still subject to considerable dispute.

LOIS N. MAGNER

Further Reading

Caplan, Eric. *Mind Games: American Culture and the Birth of Psychotherapy*. Berkeley, CA: University of California Press, 1998.

Hughes, Judith M. *Freudian Analysts/Feminist Issues*. New Haven, CT: Yale University Press, 1999.

Jones, Ernest. *The Life and Work of Sigmund Freud*. Edited and abridged by Lionel Trilling and Steven Marcus. New York: Anchor Books, 1963.

Masson, Jeffrey Moussaieff. *The Assault on Truth: Freud's Suppression of the Seduction Theory*. New York: Farrar, Straus & Girous, 1984.

Mitchell, Juliet. *Psychoanalysis and Feminism*. New York: Pantheon Books, 1974.

Sulloway, Frank J. *Freud, Biologist of the Mind: Beyond the Psychoanalytic Legend*. New York: Basic Books, 1979.

Timms, Edward, and Ritchie Robertson, eds. *Psychoanalysis in Its Cultural Context*. Edinburgh: Edinburgh University Press, 1992.

Advances in Identifying the Causes of Major Infectious Diseases

Overview

At the beginning of the twentieth century the germ theory of disease was alive and well. However, it would take time and skill to identify the specific causes of many of the infectious diseases. Robert Koch (1843-1910) and Louis Pasteur (1822-1896) had identified bacteria as the causative agents in many diseases, and scientists could stain these organisms and identify their shapes under the microscope.

Important areas developed early in the twentieth century that added to knowledge about pathogenic (disease-causing) organisms. Filterable viruses, organisms so small that they passed through a filter, were discovered in the 1890s. These viruses were so small and mysterious that many unexplained diseases were attributed to them. Also, one-celled animals, or protozoa, and a group of parasitic worms emerged as new suspects.

In 1901 the mysterious African sleeping sickness was found to be caused by a one-celled animal carried by a biting fly, the tsetse fly. Joseph Everett Dutton (1874-1905), David Bruce (1855-1931), and Aldo Castellani (1875-1971) located the trypanosome that caused this disease. In 1901 Walter Reed (1874-1905) proved that a mosquito carried the scourge of the tropics—yellow fever. In 1905 the organisms causing whooping cough, dengue fever, and measles were discovered. George Frederick Dick (1881-1967) and his wife Gladys Dick (1881-1963) found the bacteria that caused scarlet fever and developed a test for it.

These diseases caused many deaths. At the turn of the twentieth century, the average life expectancy was about 50. This meant that, of all people born in a given year, by age 50 half would be alive and half dead. The death of children and infants to a host of "childhood diseases" brought down the average life expectancy.

Background

In previous centuries the world's great plagues affected the course of history and mankind. The condition known as sleeping sickness probably originated in Africa and was noted in slaves brought from there. Thomas Winterbottom, an English doctor, first described the condition in 1803. While working in Cuba, he noted slaves with terrible fevers and enlarged glands. In 1857 the famous missionary doctor David Livingstone (1813-1873) observed how cattle died from the bite of the tsetse fly, and horses developed nagana, or stallion fever. Livingstone was the first to treat animals with arsenic compounds. These compounds eventually were used to treat humans with sleeping sickness, although the organism that caused the disease was not identified.

During the last few years of the nineteenth century, trade routes increased and people began to travel into the hot equatorial areas, In 1894 an army doctor, Sir David Bruce, and his wife went to Africa to study nagana. Drawing blood from infected horses, they looked carefully at each sample under the microscope and noted a peculiar one-celled parasite nestled among the blood cells. This parasite was later named *Trypanosoma brucei* (or *T. brucei*). Bruce also confirmed Livingstone's idea that the biting fly was the carrier.

For over 300 years yellow fever was one of the world's great plagues. The debate over whether yellow fever originated in Africa or America will probably never be solved. In the late 1800s it was discovered that the mosquito was somehow involved in the transmission of malaria. In 1881 Carlos Juan Finlay (1833-1915) of Havana proposed that mosquitoes carried yellow fever. Several scientists, using their microscopes, looked in vain.

Dengue fever is an acute, sometimes hemorrhagic fever that causes intense body pain. Also called "breakbone fever" or "Dandy fever," it makes the person very ill but is seldom fatal. In Philadelphia, during the summer of 1708, Dr. Benjamin Rush (1746-1813) first described dengue. The term "breakbone fever" refers to the fact that the joints and bones feel like they are breaking. Like malaria and yellow fever, dengue would soon be linked to the mosquito.

Among the list of major world epidemics, measles has been greatly underrated. In the last 150 years, in industrial societies, measles drew attention when the overall interest in children increased. Measles is so contagious that it is passed by casual contact, and the host may give it to someone else before the symptoms occur. It probably appeared during the Middle Ages in France. A Danish doctor, P. L. Panum (1820-1885), studied an outbreak in the Faroe Islands in 1875. He determined the exact nature of the disease, the 13-14 day incubation period, its high level of contagion, and that one attack confers lifelong immunity.

Whooping cough, also called pertussis, is characterized by a cough with a period of drawing in air or whoop. The disease was first described in 1578, but probably existed before that. About 100 years later, the Latin word "pertussis" was applied, meaning intensive cough.

Scarlet fever, also called scarlatina, was described in 1676 by English physician Thomas Syndenham as small red spots, redder than the measles.

Impact

The work of Bruce and Livingstone had described the tsetse as a carrier of sleeping sickness in animals. In 1901 Dutton, working in Gambia, examined samples of human blood from sleeping sickness victims and found a similar trypanosome. The name of this one-celled parasite was *T.b. gambiense*. Thrilled by the success, he launched an all-out investigation into other insect-born diseases. In 1902 his work ended when he died while investigating relapsing fever on the Congo River.

In 1903 a new outbreak of sleeping sickness occurred in Uganda. David Bruce and Aldo Castellani, an Italian tropical disease expert, arrived to search for a connection. Examining spinal fluid, Castellani found another tiny parasite, and Bruce realized he'd found another form of sleeping sickness. He called this strain called *T.b. rhodesiense.*

Stopping the disease then meant destroying the flies and restricting infected people and cattle. This proved to be very difficult because of African traditions.

Since there was no cure, in the 1920s people were advised to cover up completely in the hot tropical climate. In 1922 Bayer, a German pharmaceutical house, found that an arsenic compound called tryparsamide was effective, although it had side effects. Another drug, suramin sodium, was found effective for treatment in the disease's early stages. Sleeping sickness is still found in parts of sub-Saharan Africa, but aggressive programs to eradicate the tsetse and protect humans by using a preventive treatment have it under control.

Carlos Chagas (1879-1934), a Brazilian physician, found the reduviid bug, or "kissing bug," spreads a form of sleeping sickness in Central and South America. The host is the armadillo and possum. It is called the kissing bug because it bites the tender places on the face. In 1900 Carlos Finlay told Walter Reed, an army doctor, of his idea that the mosquito caused yellow fever. They joined with a group called the Yellow Fever Experiment that planned a simple experiment. One group of volunteer soldiers would live in a screened house among dirty blankets and remains of yellow fever victims. The common thinking was that the disease was spread in this way. Another group would live in clean housing without screens. The result showed that the soldiers with the screens stayed healthy; the mosquito-bitten soldiers came down with yellow fever.

By the 1900s researchers knew that agents smaller than bacteria, called viruses, could be passed through a filter. When investigators put the serum of yellow fever victims through such a filter, they found that it kept its potency. The Reed commission found that yellow fever was not only carried by a mosquito, but was a virus. After showing this connection, Reed then developed a plan that freed Havana from the disease in just a year.

When Reed died in 1902 he was succeeded by William Crawford Gorgas (1845-1920), who commanded the dramatic campaign against yellow fever during the construction of the Panama Canal. Ferdinand de Lesseps (1805-1894) had started cutting a canal through the narrow isthmus in 1881, but had failed because many workers had died of malaria and yellow fever. In 1904 Gorgas convinced the government to follow the Havana example and start an intensive war against the mosquito. Mosquito-proof buildings, netting, spraying, and draining stagnant pools were part of the plan. The last fatal case of yellow fever during the canal's construction occurred in 1906, and the canal was completed in 1914. During the 1930s, a vaccine was created and the scourge that devastated not only the tropics, but major places in the Untied States, was ended.

Dengue fever, a disease found in hot, humid climates, must be watched even today because control is most difficult. A 1906 study reports that an experiment involving human volunteers showed dengue is carried by several varieties of *Aedes* mosquitoes. In 1907 the disease was shown to be caused by a virus.

Dengue played an important part in the Pacific theater during World War II, especially in the Philippines. Even if dengue did not kill, it could incapacitate soldiers, causing them to be off-duty for weeks. Because of the four subtypes, a vaccine has never been developed, and a person must have all four to develop immunity. Late in the 1940s DDT was found to be effective in controlling the mosquito. However, during the 1970s this chemical was found to cause ill-effects, and no effective replacement has since been found.

Three infectious diseases—measles, diphtheria, and scarlet fever—had been major problems, especially for children. Deaths from these diseases declined as the twentieth century progressed and new emphasis was placed on children's health. Since no drug affects measles, the only treatment is bed rest, protection of the eyes, and steam to help breathing. A measles vaccine was later developed.

Whooping cough leads to serious complications like convulsions and brain damage. In 1906 two bacteriologists at the Pasteur Institute were studying how bacteria destroys the red blood corpuscles. Jules Bordet (1870-1961) and Octave Gengou developed tests for many diseases. They isolated the bacterium that causes whooping cough, now called *Bordatella pertussis*. Their work laid the foundation for the vaccine that would be introduced in the early 1960s.

George and Gladys Dick investigated the cause of scarlet fever and, in 1923, isolated the *streptococcus* bacterium that causes the disease. Building on the study of others in the field of toxins, they prepared a toxin used for immunization. In 1924 they developed a skin test to determine susceptibility to scarlet fever.

Infectious diseases have had a devastating effect on human beings. Identifying the insect vectors of yellow fever, dengue, and sleeping sickness has enabled those at risk to target the carriers, using eradication and good hygiene. Identifying "childhood" diseases like measles, scarlet fever, and whooping cough led to better containment through quarantine. It would not be until the latter half of the twentieth century that vaccines would cut down on incidences of such diseases in the industrial nations.

EVELYN B. KELLY

Further Reading

Bray, R. S. *Armies of Pestilence: The Impact of Disease on History.* New York: Barnes & Noble, 1996.

Cartwright, Frederick. *Disease and History.* New York: Barnes & Noble, 1972.

Humphreys, Margaret. *Yellow Fever and the South.* New Brunswick, NJ: Rutgers University Press, 1992.

Kiple, Kenneth F. *Plague, Pox, and Pestilence: Disease in History.* New York: Barnes and Noble, 1997.

The Discovery of Vitamins and Their Relationship to Good Health

Overview

Vitamins are organic substances found mainly in foods that are essential in minute quantities for growth and health. Vitamins aid in regulating the body's metabolism and serve as catalysts for important chemical processes throughout the body. Some vitamins are necessary to aid in the release of energy from digested food. Others assist in the formation of hormones, blood, bones, skin, glands, and nerves. Only vitamin D can be manufactured by the body; all others must be introduced through the diet. Lack of sufficient vitamin intake has contributed to disease and dysfunction throughout the history of the world. During the early twentieth century, the important role of diet in the prevention of disease and maintenance of health was confirmed with the discovery of vitamins. Recent advances in chemistry, physiology, and biochemistry paved the way for scientists to isolate the "accessory food factors" (vitamins) necessary for health.

Background

Since ancient times, man suspected that diet and disease shared some relationship. The Greek physician Hippocrates (c. 460-377 B.C.) encouraged those who had trouble seeing the stars at night to eat beef liver. It is now known that liver contains vitamin A, a necessary vitamin for retinal health and vision. In the early eighteenth century Native Americans realized the benefit of consuming extracts made from pine needles when afflicted with scurvy, a disease causing bleeding gums, bruising, and severe weakness. The same symptoms were observed among British sailors later in the century, and the disease was remedied by including citrus fruit at intervals in the rations of the sailors (hence the nickname "limey"). Both pine needles and citrus fruit contain vitamin C, and scurvy is caused by the lack of vitamin C in the diet.

In the nineteenth century the relationship between disease and diet diminished in prominence, as French bacteriologist Louis Pasteur (1822-1895) and German physician Robert Koch (1843-1910) established the germ theory of disease. Between 1879 and 1900 the microorganisms responsible for many of the world's dangerous diseases were being discovered at a rate of over one per year. Infectious diseases were the leading cause of death, and the prevailing assumption was that diseases such as pellagra and beriberi, later known to be caused by vitamin deficiencies, were probably the result of infectious agents.

The study of beriberi by Dutch bacteriologist Christiaan Eijkman (1858-1930) paved the way for the discovery of vitamins. Eijkman was sent to the Dutch East Indies (present-day Indonesia) in 1886 to identify the microorganism presumed responsible for beriberi, a disease common to the rice-growing cultures of Asia. Symptoms of beriberi include severe muscle weakness and, eventually, cardiac failure. As director of the medical school and research laboratory in Batavia (now Jakarta), Eijkman probed the cause of beriberi by studying the diet of the hospital chickens. Eijkman found that the chickens developed an avian (bird) form of beriberi when they were fed the polished white rice given to patients. When given their normal diet of brown rice, the chickens recovered. Eijkman and his colleagues suggested that beriberi was due to the absence of some factor in the polished rice that the brown rice hulls contained. Next, Eijkman surveyed prisoners on the Indonesian island of Java, documenting the prisoners' environment, diet, and the incidence of beriberi. Eijkman found no correlation between environment and beriberi, but he did discover that two thirds of the prisoners who were fed polished white rice had the disease, compared with only a few who were fed brown rice. Eijkman later produced an extract from the rice hulls that he was convinced contained the "antidote" for beriberi.

British biologist Frederick Gowland Hopkins (1861-1947) shared the view that certain disease states could be explained by dietary deficiencies. At Cambridge University in 1906, he conducted a series of experiments manipulating the diet of young rats. In feeding the rats starch, sugar, salt, casein, and lard, Hopkins gave the rats all the nutrients that were, at that time, presumed necessary for life. Hopkins supplemented the diet of some of the rats with milk. The rats without milk failed to thrive, while the rats given milk grew normally. Hopkins concluded that small amounts of "accessory food factors" are necessary for growth, and without them diseases of deficiency occur.

At the Lister Institute in London, Polish-born biochemist Casimir Funk (1884-1967) in 1912 isolated the "accessory food factor" in rice hulls that prevented beriberi. His name for the substance was "vitamine"—vital for life and containing nitrogen (amine). With the deficiency process shown in the laboratory, the search began to isolate and define vitamins. Later, the final "e" was dropped when it was learned that all vitamins did not contain amines.

At the time of Funk's discovery in England, the United States was plagued with pellagra, a disease whose symptoms include skin eruptions, diarrhea, dementia, and death. Pellagra was mostly confined to the southern states, with South Carolina alone reporting over 30,000 cases with a 40% mortality rate. Hungarian-American physician Joseph Goldberger (1874-1929), a surgeon with the United States Marine Hospital Service (later named the United States Public Health Service), was sent to the South in 1914 to find a cure for pellagra. During his travels throughout the southern states, Goldberger noticed that pellagra did not occur in those who consumed a well-balanced diet. Neither Goldberger nor his colleagues contracted the disease although they spent considerable time among those affected. Goldberger, convinced that the solution to pellagra lay with chemistry and nutrition, abandoned the search for an infectious agent. In a series of controlled dietary experiments, Goldberger fed children in two Mississippi orphanages and inmates at the Georgia State Asylum a diet rich in fresh meats, milk, and vegetables, instead of their usual corn-based diet. The results were dramatic. Those at the institutions recovered from pellagra. Newcomers who were fed the new diet did not develop the disease. In a repeat study, eleven Mississippi prisoners (volunteers offered pardons for their participation) were fed only a corn-based diet. Goldberger noticed pellagra rash in six of the prisoners within six months. In a final dramatic experiment to convince critics that pellagra was not an infectious disease, Goldberger and his assistant injected themselves with blood from a person suffering from pellagra. Neither developed the disease. Although Goldberger never determined the vitamin deficiency responsible for pellagra (niacin, one of the B-complex vitamins), his epidemiological studies confirmed that balanced nutritional components are necessary for healthy life and disease prevention.

Impact

Goldberger brought the politics of nutrition under scrutiny with lectures condemning the relationship between wealthy landowners of the South and the poor sharecroppers who worked

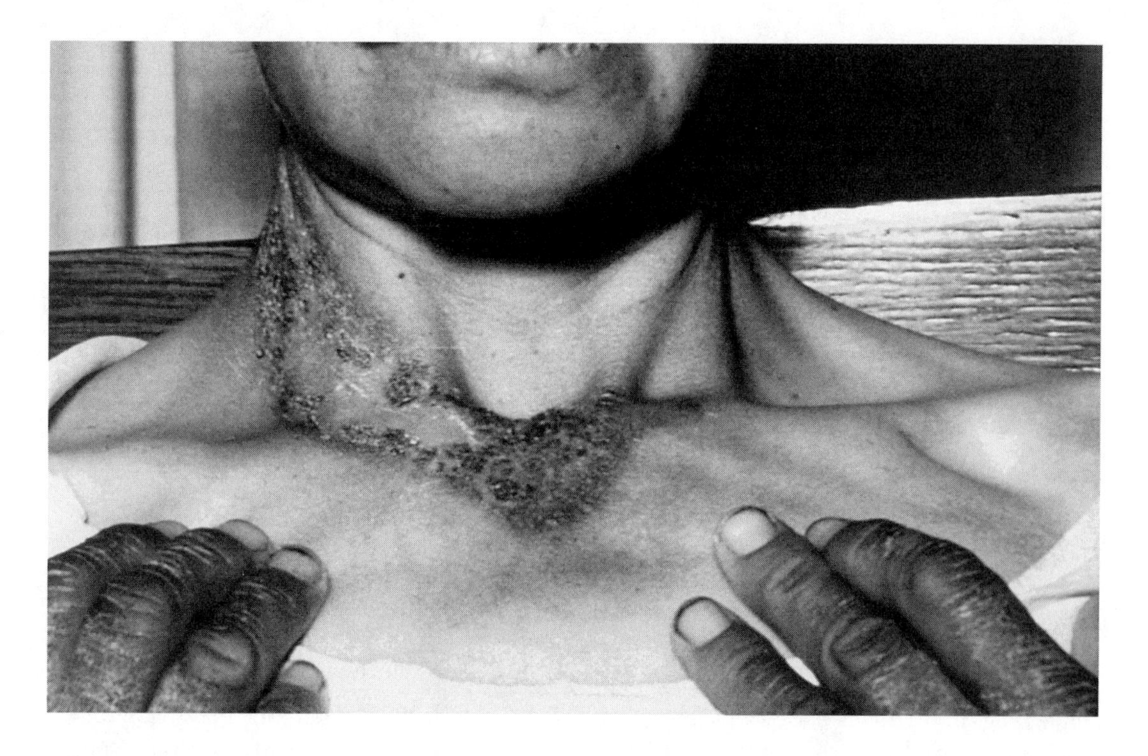

Pellagra, which results in thickening, peeling, and discoloration of the skin, is caused by a lack of niacin. *(Lester V. Bergman/Corbis. Reproduced with permission.)*

the land. Goldberger believed that the poor tenant farmers and millworkers were too destitute to eat a healthy diet and that the only permanent cure was social reform. In 1920, when a drop in cotton prices resulted in decreased income for many southerners, Goldberger predicted a rise in the number of pellagra cases to follow. Federal aid and appeals to southerners to provide local relief to the poor were inadequate, as many feared that investment and tourism would suffer if the southern states carried out a noisy public campaign against pellagra. By the end of 1922, pellagra cases had risen as Goldberger predicted. Land reform was occurring, not as a result of Goldberger's reasoning but due to the boll weevils' invasion of the cotton crop. Forced to diversify their plantings to eliminate the boll weevil, southerners also diversified their diets, and the number of pellagra cases declined.

The 1920s and 1930s brought a surge of discovery in the role of vitamins and nutrition. The anti-pellagra factor was shown to be a member of the B-group of vitamins in 1926. Eijkman won the Nobel Prize for his pioneering work in the discovery of vitamins in 1929. The vitamin deficiency responsible for beriberi, thiamine, was finally crystallized from rice bran in 1926. In 1922 vitamin D was isolated and shown to be effective against rickets, a bone-weakening disease that was estimated to affect 80% of the poor children of Boston in 1900. Nobel prize winning scientist Albert Szent-Györgyi (1893-1986), discoverer of vitamin C in 1933, frequently lectured about the importance of vitamins in an attempt to raise the nutrition consciousness of his colleagues. American biochemist Edward A. Doisy (1893-1986) and Danish biochemist Henrick Dam (1895-1976) were jointly awarded the Nobel Prize for their work in the discovery of vitamin K and its importance in metabolic processes. In 1948 the complex relationship between vitamin B_{12}, stomach secretions, and pernicious anemia was unfolding. Usually considered fatal, pernicious anemia was effectively controlled with injections of vitamin B_{12} isolated from beef liver.

World awareness of hunger and nutritional deficiency was made apparent by the economic depression of the West in the 1930s, famine in the developing world, and pockets of famine in countries devastated by World War II. In 1948 the League of Nations established the United Nations Food and Agriculture Organization, based on the belief that adequate food supply was essential to peace as well as the reduction of disease.

By 1950 vitamins became commercially available and specially formulated for the individual needs of babies, pregnant women, and adults. Fortification of foods, especially breads, cereals, flour, and other grains, with certain vitamins became standard practice. The National Academy of Sciences established recommended daily intakes for vitamins and minerals essential to health. Vitamin deficiency diseases such as pellagra and beriberi thereafter became almost non-existent in the United States and other industrial countries but remained a problem in the developing world.

BRENDA WILMOTH LERNER

Further Reading

Carpenter, Kenneth. *Beriberi, White Rice, and Vitamin B: A Disease, a Cause, and a Cure*. Berkeley: University of California Press, 2000.

Porter, Roy. *The Greatest Benefit To Mankind: A Medical History of Humanity*. New York: W.W. Norton, 1998.

Roe, Daphne A. *A Plague of Corn: The Social History of Pellagra*. Ithaca, NY: Cornell University Press, 1973.

The Discovery and Importance of Penicillin and the Development of Sulfa Drugs

Overview

The discovery of penicillin in 1928 and sulfanilamide drugs in the 1930s played a major role in treating bacterial diseases and in the creation of today's pharmaceutical industry. These chemical agents, called antibiotics, saved many lives during World War II. Though they were initially remarkable in their treatment of disease, it was soon learned that they could be harmful to humans and that the diseases they treated could become resistant to their action.

Background

Diseases have plagued human beings from the beginning of their appearance on Earth. Causes were unknown, so early humans often blamed these frightening visitations on devils or the will of the gods and frequently thought their appearance was caused by wrong behavior or portended some chaos in the future. Thousands of years later, when man had advanced enough to try to understand disease, visionary doctors suggested that disease was caused by "seeds" that were invisible. That invisibility was overcome beginning in the seventeenth century, when the microscope was invented and curious men began to look at objects through it. A man named Malpighi, an Italian, saw the movement of blood in capillaries. In the 1680s Englishman Robert Hooke (1635-1703) looked through a microscope and drew the first pictures of a cell. Anton van Leeuwenhoek (1632-1723) of Holland discovered microscopic life forms in 1676 that he called *animalcules*. He described them so clearly that they can easily be identified today as bacteria, sperm, and blood corpuscles. The structure and understanding of cells was on the verge of being discovered, but nothing more could be done until microscopes were refined two centuries later.

Bacteria were among the first living things on Earth, but their existence was not known until the sixteenth century, and their role in the cycle of life was obscure until the nineteenth century. At the time it was believed that life sprang from non-living matter, a notion called *spontaneous generation*, and no one knew what a cell was. It was known that diseases came from unseen agents, but doctors did not know what they were. They did know from observation that some diseases spread easily between people with no obvious outside agent. That was the state of affairs in the middle of the nineteenth century.

Louis Pasteur (1822-1895), a French chemist and theoretical scientist, worked on many aspects of matter that had not been understood before the microscope was improved. In 1848 he began working on bacteria and soon was able to show that they were living units. He proved that living things came only from living things. Among his other activities, Pasteur proved that many diseases are caused by living germs that multiply in the body. He invented vaccines and pioneered antiseptics and pasteurization. It was known by then that bacteria caused cholera, tuberculosis, pneumonia, and typhoid fever, among many other diseases. Air, food, and water carry bacteria from one person to another, and antiseptics were perfected to prevent the spread of bacteria.

For thousands of years humans used substances found in common molds to cure skin ailments and other infections. At the time no one knew what these molds contained, but in

A close-up of the *penicillium* mold (circle with star shape in the center), which is used to produce penicillin. *(Bettmann/Corbis. Reproduced with permission.)*

1908 a chemist discovered chemicals that had the potential of stopping infections in the same way molds had done. They were called *sulfanilomides,* or sulfa drugs, but they were ineffective against human bacterial diseases until they were refined and turned into a usable form. Many researchers engaged in trials, experiments, and testing and finally came up with a usable substance in 1930s. Sulfa drugs eliminate bacteria but do not actually kill them. When injected into a group of bacteria, sulfas prevent them from multiplying.

In 1928 Alexander Fleming (1881-1955), a British bacteriologist working at Oxford, was engaged in a program of research on staphylococci, a specific class of bacteria. He noticed a peculiar type of mold in one of his laboratory dishes and showed it to his colleagues. No one knew what it was, nor could Fleming or anyone else reproduce it experimentally. Fleming had no idea what this was nor that it would some day lead to a revolution in medicine. It was not for fifteen years that the nature and importance of penicillin was understood. Ernst Chain (1906-1979) and Howard Florey (1898-1968) came to work at Oxford in the 1930s and began to work with antibiotic substances. Eventually, they were able to identify and purify penicillin, which made it available for use as a drug. Together, Fleming, Chain, and Florey shared the 1945 Nobel Prize in medicine for their discovery.

Actually, these substances should be called *penicillins,* as nature produces many different types and more than one type has been produced in bacteriology labs. Many are active and effective against specific bacteria and many are unusable because they are destroyed in the human body. It took ten more years to prove how they did their work. It was known that penicillin was different from sulfas. Both are chemical agents, but where sulfas prevent disease cells from multiplying, penicillin damages the walls of those cells.

Impact

The effect of sulfas and penicillins on stopping diseases, infections, and their spread was electrifying. There had previously been no compound or medicine that could prevent the diseases caused by bacteria. Only after a connection was made between living cells and chemical components was it possible to produce these new medicines. Sulfas were demonstrated to be instrumental in saving many lives, and scientists finally did produce a penicillin, administered by mouth, that was used to treat soldiers during World War II. Because of their rapid effectiveness, a great deal of publicity was given to these drugs. This high visibility caused them to be hailed as "miracle drugs."

The sudden appearance of these compounds created myriad consequences in the modern world. They provided hope that many debilitating and life threatening diseases could be prevented or cured. Another effect was to revolutionize medicine by changing treatment and prognosis of many diseases. Their appearance also changed and expanded experiments into the chemical properties of natural substances. This led to the creation of chemical substitutes or synthetic replicas to act in similar ways. This visibility also created a realization that other fields could benefit from the same kind of research and that there might be substances that could affect viruses, fungi, and other disease-causing agents the same way. This greatly amplified research in microbiology and bacteriology and drew many more practitioners to these fields.

One disappointing fact that became clear by the middle of the century arose from the administration of the drugs. It was soon evident that chemical agents have side effects in the human body. Penicillins, for instance, cause allergic reactions or extreme hypersensitivity in some people. They can cause skin rashes, hives, and other allergies as well as anaphylactic shock, which can result in death. It was also discovered that those allergic to one form of penicillin are allergic to all of them. Sulfas, too, were discovered to be damaging to human kidneys. The "wonder drugs" had definite limits. Far from reducing the use of these drugs, the knowledge led to the creation of synthetic drugs. It also caused the medical profession and drug makers to take more care to determine possible side effects before the drugs were marketed. Successful synthetic penicillins are absorbed into the blood more quickly than the original while maintaining the ability to kill bacteria.

An unexpected discovery in the use of these chemical agents came when it was seen that some strains of bacteria became resistant to a particular strain of penicillin. Scientists thus had to work toward producing new substances.

The use of these drugs has been dubbed "chemotherapy"—that is, the treatment of disease by chemicals. Antibacterial chemicals are selectively toxic only to the infected cells and are designed not to harm healthy cells. The chemists who create these medicines have done so because the drugs do not exist in nature. They have been able to do so because of another exceedingly important impact of the creation of antibiotics. It has been said that it was the creation of sulfas and penicillins that created the pharmaceutical industry, one of the largest and most profitable industries in today's world. Drug companies existed before the 1930s, of course, but were much smaller and made such compounds as prescription medicines, aspirin, and antacid pills.

After the invention of the penicillins and sulfas, pharmaceutical companies expanded, hired researchers, and gave them the time, space, and money to do necessary research into refining products or searching for new ones. Because of the newly understood connection between disease and chemistry, these companies greatly expanded their ability to find the drugs that were needed and to extend their capability to produce these drugs and make them available for everyone.

Today, sulfa drugs and penicillin are no longer the premier substances they once were. Many other clinically useful antibiotics are now now produced. Most treat bacteria, while others work against fungi and protozoa. None are effective against viruses. Scientists were disappointed when they discovered that the much-praised sulfa drugs and penicillin could not cure all diseases. But the drugs were augmented by new techniques and new substances. The emergence of new antibiotics has come about because of the availability of funding for scientific testing and investigation. So, one more impact of the emergence of penicillin and sulfa drugs is that they stimulated further study and led to better understanding of the drugs. It also produced further knowledge about the physiology of the human body and its relation to chemical agents, which in turn has led to the availability of safer medicines.

LYNDALL B. LANDAUER

Further Reading

Hare, Ronald. *The Birth of Penicillin*. London: George Allen and Unwin, 1970.

MacFarlane, Gwyn. *Alexander Fleming: The Man and His Myth*. Cambridge, MA: Harvard University Press, 1984.

Wainwright, Milton. *Miracle Cure: The Story of Antibiotics*. London: Blackwell, 1990.

Wilson, David. *In Search of Penicillin*. New York: Alfred A. Knopf, 1976.

Salvarsan Provides a Cure for Syphilis and Ushers in the Field of Chemotherapy

Overview

The development of dyes that attach to specific biological tissues inspired a search for medicines that would attack disease-producing organisms while leaving healthy tissue intact. After testing over 600 arsenic compounds, German research physician Dr. Paul Ehrlich (1854-1915) announced the development of Salvarsan, a highly specific cure for syphilis, a chronic, debilitating venereal disease. While more modern antibiotic therapies have since replaced Salvarsan, Ehrlich's discovery has had a major impact on modern immunology, biochemistry, and chemotherapy.

Background

In 1900, syphilis was a much feared but little discussed disease. In over half of all cases, the disease, after first surfacing as a small, painless swelling, or sore, would enter a latent period, which could extend for many years and was void of any symptoms. The disease would then reappear as an infection of the nervous system, leading to debilitation, insanity, and death. It was not discussed in polite company because the disease was generally transmitted by sexual contact, although a pregnant woman with the disease could transmit it to her child, and during the disease's brief secondary phase it could be transmitted by more casual contact.

The nineteenth century had been a period of rapid advance in many areas of science and technology in Europe. In chemistry, the modern notion of atoms combining to form molecules took form, thanks to the work of the English chemist John Dalton (1766-1844), the Italian physicist Amadeo Avogadro (1776-1856), and many others. However, a full understanding that the atoms in a molecule had a definite arrangement in space did not develop until after 1874, when the tetrahedral arrangement of the chemical bonds around the carbon atom was proposed simultaneously by the Dutch physical chemist Jacobus van't Hoff (1852-1911) and the French chemist Joseph Le Bel (1847-1930).

In 1830, the English microscopist Joseph Jackson Lister (1786-1869) had devised a method of eliminating the distortions in earlier microscopes that had limited their usefulness. The discovery of the first aniline dye by British chemist William Henry Perkin (1838-1907) in 1856 led to the development of a vigorous dye industry, particularly in Germany. (Aniline is a colorless, poisonous liquid used to manufacture dyes, among other compounds.) With microscopes that could image microbes (microorganisms, especially those that cause disease) and dyes to stain them, a revolution in microbiology was at hand. Among the leaders in the new field was Robert Koch (1843-1910), a German physician who discovered the tuberculosis bacterium in 1882 and was named first director of the Berlin Institute of Infectious Diseases in 1891.

In the meantime, Paul Ehrlich had received his medical degree at the University of Leipzig in 1878, and as a student had developed an interest in the staining of microorganisms. Following a period of hospital practice and a convalescence of two years while he himself recovered from tuberculosis, he became Koch's assistant. There he worked on standardizing the therapeutic strength of the serums prepared by injecting disease organisms into animals. There, too, contemplating the highly specific affinity of dyes for specific tissues and the actions of serums on disease-producing organisms, he developed the beginnings of his "side chain theory." According to this theory, different types of cells were characterized by different chemical chains protruding into the surrounding medium. To stain them or to interfere with their action on a host organism, one had only to find a substance with molecules of a complementary shape that would bind chemically to the side chains; such a substance would act as a "magic bullet," affecting only the invading cell.

1899 Ehrlich became director of the Royal Prussian Institute for Experimental Therapy. To this was added, in 1906, the Georg Speyer House for chemotherapy, built and funded by a large private donation. Here, Ehrlich turned his attention to the diseases for which the immune response of animals was inadequate to produce an effective antiserum. Diseases caused by this group included sleeping sickness and yaws, diseases primarily of the tropics, and syphilis, which had afflicted Europeans at least since the end of the fifteenth century. Ehrlich hoped to cure these using not "serum therapy" but "chemotherapy"—the treatment of a disease with chemicals—using molecules specially built to attach to the pathogens. (A pathogen is any agent that causes disease.)

At the Institute, Ehrlich and his young Japanese associate, Kiyoshi Shiga (1870-1957), found an aniline dye, trypan red, that was particularly effective against trypanosomes, the parasitic protozoa that caused sleeping sickness and other diseases. In 1906, Koch introduced the use of atoxyl, an arsenic-containing compound that was thought to be related to the aniline dyes, against trypanosomes. Ehrlich was also interested in this drug, but determined that it attacked the optic nerve, and so could not be safely used on humans.

In 1905, the German microbiologist Fritz Richard Schaudinn (1871-1906) and a dermatologist, Erich Hoffman, discovered the organism, a spirochete—a spiral bacterium with some resemblance to the trypanosomes—that caused syphilis. Ehrlich began a systematic program to find an arsenic compound that would be effective, as well as safe to use, against syphilis. Over 600 compounds were synthesized and tested on rabbits infected with the spirochete, with careful attention to toxic side effects. After five years (during which time Ehrlich's work in immunology earned him a Nobel Prize in 1908), the researcher became convinced that the 606th preparation, which he named Salvarsan, was both safe and highly effective, as it attacked the disease germs but did not harm healthy cells. Thus, Salvarsan ushered in the new field of chemotherapy. He obtained a patent for the new compound and, with the assistance of a small number of physicians, began testing it on patients.

Paul Ehrlich. *(Library of Congress. Reproduced with permission.)*

Ehrlich announced his discovery on April 19, 1910, at a Congress for Internal Medicine held at Weisbaden. The demand for Salvarsan grew rapidly; the Speyer House provided about 65,000 doses to physicians before the year had ended. A few individual patients reported adverse effects, and a few patients were not cured. After standardization of the injection procedure and the dosage, most of the problems did not recur. In time, the demand became too great for Ehrlich to handle, so he reached a manufacturing agreement with the Hochst Chemical Works, which began producing Salvarsan on a commercial basis in December, 1910. Salvarsan, as well as a more soluble and more easily administered form of the drug, NeoSalvarsan, would dominate the treatment of syphilis until the advent and widespread use of antibiotics in the 1940s.

Impact

Both the book of Genesis and the Greek myths include stories of man's fall from an idyllic, disease-free state as the result of disobedience of a divine command. While the development of the art of medicine reflects an acceptance of the idea that disease could have natural causes and could be treated by natural, rather than supernatural, means, the idea that one should not interfere with the course of a disease has never been completely abandoned. Karl Wassman, an individual who shared this view and the publisher of a sensationalist paper, launched an attack against Ehrlich, claiming that he had forced patients to undergo treatment against their will. The Frankfurt Hospital began legal proceedings against Wassmann. In court, Ehrlich appeared as a witness for the hospital. Wassman was found guilty of libel and his witnesses were discredited.

Penicillin, discovered by the Scottish bacteriologist Sir Alexander Fleming (1881-1955) in 1928, quickly replaced Salvarsan as the method of choice for treatment of syphilis. Despite the availability of treatment, syphilis remained a public health problem in the United States until the 1950s, when an aggressive public health campaign was launched to identify all victims of the disease and their sexual contacts. The notion that the victims of syphilis, at least racial minority victims, were less worthy of treatment than those suffering from other diseases might have played a role in the Tuskegee experiment, begun by the United States Public Health Service in 1932, and which continued through the 1960s. In this "experiment," the course of the disease was studied in a group of African American men, who were falsely led to believe that they were being treated. Over 100 men died while participating in this study.

Ehrlich's career includes a number of features that would come to play a prominent role in determining the course of medical research in the twentieth century. Many of his discoveries were dependent on new chemical ideas. He worked not in a university or hospital, but in an institute set up by the government to advance medical science. Although a doctor of medicine, he completed, in effect, a research apprenticeship under Robert Koch and then provided research training to a new generation of researchers. He was one of the first scientists to host researchers from other countries, notably Japan. When government funding proved insufficient, he was forced to turn to wealthy individuals for grants of funding to conduct research on a larger scale. He secured a patent for his new chemical compound and had to negotiate an agreement with a major commercial company in order to make his discovery available on a large scale. To prove the value and refine the technique of Salvarsan therapy, he conducted what today would be considered a large-scale clinical trial, with numerous physicians treating thousands of patients.

Ehrlich's side chain theory is still recognizable in several areas of biomedical science. Modern immunology has identified how the presence of characteristic groups, now called haptens, on invading cells induce the generation of antibody molecules that can recognize and bind to the invading cells. Somewhat similar ideas are found in enzymology in which enzymes molecules have active sites which provide a good geometric match to the substrates on which they act. The term chemotherapy is now most often used in connection with the treatment of cancer by chemicals. Substantial research resources are currently being devoted to finding new therapeutic agents, which like "magic bullets" will attack malignant cells, while leaving normal cells undisturbed.

Paul Ehrlich had an impact on popular culture, too. Like his fellow scientists Louis Pasteur (1822-1895) and Marie Curie (1867-1934) and inventors Thomas Edison (1847-1931) and Alexander Graham Bell (1847-1922), Ehrlich was memorialized by the movie industry: *Dr. Ehrlich's Magic Bullet*, starring Edward G. Robinson, was released in movie theaters in 1940.

DONALD R. FRANCESCHETTI

Further Reading

de Kruif, Paul. *Microbe Hunters.* New York: Harcourt, Brace, Jovanovich, 1954.

Ihde, Aaron J. *The Development of Modern Chemistry.* New York: Harper and Row, 1964.

Marquardt, Martha. *Paul Ehrlich.* New York: Schuman, 1951.

Nester, Eugene W., C. Evans Roberts, and Martha T. Nester. *Microbiology: A Human Perspective.* Dubuque, Iowa: Brown, 1995.

Porter, Roy. *The Greatest Benefit to Mankind: A Medical History of Humanity.* New York: Norton, 1997.

The Development of Modern Blood Transfusions

Overview

During the nineteenth century physicians began experimenting with blood transfusions, directly from donor to patient. Most failed until Karl Landsteiner (1868-1943) discovered blood groups, which must be matched properly for a successful transfusion. The development of anticoagulant drugs around 1914 allowed blood to be stored. Blood banking began in 1937, and the system was expanded greatly during World War II.

Background

In 1628 English physician William Harvey (1578-1657) described the circulation of blood in the body. This sparked interest in understanding the functions of the blood, and physicians became interested in replacing lost blood through transfusions. Soon after Harvey published his work, transfusions between animals were attempted. In 1665 another English physician, Richard Lower (1631-1691), successfully transfused blood between dogs.

Since animals could be induced to "donate" blood more readily than seventeenth-century humans, they were eyed as possible sources of blood for human patients as well. In 1667 Lower and Jean-Baptiste Denis (1643-1704) independently reported transfusions of lamb blood into humans. However, such experiments often resulted in deadly reactions, one of which led to Denis's arrest. Animal-human transfusions were soon prohibited by law.

In 1818 an Englishwoman who was hemorrhaging after childbirth was successfully treated with a blood transfusion. Using a hypodermic syringe, physician James Blundell administered about four ounces of blood that had been extracted from her husband's arm. Blundell performed 10 additional transfusions between 1825 and 1830, of which half were successful. This was the pattern for early transfusions: sometimes they helped, but sometimes the patient had a severe, often fatal reaction. In 1875 German physiologist Leonard Landois first described how incompatible transfusions resulted in the clumping and bursting of red blood cells, a process called hemolysis. It was not until the dawn of the twentieth century that Landsteiner, an Austrian-born American immunologist, discovered why this occurred.

Landsteiner discovered three different blood groups—A, B and O—in 1901. The next year, another research team added the AB group. Red blood cells of groups A and B have A or B antigens, specific sugar-containing substances, on their surfaces. AB group blood cells carry both antigens, and O blood cells have neither. Antibodies in the blood serum react to antigens of a different group and destroy the red blood cells. Landsteiner was awarded the 1930 Nobel Prize for Medicine for his work.

Together with Alexander Wiener, Landsteiner discovered another important blood type distinction, the Rh factor, in 1940. The Rh factor accounted for the majority of the remaining transfusion reactions, especially cases in which an Rh-negative mother's antibodies attack an Rh-positive fetus' red blood cells during pregnancy, resulting in hemolytic disease in the newborn. Several other blood group systems were later identified, some as the result of a rare and unexpected hemolytic reaction after a transfusion or pregnancy.

Impact

The knowledge of the different blood groups was quickly put into practical application. The blood types of donor and recipient could be checked, and blood could be cross-matched, or mixed in the laboratory, to see if dangerous clumping occurred. The ability to type and cross-match for compatibility between blood donor and recipient greatly increased the likelihood that a transfusion would be successful.

It was soon realized that type O blood, free of A or B antigens, could be given to patients of any group; people with type O blood are universal donors. Similarly, patients with type AB blood are universal recipients, since their blood will not react to either the A or B factors. In the United States, 40-60% of the population has type O blood.

In New York, Reuben Ottenberg observed the Mendelian inheritance of blood groups. A child inherits one allele for blood type from each parent. An AA or AO pattern will result in a blood type of A. Similarly, a BB or BO inheritance will produce a B blood type. A child receiving an A allele from one parent and a B from the other will have blood type AB. Only if each parent passes along an O will the child have blood type O. This is the reason blood tests have often been employed to determine whether or not a man could have fathered a particular child. For example, if the child has a blood type of O, a man with a blood type of AB could not be the father, because he could only have passed along either an A or B allele.

With blood-typing well understood, transfusions could then be performed with greater confidence, but the technique was still difficult to implement on a large scale. Since blood clots so quickly, a donor of the right blood type had to be present on the spot to provide blood for a patient. In 1915 Richard Lewinsohn at New York's Mount Sinai Hospital used an anticoagulant, sodium citrate, to prevent donated blood from clotting. This allowed blood to be collected from the donor into a container and transported to the recipient, rather than requiring vein-to-vein transfusion. The next year, Francis Rous (1879-1970) and J. R. Turner introduced a citrate-glucose anticoagulant, allowing blood to be stored in refrigerated containers for several days. For the first time, "blood depots" could be set up, as was done by Oswald Robertson for the British military during World War I.

In the Soviet Union, thousands of quarts of blood were being collected, stored, and shipped around the country by the 1930s. However, lack of care in monitoring storage times coupled with insufficient attention to cleanliness led to a reaction rate of more than 50%. This made American physicians cautious about blood storage. American physician Bernard Fantus established the first blood bank in the United States at the Cook County Hospital in Chicago in 1937. He collected blood from donors in flasks containing sodium citrate, tested, sealed, and refrigerated it. His patients' reaction rate was far lower than experienced in Russia, and this encouraged wider use of blood transfusions.

During World War II, Charles R. Drew (1904-1950) launched a large-scale blood drive called "Plasma for Britain" involving the American Red Cross. Drew's contributions also included advocating the use of a blood component called plasma. Plasma contains important fractions of blood such as albumin to augment blood volume, fibrinogen to enable clotting and wound healing, and gamma globulin to fight infection. At the time, it could be stored longer than whole blood, making it very useful for emergency transfusions on the battlefield and elsewhere. However, it was prone to contamination, and Drew developed many strict quality control procedures to ensure its safe handling. Today, most blood donations are stored as components, including plasma or its individual fractions, platelets to control bleeding, or concentrated red blood cells to correct anemia. The improved anticoagulant citrate dextrose, introduced in 1943, reduced the volume of anticoagulant that was required and allowed storage at 4°C for three weeks.

Drew, who was African-American, protested against the racial separation of stored blood, the standard procedure in the 1940s. Although physicians were well aware that there was no scientific basis for such separation, there was concern that prejudice might discourage acceptance of the valuable technique of transfusion if white patients feared receiving "black blood." The labeling of blood by race continued in the United States until the late 1960s, when it was swept away as a result of the civil rights movement.

Tragically, Drew was killed in a 1950 automobile accident in North Carolina, at the age of 45. Perhaps understandably, a myth grew up that he had bled to death because he was refused a transfusion due to his race. In fact, although the hospital he was taken to was segregated, the emergency room did serve blacks. The white doctor on duty was acquainted with Drew, worked over him for hours, and consulted specialists at the Duke University Medical Center. A transfusion was attempted, but Drew's vena cava was severed, making it impossible to stop his bleeding. His injuries, including a broken neck and a crushed chest, were simply too severe for him to survive.

During the early part of the twentieth century, the development of successful transfusions and the ability to store and transport blood made it possible to save many other victims of trauma and hemorrhage who would otherwise have died. Later innovations, such as heart-lung machines that allowed open-heart surgery, were also enabled by progress in blood transfusion.

SHERRI CHASIN CALVO

Further Reading

Reid, Marion E., and Christine Lomas-Francis. *Blood Group Antigens Factsbook*. London: Academic Press, 1996.

Rudmann, Sally V., ed. *Textbook of Blood Banking and Transfusion Medicine*. Philadelphia: Saunders, 1995.

Speiser, Paul, and Ferdinand Smekal. *Karl Landsteiner: The Discoverer of the Blood-Groups and a Pioneer in the*

Field of Immunology. Translated by Richard Rickett. Vienna: Hollinek, 1975.

Starr, Douglas. *Blood: An Epic History of Medicine and Commerce.* New York: Alfred A. Knopf, 1998.

The First Birth Control Clinics in America and England

Overview

Primarily through the efforts of Margaret Sanger (1879-1966) in America and Marie Stopes (1880-1958) in England, deliberate family planning emerged as a social force in the early twentieth century.

Background

Until the second decade of the twentieth century, women had little choice but to bear as many children as they conceived. Rape victims, incest victims, prostitutes, sexually active unmarried women, and even wives whose husbands wanted no more children did not have any safe, readily available, or medically reliable means to prevent unwanted pregnancies. Women who used contraception were regarded as immoral, unfeminine, or abnormal.

The contraception movement began in the early nineteenth century. It drew much of its inspiration from a famous book by British political economist Thomas Malthus, *An Essay on the Principle of Population* (1798). Malthus argued that the world's population could eventually grow beyond the ability of the earth to support it. Famine, epidemics, and general poverty would result.

The first significant advocate of pregnancy prevention was Francis Place (1771-1854), one of the most revolutionary Britons of his day. His book, *Illustrations and Proofs of the Principle of Population* (1822), sought only to influence social policy and suggested no practical contraceptive methods.

The militant advocacy of birth control can be said to have originated between 1910 and 1912 when Margaret Sanger was working on Manhattan's Lower East Side as a nurse among impoverished mothers of large families. The experience radicalized her. She immediately began writing incendiary newspaper columns and staging rallies. She founded a short-lived radical journal, *The Woman Rebel*, in March 1914.

When Sanger was arrested in August 1914 for sending obscene materials through the mail in violation of the 1873 "Comstock Law," she jumped bail and fled to England. She returned from her self-imposed exile almost a year later, determined to reap the publicity of her trial, but growing public sympathy induced federal prosecutors to drop their case against her in February 1916. Deprived of this opportunity for publicity, she embarked on a nationwide birth control promotional tour, and managed several arrests along the way. Her boldness was the initial catalyst of the movement.

The two main leaders of the birth control movement were Sanger and Marie Stopes. Other important participants in their era were British psychologist Havelock Ellis, American feminist Mary Ware Dennett, American gynecologist Robert Latou Dickinson, American pediatrician Hannah Meyer Stone, British novelist H.G. Wells, Russian-American anarchist Emma Goldman, Swedish feminist Ellen Key, South African feminist Olive Schreiner, American first lady Eleanor Roosevelt, and Katharine Houghton Hepburn, the mother of the actress. Sanger's friend Otto Bobsein coined the term "birth control" in 1914.

Dennett founded the National Birth Control League (NBCL) in 1915 while Sanger was in England. When Sanger returned, she found that Dennett had attracted a wider and more sympathetic audience for birth control and had tied contraception issues to the women suffrage question. Sanger resented Dennett usurping her leadership. Their styles were different: Dennett preferred gradual legislative change while Sanger preferred confrontation and street action. Sanger's propensity for civil disobedience and radical reform soon grew into an unbridled disapproval of Dennett and the entire gradualist, progressive wing of the movement.

Stopes met Sanger in England in 1915. This encounter motivated Stopes to devote her energy to the birth control cause and to write her polemical bestseller, *Married Love* (1918). Yet

Leaders of an American birth control clinic. Margaret Sanger is third from the left. *(Bettmann/Corbis. Reproduced with permission.)*

the two women did not like each other. They differed on both philosophy and strategy. Sanger was a socialist rebel and Stopes was a class-conscious eugenicist. After 1920 they were no longer on speaking terms. Stopes allied herself with Dennett against Sanger to contest the leadership of the movement. Sanger enlisted Ellis to propagandize against Stopes.

Sanger opened America's first birth control clinic at 46 Amboy Street in the Brownsville section of Brooklyn, New York, on October 16, 1916. It advertised in English, Yiddish, and Italian that trained nurses would provide birth control information, counseling, and devices. Women of all socio-economic backgrounds stood in line for these services. In the few weeks of its existence it served 464 women. On October 26 Sanger and her sister, Ethel Byrne, were arrested for selling obscene materials and the clinic was closed. Out on bail, Sanger re-opened it. The police closed it in November under the public nuisance laws. Sanger and her sister were convicted and each spent 30 days in prison.

Stopes and her husband, Humphrey Roe, opened Great Britain's first birth control clinic in North London on March 17, 1921. This facility survived until 1977, when British reproductive health care was nationalized. Unlike Sanger, Stopes had upper-class political connections. Stopes enforced strict rules at her clinic to prevent the kind of trouble that Sanger experienced in Brooklyn. Whereas Sanger's clinic had served anyone who walked through the door, Stopes insisted that her clinic would serve only women who could prove they were already mothers. Sanger openly attacked Stopes's rules.

Stopes in England had fewer legal troubles than Sanger had in America. Stopes courted the political right and was generally content to write books and pamphlets rather than campaign in the streets. Yet she could be confrontational and even flamboyant. She took her lawsuit against a prominent Roman Catholic for libel all the way to the House of Lords, chained a copy of her book on Catholic birth control methods to the font of Westminster Cathedral, and passed around cervical caps at fancy dinner parties. On most such occasions she made sure that journalists were present to provide publicity.

Impact

History shows that when contraception is restricted, abortion rates increase, and when abortion and contraception are both restricted, infanticide rates increase. With the notable exception of Stopes, birth control advocates in the early twentieth century generally sought to dissociate themselves from advocating abortion rights. They tended to regard abortion, infanticide, and overly large families as tragedies that could be prevented by dependable, safe, and convenient contraceptive devices. Sanger and most of her colleagues mentioned the physical danger and moral undesirability of abortion in their efforts to educate the public about the reasonableness of preventing unwanted pregnancies rather than either terminating them or allowing women to bring unwanted children into the world.

Despite its standard argument that better contraception would decrease the incidence of abortion, the birth control movement earned the vigorous opposition of the Roman Catholic Church. In its campaign against birth control, and especially against the ideology that Stopes expressed in her books, the Catholic Church emphasized and sometimes exaggerated the movement's involvement with racial and class politics and with various theories of population control, eugenics, social Darwinism, and neo-Malthusianism. The opinions of Father Thomas A. Ryan and Patrick Cardinal Hayes in several official Catholic publications were a major part of this debate. They pointed out that Catholic doctrine expressly forbade the use of contraception. The most important statements of Catholic beliefs about marriage, the family, sex, parenthood, and birth control were the encyclicals of Pope Pius XI, *Casti connubii* (1930), and Pope Paul VI, *Humanae vitae* (1968).

In 1922 Sanger founded the American Birth Control League (ABCL), partly to fill the void left by the demise of the NBCL in 1919. In 1923 she founded the Birth Control Clinical Research Bureau. Its purpose was to involve physicians more significantly in the birth control movement. The Bureau encouraged doctors to monitor their patients' use of contraceptives and promoted scientific research into contraceptive methods. A raid on the Bureau in 1929 by the New York City Police backfired. The international medical community was outraged by the arrest of two physicians, the seizure of private medical records, and the violation of the confidential relationship between physician and patient. After this highly publicized raid, physicians flocked to the movement, and in 1937 the American Medical Association officially endorsed birth control as a legitimate medical enterprise.

In 1939 the ABCL and the Bureau merged to create the Birth Control Federation of America. In 1942, against Sanger's wishes, it changed its name to the Planned Parenthood Federation of America.

Birth control methods in the early twentieth century were physical barriers, not chemical or pharmaceutical preparations. Sanger's millionaire husband, J. Noah H. Slee, began manufacturing diaphragms in 1925, the first such venture in America. Stopes preferred these mechanical means, but Sanger encouraged research into spermacides and other non-mechanical contraceptives. In 1948 Planned Parenthood gave biologist Gregory G. Pincus funding to develop an oral contraceptive. This research was successful. Between 1953 and 1956 Pincus, Min Chueh Chang, and John Rock published scientific results that eventually led to the commercial production of the birth control pill.

Major legal victories for the American birth control movement occurred in 1936 when the Second U.S. Circuit Court of Appeals ruled in U.S. vs. One Package (of Japanese Pessaries) that contraceptives are not obscene, and in 1965 when the U.S. Supreme Court ruled in Griswold vs. Connecticut that states could not restrict the use of contraceptives by married couples.

ERIC V.D. LUFT

Further Reading

Books

Briant, Keith. *Marie Stopes: A Biography.* London: Hogarth, 1962.

Chen, Constance M. *The Sex Side of Life: Mary Ware Dennett's Pioneering Battle for Birth Control and Sex Education.* New York: New Press, 1996.

Chesler, Ellen. *Woman of Valor: Margaret Sanger and the Birth Control Movement in America.* New York: Simon and Schuster, 1992.

Davis, Lenwood G. *The History of Birth Control in the United States: A Working Bibliography.* Monticello, IL: Council of Planning Librarians, 1975.

Hall, Ruth. *Marie Stopes: A Biography.* London: Andre Deutsch, 1977.

Hall, Ruth. *Passionate Crusader: The Life of Marie Stopes.* New York: Harcourt Brace Jovanovich, 1977.

Kennedy, David M. *Birth Control in America: The Career of Margaret Sanger.* New Haven: Yale University Press, 1970.

Maude, Aylmer. *The Authorized Life of Marie C. Stopes.* London: Williams & Norgate, 1924.

Reed, James. *From Private Vice to Public Virtue: The Birth Control Movement and American Society since 1830.* New York: Basic Books, 1978.

Rose, June. *Marie Stopes and the Sexual Revolution.* London: Faber and Faber, 1992.

Williams, Doone, Greer Williams, and Emily P. Flint. *Every Child A Wanted Child: Clarence James Gamble, M.D. and His Work in the Birth Control Movement.* Boston: Harvard University Press, 1978.

Other

Alfred, Bruce, director. *Margaret Sanger.* Videocassette. 87 min. Princeton, NJ: Films for the Humanities & Sciences, 1998.

Sanger, Margaret. *The Papers of Margaret Sanger.* Washington, D.C.: Library of Congress, 1976. 145 microfilm reels.

Sanger, Margaret. *The Margaret Sanger Papers: Documents from the Sophia Smith Collection and College Archives, Smith College.* Edited by Esther Katz, Peter Engelman, Cathy Moran Hajo, and Anke Voss Hubbard. Bethesda, MD: University Publications of America, 1994. 83 microfilm reels plus 526-page guide.

Advances in Dentistry, 1900-1949

Overview

During the years 1900 to 1949 major advances occurred in dental development. Improvements in dental drills and filling techniques, the advent of fluoride treatment, the development of orthodontics, and new ideas about the connection between teeth and the overall health of the body enabled the profession to advance as a major segment of the health care industry. These developments would set the stage for additional advances during the later half of the century.

Nearly all major dental specialties had their inception in the first half of the twentieth century: orthodontics, the science of straightening teeth, 1901; oral surgery, 1918; periodontics, treating gums and gum disease, 1918; prosthodontics, making dentures and bridges, 1918; pedodontics, children's dentistry, 1927; public health, 1937; and oral pathology, 1946. Only endodontics (1963) did not begin during this time.

Background

From the beginning of recorded history, dental problems have plagued human beings. Ancient chronicles tell of toothaches, dental decay, periodontal disease, and tooth loss. Magic, myth, and religious experiences all played a part with some odd beliefs emerging. For example, the ancient Egyptians believed that the sun god protected mice, so for a toothache the body of a warm mouse was split and applied to the jaw to bring relief. Early Christians prayed to Saint Apollonia of Alexandria (A.D. 249) for pain relief, and she became the Patron Saint of Dentistry.

During the dark days of the Middle Ages, monks, along with barbers, became physicians and dentists. However, when a pope ruled in 1163 that this was not suitable for monks, the barbers took over completely. The barber-surgeon did extractions along with itinerant groups called tooth drawers, who operated in the public square. Many of these were spectacular showmen, such as "le grand Thomas" who appeared in Paris during the time of Louis XV (1710-1774). These practitioners developed a number of tools, although they knew very little about anatomy. It was the famous anatomist Andreas Vesalius (1514-1564) who described teeth and the chambers.

French surgeon Pierre Fauchard (1678-1761), recognized as the father of scientific dentistry, realized the condition of the teeth related to health in general. He was not just interested in extraction, but devised techniques that included drilling, filling, filing, transplanting, dentures, cosmetic tooth straightening, and surgery of the jaws and gums. He set an agenda for decades of development.

While many developments affected dentistry, the major advance was the discovery of anesthesia. Two dentists—Horace Wells (1815-1848) and William Morton (1819-1869)—are credited with introducing this development during the mid-nineteenth century. In the 1890s W. D. Miller found that teeth could be made to decay in a test tube by exposing them to bacteria and carbohydrates. Although x-rays were discovered and used by Edward Kells in dentistry in 1895, it was not until the next century that this technique was accepted and used.

Impact

The new century began with many new discoveries that would expand and lead to great improvements in dentistry.

The dental drill has been one of the most feared and hated devices, but its history reveals that ingenious investigators recognized the benefits of removing dental caries. George Washington's dentist, John Greenwood (1760-1819), used his mother's foot treadle on the spinning wheel to develop a rotating drill. In 1868 American George F. Green introduced a pneumatic drill powered by a pedal billows. Electricity was added to drills in 1874, but the devices were heavy and expensive. By 1908 almost all dental offices had electricity, though, compared to today's high speed drills, they were slow and cumbersome. However, the drills did make it possible to fill teeth and fit crowns. The use of diamond bits and carbide burrs were introduced in about 1925. The air abrasive drill was introduced in 1945. Now, modern drills are turbine-powered and rotate at speeds of 300,000 to 400,000 revolutions per minute.

Many materials have been used to fill teeth: stone chips, turpentine, resin, metals of all kinds, and gold. The improvement of drills opened up a whole new era of filling teeth. A race was on to find the best material. Several different types of amalgam (mixtures of various metals) were used, with many of them containing mercury. After a series of "amalgam wars," Chicago dentist G. V. Black (1836-1915) was able to establish a standard in 1895. In 1907 William H. Taggart successfully demonstrated the casting of gold inlays, which made possible accurate filling and inlay fitting.

Likewise, dental education was becoming more sophisticated. The first dental school had been established in the United Kingdom by the Odontological Society of London. In the United States, the first dental school associated with a university opened at Harvard in 1867. In 1899 one year of high school was established as a prerequisite for admission to dental school. Just a few years later, in 1903, a four year's course in a dental college was developed. In 1921 the requirement was changed to one year of college. The first official bulletin was published in 1913. The National Dental Association changed its name to the American Dental Association in 1922.

Dentures became an answer to the extraction of teeth. Developing satisfactory materials for dentures has a history that involves not only medical, but also legal and ethical, questions. In 1851 Nelson Goodyear, brother of Charles Goodyear (1800-1860), who first developed vulcanized rubber, found a way to make dentures from rubber that worked well and had a fairly low cost. The only problem was that the base had a dark red color and did not look desirable. These dentures were a great advance, and in 1864 Goodyear got a patent, licensed the dentists who used them, and changed a royalty for each set. Many dentists despised these patents, but always lost the legal battles. In 1879 a very respected dentist, Samuel Chalfant, shot Josiah Baron, the attorney hired by Goodyear to prosecute noncompliance. When the patents expired in 1881, Goodyear did not seek to renew the licensing of dentists. These vulcanite dentures were used until 1940, when pink acrylic dentures replaced them.

At the beginning of the twentieth century anesthesia had been introduced. The gas nitrous oxide was identified by Joseph Priestley (1733-1804) in 1772, and Humphrey Davy (1778-1829) noted that when this was breathed, people became hilarious and silly. The popular term "laughing gas" was given, but Davy's observation that it seemed to make people oblivious to pain was ignored. Horace Wells is credited with first recognizing the value of applying practical anesthesia to dentistry. Wells attended a demonstration by a traveling chemist, Gardner Q. Colton, who had people inhale the gas for the amusement of the audience. Wells noticed how the volunteers scraped their shins on the heavy benches and felt no pain. He was the first to devise the idea of inhaling anesthesia.

Cocaine was introduced as the first local anesthesia in 1884 when William S. Halstead (1852-1922) demonstrated that its injection into an area could block pain. In 1905 Professor Braun introduced novocaine (procaine hydrochloride) for dentistry. Novocaine's low toxicity and general absorption into body tissue made it a quick and relatively safe anesthesia. Novocaine is detoxified in the liver and rapidly hydrolyzed in the plasma. The drug then infiltrates around nerve endings and fibers, and its pain-killing effect is almost immediate. Now, novocaine is seldom used, for it has been replaced by primacaine or tetracaine. During World War I, Harvey Cook noted the use of cartridges in rifles and conceived the idea of putting anesthesia solution in cartridges. From this observation came the development of syringes for the delivery of drugs. In 1943 Swedish chemists Nils Lofgven and Bengst Lundquist introduced lidocaine (xylocaine hydrochloride) and tested it extensively. By 1950 it was widely accepted. Xylocaine goes quickly into the nerve sheath and rapidly deadens feeling with very few side effects.

In 1895 Karl Wilhelm Roentgen (1845-1923), a German professor of physics, was studying a cathode ray that glowed inside a spiral tube when high voltage was applied. He darkened his lab and wrapped the tube in black cardboard. He noted strange rays that produced an outline of bones in his hands through the skin. He called the beams x-rays because they were unknown. He was awarded the 1901 Nobel Prize for Physics and an honorary doctor of medicine. X-rays were accepted at once. Roentgen refused to patent x-rays because he thought it was a gift for the benefit of mankind. Dr. Edmund Kells of New Orleans, Louisiana, and Otto Walkhoff of Munich, Germany, first applied x-rays to locate impacted teeth. Soon the method of looking beneath the surface extended to locating decay and other problems in teeth.

Several nineteenth-century dentists contributed to the development of orthodontics, or straightening teeth. Norman W. Kinglet and J. N. Farrar wrote treatises on the new dental science. Edward H. Angle (1855-1930) built upon previous work to design a system of classification of malocclusion. He also designed orthodontic appliances and simplified other ones. He founded the first school or orthodontics and was an original member of the American Society of Orthodontics in 1901. He was a prolific writer and founded an orthodontic journal as well as producing a major text, *The Malocclusion of Teeth.* Some of his classifications are still used today.

In 1948 Congress created the National Institute of Dental Research as part of the National Institutes of Health, Bethesda, Maryland. The institute sup000ports about 100 research and educational centers in the United States.

EVELYN B. KELLY

Further Reading

Ring, Malvin E. *Dentistry: An Illustrated History.* New York: Abrams, 1985.

Weinberger, Bernhard Wolf. *An Introduction to the History of Dentistry.* 2 vols. St. Louis: C. V. Mosby, 1948.

Western Medicine Re-Discovers the Ancient Chinese Herb Ma Huang

Overview

For over 5,000 years Chinese physicians have used the scrubby plant ma huang to treat asthma, a severe breathing condition. Ma huang is the Chinese name of the plant of the genus *Ephedra,* the source of the modern drug ephedrine. Introduction of the drug to scientific medicine arose from the work of Carl L. Schmidt (1893-1988) and his Chinese colleague Ko Kuei Chen (1899-?). Schmidt, a pharmacologist, became fascinated with ancient Chinese herbal medicine when he was assigned to teach at Peking University in 1922. With Chen's help, he investigated the plant, extracted the essence, and found that this extract, called ephedrine, could be useful in the treatment of asthma. In the 1920s Percy L. Julian (1899-1975) synthesized the drug.

Ephedrine gave hope to a generation of asthma sufferers, although a number of side effects began to be noted. In later years of the twentieth century ephedrine was used as a street drug for its amphetamine or upper-like qualities and in diet pills. The side effects caused many deaths, warranting the U.S. Food and Drug Administration (FDA) to issue a warning about the possible dangers of using ephedrine except under very strict supervision.

Background

In ancient medicine many ways—including magic and incantations—were used to treat people with a multitude of conditions. More active therapies included tonics to strengthen the person or "poisons" to drive off the evil spirits. Choice of the right herbal medicine depended on the symbolic use of a plant, which could be chosen for its shape or sound.

The Chinese have a long tradition of medicinal herbs. Ma huang, the plant from which ephedrine comes, is part of ancient history. The Chinese book of medicine *Nei Jing*, or *Book of Medicine of the Yellow Emperor*, written about the third century B.C., established the principle of yin and yang. The idea of yin and yang is based on balance—between the active and passive, hot and cold, male and female. Illnesses occur when these qualities are out of balance. The role of Chinese physicians was to restore this balance, and the use of herbs was only one of the ways adopted by Chinese physicians. *Shen Nong Ben-*

caojung (Classes of Roots and Herbs of Shen Nong), a book that appeared in the third century A.D., mentioned ma huang.

Li Shih-chin (fl. 1500s), a physician of the Ming dynasty, gathered information on drugs in a giant encyclopedia of medicine called the *Pen-ts'ao kang-mu* or "Great Pharmacopoia." These volumes described more than 2,000 drugs and gave directions for more than 8,000 prescriptions. Li included modern processes such as distillation. Ephedra and other plants, as well as mercury, iodine, and even information on smallpox inoculations were also discussed.

For over 5,000 years ma huang was used to treat asthma and hay fever. From the genus *Ephedra,* the Chinese species is *E. sinica.* The plant itself is a straggling gymnosperm or part of the evergreen family. It appears as a low or climbing desert shrub. North American kin include a small fir and a species called the Mormon tea bush.

The dried *Ephedra* stems look very much like the end of a broom. Users put a handful of crumbled stems in a glass or enamel pot with a quart of cold water. After boiling for 20 minutes, the tea is strained. The person is advised to drink 1-2 cups every 2-4 hours as needed. Chinese *Ephedra* is also found as tinctures and capsules.

In the 1800s concern over the quality of herbs and herbal preparations received attention. The new scientific knowledge from chemistry gave impetus to isolate "essences" from the plants, and among the new products developed was a group of plant constituents called alkaloids. Between 1817 and 1898 42 alkaloids, including morphine and quinine, were found. These preparations were presented as pills, infusions, tinctures, and extracts.

Folk medicine and treatments were popular in America in the 1800s, but toward the end of that century the production of medicines came into being. Small manufacturing companies emerged that later became household names like Parke-Davis, Burroughs-Wellcome, and Squibb. The new pharmaceutical houses became concerned with quality control and producing single doses in tablet form. The role of herbal extracts generally declined. However, some new herbals did appear in the 1920s.

Scientific medicine, as it was developing in the early twentieth century, basically denied and dismissed ancient herbal medicine as superstition. However, some researchers, studying ethnobotany, saw that certain herbs were effective painkillers, anesthetics, and narcotics. These drugs included salicylic acid (aspirin), ipecac, quinine, cocaine, digitalis, and later ephedrine.

Ma huang was introduced to modern medicine by a University of Pennsylvania researcher named Carl Frederick Schmidt. Schmidt graduated with an M.D. from the University of Pennsylvania in 1918 and joined the faculty at Penn in 1920. He became interested in the emerging field of pharmacology as it applied to clinical medicine and dentistry. In 1922 his life and outlook changed when he went to teach at Peking Union Medical College in China. His China experience was a powerful eye opener. He had always assumed that the scientific medicine of the West had all the answers and that Chinese folk medicine was little more than old-time myth.

He became enthralled with how effective certain Chinese medicinal herbs were and was impressed with a Chinese colleague, Ko Kuei Chen, who pointed out the uses of many of these ancient herb remedies. When Western medicine had been imported to China by missionary doctors and others, the Chinese physicians combined the best of both worlds. They practiced Western medicine but also used the herbal remedies when necessary or advantageous.

The investigations of Schmidt and Chen led them to look at many plants, including the herb ma huang from the *Ephedra* plant. Returning to Penn in 1924, the two continued their collaboration—isolating the medicine from plants and investigating the effects on patients who suffered from bronchial asthma. The research was done in the late 1920s and published in *Ephedrine and Related Substances* in 1930.

Another step in the development of the use of ephedrine was made by Percy L. Julian (1899-1975). Julian first synthesized both ephedrine and nicotine. An African-American chemist, he sought to isolate simple compounds in natural products, understand how the body used them, and then create the synthetic compound.

Impact

Chemically, ephedrine is an alkaloid closely related to epinephrine or adrenaline. The two compounds have similar effects on the body. Ephedrine stimulates the central nervous system. In sufficient doses, it acts to constrict small blood vessels, elevates the blood pressure, and accelerates the heart rate. Ephedrine relaxes the bronchioles, the small branches of the bronchial tubes, and dilates the pupils of the eyes. Ephedrine has numerous applications in medicine. The primary treatment is for asthma, but physicians have also used it to treat low blood pressure and as an antidote to poisons that depress the central nervous system. Eye physicians may use it to dilate the pupils, but it seems to work only on light-colored eyes. The drug does have some undesirable side effects of restlessness, palpitations, and insomnia.

Schmidt continued his influence and interest in herbal research and ultimately became head of the Department of Pharmacology at Penn from 1936-59. In 1948 he was appointed to the Board of Directors of the American Bureau for Medical Aid to China. He went on to hold the office of Secretary of the International College of Pharmacologists (1954) and research director of the Aviation Medical Acceleration Laboratory in the Navy Aid Development Center.

Asthma has been a condition of age-old concern for children as well as adults. A sudden constriction of the bronchial tubes results in wheezing when one breathes out and difficulty breathing air in. Such an attack is called an "episode." The condition is often very mysterious and difficult to treat. The actual attack is caused by a tightening of the muscles that regulate the size of the bronchial tubes. Nerves normally control these muscles, but the reasons why the nerve ceases to function properly are not clear. Asthma may be an allergic reaction. Emotional stress may trigger an attack, as may exercise or respiratory infection. A case that appears with no obvious cause is called "intrinsic." Striking without warning, an intrinsic episode may require emergency treatment, without which the patient may die. Toward the end of the twentieth century, incidence of asthma in the United States was increasing, perhaps due to increased air pollution and other environmental problems.

In the last decades of the twentieth century, some herbals re-emerged. The argument that plant products were safer and more pleasant than synthetics became popular. Ephedrine received a great deal of attention as a type of street drug with "upper"-like properties, similar to amphetamine. It was also put into diet drugs. In 1997 the U.S. Food and Drug Administration (FDA) proposed strong safety measures for ephedrine as a weight-loss source. The FDA warned that ephedrine alkaloids are amphetamine-like compounds with powerful stimulant-like effects on the nervous system and heart. However, under the supervision of a physician and responsible use, ephedrine is very useful and effective for the treatment of asthma.

EVELYN B. KELLY

Further Reading

Chen, Ko Kuei and Carl F. Schmidt. *Ephedrine and Related Substances.* Baltimore: Williams and Wilkins, 1930.

Porter, Roy. *Medicine: A History of Healing.* New York: Barnes and Noble, 1997.

Porter, Roy. *The Greatest Benefit to Mankind.* New York: W.W. Norton, 1997.

New Diagnostic Tests Are Developed

Overview

Advances in medical diagnostics in the early twentieth century armed physicians with new tools and methods for diagnosing the maladies of their patients. Scientists borrowed from recent advances in physics, chemistry, and microbiology to create valuable technologies for the diagnosis of disease and injury. A half-century of war and social change prompted the development of some new tests, while meticulous planning in disease prevention resulted in the development of other diagnostic methods.

Background

Prior to 1900 the physician's main diagnostic tools were his assessment skills and his microscope. The relatively new science of bacteriology offered confirmation of some infectious diseases, although most often diagnosis was made according to the patient's symptoms and the course of the disease. The Wasserman test, designed in 1906 by German bacteriologist August von Wasserman, provided a definitive diagnosis for syphilis, a sexually transmitted disease known since ancient times. Wassermann built on newly discovered basics of immunity to create a test that showed the presence of antibodies in the blood or spinal fluid in persons infected with syphilis.

Before the Wasserman test, physicians often had difficulty making a diagnosis of syphilis until its late stages, when the patient was no longer infectious to others. The early stages of the disease caused mild symptoms for which many did not seek treatment. Early syphilis may last up to two years, and throughout this phase the disease can be transmitted, often unknowingly. This resulted in periodic epidemics of the disease throughout the world. The Wassermann test proved valuable until after World War II, when it was largely replaced with the VDRL (Venereal Disease Research Laboratory) test in 1946. Similar in technique, the VDRL test used a more specific antigen (the substance capable of inducing the formation of the antibody), and was less difficult and time consuming to perform.

German physics professor Wilhelm Konrad Roentgen (1845-1923) made one of the most important and long-lasting contributions to medical diagnostics with his discovery of X-rays. Although discovered in 1885 and instantly recognized for its potential medical benefits, the practical matters of confirming Roentgen's work, and designing, manufacturing, and installing equipment delayed X-ray's availability in most hospitals until after the turn of the century. The first X-ray facilities in the United States were located in Boston, where physicians were soon overwhelmed with patients. By the end of World War I, X-ray machinery in hospitals throughout the United States aided physicians to visualize fractures and disorders of internal structures. Roentgen was awarded the 1901 Nobel Prize for Physics for his illuminating discovery.

Electrocardiography, first introduced in 1901, provided physicians with valuable information on the conduction and performance of the heart. Dutch physician Willem Einthoven (1860-1927) developed the technique, based on the previous work of British physiologist Augustus D. Waller (1856-1922). Einthoven modified a string galvanometer (at the time galvanometers were used for underwater telegraph lines) to record the deflections that represented the contractions of the upper and lower chambers of the heart. By applying electrodes to the body, minute amounts of electrical current from the heart led the deflections to the galvanometer, where they were represented as tracings called electrocardiograms. Einthoven's first machine weighed over 600 lb (272 kg).

By 1950, electrocardiographs were small, portable, and capable of pinpointing irregularities in heart rhythm, as well as aiding in the diagnosis of heart enlargement, thyroid disease, hypertension (high blood pressure), and myocardial infarction (heart attack). For his invention of the electrocardiogram and further research on the phases of the electrical activity of the heart, Einthoven was awarded the 1924 Nobel Prize for Physiology or Medicine.

Electroencephalogram (EEG) tests provided insight into the electrical activity of the brain. First introduced in the 1920s, EEG interpretation was likened to a scientific art, requiring many years of comparative experience. Thus, other than by a few pioneering experts, EEG was not widely used until the early 1940s. EEG records the changes in electrical potential in various aspects of the brain by electrodes placed on the scalp, which are attached to an amplifier. The amplified impulses move an electromagnetic pen that records the brain waves. Physicians then screen the wave tracings and the spectral background of the EEG for particular features, such as spikes, that may indicate the presence of injury or illness. EEG was particularly helpful in the diagnosis of epilepsy, and was also used to aid in the diagnosis of brain tumors and brain dysfunction.

Greek physician Nicholas Papanicolauo (1883-1962) developed the Pap test in 1939, which assists in the diagnosis of cervical cancer. During his research, Papanicolauo observed that smears prepared from secretions and cells scraped from the area near the cervix in women with cervical cancer showed cellular abnormalities. Papanicolauo reasoned that cancerous cervical lesions could be detected by observable and measurable cellular changes while the cells were still in a pre-invasive phase. The Pap test made it possible to diagnose cervical cancer in patients with no symptoms of the disease. Moreover, such early diagnosis enabled physicians to treat patients while they were in the earliest and most treatable stages of cancer. The Pap test also meant that cervical cancer could be detected and treated before it made the deadly leap (metastasis) to other sites, most commonly the uterus.

The Minnesota Multiphasic Personality Inventory, or MMPI, was developed in the late 1930s to provide an objective measure of personality and to aid in the diagnosis of mental illness. The test consisted of a set of 550 questions, interrelated to show basic personality traits. The emphasis of the test is psychiatric, and the scores were classified in terms of psychiatric categories. The examiner determines the degree to which the subject's pattern of responses to the questions resemble that of schizophrenic patients, depressed patients, and patients with other psychiatric and major personality disorders. The test was originally designed for adults. It required at

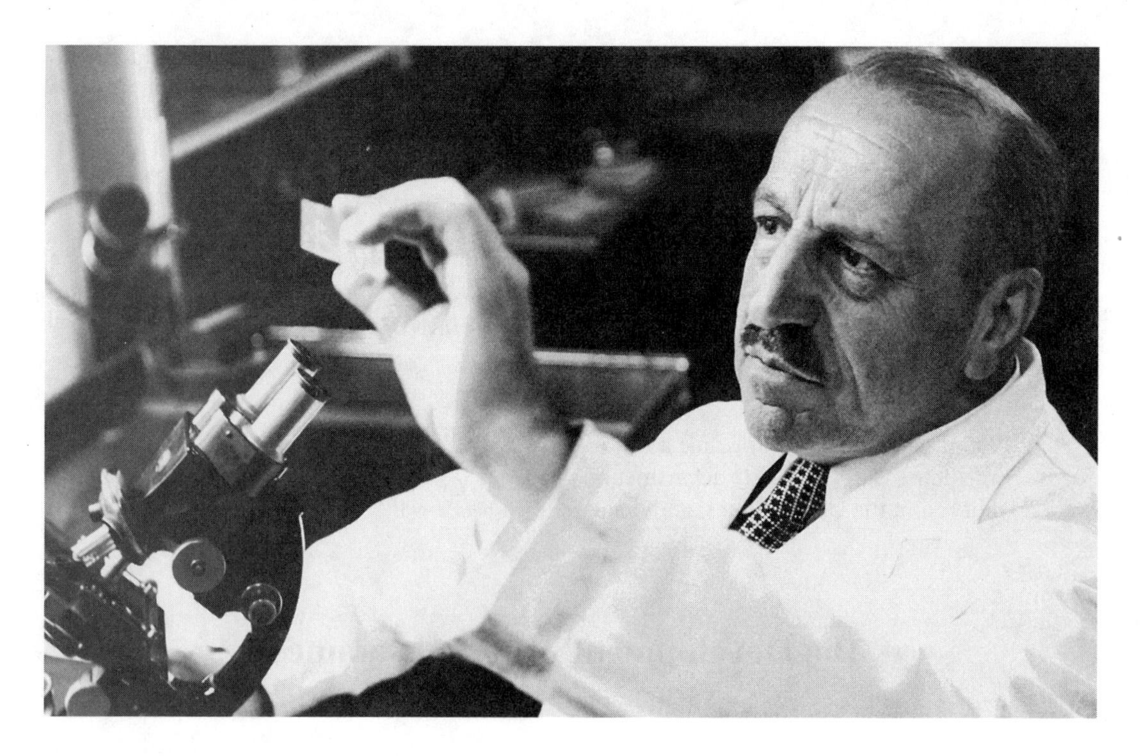

George Papanicolaou. *(Bettmann/Corbis. Reproduced with permission.)*

least a sixth grade reading level, and was not applicable to most children below the age of 13 or retarded persons. Occasionally the test was administered to teenagers, but the results were interpreted with caution. The MMPI became one of the most widely used diagnostic tests, and was revised several times.

Impact

The advent of the MMPI and its widespread popularity among clinicians reflected the climate of social upheaval and change that was characteristic of the first half of the twentieth century. The outbreak of World War I produced large numbers of psychiatric casualties. In 1904, during the Russo-Japanese War, the Russians became the first to post psychiatrists at the front lines of battle. Although Russian clinicians were the first to determine that the mental collapse associated with the stress of war was a true medical condition, Americans and western Europeans saw "shell shock" as a new phenomenon in World War I. Military psychiatry became a legitimate field of study during this period.

By World War II psychiatric testing, including tests modeled after the MMPI, was used to screen potential victims of mental collapse. During the war, however, American forces still lost over half a million men from the fighting effort due to psychiatric collapse. The rate at which soldiers became debilitated led psychiatrists to conclude that psychiatric breakdown during the extreme stresses of war was the consequence of a normal human reaction, not of cowardice. After World War II, the National Mental Health Act of 1946 led to the creation of the National Institute of Mental Health (NIMH). In 1949 NIMH was established in response to mental health problems suffered by American veterans of both World Wars.

At the beginning of the century, the Industrial Revolution brought about technological change at a rate many Americans found difficult to accept. Although Americans left rural areas by the thousands and poured into the cities to find work in the new industries, many remained distrustful of the new technologies. X-rays were of particular concern in the post-Victorian era. Companies advertised protective garments for women (at that time it was considered daring to expose an ankle) designed to protect them from the peering indecency of misused x-rays. Other new medical technologies, such as the electrocardiogram, helped to reassure Americans that industrialization contributed to helpful technology. By 1950 new technologies were eagerly anticipated and discussed by a less wary American public that viewed technological invention as further evidence of American superiority in the postwar world.

Diagnostic tests such as the Pap test ushered in a new era of preventive medicine. Since its inception, cervical cancers dropped dramatically in countries where Pap smears are routine medical practice, and the test is considered one of the greatest life-saving techniques in medical practice. The Pap test was one of the first screening tests (tests that are performed in the absence of known disease and negative results are the norm) widely used in medicine. In contrast to other medical innovations of the early twentieth century, the Pap test was inexpensive and required simple technology.

By 1950 disease prevention became a major focus in medicine. Public health departments were established in the United States, and diagnostic screening tests were incorporated into routine medical practice throughout much of the industrialized world.

BRENDA WILMOTH LERNER

Further Reading

Dejauregui, Ruth. *100 Medical Milestones That Shaped World History.* San Mateo, CA: Bluewood Books, 1999.

Handbook of Diagnostic Tests. 2nd ed. Springhouse, PA: Springhouse Publishers, 1999.

Hoffstaetter, Henrick Hooper. *Public Health: Index of New Information of Advances, Practices, Problems, Risks, and Research Development.* Washington, DC: ABBE Publishers, 1995.

The Development of Antihistamines

Overview

An antihistamine is a drug used to counteract the effects of histamine, the chemical released by certain cells in the body during an allergic reaction. Although antihistamines do not change the cause of the allergic reaction, they do suppress the symptoms associated with allergy. The groundwork for the development of antihistamines was made in the first half of the twentieth century by Swiss-Italian pharmacologist Daniel Bovet (1907-1992). Bovet's work led to the discovery and production of antihistamines for allergy relief and earned him the Nobel Prize for physiology or medicine in 1957.

Background

In some humans, the immune system perceives irritants such as pollen or animal dander to be foreign substances dangerous to the body. When these are inhaled into the body, antibodies seek out the irritant and combine with it. A large blood cell known as a basophil, or mast cell, then releases the compound histamine, which attaches itself to receptor cells in mucous membranes. Histamine then causes the local blood vessels to dilate. A drop in blood pressure and increased permeability of the vessel walls also occurs, allowing fluids from the blood to escape into surrounding tissues. These reactions are responsible for the itching, "runny nose," and "watery eyes" of a cold or hay fever, as histamines attempt to rid the body of the irritant.

While working in the therapeutic chemistry department of the Pasteur Institute in Paris, Bovet discovered that certain chemicals counteracted the reactions of histamine in guinea pigs. The first antihistamine identified was named compound F929 (thymo-ethyl-diethylamine), described by Bovet and his colleagues in 1937. Bovet observed that compound F929 protected the guinea pigs against histamine in doses high enough to be presumed lethal. F929 also lessened the symptoms of anaphylaxis, a severe allergic reaction often resulting in airway obstruction, shock, and death. Although compound F929 proved unsuitable for clinical use in humans due to weakness and toxicity, its discovery opened the door for the pursuit of a histamine-blocking agent suitable for clinical use.

By 1942 the first antihistamine successfully used to treat humans, Antergan, was developed in France. Antergan was revised to Neo-Antergan in 1944. Scientists in the United States introduced diphenhydramine and tripellinamine in 1946, both of which remained in use through the end of the century. A flurry of antihistamine development ensued, with virtually all derived from the original compound F929.

Impact

By 1950 antihistamines were mass-produced and prescribed extensively as the drug of choice for those suffering from allergies. Hailed as "wonder drugs," antihistamines were often mistakenly perceived by the public as a cure for the common cold. Although not a cure, antihistamines provided the first dependable relief for some of the cold's symptoms.

Physicians prescribed antihistamines for conditions associated in any way with allergic response, including asthma. Thus began a debate continuing long after 1950 on the benefits and risks involved with asthma and antihistamines. In 1948 medical literature showed that histamine could induce bronchial constriction in asthmatics. Antihistamines were recommended to block this effect. In 1950 asthma had proved unresponsive to the antihistamines available at that time. In 1951 scientists suggested that antihistamines would be beneficial to asthmatics, since histamine release and bronchial constriction occurred simultaneously when tissue from the lungs of allergic asthmatics was exposed to specific antigens in vitro (i.e. in a laboratory environment). By 1955 the prevailing thought was that antihistamines may actually be harmful to asthmatics by drying their lung secretions and making the secretions more viscous (thick). After years of indecision by the medical establishment, medical students were taught after 1955 not to prescribe antihistamines to patients with asthma. The debate and research into the potential benefits of antihistamines for asthmatics continued.

As different formulations were developed, scientists eventually found additional indications for the use of antihistamines. Diphenhydramine became available as a sleeping preparation and a remedy for itching. The piperazine group of antihistamines were used to prevent motion sickness. Dimenhydrinate was used as an anti-nausea medicine. The phenothiazines have a pronounced sedative effect, and hydroxizine was used as a tranquilizer. These wide-ranging properties of available drugs led to the potential for overuse and, at times, abuse of antihistamines.

Patients taking antihistamines experienced side effects that affected their daily activities of life. Studies indicated that one third of patients receiving antihistamines experienced drowsiness substantial enough to impair their concentration. While this was perhaps a desirable consequence at nighttime, it was a potential serious complication during waking hours. Antihistamine therapy became one of the first opportunities for physicians to educate patients and enlist their involvement in managing an ongoing drug therapy, rather than administer the therapy in a hospital or confine the patient to home. Patients were advised not to drive a vehicle or operate equipment requiring fine motor coordination while taking antihistamines. Patients were also encouraged to time doses according to their own schedules and body rhythms to minimize side effects, rather than adhering to a rigid schedule.

Daniel Bovet (right). *(Library of Congress. Reproduced with permission.)*

Despite the side effects, antihistamines have become widely used for a variety of medical problems. Their broad use in medicine at the turn of the twenty-first century is a testament to the importance of Bovet's discovery.

BRENDA WILMOTH LERNER

Further Reading

Bender, G. A. *Great Moments in Pharmacy*. Detroit, MI: Northwood Institute Press, 1966.

Higby, G. J. and E. C. Stroud, eds. *The Inside Story of Medicines: A Symposium*. Madison, WI: American Institute of the History of Pharmacy, 1997.

Johannes Fibiger Induces Cancer in Lab Animals and Helps Advance Cancer Research, in Particular Leading Directly to the Study of Chemical Carcinogens

Overview

Johannes Andreas Grib Fibiger (1867-1928), Danish physician, pathologist, and bacteriologist, was awarded the 1926 Nobel Prize in physiology or medicine for his research on the etiology of cancer and for his discovery of a parasite that he claimed was the cause of cancer. Fibiger called the parasite, which was a nematode worm, *Spiroptera carcinoma,* or *Spiroptera neoplastica.* The organism is now regarded as a member of the genus *Gongylonema.* Although Fibiger's work seemed to show that nematodes caused carcinoma in rodents, other researchers were unable to confirm his results. Unfortunately, the great hope expressed in the 1920s that Fibiger's research would lead to a practical solution for the cancer puzzle proved to be unfounded. Although for many years other cancer investigators rejected his work on *Spiroptera carcinoma* and his general ideas about cancer, he is regarded as one of the pioneers of randomized controlled clinical trials for his work on diphtheria serum. Changing concepts of disease and recent studies of the relationship between other pathogens and cancer has led to a reevaluation of Fibiger's place in the history of medicine.

Background

Johannes Andreas Grib Fibiger was born in Silkeborg, Denmark. His father, C. E. A. Fibiger, was a local medical practitioner and his mother, Elfride Muller, was a writer. Although his father died when Johannes Fibiger was very young, Fibiger hoped to become a physician like his father. While an undergraduate at the University of Copenhagen, Fibiger became interested in bacteriology. He was awarded his bachelor's degree in 1883 and his M.D. in 1890 from the University of Copenhagen. After a period of working in hospitals and studying under Robert Koch (1843-1910) and Emil von Behring (1854-1917) in Germany, he returned to Copenhagen. From 1891-1894 he was assistant to Professor Carl Julius Salomonsen in the Department of Bacteriology of Copenhagen University. Fibiger served as an army reserve doctor at the Hospital for Infectious Diseases in Copenhagen from 1894-1897 while completing his doctoral research on the bacteriology of diphtheria. In 1895 he received his doctorate from the University of Copenhagen for a thesis on diphtheria and returned to Germany to work at the pathological institute of Johannes Orth. In 1897 Fibiger was appointed dissection assistant for the University of Copenhagen's Institute of Pathological Anatomy. In 1900 he became professor of pathology at the University of Copenhagen. He also served as Principal of the Laboratory of Clinical Bacteriology of the Army from 1890-1905. In 1905 he became Director of the Central Laboratory of the Army and Consultant Physician to the Army Medical Service.

Although Fibiger is often referred to in cautionary terms to demonstrate how researchers can leap to incorrect conclusions, he was a careful researcher and eminent scientist. His pioneering studies of randomized controlled clinical trials demonstrate this aspect of Fibiger's legacy. The British Medical Research Council's trial of streptomycin for pulmonary tuberculosis (1948) is often referred to as the first modern randomized clinical trial. However, the study of the effect of serum treatment on diphtheria published by Fibiger in 1898 was actually the first clinical trial in which randomization was recognized as a fundamental methodological principle. The Copenhagen diphtheria trial was conducted in 1896-1897. Although the results of previous trials were ambiguous and did not seem to justify the side effects caused by serum treatment, Fibiger discovered methodological deficiencies in previous trials. Fibiger believed that a rigorous new trial was necessary: his objective was to investigate the effects of serum treatment on the mortality and morbidity of patients suffering from diphtheria. From May 1896 to May 1897 patients with diphtheria were treated with diphtheria serum or with standard treatment alone. Patients were assigned to the two groups on the basis of the day of admittance to the hospital. Serum sickness occurred in 60% of the patients. Only 8 out of 239 patients in the serum treated group died; 30 out of 245 in the control group died. Epidemiologists who have examined Fibiger's trial believe that its quality was high, even when judged by modern standards. Although few physicians followed Fibiger's excellent model in future clinical trials, his results did lead to a high demand for the new serum treatment for diphtheria.

After he was appointed Professor of Pathological Anatomy and Director of the Institute of Pathological Anatomy in 1900, Fibiger initiated a new research project in tuberculosis. He was particularly interested in the relationship between human and bovine tuberculosis. Although he was also interested in cancer, he had no experimental model that could be used to solve the problem of the origins of the cancer cell. At the time, scientists could transfer cancerous tissue from one animal to another, but they lacked techniques that could reliably induce cancer in a healthy animal. Influenced by the success of bacteriology in finding the cause of so many diseases, Fibiger hoped that similar methods might provide insights into the genesis of cancer.

Johannes Fibiger. *(Library of Congress. Reproduced with permission.)*

In 1907, while dissecting three rats that had been injected with tubercle bacilli, Fibiger found that all of these animals had papillomatous stomach cancers. Microscopic examination of the tumorous tissue revealed traces of eggs and peculiar little worms. Intrigued by these observations, Fibiger abandoned his research on tuberculosis and joined the legions of scientists who were trying to solve the problem of cancer through experimental research. His initial attempts to find similar parasites in rats failed after he had tested almost 1,000 animals. Based on reports that a tiny worm known as a nematode infested rats and cockroaches, Fibiger attempted to collect rats that subsisted largely on roaches. He found that the rats and roaches at a Copenhagen sugar refinery were infested by nematodes. About 75% of the rats were infested with the nematodes he had identified previously and 20% had stomach tumors. After isolating the parasitic nematode in question and determining that it was a new species, he passed it repeatedly from infested rats to cockroaches and then to healthy rats. Fibiger reported that these rats eventually developed tumors of the stomach. In 1913 Fibiger reported on his discovery to the Royal Danish Academy of Science. He called the nematode *Spiroptera neoplastica.* Fibiger thought that roaches became infested by eating rat excrement that contained the eggs of the parasite. Rodents became infected, or re-infected, by eating larvae-infested roaches. According to Fibiger, his work proved that cancers were caused by the nematodes or by chemicals released by the nematodes. In other words, chronic irritation by a foreign agent or toxin induced tumor formation.

Many investigators were skeptical, but some considered Fibiger's work quite ingenious and convincing. Fibiger was not the first to suggest that some parasite was the cause of cancer, but most of these other claims were based on poorly designed experiments and were rapidly discredited. Several scientists agreed with Fibiger's conclusion and published reports of other parasite-induced cancer. For example, in 1920 Frederick Dabney Bullock, M. R. Curtis, and G. L. Rohdenberg reported that liver sarcoma in rats was caused by the parasite *Cysticercus fasciolaris,* the larval stage of *Taenia Crassicolis.* As a new research assistant, future Nobel Laureate Gerty Cori (1896-1957) was assigned to work with a researcher in the Pathology Department of the State Institute for the Study of Malignant Diseases, in Buffalo, New York, who claimed that encapsulated *Amoeba histolytica* was the cause of Hodgkin's disease. By the time Fibiger was awarded the Nobel Prize, he realized that his parasites probably were not the key to the etiology of cancer. Other researchers had little or no success in duplicating Fibiger's results in their experimental animals. Some investigators found that tumors similar to those reported by Fibiger also occurred in rodents when their diet was deficient in vitamin A.

Impact

Despite the optimism and high praise expressed when Fibiger was awarded the 1926 Nobel Prize in Physiology or Medicine, his work soon subjected to sharp criticism and was then all but forgotten. Eventually, Fibiger himself suggested that some external agents might generate cancer in genetically susceptible individuals. He believed that the lasting value of his work was the demonstration that "spontaneous" cancer (tumors that were not transplanted) could be routinely produced in laboratory animals under appropriate conditions. Certainly, the hope and example generated by Fibiger's work helped open up new approaches to the experimental study of cancer. Although many lines of research had been attempted in the early twentieth century, few observers were willing to claim that any results had been of significant practical value. During the early twentieth century, older clinical observations of the development of skin cancer, such as Percival Potts's (1714-1788) classic report of an occupational cancer, i.e., cancer of the scrotum in chimney sweeps, were confirmed and extended by laboratory research. During the nineteenth century several studies had linked cancer to external causes. Skin cancers had been associated with tar and paraffin, cancer of the lung was thought to be an occupation disease of miners, and aniline was found to cause various cancers. In 1916 Katsusaburo Yamagiwa (1863-1930) demonstrated that repeated applications of tar led to the production of skin cancer in laboratory rabbits. This extended knowledge of the etiological relationship between the incidence of tumors and exposure to certain chemicals. Exposure to x rays and radium among researchers was soon associated with the induction of cancers. Fibiger died of a heart attack shortly after he was diagnosed with colon cancer and only six weeks after participating in the Nobel Prize ceremonies.

Perhaps somewhat embarrassed by the lack of confirmation of Fibiger's cancer hypothesis, the Nobel Foundation made no other awards for cancer research for 40 years after honoring Fibiger. In 1966 the Nobel Prize was awarded to Francis Peyton Rous (1879-1970), who discovered that a virus might cause cancer, and Charles Huggins (1901-1997), who introduced hormonal treatment for prostate cancer. Subsequent recognition of the relationship between infections and certain malignant diseases has created new interest in Fibiger's work. Some researchers suggested that the tumors observed in Fibiger's experimental animals might have been caused by a virus carried by the parasite worm, even though the worm itself was not the cause of cancer. Nematodes are known to serve as carriers of various disease agents. Pathogens have been linked to other forms of cancer; for example, *Schistosoma haematobium* is associated with bilharzia cancer of the bladder, *Spirocerca lupi* appears to cause malignant neoplasms in dogs, and *Helicobacter pylori* has been associated with gastric ulcers and stomach cancer in humans.

LOIS N. MAGNER

Further Reading

Periodical Articles

Campbell, William C. "The Worm and the Tumor: Reflections on Fibiger's Nobel Prize." *Perspective in Biology and Medicine* 40, no. 4 (summer 1997): 498-506.

Clemmesen, J. Johannes Fibiger. "Gongylonema and Vitamin A in Carcinogenesis." *Acta Pathologica Microbiology Scandinavia* supplement 270 (1978): 1-13.

Hrobjartsson, Asbjorn; Gotzsche, Peter C; Gluud, Christian. "The Controlled Clinical Trial Turns 100 Years: Fibiger's Trial of Serum Treatment of Diphtheria." *British Medical Journal* 317 (October 31, 1998): 1243-1246.

MacKenzie, Debora. "Worming a Way In." *New Scientist* 161 (Jan. 30, 1999): 14.

Raju, Tonse N. K. "The Nobel Chronicles 1926: Johannes Andreas Grib Fibiger (1867-1928)." *The Lancet* 352 (November 14, 1998): 1635.

Stolley, P. D., Lasky, T. "Johannes Fibiger and His Nobel Prize for the Hypothesis That a Worm Causes Stomach Cancer." *Annals of Internal Medicine* 116 (1992): 765-769.

Weise, Allen B. "Barry Marshall and the Resurrection of Johannes Fibiger." *Hospital Practice* 31, no. 9 (Sept. 15, 1996): 105-113.

Other

Fibiger, Johannes. "Investigations on *Spiropter carcinoma* and the Experimental Induction of Cancer." Nobel Lecture, December 12, 1927. Reprinted in *Nobel Lectures in Physiology or Medicine*, vol. 2 (1922-1941). Amsterdam: Elsevier, 1965: 122-152.

Introduction of Electroshock Therapy

Overview

Electroshock was one of a number of "antagonist therapies" introduced in the early part of the twentieth century. Based on a belief that epilepsy and schizophrenic disorders had opposite effects on human brain anatomy, a number of physicians tried to relieve the symptoms of serious mental illness by inducing convulsions, first by chemical means and then by passing a current through the brain. Electroshock was opposed by the advocates of "talk" psychotherapy, and was largely supplanted by new tranquilizing drugs after 1950. In its modern form, drugs are used to eliminate the actual convulsion, and memory loss is minimal, so that it has become more acceptable for patients who do not respond to other treatments. The mechanism by which electroshock acts on the brain is still poorly understood.

Background

The humane treatment of the mentally ill made great strides in the nineteenth century, representing both the efforts of social reformers and a degree of progress in medical knowledge. By the year 1900, there were asylums for the mentally ill in most industrialized countries, and a number of physicians were restricting their practice to psychiatry, the treatment of mental illness. Early psychiatry recognized three major mental illnesses: dementia paralytica, now called neurosyphilis, the third stage of the venereal disease syphilis; dementia praecox, now called schizophrenia; and manic-depressive disorder, which today would include bipolar disorder and severe clinical depression. In 1900, the largest portion of hospitalized mental patients were being treated for neurosyphilis.

The microorganism causing syphilis, the spirochete *Treponema pallidum*, had been discovered in 1905, and a treatment effective against the early stages of the disease—injection of the arsenic-containing drug salvarsan—had been introduced in 1910 by the German medical researcher Dr. Paul Ehrlich (1854-1915). Prospects for those patients in which the disease had advanced to the third stage, however, remained bleak. Drawing on the discovery by the French chemist Louis Pasteur (1822-1895), that heat could kill many kinds of bacteria, and reports that mental patients sometimes improved after recovering from fever-producing illness, there was speculation that neurosyphilis patients could be cured by deliberately inducing fever.

With the discovery that quinine provided a cure for malaria, Professor Julius Wagner-Jauregg (1857-1940), a member of the psychiatric faculty of the University of Vienna, began injecting neurosyphilis patients with the malaria parasite, allowing the patients to experience recurring fevers for about a month before treating them with quinine. In 1918, he was able to report that of an original experimental group of patients, two-thirds experienced improvement, while none deteriorated. With this work, for which Wagner-Jauregg received the Nobel Prize in 1927, began a major effort to discover other "antagonist therapies" in which one disease might cure another.

Anecdotal reports also existed about schizophrenic patients who improved after experiencing epileptic-like seizures caused by head injury or infection. Experiments were done on epileptic patients, injecting them with the blood of schizophrenics, but without success.

Ladislas Meduna (1896-1964), a Hungarian physician, believed that he had discovered differences in brain anatomy between schizophrenics and epileptics in the 1930s. The brains of schizophrenic patients were found to have fewer cells of a certain cell type, called neuroglia, than normal brains, while the brains of epileptics were found to contain an excess of these same cells. Meduna reasoned that inducing seizures might stimulate the brain to form additional neuroglia. After experimenting on animals to find a safe method of producing seizures, he began inducing seizures in his patients by the injection of a solution of camphor in oil. He later chose the drug Metrazol to induce seizures. In 1937 he was able to report improvement in roughly half of the 110 severely ill patients on whom he had tried the procedure.

Drug-induced seizures were, however, a terrifying experience for the patient, who experienced extreme anxiety prior to loss of consciousness. The use of electrical shocks, with current passing through the brain, to induce convulsions was introduced on April 28, 1938, by Italian psychiatrists Ugo Cerletti (1877-1963) and Luigi Bini (1908-). These physicians sought to eliminate the unpleasant preliminary effects found with drugs. Like Meduna, they did not begin experiments on humans until they had a seizure-inducing procedure that worked safely when applied to animals. A 39-year-old man suffering from a psychotic episode was the first patient. After three weeks of treatments on alternate days, the patient was considered to have recovered. By 1940, both electroshock and drug-induced shock therapies were being used on hundreds of patients a year.

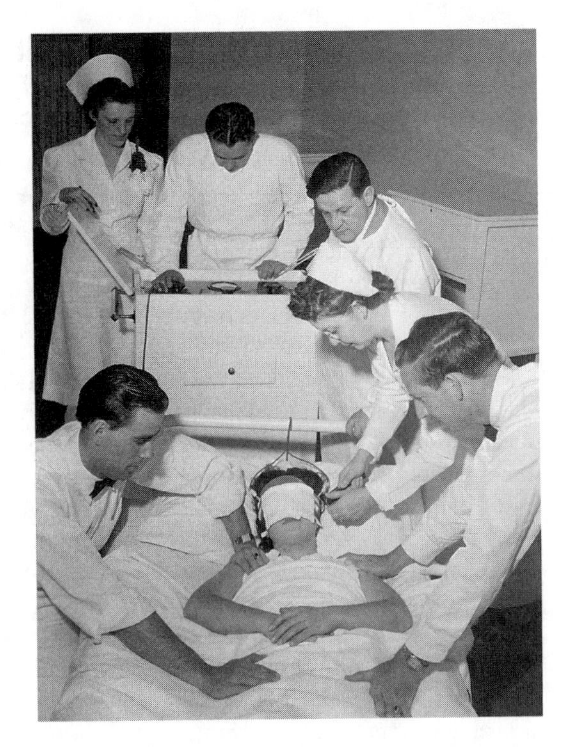

In this 1942 photograph doctors attempt to "cure" a patient suffering from manic depression using electroshock therapy. *(Bettmann/Corbis. Reproduced with permission.)*

Impact

The early twentieth century was indeed a period of experimentation with drastic therapies for the seriously mentally ill. Also originating in this period were insulin shock or insulin coma therapy, in which the patient is rendered unconscious by a high dose of the hormone insulin, and the operation called prefrontal lobotomy, in which some of the connections between the brain's frontal lobe and the remainder of the brain are severed. While lobotomized patients might cease to exhibit some undesirable behaviors, they would often exhibit personality changes and a reduced ability to cope with social situations. As a result, such operations have largely been abandoned today.

An alternative to the drastic physical interventions of shock therapy and lobotomy appeared to be offered by psychoanalysis, a technique introduced by the Austrian neurologist Sigmund Freud (1856-1939), and the many variations of "talk" psychotherapy that followed. Freud theorized that mental illness, except that resulting from infection or injury to the brain, was caused by the patient having repressed memories or desires, often of a sexual nature. Using a process of free association, the patient would be guided to make those repressed thoughts conscious, whereupon the troubling symptoms would often disappear. Having demonstrated that many serious psychological problems could be treated without medical intervention, psychotherapists, even those with medical training, tended to consider the more drastic treatments primitive and unnecessary. In addition, the potential of abuse inherent in the drastic therapies and brain surgery was increasingly appreciated after the treatment of political dissidents as insane in Nazi Germany and Stalin's Soviet Union became known.

The nature of the relationship between mental phenomena and the physical structure that exhibits them, the so-called mind-body, or mind-brain, problem, has stimulated philosophical discussions since the ancient Greeks. For several centuries the dominant position in Western thinking had been that of Cartesian dualism, first proposed by the great French Philosopher Rene Descartes (1596-1650), in which the mind was of a separate nature yet connected to the brain and body in some fashion. As understanding of the brain improved through experiments on animals and the observation of individuals with injured brains, more, but not all, scientists and philosophers have tended to accept the position that "the mind is what the brain does." The psychotherapists generally followed the dualist position. The advocates of the more drastic therapies, of course, were less inclined to the dualist view.

A third alternative for the treatment of serious mental illness was provided by the development of psychotropic drugs in the years following the Second World War. With the proper medications many mental patients became able to function in society. During the 1960s and 1970s many state governments passed laws restricting the practice of electroshock, which was opposed by therapists and the pharmaceutical industry. Opposition to the drastic therapies was also expressed in the popular media. In the 1975 film *One Flew over the Cuckoo's Nest,* the main character is subjected to repeated electroshock therapies and then surgery by the sadistic medical staff.

Since the 1970s, however, electroshock, more often called electroconvulsive therapy, has become a bit more acceptable in the United States, particularly in hospitals affiliated with medical schools. Its use outside the United States varies from one country to another. The fourth edition of the *Diagnostic and Statistical Manual*, published by the American Psychiatric Association in 1994, lists 18 diagnoses in which electroconvulsive therapy is believed to be effective. The illnesses listed are mainly forms of depression and schizophrenia, although a category of atypical psychosis is included. Generally, electroshock is not given unless a signed consent is obtained from the patient or (rarely) after a court has given permission. The consent form includes a description of the procedure and the likely number of repetitions, and informs the patient that he or she may discontinue treatment at any time. In modern practice the patient is sedated and given pure oxygen to breathe for the period preceding surgery. He or she is also given a muscle-relaxing drug so that the seizure occurs without violent movements. Vital functions, including brain wave activity, are continuously monitored. Modern practitioners claim that there is no significant memory loss or other injury with the current procedure.

The mechanism of electroshock therapy is not well understood, and there is little financial support for research on its use. Meduna's theory is no longer taken seriously. Current speculation centers on the release of hormones from the hypothalamus and pituitary gland, which have been shown to appear in the fluid surrounding the brain within a few minutes following the induced seizure. With a better understanding of the effect, it is possible that even less drastic forms of the therapy will be developed.

DONALD R. FRANCESCHETTI

Further Reading

Books

Fink, Max. *Electroshock: Restoring the Mind.* New York: Oxford University Press, 1999.

Porter, Roy. *The Greatest Benefit to Mankind: A Medical History of Humanity.* New York: Norton, 1997.

Shorter, Edward. *A History of Psychiatry.* New York: Wiley, 1997.

Advances in Surgical Techniques

Overview

Surgeons of the twentieth century inherited many ideas and techniques from earlier physicians, which they continued to investigate and improve. Three age-old problems that plagued surgeons—pain, infection, and shock—were beginning to be conquered. Surgeons also ventured into areas of the body, like the heart, that had previously been off-limits and developed unique approaches for their particular specialty.

During the first half of the twentieth century, many doctors played critical roles in developing surgical techniques. At the beginning of the century, Alexis Carrel (1873-1944), practicing on paper with fine needles and silk thread, introduced a method for linking blood vessels. Harvey Cushing (1869-1939) was a pioneer in neurosurgery. Ernst Sauerbruch (1875-1951) and Louis Rehn (1849-1930) first worked with the chest and esophageal surgery, with Rehn performing the first pericardectomy. Evarts Graham (1883-1957) and Jacob Singer (1885-1954) specialized in lung surgery.

Many unnamed surgeons worked to perfect anesthesia, asepsis against infection, pain management, as well as sophisticated blood transfusions. By mid-century, medicine was at a high peak with hope for many additional breakthroughs.

Background

Though surgeons have become highly respected members of the medical community, many doctors were involved in conquering the great enemy of invasive surgery—infection. Joseph Lister (1827-1912) had found that antisepsis, using carbolic acid or phenol, sterilized wounds and equipment. Instruments were put into the solution, and the surgeon may also don a gown soaked in the chemical then wrung out. In a Berlin clinic in 1886 Ernst von Bergmann (1836-1907) introduced the steam sterilizer. The idea that everything that comes into contact with the wound must be sterilized was being developed. Hermann Kummell of Hamburg invented the process of scrubbing up while

EARLY FEMALE ANESTHESIOLOGISTS IN THE UNITED STATES

Women moved into positions of prominence in anesthesiology in the early twentieth century, just as the specialty began taking shape in the United States. In 1899 Dr. Isabella Herb joined the staff at the Mayo Clinic in Rochester, Minnesota, as the first physician anesthetist at that institution; previously, anesthesia had been administered by nurses or other staff. By 1905 she had moved to Chicago, and four years later became chief anesthetist at Presbyterian Hospital and Rush Medical College, remaining in that position until her retirement in 1941. In addition to holding various other offices, she served in 1922 as the tenth overall and first female president of the American Association of Anesthetists. Another female anesthesiologist in the Chicago area was Dr. Huberta M. Livingstone. After completing medical school and a year of internship, she became Director of the Department of Anesthesiology at the University of Chicago's Pritzker School of Medicine in 1929 and held that position until resigning in 1952.

Dr. Alice McNeal spent the first part of her anesthesiology career in Chicago, working at Presbyterian Hospital from 1925 until 1946. In that year she moved south to Birmingham, Alabama, and became Chief of the Division of Anesthesia in the Surgery Department of the newly created University of Alabama School of Medicine. Two years later she was named Chairman of the new Department of Anesthesiology in the medical school and retained that position until her retirement in 1961. Dr. McNeal's case load at the University Hospital in Birmingham is typical of how busy physician anesthesiologists of this era—both male and female—could be. She was the only full-time faculty member in her department for many years, and had just a few nurses, medical students rotating through the department, and occasional assistance from local private-practice anesthesiologists to help her administer anesthesia for thousands of cases each year.

On the West Coast, Dr. Mary Botsford joined the staff of Children's Hospital in San Francisco in the late 1890s. She later was appointed the first Professor of Anesthesia at the University of California at San Francisco, where she trained more than 40 female anesthesiologists. Dr. Botsford remained in those clinical and academic positions until her retirement in 1920. She also served a term as President of the Associated Anesthetists of the U.S. and Canada, as did one of her trainees, Dr. Eleanor Seymour of Los Angeles.

A. J. WRIGHT

William Stewart Halstead (1852-1922) of Johns Hopkins ordered special rubber gloves just for operating. In 1896 the gauze mask was used.

At the end of the nineteenth century the master surgeon often performed in an arena. Students sat in rows and observed as he operated on the patient—explaining exactly what he was doing as he proceeded.

Theodor Billroth (1829-1894), the great nineteenth-century surgeon, laid the foundation for many improvements. He developed the concept of experimental surgery by taking problems to the laboratory before going to the operating theater. He developed procedures for recording and reporting the operative procedure, studying complications and mortality, and conducting five-year follow-ups. As a great teacher, he influenced a large number of German and Austrian surgeons. However, the aseptic tradition caught on slowly.

Impact

In 1894, in Lyons, France, Alexis Carrel witnessed the stabbing of the French president, who subsequently bled to death because there were no procedures to repair blood vessels. The son of a silk merchant, Carrel practiced sewing on paper with a fine needle and silk thread until he had perfected the technique to repair blood vessels. In 1906 Carrel moved to New York to the Rockefeller Institute, where he worked to perfect his blood vessel suturing techniques. Carrel's innovative and patient thinking on suturing blood vessels led to a time when blood transfusions and transplants would be a reality. He even did a successful kidney transplant from one dog to another.

Billroth, the famous surgeon of Vienna, influenced many students, including American William Stewart Halstead (1852-1922). Halstead became a top surgeon until he began to experiment with cocaine as an anesthetic and became addicted. However, he had a remarkable career during which he trained surgeons in the philosophy of "safe surgery." His great pupil was neurosurgeon Harvey Cushing.

The science of neurosurgery was revolutionized by the work of Cushing. In 1902 Cushing became a surgeon at Johns Hopkins and began to specialize in neurosurgery. He developed many procedures and techniques basic to brain surgery and greatly reduced mortality connected with such serious treatments.

Cushing probed deeper into the brain. He developed a technique used in the battlefields of

World War I in which he applied gentle suction to injured the brain and even developed a magnet to remove bits of metal embedded in the brain. He became an expert on brain tumors and pioneered techniques for their diagnosis and treatment. He was the first to find and understand the role of the pituitary gland, located at the base of the brain. His efforts and numerous writings on surgery won him international recognition in 1912.

The use of anesthesia at the turn of the twentieth century progressed slowly. Since there were no specialists in the field, a surgeon would have to get a nurse, a student, or anyone around to put chloroform over a rag to knock the person out. Chloroform and ether were both popular. Chloroform was easier to administer, although it had a deadly effect on the heart. By 1910 nitrous oxide (laughing gas), combined with ether, was replacing chloroform. With the discovery of other agents, anesthesia came into its own.

The 1920s brought great improvements in anesthesia, which broadened the possibilities of surgery. Ralph Waters of Madison, Wisconsin, introduced the general anesthesia cyclopropane in 1933. John Lundy of the Mayo Clinic, building upon the trials of many others, successfully introduced pentothal or thiopental sodium to put patients peacefully to sleep. Howard Griffith and G. Enid Johnson of Montreal purified curare to inject it into the muscles for a local anesthesia. The development of the specialty of anesthesiology assisted the surgeon, as now he or she could concentrate on operative skills with another specialists taking care of respiration and vital signs.

The progress in anesthesia certainly aided the chest or thoracic surgeon. When the chest cavity was opened, the lung collapses. Tuberculosis, or the White Plague, was the scourge of lungs. Carlo Forlianini (1847-1918) of Pavia, Italy, had attempted the first pneumothorax surgery in 1888. The surgery was unsuccessful because of the collapse of the lungs when the chest was opened. To overcome the problem, the Prussian surgeon Ernst Sauerbruch designed a negative pressure cabinet to control the collapse. His father had died of tuberculosis, and he was determined to conquer the technique.

Experimenting with animals in pressurized cages, he found a way to operate while the animal continued to breathe. Sauerbruch built and demonstrated this cabinet in 1904 at a clinic in Breslau. The devise held patient, table, and a full operating team. By the 1920s he found a new way to establish a temporary collapse by nitrogen displacement or pneumothorax. Although it was a skillful operation, the pneumothorax became obsolete when inhalation under pressure was introduced after World War II.

Thoracic surgery still had many problems to overcome. In San Francisco surgeon Harold Brunn successfully used suction after surgery to keep the lung cavity free of fluid until the remaining lobes could expand.

When Lister first started his operations along with his protegee William Watson Cheyne (1852-1932), about 60% of their practice involved accidents, broken bones, and superficial tumors. They had excellent results, but up until 1893 they tried few intestinal operations. Then Cheyne expanded his practice into a relatively unknown area of abdominal surgery. As a result, the surgeon's status increased.

William Mayo (1861-1939) and Charles Mayo (1865-1939) became masters of abdominal and thyroid surgery, making the Mayos household names and also millionaires. The Mayo Clinic in Rochester, Minnesota, is world renowned.

Shock was one of the most difficult problems to conquer. Understanding of shock is very elusive, but it was known to be caused by a loss of blood. Carrel, with his surgical techniques of sewing blood vessels, enabled transfusions. By developing knowledge of blood typing and blood factors, Karl Landsteiner (1868-1943) added to making successful transfusions.

The heart had been an off-limits area among medical practitioners. Ludwig Rehn of Frankfurt performed the first cardiac surgery in 1896. He showed that wounds of the heart could be treated successfully by draining the infection. In 1912 James Herrick (1861-1954), a Chicago physician, insisted that heart attacks were caused by blood clots and were indeed survivable. In 1913 Rehn and Sauerbruch performed surgery on the pericardium for relief of pericarditis or inflammation. However, these procedures did not gain wide acceptance.

Experimental work on the heart was being done in the first two decades of the twentieth century. Theodore Tuffier operated on the aortic valve and Carrel had worked diligently to perfect the technique. In 1925 Henry Souttar used a finger to dilate a mitral valve, a feat that was 25 years ahead of its time. There was still resistance to heart surgery. It took the experience of World War II, and the remarkable work of Dwight

Harden, who removed 134 missiles from the chests of wounded soldiers without losing a single patient. The surgery was performed "blind," as it was not until 1953 that the heart-lung machine, which could stop the heart so the surgeon could see the procedure, was realized. However, the machine had been the dream of John Gibbon Jr. (1903-1973), who began the research on this machine in 1937.

Operations were developed for many forms of cancer. Evarts Graham did a lung resection in 1933. The patient, a doctor, was alive at the time of Graham's death in 1957. He also tackled a lobectomy (removing a lung lobe), a segmental resection (removing only part of a lobe), and a pulmectomy (taking out a whole lung).

The pioneers of surgery during the first half of the twentieth century opened the door for many of the procedures used today.

EVELYN B. KELLY

Further Reading

Klaidman, Stephen. *Saving the Heart: The Battle to Conquer Coronary Disease.* New York: Oxford Press, 2000.

Wolfe, Richard J., and Leonard F. Menczer, eds. *I Awaken to Glory.* Boston: Boston Medical Library, 1994.

Zimmerman, Leo, and Ilza Weith. *Great Ideas in the History of Surgery.* San Francisco: Norman, 1961.

The Treatment of Blue Baby Syndrome

Overview

In 1944 a small, frail child was wheeled into the operating room at Johns Hopkins Hospital in Baltimore. As the doctors prepared for surgery, the team in the operating room looked at the cyanotic (blue) 15-month-old girl, who hovered close to death. The operation she was about to undergo would be the first attempt to treat her congenital heart condition, which was called the tetralogy of Fallot or blue baby syndrome. The team consisted of surgeon Alfred Blalock (1899-1964), pediatric cardiologist Helen B. Taussig (1898-1986), and surgical technician Vivien T. Thomas (1910-1985).

The groundbreaking surgery took place on November 29, 1944. The operation joined an artery leading from the heart to an artery leading to the lungs, giving the sick child a vital oxygen supply and taking the necessary first step toward a complete surgical cure. The success of this procedure, known as the Blalock-Taussig shunt, made medical history.

Background

Even in the early 1900s the heart itself was thought to be inoperable, although medical procedures had improved to allow surgery on the arteries surrounding it. Alexis Carrel (1873-1948) had shown that a piece of aorta wall could be replaced with a fragment of another artery or vein. Carrel's observation of a lacemaker had helped him develop an effective way to sew vessels together, a process called anastomosis; in 1920 he transplanted an entire vessel by sewing the ends together. Robert Gross (1905-1988) of Boston performed the first successful surgery on one of the large vessels. All of these could be carried out while the heart was beating.

Congenital (present at birth) heart problems originate early in fetal development, probably within the first two months. The defect has little effect while the baby is in the womb because circulation and respiration are accomplished thorough the umbilical cord attached to the mother's placenta. However, at birth, when the child must use its own respiratory and circulatory systems, the deformities become life threatening. Before surgical intervention was possible, children who were born with heart abnormalities usually died or suffered severe disabilities.

The tetralogy of Fallot (ToF) or blue baby syndrome is a congenital condition. It was named for Etienne-Louis-Arthur Fallot, who first described the defect and its four components in 1888. The condition is outwardly characterized by a bluish discoloration of the skin called cyanosis, and accompanied by difficult breathing, and digital clubbing (short, blunt fingers).

Dr. Fallot identified the set of four (tetralogy) anatomical defects that comprised the syndrome after studying autopsies and postmortems performed on blue babies:

> The pulmonary valve, between the heart and pulmonary artery, which enables the blood to pick up oxygen, was narrowed. This is called infundibular pulmonary stenosis.

The right ventricles were swollen or enlarged (from trying to pump blood through the narrow valve).

The heart had a large ventricular septal defect (VSD), literally a hole in the wall between the two ventricles of the heart; this allowed oxygenated and deoxygenated blood to mix.

The aorta (the main artery that sends blood to the rest of the body), normally on the left, had shifted to the right, sending blood from both sides into systemic circulation.

To understand why ToF is so lethal, you must first understand how a normal heart works: The heart muscle has an atrium and a ventricle on each side. The atria are the receiving chambers, the ventricles are pumps. The atria and ventricles are each divided into two chambers by a partition or wall called a *septum*. The wall between the two chambers of the atria is called the interatrial septum; the wall between the ventricles is called the interventricular septum.

The right atrium receives blood that has circulated through the body. Because this blood is deoxygenated (without oxygen) and also contains waste and carbon dioxide, it appears dark. (This is why veins appear bluish when viewed through the skin.) The blood then goes into the right ventricle where it is pumped to the lungs through the pulmonary artery. Here it picks up oxygen and gives off carbon dioxide. It travels through the pulmonary vein to the left atrium. This is called *pulmonary circulation*. The blood then flows from the left atrium to the left ventricle where, now rich with oxygen and bright red, it goes through the aorta, into the arteries, and out to the cells of the body. This is called *systemic circulation*.

Babies born with ToF are cyanotic, an indication their blood is not sufficiently oxygenated. Before surgical intervention became possible, one-fourth of children born with ToF died before their first birthday. Within 10 years 70% had died; only 5% survived past the age of 40. When the careers of three physicians came together in 1944, the procedure they performed would save many lives. The unique preparation of Blalock, Taussig, and Thomas allowed them to each contribute their talents.

Impact

After Blalock graduated from Johns Hopkins Medical School in 1922, he went to Vanderbilt University Hospital in Nashville as a resident in

Helen Brooke Taussig. *(AP/Wide World Photos. Reproduced with permission.)*

surgery. His research led him into pioneering work on how to overcome shock when much blood is lost. While working on shock, he created different conditions in dogs. In 1938 he did an experiment in which he joined the left subclavian artery to the left pulmonary artery to produce pulmonary hypertension. (The subclavian artery, at the back of the neck, supplies blood to the arm. The joining of the arteries is referred to as anastomosis.) The experiment failed and Blalock put aside his idea.

In 1941 he returned to Johns Hopkins as surgeon in chief and director of the department of surgery at the medical school. While there, he continued to work on a shunt technique that could be used to bypass an obstruction (coarctation) in the pulmonary artery. While at Vanderbilt in 1930, Blalock had hired Vivien Thomas, a young African-American who had been forced to leave college by a lack of funds. Thomas got a job in Blalock's surgery laboratory and soon became his right-hand surgical assistant. He learned to perform surgery, make chemical determinations, and keep precise records. When Blalock went to Baltimore, he took Thomas with him and the two continued their collaboration in research for 30 more years.

In 1930, Helen B. Taussig was appointed to head the pediatric cardiac unit at John Hopkins, the Harriet Lane House. She had been denied entry into Harvard Medical School because of her gender, but had been accepted at Johns Hopkins. Her studies of congenital heart conditions led her to discover the single-biggest problem with ToF: the lack of blood flow to the lungs.

One day in 1943, Taussig overheard a conversation between Blalock and a colleague about the difficulty of cross-clamping the aorta to repair a coarctation. The colleague wondered aloud if it would be possible to anastomose the carotid artery to the aorta below the obstruction. Taussig, inspired, then joined the conversation, asking, if that were possible, why couldn't you also put the subclavian artery into the pulmonary artery, and repair the biggest physical defect in blue babies?

Bells went off in Blalock's mind. He remembered the experiment he had done at Vanderbilt. He and Thomas went right to work to retry the experiment. Taussig believed that the pulmonary artery could be repaired and attached to a new vessel. She persuaded Blalock to work on the problem. During the 18 months, Blalock and Thomas worked on technical procedures using 200 experimental dogs.

Eileen Saxon was a desperately ill child in 1944, and it was obvious she would die if nothing were done. For Thomas the surgery came sooner than he wished because he felt the procedure was not yet perfected. But they had a dying child who could not wait. They wheeled her into surgery on November 19th. With Taussig as consultant and Thomas giving advice on technicalities and suturing, Blalock did the surgery. He took a branch of the aorta that normally went to the arms and connected it to the lungs. The child who had been so much at risk survived and gradually regained the normal color. After two weeks it was obvious the procedure was a success. Second and third operations were performed, and the success of each was apparent. Taussig described the third patient as "an utterly miserable six-year-old who could no longer walk." What a thrill it was for her to see the little boy with deep purple lips return to a happy, normal child.

In 1945 Blalock and Taussig published reports of their three successful operations in the *Journal of the American Medical Association*. The news caused a worldwide stir in the scientific community. Doctors from all over the world came to Johns Hopkins to learn their techniques. The operation became known as the Blalock-Taussig Shunt. While further operations were still necessary to repair the remaining ToF defects (the hole in the ventricular wall and narrowed pulmonary valve) patients were well on the road to a normal childhood.

The shunt is now widely performed and has saved thousands of children. The isolation and division of the subclavian artery, a time-consuming and difficult procedure, is rarely performed today. A polytetrafluoroethylene (PTFE) tube is most often used to create the shunt. The tube is sutured to the subclavian artery on one side, and the pulmonary artery on the other. The same effect is achieved but neither artery is interrupted; the risk of damaging nerves and disrupting blood supply to the arm is also lessened.

The feat of Blalock, Taussig, and Thomas is heralded as one of the great achievements of twentieth-century medicine. The three were recognized throughout the world and had numerous honors bestowed upon them.

EVELYN B. KELLY

Further Reading

Taussig, Helen, with Joyce Baldwin. *To Heal the Heart of a Child: Helen Taussig, M.D.* New York: Walker, 1992.

Thomas, Vivien T. *Partners of the Heart: Vivien Thomas and His Work With Alfred Blalock*. Philadelphia: University of Pennsylvania Press, 1998.

The Development of Modern Hearing Aids

Overview

Artificial aids to hearing may have been developed in ancient times to help hunters, warriors, and sailors send messages over long distances. Eventually, similar devises were adapted to help individuals who had suffered hearing loss. Deafness can be caused by accidents, disease, or aging and the loss of sensitivity to sound can vary from partial to severe. A hearing aid can be thought of as a miniature amplifier, that is, a device that increases the level of sound for the user. The history of aids to hearing includes various kinds of ear trumpets, tubes, bone-conduction devices, and electric aids. Modern hearing aids use electric amplifiers to boost the original sound.

Background

The technology of modern medicine, including diagnostic devices and prosthetic devices, reflects the profound changes that have occurred in nineteenth- and twentieth-century medicine. By 1900, diagnostic instruments were being used on apparently healthy people to detect early signs of disease or as a means of detecting hidden disorders, such as impairment of vision or hearing. The foundations of the modern science of acoustics were established by Herman Ludwig Ferdinand von Helmholtz (1821-1894) in his classic treatise, *On the Sensations of Tone* (1862). Important twentieth-century contributions to understanding and measuring hearing loss were made by Robert Bárány (1876-1936) and Georg von Békésy (1899-1972). Bárány won the Nobel Prize for Physiology or Medicine in 1914 for his contributions to otology, the field of medicine that deals with the study, diagnosis, and treatment of diseases of the ear and related structures. His research on the equilibrium system, based in the vestibular apparatus of the ear, led to the establishment of valuable clinical techniques and tests, such as the pointing test for the localization of circumscribed cerebellar lesions and the caloric test for labyrinthine function. Bárány's syndrome was named in honor of his observations of a condition characterized by unilateral deafness, vertigo, and pain in the occipital region. Békésy was awarded the Nobel Prize in 1961 for research in physiological acoustics, especially his discoveries about the physical mechanism of stimulation within the cochlea. As a result of his basic research, he developed a new audiometer, the semi-automatic Békésy audiometer, in 1947. Also important during the early twentieth century were improvements in tests for deafness and hearing loss, including early assessment of deafness in infants.

Alterations in the functioning of the ear caused by accidents, disease, or aging can result in deafness. The loss of hearing can be partial, but in some cases sensitivity to sound is totally lost. Compensation for some forms of hearing loss can sometimes be accomplished by decreasing the distance between the hearer and the speaker. The oldest method of increasing hearing—and reducing the distraction of background noises—is to cup the hand behind the ear. This primitive approach actually increases sound energy by about 5-10 decibels. The speaker can increase the level of sound by speaking more loudly or by using a mechanical device. Artificial aids to hearing may have been developed in ancient times to help hunters, warriors, and sailors send messages over long distances. Although speaking tubes of various sizes and shapes are apparently quite old, the first written description of the use of ear trumpets and other devices as aids to the deaf appeared in the seventeenth century. By the eighteenth century, some surgeons were attempting surgical interventions for diseases of the ear.

A hearing aid is a device that increases the level of sound for the user. Various devices can be used to collect more sound energy and direct it to the auditory canal or sound vibrations can be directed into the skull. The history of hearing aids includes various kinds of trumpets, tubes, bone-conduction devices, and electric aids. Speaking-listening tubes channel sound directly from the speaker to the listener. These devices were shaped like bells and trumpets. Depending on their size and shape, ear trumpets could amplify sound by about 10-20 decibels, but much of the normal range of human speech is lost. Even the best ear trumpets could only help people with fairly mild hearing impairments. Large trumpets, which were heavy and difficult to use, could provide more amplification. Hearing impaired individuals, however, often tried to disguise or conceal their hearing aids and preferred small devices that could be hidden behind the ear, under long hair, under a beard, or even as part of the furniture. Most early trumpets were custom made until specialized business firms that manufactured and sold hearing aids were established in the nineteenth century.

Conversation tubes sometimes resembled and even functioned as monaural stethoscopes. Conversation tubes had a long, flexible rubber tube, 2-3 ft (0.7-1m) long, and hard rubber mounts on each end. The devices had an ear peg on one end and a removable, conical chest-piece on the other. The cone-shaped end could be given to the speaker or placed on the chest if a physician wanted to use the device as a stethoscope. The ear peg was inserted into the ear of the deaf person. Such tubes were quite common at the beginning of the twentieth century.

Sounds, of course, are transmitted to the ear by vibrations in the air, but also by bone conduction, that is, by the vibration of the bones in the skull. Sound energy causes hard, thin materials to vibrate. For some people, aids to hearing based on bone conduction provided the best way to transmit amplified sound. Appropriate devices could conduct vibrations to the internal mechanism of the ear via the teeth or skull. For example, a woman could hold an acoustic fan against her teeth and a man could use a pipe with a vibrating stem. Bone-conduction devices can be traced back to the sixteenth century, but the first practical one seems to have been the 1879 Rhodes Audiophone, which used a fan to pick up air vibrations and transmit them to the teeth. Electric bone-conduction hearing aids appeared in the 1920s and were a major improvement. Today, most hearing losses related to bone conduction are corrected surgically.

Impact

By the end of the nineteenth century many different hearing devices were available. These devices ranged from cheap models made of tin or hard rubber to expensive ones made from precious materials. Expensive devices, however, did not necessarily work any better than their less elegant counterparts. Effectiveness varied depending on the characteristics of the device and the specific impairment of the user, but, until the twentieth century, choosing a hearing aid was a matter of trial and error. During the nineteenth century, inventors found that electrical energy could be used to amplify speech. Alexander Graham Bell (1847-1922), the inventor of the telephone, was one of the early pioneers in this field. Although Bell did not invent the electrical hearing aid, he seems to have devised a simple transmitter, receiver, and battery system for his hearing-impaired mother. He also attempted to build an amplification system for his future wife, Mabel Hubbard, who had lost her hearing when only four years old. His experiments on what he called "speech reading" led to his invention of the telephone.

A few hearing aids that used the principle of electrical amplification appeared as table models at the turn of the century, but more practical and wearable instruments were not available until about 1902. The first electric hearing aids were essentially miniature telephones composed of a power source, a magnetic earphone/receiver, and a carbon microphone. A microphone is an energy converter that changes acoustic energy into electrical patterns. Electric hearing aids generally used a large 3-volt or 6-volt battery as the power source. Electrical current can be more easily amplified than acoustic or mechanical energy. Therefore, the electrical hearing aids, also known as carbon hearing aids, that were introduced at the turn of the century provided useful amplification for a broader range of hearing impairments.

The earliest electric aids offered the same amplification ear trumpets had but covered a wider range of sound frequencies. The amplification of sound energy provided by the early microphone-battery-receiver hearing aids was not significantly greater than that of some of the better pre-electric aids, but the range of the frequencies that were amplified was substantially extended. Later models with multiple microphones provided 25-30 decibels of amplification. During the 1920s, amplifiers were incorporated into carbon hearing aids and the amplification increased by 45-50 decibels. Volume controls that varied the amount of current flowing into the system from the battery were incorporated into the carbon hearing aids.

At first, the receivers for carbon hearing aids were large hand-held earphones. The size was eventually reduced so that the earphone could be supported by a headband. Eartips and earmolds carried the acoustic energy into the auditory canal. Custom earmolds were introduced in the 1930s. For cosmetic reasons, many devices were designed to be camouflaged as purses, camera cases, or headbands. Carbon hearing aids remained common into the 1940s and helped people with moderate hearing loss, but they did not have enough power to assist people with severe problems. More modern devices were made possible with the development of small dry-cell batteries and the electron tube.

In the early 1930s, the carbon hearing aids were being replaced by table vacuum-tube hearing aids, but, even in 1944, some 50,000 carbon hearing aids were still being used in the United States. The technology used in the vacuum-tube hearing aids can be traced back to experiments conducted by Thomas A. Edison (1847-1931) in the 1880s. More practical modifications of the vacuum tube appeared early in the twentieth century. Unlike the previous generation of carbon instruments, the new vacuum-tube hearing aids had adequate power for severe hearing losses. This type of hearing aid first appeared in 1921, but vacuum-tube devices were expensive well into the 1930s.

Electronic hearing aids, also known as transistor hearing aids, appeared in 1952 and essentially replaced vacuum-tube hearing aids by the end of 1953. Because transistors only needed one battery, the size of hearing aids could be significantly reduced. The body aid (or pocket aid) and the behind-the-ear (or over-the-ear) version were useful for individuals with severe hearing loss. After Eleanor Roosevelt endorsed the use of eyeglass hearing aids, they became quite popular. Further improvements and miniaturization led to various in-the-ear models, canal aids, and aids that fit entirely within the ear canal. Further improvements included hybrid hearing aids with a combined digital/analog circuitry (1970s) and aids with a digital chip to be integrated into an analog hearing aid (1980s). The digital signal processing (DSP) chips that became available in 1982 were incorporated into digital hearing aids. The early models were quite large and expensive, but, by the end of the century, fully digital behind-the-ear and in-the-ear hearing aids were commercially available. By the end of the twentieth century, amplification had increased substantially, but, more importantly, hearing aids could be individually designed to accommodate specific hearing impairments.

A series of experiments that began in the 1950s led to the development of the cochlear implant, an electronic device that restored partial hearing to the deaf by bypassing the mechanisms of the middle and inner ears and directly stimulating the auditory nerves. The device was granted FDA approval for general use in 1984. Cochlear implants are surgically implanted in the inner ear and activated by a device worn outside the ear. Traditional hearing aids make sounds louder or clearer, but the cochlear implant works on an entirely different principle. A microphone in the cochlear implant system transforms sound into an electric signal. Electrodes in the system then use these electric signals to stimulate the auditory nerves. If the auditory nerves are healthy, cochlear implants can restore some measure of hearing in cases in which the inner ears are badly damaged.

During the early twentieth century, several surgeons introduced operations that could cure or ameliorate certain kinds of deafness. Sir Charles Alfred Balance (1856-1926) and Charles David Green (1862-1937) published their landmark two-volume treatise, *Essays on the Surgery of the Temporal Bone,* in 1919. The operating microscope brought the benefits of magnification to neurosurgery and to ear surgery, which made it possible for surgeons to perform novel operations on the eardrum and within the middle ear. In 1928 Walter Edward Dandy published an account of the diagnosis and surgical treatment of Menière's disease. George John Jenkins's experimental operations for otosclerosis, reported in 1917, led to the modern fenestration operation for otosclerosis. (This condition occurs when a formation of spongy bone impedes the movement of the small bones in the middle ear, causing severe and progressive hearing loss.)

During the 1930s and 1940s, many surgeons devised operations for various diseases of the ear. In the 1930s, Julius Lempert (1890-1968) introduced the one-step fenestration operation to treat otosclerosis. Lempert is considered a pioneer in the field of otology. His innovative operations led many medical schools to endorse otolaryngology as an area of specialization. Further advances in the treatment of otosclerosis occurred in the 1950s, when Samuel Rosen rediscovered stapes mobilization. In patients with otosclerosis, the stapes (a stirrup-shaped bone of the middle ear) often becomes rigid. By carefully applying pressure to the stapes, the surgeon can sometimes free the bone and restore hearing. Surgeons had attempted to treat otosclerosis by stapes mobilization in the late nineteenth century, but, at the time, the medical community regarded this as an unsafe operation of questionable value. Because of Rosen's success, however, the operation became widely accepted. Unfortunately, although stapes mobilization benefits about 70% of patients with otosclerosis, the improvement is often only temporary. In an attempt to improve on stapes mobilization, John Shea Jr. reintroduced and improved upon a procedure known as stapedectomy. In this operation, the oval window in the ear is covered with a living membrane and a prosthesis is used to reconstruct the bones of the middle ear. In the 1960s, Shea developed an improved stapedectomy procedure in which the stapes was completely removed and replaced with a prosthesis made of Teflon.

Insights into the struggles to overcome the obstacles caused by deafness can be gained by reading the autobiography of Helen Keller (1880-1968), who became blind and deaf as the result of an acute illness when she was only 19 months old, and Ruth Elaine Bender's book *The Conquest of Deafness: A History of the Long Struggle to Make Possible Normal Living to Those Handicapped by Lack of Normal Hearing.* Hearing aids have been collected by several museums, including the Hearing Aid Museum and Archives at Kent State University, the Central Institute for the Deaf in St. Louis; the Smithsonian Institution in Washington, D.C.; and the Oticon A/S in Denmark.

LOIS N. MAGNER

Further Reading

Balance, Sir Charles Alfred, and Charles David Green. *Essays on the Surgery of the Temporal Bone.* 2 vols. London: Macmillan, 1919.

Békésy, Georg von. *Experiments in Hearing.* Translated and edited by E. G. Wever. New York: McGraw-Hill, 1960.

Bender, Ruth Elaine. *The Conquest of Deafness: A History of the Long Struggle to Make Possible Normal Living to Those Handicapped by Lack of Normal Hearing.* Cleveland, OH: Western Reserve University Press, 1960.

Bennion, Elisabeth. *Antique Medical Instruments.* Berkeley, CA: University of California Press, 1979.

Berger, Kenneth W. *The Hearing Aid: Its Operation and Development.* Livonia, MI: National Hearing Aid Society, 1974.

Davis, Audrey B. *Medicine and Its Technology: An Introduction to the History of Medical Instrumentation.* Westport, CT: Greenwood Press, 1981.

Donnelly, Kenneth, ed. *Interpreting Hearing Aid Technology.* Springfield, IL: C. C. Thomas, 1974.

Keller, Helen Adams. *The Story of My Life.* New York: Grosset & Dunlap, 1905.

Watson, Leland A., and Thomas Tolan. *Hearing Tests and Hearing Instruments.* Baltimore, MD: Williams & Wilkins Company, 1949.

Hormones and the Discovery of Insulin

Overview

Hormones are chemical messengers secreted mainly by the body's complex network of glands and glandular tissue that comprise the endocrine system. Each hormone has a unique function and travels via the bloodstream to exert influence upon its targeted cells, tissues, or organs. Hormones affect many vital life functions, stimulating certain life processes, while retarding others. Among other functions, hormones regulate reproduction, the body's fluid balance, growth, and metabolism. The nervous system also acts in conjunction with hormones to enable the body to react to sudden internal or external changes in environment.

After almost 50 years of experimentation, the fledgling science of endocrinology (the study of hormones and their effects on the body) exploded in the 1920s with the discovery of the hormone insulin. Insulin, necessary for the regulation of carbohydrate metabolism, offered the first effective treatment for diabetes, a metabolic disorder documented since ancient times. After the discovery of insulin, the isolation and characterization of other hormones led to a new understanding of the body's physiology and new methods of treating disease.

Background

Prior to the 1920s, scientists studied hormones primarily through experimentation with hormone-producing glands. In search of youthful regeneration, French-American physiologist Charles Brown-Sequard (1817-1894) experimented on himself by injecting testicular extracts in 1889. He lived six years afterward, and although he failed to reverse the effects of aging, his glandular extracts attracted attention among scientists. Crude extracts containing iodine and hormones from the thyroid gland were prepared in the late nineteenth century with no correlation recognized at that time for its relevance in the treatment of disease.

In 1902 two British scientists, William Bayliss (1860-1924) and Ernst Starling (1866-1927), first proposed the theory that internal secretions control the functions of the body. By the end of World War I, scientists knew of a relationship between the secretions of the pancreas (an endocrine gland) and diabetes. When that connection was narrowed down to a specific region of the pancreas, the islets of Langerhans, the search was on to find and extract the hormone that lowered the characteristically and abnormally high concentrations of glucose (sugar) in the blood of diabetics.

In 1920, while reading medical journal articles in preparation for delivering a lecture on the functions of the pancreas, Canadian physician Frederick Banting (1891-1941) jotted down an area of interest and potential research—isolation of the secretion that lowers glucose in the blood. Earlier, German physician Charles Minchowsky (1858-1931) suggested that if the pancreas was minced, the fluid extracted, then injected into diabetic animals, it might help control diabetes. Minchowski's efforts failed when the enzymes of the minced pancreas destroyed the insulin before it could be injected. Banting proposed a different approach. By ligating (tying off) the pancreatic duct and allowing the pancreas to degenerate slowly within the animal, Banting reasoned the pancreatic enzymes would not destroy the insulin, and would remain viable in an extract.

Banting approached Scottish physiologist John Macleod (1876-1935) with his proposal. Macleod, an expert in carbohydrate metabolism at the University of Toronto, was initially skeptical, but agreed to provide Banting with laboratory space at the university during the summer break. Macleod also furnished 10 laboratory dogs and a medical student assistant, fellow Canadian Charles Best (1899-1978). In May 1921, the two began their research as Macleod traveled on summer holiday. Banting performed most of the surgeries, first ligating the pancreas of the dog, then later removing it to produce diabetes in the animal. Best performed most of the biochemical tests for measuring sugar in the blood and urine, and occasionally assisted Banting with the surgeries.

By the end of July, the pair had their first conclusive results. After removing the ligated, shrunken pancreas from of the dogs, the pancreas was ground in a mortar with a saline solution, strained, and a small amount was injected into a diabetic dog. The dog's blood sugar levels were significantly reduced, and the dog became more active. After repeating the results with other dogs in the laboratory, Banting and Best were confident they had isolated the anti-diabetic hormone from the islets of Langerhans in the pancreas. They named their discovery "Isletin" and later changed the name to "Insulin" (from the Latin for "Island"). Macleod requested that the pair validate their results by repeating the experiments two more times. Although in the repeated experiments the results varied due to the quality and concentration of the insulin extraction, the discovery of the role and extraction of insulin was confirmed.

Before insulin could be tested on a human it was necessary to eliminate the variations within the extraction. Canadian biochemist James D. Collip (1892-1965) joined the research team in December 1921, and worked to purify the quality and concentration of insulin. When Banting suggested the use of fetal calf pancreas, which contained few digestive enzymes, Collip was able to produce greater amounts of insulin of sufficient purity that the injection site no longer formed an abscess, as did earlier extractions. Collip's efforts had produced insulin in a form suitable for human clinical use.

In January 1922, Leonard Thompson, a 14-year-old patient dying of diabetes in Toronto General Hospital, was the first human to receive insulin. Thompson's blood sugar levels, initially high enough to produce diabetic ketoacidosis and coma, dropped dramatically after insulin was injected. Two weeks were necessary to prepare additional extract, and with a second insulin injection, Thompson's blood sugar levels fell further. After several injections Thompson recovered from his bout with hyperglycemia (abnormally high blood sugar associated with diabetes) and was discharged home from the hospital in May 1922.

Earlier, in February 1922, Banting and Best published their famous paper, "The Internal Secretion of the Pancreas," in the *Journal of Laboratory and Clinical Medicine*. Ironically, Banting and Best later learned that their original premise was not entirely correct; insulin could, in fact, be obtained from an intact pancreas. Nevertheless, by the spring of 1922, the group produced quantities of insulin sufficiently purified to prove the practical value of insulin in treating diabetic patients. The Eli Lilly and Connaught companies of North America offered laboratory space and assistance to help improve the concentration, stability, and purification of insulin. Soon the supply of insulin dramatically increased, with Connaught as its primary manufacturer.

The 1920 Nobel Prize for Physiology or Medicine was awarded to Banting and Macleod for the discovery of insulin. Banting felt strongly that Best should have been recognized by the Nobel committee as well. Banting shared half of his prize money with Best, and upon learning this, Macleod shared his prize with Collip. For some years afterward the four men privately feuded over the essential nature of each of their roles in the discovery. Macleod's role perhaps caught the attention of the Nobel committee because, as the only member among the group belonging to the American Association of Physicians, only Macleod was eligible to present the paper at the association's meeting. Macleod also presided over another professional meeting where the paper was presented by Banting and Best. On this occasion, Macleod introduced Banting and Best as his assistants, and reporters covering the event gave Macleod heightened credit and publicity. Nevertheless, the contributions of all four men are fully recognized by history and related professional literature.

Impact

Insulin was hailed as a miracle drug and, incorrectly, initially as a cure for diabetes. No longer was diabetic coma assumed a death sentence. At first, insulin was reserved for those suffering the most severe effects of diabetes. Later, as supplies became more abundant and physicians learned the intricacies of insulin dosages and reactions, insulin therapy became standard treatment for diabetics. Before 1920, diabetics were treated with starvation diets of proteins considered easily digested. Insulin enabled diabetics to gain the benefits of a varied diet rich in nutrients.

By 1950 there were an estimated seven million diabetics in the United States alone, many of whom were unaware they had the disease. Armed with simplified lab procedures to detect abnormal blood sugar levels, physicians began to incorporate diabetic screening in their patient's routine health examinations. With insulin available to treat diabetes, victims of one of the world's ancient and deadly diseases were given hope to live longer, more productive lives.

After the discovery of insulin, research on the function of other endocrine glands and their hormone secretions burst into scientific prominence. The discovery of cortisone, an anti-inflammatory hormone released by the adrenal cortex, offered relief for patients suffering from the debilitating effects of rheumatoid arthritis. First tested on humans in 1947, initial results showed the beneficial effects of cortisone were temporary and patients suffered significant side effects. With further research, other hormones with anti-inflammatory properties would eventually become mainstream treatment for many varied ailments, including autoimmune diseases, asthma, and some traumatic injuries.

In 1945 American endocrinologist Fuller Albright (1900-1969) proposed the possibility of "birth control by hormone therapy," an idea that planted the seeds for a subsequent social revolution. After hearing of what became known as "Albright's Prophecy," American Planned Parenthood founder Margaret Sanger (1879-1966), then 70 years old, set out to inspire the development of an oral contraceptive. Sanger and her wealthy American friend Katharine McCormick (1875-1967), also in her seventies, approached research biologist George Pincus (1905-1967) of the Worcester Foundation for Experimental Biology with the task in 1948. Two years later, funded by McCormick, Pincus began the research that led to the introduction of a hormonal oral contraceptive (birth control pill). Ultimately, the birth control pill would lead to the first generation of women able to separate sexuality from reproduction and to plan the size of their families. This new freedom precipitated the sexual revolution of the 1960s, the women's movement of the 1970s, and the influx of women into the workplace.

BRENDA WILMOTH LERNER

Further Reading

Bliss, Michael. *The Discovery of Insulin.* Edinburgh: Paul Harris, 1983.

Hardie, D. G. *Biochemical Messengers: Hormones, Neurotransmitters, and Growth Factors* London: Chapman & Hall, 1991.

Medvei, Victor C. *The History of Clinical Endocrinology: A Comprehensive Account of Endocrinology From the Earliest Times To Present Day.* New York: Parthenon, 1993.

Oudshoorn, Nelly. *Beyond the Natural Body: An Archaeology of Sex Hormones.* New York: Routledge, 1994.

Therapeutic Innovations Related to Respiration

Overview

During the first half of the twentieth century, several great innovations designed to treat respiratory failure were developed. The emerging fields of bioengineering and biomaterials developed devises to assist in medicine and therapy. One of the great bioengineering feats was the development of the "iron lung" to assist breathing. Invented by Philip Drinker (1894-1972) in 1927, this device provided respiration for those whose lungs could not breathe on their own. Drinker's machine, originally made from a household vacuum cleaner, some materials from the laboratory, and a large steel box, was used to treat one of the most frightening conditions of the day—infantile paralysis, now called poliomyelitis.

With the aim of treating people who were dying from drowning or other interruptions of breathing, Edward Sharpey-Schafer (1850-1935), an English physiologist, decided that a

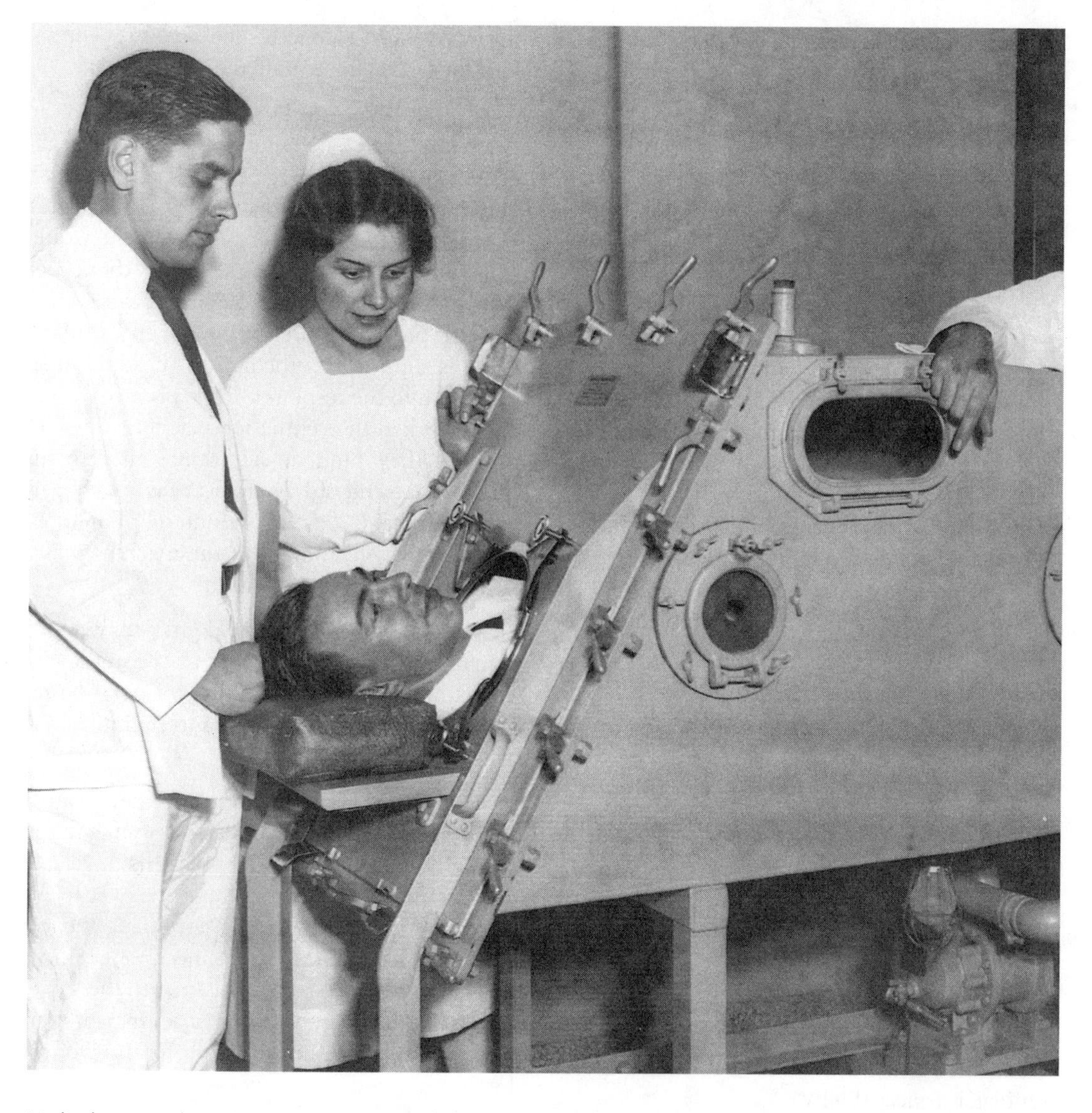

Medical personnel monitor a patient in an iron lung. *(Minnesota Historical Society/Corbis. Reproduced with permission.)*

mechanical method or artificial respiration could start breathing until the lungs of the person could take over. He described the first prone-pressure method of artificial respiration that became known as the Schafer method. In addition, John Scott Haldane (1860-1936), a British physiologist known for his work on respiration, first introduced modern oxygen therapy.

These pioneering innovations have been refined in the last part of the twentieth century. Some of the inventions and methods are no longer used, however, they once saved many lives and lead to new improvements in the treatment of respiratory failure.

Background

The study of respiration developed as part of the physiology of chemistry. Gases necessary for life were studied by many chemists and physiologists who formed the basis of the science of biochemistry. Gases necessary for life—mainly oxygen and carbon dioxide—were known in earlier centuries.

Asphyxia, unconsciousness caused by oxygen deprivation, had long been recognized as the most common cause of death. Oxygen can be deprived in several ways, each occurring when respiration is slowed or stopped.

In the early days of the twentieth century, poliomyelitis was a disease that caused paralysis to the lungs. Also called, infantile paralysis, the condition is caused by a virus that injures the upper part of the spinal cord and then causes difficulty in breathing. When breathing is affected, the person requires mechanical help or will die.

Asphyxiation also may occur when normal respiration is ineffective because the person is in an atmosphere where there is little oxygen. Examples are unventilated or poorly ventilated enclosed spaces such as mine tunnels, sewers, and industrial tanks where oxygen may be replaced by other gases. Many of these gases are odorless. Methane, for example, may accumulate in coal mines or come from decomposing sewage. People may or may not be aware of breathing these poisonous gases.

Another form of asphyxiation occurs when respiration is interfered with by gases like carbon monoxide that impede the oxygen-carrying ability of the cells. Carbon monoxide combines with hemoglobin, the substance in red blood cells that carries oxygen to the body tissues. The carbon-monoxide-hemoglobin combination does not allow the red blood cells to carry oxygen, so the body does not get the oxygen it needs. This condition may cause problems even though normal oxygen is present.

Impact

The idea of an apparatus to help breathing had been the dream of many people. In 1670 a scientist named John Mayow (1641-1679) built a model consisting of bellows and a bladder to pull in and expel air. Inspired by the death of his son from respiratory failure, Alexander Graham Bell (1847-1922), inventor of the telephone, designed a vacuum jacket in 1892. Throughout the nineteenth century, physicians and engineers experimented with External Negative Pressure Ventilator devices (ENPV).

Few people can imagine the terror of polio during the early days of the twentieth century. Patients with anterior polio affecting cervical and thoracic areas of the spinal cord might only have a few hours between the onset of the disease and death. Children who were afflicted would smother and drown in their own saliva. In 1926 Philip Drinker, a chemical engineer, was employed to teach applied physical science at Harvard. In 1926 he was asked by a committee of the Rockefeller Research Institute to study methods of resuscitation. He happened to observe an experiment conducted by his brother Cecil, and another physiologist Louis Shaw. The two measured the respiration of a cat by putting it in a sealed box with only its head sticking out.

Drinker reasoned that perhaps a similar design could help one breath. He injected a cat with curare, a poison and muscle relaxant, to slow down breathing. He modified the box so that he could manipulate a syringe to increase or lower the air pressure. He successfully ventilated the animal for several hours until the drug wore off.

Drinker concluded that if it worked for a cat, it could also work for a human being. He recruited a tinsmith to build a metal cabinet, to which he added a vacuum cleaner blower and some valves from the laboratory. The patient would slide into the respirator on a creeper, the device that wheels a mechanic under a car to work on it. The end point is secured at the feet and a rubber collar is slipped over the person's head. Drinker climbed into the respirator while his brother and Shaw worked the ventilation machine. For four minutes they pumped additional air into his lungs. At the end of 15 minutes he was so hyperventilated that it took four minutes for him to regain normal breathing. But the machine was a success; it had breathed for him.

In 1928 an eight-year-old girl who was in a coma from lack of oxygen was placed in the machine, and it started breathing for her. She regained consciousness and in a few minutes was asking for ice cream.

Another opportunity to use the device arose when Harvard senior Barret Hoyt was dying in the Peter Bent Brigham Hospital. His physician called Drinker to bring the machine over to the hospital. The machine would not fit into the taxi he had called, and there was no time to find a truck; the machine was finally tied to the top of the taxi. After a stay in the machine for four weeks, Hoyt lived, finished college, and worked for 30 years as a successful insurance agent.

An unknown journalist dubbed the machine the "iron lung." An improved version went into production and by 1931, 70 were in hospitals throughout the United States.

The iron lung also became an extravaganza. Department stores—wanting to show their support for polio research—would display the instrument in their windows. Machines were invented that would hold a group of people, and these would be shown with several of the large iron boxes stacked on top of each other with multiple heads sticking out. The machine was then sent to Europe, and over the years was improved and imitated. Actually, it was a great tool against the dreaded disease. When the polio vaccine was developed in 1955, the machine—no longer needed—was relegated to museums.

Edward Albert Sharpey-Schafer was an English physiologist and inventor of the Schafer method of artificial respiration. He was the first student to hold a scholarship under Dr. William

Sharpey; to honor his teacher, he attached his surname to his own to become Sharpey-Schafer.

A professor named Sylvester had devised a method of reviving a person who was drowning by laying the patient on his or her back and working the arms sideways from straight down to up above the patient's heart to pump air into the lungs. Schafer turned the patient face down to let the water drain out of the lungs. The operator kneeled across the back pressing down and counting "one two three" then relaxing counting "one-two-three."

The Schafer method became generally accepted by the medical profession and was taught to Scouts, first aid professionals, and people throughout the world. Although the procedure has been replaced by cardiopulmonary resuscitation (CPR), it is still credited to saving many lives.

British physiologist John Scott Haldane discovered in 1905 that breathing is regulated by the concentration of carbon dioxide in the respiratory center of the brain. He devised an apparatus for the analysis of blood gas called a hemoglobinometer and an apparatus for the analysis of a mixture of gases. In 1911 he led an expedition to Pike's Peak to study the effect of low barometric pressure on breathing.

Going into the depths of coal mines, he found that carbon monoxide and other dangerous gases were causing the deaths of miners. Taking a clue from French physiologist Claude Bernard (1813-1878), who showed that carbon monoxide combined with hemoglobin, Haldane found that rats survived carbon monoxide poisoning when placed at 2 atmospheres of pressure. Carbon monoxide deaths were occurring in many places because of illuminating gas and gas heaters. In 1927, for example, 611 carbon-monoxide related deaths occurred in New York City.

Haldane's innovative thinking related to his study of blood gases. He noted that divers who ascended in stages could avoid the nitrogen accumulation in the blood commonly called the "bends." Haldane used experimental animals, especially goats because their respiratory systems are similar to humans. Haldane also carried out an experiment in which a goat was decompressed to a low pressure using a mixture of helium and oxygen, then given pure oxygen to speed up recompression. The pioneering work of Haldane would lead to the first clinical use of the hyperbaric chamber in 1960.

The work of these three innovative pioneers—Drinker, Sharpey-Schafer, and Haldane—proved essential to the development of a variety of treatments for respiratory ailments in the second half of the century.

EVELYN B. KELLY

Further Reading

Porter, Roy. *The Greatest Benefit to Mankind: A Medical History of Humanity.* New York: W. W. Norton, 1997.

Porter, Roy. *Medicine: A History of Healing.* New York: Barnes & Noble, 1997.

The Impact of Radioactivity on Medicine between 1900 and 1949

Overview

In 1896 Antoine Henri Becquerel (1852-1908), a professor at the Ecole Polytechnique in Paris, France, discovered radioactivity in uranium when he was investigating fluorescent crystals for x rays. He shared the Nobel Prize for physics in 1903 with Marie and Pierre Curie (1867-1934 and 1859-1906, respectively). It was mathematician Henri Poincaré (1854-1912) who told Becquerel that the x rays were emitted from a fluorescent spot on the glass cathode-tube used by Wilhelm Röntgen (1845-1923) when he first detected x rays in 1895. The era of artificial radioactivity—in which humans make more radioactivity than is naturally existing in the Earth and its atmosphere—had begun in earnest by the start of the twentieth century. But despite the fact that radioactivity devices such as the early "fluoroscope" x-ray machines developed by Thomas Edison (1847-1931) were becoming rapidly available to the general public, scientists did not actually understand what was making them work so that they could take photographs of the bones inside the human body, or indeed of bullets that had been shot into people.

Background

X-ray machines were offered as entertainments and curiosities throughout the United States in department stores such as Bloomingdale's at the turn of the century, and people lined up to have their hands or their feet x rayed so that they could see their bones. When Wilhelm Röntgen had first detected x rays, he showed his fellow scientists what he could do by sending photographs of his x ray of his wife's hand, showing only her bones and wedding ring but no tissues. When she had seen this photograph, Frau Röntgen said she felt she was looking at her own death. It took several years for the scientists and the public to realize that the burns many people working with x rays were suffering were not from the electrical charge in the apparatus but from the actual rays themselves, the radioactivity. This also caused people's hair and teeth to fall out because there was no realization in the first years of the twentieth century that it was necessary to using shielding materials such as lead to protect both the patient and the operator (increasingly known as the "Röntgenologist" and then the "radiographer") from getting too much radioactivity for their bodies to cope with.

But even with this major disadvantage, the use of x rays was becoming widespread in medicine, particularly when metal objects such as bullets had entered the patient's body and had to be located in order to be removed by surgery. Within ten years of Röntgen's discovery, primitive x-ray machines were being used by British, Italian, and American military doctors in conflicts as far afield as Abyssinia, the Balkans, Afghanistan, and Egypt. They were widely used in the Spanish-American War in 1898, and many soldiers' lives were saved by the operations the new x-ray technology made possible, together with other new developments such as antiseptics and anesthesia.

Still, however, it was not understood that the very use of the x rays might be killing some of the patients, and even the operators. Thomas Edison was one of the first experimenters with x-ray technology to realize that his own burns, particularly to the hands and fingers that had handled the radioactive materials, were due to the rays themselves. One of his assistants, Clarence Dally, lost his mustache and beard together with his eyebrows and eyelashes—and then all his fingers and his left hand, which had held the substances causing the x-ray effects. Edison said Dally "died by inches," finally succumbing to his dreadful injuries in 1904. He was the first of many radiographers (such as the Army radiographer Elizabeth Fleischmann) to suffer such serious injuries in the first third of the century before the cause was realized and strict protection measures instituted. But before then, the very burning effects of radiation—which often begin only slightly with a reddening of the skin like sunburn—were used to treat acne and excema, with dreadful effects. The trouble was that weeks later deep ulcers developed that could not be treated. Amputation was often the only solution to stop the intense pain. Even so, x rays were widely used in both medicine and dentistry, particularly in the United States, within a decade of Röntgen's discovery. Even U.S. Customs officials started using x-ray machines at the turn of the century. X-ray photographs were already being used as evidence in court in shooting cases and similar incidents of personal injury in the first years of the century.

Impact

Some scientists working with radiation also began to realize that it could make people and animals sterile. To some this was seen as an advantage in the early years of the century when the "eugenicists"—those who believed that human reproduction should be managed so that only the fit, clever, and healthy reproduced themselves—were considered respectable. Still, scientists such as Marie and Pierre Curie went on working with uranium to understand its radioactive properties. Pierre Curie died in 1906 in a road accident, but his wife suffered many burns to her hands before succumbing to anemia and leukemia like so may other of the early radiation workers. But she had been awarded two Nobel prizes, one for discovering "polonium" (named after her native Poland) and radium.

The realization that different substances had different levels of radioactivity, and indeed different types of radioactivity, enabled their adaptation for different medical uses. The field of medicine called "nuclear medicine" arose from the ability of doctors to feed patients radioactive substances that could then be traced in their bodies with x-ray machines to detect if there were cancerous blockages or other medical problems. Eventually, the radioactive substances would themselves be developed as therapies and medicines, relying on their capacity to target cancerous cells and kill them. While surrounding healthy cells would also be in jeopardy, they would be able to "repair" themselves more quickly than the cancers, and so people would survive and benefit from radiotherapy.

The administration of these therapies in twentieth-century medicine was made safer by the development of lead suits and aprons to shield the health workers and the patient from excess exposures to the radiation. The "scattering" effect of radiation rays also came to be understood and devices were developed to "collimate" the rays to the desired focus and to protect the people and equipment surrounding the patient from receiving too much radiation. The film badge dosimeter was developed utilizing the known effects of radiation on photographic film to serve as a measure and monitor of how much radiation individual nurses, doctors, radiologists, and radiographers were receiving.

Radiographic equipment manufacture was big business by the 1920s, and major companies such as Siemens, General Electric Company, and Kodak were constantly refining the devices as the safety issues were increasingly understood. The professions of radiographer and radiologist grew with great rapidity, especially after the First World War when radiation scientists such as Marie Curie contributed greatly to the war efforts of their countries by developing more mobile and effective x-ray units that could be taken to the battlefronts. Training schools for people to operate these units were set up during the War, and many women found it possible to embark on a new career in this field. The soldiers who were the beneficiaries of these x-ray units came to expect them to be available to them in civilian life, and this also hastened the acceptance of this new field of medicine in the first half of the century. Two thirds of U.S. hospitals had their own x-ray departments by the 1920s, and the rest depended on private practitioners who were offering the service from their own offices and clinics.

William David Coolidge had developed smaller, quieter, more efficient systems of generating x rays, and these came into wide use particularly by the U.S. Army when it entered the First World War. By now the technology was useful not only for locating metal objects such as bullets but also for examining the state of patients' chests. The chest x ray became a widespread technique in both civilian and military medicine by mid-century, particularly to detect tuberculosis. If anything, it was over-used (some servicemen being x rayed annually), as there was still a reluctance on the part of enthusiasts to acknowledge that the technology that was providing such a marvelous view into the human body was itself very dangerous. It even became common practice to x ray fetuses in the womb to check that they were developing properly, and it was not until the 1950s that scientists determined that this was extremely hazardous to the fetuses, as the youngest tissues are those most vulnerable to radiation damage. Since infants

EARLY USES (AND MISUSES) OF RADIATION AND RADIOACTIVITY

The years immediately following the discovery of radiation, radioactivity, and radioactive elements such as uranium, thorium, and radium were heady ones for the practitioners of radiation science. They had given medicine the ability to see within the body and, it seemed, to heal many diseases, because it was soon found that high doses of radiation could kill cells, including cancers. To be sure, in high doses that were applied inappropriately, radiation caused injury, but this was true of many phenomena, including electricity and steam.

In those days of excited experimentation, a number of things seem, in retrospect, particularly ill-advised. However, it should be remembered that, at that time, radiation did not carry the stigma it now does and most of these experiments and uses were well-intentioned efforts to help people.

One use of radiation was to remove unwanted facial hair, primarily in women. This was discontinued when it was found that, improperly applied, scarring could result that was even more unsightly. Radiation in the form of radium inserts were used to try to clear up sinus problems, and many companies sold devices that "invigorated" water by infusing radium or radon into it. Thomas Edison tried to x ray the brain to see if any insights into thought could result, and others tried to locate or cure criminal behavior in a similar manner.

One of the more interesting anecdotes, however, came from the World War II years when Allied intelligence agents discovered that a German company was trying to buy the world's stocks of thorium ore. At first thinking that the Germans had somehow discovered how to make a nuclear weapon from thorium, it was later discovered these stocks were purchased by a toothpaste company that used thorium for putative extra whiteness—the company was trying to corner the post-war market to keep competition at bay.

P. ANDREW KARAM

and children are also undergoing more rapid cell division than the grown adult, it is now also understood that they are more susceptible to radiation damage than a mature human. X rays were also used to treat excessively heavy bleeding in women's menstrual cycles before it was realized that the uterus and ovaries are particularly susceptible to radiation damage.

Radiology was declared a specialty within American medicine in 1934. Various Röntgen societies existed around the world, and they began to realize the need to reach agreed safety standards for the practice of nuclear medicine. This involved the definition of units of measurement, the first of which was named the *Röntgen*. The concept of a safe dose of radiation was increasingly debated, and doctors and scientists around the world tried to agree on where the limits for human exposures of both patients and health workers should be set. When the Second World War broke out in 1939, the technology of the x-ray industry again went into fast gear, and most of the fighting forces were well served by increasingly effective and more mobile equipment. It is one of the great ironies of the twentieth century that the use of radioactivity, which had been such a dangerous but wonderful resource in medicine, was adapted in the next four years to become a weapon of mass destruction when it was used as part of the atom bombs dropped on Hiroshima and Nagasaki in Japan in 1945.

SUE RABBITT ROFF

Further Reading

Kevles, Bettyann Holtzmann. *Naked to the Bone: Medical Imaging in the Twentieth Century*. New Brunswick, NJ: Rutgers University Press, 1997.

Quinn, Susan. *Marie Curie: A Life*. London: Heinneman Publishing Co., 1996.

Medical Advances During War

Overview

Several great military conflicts occurred during the first half of the twentieth century. With improved weapons came great destruction and mayhem. However, in the backdrop of arenas of war, physicians and scientists learned valuable medical lessons, later applied to civilian care, from the agony of the battlefields and the horrors of war crimes.

World War I (1914-18), including American involvement in 1917-18, took place in the trenches of Western Europe. World War II (1941-45) was a worldwide conflict with many casualties. It has been said that the only benefit from these conflicts was surgery. This statement can well be extended to medicine in general.

Throughout the world, many changes in postwar civilian medicine were a direct result of war medicine. For example, the idea of centralized medical administration and a system of free health care for all emerged, especially in Great Britain and other European countries.

Nazi medical experiments, Japanese atrocities, along with a few Allied abuses, offered a chilling view of medical science that could not be tolerated. The revelation at the Nuremberg trails that Nazi doctors had performed horrendous experiments on living prisoners—presumably conducted in the name of science—led to an international outcry. The ethical principle that there should be no human experimentation without full disclosure and consent has been a lasting legacy from this experience.

Background

Before firearms were introduced, hand-to-hand combat with swords and spears produced little tissue damage. If the soldier survived the initial battle, generally he would live. That changed with gunpowder and firearms. High-powered projectiles could pierce the flesh and cause major tissue destruction. Puncture wounds introduced a group of bacteria called anaerobes. The term "anaerobe" means "without air," and the spores, the inactive state of these organisms, are present in dirt and other places. When objects such as bullets or other debris penetrate the flesh, spores of the anaerobe are sealed in, then the flesh closes around it. The spore's coat dissolves and bacteria start to grow. The major diseases caused by anaerobes and puncture wounds are tetanus and gangrene. Both are fatal if not treated.

In the sixteenth century an army surgeon named Ambroise Pare (1500-1590) was appointed to the Duke of Montejan. At that time, surgeons were attached only to generals and helped the common soldiers only when they had spare time. The common soldiers were treated by their companions, farriers, female camp followers, or anyone around. The best known treatment of the time was pouring boiling oil on the wound. Pare changed the treatment of battlefield wounds from boiling oil to sewing together blood vessels.

For two and a half centuries no more improvements were made until Dominique-Jean Larrey (1766-1842) improved the rapid evacuation of the wounded with his "flying ambulance," a horse-drawn box on wheels. This established the concept that the wounded should be rapidly removed from the battlefield and taken to special surgery areas.

Medicine during the American Civil War was a disaster. More soldiers died from disease and poor care than from actual combat. Previous medical and procedural advances during the Napoleonic Wars were forgotten. The Civil War was indeed a dark time for medicine and warfare.

Impact

One great result of the two world wars was the improvement of surgery. The battlefields taught surgeons important lessons that were then translated into civilian life. While initial principles of casualty evacuation and surgery advanced during the Napoleonic Wars had to be relearned, some of the new investigations led to advances during wartime.

World War I broke the tradition by which surgeons had to study under a master surgeon for years. Suddenly, young surgeons were pressed into dealing with situations that would have daunted their masters. The surgeon's training had been in clean, or antiseptic, surgery of the carbolic acid-sprayed hospital operating rooms, introduced by Joseph Lister (1827-1912). Now, high-speed bullets and missiles were tearing flesh in such a way that bones were shattered and human flesh destroyed. The fertile farmland of Flanders was rife with bacteria that produced infection. When the men were wounded, skin would be punctured and the individual would get the anaerobes that cause tetanus and gangrene. Never had gangrene been seen with such a high incidence. Tetanus was found in 8.8 per 1000 wounds.

Surgical units, called Casualty Clearing Stations, went back to the days of Ambroise Pare, but now frontline surgeons found that they must remove all dead tissue, take out foreign materials, and leave the wound open. This process, called debridement, removes dead, infected and decaying tissue. The wound was lightly packed with gauze so that the well-oxygenated areas would have a chance of overcoming infection. About five days later, a technique called "delayed primary suture" would close the wound. The discovery of debridement and surgical excision of dead and dying tissue became a standard of care during warfare.

There was still a search for antiseptic treatment. Alexis Carrel (1873-1948) and Henry Dakin (1880-1952) developed the Carrel-Dakin treatment, in which a sodium hypochlorite or bleach solution was used to clean the wound. However, this very complicated process was effective only after the wound had been debrided. Other controls included antitoxin for tetanus and antiserum injections for gangrene.

World War I also established some first aid procedures like the early splinting of fractures. Harvey Cushing (1869-1939) treated gunshot wounds to the head with gentle suction and used powerful magnets to remove metal objects from the brain.

Since the days of the South African War, or Boer War (1899-1902), army surgeons left abdominal wounds to heal on their own. By 1915 it became evident that early surgery for abdominal wounds was important, since missile injuries to that area are usually fatal. Chest wounds were opened up to drain pus from the pleural cavity.

An enduring advance from World War I was the treatment of injuries that disfigure the face and other body parts. Harold Gillies (1882-1960), a pioneer in the field of reconstructive surgery, developed the pedicle flap to move skin from one part of the body to another. In this graft, subcutaneous tissues are temporarily attached to the site where the graft was taken. Plastic or cosmetic surgery would later become an important branch of medicine.

World War I gave doctors and medicine a voice. Working previously on their own, doctors were now forced to work together and began to see the value of centralized planning. Nurses became an integral part of medicine and all personnel saw the advantages of a coordinated, well-organized system. World War I confirmed that health was a national concern, so much so that in Great Britain liberal politicians like Lloyd George began to push for a national health service—free to all. The service would come to fruition after World War II when Nye Bevan and the Labor Party passed the National Service into law in 1948.

When World War II began, military doctors were again pressed into service. The urgency of tending to casualties at the earliest possible time was paramount. Air lifting, which had only been hinted at during World War I, helped to evacuate the wounded. Anesthetics were available to help alleviate pain. Blood transfusions, which were enabled by blood typing and understanding of related factors, helped to treat blood loss and shock. World War II marked the first use of blood transfusions on a massive scale.

Sir Alexander Fleming (1888-1955), while experimenting in his laboratory, found that bacterial plates that had mold did not have bacteria growing in the area. The search for an agent that would kill bacteria inside the body soon met with success. At Oxford, Howard Florey (1898-1968) and Ernst Chain (1906-1975) investigated a type of fungi that might have anti-bacterial properties and found *Penicillium notatum*. By 1941 penicillin was sent to wounded troops in North Africa to fight infection.

Medical research took an unusual twist in Germany. In 1883 Chancellor Otto von Bismarck implemented a national health service for all. During the Weimar Republic (1918-1933), the period between World War I and II, clinics continued to provide services for mothers and children. German pride was intense, and the passion for national fitness meant building the strong and eliminating the weak. When Adolf Hitler became chancellor in 1933, the sentiment that certain groups were weak and inferior had the backing of many doctors and psychiatrists. Physicians and scientists supported sterilization of people with mental disabilities and "mercy deaths" for those with physical handicaps. Between January 1940 and September 1942 physicians sent 70,723 mental patients—whose "lives were not worth living"—to their deaths in the gas chambers. Nazi doctors similarly dictated who would live and die at concentration camps like Auschwitz and others.

German medical scientists also saw the opportunity for human experimentation "for the advancement of science." To study the effects of mustard gas—or phosgene—used in World War I, gangrene, freezing, and typhus, camp doctors used inmates as their test subjects. Children were brutally injected with petrol, frozen, drowned, or slain for dissection.

The leading doctor at Auschwitz was Josef Mengele (1911-1979). A highly trained physician with doctorates in medicine and anthropology, Mengele was dedicated to human experimentation. He chose over 100 pairs of twins, injecting them with typhus and tuberculosis, and after their deaths sent the organs to other doctors for study. Through autopsy, he investigated the different reactions of various races to infectious disease. The projects were financed by the German Research Facility headed by Ernst Sauerbruch (1875-1951), an eminent surgeon.

In the Pacific area, Japanese doctors created "Unit 731" in 1936. In Northern Manchuria, Dr. Shiro Ishii set up studies of lethal microbes—anthrax, dysentery, typhus, cholera, and bubonic plague. Using prisoners, he determined the quantity of lethal bacteria necessary to ensure epidemics and even sent disease bombs into China. Dr. Ishii was never prosecuted because he gave the United States his trade secrets for manufacturing anthrax and other biological weapons. American citizens were shocked to learn that the United States had subjected troops to secret radiation tests as part of the atomic studies program.

Reactions against such tests spawned an ethical movement in medicine drawn up as part of the Nuremberg Code and refined in the Declaration of Helsinki in 1964. The prevailing principle states that "fully informed and voluntary consent of the subject is essential for any human experimentation." Medicine in the early part of the twentieth century was defined by war. By 1950, medicine was the hope to answer all ills. Unfortunately, the era of optimism would not last.

EVELYN B. KELLY

Further Reading

Gabriel, Richard A., and Karen S. Metz. *A History of Military Medicine: From the Renaissance Through Modern Times*. Westport, CT: Greenwood, 2000.

Gold, Hal. *Unit 731 Testimony: Japan's Wartime Human Experimentation and the Post-War Cover-up*. Rutland, VT: Charles E. Tuttle, 1996.

Lifton, Robert Jay. *The Nazi Doctors: Medical Killing and the Psychology of Genocide*. New York: Basic Books, 1986.

Fluoridation and the Prevention of Tooth Decay

Overview

Early in the twentieth century dentists became aware that tooth decay, or dental caries, was a public health problem. Several scientists had investigated tooth decay and connected it to bacteria residing in the plaque or build-up on the teeth. Proper hygiene and care were being encouraged and research led to an unusual source of prevention—fluoride. Fluoride is a chemical found in water in some areas of the country. At very high levels it is toxic. Because of this fact, many groups objected to the addition of fluorine to the public drinking supplies of a town. The pros and cons of adding fluoride continue today.

Background

As the nineteenth century ended, researchers in the tradition of Louis Pasteur (1822-1895) and Robert Koch (1843-1910) proposed that bacteria might be a cause of tooth decay. The processing of food that became popular between 1860 and 1960 doubled the average consumption of sugar. This led to an increase in tooth decay as well as other health problems like diabetes. In 1890 Willoughby Miller, a student of Dr. Koch, described his findings in the book *Microorganisms of the Human Mouth.* Miller stated how food bits would decay between the teeth, attack the enamel of the tooth, and cause cavities. Another scientist, J. Leon Williams, showed that dental decay results from acid acting on a particular spot on the tooth. That acid is produced by bacteria in an area of plaque buildup. A gelatinous "gunk" prevents that area from coming in contact with the saliva, which has a germicidal effect. Williams advocated frequent and thorough cleaning of the teeth. In 1896 he introduced the slogan, "A clean tooth never decays."

The connection between dental decay and oral hygiene was the subject of much research at the turn of the twentieth century. In 1910 Sir William Hunter pointed out that infections around the teeth affect general health. He wrote an article called "The Role of Sepsis and Antisepsis in Medicine." He also advocated teaching medical subjects to dental students, as heart disease and other problems are related to the condition of the teeth.

When W. G. Ebersole became chairman of the Oral Hygiene Committee of the National Dental Association in 1909, he campaigned for dental colleagues to wage war for good dental hygiene. He designed a plan to teach school children about good dental care. The discovery of fluoride added perfectly to this idea of prevention.

Impact

Fluoride propelled dentistry into the area of prevention rather than just correcting problems. In 1901 Frederick Sumter McKay left the East coast to set up a dental practice in Colorado Springs, Colorado. He was intrigued by the large brown, chocolate stains on the teeth of many area residents. Many people had stains so severe that the entire tooth was splotched the color of chocolate candy. No one could explain the condition, and there was nothing written about it in the literature of the day. The natives attributed it to such things as too much pork, inferior milk, and calcium-rich water. McKay decided to launch his own investigation. He sought the help of local dentists, but they were not interested in the problem and scoffed at McKay for undertaking such studies.

McKay suspected that the stains, which became known as the Colorado Brain Stain, might have something to do with the water. A big break came in 1909 when McKay posed the problem to renowned dental researcher Greene V. Black (1836-1915) of Northwestern University. Black made his way to Colorado Springs and, likewise, was amazed that such a disorder would not be reported in dental literature. The Colorado Springs Dental Society did issue a report that 90% of the city's locally born people had signs of the brown stain. While watching the children of Colorado Springs at play, Black lamented that it was not just a problem of childhood but a deformity for life.

When Black died in 1915 he had spent six years investigating the condition. He and McKay had determined: (1) the mottled teeth resulted from a developmental imperfection to teeth; (2) people with the stain had fewer cavities. There were still no clues as to the cause. A big break came in 1923 when McKay was invited to Oakley, Idaho, across the Rocky Mountains to talk with parents who had noticed strange brown stains on their children's teeth. They noted that the stains appeared after the community had installed a pipeline coming from a warm spring five miles away. McKay suggested they abandon the pipeline for another. It worked; the stains began to disappear.

McKay and Dr. Grover Kempf of the United States Public Health Service were invited to Bauxite, Arkansas, a town owned by the Aluminum Company of America (Alcoa). They analyzed the water but found nothing. At Alcoa's head office in Pennsylvania, their report came across the desk of head chemist H. V. Churchill. He brought samples from Bauxite and used a very sophisticated technique of water analysis called photospectrographic analysis. He could hardly believe it when he found the element fluorine in the water. He wrote a letter to McKay about his revelation and then asked him to send water samples from all the communities with the brown stain problem. McKay had his answer. The connection was made between the mottling and the presence of fluorides in the water.

Fluorides are compounds of the element fluorine. The element is in the same chemical family as chlorine, a group called halides, and in many substances in nature. Actually, by itself fluorine is a strong, brownish-yellow, poisonous gas. It is very reactive and combines with most elements. The term fluoridation began to be used with the addition of the chemical to the water supply.

Armed with the initial studies, health scientists started looking seriously at fluoride. H. Trendley Dean, head of the Dental Hygiene Unit at the National Institute of Health, became interested in fluorine in 1931. He wanted to know how high the level of fluoride must be before staining began. He assigned Dr. Elias Elvove, a chemist, the task of determining this level. He had state-of-the-art equipment that could measure an accuracy of 0.1 parts per million (ppm). He traveled the Untied States measuring fluorides in drinking water and determined the safe threshold was 1 ppm. Above this level, fluorosis—the term given to the accumulation of fluoride in teeth enamel—began to occur.

In 1938 the Public Health Service was involved in a study of two communities. Galesburg, Illinois, a community with a high degree of tooth mottling, was compared to Quincy, Illinois, where the water was free of fluorides. The study showed children of Galesburg had fewer cavities than the children of Quincy.

The Public Health Service then went in another direction. Having determined that one part of fluoride per million of water was the ideal level to prevent tooth decay but not cause mottling, they reasoned that adding fluoride to a community's water would reduce decay. Extensive tests were done on animals, and in 1945 the Public Health Service concluded it was safe to begin tests. Water systems in Newburgh, New York, and Grand Rapids, Michigan, became the first to add the chemical sodium fluoride to water. The city of Kingston, New York, was a control group. During a 15-year study, researchers monitored the rate of tooth decay among Grand Rapids' 30,000 school children. The conclusion was that fluoride dropped decay by about 60%. A group of dentists in Wisconsin also added fluoride in their state.

Results from these initial studies indicated a reduction of dental caries by two-thirds. The results encouraged the Public Health Service to recommend in 1950 that fluorides be added to the public water supplies of all communities. The next year the American Dental Association (ADA) also endorsed it. The medical, dental, and public health communities were solidly behind adding fluoride to water.

A large outcry arose among some groups who believed that adding fluorides might be unsafe and risky. Some groups decried the addition of a chemical that, in large amounts, was toxic. Others objected that people did not have a choice if they wanted the chemicals or not. Some communities resisted and turned down the proposal in a referendum. Today, about 60% of people live in communities with fluoridated water. About 30 other countries practice fluoridation.

Other ways of applying or getting fluorides emerged. The fluoride may be taken in tablet form, in a mouthwash solution, or painted onto the teeth. In the early 1950s Proctor and Gamble (P&G) had the idea of adding the chemical to toothpaste. Researchers at the University of Indiana, working with stannous fluoride, found a way to bond the substance to teeth. Then, in 1956, P&G introduced its "Crest" toothpaste with fluoristan and launched an immediate media blitz with the refrain, "Look mom, no cavities." The timing was right because television was just coming into households throughout the country. Later, the Council on Dental Therapeutics gave "Crest" their seal of approval.

The "Crest" claim that cavities could be reduced by two-thirds has since been modified to about 20-25%. However, the addition of fluoride, combined with education, has been an effective tool for dental care and prevention of cavities.

EVELYN B. KELLY

Further Reading

Meinig, George E. *Root Canal Cover-up*. Ojai, CA: Bion, 1994.

Porter, Roy. *The Greatest Benefit to Mankind: A Medical History of Humanity*. New York: W. W. Norton, 1998.

Travers, Bridget, ed. *World of Invention*. Detroit: Gale Research, 1994.

Founding of Major Health Organizations

Overview

Three major heath initiatives were founded in 1948—the National Institutes of Health (NIH), the National Health Service in the United Kingdom, and the World Health Organization (WHO). Each of these organizations has a background with different missions, and each would develop in completely different directions. While the National Health Service administers health care for all citizens in the United Kingdom, NIH provides for research into major areas of medicine and health, and WHO seeks not only the elimination of worldwide disease, but the well-being of all nations.

Background

The history of the medical profession up to the end of the nineteenth century was a jumble of patient-doctor relationships. Doctors were self-employed with little support, back-up, or help. They may have met briefly at a college or social gathering, but basically it was each unto himself. A major change occurred in 1883. Chancellor Bismarck of Germany set up state-run medical insurance, and politicians immediately seized upon the issue as a service appealing to an entire electorate.

Before 1900, organized medical practice was just beginning. Individual doctors and hospitals operated without much direction and with little government intervention. But as medicine became more complex and new developments, techniques, and technologies grew, it became evident that governing policies and initiatives were necessary.

In 1887 a small, one-room laboratory was set up with the Marine Hospital Service (MHS) in Staten Island, New York. The MHS had been created in 1798 to provide for the medical welfare of merchant seamen; 20 cents per month was deducted from the salaries of the seamen to pay for the services. In the 1880s Congress expanded the role of MHS to examine passengers arriving in the United States for dreaded diseases like cholera and yellow fever. With developing evidence that microorganisms caused disease, European scientists, especially in France and Germany, set up laboratories for the public's health.

The MHS authorized Joseph J. Kinyoun, a physician trained in the new bacteriology at the Marine Hospital in Stapleton, Staten Island, to form a Hygienic Laboratory. In 1884 Robert Koch (1843-1910) found the cause of cholera, the most feared communicable disease of the nineteenth century. Kinyoun almost immediately found this bacteria in suspicious cases and confirmed the diagnosis. Kinyoun and the Hygienic Lab initiated what would become the NIH. In 1891 the Hygienic Laboratory was moved to Washington, D.C., near the capitol, and for 10 years Kinyoun was its only full-time employee.

Before 1900, in Great Britain doctors practiced where they could find patients, and the development of hospitals was usually associated with religious groups or orders.

Epidemics and plagues cut across national borders. The hope of medical internationalism first emerged in 1859 at the International Sanitary Conference in Paris. Concern for the health of the world and developing nations grew out of the missionary tradition. The nineteenth century was known for the great missionary movement, and missionaries were responsible for bringing Western medicine to non-Western people throughout the world. They evoked the idea of compassion and caring for the world as well as its health.

The basic mission of organizations like the NIH, National Health Service, and WHO was not altogether new, however, it was not until the mid-twentieth century that conditions were ripe for the development of these organizations.

Impact

The 1948 establishment of the National Health Service in Great Britain can be traced to earlier developments. In 1911 the liberal politician

Lloyd George launched a National Insurance scheme, modeled after Bismarck. Working wage earners would contribute four pence per week from their paychecks, the employer would give two pence, and the state three. The workers in turn would receive approved medical treatment from a "panel doctor" of their choice. The electorate loved the plan, as they received much for quite a little. However, National Insurance was a divisive issue; the conservatives were generally against it, while the labor party supported it. The first half of the twentieth century saw two major wars that devastated Great Britain. Despite the distractions of basic survival, the wars had shown British physicians that an organized, well-planned effort was much more efficient. The blueprint for total reform was written in 1942 by Sir William Beveridge (1879-1963), proposing a new health service be available to everyone.

In postwar Britain, the conservative party of Winston Churchill was defeated and Clement Attlee, a member of the Labor Party, became prime minister in 1945. Attlee appointed as minister of health Aneurin (Nye) Bevan (1897-1960), the son of a miner, who became leader of the left-wing group of the labor party called the Bevanites. A bill was introduced in April 1946, for the national health program. It was signed by the queen and began July 5, 1948.

Bevan became the architect of the National Health Service. The law provided free medical services to everyone. The person registers with a general practitioner, and the government pays a capitation, or head fee, according to the number of people enrolled with him or her. A physician may operate free of the National Health Service but few do. The maximum number of patients allowed for a physician is 3,500; the average is 2,500. Before about 1950 doctors visited patients in their homes, making as many as 15-20 calls a day, as well as seeing patients at the office in the evenings . This practice changed after the midpoint of the century.

Likewise, Bevan nationalized both municipal and charity hospitals under National Health, with government-paid specialists providing the services. The National Health Service is financed by general taxes, with some local and private fees. When implemented, NHS was powerful, popular, and cheap compared to other standards. However, with increased technology and more expensive and longer hospital stays, the system has recently become financially stressed.

A world health initiative has been the dream of many people. In 1909 twenty-three European nations established a permanent office called the International Office of Health. It established a permanent base in Paris. The mission was to collect knowledge about infectious diseases. After World War I, the Treaty of Versailles created the League of Nations, which, in 1923, set up a Health Organization to promote the "highest possible level of health" for all peoples. When Hitler invaded Poland in 1939, the League of Nations failed, but some retained the idea of an organization dedicated to helping every nation. In 1948 the French Organization Mondiale de la Santa, a specialized agency of the United Nations, was established. The organization, called the World Health Organization, now has headquarters in Geneva, Switzerland. Its three main purposes are: 1) to serve as a central clearinghouse for research (member countries are informed of vaccines, cancer research, and other discoveries); 2) to control epidemics and endemic diseases by promoting mass education programs; and 3) to expand public health areas.

WHO is administered by a World Health Assembly, which meets annually and has a Secretariat with offices and field staff throughout the world. After 1951 the technical assistance program of the Untied Nations allotted money to WHO.

In the United States, the Hygienic Laboratory received recognition in 1901, along with $35,000 from Congress to build a new laboratory. The agency was renamed the Public Health and Marine Hospital Service (PH-MHS). Completed in 1904, the new building was located in Washington and provided for the investigation of infectious and contagious disease and matters pertaining to public health. As Congress was skeptical of the values of such bureaucracies, they reserved the option not to renew the funding if it did not work.

In 1912 another law shortened the name from PH-MHS to Public Health Service (PHS) and, in addition to contagious diseases, authorized the organization to study non-contagious diseases and the pollution of lakes and streams. One PHS researcher, Dr. Joseph Goldberger (1874-1929), identified pellagra as a dietary deficiency of poor Southerners and recommended a cure.

The delivery of medicine in the United States took a different path. Emphasis remained on the free market, not on the state. An organization of doctors, the American Medical Association, became a powerful force championing health causes. In 1912 a short-lived Progressive Party championed national health care but the free market idea prevailed and the national impetus became a system of supporting research.

In 1930 the PHS was renamed the National Institute of Health; the singular version of "Institute" was then used. Louisiana Senator Joseph E. Ransdell sponsored a bill that changed the name from the Hygienic Laboratory, and made its purpose to establish fellowships for research into basic biological and medical problems. Although the country was deep in the Great Depression at that time, this legislation, which became known as the Ransdell Act, marked a great change in attitude toward funding medical research.

The problem of cancer was attracting growing attention and, in 1937, every senator voted to fund the National Cancer Institute (NCI). A new campus was built in Bethesda, Maryland, to house the NCI and, in 1944, it was designated that the NCI would be part of the NIH.

World War II prompted the NIH to focus on war-related efforts and to improve conditions in the defense industry. After the war, Congress appropriated money to expand the NIH and, in 1948, the name changed to the plural National Institutes of Health. By 1998 the NIH had 24 institutes and centers.

Founded in the aftermath of World War II, all three major organizations—the NIH, National Health Service, and WHO—reveal the importance of cooperative medical care and a new commitment to public health on both a national and international scale.

EVELYN B. KELLY

Further Reading

Beigbeder, Yves, et al. *The World Health Organization.* Boston: M. Nijhoff, 1998.

Harden, Victoria. *Inventing the NIH: Federal Biomedical Research Policy, 1887-1937.* Baltimore: Johns Hopkins University Press, 1986.

Porter, Roy. *The Greatest Benefit to Mankind: A Medical History of Humanity.* New York: W. W. Norton, 1997.

Aspirin

Overview

Aspirin is a trade name for acetylsalicylic acid (ASA), a mild, nonnarcotic analgesic that was first marketed by the Bayer Company in 1899. Until the end of World War I, the name "Aspirin" (with a capital *A*) was a trademark of the German firm Bayer. Aspirin quickly became one of the most widely used drugs in the world, but disputes over the discovery of aspirin were still raging at the end of the twentieth century. Often called a miracle drug, aspirin was originally prescribed for the relief of inflammation, headaches, muscle pains, and fever. Before the introduction of antibiotics, aspirin was also used in the treatment of rheumatic fever, rheumatoid arthritis, and infections. Although aspirin generally does not modify or shorten the duration of these diseases, it does relieve some of their symptoms. One hundred years after it was first marketed, scientists continue to find new uses for aspirin. Indeed, statistical evidence indicates that taking aspirin can reduce the risk of a second heart attack and may reduce the risk of heart disease, colon cancer, and preeclampsia during pregnancy.

Background

During the time of Hippocrates (460-377 B.C.), a preparation made from the bark and leaves of the willow tree was prescribed as an analgesic and antipyretic. (Analgesics are drugs that alleviate pain without affecting consciousness and antipyretics are used to reduce fevers.) The antipyretic properties of willow bark were rediscovered in the eighteenth century, but the active ingredient, salicin, was not isolated until 1828. Many other important drugs were first isolated from plant sources during the nineteenth century. Among the alkaloids purified during this time period were narcotine, morphine, emetine, strychnine, brucine, piperine, colchicine, quinine, nicotine, atropine, cocaine, and physostigmine. Chemical purification made it possible to study the active factors of these substances and to eliminate the toxic effects of impurities. Progress in pharmacology also made it possible to prescribe accurate doses of purified compounds. As scientists elucidated the chemical structures of pharmacologically active compounds, they were able to synthesize important derivatives of existing drugs and create entirely new drugs in the laboratory.

Although salicin, salicylic acid, and sodium salicylate were introduced as pain relievers in the 1870s, the salicylates caused nausea, ringing in the ears, and irritation to the stomach lining. Based on experience with other drugs, Bayer chemists hoped that adding an acetyl group to salicylic acid would make it less irritating to the stomach. (Indeed, adding acetyl groups to the morphine molecule led to the production of another highly successful Bayer product that was sold under the trademarked name "Heroin" as a new and superior cough suppressant.)

In 1897 Felix Hoffman (1868-1946), a German chemist working for Bayer, synthesized and purified acetylsalicylic acid. Two years later, the drug was given the trademark name "Aspirin" (*A* for acetyl chloride, *spir* from the plant *Spiraea ulmaria,* and *in* being a common ending for names of medicines) and was distributed to physicians and hospitals for testing. Pleased by the enthusiastic response of patients and physicians, Bayer marketed the analgesic as a safe remedy for the treatment of arthritis, joint pains, muscle pains, and fever. Aspirin was sold as a powder until 1915, when Bayer began to market Aspirin tablets. Until 1915, Aspirin, which would become one of the most popular over-the-counter drugs for self-medication, was not available to patients unless they obtained a prescription from their doctor.

Within a few years of each other, Bayer had named and marketed both Aspirin and Heroin, two drugs that were to remain among the most widely used and abused in Western society over a hundred years later. Until the end of World War I, the Bayer Company maintained its valuable trademark protection for Aspirin and Heroin. After Germany's defeat, however, Bayer was forced to give up these trademarks as part of the country's war reparations. In 1919 the Treaty of Versailles assigned Bayer's trademarks to France, England, Russia, and the United States. Eventually, "aspirin" and "heroin" became the common names of these drugs. Bayer Aspirin, however, remained the most familiar brand in what would become a highly competitive analgesic market. Advertisements touted Bayer's product as "pure aspirin, not part aspirin" and claimed that "nine out of ten doctors recommend Bayer's aspirin." Other pharmaceutical companies sold aspirin in combination with caffeine, antacids, and special coatings, but no medical studies have found any version of the drug to be better than plain aspirin for headaches, fever, and inflammation.

Bayer had established its presence in the United States as early as 1903, when Friedrich Carl Duisberg Jr. (1861-1935) came from Germany to Rensselaer, New York, to build a chemical plant. The most important product of this plant was Bayer Aspirin. Duisberg was a chemist and the founder of the industrial empire known as I.G. Farbenindustrie (or I.G. Farben), the chemical empire that became identified with Nazi Germany. After World War II, the directors of I.G. Farben were tried along with Nazi war criminals at Nuremberg.

Impact

The Bayer Company had begun as Friedrich Bayer & Company, one of Germany's first dye factories. Friedrich Bayer and Johann Friedrich Weskott established the firm in 1863. After Weskott and Bayer died, the firm was managed by Carl Rumpff, Bayer's son-in-law. The firm then became officially known as Farbenfabriken vormals Friedrich Bayer & Company ("the Dye Factory formerly known as Friedrich Bayer & Company"). When Rumpff died in 1890, Duisberg assumed leadership of the company. As the director of Bayer's research and patent program, part of Duisberg's job was to diversify production and create new products. Duisberg thought that the market for antipyretics—then limited to salicylic acid, quinine, antipyrine, and acetanilid (Antifebrin)—was promising. Acetanilid, a coal-tar derivative whose antipyretic properties were discovered in 1886, was the first German drug to be marketed by a commercial name rather than its chemical or generic name. Seeing the successful marketing of Antifebrin, Duisberg decided to enter the market for fever and pain relievers.

Ernst Arthur Eichengrün (1867-1949), a research chemist who ran the pharmaceutical section of Bayer's drug laboratory, assigned the task of producing a form of salicylic acid with fewer side effects to Felix Hoffmann. According to Hoffmann's laboratory notes, he isolated acetylsalicylic acid (ASA), a derivative of salicylic acid, on October 10, 1897. Although the same compound had been produced by the French chemist Charles Frédéric Gerhardt in 1853 and by the German chemist Karl Johann Kraut in 1869, Hoffmann's methods were an improvement over those of his predecessors.

When the drug was tested by Heinrich Dreser (1860-1924), who, as the director of the pharmacological section of Bayer's drug laboratory, had to approve all the drugs prepared by Eichengrün and his associates, it was initially rejected. Apparently, Dreser thought that ASA might "enfeeble" or exhaust the heart, much like salicylic acid. Because these drugs were often given in large doses for the pain and swelling of rheumatic fever, many patients experienced severe reactions, including shortness of breath, panting, and the feeling that their hearts were racing. Thus, Bayer ignored ASA for over a year while the company promoted its successful new drug, Heroin. In 1898, at the Congress of German Naturalists and Physicians, Dreser announced that Heroin was ten times as effective as codeine as a cough medicine and only one-tenth as toxic. Heroin was promoted as a safe, non-habit-forming drug that could be used to treat colic in babies and to cure morphine addiction.

Eichengrün, who resented Dreser's veto power, tested ASA himself and passed it on to doctors, who reported that it relieved fevers, aching joints, and headaches. When Dreser was finally forced to accept ASA, he wrote a paper on its beneficial properties but neglected to mention either Hoffmann or Eichengrün. Dreser earned royalties on sales of ASA and became wealthy, but Hoffmann and Eichengrün did not share in this wealth because they could only receive royalties for drugs that had been patented. The German patent office refused to patent ASA because it was not a new product nor did its synthesis involve a new process. Bayer fought for patent protection in other nations—with varying success. Notably, Bayer enjoyed both trademark and patent protection on Aspirin only in the United States. In 1899 Bayer began distributing Aspirin to doctors, medical professors, and hospitals. By 1902 about 160 scientific studies had been reported and a voluminous literature on Aspirin had been published. Most reports were very positive. The drug was taken for long periods of time in large doses for inflammation and arthritis. Smaller doses of Aspirin were prescribed for fevers and headaches.

One hundred years after Bayer introduced Aspirin, Walter Sneader, a Scottish scientist and an expert on the history of pharmacology, announced that the man who deserved credit for the discovery of aspirin was not Hoffmann but Eichengrün. Because Eichengrün was Jewish, he had been incarcerated in a concentration camp during World War II and his role in the success of Bayer Aspirin concealed by the Nazis. Although Hoffmann did indeed synthesize and purify acetylsalicylic acid in 1897, the project had been initiated and supervised by Eichengrün. It was not until 1934 that Hoffmann began to assert that he had discovered Aspirin. By that time, however, the Nazis were in power and Eichengrün was unable to challenge Hoffmann's story. After the war, Eichengrün attempted to correct the commonly accepted story of the discovery of aspirin, but he died in 1949, shortly after his account was published. While examining Hoffmann's laboratory notes, Sneader found that Hoffman had adapted methods previously developed by Eichengrün.

Of all the products sold by Bayer in the United States, Aspirin quickly became the most important. By 1909 it accounted for over 31% of Bayer's total sales in the United States. Because Bayer did not have patent protection for Aspirin in Canada, lower-priced Aspirin was smuggled into the United States from Canada, where it sold for about one-third the price. In 1910 the Bayer Company successfully sued a pharmaceutical dealer engaged in smuggling and its American patent was upheld. Bayer, however, estimated that counterfeit Aspirin held about 75% of the American market.

In anticipation of the expiration of patent protection in 1917, Bayer tried to boost sales by bringing the trademarked Bayer Aspirin name to the public. Aspirin tablets were stamped with the Bayer Cross and tablets were enclosed in Bayer packages. The change from powder to tablets had begun in 1914. In 1916 the company began placing ads for Bayer Aspirin in American newspapers. The American Medical Association, which was opposed to the advertising of trademarked or patented products, argued that ASA should be sold under its generic name and that Bayer should not have a monopoly on the drug. At the end of World War I, the U.S. government sold the American assets of Farbenfabriken Bayer to American businesses.

The mechanism by which aspirin works was discovered in the 1970s by the British pharmacologist John Vane (1927-), a 1982 Nobel laureate, who found that aspirin inhibited the release of the hormone-like substances called prostaglandins. Prostaglandins were discovered in 1935 by the Swedish physiologist and 1970 Nobel laureate Ulf Svante von Euler (1905-1983). Prostaglandins are modified fatty acids that have diverse functions in the body; they affect blood-vessel elasticity, blood platelets and clotting, the functioning of the digestive tract, sensitivity to pain, and inflammation. The studies by these researchers have provided a molecular explanation for using aspirin in the treatment of cardiovascular disease and inflammatory diseases, including rheumatoid arthritis.

Of course, taking aspirin is not without risk. Aspirin can cause allergic reactions and gastrointestinal problems. In children with viral infections such as influenza or chicken pox, the use of aspirin has been linked to the onset of Reye's syndrome, an acute disorder of the liver and central nervous system.

Many other drugs, such as acetaminophen, ibuprofen, naproxyn sodium, and ketoprofen, have challenged aspirin's place as an analgesic. Other pain relievers, however, do not seem to mimic aspirin's beneficial impact on cardiovascular health. Although in the 1920s Bayer's advertisements assured consumers that Aspirin would not affect the heart, clinical studies in the 1940s indicated that men who took aspirin were less likely to suffer heart attacks. Some physicians began to suggest that all their patients should take an aspirin tablet every day to reduce the risk of a heart attack. By the end of the twentieth century, aspirin was being prescribed by physicians as a means of preventing heart attacks. Most of the 80 million aspirin tablets taken by Americans every day in the 1990s were taken to reduce the risk of heart disease and stroke. More medical uses for aspirin are still under study, including the treatment of migraine headaches, the prevention of various cancers, and the improvement of memory and brain function.

LOIS N. MAGNER

Further Reading

Ackerknecht, E. H. *Therapeutics: From the Primitives to the 20th Century.* New York: Hafner Press, 1973.

Barnett, H. J. M., and J. F. Mustard, eds. *Acetylsalicylic Acid: New Uses for an Old Drug.* New York: Raven Press, 1982.

Borkin, J. *The Crime and Punishment of I.G. Farben.* New York: The Free Press, 1978.

Fields, W., and N. Lemak. *A History of Stroke.* New York: Oxford University Press, 1989.

Irwin, Michael H. K. "Aspirin: Current Knowledge about an Old Medication." (Public Affairs Pamphlet, No. 614.) New York: Public Affairs Committee, 1983.

Jones, Charlotte Foltz. *Mistakes That Work.* New York: Doubleday, 1991.

Mann, Charles C., and Mark L. Plummer. *The Aspirin Wars: Money, Medicine, and 100 Years of Rampant Competition.* New York: Alfred A. Knopf, 1991.

Vane, John R., and Regina M. Botting, eds. *Aspirin and Other Salicylates.* New York: Chapman & Hall Medical, 1992.

Young, James Harvey. *The Toadstool Millionaires: A Social History of Patent Medicines before Federal Regulation.* Princeton, NJ: Princeton University Press, 1961.

Biographical Sketches

Sir Frederick Grant Banting

1891-1941

Canadian Physician and Physiologist

Frederick Banting, along with Charles H. Best (1899-1978), discovered insulin, a hormone secreted by the pancreas that regulates sugar in the blood. This discovery led to the treatment of diabetes, a fatal disease. For the discovery of insulin, Banting won the 1923 Nobel Prize for physiology or medicine, the first to be awarded to a Canadian.

Born November 14, 1891, near Alliston, Ontario, Canada, Banting was the youngest of five children of William Thompson Banting, a farmer whose parents were Irish immigrants. Banting attended the University of Ontario with the goal of becoming a Methodist minister, but changed to medicine and completed his degree in 1916.

After graduation, he was immediately drafted into the Canadian Medical Corps as a lieutenant during World War I. In 1918, while on the battlefield in France, his right arm was severely wounded by shrapnel, but he continued to attend to wounded men despite his own injury. In 1919 the British Government awarded him the Military Cross for bravery.

Returning to London, Ontario, in 1920, Banting set up a surgical practice. During the first year, however, he had only one patient. To help supplement his income, he took a job as lecturer in surgery and anatomy at the University of Western Ontario and became fascinated with the whole concept of medical research.

One sleepless night, while preparing for a lecture on how the body metabolizes carbohydrates, he came upon the work of two early researchers who found that when the pancreas of a dog was removed, the animal developed diabetes. However, when a small area of cells, called the islets of Langerhans, was preserved, the animals were normal. The studies gave Banting an idea of how he could establish the role of the pancreas in diabetes. He would tie off the pancreatic ducts of the dogs, preserving the cells of the islets. After 6-8 weeks, he would make an extract of the cells and inject it into dogs made diabetic by complete pancreas removal.

Banting traveled to Toronto to talk to J.J.R. Macleod (1876-1935), an expert of carbohydrates. Macleod was not impressed, but Banting persisted and convinced him that he could prepare an extract and give it to dogs that had been made diabetic through surgery. After six months of persuading, Macleod gave Banting a laboratory and provided him with an assistant who knew how to do analysis. That assistant, 22-year-old Charles Best, had just finished an undergraduate degree in chemistry.

Banting and Best performed the experiment. They chopped up the cells of the islets of Langerhans and injected the extract, which they called isletin—later insulin—into dogs with surgically removed pancreases. The dogs recovered. To avoid surgery to get insulin, they used the glands of fetal calves from the slaughterhouse and injected the dogs that were made diabetic by removal of the pancreas. Macleod, who had been on vacation during these experiments, was pleased with the work and gave Banting a position and enlisted James Collip (1892-1965), a Ph.D. in biochemistry, to assist.

In 1922 the researchers tested insulin on 14-year-old Leonard Thompson, who had suffered with diabetes for two years. The use of insulin and a regulated diet enabled Thompson to live for 13 more years, when he died in an accident. The university contracted with the pharmaceutical firm Eli Lilly to produce insulin.

In 1923 the Nobel Prize for physiology or medicine was awarded to Banting and Macleod. Banting was furious that Macleod had a part in the award and not Charles Best, who had worked so diligently with him. The reason given was that Best did not have a doctorate at the time and Nobel prizes are only given to credentialed scientists. Banting gave half of his prize to Best. Macleod then gave half of his prize to Collip.

Banting received much praise and was given an annuity by the Canadian government. In 1923 he became head of the Banting and Best Department of Medical Research at the University of Toronto. He was knighted by the British Crown in 1934.

Frederick Banting. *(Library of Congress. Reproduced with permission.)*

Not content with these successes, Banting launched into new areas of research on cancer, silicosis, and the effect of flying at high altitudes on the body.

Banting married in 1924 and had one son. The unhappy marriage ended in divorce in 1934, and in 1939 he married a technician from his laboratory at the university. He became very interested in art and collected paintings by Canadian artists.

When Canada entered World War II in 1939, Banting volunteered for service. He was killed in an airplane crash off the coast of Newfoundland on February 21, 1941, while on a medical military mission.

EVELYN B. KELLY

Alfred Binet
1857-1911
French Psychologist

Alfred Binet played an important role in the development of experimental and child psychology. Together with Theodore Simon, he developed the first tests designed to measure intelligence and compute *mental age*. Many present-day intelligence tests are based on the original Binet-Simon tests.

Binet was born on July 18, 1857, in Nice, France. His father was a physician, and his mother an artist. Although trained as a lawyer, he found that the legal profession did not suit him; he was more interested in understanding the workings of the human brain. In 1878 he went to study and work at the Salpetriere Hospital in Paris under the neurologist Jean Charcot (1825-1893). He married Laure Balbiani, the daughter of an embryologist at the College de France, in 1884. He remained at the Salpetriere for 13 years, and also worked in his father-in-law's laboratory.

In 1891, Binet began working in an experimental psychology clinic at the Sorbonne. He was appointed its director four years later, and held the position for the rest of his life. Binet was interested in measuring individual differences in mental characteristics, believing that understanding normal variations was key to developing a general theory of intellectual development. He was not convinced of the importance of research on human perception, which was stressed in Germany at that time. Instead, he attempted to measure reasoning ability using paper and pencil tests, often involving pictures. He undertook a careful study of his own two daughters as they grew, and published his observations in *The Experimental Study of Intelligence.* He also founded the first French psychology journal, called *L'Année Psychologique*, which still exists today.

In order to research practical problems regarding the education of children, Binet and Ferdinand Buisson (1841-1932) established an organization called "Societe Libre pour l'Etude Psychologique de l'Enfant." This prompted the French government to appoint Binet to a commission charged with designing a mechanism for early identification of slow learners so that they could be afforded remedial assistance. The test he developed with Theodore Simon was intended to measure the mental age of the child. It evaluated the degree to which the child could perform such tasks as following commands, copying patterns, and sorting objects. The test was administered to Paris schoolchildren so that a standard scale of expected performance could be established. The ratio of a child's measured mental age to his or her chronological age was called the *intelligence quotient*, or IQ.

The concept of the IQ became enormously influential, with standardized tests like the Stanford-Binet being widely administered for generations. Today the entire idea of intelligence testing is controversial. Many argue that the tests are discriminatory, or that they measure knowledge rather than inherent intelligence. In fact, there is no real agreement on the definition of general intelligence. The notion of "multiple intelligences," that is, different but equally valuable types of abilities, has become more popular.

Probably due to the rather unusual educational path he had taken in his career, Binet was never offered a professorship in France, and this bothered him greatly. After visiting Bucharest to great acclaim he was offered a chair in psychophysiology, but was not interested in moving away from Paris. He died there on October 18, 1911.

SHERRI CHASIN CALVO

Daniel Bovet

1907-1992

Swiss-French-Italian Pharmacologist

Daniel Bovet received the 1957 Nobel Prize in physiology or medicine for his discoveries of important chemotherapeutic agents, including the first antihistamine and various muscle relaxants. His research on structure-activity relationships made it possible to block the effect of biologically active amines with pharmacological agents. The naturally occurring biogenic amines—epinephrine (adrenaline), histamines, and acetylcholine—exert important effects on several organ systems, including the vascular system and the skeletal muscles. In addition to his work on the effects of biologically active amines, he made significant contributions to psychopharmacology and chemotherapy.

Bovet was born in Neuchâtel, located in the French-speaking region of Switzerland. He was the son of Pierre Bovet, a psychologist and professor of pedagogy at the University of Geneva, and his wife, Amy Babut. Bovet studied biology at the University of Geneva and graduated in 1927. He remained at the university for postgraduate work in physiology and zoology and wrote a thesis on the influence of the nervous system on organ regeneration. After obtaining his doctorate in 1929, he began working in pharmacology at the Pasteur Institute in Paris, where he was greatly influenced by Ernest Fourneau. In 1939 Bovet became head of the laboratory of therapeutic chemistry. He married Filomena Nitti, the sister of Bovet's colleague, the bacteriologist Federico Nitti. Filomena was closely and continuously associated with Bovet's research and publications. Their son Daniele became a professor of information science at the University of Rome.

In 1947 Domenico Marotta, director of the Superior Institute of Health in Rome, invited Bovet to organize a chemotherapeutics laboratory in Italy. In 1964 Bovet left the Superior Institute to become a professor of pharmacology at the University of Sassari Medical School in Sardinia. He was a visiting professor at the University of California in Lost Angeles in 1965. From 1969-1971 he was the head of the new psychobiology and psychopharmacology laboratory of the National Research Council in Rome. He also served as a professor of psychobiology at the University of Rome until he reached age 75, when the appointment was turned into an honorary professorship.

When Bovet began work at the Pasteur Institute, he joined Fourneau in the search for antimalarial drugs. This work led to Bovet's discovery of Prosympal, the first synthetic sympatholytic (or antiadrenergic) drug, and other antiadrenergic compounds. Although these drugs had little impact on clinical medicine, they did point the way to the eventual development of drugs for the treatment of hypertension. Moreover, Bovet realized that apparently unrelated compounds might reveal common structural requirements for specific physiological and pharmacological activities. Bovet and his associates at the Pasteur Institute studied the work of Gerhard Domagk (1895-1964) and proved that the antistreptococcal action of Prontosil was due to the sulfanilamide portion of the molecule.

Within 20 years of Henry H. Dale's (1875-1968) 1910 demonstration that histamine provoked anaphylactic shock, researchers had generally accepted the idea that histamine played a key role in the allergic response. Bovet reasoned that if ergot and synthetic antiadrenergic drugs could block the effects of epinephrine, it should be possible to discover drugs that could block the effect of histamine. In 1937 he began the search for antihistaminic compounds that could be used to treat urticaria, hay fever, edema, and other allergic diseases. No naturally occurring compounds with antihistaminic action were known. In 1942 Bovet discovered Antergan, a weak antiadrenergic and antihistaminic compound. Two years later, he discovered pyrilamine (mepyramine, or Neo-antergan), which effectively counteracted the histamine reaction and is still used in the treatment of allergies. By the 1950s hundreds of synthetic antihistamines had been synthesized and used in the treatment of allergies, motion sickness, nausea, and vomiting.

In 1947 a search for a synthetic substitute for curare led Bovet to the discovery that gallamine and various derivatives of succinsylcholine could mimic the effects of curare. Since 1942 curare, a South American arrow poison, had been used in conjunction with light anesthesia during surgery to induce muscle relaxation. Researchers knew that tubocurarine, the active agent in curare, antagonizes acetylcholine at the neuromuscular junction. Bovet's search for synthetic cholinergic and anticholinergic drugs produced diethazine, one of the first synthetic drugs for Parkinson's disease, and curare-like drugs, which produced fewer unwanted side effects than tubocurarine. Moreover, Bovet's sophisticated approach to structure-activity relationships helped guide further research in pharmacology.

LOIS N. MAGNER

Alexis Carrel
1873-1944
French Physician and Surgeon

Alexis Carrel is recognized for making advances in surgery and promoting interest in organ transplants. He received the 1912 Nobel Prize for physiology or medicine for his development of a technique to sew blood vessels together end-to-end.

Born in Ste-Foy-les-Lyon, France, on June 28, 1873, he was the son of a silk merchant. He was interested in many subjects and received two degrees before getting his medical degree from the University of Lyons in 1900. In 1894, while in Lyons, the French president was stabbed and bled to death because there were then no procedures for repairing blood vessels. The death had a profound effect on Carrel, who sought to find ways to sew blood vessels. Using a very fine silk thread and needle, he practiced on paper to perfect the technique. He then turned to ways of keeping cells around the area alive and well. He published an account of his successes in a French medical journal in 1902, but the French establishment was not impressed.

Carrel moved briefly to Canada, then to the University of Chicago. In 1906 he joined the Rockefeller Institute for Medical Research, New York, where he worked to perfect his technique for suturing blood vessels. Carrel envisioned a time when blood transfusions and organ transplants would become reality. He even performed successful kidney transplants on dogs. For his initiatives, he was given the 1912 Nobel Prize for Medicine or Physiology.

In 1912 Carrel began a fascinating experiment in which he kept heart tissue from a chick embryo alive and beating by maintaining the proper nutrient culture and carefully removing waste accumulation. The monstrous heart became a media sensation, with one New York newspaper reporting each year on the "birthday" of the chick heart. The heart was maintained for 34 years, even outliving Carrel, before it was deliberately destroyed.

Although he lived in New York, he never became a United States citizen and maintained a home in France off the coast of Brittany. When World War I broke out, the French government called Carrel into service and placed him in a front line hospital. He was assisted by his wife of one year, who was a surgical nurse. Carrel worked with biochemist Henry Dakin (1880-1952) to develop a method for cleaning wounds using sodium hypochlorite. The complicated Carrel-Dakin method was very important in its time, but now has been replaced by antibiotics.

After his discharge in 1919, Carrel returned to Rockefeller Institute to continue his work with tissue cultures and began to focus on causes of cancer. He won several awards, including the Nordhoff-Jung Prize in 1931, for his study of malignant tumors.

Carrel was always fascinated with keeping organs alive outside the body. He and famed aviator Charles A. Lindbergh (1902-1974) designed a special pump made of glass that would circulate fluids around organs and keep them viable for a period of time. Called the perfusion pump, the device laid the foundation for the future development of the heart-lung machine. Carrel and Lindbergh became good friends. They appeared together with their mechanical heart in the July 1, 1935, issue of *Time* magazine and published a book together called *The Culture of Organs.*

Carrel had many personal interests and was deeply religious. He visited Lourdes in France as a young man and made a spiritual pilgrimage there each year. He wrote several books on the topic, including *Man the Unknown,* which became a bestseller in 1935. Some considered his mixing of religion and science inappropriate. One such person was the new director of the Rockefeller Institute, who encouraged Carrel's "retirement" and closed the division of experimental surgery.

Alexis Carrel. *(Library of Congress. Reproduced with permission.)*

Although France had fallen to Germany, Carrel returned to his homeland in 1941 by making his way through Spain. The Germans had set up a puppet French government at Vichy. Carrel refused to serve as director of health under the new regime, but did accept a position with the Foundation for the Study of Human Problems. He brought together young intellectuals for philosophical discussions.

When the Germans were defeated in 1944, the French government accused Carrel of collaborating with the Nazis and intended to prosecute him. Shortly before the trial, Carrel had a heart attack and died in Paris on November 4, 1944. He was buried in Saint Yves Church near his home on Saint Gilder.

EVELYN B. KELLY

Boris Ernst Chain
1906-1979
German Biochemist

In 1945 Boris Ernst Chain shared the Nobel Prize in physiology or medicine for 1945 with Alexander Fleming and Howard Walter Florey "for the discovery of penicillin and its curative effect in various infectious diseases." Fleming had discovered the antibacterial action of the *penicillium* mold in 1928, but Chain and Florey recognized its therapeutic powers in 1940 and went on to isolate and purify penicillin. Although Fleming noted that his crude penicillin preparation was nontoxic when injected into mice, he did not carry out experiments to determine whether it would actually cure mice that had been infected with a virulent bacterium, such as streptococcus.

Chain, the son of Michael Chain, a chemist and industrialist, was born in Berlin. Chain became interested in chemistry while he was a student at the Luisengymnasium in Berlin. Both his teachers and his visits to his father's workplace stimulated his interest in chemistry and biochemistry. He graduated from the Friedrich-Wilhelm University in 1930 and spent the next three years carrying out research on enzymes at the Charité Hospital in Berlin. In 1933, when the Nazi regime assumed power in Germany, he immigrated to England. (Unfortunately, his mother and sister were not able to follow him and both died in concentration camps.) He worked on phospholipids at the School of Biochemistry at Cambridge University under the direction of Sir Frederick Gowland Hopkins for two years before transferring to Oxford University in 1935. Here, he worked as a demonstrator and lecturer in chemical pathology at the Sir William Dunn School of Pathology. In 1948 he was appointed the scientific director of the International Research Center for Chemical Microbiology at the Superior Institute of Health (Istituto Superiore di Sanità) in Rome. In 1961 he became a professor of biochemistry at Imperial College in London.

In 1939 Chain and Florey began a systematic study of antibacterial substances produced by microorganisms. This work led to the discovery of the chemotherapeutic action of penicillin. Chain's work was critical to the next stage of this research, the isolation and elucidation of the chemical structure of penicillin and other natural antibiotics. Chain and Florey began their work on penicillin in order to find out whether penicillin preparations contained enzymes that could break down (lyse) bacterial walls. They thought that penicillin might actually be similar to lysozyme, a lytic enzyme that Fleming had discovered before he discovered penicillin. Chain and Florey later said that they had not considered the possibility that penicillin could have practical uses in clinical medicine when they began to work on penicillin. Their research on penicillin began in 1938, before the outbreak of World War II. Although penicillin proved to

Ernst Chain. *(Library of Congress. Reproduced with permission.)*

be a valuable chemotherapeutic agent for the treatment of infected war wounds and venereal disease, Chain emphasized that only purely scientific curiosity had motivated him to begin studying penicillin.

Although he is primarily remembered for his landmark work on penicillin, his research interests were very broad and he made significant contributions to the knowledge of snake venoms, tumor metabolism, the mechanism of lysozyme action, the carbohydrate-amino acid relationship in nervous tissue, the mode of action of insulin, fermentation technology, 6-aminopenicillanic acid and penicillinase-stable penicillins, lysergic acid production in submerged culture, the isolation of new fungal metabolites, and biochemical microanalysis.

Chain was author or co-author of many scientific papers and classic monographs on penicillin and antibiotics. In addition to the Nobel Prize, he received many honorary degrees and awards, including the Silver Berzelius Medal of the Swedish Medical Society, the Pasteur Medal of the Pasteur Institute and the Societé de Chimie Biologique, the Paul Ehrlich Centenary Prize, and the Gold Medal for Therapeutics of the Worshipful Society of Apothecaries of London. He was also a member or fellow of many learned societies in various countries. Chain married Anne Beloff in 1948. They had a long and happy marriage that was productive both in terms of collaborative research in biochemistry and a supportive family life.

LOIS N. MAGNER

Harvey Williams Cushing
1869-1939
American Neurosurgeon

Harvey Cushing was generally acknowledged during his lifetime as the world's greatest brain surgeon. He invented many basic neurosurgical procedures, operated on tumors previously considered inoperable, and worked closely with pathologists, endocrinologists, and other medical specialists to understand and classify various brain lesions and their outcomes.

Cushing was born in Cleveland, Ohio, the youngest of ten children of Betsey Maria Williams Cushing and Henry Kirke Cushing. His father, grandfather Erastus Cushing, and great-grandfather David Cushing were all physicians.

He received his A.B. from Yale University in 1891 and his M.D. from Harvard Medical School in 1895. After a one-year internship at Massachusetts General Hospital, he became a resident surgeon at the Johns Hopkins Hospital and an instructor in surgery at the Johns Hopkins School of Medicine. He studied under William Stewart Halsted (1852-1922) at Hopkins from 1896-1900, then under Emil Theodor Kocher (1841-1917) and Karl Hugo Kronecker (1839-1914) in Berne, Switzerland, and Sir Charles Scott Sherrington (1857-1952) in Liverpool, England, from 1900-1901.

Cushing returned to Hopkins in 1901. The following year he married Katharine Crowell in Cleveland. He was Associate Professor of Surgery at Hopkins from 1903 until 1912, when he became jointly the Moseley Professor of Surgery at Harvard and Surgeon-in-Chief at the Peter Bent Brigham Hospital in Boston. He held both these posts until 1932, with the exception of the time he spent in volunteer surgical service in World War I. He commanded the U.S. Army Base Hospital No. 5 in France and was discharged in 1919 at the rank of colonel.

Among Cushing's major books are *Surgery of the Head* (1908), *The Pituitary Body and its Disorders* (1912), *Tumors of the Nervus Acusticus and the Syndrome of the Cerebellopontile Angle* (1917), *Studies in Intracranial Physiology and Surgery: The Third Circulation, The Hypophysis, The Gliomas* (1926), *Consecratio Medici* (1928), *Intracranial Tumours* (1932), *Papers Relating to the Pituitary Body, Hypothalamus, and Parasympathetic Nervous System* (1932), and several memoirs. With neurosurgeon Percival Bailey (1892-1973) he wrote *A Classification of the Tumors of the Glioma Group on a Histogenetic Basis with a Correlated Study of Prognosis* (1926) and *Tumors Arising from the Blood-Vessels of the Brain: Angiomatous Malformations and Hemangioblastomas* (1928). With neuropathologist Louise Charlotte Eisenhardt (1891-1967) he wrote *Meningiomas: Their Classification, Regional Behaviour, Life History, and Surgical End Results* (1938).

His pet project was the "Brain Tumor Registry." His idea was to preserve and document every brain lesion that he personally obtained either in the operating room or at autopsy and to chart the life expectancies of all his surviving patients. After workers in the pathology laboratory at Hopkins lost a pituitary cyst that Cushing had removed in 1902, he always demanded to manage his pathological specimens himself. By the early 1930s the result was a collection of over two thousand brain lesions and a year-by-year account of over a thousand brain surgery survivors. Shortly after Cushing became Sterling Professor of Neurology at Yale in 1933, Yale appointed Eisenhardt to oversee the registry. She continued this task long after Cushing's death. In 1945 she was still tracking about 800 of his survivors.

Besides being a skilled clinician, Cushing was also an expert on medical history. He lectured frequently on that subject; directed history of medicine studies at Yale from 1937 until his death; wrote the standard bibliography of the founder of the modern science of anatomy, Andreas Vesalius (1514-1564); and won the Pulitzer Prize in 1926 for his biography of Sir William Osler (1849-1919).

Even though Cushing was universally respected and sometimes deified, he was not easy to like. He was a bad-tempered, self-certain show-off who displayed nearly constant scorn for his colleagues and subordinates. Nevertheless, he could occasionally be gracious and even charming. Bailey expressed the thoughts of many when he said of Cushing: "One forgave him much because of his accomplishments." Evidence of the extent of this forgiveness is that the American Association of Neurological Surgeons was known from its founding in 1931 until 1967 as the Harvey Cushing Society.

ERIC V.D. LUFT

Philip Drinker

1894-1972

American Chemical Engineer

Philip Drinker was a pioneer in the fields of bioengineering and industrial safety. His famous invention, the "iron lung," saved many lives, especially those afflicted with polio who could not breathe.

Born in Haverford, Pennsylvania, on December 12, 1894, Drinker received a B.S. degree from Princeton in 1915 and a degree in chemical engineering from Lehigh University in 1917. During World War I, he was sent to France to inspect the coating fabrics of airplane wings and to investigate toxic conditions in munitions plants. Returning to civilian life, he became a chemical engineer for the Buffalo Foundry and Machine Company. In 1921 he went to Harvard Medical School as an instructor in applied physiology and helped create the first industrial hygiene program at any institution.

During this time, polio affected thousands of people and caused many deaths when the lungs of those with the disease were paralyzed by the virus. In 1926 Drinker was appointed to a Rockefeller Institute commission to investigate methods of resuscitation. Respiratory problems of all kinds were the result of a society modernizing too quickly without regard to the safety conditions for workers.

Philip observed an experiment in which his brother Cecil and physiologist Louis Shaw studied respiration in cats. They sealed a cat inside a small chamber with its head sticking out of a rubber collar and used a pressure gauge to measure respiration. Philip Drinker pondered the question of whether such a device could breathe for the animal. He temporarily paralyzed the cat and simulated breathing for hours by using a hand-held syringe.

Then Drinker had an idea. Using two vacuum cleaner blowers, valves salvaged from the laboratory, and a mechanics creeper (the device which rolls repairman under cars), he enclosed it all in a large iron chamber. Drinker himself climbed into the machine with only his head sticking out. Cecil Drinker and Shaw monitored as the machine sucked additional air in and out of his lungs for 15 minutes. At the end of the test Philip was so hyperventilated that it took four minutes for him to resume normal breathing. The machine was a success; it had breathed for him.

On October 14, 1928, an unconscious eight-year-old girl was rushed to Drinker in the wee hours of the morning. The machine pumped life-giving oxygen into her lungs and within a minute or two she was awake and asking for ice cream. She was kept breathing for five days but later died of pneumonia, a complication not related to the machine.

Another chance soon came when Barret Hoyt, a Harvard senior, was dying from polio because his lungs were paralyzed. The student's physician begged Drinker to bring the machine to the Peter Bent Brigham Hospital. The big machine would not fit in a taxi, and they had no time to get a truck. They finally tied the iron lung to the top of the cab and made it to the hospital just before the patient arrived. Hoyt was barely breathing, but the machine forced air in and out of his lungs for four weeks. The "iron lung" never faltered and immediately became standard equipment to help people not only with polio, but with all types of respiratory failure, including gas poisoning and acute alcoholism. The iron lung became known as the "Drinker Respirator."

Today, it is difficult to fully understand the fear inspired by the dreaded disease polio and how the iron lung became the hope for many who were paralyzed. When the Salk vaccine was developed in 1955, the need for the iron lung was greatly reduced, but it was hailed as one of the great medical feats of its time.

In 1925 Drinker proposed the first air-conditioned ward for the treatment of premature infants. He found the use of a helium-air mixture could assist divers and made possible the dramatic escape of the crew of a disabled submarine, the *Squalus,* in 1939.

Drinker was interested in many problems of occupational medicine and used himself as a guinea pig, breathing poisonous fumes and dust. He campaigned for the use of protective devices, such as dust and fume masks, for workers.

During World War II, Drinker was appointed Director of Industrial Health for the United States Maritime Commission and Navy shipyards. He organized seminars on industrial hygiene and founded the American Industrial Hygiene Association. The new United States Atomic Energy Commission called upon him to study safety related to the nuclear industry. This led him to develop an early leadership position in researching air pollution.

After World War II Drinker went to Belfast, Ireland, as a Fulbright Visiting Professor and maintained close ties with British colleagues. While investigating the toxic effects of sulfuric acid mist and sulfur dioxide, he was called upon for advice when the disastrous London smog hit in 1952. He determined that the smelting and release of these chemicals in the air had caused the many fatalities associated with the smog. He also began to formulate United States air quality standards.

Drinker was an excellent writer and produced many articles from his research. His textbooks, *Industrial Medicine* and *Industrial Dust,* are classics on their subjects. He retired from Harvard in 1961 after 40 years. Drinker received many honors, including honorary doctorates from Norwich University and Hahnemann Medical School.

A pioneer in the field of bioengineering, Drinker was dubbed "Mr. Industrial Hygiene." His son Philip A. Drinker (1932-) followed his father in the bioengineering field and designed a neonatal breathing unit. The elder Drinker died in Fitzwilliam, New Hampshire, on October 19, 1972, after a brief illness at the age of 78.

EVELYN B. KELLY

Paul Ehrlich

1854-1915

German Bacteriologist

Paul Ehrlich was a German bacteriologist who made tremendous contributions to medical science in the areas of immunology, hematology, and pathology. Although he was awarded the Nobel Prize for Physiology or Medicine in 1908 in conjunction with Elie Metchnikoff (1845-1916) in recognition of their work in immunology, he is best known as the pioneer of chemotherapy. He created the first successful treatment for syphilis. In another noteworthy accomplishment, Ehrlich put forth the side-chain theory, becoming the first person to attempt to correlate the chemical structure of a synthetic drug with its biological effects.

Paul Ehrlich was born in Germany on March 14, 1854, into a family of Jewish faith. His father was prominent in business and industry, while his mother was the aunt of exceptional bacteriologist and scientist Karl Weigert. His early education was at the Gymnasium at Breslau. Following that, he attended various institutions, which culminated in his completion of the doctorate of medicine degree at the University of Leipzig in 1878. The subject of his dissertation concentrated on the theory and methods of staining animal tissues, a practice he revolutionized during his lifetime.

After graduation, he was given an appointment at the Berlin Medical Clinic, where he continued researching tissue staining. During this time, Ehrlich demonstrated that all the stains could be classified as being acidic, basic, or neutral, depending on its properties. He also studied the effect dyes had on cells in the blood, which laid the foundation for hematology. Ehrlich studied the tuberculosis bacterium and published a staining method in 1882 that is still used today with only minor modifications. This paper was also significant in that it paved the way for the Gram method of staining bacteria that is widely used in modern medicine to help differentiate between certain types of microorganisms. In 1883 Ehrlich married Hedwig Pinkus, and they subsequently had two daughters.

Ehrlich's reputation as a skilled researcher allowed him to move up quickly through the hierarchy of German science, despite a bout with tuberculosis that forced Ehrlich to interrupt his work and travel to Egypt for treatment. He returned to Berlin in 1889 when the disease had arrested permanently. Ehrlich was appointed as assistant to Robert Koch (1843-1910), who was the director of the newly established Institute for Infectious Diseases in 1890. It was there that Ehrlich began a series of immunilogical studies for which he will always be remembered. The working hypothesis that Ehrlich developed was called the side-chain theory. He believed each cell consisted of a protein center surrounded by a series of side chains that are responsible for absorbing nutrients as well as toxins. If a toxin were to attach to a cell, the cell would produce large amounts of immune side chain bodies gauged to prevent a new infection.

Ehrlich recognized the limitations of the treatment therapies at that time and started experimenting with substances not normally found in nature that could kill or retard growth in parasites. For this, Ehrlich is often considered to be the father of chemotherapy. He worked primarily with arsenic compounds that could be effective against disease, yet have little side effects. After testing nearly 1,000 compounds, Ehrlich directed his colleague, Sahachiro Hata, to go back and retest the effectiveness of compound #606 against syphilis. It was found effective and was eventually marketed as a cure for syphilis as Salvarsan. This drug had a profound effect on Alexander Fleming (1881-1955), who, after administering it to patients, set out to find similar cures for bacterial diseases and subsequently discovered penicillin.

Ehrlich suffered a stroke in December of 1914. The following year, Ehrlich died of a second stroke, leaving a legacy of tremendous scientific achievement.

JAMES J. HOFFMANN

Christiaan Eijkman
1858-1930
Dutch Physician and Pathologist

Christiaan Eijkman was a Dutch physician who shared the 1929 Nobel Prize in Physiology or Medicine with Sir Frederick Hopkins (1861-1947) for their demonstration that beriberi is caused by poor diet. This work subsequently led to the discovery of vitamins. Eijkman helped to establish the first laboratory dedicated to the physiological study of people living in tropical regions. His work helped to show that there were few physiological differences attributable to the change in geography. Eijkman also developed a well-known fermentation test to readily establish if water has been polluted by human and animal feces containing *E. coli*.

Christiaan Eijkman was born on August 11, 1858, in Nijkerk, the Netherlands. A year later his family moved to Zaandam, where his father was appointed head of a newly founded school for advanced elementary education. It was here that Christiaan received his early education and training. In 1875, he became a student at the Military Medical School of the University of Amsterdam, where he was trained as a medical officer for the Netherlands Indies Army, graduating with honors. He finished his doctor's degree, with honors, in 1883. That same year he left Holland for the Indies, where he served as medical officer of health in Java. He developed a case of malaria, which significantly impaired his health to the extent that he had to return to Europe in 1885.

This disease turned out to be a significant event for Eijkman in that it enabled him to return to Amsterdam and get involved with a Dutch research team that was returning to the Indies to conduct investigations into beriberi. At that time, beriberi was very prevalent in the region and was responsible for many medical problems. Although the mission ended prematurely, it was proposed that Eijkman stay behind and be appointed director of the newly established "Geneeskundig Laboratorium" (Medical Laboratory) in Java. He served at this post for over eight years and made many important contributions during this time.

This laboratory was the first to be devoted exclusively to studying the physiology of people living in tropical regions. Eijkman was able to disprove a number of inaccurate theories regarding life in tropical regions. Among them, he clearly demonstrated that the blood of Europeans living in the tropics unaffected by disease show no measurable differences when compared to the blood of Europeans living in Europe. He also compared the metabolism, perspiration, respiration and temperature regulation of Europeans to natives and found that there were no differences. Thus, Eijkman put an end to a number of speculations on the acclimatization of Europeans in the tropics, which had previously necessitated the taking of various precautions.

Eijkman began to study beriberi and believed that a bacterial agent was responsible for the disease. In 1890, a disease called polyneuritis broke out among his laboratory chickens. Eijkman noticed the disease's striking resemblance to beriberi, and he was eventually able to show that the condition was caused by a diet of polished, rather than unpolished, rice. Although he still believed that the disease was related to some toxin produced in the intestine by the action of bacteria on boiled rice, Eijkman demonstrated that the problem was a nutritional deficiency. It was later determined to be a lack of vitamin B_1 (thiamin), which helped lead to the discovery of vitamins.

Eijkman returned to the Netherlands in 1896 to serve as a professor at the University of Utrecht. In 1883, before his departure to the Indies, Eijkman married Aaltje Wigeri van Edema, who died in 1886. Professor Eijkman later married Bertha Julie Louise van der Kemp in 1888. Christiaan Eijkman died in Utrecht in 1930.

JAMES J. HOFFMANN

Willem Einthoven
1860-1927
Dutch Physiologist

Willem Einthoven was a Dutch physiologist who, in 1924, was awarded the Nobel Prize in Physiology or Medicine for his invention of a string galvanometer that he used to produce the electrocardiogram (EKG), a physical recording of the electrical activity of the heart. He originated many of the ideas that govern both the technical aspects and clinical interpretation of EKG. This instrument has been vital in assessing certain types of heart disease because it records the electrical action of the heart in a noninvasive fashion. He also developed "Einthoven's Triangle," which is a central theoretical aspect in the administration and interpretation of EKG records.

Einthoven was born on May 21, 1860, in Semarang, Java, Dutch East Indies. His father, although born in the Netherlands, served as an army medical officer in the Indies. Later, he set up a medical practice in Semarang, but passed away when Einthoven was 10. His mother decided to return the family back to Holland. He entered the University of Utrecht in 1878 as a medical student, intending to follow in his father's footsteps. Einthoven graduated with a degree in medicine and was appointed professor of physiology at the University of Leiden. He served at this post from 1886 until his death in 1927. Although he was trained in medicine, Einthoven discovered that he excelled in physics and worked to measure electrical signals in humans.

He showed interest in recording the heartbeat as early as 1891. At that time, a device called the capillary electrometer was used to register the changes of electrical potential caused by contractions of the heart muscle. The apparatus was not accurate or practical for scientific and diagnostic applications. Einthoven was interested in developing an instrument that would directly record the electrical changes in the heart. In 1903, he designed the string galvanometer, which measured the changes in the electrical activity of the heart caused by muscular contractions and recorded them graphically. It consisted of a very thin wire of quartz held in a magnetic field. When the wire was magnified and its movements recorded on film, Einthoven could make precise measurements of the heart's electrical activity. This device revolutionized the field of medicine. It was originally referred to in German as "Elektrokardiogramm" or EKG. In most Western countries it is now called ECG (electrocardiogram).

The instrument became an important tool in the diagnosis of heart disease, with Einthoven spearheading the investigation into the applications of the technology. Einthoven recognized differences in the tracings obtained from various types of heart disease. In 1906, he published the first organized presentation of normal and abnormal electrocardiograms recorded with a string galvanometer. In his treatise, Einthoven identified pathologies such as hypertrophy, ventricular premature beats, and complete heart block. This continued a previous pattern of Einthoven's, for as early as 1895, he had distinguished the five deflections normally seen in an EKG recording which he designated P, Q, R, S, and T. Subsequently, Einthoven refined models of the string galvanometer, including the development of an electrophonocardiogram to measure heart sounds in 1915. He also applied this technology to other devices used to measure electrical currents in muscles, nerves, and the eye. Modern EKG machines are still based on Einthoven's original invention and allow physicians to monitor heart function and to detect damage from heart attacks and other causes.

In 1886, Einthoven married Frédérique Jeanne Louise de Vogel. Together, they had four children, some of whom had distinguished scientific careers themselves. Einthoven died on September 28, 1927, leaving a remarkable legacy in the field of electrocardiography.

JAMES J. HOFFMANN

Johannes Andreas Grib Fibiger
1867-1928
Danish Physician, Pathologist and Bacteriologist

Johannes Andreas Grib Fibiger was awarded the Nobel Prize in Physiology or Medicine for his research on the etiology of cancer and for his discovery of a parasite that he claimed was the cause of cancer, the *Spiroptera carcinoma*. Unfortunately, the great hope that Fibiger's research on cancer would solve the cancer puzzle proved to be unfounded.

Fibiger was born in Silkeborg, Denmark. His father, C. E. A. Fibiger, was a local medical practitioner and his mother, Elfride Muller, was a writer. His father's career stimulated Fibiger's interest in medicine, although Dr. Fibiger died when his son was very young. Fibiger became interested in bacteriology as an undergraduate at the University of Copenhagen. He was awarded his bachelor's degree in 1883 and his M.D. in 1890. After a period of working in hospitals and studying under Robert Koch (1843-1910) and Emil von Behring (1854-1917) in Germany, he returned to Copenhagen. From 1891-1894 he was assistant to Carl Julius Salomonsen in the Department of Bacteriology at Copenhagen University. Fibiger served as an army reserve doctor at the Hospital for Infectious Diseases in Copenhagen from 1894-1897, while completing his doctoral research on the bacteriology of diphtheria. In 1895 he received his doctorate from the University of Copenhagen. He returned to Germany for six months to work at the pathological institute of Johannes Orth. In 1897 Fibiger was appointed dissection assistant at the University of Copenhagen's Institute of Pathological Anatomy. He also served as principal of the Laboratory of Clinical Bacteriology of the Army from 1890-1905. In 1905 he became director of the Central Laboratory of the Army and a consultant physician to the Army Medical Service.

In 1900 Fibiger initiated a new research project, studying tuberculosis. He was particularly interested in the relationship between human and bovine tuberculosis. Although he was also interested in cancer, he had no experimental model that could be used to solve the problem of the origins of the cancer cell. Transferring cancerous tissue from one animal to another was possible, but initiating cancer in a healthy animal was not. Influenced by the success of bacteriology in finding the causes of many diseases, Fibiger hoped that similar methods would provide insights into the genesis of cancer.

In 1907, while dissecting three rats that had been injected with tubercle bacilli, Fibiger found that these animals all had stomach cancers. Microscopic examination of the tumorous tissue revealed traces of eggs and worms. Intrigued by these observations, Fibiger abandoned his research on tuberculosis and focused on cancer. Based on reports that a tiny worm known as a nematode infested rats and cockroaches, Fibiger attempted to collect rats that subsisted largely on roaches. He found that the rats and roaches at a Copenhagen sugar refinery were infested by nematodes. After isolating the parasitic nematode in question and determining that it was a new species, he passed it repeatedly from infested rats to cockroaches and then to healthy rats. Fibiger reported that these rats eventually developed tumors of the stomach. In 1913 Fibiger reported his discovery to the Royal Danish Academy of Science. He called the nematode *Spiroptera neoplastica*. According to Fibiger, his work proved that cancers were caused by chronic irritation from a foreign agent or toxin. By the time he was awarded the Nobel Prize, he realized that his parasites probably were not the key to the etiology of cancer. Other researchers had little or no success in duplicating his results in experimental animals. Later, researchers suggested that the tumors observed in his experimental animals might have been caused by a virus carried by the parasite and that Fibiger's *Spiroptera neoplastica* had no relationship to human cancer at all. Eventually, Fibiger himself suggested that some external agents might generate cancer in genetically susceptible individuals. Fibiger died of a heart attack shortly after he was diagnosed with colon cancer.

LOIS N. MAGNER

Sir Alexander Fleming
1881-1955
Scottish Bacteriologist

Alexander Fleming discovered penicillin, a "wonder drug" that ushered in the era of antibiotics. This new weapon against bacterial disease offered hope in fighting many infections that could not be treated effectively in the past. Together with Sir Howard Walter Florey (1898-1968) and Ernst Boris Chain (1906-1979), Fleming was awarded the Nobel Prize in 1945. In addition to penicillin, he also discovered lysozyme, an antibacterial agent found in tears and saliva.

Fleming was born on August 6, 1881, on a farm in Lochfield, Ayrshire, Scotland. He began his education in Scotland and then went on to London, where he received his medical degree in 1906 from St. Mary's Hospital Medical School at the University of London. He lectured there until World War I, during which he served in the Army Medical Corps. After the war he returned to St. Mary's, and in 1921 identified lysozyme. He was appointed a professor in 1928.

That same year, Fleming was experimenting with the staphylococcus bacteria, and had the good luck to have one of his culture plates contaminated by a mold called *Penicillium notatum*. Around the mold, there was a ring in which no bacteria grew. Further experiments confirmed that the effect was due to a substance produced by the mold, and it not only inhibited the growth of many types of bacteria, it killed existing growth. The effect was specific to *Penicillium notatum*; other molds did not work the same way.

Fleming called the active substance penicillin. He diluted it hundreds of times, and it was still effective. He showed that, unlike harsh antiseptics such as phenol, penicillin did not destroy white blood cells. He injected it into animals without negative effects. He published the results of his experiments in 1929, suggesting that the substance might have therapeutic uses, but the mold extract was too perishable to allow extensive tests. Penicillin was largely neglected for a decade.

Alexander Fleming. *(Library of Congress. Reproduced with permission.)*

In 1939, the sulfanilomide derivatives, or sulfa drugs, were introduced by the Italian pharmacologist Daniel Bovet (1907-1992). Sulfa was effective against streptococcus and other disease-causing bacteria, and saved many lives during World War II. However, it was somewhat toxic, tending to cause kidney damage if not used very carefully. Its promise, coupled with the needs of the war, interested researchers in finding other valuable medicines with fewer toxic effects.

It was at this point that Chain and Florey obtained a sample of Fleming's *Penicillium notatum* culture. Isolating the active substance, penicillin, they were able to purify it so that it was more stable and produce it in larger quantities. Then they performed clinical trials that demonstrated its effectiveness for use as an antibiotic. Like sulfa, penicillin was employed in what amounted to a massive field trial during World War II.

Honors soon followed for all three men. In addition to the Nobel Prize, Fleming was elected a Fellow of the Royal Society in 1943 and knighted in 1944. In 1948, he became emeritus professor of bacteriology at the University of London. He died in London on March 11, 1955.

SHERRI CHASIN CALVO

Sir Howard Walter Florey

1898-1968

English Pathologist

Sir Howard Walter Florey was also known as Baron Florey of Adelaide. He was an Australian-born British pathologist who worked with Ernst Boris Chain (1906-1979) isolating and purifying penicillin. They further demonstrated its effectiveness against harmful bacteria, developed methods for mass production, and helped introduce it into general clinical use. Together, they, with Alexander Fleming (1881-1955), the discoverer of penicillin, shared the 1945 Nobel Prize in Physiology or Medicine. Because of his contributions to science, Florey was knighted in 1944.

Howard Florey was born in 1898, at Adelaide, South Australia. His early education was at St. Peter's Collegiate School in Adelaide. He subsequently attended Adelaide University, where he graduated in 1921. He accepted a Rhodes Scholarship to Magdalen College and moved to England, never to return to his native Australia. Following that experience, he went to Cambridge and eventually received his Ph.D. in 1927. He was then appointed Huddersfield Lecturer in Special Pathology at Cambridge. In 1931 he became Chair of Pathology at the University of Sheffield. Leaving Sheffield in 1935 he became Professor of Pathology and a Fellow of Lincoln College, Oxford. In 1962 he was made Provost of The Queen's College, Oxford.

In 1924 Florey joined the Oxford University Arctic Expedition as its medical officer. This expedition undertook various scientific investigations, including the study of topography and ice crystals. This expedition was the first arctic expedition to use an airplane as a mode of transport. However, the plane crashed, and Florey treated the slight injuries of the pilot and the team leader.

Florey's best-known research is from his collaboration with Chain. In 1938, they investigated tissue inflammation and secretion of mucous membranes. Florey was convinced that there had to be a naturally occurring substance that would kill bacteria, so they conducted an intensive investigation of the properties of naturally occurring antibacterial substances. Their original subject of investigation was lysozyme, an antibacterial substance found in saliva and human tears. However, they did not produce any significant findings, and their interest shifted to a class of substances now known as antibiotics.

Howard Florey. *(Library of Congress. Reproduced with permission.)*

They were particularly interested in a paper published by Fleming 10 years prior.

Fleming had discovered penicillin in 1928 as a result of observations on bacterial culture plates where a mold had developed in some areas and inhibited germ growth at those points. However, the active substance was never isolated. In 1939, Florey and Chain headed a team of British scientists that attempted to isolate and purify that substance. In 1940 they issued a report describing their success. The results indicated that penicillin was capable of killing germs inside the living body. Thereafter, great efforts were made to develop methods for mass production that would enable sufficient quantities of the drug to be made available for use in World War II to treat war wounds.

Despite his success as a scientist, Florey's personal life was not fulfilling to him. He married Ethel Reed, a fellow Australian, who had studied medicine with Florey at the University of Adelaide in 1926. Almost immediately, there was friction and it seemed that the only emotional outlet for Florey was his work. Between them they had two children and Ethel assisted her husband with the clinical trials of penicillin. Despite the difficulties, they remained married. Ethel Florey died in 1966. In June 1967, Florey married Margaret Jennings, a long-time colleague and friend. Here he found the happy marriage that had previously eluded him, but it was tragically brief. Florey died suddenly, less than a year later in 1969.

JAMES J. HOFFMANN

Sigmund Freud

1856-1939

Austrian Neurologist and Psychiatrist

Between 1895 and 1939 Sigmund Freud created psychoanalysis, a therapy for the mentally ill, a tool for cultural criticism, and a field of psychiatric and philosophical research. His theories of human behavior and motivation amounted to no less than a revolution in psychiatry and social mores. Few thinkers affected the arts and sciences of the twentieth century as profoundly as did Freud. His thought has influenced the development of the fine arts (art, literature), the social sciences (history, philosophy, anthropology, sociology, education), and the health sciences (psychiatry, social work, psychology). Freud's ideas constitute a core element in the social and political transformations of the century.

Sigmund Freud was born in Freiberg, Moravia (then in Austria, now in the Czech Republic) in 1856, the son of a poor, Jewish merchant. The family relocated to the capital of the Austrian Empire, Vienna, at an early age, and Freud spent almost his entire life in that German-speaking city. He received a classical education and earned his M.D. degree in 1881. The combination of Austrian, German, Hellenic, Jewish, and Roman Catholic cultures created an intellectual and social environment that catalyzed the synthesis of nineteenth-century neurology and psychiatry with Freud's creativity and productivity. In 1885 he spent several months in Paris with J. M. Charcot (1825-1893), an expert in neurology and hysteria, which led him to theorize new approaches to treating hysteria.

In the century since the publication of Freud's landmark book *The Interpretation of Dreams* in 1900, his ideas have so fundamentally transformed our treatment for mental illness, our language, and the way we view ourselves and our society that, as W. H. Auden wrote in elegy, "he is no more a person now, but a whole climate of opinion." Freud brought the scientific study of the unconscious to the forefront of academic psychology, psychiatry, and neurology as well as to the public consciousness.

Before 1900, Freud was a neurological researcher with an interest in psychopathology. In

1895 he combined with Josef Breuer (1842-1925; Breuer used a "talking cure" to discover the roots of his patients' neuroses) on *Studies on Hysteria*, which theorized that hysterical patients suffered from the undischarged emotional energy of repressed (i.e., consciously forgotten but unconsciously active) memories of infantile seduction in which fathers sexually abused their children.

In *The Interpretation of Dreams,* Freud theorized that the Greek mythological figure Oedipus (abandoned at birth, miraculously surviving to unwittingly kill his father and marry his mother) symbolized the primal wish of every boy to supplant the father and claim the mother. He later called his discovery of this, the Oedipus complex, his most important theoretical contribution. He then rejected infantile seduction as untenable and hypnosis as unreliable, developing instead free association, a technique that permitted unconscious emotions to surface. Freud continued to emphasize the sexual nature of repressed memories.

After 1905 Freud's ideas began to gather converts across Europe and in the United States; in 1910 he established the International Psychoanalytical Association. He further elaborated his theories of the unconscious and sexuality, adding aggression during the First World War as an instinctual force as powerful as sexuality. "The Ego and the Id" (1923) proposed a structural model to contrast the earlier topographic model of instinctual energies passing between unconscious, preconscious, and consciousness. Many of Freud's most influential papers on culture, literature, and art appeared between 1913 and 1930, such as "Totem and Taboo" (1913), "Group Psychology and the Analysis of the Ego" (1921), "The Future of An Illusion" (1927) and "Civilization and Its Discontents" (1930).

To clinical mental health Freud contributed the concept of psychic determinism, the lawfulness of all psychological phenomena, even the most trivial such as dreams, fantasies, and "Freudian slips." He furthermore demonstrated that the subject matter of an investigation into psychosexual development and character formation consists of life histories. The concept of "psychodynamics" and the awareness of the importance of sexuality and aggression are now omnipresent in psychological and psychiatric treatment. The ideas of conflict, anxiety, and defense are Freudian, although much is owed to other dynamic theoreticians such as Carl Jung (1875-1961), Alfred Adler (1870-1937), and Sigmund's daughter Anna Freud (1895-1982).

Sigmund Freud. *(Library of Congress. Reproduced with permission.)*

Freud's research into irrational human behavior was underway while warfare on scales hitherto unimaginable and the mass extermination of entire populations and cultures based on racist theory and industrial efficiency belied the hope that a greater understanding of our minds would lead to greater peace and prosperity. His theories permit clinicians, artists, and social scientists to comprehend these, and many other, phenomena.

DAVID D. LEE

Alice Hamilton
1869-1970
American Pathologist and Industrial Toxicologist

Alice Hamilton was a social crusader who fought for safe conditions in industry and for workers' compensation laws. She laid the foundation for the new field of industrial medicine and became the first female professor at Harvard Medical School.

Born on February 27, 1869, in New York City, Hamilton was second in the family of five born to Montgomery Hamilton, a wholesale grocer, and Gertrude Hamilton. In this very privileged family the children were encouraged to develop their minds and were taught at home and in private schools. Her sister, Edith Hamilton, became a well-known scholar of myth and culture in ancient Greece.

Alice became very interested in medicine and first attended a small, uncertified school before transferring to the University of Michigan, where she received a M.D. degree in 1893. After interning in Minneapolis and Boston, she decided that research in bacteriology and pathology was what she would pursue. She went to the Universities of Leipzig, Munich, and Frankfurt in Germany. She then had the very best training at Johns Hopkins, the University of Chicago, and the Paris Pasteur Institute. In 1897 she was appointed professor of pathology at the Women's Medical College in Chicago, which closed in 1912. She was subsequently hired by the McCormick Institute for Infectious Diseases and stayed until 1909.

While in Chicago, Hamilton lived in the settlement called Hull-House, developed by social worker Jane Addams (1860-1935) to care for the poor. For the first time, Hamilton experienced poverty and disease firsthand and was truly shocked by the conditions. She persuaded the local health department to make an all-out effort to improve these sanitation conditions. She was appalled at how immigrant workers had been afflicted by the fumes and chemicals they were exposed to in their jobs in the steel mills, foundries, and factories. Later, while traveling the country, she found the same hazardous conditions everywhere. A book called *Dangerous Trades,* by Englishman Sir Thomas Oliver, further inspired Hamilton to begin her lifelong crusade against industrial hazards.

In 1910 she persuaded the governor of Illinois to set up a commission on occupational diseases, in which she was assigned to research the dangers of lead and phosphorus in industry. The next year she became an investigator of industrial poisons for the newly formed Department of Labor. During World War I she exposed the hazards of nitrous fumes that were causing deaths among workers in the explosives industry, but were being passed off as deaths from natural causes.

In 1919 Harvard University appointed Hamilton professor of industrial medicine. She was the first female professor at that institution. She then took her crusade worldwide, serving as the only female delegate to the League of Nations Health Commission in 1924. Her book, *Industrial Poisons in the United States,* was the first textbook written on industrial medicine. She convinced the Surgeon General to investigate the effects of tetraethyl lead and radium. In 1934 she wrote the classic work, *Industrial Toxicology,* which led to the passage of workers' compensation laws as part of President Franklin Roosevelt's New Deal.

Alice Hamilton. *(Library of Congress. Reproduced with permission.)*

Retirement from Harvard in 1935 enabled Hamilton to further pursue her mission to influence legislation on health and safety, which she continued until into her late eighties. As a consultant in the United States Department of Labor, she launched a major investigation into the toxic practices of the viscose rayon industry. This led to the first workers' compensation laws in Pennsylvania. The story of her crusade and mission is described in her autobiography *Exploring the Dangerous Trades,* published in 1943.

Hamilton never married and continued to work for the causes of social justice and pacifism. She died in Hadlyme, Connecticut, in 1970 at the age of 101.

EVELYN B. KELLY

William Augustus Hinton
1883-1959

American Bacteriologist and Pathologist

William Augustus Hinton was an American bacteriologist and pathologist who made significant contributions to the areas of public health and medicine. Dr. Hinton overcame poverty and racial prejudice to become the foremost investigator of his time in the area of venereal disease, specifically syphilis. He developed a blood serum test called the Hinton test that accurately diagnosed the presence of syphilis. He is also noteworthy because he was the first black professor at Harvard University Medical School, and he authored the first medical textbook by a black American to be published, *Syphilis and Its Treatment* (1936).

William Hinton was born in Chicago, Illinois, on December 15, 1883. He was the son of former slaves who imparted a strong belief in the importance of equal opportunity for everyone. Hinton spent his younger years in Kansas and was the youngest student to ever graduate from his high school. He initially attended the University of Kansas, but had to leave after two years so that he could earn enough money for school. He then attended Harvard University and graduated in 1905. Hinton delayed medical school and accepted a job as a teacher to earn money. Because of his belief in equal opportunity without special treatment, he refused a scholarship reserved for black students. This was to be a common theme throughout his life.

Hinton graduated with a medical degree from Harvard Medical School in 1912, completing his degree in only three years. After graduation, he worked for the Wasserman Laboratory, a biological laboratory that was associated with the Medical School at that time. Hinton was named chief of the laboratory when it was transferred to the Massachusetts Department of Public Health in 1915. Hinton was an excellent teacher and was appointed as an instructor in preventive medicine and hygiene at Harvard Medical School in 1918. He taught at Harvard for over 30 years until 1949, when he was promoted to the rank of clinical professor. Hinton was the first black person to become a professor at Harvard Medical School in its 313-year history.

An expert in the study of disease detection and the development of medications to combat those ailments, Hinton is best known for his advances in specific tests used to detect syphilis. His "Hinton Test" greatly enhanced syphilis screening by reducing the number of people who were falsely believed to have the disease. These false positives were common with standard tests of that time. This breakthrough significantly reduced the number of patients who had to undergo needless treatment. In 1934, the U.S. Public Health Service reported that the Hinton test was the most effective test for syphilis at that time. He later developed the Davies-Hinton test for detection of syphilis in blood and spinal fluid.

During his professional career, Hinton published many important articles and books in the field of medicine. He was a dedicated scientist who believed in his principles. He declined the Spingarn Medal from the NAACP in 1938 because he wanted to be rewarded for the quality of his work, not his race. He was also concerned that his productivity as a researcher would be compromised if his colleagues knew he was black. It is believed that Hinton could have had a very successful private practice, but he chose to serve in the field of public health. Hinton had diabetes, which strained his eyesight and strength. This caused him to retire completely from active service in 1953. He died in 1959.

JAMES J. HOFFMANN

Frederick Gowland Hopkins

1861-1947

English Biochemist

Frederick Gowland Hopkins won the Nobel Prize in Physiology or Medicine in 1929 for his research on the chemistry of nutrition. Hopkins proved that, even if a diet was adequate in terms of total calories and protein content, small quantities of specific "accessory factors" were essential to health. The chemical studies of these dietary factors carried out by Hopkins and his co-workers stimulated the isolation and characterization of vitamins.

Hopkins was born in Eastbourne, England. His father, a bookseller in London, died when Hopkins was only an infant. His mother, Elizabeth Gowland Hopkins, gave him a microscope with which he studied marine life. As a child, however, Hopkins seemed more interested in poetry and literature than science. When Hopkins was 10 years old, he was sent to the City of London School, where his interest in science was stimulated and he received honors in chemistry. Bored by his coursework, however, he left school when he was seventeen. Shortly after leaving school, he published his first scientific paper, an article on the bombardier beetle in *The Entomologist.*

After working as an insurance clerk for six months, Hopkins became an assistant to a chemist in a commercial analytic laboratory. A small inheritance allowed Hopkins to take a chemistry course at the Royal School of Mines in London. The lectures on chemistry given by Edward Franklin led him to join a private laboratory run by Franklin's son. Finding the work routine and boring, however, Hopkins enrolled in coursework at University College London in order to prepare for the associateship examination of the Institute of Chemistry. His outstanding performance on the examination led to a position with Sir Thomas Stevenson at Guy's Hospital. Stevenson was a distinguished expert on poisoning and a forensic specialist for the Home Office. While working for Stevenson and participating in several important legal cases, Hopkins was able to study for his B.Sc. degree. In 1888 he became a medical student at Guy's Hospital Medical School, where he was awarded the Sir William Gull Studentship, earning his degree in 1894. In 1898 Hopkins married Jessie Anne Stevens, with whom he had two daughters. In 1891 he published in *Guy's Hospital Reports* a paper on an assay method for uric acid in urine. He also published work on the pigments in butterfly wings.

From 1894-1898 he taught physiology and toxicology at Guy's Hospital, conducted research on blood proteins, and helped organize the Clinical Research Association. In 1898 Sir Michael Foster invited him to move to Cambridge to establish a lectureship in chemical physiology. Because the stipend connected to the lectureship was inadequate, Hopkins had to supplement his income by supervising undergraduates, giving tutorials, and conducting part-time work for the Home Office. In 1902 he was appointed to a readership in biochemistry; in 1910 he became a Trinity College fellow and an honorary fellow of Emmanuel College. In 1914 he became the chair of the biochemistry department at Cambridge University. In 1925 he moved his laboratory into the new Sir William Dunn Institute of Biochemistry.

Among the many areas that interested Hopkins was the puzzle of nutritional-deficiency diseases, such as scurvy or rickets. At first, he assumed that dietary proteins would provide the key to understanding nutrition. This approach led to investigations on uric acid output, the purification of proteins, and the isolation and characterization of the amino acid tryptophane. Hopkins and his associates developed synthetic diets for experimental animals and discovered that a diet that consisted of purified proteins, carbohydrates, fats, and salts was inadequate. Hopkins concluded that small quantities of "accessory food factors" were essential for growth and health. When he first described the work in 1906, other scientists were skeptical, but Hopkins continued to gather evidence. His experiments, published in 1912, are considered classics in the history of nutrition. Inspired by his ideas, other researchers subsequently identified the fat-soluble vitamins A and D and the water-soluble vitamin B. Hopkins also conducted research on the metabolic changes occurring in muscular contractions and rigor mortis, developed analytical tests for lactic acid, isolated glutathione, and discovered the enzyme xanthine oxidase.

Frederick Hopkins. *(Library of Congress. Reproduced with permission.)*

In addition to the Nobel Prize, Hopkins received many other honors, including knighthood, the Order of Merit, the Royal Medal of the Royal Society of London, and the Copley Medal. From 1930 until 1935 he was president of the Royal Society.

LOIS N. MAGNER

Karl Landsteiner

1868-1943

Austrian-American Immunologist and Pathologist

Karl Landsteiner is recognized for his pioneering research into the workings of the human immune system. He was awarded the 1930 Nobel Prize for Physiology or Medicine for his discovery of human blood groups, which he classified in the ABO blood type system. He also discovered the polio virus and developed a test for syphilis.

Born in Vienna on June 14, 1868, Landsteiner was the only child of Dr. Leopold Landsteiner, a noted journalist. His father was the Paris correspondent for several German newspapers and later founded his own newspaper, *Presse.* When Karl was six, his father died of a massive heart attack, and he was placed under the guardianship of a family friend.

At age 17, Landsteiner entered the medical school at the University of Vienna, where he developed an interest in organic chemistry. He remained close to his mother, and in 1889 they both converted from Judaism to Catholicism. He graduated from medical school at the age of 23 and immediately went into a research laboratory, where he applied his passion for organic chemistry to the field of medicine.

While at the Vienna Pathological Institute (1898-1908), he became interested in the problem of why some people could get blood from others but some died. While doing autopsies at the institute, he noted that some blood would clump or agglutinate when mixed with other blood. He proposed there were factors in the blood that were compatible and others that were not. He called these A, B, and C (C was later renamed O). Two other researchers added AB. He showed that a kind of sugar-containing substance, called an antigen, was attached to the plasma membrane of the red blood cells that determined these factors. Further investigation showed that A would always clump with type B. Type O was a universal donor and did not clump with either. The rare form AB was identified as the universal recipient.

Testing the research of Landsteiner in 1907, Dr. Reuben Ottenberg of Mt. Sinai Hospital, New York, performed the first successful blood transfusion. The discovery saved many lives during World War I, the first time that transfusions were performed on a large scale.

Using the information on blood types, Landsteiner and a colleague, Max Richter of the Institute of Forensic Medicine, also devised a plan to use dried blood left at crime scenes to help investigators solve the crime.

From 1908-1920 Landsteiner was head of the pathology department at the Vienna Wilhemina Hospital. There, he wrote numerous papers on bacteriology and immunology and made several important discoveries, including how antigens and antibodies are related. He also developed methods for purifying antibodies and laboratory techniques related to immunology. Although refined, many of the techniques are still used today.

Karl Landsteiner. *(Library of Congress. Reproduced with permission.)*

Landsteiner was one of the first to use animal models to study disease. He successfully transferred the sexually transmitted disease syphilis from humans to apes. Later, he worked to develop a technique called dark-field illumination to identify the bacteria that cause syphilis.

While at the institute, a young boy who had died of polio was brought in for autopsy. Using part of the spinal cord from the boy, Landsteiner injected it into rabbits, guinea pigs, mice, and monkeys. He was puzzled that only the monkeys developed the disease. It was thought that the condition was caused by a bacterium, and Landsteiner went to Paris to collaborate with Romanian scientist Constantin Lefaditis of the Pasteur Institute. The two realized that polio was not caused by a bacterium, but a virus. They traced how it was transmitted, how long the exposure must be, and how the serum of another patient could neutralize the virus in the laboratory. In 1912 they predicted that a vaccine against this virus could be developed, however, this did not occur until 1955, when Jonas Salk developed the first successful polio vaccine.

Dedicated completely to his work, Landsteiner did not marry until age 48, when he met his wife, Helene, at a war hospital. They married in 1916 and had one son, Ernst Karl, in 1917. Postwar Austria was very chaotic, with shortages of food and supplies. Fortunately, Landsteiner was able to get a job at a small Catholic hospital in the Netherlands. Although he was assigned to do routine blood and urine work, he made a major discovery with haptens, small organic molecules that determine antigen-antibody reactions in the presence of a protein. He showed that certain inflammations of the skin caused contact dermatitis and launched into the study of allergic reactions.

Landsteiner was fortunate to be offered a position at the Rockefeller Institute in New York. Refining his previous work, he found new blood groups M, N, and P. He and Philip Levine published a work in 1927 that described the use of blood groups in paternity suits.

In 1929 Landsteiner became a United States citizen, but did not like the crowds of New York and did not care to be a celebrity. When he won the Nobel Prize in 1930, he shunned publicity. His Nobel lecture, describing differences in blood between individuals, as well as species, was at odds with the medical community of the day, although it is well accepted now.

He officially retired in 1939, but continued to work on another blood factor called Rh, because it was first discovered in rhesus monkeys. The incompatible blood factor is responsible for a condition that occurs when a mother without the blood factor carries a fetus with the factor. The infant may have a deadly condition known as erythroblastosis fetalis.

Toward the end of his life, Landsteiner became increasingly worried that the Nazis would take over the world. He died in 1943, at age 75, of a massive heart attack just after completing the manuscript of a new book and seeing his son finish medical school. Tributes were paid to Landsteiner, but there was no mention of his death in Germany or Austria, his home, until 1947, after the Nazis had been defeated.

EVELYN B. KELLY

George Nicholas Papanicolaou
1883-1962
Greek-American Physician

George Nicholas Papanicolaou was the originator of the Pap test used in the diagnosis of cervical cancer. Papanicolaou's test (known as a Pap smear) became the most effective cancer prevention method ever devised.

Papanicolaou, born in Greece, undertook his medical training at the University of Athens, where he earned his medical degree in 1883. After serving in the army, he joined his father's medical practice for awhile before pursuing a career in academic medicine. In 1910, Papanicolaou earned a Ph.D. degree in zoology from the University of Munich, and moved on to participate in a long oceanographic expedition. When the Balkan War broke out, Papanicolaou once again served his country in the Medical Corps. In 1913—with a new wife and no firm prospects—Papanicolaou set sail for America.

After working briefly as a salesman in a department store and playing the violin in restaurants, Papanicolaou secured a research position at Cornell Medical College, where he quickly rose to the rank of instructor. Papanicolaou's early work studied the role of chromosomes in sex determination. During these studies Papanicolaou noted cyclical changes in various vaginal discharges from test animals that Papanicolaou linked to the ovarian and uterine cycles. Papanicolaou's findings were of great benefit to the fledgling field of endocrinology. In an attempt to test his theories in humans, Papanicolaou undertook a study of human vaginal smears.

During his research Papanicolaou observed that smears from women who had been diagnosed with cervical cancer showed cellular abnormalities (enlarged, deformed, or hyperchromatic nuclei). In 1939, Papanicolaou and clinical gynecologist Herbert F. Traut published a paper titled *Diagnosis of Uterine Cancer by the Vaginal Smear*. Papanicolaou and Traut argued that cancerous cervical lesions could be detected by observable and measurable cellular changes while the cells were still in a preinvasive phase. Accordingly, Papanicolaou's diagnostic technique made it possible to diagnose asymptomatic patients. Moreover, such early diagnosis enabled physicians to treat patients while they were still in the earliest, and most treatable, stages of cancer. Papanicolaou's findings meant that cervical cancer could be detected and treated before it could metastasize to other sites. The paper and the Pap test proved to be a fundamental milestone in the treatment of a deadly cancer in women.

Ultimately, the Pap smear became a routine, clinical diagnostic test. The Pap smear for cervical cancer was designed as an inexpensive screening test. The idea was that the test should be repeated frequently. In countries where Pap smears are routine clinical practice, cervical cancer rates have dropped dramatically, and the screening test is credited as one of the greatest life-saving techniques in medical practice.

Other diagnostic tests based upon Papanicolaou's methodology (exfoliated cytology, the scraping, staining, and examination of cells from the test site) proved effective in screening for abnormalities in cells from other organs and systems.

Papanicolaou's colleagues often cited Papanicolaou's strict work regimen as evidence of his meticulous dedication to science. Despite success and acclaim, Papanicolaou continued to work at least six days a week and went years between vacations. After nearly a half-century of research at Cornell, Papanicolaou retired with his beloved wife to Florida. Even in retirement, Papanicolaou proved restless: shortly before his death, he founded the Papanicolaou Cancer Research Institute in Miami.

BRENDA WILMOTH LERNER

Florence Sabin. *(Library of Congress. Reproduced with permission.)*

Florence Rena Sabin
1871-1953
American Physician

Florence Sabin was born in Central City, Colorado, and, after earning her baccalaureate from Smith College in Northampton, Massachusetts, went on to became the first woman to graduate from Johns Hopkins Medical School. Sabin is honored as one of the leading women scientists of her time.

While still in medical school, Sabin studied anatomy and helped transform the field of anatomy from a purely descriptive science to an academic discipline concerned with the relationships between form and function. Sabin's text, *An Atlas of the Medulla and Mid-Brain* (1901), became a classic.

In admitting Sabin, Johns Hopkins fulfilled a charter promise to admit and train both men and women in medicine. After graduation, Sabin performed her medical internship at Johns Hopkins Hospital in Baltimore and subsequently accepted a postgraduate fellowship in the Department of Anatomy. Sabin started research on the embryological development of blood cells and lymphatic system. Sabin's work eventually demonstrated the development of systems from early embryonic "buddings" within the veins of the developing embryo.

Sabin consistently broke ground for women in the sciences. In 1917, Sabin became the first woman to be appointed to a full professorship at Johns Hopkins. In 1924, Sabin was elected the first woman president of the American Association of Anatomists.

After becoming the first lifetime woman member of the National Academy of Science, Sabin accepted a post at the Rockefeller Institute for Medical Research as the head of the Department of Cellular Studies.

Sabin expanded her work to include research on understanding the pathology and immunology of tuberculosis. Although technically tuberculosis refers to any of the infectious diseases of animals and humans caused by the microorganism *Mycobacterium tuberculosisan*, Sabin's work concentrated on the damage inflicted on the lungs and other parts of the body through the formation of tubercles.

Sabin's methods for determining variability in white corpuscle count became important indictors of various disease states. Sabin's findings took on urgent importance in later attempts to combat tuberculosis. Sabin's work continued until her retirement in the late 1930s.

Near the end of World War II, Sabin was called upon to chair a study on public health practices in Colorado. As part of her study Sabin conducted studies on the effects of water pollution and the prevention of brucellosis in cattle. Sabin concentrated her efforts on identifying livestock brucellosis because it is a contagious disease that affects both humans and domesticated animals. Sabin's efforts resulted in the passage of the Sabin Health Laws that signaled an important change in public health policy. The Sabin Health Bills mandated stringent regulations regarding infectious disease, milk pasteurization, and sewage disposal.

Sabin was a refined and distinguished teacher and humanitarian. From the mid-1940s until her death, Sabin accepted various positions related to public health management, including a position as Chairman of the Interim Board of Health and Hospitals of Denver. Sabin also consistently donated her salary to medical research.

The University of Colorado School of Medicine named its principal facility in Sabin's honor, and a statue of Sabin represents the State of Colorado in the United States Capitol's Statuary Hall.

BRENDA WILMOTH LERNER

Margaret Sanger. *(Planned Parenthood Federation of America, Inc. Reproduced with permission.)*

Margaret Louisa Higgins Sanger
1879-1966
American Nurse and Birth Control Advocate

Margaret Sanger was an early champion in the struggle for women to gain control of their own reproductive systems. She opened the first birth control clinic in America, braved frequent arrest and censorship by civil authorities and lifelong persecution by the Roman Catholic Church, and founded the two organizations that later merged into the Planned Parenthood Federation of America.

She was born into an Irish Catholic family in Corning, New York, the sixth of eleven children. Her father, Michael Hennessy Higgins, was an immigrant stonecutter and a decorated veteran of the American Civil War. Her mother, Anne Purcell Higgins, was a sickly, passive, obedient woman, completely devoted to the idea that the husband was the natural lord and master of the house. Margaret never accepted that idea.

The family fortunes declined after 1894 when Michael, a political radical and iconoclast, alienated the local Catholic constituency by engaging the atheist socialist Robert Ingersoll to speak at a public meeting in Corning. A Catholic priest would not let Ingersoll into the meeting hall and the Higgins family was pelted with rotten fruit. Thereafter, a local Catholic boycott of Michael's business was in effect, and the family was ostracized.

Margaret Higgins always hated the provincialism of Corning and longed to escape that environment. She wanted to become a physician, but finances prevented it. Her second choice was to become a nurse. With financial help from her sisters, Margaret enrolled at the Claverack College and Hudson River Institute in 1896 to finish her secondary education, then began the nurse training program at White Plains (New York) Hospital in 1900. She received her nursing credentials in 1902 and the same year married architect William Sanger.

The Sangers lived in suburban Westchester County, New York, until 1910, when they moved to Manhattan. There they became involved in a prominent socialist circle including activists Emma Goldman and John Reed and author Upton Sinclair. Margaret worked as a visiting nurse midwife on the Lower East Side. Her patients were women suffering from too many children, inadequate reproductive health care, frequent miscarriages, sexually transmitted diseases, and abortion. These conditions were much worse than any she had experienced in Corning. She had watched her own mother deteriorate into an early grave from the strain of too many pregnancies. She blamed her father for her mother's death and never forgave him for his sexual tyranny. Incensed by the misery of impoverished, sick, and downtrodden mothers in New York City, Sanger quit nursing in 1912 to devote full time to the cause of freeing women from the medical and economic afflictions of unwanted pregnancy.

In 1912 she began writing a sex education column for *The New York Call.* Her first monthly issue of *The Woman Rebel,* a journal subtitled "No Gods, No Masters," appeared in March 1914, but it was soon suppressed. In 1917 she founded *Birth Control Review.* In the next twenty years she wrote dozens of political pamphlets and books.

With her sister, Ethel Higgins Byrne, and Fania Mindell she opened a clinic in Brooklyn, New York, in October 1916 to distribute family planning literature and contraceptive devices. Because sex education was considered legally obscene, the clinic was almost immediately raided and both sisters spent time in jail.

She divorced William Sanger in 1920 and two years later married J. Noah H. Slee, who gave significant financial support to the birth control movement. In 1922 she founded the American Birth Control League and the next year the Birth Control Clinical Research Bureau.

Sanger worked tirelessly for pro-contraceptive legislation and the social acceptance of women who use contraception. She retired to Tucson, Arizona, in 1942. Some historians claim that she was forced out of the movement by those who thought she was too radical to attract mainstream citizens.

ERIC V.D. LUFT

Ernst Ferdinand Sauerbruch
1875-1951
German Surgeon

Ernst Ferdinand Sauerbruch was one of the most outstanding surgeons of the first half of the twentieth century. His inventiveness and technical brilliance solved a problem that had stumped physicians for centuries—how to open the chest without losing respiratory function. In addition to this breakthrough, Sauerbruch introduced many advances in surgery of the lung and was the first to devise an artificial hand worked by the muscles of the amputated arm.

Born in Barmen, Germany, on July 3, 1875, Sauerbruch was raised in humble circumstances by his mother and grandfather after his father's death in 1877. His family sacrificed to send him to medical school at Marburg, Jena, and Leipzig, where he qualified in 1902. His highly original essay on cerebral pressure attracted the attention of the renowned surgeon Johann von Mikulicz-Radecki (1850-1905), who invited Sauerbruch to his surgical clinic at the University of Breslau. Mikulicz immediately challenged him to research the problem of pneumothorax, inrushing air leading to collapse of the lung, which occurred whenever the chest cavity was opened.

Sauerbruch realized that respiratory function depends on maintaining a lower pressure inside the chest than in the surrounding atmosphere, and conceived the idea of building a low pressure chamber within which the chest might be safely opened. After repeated experiments on rabbits and dogs, the chamber was successfully used on human subjects. At the age of 28, Sauerbruch was invited to demonstrate his device at the prestigious German Surgical Congress in Berlin in 1904.

Following brief periods at Griefswald and Marburg, Sauerbruch was invited in 1910, at age 35, to fill the Chair of Surgery at the University of Zurich, near the Swiss mountain region that had become a haven for patients with lung diseases, especially tuberculosis. Here he developed the two-stage thoracoplasty—removal of portions of the ribs—as a treatment for tuberculosis, which became a classical operation practiced throughout the world. Also significant were the reliable techniques he developed for lung resection—removal of a lobe or an entire lung—which were fundamental to successful treatment for lung cancer.

During his World War I service as a military surgeon, Sauerbruch witnessed the suffering of soldiers who had lost arms or hands. In response he designed the "Sauerbruch hand," an artificial limb controlled by the muscles of the amputated stump.

After the war Sauerbruch was called to the University of Munich and ten years later to the Charite Hospital in Berlin, where his operating theater became a mecca for thoracic surgeons from around the world. Would-be entrants, however, were required to sign their names in support of the restoration of German surgeons to the membership of the International Society of Surgery, an unacceptable condition to many who had fought against Germany. During World War II he was appointed Surgeon-General of the Armed Forces. His relations with the Nazis remain uncertain; though he accepted its rewards he professed opposition to the regime, and in 1949 he was cleared by a denazification court.

Sauerbruch's later years were marred by dismissal in 1949 from his Berlin post: due to the debilitating effects of cerebral sclerosis he had become dangerous to patients, a fact that others could recognize but he could not. Increasingly incapacitated from that point on, Sauerbruch died on July 2, 1951, a day short of his seventy-sixth birthday.

In this half century no other single individual so decisively influenced the direction of surgery. His innovations were known and sought after not only in Europe but throughout the world. Although his low pressure chamber was soon superseded by less cumbersome techniques using inhalation anesthesia under pressure, the underlying principle of pressure differential remains valid. His definitive work, *Surgery of the Organs of the Chest* (1920-1925), was the first to treat this subject systematically and became the foundation for future developments in the field. Because of his work, surgeons gained the knowledge and confidence to attempt life-saving operations involving not only the lungs but also the heart and esophagus.

DIANE K. HAWKINS

Henry E. Sigerist
1891-1957
Swiss Physician and Medical Historian

Henry Sigerist was a medical man who focused on illuminating the history and development of medicine and its effect on patients, practitioners, and society. He was the first medical historian to emphasize the impact of culture and conditions on medicine and the importance of medicine in modern society. He was one of the first respected medical scholars to take up the field of the history of science, a brand-new discipline in the early twentieth century.

Born in Switzerland in 1891, Henry E. Sigerist graduated from local schools unsure of his intended profession. In 1910 at the University of Zurich, he took courses in medicine, history, and art and eventually decided to combine them. He learned about oriental philology, the study of historic and comparative language, and transferred to University College London for a year. He spent 1914 at the University of Munich, then returned to Zurich to earn a degree in medicine. He graduated in 1917 at the height of World War I. Although Switzerland was neutral in the war, Henry spent two years in the Swiss Army, as all Swiss citizens did. As a medical student, he served in the Medical Corps, after which he returned to his graduate medical studies. He was 27 when the armistice ending the war was signed in 1918.

He was interested in social problems and their affect on people and on medicine. He was also interested in the history of medicine, but this was considered just a "delightful hobby" and not a profession. Still, he persisted. In 1919 at the University of Leipzig in Germany, he studied medicine during the Middle Ages. By 1925 he was teaching medical history there and became the director of the Institute of the History of Medicine. This institute became the European focus of medical studies in historical context. His first book, of which he was an editor, was published in 1927.

After a disagreement with the administration over the expansion of his institute, he took a leave of absence and spent the winter of 1931-32 as a lecturer at Johns Hopkins University. He gave a number of lectures and in 1940 toured the United States. He spent the years of World War II in the U.S. teaching at Johns Hopkins and never went back to Germany. He counted himself lucky to have gotten out before Adolf Hitler came to power. He toured South Africa and India with a commission surveying medical conditions in those places. In his books and articles for professional journals and magazines, he concentrated on history, medicine, and sociology, areas seldom previously combined.

He was highly effective as a teacher and respected as a writer and scholar of medical history. He thought the structure of medicine should be reorganized to serve the needs of all people equally. He influenced students whose medical school curricula had not previously included history or sociology and helped to shape the philosophy of many future doctors.

He returned to Switzerland after the war in 1947 and continued to write on medicine and society. In 1954 he had a stroke and was recovering when a second stroke hit him in 1957 from which he did not recover. He died at the age of 66. He is remembered today for his monumental *History of Medicine* and for works like *Medicine and Human Welfare*, *The Sociology of Medicine*, and *Socialized Medicine in the Soviet Union*. He was a prolific writer and a pioneer in combining humanistic ideas with science in general and medicine in particular.

LYNDALL BAKER LANDAUER

Marie Charlotte Carmichael Stopes
1880-1958
British Social Reformer, Botanist, and Writer

In 1921 Marie Stopes opened the first birth control clinic in Great Britain. She is best known for her crusade to gain reproductive freedom for women, but that was only a third of her career. Until about 1914 she was a scientist, until the late 1930s a social reformer, and for the rest of her life a poet.

Stopes was born in Edinburgh, the daughter of feminist Charlotte Carmichael, one of the first women to attend university in Scotland, and geologist Henry Stopes. She won a science scholarship to University College London. In 1902 she was graduated B.Sc. with a double first in botany and geology. Specializing in paleobotany, she received her Ph.D. in 1904 from the University of Munich with a dissertation on fossil plants. The same year she became the first woman scientist hired by the University of Manchester. When University College, London, awarded her the D.Sc. in 1905, she became the youngest Briton of either gender to achieve that degree.

She married botanist and geneticist Reginald Ruggles Gates in 1911, but in 1916 obtained an annulment on the grounds of his impotence. Humphrey Verdon Roe became her second husband in 1918. His wealth paid for much of her campaign for birth control. Together they opened "The Mothers' Clinic" at 61 Marlborough Road, Holloway, North London, on March 17, 1921. This pioneering facility survived public challenges and moved to larger quarters at 108 Whitfield Street, London, in 1925. Even two bomb hits during World War II did not stop its operation. It continued at Whitfield Street until 1977, when the British National Health Service assumed responsibility for providing birth control.

Stopes wanted women to be able to enjoy sex without fear of conception. In many ways she was a voluptuary. After she grew tired of physical intimacy with Roe in the 1930s, she, with his consent, took on a series of younger lovers.

Her opponents included the medical and legal communities, religious leaders, and other birth control advocates, but her strongest opposition came from the Roman Catholic Church. Pope Pius XI's encyclical, *Casti connubii* (1930), codified the traditional Catholic views that artificial contraception is immoral and the purpose of marriage is to procreate. This doctrine, written

Marie Stopes. *(Library of Congress. Reproduced with permission.)*

partially in reaction to Stopes, was reaffirmed in Pope Paul VI's encyclical, *Humanae vitae* (1968). Stopes further irked Catholics because, unlike most other early advocates of birth control, she did not oppose abortion.

She never succeeded against Catholicism but did manage to influence other churches. The Lambeth Conference of the Anglican Communion officially accepted the idea of birth control in 1958, just before her death.

Stopes was a prolific author. Among her books are *Married Love: A New Contribution to the Solution of Sex Difficulties* (1918), *Wise Parenthood: A Sequel to "Married Love"; A Book for Married People* (1919), *Radiant Motherhood: A Book for Those Who Are Creating the Future* (1920), *Contraception (Birth Control): Its Theory, History and Practice* (1923), *The Human Body* (1926), *Sex and the Young* (1926), *Enduring Passion: Further New Contributions to the Solution of Sex Difficulties* (1928), *Mother England: A Contemporary History* (1929), *Roman Catholic Methods of Birth Control* (1933), *Birth Control To-Day* (1934), *Marriage in My Time* (1935), *Change of Life in Men and Women* (1936), and *Your Baby's First Year* (1939).

Her interest in eugenics was extreme. She founded the Society for Constructive Birth Control and Racial Progress in 1921 and after 1937 was a Life Fellow of the British Eugenics Society. She advocated the involuntary sterilization of anyone she deemed "unfit" for parenthood, including lunatics, idiots, addicts, subversives, criminals, and half-breeds. She disinherited her son, Harry Stopes Roe, because he married a woman with bad eyesight.

She was plagued throughout her public life by charges of anti-Semitism, political conservatism, and egomania. Many of these charges came from left-leaning rivals within the birth control movement, such as Margaret Sanger (1879-1966) and Havelock Ellis (1859-1939). Despite the ulterior motives of her accusers, a kernel of truth exists in what they said about her.

ERIC V.D. LUFT

Helen Brooke Taussig
1898-1986
American Physician

Helen Brooke Taussig is recognized as the founder of pediatric cardiology. Along with Alfred Blalock (1899-1964), she devised a procedure for the treatment of newborns afflicted with "blue baby" syndrome.

Born May 24, 1898, in Cambridge, Massachusetts, Taussig was the youngest of four children of famed Harvard economist Frank Taussig. Her mother had attended Radcliffe College but died of tuberculosis when Taussig was 11. Taussig began her studies at Radcliffe in 1917, where she was a member of the tennis team, but finished her B.A. at the University of California, Berkeley, in 1919.

Though medicine was her primary interest, she enrolled at the Harvard School of Public Health. At the time, Harvard permitted women to take classes, but did not allow female students to earn a degree in medicine until 1945. Thus, she studied medicine at Boston College, where her professor, Alexander Begg, suggested she study the heart. Begg also helped Taussig gain acceptance to Johns Hopkins Medical School in Baltimore, Maryland.

Taussig especially enjoyed work at the heart station, and after receiving the M.D. in 1927, she completed additional studies in cardiology and pediatrics. In 1930 she became head of the Pediatric Cardiac Clinic called the Harriet Lane Home, a position she held until her retirement in 1963. She was appointed associate professor of pediatrics at the Johns Hopkins Medical School in 1946 and became the first female full professor in 1959.

Taussig began a study of babies who had congenital heart defects and hearts damaged by rheumatic fever, a condition caused by complications of a streptococcal infection. She became proficient in the use of a new instrument called the fluoroscope, which passes x-ray beams through the body and displays the image on a screen. This device showed the exact location of the malformation. She also used another newly developed device, the electrocardiograph (EKG), which records heart beat patterns on a graph.

In the early 1940s Taussig began to study a birth deformity known as the tetralogy of Fallot or "blue baby" syndrome. The skin of babies suffering from this condition appears blue because a defect in the heart prevents their blood from getting enough oxygen.

Taussig developed an idea for a procedure to repair the artery or attach a new artery. She sold the idea to Alfred Blalock, a Johns Hopkins surgeon who had experimented with artificial arteries. Along with Vivien Thomas (1910-1985), the two developed the procedures after using about 200 experimental dogs. The first human trial came sooner than expected when, in 1944, one of Taussig's patients was born with a serious deformity. Blalock reconnected a branch of the child's aorta, which normally went to the arm, to her lungs. The successful procedure became known as the Blalock-Taussig procedure and since that time has saved the lives of thousands of children.

News of the success spread worldwide, and Taussig was flooded with patients as well as students who wanted to study the procedure. She developed a team approach to cardiology care, which has been adopted in clinics today. Her 1947 textbook, *Congenital Malformations of the Heart,* was a classic text that described heart defects and the techniques and tools to treat them.

Taussig served on several national and international committees. In 1963, one of her students from Germany reported an increase in the prevalence of a birth defect call phocomelia, in which the arms and legs develop a flipper-like appearance. The student thought the drug Contergan, used to counter morning sickness among pregnant women, may be the cause. Taussig spent six weeks in Germany, during which she investigated the problem. Noting the absence of the defect in the United States military camps, where the drug was banned, she was able to make the connection between the drug and defect. She was also able to halt usage of the drug, renamed thalidomide, in the United States.

When Taussig retired in 1963, she turned her attention to causes involving the treatment of laboratory animals and a woman's right to terminate pregnancy. During her career, she wrote over 100 papers, 40 of which were published after her retirement. She became very interested in the embryological development of heart defects and had launched into a major study shortly before her death at age 87. She was killed in an automobile collision while driving home after voting in a political primary.

Taussig was recognized for her many achievements with 20 honorary degrees and numerous awards, including the 1984 Medal of Freedom.

EVELYN B. KELLY

Julius Wagner-Jauregg
1857-1940
Austrian Psychiatrist

In 1927 Julius Wagner-Jauregg was awarded the Nobel Prize for medicine or physiology for his attempts to treat patients who had syphilitic paresis, or dementia paralytica, by inducing fevers using malaria inoculations. His early research on shock therapy led to experiments on the effect of fever on mental illness.

Julius Wagner, the son of Ludovika Ranzoni and Adolf Johann Wagner, an Austrian government official, studied at the Schottengymnasium in Vienna. He began his medical studies in 1874 at the University of Vienna, where he became interested in microscopy and experimental biology. To immerse himself in these new areas of research, he served as assistant to Salomon Stricker at the Institute of General and Experimental Pathology from 1874-1882. Additionally, Wagner and Sigmund Freud (1856-1939) shared many scientific interests and became lifelong friends. In 1880 Wagner was awarded a Ph.D. for his thesis on the heart. His father was granted the title "Ritter von Jauregg" in 1883, at which point Julius became Wagner von Jauregg. Titles of nobility were eliminated with the breakup of the Hapsburg Empire in 1918 and Julius changed his name to Wagner-Jauregg. He married Anna Koch in 1899; together they had a son and a daughter.

After working in the Department of Internal Diseases at the University of Vienna, Wagner-Jauregg became an assistant to Max von Leidesdorf in the psychiatric clinic. Despite his lack of training in either psychiatry or neurology, Wagner-Jauregg became very interested in these fields. In 1885 he was considered qualified to teach neurology. By 1888 he was also teaching psychiatry. When Leidesdorf became ill in 1887, Wagner-Jauregg served as director of the clinic. Two years later, he was appointed professor of psychiatry and neurology and director of the neuropsychiatric clinic at the University of Graz. After his appointment to the staff of the State Asylum, he returned to Vienna as "extraordinary professor of psychiatry and nervous disease" and director of the university clinic.

One of Wagner-Jauregg's lesser known, but very valuable, studies involved the relationship between cretinism and goiter. He attributed these conditions to a malfunction of the thyroid gland and investigated the benefits of glandular extracts and the addition of iodine to table salt. The importance of this work was not fully recognized until the 1920s, when the Austrian government began formally requiring the use of iodized salt.

Wagner-Jauregg was also interested in trauma, nerve action, and shock therapy. His experiments on shock therapy led him to consider the possible value of fever in the treatment of mental illness. In 1887 he published his first paper on the effects of fevers on psychosis. According to Wagner-Jauregg, the mental state of his patients seemed to improve after they recovered from a febrile illness. He thought patients with paresis, or dementia paralytica associated with advanced syphilis, particularly benefited from episodes of high fever. At first, he preferred to induce fevers with tuberculin, which Robert Koch (1843-1910) had recently introduced as a possible remedy for tuberculosis, but this substance had dangerous side effects. By 1917 he concluded that the pyretic agents he had been using did not produce a permanent cure. He then decided to test malaria therapy, confident that, after he had cured the patient of syphilitic paresis, he could cure the malaria with quinine. Blood from a patient suffering from malaria was used to inoculate nine syphilitic patients; six seemed to improve significantly. Until penicillin became widely available after World War II, malaria induction remained one of the standard treatments for neuro-syphilis. Physicians generally reported improvement in 30-40% of their patients, especially patients in the early stages of the disease.

LOIS N. MAGNER

Biographical Mentions

John Jacob Abel
1857-1938

American physician known as an authority on the chemistry of the ductless glands. Born in Cleveland, Ohio, he received a pharmacy degree from the University of Michigan in 1883 and went abroad to study physiology and medicine. He received his M.D. degree at Strausbourg in 1888. He came to Johns Hopkins in 1893 and remained there as director of the Laboratory of Endocrine Research until his death. He made numerous discoveries on the chemistry of tissues, including isolating insulin and adrenaline in crystalline form.

Edward Hartley Angle
1855-1930

American dentist who published a landmark treatise on the treatment of malocclusion of the teeth. He was a pioneer in the evolution of orthodontics as a specialty within the profession of dentistry. Angle established classifications that organized the various abnormalities of the teeth and jaws and developed methods for treating them. Angle's classification system is used to define three major types of malocclusion, based on the relationship of the permanent molars as they erupt and lock into position.

Percy Moreau Ashburn
1872-1940

American physician, army medical officer, and historian of the army medical corps. A native of Texas, in 1929 he wrote *A History of the Medical Department of the U.S. Army* and many treatises on the role of medicine in the battles of the Civil War, the Indian Wars, and others. In May of 1943, a hospital housing veterans in McKinney, Texas, was renamed the Ashburn General Hospital in his honor. When the last patient was discharged in December 1945, the center was remodeled as a U.S. Veteran's Hospital.

Robert Barany
1876-1936

Austro-Swedish physician recognized for his work on the physiology of the inner ear. Born in Vienna, Austria, he received his doctor of medicine degree from the University of Vienna in 1900. His interest in the structure of the inner ear and its effect on balance led to a 1914 Nobel Prize in physiology or medicine. He also related the cerebellum of the brain with balance and coordination. Barany went to the University of Uppsula to teach otology and practice clinical medicine. He won several national prizes in Sweden and died in Uppsula.

Wlliam Harry Barnes
1887-1945

African-American physician who gained recognition in the field of otolaryngology, or study of ear and throat. Born to poor parents, Barnes did so well on an entrance exam to the University of Pennsylvania that he won a scholarship. He received his M.D. degree in 1912 and after his residency, went to France in 1924 for additional study. In 1927 he was the first African-American to become a Diplomat of the Medical Board of Otolaryngology. Barnes was an active social leader in Philadelphia and published over 23 scientific studies.

Emil Adolf von Behring
1854-1917

German microbiologist recognized as one of the founders of immunology and serum therapy. Born in Hansdorf, Germany, Behring studied medicine in Berlin before working at the Robert Koch Institute. While investigating toxin-producing bacteria, he injected dead or weakened diphtheria bacilli into guinea pigs and showed how the toxin was neutralized by the injection. He coined the term "antitoxin." He found the same results with tetanus. Behring tested with the first human patient in 1880 and in 1892 widespread immunization began. His work set the stage for antitoxins for a number of diseases. He received the Nobel Prize in Physiology or Medicine in 1901.

Vladimir Bekhterev
1857-1929

Russian physiologist known for his research on brain formation and conditioned reflexes. He developed a behaviorist theory of conditioned response, independent of Ivan Pavlov. Bekhterev's greatest work was on brain structure and specific brain diseases. Born in Kirov, Russia, he received his doctorate from Medical-Surgery Academy of Saint Petersburg in 1881 and worked at the University of Petrograd until his death. His works greatly contributed to the behaviorist psychology movement in the United States.

Charles H. Best
1899-1978

Canadian physiologist who worked to find insulin for the control of diabetes. Born in West

Pembroke, Maine, he became an assistant to Sir Frederick Banting at the University of Toronto. The two scientists extracted insulin from the pancreas and showed how it was effective in treating diabetic dogs. Success with humans followed. Best did not receive his M.D. until 1925, and for this reason could not share the 1923 Nobel Prize in physiology or medicine with Banting. He succeeded Banting as director of the Banting and Best Department of Medical Research in 1941 and discovered the vitamin choline, the enzyme histamine, and anticoagulants for treatment of blood clots.

Francis Gilman Blake
1887-1952

American physician recognized as a specialist in the field of internal medicine and later dean of the Yale School of Medicine. Born in Mansfield Valley, Pennsylvania, he graduated from Harvard in 1913. Blake was very influential in wartime medicine, especially in the control of influenza and other epidemics in the army. His research contributed to treatment of scarlet fever with antitoxin and to the treatment of pneumonia with serums and sulfapyridine and related compounds. The Francis Gilman Blake award is given at Yale to the outstanding teacher in the College of Medicine.

Alfred Blalock
1899-1964

American surgeon who performed the first surgery for the "blue baby syndrome," a condition known as tetralogy of Fallot. Born in Cullonden, Georgia, he received his doctor of medicine degree from Johns Hopkins University School of Medicine and did a residency in surgery at Vanderbilt. His research into trauma and shock led to treatment that saved many lives during World War II. Returning to Johns Hopkins in 1941, he collaborated with Helen B. Taussig on the procedure to correct the "blue baby" and performed the operation in 1944.

Jules Jean Baptist Vincent Bordet
1870-1961

Belgian microbiologist known for his discovery of immune factors in blood and for the bacterium that causes whooping cough. Born in Soignies, Belgium, he studied and conducted research at the Pasteur Institute, Paris. In 1885 he found that the bacterial wall will rupture if exposed to a heat-stable antibody and a heat-sensitive complement. He founded the Pasteur Institute in Brussels and along with Octave Gengou developed tests for typhoid, tuberculosis, and syphilis. He received a Nobel Prize in physiology or medicine in 1919.

George Sidney Brown
1873-1948

American electrical engineer and inventor who designed an electromagnetic bone-conducting device as an aid for the hearing impaired. Called an Ossiphone, the name recalled a popular diaphragm earpiece trumpet that had been marketed under the name Otophone since 1877. Brown's invention was the first of the electrically amplified bone-conductor type. Brown, who was born in Chicago, invented several devices and was awarded patents in the United States and Great Britain. In 1915 Brown received a patent in the United States for a telephone relay. In 1921 he was awarded a British patent for a carbon hearing aid and three years later a patent for a loudspeaker improvement. The trade name of S. G. Brown Ltd. was Weda, which in 1949 marketed a vacuum-tube, one-piece earphone-type hearing aid.

Frank Macfarlane Burnet
1899-1985

Australian physician and biologist who made major contributions to immunology, virology, and the ecology of diseases. He was awarded a Nobel Prize in medicine in 1960 for developing two key concepts in immunology: acquired immunological tolerance and the clonal selection theory of antibody production. His investigation into the deaths of children who had received diphtheria vaccinations led to safer methods for the storage of vaccines. Burnet developed methods for culturing viruses, established a classification for bacteriophages, and investigated such diseases as influenza, Q fever, psittacosis, and encephalitis.

Léon Charles Albert Calmette
1863-1933

French physician who is best known for establishing the live, attenuated strain of the tubercle bacillus that was used as a vaccine against tuberculosis. Bacillus Calmette-Guérin (BCG) was originally produced by Calmette and Camille Guérin in 1906. Since the 1920s BCG has been used as a vaccine against tuberculosis in children. Calmette also conducted studies of venoms, immunized animals in order to create therapeutic serums, and investigated vaccines for bubonic plague and water purification.

Ugo Cerletti
1877-1963

Italian psychiatrist and neurologist who developed the method of electroconvulsive shock (electroshock) therapy (ECT) to treat certain mental pathologies. While chair of the Department of Mental and Neurological Diseases at the University of Rome, Cerletti began to use electroshock on animals while studying the neuropathological effects of recurrent epilepsy attacks. He first used ECT on a human patient with schizophrenia in 1938. Through experimentation, Cerletti and colleagues found ECT to be useful in treating some forms of schizophrenia, manic depression, and severe depression, and ECT quickly became a popular therapeutic tool.

William Montague Cobb
1904-1990

African American physician who was a longtime professor of anatomy at Howard University in Washington, D.C., and who served as editor of the *Journal of the National Medical Association* from 1949-77. Cobb also was National President of the National Association for the Advancement of Colored People (NAACP). Working with other black leaders in science in medicine, Cobb sought to change the separate-but-equal status of medical schools and hospitals and instead promote racial integration of the medical establishment. The first permanent building built at Charles R. Drew University of Medicine and Science in Los Angeles, California, was named the W. Montague Cobb Medical Education Building to honor Cobb's work.

Carl Ferdinand Cori
1896-1984

Gerty Theresa Radnitz Cori
1896-1957

Czech-American husband and wife physicians who won the Nobel Prize in medicine in 1947 for their discoveries concerning the enzymes involved in the metabolism of glycogen and glucose. Carl and Gerty Cori worked in close collaboration for 35 years as pioneers in the development of "dynamic biochemistry." Their work also established the effects of such hormones as insulin and epinephrine on carbohydrate metabolism.

George Washington Crile
1864-1943

American surgeon whose research into the causes and control of hemorrhage saved countless lives. Three developments were necessary for modern surgery to arise: control of pain (anesthesia), control of infection (antisepsis), and control of surgical shock due to loss of blood. Crile provided this third development in the 1890s. Among his books are *An Experimental Research into Surgical Shock* (1899), *Blood-Pressure in Surgery* (1903), and *Hemorrhage and Transfusion* (1909).

Henry Hale Dale
1875-1968

English physiologist who shared the 1936 Nobel Prize in Physiology or Medicine with Otto Loewi. Dale conducted pioneering research in experimental pharmacology and chemotherapy. His work included studying pituitary hormones, the effect of secretin on pancreatic cells, the effect of ergot on the activity of epinephrine, the effect of epinephrine on the sympathetic branches of the autonomic nervous system, the role of acetylcholine in neurotransmission, and other regulatory mechanisms, providing insights into the chemical mechanisms that direct physiological functions.

Walter Edward Dandy
1886-1946

American neurosurgeon known for his contributions to neurosurgery. He pioneered successful operations for Ménière's disease, glossopharyngeal neuralgia, intracranial aneurysm, and various tumors of the central nervous system. In 1918 he invented pneumography, a diagnostic procedure whereby x rays are taken after air is injected into the cerebral ventricles (ventriculography). The following year he modified this procedure by injecting air through a lumbar puncture needle (pneumoencephalography). This was more dangerous, and ventriculography was preferred.

George Frederick Dick
1881-1967

Gladys R. Henry
1881-1963

American physician and pathologist, respectively, who discovered the cause of scarlet fever and developed methods for preventing this disease. In 1923 George and Gladys Dick identified and isolated the hemolytic streptococcus bacterium that causes scarlet fever. They used the Dick toxin, which is secreted by the causative agent, in a series of immunization experiments. Injections of the toxin-antitoxin were used to immunize children against the disease. The Dick skin test for susceptibility to scarlet fever was successfully introduced in 1924.

Gerhard Domagk
1895-1964

German pathologist and chemist who won the Nobel Prize in 1939 for his discovery that a red, sulfur-containing dye called Prontosil was a safe and effective treatment for streptococcal infections in mice. Researchers at the Pasteur Institute later proved that Prontosil itself was not antibacterial; instead, the dye substance was broken down in the body, releasing the active sulfonamide portion of the molecule. This revelation led to the development of a series of related drugs called the sulfonamides, or sulfa drugs.

Charles R. Drew
1904-1950

African-American physician and surgeon who made major contributions to the preservation of blood for transfusion in "blood banks." Drew was the first black surgical resident at the Presbyterian Hospital of New York City. During World War II, he organized blood-banking programs for the United States and Great Britain. Drew fought against policies that allowed blood banks to exclude or segregate blood contributed by blacks. The Charles R. Drew University of Medicine and Science in Los Angeles was established in the 1960s.

Joseph Everett Dutton
1874?-1905

English physician whose research helped describe the causes and effects of African sleeping sickness. While working at the Liverpool School of Tropical Medicine, Dutton and Robert Mitchell Forde isolated a parasitic flagellate protozoan (a trypanosome) from the blood of an English shipmaster. Dutton named the parasite *Trypanosoma gambiense.*

Guy Henry Faget
1891-1947

American physician who discovered the first practical treatment for leprosy in 1941. While working at a hospital in Carville, Louisiana, that was home to many lepers, Faget found that a sulfone drug similar to that used to treat tuberculosis controlled leprosy and helped put the disease into remission. Faget, a specialist in tropical diseases, worked in Central America as an employee of the British government during World War I.

Werner Theodor Otto Forssmann
1904-1979

German physician who, by experimenting on himself, was the first to develop a technique for catheterization of the heart. Forssmann was awarded the 1956 Nobel Prize for physiology and medicine (along with two other cardiology researchers) for his pioneering work in the field of cardiology. Despite the significance of Forssmann's cardiac research, he was scorned by the medical community for his unorthodox method of self-experimentation. Forssmann ultimately switched his focus to urology.

Ernest François Auguste Fourneau
1872-1949

French pharmacist who developed the first successful substitute for the active anesthetic ingredient of cocaine. Fourneau was one of many doctors looking for a substitute for the addictive drug cocaine, and in 1904 he discovered stovaine. For 30 years Fourneau was the head of chemical laboratories at the Institut Pasteur in Paris, and he was successful in developing new chemical compounds to fight specific pathogenic organisms. Fourneau's research led to better treatment for syphilis and paved the way for the development of sulfonamide drugs and antihistamines.

Casimir Funk
1884-1967

Polish-American biochemist who demonstrated in 1912 that chickens with beriberi recovered when their test diet was supplemented by a concentrate made from rice polishings. Beriberi was endemic in Asian regions, where white rice was a major component of the human diet. Funk suggested the name "vitamine" for the antiberiberi factor (vitamin B_1, or thiamine) found in the supplement and other "accessory substances" because he thought that they were amines. Eventually, chemists demonstrated that not all accessory factors were amines and the term "vitamin" was adopted instead.

Archibald Edward Garrod
1857-1936

English physician who conducted pioneering studies of genetic diseases, which he referred to as "inborn errors of metabolism." Garrod concluded that these disorders were the results of an inherited alteration in some metabolic process. He studied the characteristics and hereditary patterns of alcapatonuria, albinism, cystinuria, and pentosuria, arguing that these disorders resulted from "inborn errors of metabolism." The investigation of the family trees of affected individuals suggested that these conditions were inherited as Mendelian recessive genes.

Walter Holbrook Gaskell
1847-1914

English physiologist whose research in experimental physiology contributed to a greater understanding of the autonomic nervous system. In particular, Gaskell studied the vasomotor nerves of striated muscle, the central sympathetic nerves, and innervation of the heart. He described his findings in his book *The Involuntary Nervous System.*

Octave Gengou
1875-1957

French bacteriologist who with Jules Bordet described the bacterium responsible for causing whooping cough (Pertussis). While working at the Pasteur Institute in Brussels, the two researchers successfully cultured the bacterium in 1906. In 1912 they developed the first whooping cough vaccine by growing Pertussis bacteria, killing them with heat, and mixing them with formaldehyde. The pair's research led to the development of tests for the presence of many disease organisms, including those that cause typhoid fever, tuberculosis, and syphilis (the Wassermann test).

John Heysham Gibbon, Jr.
1903-1974

American physician who is known for his invention of the heart and lung apparatus that was utilized to prevent heart and lung failure during open chest surgery. He also invented the Jefferson ventilator. He was a professor of surgery and director of surgical research at Jefferson Medical College from 1946-56 and was the recipient of the Research Achievement award from the American Heart Association in 1965.

Joseph Goldberger
1874-1929

American physician and epidemiologist who is best known for demonstrating that pellagra is a nutritional deficiency disease. Pellagra is characterized by dermatitis, diarrhea, and dementia. The disease was endemic in areas where the diet was based almost exclusively on corn. Goldberger proved that the disease could be cured and prevented by enriching the diet with milk, eggs, or meat. Eventually, scientists discovered that pellagra is caused by a deficiency of niacin, a member of the vitamin B complex. Goldberger's research also included studies of yellow fever, typhoid fever, dengue fever, typhus fever, and diphtheria.

Evarts Ambrose Graham
1883-1957

American surgeon who in 1933 with J. J. Singer performed the first successful total lung removal. In 1924 Graham and W. H. Cole developed the Graham-Cole test, which is an x ray study of the gall bladder. He was also responsible for a new treatment of the chronic occurrence of an abscess in the lung and contributed to the understanding and treatment of tumor growth in the bronchus of the lung.

Sir Norman McAlister Gregg
1892-1966

Australian ophthalmologist who discovered the relationship between rubella and babies born with cataracts. In 1941 Gregg examined two babies for cataracts, which is a problem in the eye regarding the lens. Gregg was struck with the fact that the mothers of two babies stated that they had had rubella while pregnant. Following this, he studied the possibility and discovered that there was a direct connection. Gregg was knighted for this discovery in 1953.

Camille Guérin
1872-1961

French veterinarian who developed bacillus Calmette-Guérin, or BCG, in association with Albert Calmette at the Pasteur Institute in Lille. BCG was widely used as a vaccine to prevent childhood tuberculosis. Guérin and Calmette spent 13 years developing a weak strain of bovine tuberculosis bacteria. In the 1920s, convinced that BCG was harmless to humans but could induce immunity to the tubercle bacillus, they began a series of experimental inoculations of newborn infants at the Charité Hospital in Paris.

Granville Stanley Hall
1844-1924

American psychologist who was a pioneer in the development of psychology in the United States and a founder of child and educational psychology. Hall incorporated the ideas of Charles Darwin, Sigmund Freud, Wilhelm Wundt, and Hermann von Helmholtz into American psychology. He helped establish a scientific base for experimental psychology. Hall founded one of the first psychological laboratories in the United States, the first American journal devoted to research in psychology, and the first journal for child and educational psychology.

Philip Showalter Hench
1896-1965

American physician who is known for his contribution in the treatment of rheumatism, an umbrella term that includes arthritis and other pains associated with the joints and muscles. His research with arthritis led him to discover that in some cases the symptoms can be reversed. He also successfully used cortisone and extracts of the pituitary and adrenal cortex to combat arthritis. He shared the Nobel Prize for physiology or medicine with E. C. Kendall and Tadeus Reichstein in 1950.

James Bryan Herrick
1861-1954

American physician and cardiologist who published the first description of sickle cell anemia, including observations of the abnormal red blood cells that characterize this disorder. Eventually, scientists discovered that the disorder is a genetic disease caused by an abnormal form of hemoglobin, called hemoglobin S. The disorder is inherited and occurs predominantly in blacks. Herrick's medical practice and clinical research program developed a specialization in cardiovascular disease. Herrick published classical accounts of cardiovascular disease, including myocardial infarctions (heart attacks).

Russel Aubra Hibbs
1869-1932

American surgeon who was probably the first to perform a posterior spinal fusion for the treatment of scoliosis. Scoliosis (from the Greek word for "crookedness") refers to various forms of abnormal lateral curvatures of the spine. Hibbs performed the operation on January 9, 1911, at the New York Orthopedic Hospital. He published an account of this operation for progressive spinal deformities in the *New York Medical Journal.*

William Henry Howell
1860-1945

American physiologist and biochemist who isolated and characterized heparin, a substance that inhibits blood clotting, and conducted pioneering studies of the biochemistry of the blood. The activity of heparin preparations is now measured in terms of Howell units. He was the first to describe the "Howell-Jolly bodies" that are sometimes seen in erythrocytes, especially after a splenectomy. Howell demonstrated the presence of amino acids in the blood and lymph and conducted pioneering studies into the preparation and properties of thrombin, antithrombin, and prothrombin.

Carl Gustav Jung
1875-1961

Swiss psychoanalyst, at first a Freudian, but who broke with Freud in 1914 to found "analytic" psychoanalysis. Freud believed that all neurosis is caused by sexuality, but Jung blamed it on disharmony between the conscious self and the "collective unconscious." Jung defined two basic psychological types—introvert and extrovert—and four basic psychological functions—thinking, feeling, sensing, and intuiting. His psychology has profound religious aspects. He took myths and dreams seriously as keys to the memories, symbols, and archetypes of the collective unconscious.

Willem Johan Kolff
1911-

Dutch-American physician who is best known for inventing the artificial kidney, or dialysis machine, in Nazi-occupied Holland. His original dialysis machine was ingeniously constructed from improvised materials, including a washing machine and sausage-link casings, which served as semipermeable membranes. Improved versions of this machine have kept thousands of people with end-stage renal disease alive. Kolff has been involved in various artificial organ programs. He directed a team that invented an artificial heart.

Emil Kraepelin
1856-1926

German psychiatrist, a specialist in diagnosis, who developed the classification of mental disorders that still serves as the basis of the American Psychiatric Association's *Diagnostic and Statistical Manual of Mental Disorders* (*DSM-IV*) and the World Health Organization's *International Statistical Classification of Diseases and Related Health Problems* (*ICD-10*). He discovered and named several psychoses and neuroses, including manic depression (now called bipolar disorder) and dementia praecox (now called schizophrenia).

Emanuel Libman
1872-1946

American physician who is known for his contributions to clinical medicine. He was the first to show that an aneurysm, which is a weakening of the blood vessels walls that leads to the formation of a sac in the vessel, can be caused by impact. He also made the first clinical study of blood transfusions in 1915 with R. Ottenberg. He studied the blood clotting of the coronary artery in the heart from 1917-28.

John James Rickard Macleod
1876-1935

Scottish physiologist who shared the Nobel Prize for physiology or medicine in 1923 with Frederick Banting for the discovery of insulin. Macleod was already well known for his research on carbohydrate metabolism when Banting approached him with a plan to isolate the anti-diabetes factor from the pancreas. Macleod arranged laboratory facilities at the University of Toronto for the research carried out by Banting and Charles H. Best. Insulin was successfully isolated and prepared in 1921.

Jay McLean
1890-1957

American physician who is known for his discovery of the anticoagulant, or non-blood clotting substance, that later became known as heparin. As a medical student at Johns Hopkins University, McLean studied under William Howell, who was studying blood coagulation with a clotting substance from the brain. McLean, while comparing similar substances from the liver, discovered that the liver substance did not clot blood. This substance is now widely used in surgery and medicine for its anticoagulant properties.

Antonio Caetano de Abreu Freire Egas Moniz
1874-1955

Portuguese physician, neurologist, and politician whose introduction of prefrontal lobotomy as psychosurgery in 1936 won him the Nobel Prize for physiology or medicine in 1949. Of more lasting value was his work in neuroradiology, diagnostic techniques that allow doctors to see into the living brain. Around 1926 he invented cerebral angiography. This involved inserting an opaque contrast medium such as thorium dioxide into both carotid arteries, then using x rays to find brain tumors and other lesions.

Thomas Hunt Morgan
1866-1945

American zoologist and geneticist who is best known for his genetics research on the fruit fly (*Drosophila*), which established the chromosome theory of heredity. Morgan's work demonstrated that genes are linked in a linear sequence on the chromosomes. Morgan was awarded the Nobel Prize for physiology or medicine in 1933 for his contributions to genetics. He also contributed to the fields of experimental embryology and cytology. His laboratory at Columbia University was a major center of research into heredity. Morgan and his associates established techniques for chromosome mapping.

John Benjamin Murphy
1857-1916

American surgeon who was a pioneer in surgical procedures. He began a new era in intestinal surgery by developing instruments to connect the severed ends of the intestines. He also made advances in surgery of the gall bladder as well as developing methods with which to repair damaged blood vessels. Later in his career he studied joint diseases and was president of the American Medical Association.

Hideyo Noguchi
1887-1928

Japanese bacteriologist who was the first to demonstrate the presence of *Treponema pallidum*, the causative agent of syphilis, in the brains of persons suffering from paresis (tertiary syphilis). Most of Noguchi's research in bacteriology and virology was carried out at the Rockefeller Institute for Medical Research in New York City. Noguchi also contributed to studies of poliomyelitis, trachoma, yellow fever, and Carrion's disease. He died of yellow fever while conducting research on that disease in Africa.

Jean Piaget
1896-1980

Swiss psychologist who identified a universal sequence of four stages in the cognitive, linguistic, and social development of children, from egocentric, imitative, and concrete reaction to constructive, logical, and abstract thought. These stages are: (1) sensorimotor; (2) preoperational; (3) concrete operational; and (4) formal operational. Because teachers' understanding of these stages has been shown to affect students' progress, Piaget's theory of developmental psychology, "genetic epistemology," has had tremendous impact on educational philosophy and policy.

Clemens von Pirquet
1874-1929

Austrian pediatrician who is known for his development of the term "allergy" and for beginning to scientifically study allergies. Pirquet noticed in 1906 that some of the young patients he treated for diphtheria—with an antitoxin based on a horse serum—developed symptoms not related to the original disease. He used the term "allergen" (from where we get allergy) to describe an "altered response" some patients had to such items as medications, food, or by breathing.

Armand James Quick
1894-1978

American hematologist who specialized in the study of blood. He is best known for his study of the coagulation of blood and his subsequent discovery of coagulation factor V. He is also known for his work on the mechanisms of detoxication, the functioning of the liver, and for tests regarding hippuric acid, an acid that is produced and excreted by the kidneys. Quick, in 1944, contributed to Sayhun's outline of the amino acids and proteins.

Walter Reed
1851-1902

American physician, military surgeon, and epidemiologist whose discovery in 1900 that the aedes aegypti mosquito is the vector of yellow fever soon led to the control and conquest of this disease. Yellow fever is a tropical viral hepatitis. For centuries it was among the most feared plagues in the world. Carlos Juan Finlay expressed the mosquito vector theory in 1881, but only after Reed's work could William Crawford Gorgas institute the effective public health policies that eliminated yellow fever from Havana, Cuba, by 1902.

Ludwig Mettler Rehn
1849-1930

German physician and surgeon who was the first to operate successfully on the heart in 1896, an operation that included the suture of a stab wound of the heart. In 1880 Rehn was the first surgeon to remove an enlarged thyroid in a patient that was protruding the eyeball out of its socket, a condition known as Basedow's disease. He was also the first to operate on the esophagus to combat diseases that were regarded as inoperable in 1884.

Tadeus Reichstein
1897-1996

Polish-born chemist educated in Switzerland who received the 1950 Nobel Prize for Physiology or Medicine for his research on adrenal cortex hormones. Reichstein and his team isolated and characterized about 29 hormones. Their work led to the isolation of the hormone cortisone, which was later discovered to be an anti-inflammatory agent. Reichstein is also known for his synthesis of vitamin C.

Cornelius P. Rhoads
1898-1959

American physician who helped found the Memorial Sloan-Kettering Cancer Research Institute. Rhoads's research into the causes and cures for cancer in the 1930s and 1940s helped to set the stage for many of the advances of recent years. However, Rhoads also generated a great deal of controversy in 1931 when he injected cancer cells into Puerto Rican patients to research the causes of cancer.

Hermann Rorschach
1884-1922

Swiss physician and psychiatrist who developed the inkblot test, now known as the Rorschach test, which is widely used for psychological evaluation and the diagnosis of psychopathology. His experiments, which began in 1918, involved asking patients to describe their subjective responses to a series of accidental inkblots. The results of hundreds of tests of patients and normal subjects were published in his *Psychodiagnostics* (1921). Rorschach concluded that he could distinguish his subjects in terms of their personality traits, emotional characteristics, perceptive abilities, impulses, and intelligence on the basis of their responses to these ambiguous stimuli.

Bela Schick
1877-1967

Hungarian physician who is best known for his contributions to the battle against diphtheria, a childhood disease that killed thousands. The Schick test, which was developed in 1913, reveals susceptibility to diphtheria, which is caused by a toxin produced by *Corynebacterium diphtheriae* (Klebs-Loeffler bacillus). Injections of toxin-antitoxin (or, later, toxoids) produced immunity; by using the Schick test, physicians could avoid subjecting patients to unnecessary treatment with antitoxins.

Carl Frederick Schmidt
1893-1988

American physiologist who carried out experiments with Alfred Newton Richards in the 1920s concerning the secretion of urine. Richards and Schmidt published a description of the glomerular circulation in the frog's kidney as well as observations of the effect of adrenalin and other substances on the circulation. They collected and analyzed the fluid from a single glomerulus, confirming the theories of Carl Ludwig and Arthur Cushny. In 1918 Schmidt and Ko Kuei Chen published an influential treatise on ephedrine and related substances that summarized the current literature and provided an extensive bibliography.

Sir Charles Scott Sherrington
1857-1952

British physician and neurologist who made important contributions to understanding the relations between brain and spinal cord, efferent and motor nerves, and nerves and muscles. After he published *The Integrative Action of the Nervous System* (1906), he was acknowledged as the world's foremost neuroanatomist and neurophysiologist. He shared the 1932 Nobel Prize for physiology or medicine with Lord Edgar Douglas Adrian for their work on the structure and function of neurons.

Jacob Jesse Singer
1882-1954

American physician who, with Evarts Graham, was the first to successfully remove an entire lung in the treatment of lung cancer. This procedure, radical at the time, later proved one of the more successful treatments for lung cancer and is still used in many cases today. Most people can survive with a single lung, although they cannot exercise vigorously.

Ernest Henry Starling
1866-1927

British physiologist who, in collaboration with William Maddock Bayliss, discovered secretin, a hormone produced by the intestinal mucosa. Staring and Bayliss demonstrated nervous control of the peristaltic wave (the muscle action that causes the movement of food through the intestine). In 1902 they isolated secretin, a substance that stimulates the secretion of pancreatic digestive juice into the intestine. In 1904 Starling coined the term "hormone." Starling also investigated the mechanisms that maintain fluid balance, the regulatory role of endocrine secretion, and the action of the heart and kidneys.

Albert Szent-Györgyi
1893-1986

Hungarian-American biochemist who was awarded the Nobel Prize for physiology or medicine in 1937 for his research concerning the roles played by vitamin C in cellular metabolism. Szent-Györgyi isolated a reducing agent he called "hexuronic acid" and demonstrated that hexuronic acid (ascorbic acid) was identical to vitamin C, the antiscorbutic factor. Szent-Györgyi also investigated carbohydrate metabolism, the biochemistry of muscular action, the production of adenosine triphosphate (ATP, which serves as the immediate source of the energy needed for muscle contraction), and proposed theories concerning the causes of cancer.

Lewis Madison Terman
1877-1956

American psychologist who developed the Stanford-Binet intelligence test. In 1916, while a professor of education at Stanford University, Terman published *The Measurement of Intelligence,* which served as a guide to his version of the Binet-Simon intelligence scale. Stanford-Binet test scores were expressed as the intelligence quotient (IQ). Test scores were based on both chronological age and mental age so that the average person of any age would have an IQ of 100. The test was used extensively by the U.S. Army during World War I. Terman also carried out research on gifted children.

Max Theiler
1899-1972

South African-American microbiologist who was awarded the 1951 Nobel Prize for physiology or medicine for his research on yellow fever. Theiler's discovery that mice are susceptible to yellow fever expedited research on the disease and made possible the development of an attenuated strain of the virus. Theiler and his associates developed an improved vaccine that is widely used to protect humans against yellow fever. Theiler made many other contributions to tropical medicine and the study of infectious diseases.

John Lancelot Todd
1876-1949

English microbiologist who demonstrated that relapsing fever in monkeys is transmitted by infected ticks (*Ornithodorus moubata*). Philip Hedgeland Ross and Arthur Dawson Milne also made this discovery independently at about the same time. The causative agent was named *Borrelia duttoni.* In the course of their work, in what was then the Congo Free State, both Todd and his research partner, Dutton, contracted the disease. Dutton died of "tick fever" before their landmark paper was published in the *British Medical Journal* in 1905. In collaboration with Simeon Burt Wolbach and F. W. Palfrey, Todd proved that *Rickettsia prowazeki* is the causative agent of typhus fever.

James Dowling Trask
1821-1883

American physician who helped found the American Gynecological Society. Trask was also a professor of obstetrics and the diseases of women and children at the Long Island College Hospital for several years, and he practiced medicine for nearly 40 years.

Edward Bright Vedder
1878-1952

American physician, pathologist, and nutrition scientist who received his M.D. from the University of Pennsylvania in 1902, then spent the next 31 years as an Army medical officer. He discovered that emetine, the active ingredient of the ancient emetic ipecacuanha, is an amoebicide and therefore effective against amoebic dysentery. He proved that beriberi and pellagra are deficiency diseases and developed diet therapy for both. His other research concerned typhoid, whooping cough, leprosy, syphilis, sprue, scurvy, sanitation, and public health.

Selman Abraham Waksman
1888-1973

Ukrainian-born American biochemist who won the 1952 Nobel Prize for physiology or medicine for his discovery of the antibiotic streptomycin, the first specific agent effective in the treatment of tuberculosis. Waksman coined the term "antibiotic" in 1941. After the discovery of penicillin, Waksman initiated a systematic search for other antibiotic-producing microbes. He was able to extract various antibiotics from the microorganisms known as actinomycetes. Actinomycin proved to be very toxic to animals, but streptomycin was relatively nontoxic and was effective against gram-negative bacteria, including the tubercle bacillus (*Mycobacterium tuberculosis*).

Otto Heinrich Warburg
1883-1970

German biochemist who won the Nobel Prize for physiology or medicine in 1931 for his research on respiratory enzymes. Warburg established the value of mannometry (the measurement of changes in gas pressure) as a way to measure the rates of oxygen uptake in living tissue. He discovered the role of the cytochromes, enzymes that, like hemoglobin, contain heme groups. Warburg isolated the first of the flavoproteins, enzymes involved in dehydrogenation reactions, and demonstrated that these enzymes contain a crucial nonprotein component, the coenzyme. He discovered the coenzyme now called nicotinamide adenine dinucleotide, investigated photosynthesis, and established differences between the oxygen requirements of cancer cells and normal cells.

August von Wassermann
1866-1925

German physician who made major contributions to immunology and in 1906 developed the test for syphilis that bears his name. The "Wassermann reaction" is a specific and sensitive test for infection by the microbe that causes syphilis. Wassermann also investigated tetanus, diphtheria, tuberculosis, and cerebrospinal fever. He immunized horses to meningococcus and used the resulting immune serum to treat patients with cerebrospinal fever. He served as editor of an important multi-volume work called *Handbook of Pathogenic Microorganisms* (1903-1909).

John Broadus Watson
1878-1958

American psychologist whose work involving the experimental study of the relations between environmental events and human behavior became the dominant psychology in the United States in the 1920s and 30s. Watson's first major published work, *Behavior: An Introduction to Comparative Psychology*, argued forcefully for the use of animals in psychological study. Watson established a laboratory for comparative, or animal, psychology at Johns Hopkins University in 1908.

Ernst Wertheim
1864-1920

German physician who performed the first radical mastectomy in the early part of the twentieth century. This surgical procedure, which involves removing the entire breast and portions of surrounding tissue, remains one of the preferred treatments for breast cancer because it removes not only the affected tissue, but also removes nearby tissues that often become the sites of metastases, leading to further complications. By pioneering this technique, Wertheim took one of the first steps towards an effective breast cancer treatment.

Max Wertheimer
1880-1943

Czechoslovakian-born psychologist who founded the Gestalt school of psychology. Gestalt psychology examined psychological phenomena as structural wholes, rather than components. While Wertheimer's early work focused on the perception of complex and ambiguous structures, his school later delved into other areas of psychology. In the 1930s, Wertheimer fled from Europe to the United States, shortly before the Nazis came to power.

Robert Mearns Yerkes
1876-1956

American psychologist who is regarded as one of the founders of comparative animal psychology. Yerkes was interested in the great apes as well as psychological testing and the measurement of human mental ability. During World War I, Yerkes directed the first mass-scale testing program, which administered psychological tests to almost two million men. The Yale Laboratories of Primate Biology (later, the Yerkes Laboratories of Primate Biology) became a major center for the study of primate behavior.

Hugh Hampton Young
1870-1945

American surgeon and physician who developed important new techniques for the diagnosis and treatment of genito-urinary disorders. For carcinoma of the prostate, he developed new approaches to the prostatectomy, including the perineal prostatectomy and the punch prostatectomy operations. Young and his associates introduced the use of mercurochrome as a germicide for the treatment of infections of the genito-urinary tract and the use of vesiculography (involving thorium and x rays). In 1937 Young published a treatise entitled *Genital Abnormalities, Hermaphroditism, and Related Adrenal Diseases.* He had a long and distinguished career as chief of the Department of Genito-Urinary Diseases at Johns Hopkins University Hospital.

Hans Zinsser
1878-1940

American microbiologist, epidemiologist, and immunologist whose main research concerned epidemic louse-borne typhus and other rickettsial diseases. His recognition that a form of typhus, described by Nathan E. Brill as endemic, was actually epidemic but recrudescent (reactivating after recovery) led to its being named Brill-Zinsser disease. In the 1930s he and his associates at Harvard developed killed-bacteria vaccines for typhus. He is best known as the author of *Rats, Lice, and History: ... the Life History of Typhus Fever* (1935).

Bibliography of Primary Sources

Books

Adler, Alfred. *Practice and Theory of Individual Psychology* (1924). This work is the landmark treatise for the school of thought known as individual psychology.

Balance, Charles Alfred and Charles David Green. *Essays on the Surgery of the Temporal Bone* (2 volumes, 1919). A classic work by two physicians who were instrumental in developing operations that could cure or ameliorate certain kinds of deafness.

Chen, Ko Kuei, and Carl F. Schmidt. *Ephedrine and Related Substances* (1930). In this book the authors described their work with the ancient Chinese herb ma huang. The investigations of Schmidt and Chen resulted in the isolation of ephedrine, used to treat patients with asthma.

Crile, George Washington. *Blood-Pressure in Surgery* and *Hemorrhage and Transfusion* (1903) (1909). Classic works by the American surgeon whose research into the causes and control of hemorrhage during surgery saved countless lives.

Freud, Anna. *The Ego and Mechanisms of Defense* (1937). Freud, one of the founders of child and adolescent psychology, here offered an important contribution to ego psychology.

Freud, Sigmund. *Three Essays on the Theory of Sexuality* (1905). This work established Freud as a pioneer in the serious study of sex, mind, and behavior, both normal and pathological.

Freud, Sigmund. *The Origin and Development of Psychoanalysis* (1910). This book contains lectures given by Freud in 1909 at Clark University in Worcester, Massachusetts.

Jung, Carl Gustav. *Psychology of the Unconscious* (1916). Jung is best known for his concepts of the extroverted and introverted personality types, archetypes, and the collective unconscious. This 1916 work is his major treatise about the collective unconscious.

Klein, Melanie. *The Psychoanalysis of Children* (1932). A landmark work of child analysis by Klein, whose insights and methods are now widely used to help troubled children.

Rorschach, Hermann. *Psychodiagnostics* (1921). This work contains the results of the author's inkblot experiments, which began in 1918. They involved asking patients to describe their subjective responses to a series of accidental inkblots.

Sauerbruch, Ernst. *Surgery of the Organs of the Chest* (1920-1925). Sauerbruch realized that maintaining respiratory function during open-chest surgery depends on maintaining a lower pressure inside the chest than in the surrounding atmosphere, and conceived the idea of building a low pressure chamber within which the chest might be safely opened. This book was the first to treat this subject systematically and became the foundation for future developments in the field.

Sherrington, Charles Scott. *The Integrative Action of the Nervous System* (1906). After the publication of this book, Sherrington was acknowledged as the world's foremost neuroanatomist and neurophysiologist.

Terman, Lewis. *The Measurement of Intelligence* (1916). This book is Terman's guide to his version of the Binet-Simon intelligence scale.

Zinsser, Hans. *Rats, Lice, and History: ... the Life History of Typhus Fever* (1935). An important work by the American microbiologist, epidemiologist, and immunologist whose main research concerned epidemic louse-borne typhus and other rickettsial diseases.

Periodical Articles

Freud, Sigmund. *The Ego and the Id* (1923). This 1923 work proposed a structural model to contrast the earlier topographic model of instinctual energies passing between unconscious, preconscious, and consciousness.

Freud, Sigmund. *Civilization and Its Discontents* (1930). Written after Freud's battle with cancer of the jaw, this pessimistic tract examines human guilt and notes the impossibility of achieving full happiness. In the end Freud concluded that there was no possible solution to civilization's discontents.

NEIL SCHLAGER

Physical Sciences

Chronology

1900 German physicist Max Planck publishes an epoch-making paper in which he establishes the principles of quantum physics, ideas that will have a profound impact on Albert Einstein, Niels Bohr, and others.

1905 Albert Einstein publishes his first papers on the special theory of relativity (including the famous equation $E=mc^2$), which he will follow up with his general theory of relativity in 1916.

1911 Dutch physicist Heike Kamerlingh Onnes discovers the principle of superconductivity when he finds that certain metals lose all electrical resistance at very low temperatures.

1926 German chemist Hermann Staudinger discovers a common structure to all polymers, thus paving the way for advances in polymer plastics technology.

1927 Building on ideas postulated five years earlier by Russian mathematician Alexander Friedmann, Belgian astronomer and Catholic priest Georges-Edouard Lemaître formulate what comes to be known as the Big Bang theory of the universe's origins.

1927 Werner Karl Heisenberg, a German physicist, postulates his principle of uncertainty, which states that it is impossible to determine accurately and simultaneously two variables of an electron.

1929 American astronomer Edwin Hubble formulates a law, named after him, that initiates the idea of the expanding universe.

1931 Radio astronomy, which allows astronomers to see far beyond the range of ordinary telescopes, is born when American radio engineer Karl Jansky first detects radio waves coming from outer space.

1934 Frédéric Joliot and Irène Joliot-Curie, a husband-wife team of French physicists, discover that any element can become radioactive if the proper isotope is used.

1935 American engineer Charles Richter introduces the Richter scale, which becomes a widely used method for measuring earthquakes.

1942 At the University of Chicago, the Manhattan Project team under the direction of Enrico Fermi creates a self-sustaining chain reaction, thus inaugurating the atomic age.

1947 American chemist William Frank Libby develops the carbon-14 dating technique for measuring the age of fossils.

Overview: Physical Sciences 1900-1949

By the dawn of the twentieth century, more than two centuries had elapsed since the publication of Isaac Newton's (1642-1727) *Principia,* which set forth the foundations of classical physics. During those intervening centuries, scientists embraced empiricism and sought new and ingenious ways to understand the physical world. In addition to fueling industrial revolutions in Europe and the United States, that same persistence and technological inventiveness allowed scientists to make increasingly exquisite and delicate calculations regarding physical phenomena. Advances in mathematics, especially during the nineteenth century, allowed the development of sophisticated models of nature that became accepted as a common language of science.

More tantalizingly, many of these mathematical insights pointed toward a physical reality not necessarily limited to three dimensions and not necessarily absolute in time and space. On top of a steady tempo of refinement and discovery there emerged a new and uncharted harmony of mathematics, experimentation, and scientific insight. During the first half of the twentieth century, these themes found full expression in the intricacies of quantum and relativity theory. Scientists, mathematicians, and philosophers united to examine and explain the innermost workings of the universe—both on the scale of the very small subatomic world and on the grandest of cosmic scales.

Nineteenth-century experimentalism culminated in the unification of concepts regarding electricity, magnetism, and light, formulated by Scottish physicist James Clerk Maxwell (1831-1879) in his four famous equations describing electromagnetic waves. Moreover, at the start of the twentieth century, the speed of light was well known and precisely determined. The ingenious experiments of Albert Michelson (1852-1931) and Edward Morley (1838-1923), however, demonstrated an absence of a propagation medium or "ether" and, as a consequence, cast doubt on the existence of an absolute frame of reference for natural phenomena. In 1905, in one grand and sweeping theory of special relativity, Albert Einstein (1879-1955) provided an explanation for seemingly conflicting and counterintuitive experimental determinations of the constancy of the speed of light, length contraction, time dilation, and mass enlargements. A scant decade later, Einstein again revolutionized concepts of space, time, and gravity with his general theory of relativity.

Prior to Einstein's revelations, German physicist Maxwell Planck (1858-1947) proposed that atoms absorb or emit electromagnetic radiation in discrete units of energy termed quanta. Although Plank's quantum concept seemed counterintuitive to well-established Newtonian physics, quantum mechanics accurately described the relationships between energy and matter on an atomic and subatomic scale and provided a unifying basis to explain the properties of the elements.

Concepts regarding the stability of matter also seemed ripe for revolution. Far from indivisibility, advancements in the discovery and understanding of radioactivity culminated in a renewed quest to find the most elemental and fundamental particles of nature. In 1913 Danish physicist Niels Bohr (1885-1962) published a model of the hydrogen atom that, by incorporating quantum theory, dramatically improved existing classical Copernican-like atomic models. The quantum leaps of electrons between orbits proposed by the Bohr model accounted for Planck's observations and also explained many important properties of the photoelectric effect described by Einstein.

More mathematically complex atomic models were to follow based on the work of the French physicist Louis Victor de Broglie (1892-1987), Austrian physicist Erwin Schrödinger (1887-1961), German physicist Max Born (1882-1970) and English physicist P.A.M. Dirac (1902-1984). More than simple refinements of the Bohr model, however, these scientists made fundamental advances in defining the properties of matter—especially the wave nature of subatomic particles. Matter became to be understood as a synthesis of wave and particle properties. By 1950 the articulation of the elementary constituents of atoms grew dramatically in numbers and complexity.

Advancements in physics also spilled over into other scientific disciplines. In 1925 Austrian-born physicist Wolfgang Pauli (1900-1958) advanced the hypothesis that no two electrons in an atom can simultaneously occupy the same quantum state (i.e., energy state). Pauli's exclusion principle made completely understandable the structure of the periodic table and was a major advancement in chemistry.

Profound scientific advances were also made possible by the advent and growth of commercial research labs that provided additional outlets and facilities for research beyond traditional academia. In contrast, radio astronomy had more humble (literally backyard) beginnings as American engineer Karl Jansky (1905-1945) discovered the existence of radio waves emanating from beyond Earth. Subsequently, radio astronomy advanced to become one of the most important and productive means of astronomical observation.

A greater understanding of nature was not limited to the subatomic or astronomical worlds. Advances in geology and geophysics allowed scientists a deeper understanding of Earth processes. As details regarding the inner Earth yielded to seismic analysis, the enigmatic patterns of weather and climate yielded to meteorological analysis. Geologists and astronomers began to understand the dynamic interplay of gradual and cataclysmic geologic processes both on Earth and on extraterrestrial bodies.

Against a maddeningly complex backdrop of politics and fanaticism that resulted in two World Wars within the first half of the twentieth century, scientific knowledge and skill became more than a strategic advantage. The deliberate misuse of science scattered poisonous gases across World War I battlefields at the same time that advances in physical science (e.g., x-ray diagnostics) provided new ways to save lives.

Countries struggled to gain both offensive advantages and defensive capabilities by the development of radar and sonar. As the century progressed, the balance of scientific and technological power often swayed between adversaries. This balance eventually tipped toward the United States when large numbers of continental European scientists, particularly German Jews, emigrated to America during 1930s and 1940s to avoid Adolf Hitler's cruel and murderous Third Reich. Without such an exodus, these brilliant scientists may have perished in the Holocaust along with their great contributions to both wartime and postwar science. The dark abyss of World War II brought the dawn of the atomic age. In one blinding flash, the Manhattan Project created the most terrifying of weapons, which—in an instant—could forever change the course of human history and life on Earth.

Einstein's theories of relativity brought about a revolution in science and philosophy, rivaled only by the contributions of Newton. Although the development of relativity and quantum theory mooted the quest—spurred by prominent philosophical and religious thought—to find an absolute frame of reference for natural phenomena, the insights of relativity theory and quantum theory also stretched the methodology of science. No longer would science be mainly an exercise in inductively applying the results of experimental data. Experimentation, instead of being only a genesis for theory, became a testing ground to falsify the apparent truths unveiled by increasingly mathematical models of the universe. With the formulation of quantum mechanics, physical phenomena could no longer be explained in terms of deterministic causality, that is, as a result of at least a theoretically measurable chain causes and effects. Instead, physical phenomena were described as the result of fundamentally statistical, unreadable, indeterminist (unpredictable) processes.

The development of quantum theory, especially the delineation of Planck's constant and the articulation of the Heisenburg uncertainty principle, carried profound philosophical implications regarding the limits on knowledge. By mid-century, scientists were able to substantially advance the concept that space-time was a creation of the universe itself. This insight set the stage for the development of modern cosmological theory (i.e., theories regarding the nature and formation of the universe) and provided insight into the evolutionary stages of stars (e.g., neutron stars, pulsars, black holes, etc.) that carried with it an understanding of nucleosynthesis (the formation of elements), which linked mankind to the lives of the stars.

K. LEE LERNER

Mass Migration of Continental European Scientists to the U.S. and Elsewhere

Overview

Hans Bethe (1906-), Felix Bloch (1905-1983), Albert Einstein (1879-1955), Niels and Aage Bohr (1885-1962 and 1922-), Enrico Fermi (1901-1954), Emilio Segrè (1905-1989), Eugene Wigner (1902-1995). All won the Nobel Prize in physics, all were involved with the Manhattan Project in some manner, and all were born in Europe, driven out by the rise of Fascist governments in the 1930s and 1940s. These were the proverbial tip of the iceberg; many more distinguished scientists fled Europe to escape the Fascists and more were captured by Allied armies. Of these scientists, most came to the United States, where they became prominent in physics, mathematics, and engineering. These scientists not only helped the Allies win World War II, but also helped the U.S. achieve and maintain technological superiority over its foes for over 50 years. They accomplished this by bringing their intelligence, ambitions, and hatred of Hitler and Communism to the U.S. and by taking their talents *from* the countries they fled in what may well have been the greatest transfer of intellectual power the world has known.

Background

Central Europe at the end of the nineteenth century was the home of great universities and a great intellectual tradition. Vienna, Berlin, Budapest, Rome, Paris, and the other great cities of Europe hosted prestigious universities and some of the greatest thinkers of the time. The people born into this time, attending university, had the advantage of a relatively stable government and economy and access to intellectual leaders. Many of the brightest and most inquisitive students were Jewish, drawing on the long Jewish tradition that values education. Some of the best and brightest of this generation of scientists were born in Hungary, near the Carpathian Mountains. These included Edward Teller (1908-), John von Neuman (1903-1957), Theodor von Kármán (1881-1963), Eugene Wigner, George de Hevesy (1885-1966), Leo Szilard (1898-1964), and Michael Polanyi (1891-1976). So remarkable was this group, in fact, that other scientists joked they could not possibly be from Earth but were, instead, from Mars.

Outside of Hungary, other great thinkers were being trained and were coming of age at the time that classical physics was being stood on its head. Einstein, Bohr, and Bethe were all important players in a field that was suddenly seemingly wide open, where little could be taken for granted. In such an atmosphere, the brightest and most creative scientists flourished, partially because their creativity was rewarded during the revolution that physics was undergoing.

In the wake of World War I, the Austro-Hungarian Empire dissolved, replaced by a number of smaller nations, each with their own government. While this was occurring, Germany was humiliated and impoverished by the terms of surrender, as were many of the new nations that now existed. In this unsettled environment, anti-Semitism flourished and strong, fascist governments came to power, including those of Hitler and Mussolini. This combination of events, the rise of Fascism and rampant anti-Jewish actions, drove many of Europe's best and brightest to the only havens they could see, England and the United States. As Germany invaded Austria, Czechoslovakia, and most of the rest of Europe, those scientists who were able fled Europe, partly for self-preservation and partly to be able to use their intellects to help restore peace and sanity to Europe. During the closing days of World War II, even more scientists came to the U.S. and England. These were German scientists who were "collected" during special operations aimed at capturing German scientists or, in some cases, scientists who arranged to be captured by the U.S. to avoid capture by the Soviet Army.

As mentioned above, a large number of these scientists had or went on to win the Nobel Prize in Physics. But the others were brilliant, too. Edward Teller helped invent the hydrogen bomb, John von Neuman helped design the earliest computers, and Leo Szilard developed the concept upon which nuclear reactors operate. Of the German scientists, perhaps the best known was Wernher von Braun (1912-1977), whose team of engineers and scientists were crucial to America's space program in the 1950s and 1960s.

Impact

The effect of these scientists moving to England and America can hardly be overstated. Whether

Albert Einstein was among the prominent scientists who were forced to flee their homes in continental Europe during the 1930s and 1940s. Here, Einstein (second from left) is shown in England after fleeing Nazi Germany in 1933. *(Bettmann/Corbis. Reproduced with permission.)*

they left voluntarily, were forced to leave, or were captured by the American Army, their chief impact was in the transfer of their intellect from Europe to the United States, which is where the majority of them ended up. As a result, Germany never developed an atomic weapon, the United States became a military, economic, and intellectual powerhouse, and European science (particularly German science) suffered a profound loss of leadership for several decades.

The most obvious impact of these events was, of course, on the military. Without these expatriate scientists it is very likely that America would not have developed the atomic bomb, and may never have launched the Manhattan Project. The scientists fleeing Europe not only brought with them their knowledge of atomic physics and the intelligence to turn that knowledge into a working atomic weapon, they also had first-hand knowledge of life under a dictator, a hatred of Fascism, and a fear of a nuclear-armed Hitler. They convinced Albert Einstein to write to President Roosevelt a letter urging the design of atomic weapons. Without Einstein's prestige, this letter would never have received the careful consideration it did, and it is quite likely the Manhattan Project would never have been launched. This, in turn, would have greatly delayed the development of nuclear weapons, although it is likely that they would still have been developed at some time in the post-war years. However, without the examples of Hiroshima and Nagasaki, it is also possible that nuclear-armed nations would have been less unwilling to use their weapons, although this can never be known with certainty.

Perhaps the most significant post-war weapon development was the hydrogen bomb. Although fission weapons (such as the bombs dropped on Hiroshima and Nagasaki) have an upper size limit, based on the nuclear properties of uranium and plutonium, there is no theoretical limit to the size of a hydrogen bomb. America's H-bomb was designed almost entirely by expatriate Europeans; Edward Teller (Hungary), Stanislaw Ulam (Poland, 1909-1985), Enrico Fermi (Italy), John von Neumann (Hungary), and Hans Bethe (Germany) all made important contributions to this new weapon, and the major conceptual breakthroughs were developed by Ulam and Teller. In an interview in 1992, Teller said it was his fear of Russia that prompted him to develop this weapon.

Other technological developments that arose from the contributions of European scientists are literally too numerous to mention here. However, it is worth noting that our space program, too, was launched on rockets designed by German scientists, many of whom had earlier worked on the V1 and V2 rockets of Nazi Germany. Although much of the work was performed by Americans, the German scientists and engineers provided crucial leadership and expertise that the Americans lacked. Every rocket through the Saturn V, the moon rocket, owed its design to the Germans.

Nuclear weapons helped America retain its place as a victorious superpower in the post-war years. Space exploration, including the lunar landing, made America prominent and was evidence of American technological superiority over the rest of the world, and other technological advances ensured that this technological and military superiority did not wane with the passing years. In a very real way, these scientists contributed mightily towards American military and economic power during the last half of the twentieth century, and they, in turn, had enormous political implications. In fact, because of these contributions, America was able to "win" the Cold War, resulting in a world with far fewer repressive governments than might otherwise have been the case.

Finally, these scientists helped launch major social changes in the U.S. By their very presence and prestige, they helped to shape American science, giving the U.S. a very strong showing in Nobel Prize awards and other awards for decades. By so doing, they also raised the prestige and status of science in the U.S. Although scientists and engineers are not revered in the U.S., they are respected, in some cases inordinately so. A great deal of this prestige is due to the important role science and technology have played in American society over the years as well as the realization that scientific and technological advances have been enormously valuable to the U.S. for the past 50 years. This, in turn, has led to continuing funding for scientific research, including at national laboratories, that continues to churn out important scientific discoveries, inventions, and military weapons that help the U.S. to maintain its status as a world leader.

There is a great deal of irony in this. The governments of Central Europe raised one of the most gifted generations of scientists that ever lived. Just when they were reaching the peaks of their intellectual powers, Central European dictators threw or drove them away. In their new homes, these same scientists developed the weapons that helped ensure the defeat of their former oppressors. After the war, instead of returning home, these scientists remained in the U.S., because of the devastation in their former homelands and because many of those nations were now Communist. In so doing, they changed the course of history.

P. ANDREW KARAM

Further Reading

Books

Powers, Thomas. *Heisenberg's War.* Little, Brown, and Company, 1993.

Rhodes, Richard. *The Making of the Atomic Bomb.* New York: Simon & Schuster, 1986.

Rhodes, Richard. *Dark Sun.* New York: Simon & Schuster, 1995.

The Manhattan Project and the Decision to Drop the Bomb

Overview

The Manhattan Project ushered in the nuclear age, and with it the concept of international relations and war changed forever. Humankind now had the ability not only to destroy nations and civilizations but to end life as we know it on the planet.

Background

During the years immediately following the First World War, Germany became the center for the study of physics. The most important work in the area of atomic energy was carried on in German universities, especially at the University of Berlin. The most important people in the field traveled to Germany to study, teach, and attend symposiums. Among the many scientists who were to have an impact on the study of atomic energy with this German connection was the Danish physicist Niels Bohr (1885-1962), who would create the scientific model of the hydrogen atom. His extensive work in this area was largely based upon the research of the two famous German physicists Albert Einstein (1879-1955) and Max Planck (1858-1947). The Hungarian Leo Szilard (1898-1964) and the Italian Enrico Fermi (1901-1954), who both played an important role in the development of the atomic bomb, spent time in Germany. Robert Oppenheimer (1904-1967), the great American theoretical physicist and leader of the Manhattan Project, and Edward Teller (1908-), the father of the American hydrogen bomb, both studied physics at German universities. Werner Heisenberg (1901-1976), one of the founders of quantum theory and Teller's dissertation advisor, would play a major role in the German bomb program. As political events of the 1920s and 1930s began to unfold, German preeminence in this field would be an important factor in the history of the twentieth century.

In the 1920s, historical forces were changing the international landscape of the world. Totalitarian states of both the left and the right began to have an impact on the nations of Europe and Asia. In 1917, the largest country on the Eurasian landmass, Russia, became the world's first Communist state as a result of the Bolshevik Revolution. This event was based upon the scientific principles of dialectical materialism found in the writings of Karl Marx. Both Lenin and Stalin viewed science and technology as essential to the development of the modern socialist state. The same held true for the emerging Fascist states of Italy and Germany.

Allegiance to the state was seen as the modern secular replacement of the earlier European quest for religious immortality. Leaders and generations would come and go, but the socialist workers' paradise and the thousand-year Reich would last forever. Science and technology would be two of the tools used for the betterment of the state. They would increase and strengthen both economic productivity and the nation's military capacity, thus allowing the new secular ideology to dominate the world. The same model was present in Japan, which, during the Meiji Restoration, had adopted Western-based science and technology and by the end of the 1920s moved toward a militaristic right-wing government.

The Western democratic industrialized nations such as Britain, France, and the United States had also embraced this new scientific model. However, unlike the totalitarian states, they did not focus much of their attention on weaponry. As a result of the Great Depression, these countries continually chose societal needs in the "guns and butter" debate. Since these states were influenced by political pressure, they had to respond to the pressing needs of a democratically active populous suffering the pain of economic uncertainty. This is not to say that there was an absence of interest in nuclear physics in these nations. Great and important work was done in the Western democracies, especially at Berkeley under the direction of Oppenheimer, but this was theoretical in nature and not aimed at an emerging military industrial complex.

Unfortunately, these great accomplishments in physics did not occur in isolation but coincided with a series of political events that would eventually end with the outbreak of war and the development of the atomic bomb. The first two years of the 1930s saw Western science split the atom and discover the neutron. It was also at this time that the Japanese invaded Manchuria and created the puppet state of Manchukuo. By 1934, the discovery of artificial radioactivity showed that humankind had the potential of creating a new and very powerful source of energy. Radioactivity is the energy produced from the destruction of atomic nuclei in certain elements such as uranium. This was followed in 1936 with the formation of the Rome, Berlin, Tokyo Axis. In 1939, the basic model of atomic fission was constructed, and this was quickly followed by the first chain reaction. A chain reaction is the self-sustaining process of energy release from atomic fission. By the end of the same year many scientists believed that the process of nuclear fission could be used to create the world's most powerful weapon. Any country that had a monopoly on this technology could impose its will upon the entire world. By September of that year Hitler had invaded Poland and the Second World War had begun.

When the war started, the United States had no intention of participating in the conflict. Public opinion was against any involvement in international disputes. This was the result of two decades of an isolationist foreign policy and 10 years of economic depression. Americans were more concerned with pressing problems at home and were concentrating on the programs of Roosevelt's Second New Deal. However, the President knew that the United States could not remain isolationists in the face of a growing Fascist menace, and that it was only a matter of time before we were drawn into the war.

Ironically, America would be the beneficiary of the racist ideology of the Nazi regime. In 1936, Hitler set into motion the racial purity program that he described in detail in *Mien Kampf*. Hitler, through his racist anti-Semitic ideology, used the Jewish people as scapegoats for the economic, social, and political problems facing Germany. Many Jewish intellectuals who were lucky enough to have the ability to leave Germany after 1933 settled in the United States. Among the more notable scientists were Albert Einstein, the most famous German physicist, and two Hungarians, Leo Szilard and Edward Teller, both of whom had studied and taught in Germany. The free and open environment of the United States allowed our nation to "inherit" some of the best minds working in the field of nuclear physics.

In 1939, the Nazis stopped exporting uranium from Germany and Czechoslovakia. This convinced Einstein and his colleagues that Germany in fact was in the process of constructing an atomic bomb. A group of concerned nuclear scientists composed a letter, and Einstein delivered it to President Roosevelt. The communique dealt with four major areas. Roosevelt was informed that Enrico Fermi and Leo Szilard had proved that uranium could be turned into an important power source. He was also notified that it was possible to create a chain reaction that would unleash a considerable amount of energy, and that with this process a device or bomb could be developed to be used against a military target. The letter also stated that the largest available source of uranium was located in the Belgian Congo and that Roosevelt would be well advised to begin steps to assure Allied control of that area. Finally, the scientists strongly suggested that open lines of communication be established between university professors working on this new technology and the government of the United States.

Impact

Roosevelt and Congress finally agreed on the importance of the situation, and they began to develop a program to investigate the potential of atomic energy. The initial location of the program was in the Manhattan district of the Army Corps of Engineers, and because of this it was designated the Manhattan Project. The first major accomplishment was the signing of the Quebec Agreement in August of 1943. This created broad guidelines for American/British cooperation. Both countries agreed that this new weapon would never be used against each other and that each would notify the other before it was used against an enemy. Guidelines were also created to control the flow of information and to cooperate on research for the peaceful application of this new-found energy source. Finally, an agreement was reached to create a policy board made up of representatives from both countries.

Within a very short period of time an oversight committee was formed to handle the major problems concerned with the construction of the bomb. This group dealt with a number of topics, including research, military applications, control of secret information, the international community, (including the Soviet Union), and the destructive capability of the weapon. The leader of this committee was Secretary of War Henry Stimson. In his opening remarks he emphasized that this program was not just for military purposes. He believed it would provide humankind with a clearer view of the universe and should be compared to the work of Nicolaus Copernicus's (1473-1543) heliocentric theory and Isaac Newton's (1642-1727) universal law of gravity. Most importantly, he believed this new science should be controlled for peaceful means.

Robert Oppenhiemer (left) looks at a photograph of the mushroom cloud that formed after an atomic bomb was dropped on Hiroshima, Japan. Also pictured (left to right) are H.D. Symthe, General Nichols, and G. T. Seaborg. *(Bettmann/Corbis. Reproduced with permission.)*

The leading scientist and manager of the Manhattan Project, Dr. J. Robert Oppenheimer, believed consensus had to be reached concerning the regulation of fundamental research. The debate revolved around how much academic freedom the scientists should be given. Most of the scientists believed that the research process should be as open as possible with each person working on the project being allowed total access to the work of every other individual doing research in the area. Furthermore, the opportunity to return to the university environment should be accorded to everyone. Oppenheimer believed that a freer mode of inquiry would ultimately accelerate research, which would result in shortening the war. Most of the scientists knew that the basic knowledge related to nuclear energy was widely known throughout the world. Oppenheimer and his colleagues believed that if all the new research were made readily available, emphasis would be placed upon peaceful applications, which in turn would benefit the entire human community. Other scientists argued that the success of recent industrialization was in great part the result of the free exchange of information among Western technologists. It was an accepted fact that modern science and technology was based upon teamwork, and that it was virtually impossible to keep the findings of modern research secret for long periods of time. Government representatives urged caution and proposed the creation of an international committee that would regulate both the free exchange of information and also enforce strict inspection to insure that potential aggressors would not use the knowledge for the benefit of their military.

The Allied nation that created the most concern was the Soviet Union. Since its inception in 1917, the Western democracies regarded it as a major threat to their existence. The United States did not recognize it as a member of the community of nations until 1933, when Franklin Roosevelt opened the first United States embassy in Moscow. In August 1939, relations hit an all time low when Stalin signed a nonaggression pact with the Nazis. This relieved Hitler from the problem of fighting a two-front war, and during September 1939 the combined forces of Fascist Germany and Communist Russia invaded Poland, beginning the Second World War. According to the agreement, the Soviet Union was also allowed to invade the Baltic nations of Estonia, Lativa, and Lithuania. These events revealed Stalin as a premier political opportunist who was more concerned with personal power than with political philosophy. Much of the petroleum used by the German Luftwaffe to bomb London during the "Blitz" came from Soviet-controlled oil fields in eastern Poland. Ironically, Stalin became "Uncle Joe," our trusted ally in June of 1941, after Hitler betrayed the Soviets and invaded their homeland. It was obvious from the beginning of the relationship that Stalin was untrustworthy and ready to betray any agreement in order to advance his personal agenda.

Oppenheimer again argued that it was virtually impossible to stop the information about the bomb from reaching the Soviet Union. This was a very delicate subject for him to deal with. His wife, younger brother, and mistress had all been active members of the American Communist Party, and on more than one occasion he had given financial aid to the party. Early on in the Manhattan Project he had been approached by a Soviet agent seeking his cooperation. Even though he refused, the specter of treason would follow him for the rest of his career. Secretary of War Stimson did not trust the Russians and believed the only solution was to create a coalition of powerful democratic nations which would force the totalitarian countries to act in a peaceful fashion.

Not long after the project began, extensive work was done on the possible effects of the bomb on a military target. The scientific community warned that the impact would be terrifying. They stated that the device would create a luminous mushroom cloud that would extend to an elevation of between 10,000 and 20,000 ft (3,048-6,096 m). The scientists predicted the explosion would create such a strong neutron effect that it would kill all human life within a radius of two-thirds of a mile.

Most of the important actions of the Manhattan Project were based upon the ultimate goal of shortening the war. This resulted in the decision not to give the target any advanced warning but also not to hit a civilian area if possible. Most of the military men wanted a strategic target,

THE NAZI BOMB

One of the driving factors behind the Manhattan Project was the fear that German scientists were working as feverishly on an atomic bomb that would let Hitler dominate the world. After the war it was discovered that, while the Nazis had an active nuclear weapons program, they had failed to make many of the breakthroughs necessary to design a working nuclear weapon. In fact, German scientists became so discouraged by their lack of progress that they believed nuclear weapons to be impossible to build and had largely turned their attentions towards designing nuclear reactors.

The German atomic weapons efforts were led by the great German physicist, Nobel prize winner Werner Heisenberg. Heisenberg's leadership was precisely what worried expatriate European scientists because they knew of his brilliance firsthand. What puzzled them in later years, especially after the war, was how little progress the Germans made, in spite of Hitler's enthusiastic support of their efforts.

After the war, many of these scientists were captured by the Allies and taken to England, where they were interned in a house that was bugged to record their conversations. Of particular interest were their comments upon learning of the atomic bombing of Hiroshima and Nagasaki. Of the scientists there, the only one who seemed truly unsurprised was Heisenberg. Disappearing for a short time, he returned and announced he had just calculated the amount of fissionable material required for a weapon, and that his previous calculations had been in error by a factor of ten or so. To some Manhattan Project scientists this could mean only one thing; that Heisenberg had purposely mislead Hitler and his scientists to keep from inventing a Nazi bomb. Heisenberg's actions—and intentions—have never been clarified.

P. ANDREW KARAM

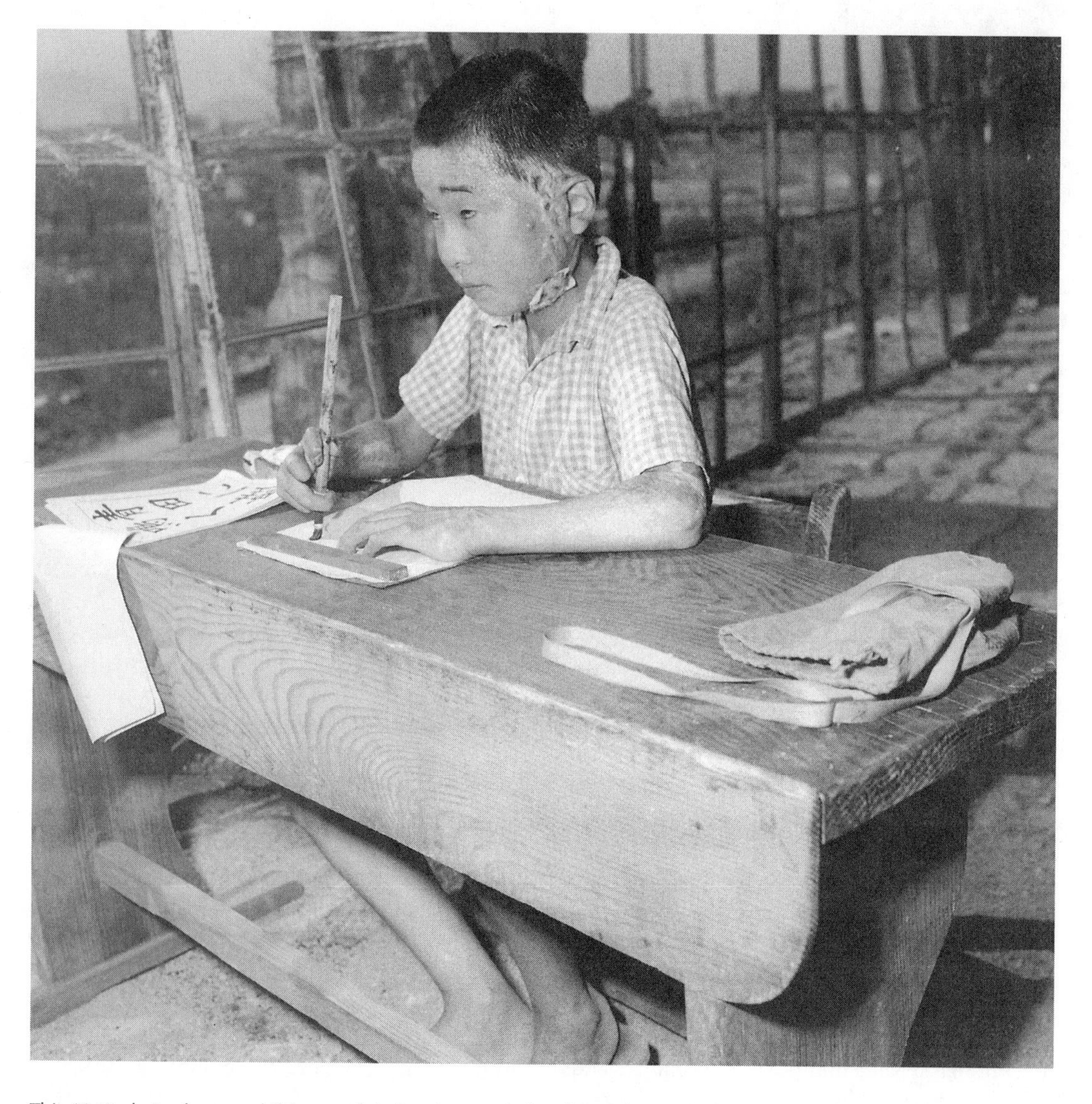

This 1946 photo shows a child scarred during the atomic bomb attack on Hiroshima, Japan. *(Bettmann/Corbis. Reproduced with permission.)*

such as a military or naval base or an important war plant. There was also discussion of simultaneous strikes, but this was rejected because the scientists believed the collection of meaningful data would be easier if there were individual detonations. The engineers also cautioned that an attempt to have multiple strikes would reduce quality control and cause a possible disaster.

In the early morning hours of July 16, 1945, near Alamogordo, New Mexico, the first nuclear device, designated Trinity, was detonated. The force of the blast equaled 20,000 tons (18,160,000 kg) of TNT and created a fireball 10,000 times hotter than the Sun. A deadly radioactive mushroom cloud rose 8 miles (12.9 km) into the sky, and at the same time back on Earth a 1,200-foot (366 m) crater was dug out of the surface of the planet. Members of the Manhattan Project, along with important government dignitaries, viewed the explosion from a fortified location approximately 6 miles (9.6 km) from ground zero. Most of the witnesses stood in stunned silence, knowing that humankind had crossed over into a new age, the nuclear age. Everyone present knew that the world had reached the point where it could not only destroy nations and civilizations but all life on Earth. This was stated best by Oppenheimer when he spoke these chilling words from the *Bhagavad-Gita*, "I am become death, the shatterer of worlds."

Almost immediately, a debate began over whether the bomb should be used. Most of the members of the Manhattan Project were now against using the weapon, even if it meant shortening the war. Many scientists believed that dropping the bomb on Japan would ultimately place the country in grave danger because it would initiate an arms race. Most experts believed that any number of countries would in a relativity short time be capable of constructing a bomb. The knowledge to build such a weapon was now known in great detail by many scientists around the world. Both the French and the Russians had acquired enough information to catch up with the United States in fairly rapid fashion. Not only was it impossible for the United States to control the vital information concerning weapons development, but it was just as impossible to control the strategic materials used to construct the bomb. Everyone knew that the Soviet Union had considerable deposits of uranium, and once they acquired the necessary knowledge they would be able to build a weapon that would rival the ones of the United States.

At the same time, American military analysts were becoming very concerned about the invasion of the Japanese homeland. The war in the Pacific had become a very bloody campaign. A year before Pearl Harbor, Churchill and Roosevelt had decided on a Europe-first strategy. Both leaders agreed that Hitler and his Nazi military machine were the greatest threat facing the Allies. Despite this agreement, the United States was still able to develop a strong and active military campaign in the Pacific Theater, scoring two important naval victories in early 1942, at Coral Sea and Midway. By mid-year, the military decided on the tactical strategy known as "island hopping." The plan was to control a specific number of important islands that would eventually put the Allies in the position to mount an invasion against the Japanese homeland. Locations such as Guadalcanal, Iwo Jima, and Okinawa quickly became part of United States military history. The combat on these islands was always very intense; both the Japanese and the Allies suffered a great number of casualties. Time and again the Japanese soldiers chose death over surrender. As the war progressed, Japan finally resorted to the use of kamikaze tactics. They would take young men, give them very little training, and send them on suicide missions against American ships. They were given just enough fuel to reach their targets. They were then expected to crash their planes loaded with explosives into the American fleet. Military intelligence believed that if the Japanese continued to exhibit this fighting spirit it would cost the Allies a million casualties to invade the Japanese homeland.

President Truman received the news of the successful test at Trinity while he was attending the Potsdam Conference. When Truman informed Stalin that the United States had successfully tested this new super weapon, he reacted as if he knew nothing about the Project. In fact, however, the Soviets had a spy deep within the Manhattan Project, a British theoretical physicist by the name of Klaus Fuchs (1911-1988). The Russians had been conducting research in this area as early as 1941.

Recently much has been written about an extensive Japanese bomb project. There is a growing amount of evidence that suggests that the Japanese had constructed a research facility in Manchuria. The historical record also suggests that their military had plans of attaching small nuclear bombs to the Kamikaze planes and using them against the Allied invasion force.

President Truman faced a monumental decision about whether to use the bomb. His final decision was based upon the overwhelming desire to save American lives and to bring the war to a speedy conclusion. In typical Truman style, he not only made the decision to drop the bomb but also took full responsibility for the action. On his orders, the *Enola Gay* delivered the first atomic bomb on the city of Hiroshima on August 6, 1945. Within an instant 50,000 people were incinerated. Almost 30,000 more would eventually die of radiation sickness and leukemia within a few short months. The head of the Japanese bomb project, Yoshio Nishina, was flown to Hiroshima to inspect the damage. It was his report, along with a second attack on Nagasaki on August 9, that convinced the military government to accept the plan of unconditional surrender. Yoshio Nishina died of leukemia as a result of the radiation poisoning he received during his inspection of Hiroshima. The Japanese government surrendered on September 2, 1945.

RICHARD D. FITZGERALD

Further Reading

Books

Frank, Richard B. *Downfall: The End of the Imperial Japanese Empire.* New York: Random House, 1999.

Goodchild, Peter. *Oppenheimer: Shatterer of Worlds.* New York: Fromm International, 1985.

Macrakis, Kristie. *Surviving the Swastika: Scientific Research in Nazi Germany.* New York: Oxford University Press, 1993.

Rhodes, Richard. *The Making of the Atomic Bomb.* New York: Simon & Schuster, 1988.

Wilcox, Robert. K. *Japan's Secret War: Japan's Race Against Time To Build Its Own Atomic Bomb.* New York: Marlowe & Company, 1995.

National Jewels: The Beginnings of Commercial Research Labs

Overview

Commercial research labs have given us the inventions that have changed our lives, including transistors, microchips, nylon, floppy disks, and the laser. At the same time, they've provided fractals, evidence for the Big Bang theory, and detailed pictures of atoms and molecules. Funding for these labs emerged from competition for new markets and prestige. Ultimately, research labs became an integrated part of the modern system of industrial production. Their findings are the engines of a new technology-driven economy.

Background

During the twentieth century, Bell Labs (formerly part of AT&T, now part of Lucent Technologies, Inc.) became the model for the modern commercial research lab. Incorporated in 1925 as Bell Telephone Laboratories Inc., Bell Labs has patents for lasers, light-emitting diodes, and solar cells to its credit. Its scientists have been awarded four Nobel prizes over a 60-year period for work that demonstrated the wave nature of matter, the invention of the transistor, the discovery of cosmic background radiation (proof for the Big Bang theory), and the fractional quantum Hall effect. Bell scientists created UNIX, the software operating system that runs much of the World Wide Web, and the C programming language.

The origins of Bell Labs are unusual. Its mandate was shaped by the status of AT&T as a monopoly, which was made explicit in the Graham-Willis Act of 1921. This assured the lab of a steady source of funding and made research for the public good part of its reason for being. This set it apart from other commercial labs, which generally had to justify all projects based on market potential. For Dow Corning, Westinghouse, and other contemporary labs, the model was set by Thomas Edison (1847-1931).

Edison's first patent, in 1868, was for a speedy vote-counting machine. It was a clever device, but he couldn't sell it, and, after that, he did not believe in inventing anything for which there was not a market. His lab in Menlo Park, New Jersey, was a commercial enterprise, dedicated to putting out a new invention every 10 days. Edison and his associates were responsible for an unprecedented, regular flow of patents (1,093) and inventions, including the phonograph, the incandescent light, and motion pictures. General Electric, which maintained a relationship with Edison through his patents and consulting, established its research laboratory in 1900 in Schenectady, New York. A genius named Irving Langmuir (1881-1957) came to work there in 1909. By 1913, he had invented the gas-filled incandescent lamp, which, to this day, lights our homes. The invention was so important that Langmuir was allowed to do basic research. He won the 1932 Nobel Prize in chemistry for his work in surface chemistry.

DuPont took a different direction from other businesses. In 1927, it established a laboratory for "pure science or fundamental research work." This attracted an energetic, young Harvard chemist named Wallace Carothers (1896-1937), and in 1931 his team introduced synthetic rubber. A more important challenge was in front of them, however—finding a substitute for silk. By 1934, Carothers was producing synthetic fibers and in the following year nylon was patented. Basic research had paid off.

During World War II, the pressure was on for laboratories to contribute to the war effort. Management techniques and process engineering recognized the role of research labs, and the labs began to become incorporated into the industrial system. Many scientists made a good living at commercial labs, but after the war, the labs faced competition from government labs and National Science Foundation-sponsored university research programs. During the war, the physics community had proven that militarily important discoveries could come from unexpected places. The U.S. was competing with the Soviets, and it was willing to put money into pure research.

Bell, with its dual mandate, could attract the best and the brightest who often were more interested in extending the frontiers of knowledge than in helping create new, improved products. The results were the right chemistry. In the second half of the twentieth century, Bell regularly burnished their image with noncommercial discoveries. They attracted keen minds, and led innovations both in pure and applied science. Other labs watched and learned as Bell Labs's discoveries were publicized and its scientists collected honors.

Impact

Today, we live in a world that has largely been invented in commercial laboratories. The artificial fibers for clothing and furnishings—including nylon, rayon, and polyester—have their origin in the labs of DuPont. Plastics and other polymers are now commonplace, found in toys, garbage bags, and artificial heart valves. In software, C and UNIX came from Bell Labs, while FORTRAN and relational databases came from IBM Research. Computers are unimaginable without the transistor, the multiprocessor, and magnetic storage, from researchers at Bell, Intel, and IBM, respectively. We depend on communications satellites (Bell), power plants (Westinghouse), and jet engines (GE) that were created in commercial laboratories.

Perhaps the most significant laboratory-related invention didn't come from the labs. It was the integration of research and development into the industrial process by managers and operations research analysts. The modern industrial system begins with an idea, proceeding onto pilot to prototype to manufacturing start-up to production and marketing. This is true whether the industry is telecommunications, paper, food, chemicals, aerospace electronics, or automobiles. And it works, regularly producing new and better products and services while increasing industrial productivity. One of the most intensively research-driven industries is pharmaceuticals. Most of the wonder drugs of the second half of the twentieth century emerged from a systematic approach of investigation, development, and testing. The pharmaceutical industry invests heavily in its research laboratories. In 1999, the top 10 drug companies worldwide accounted for over 2,000 new patents.

Innovation has been recognized by the financial community, and the market valuation of companies is increasingly related to the intellectual capital that is coming out of their labs. In fact, the cycle of innovation has sped up, and one of the management challenges is to place the right bets on technologies earlier in their development. Often the best bet is on the talent, and, just as DuPont attracted Carothers and other innovators to its labs by explicitly offering the opportunity to do basic research, many of the top commercial laboratories lure top scientists to their companies with a measure of freedom.

IBM Research has operated in the tradition of Bell Labs, with he explicit goal of doing things to make itself "famous and vital." Thus, while practical inventions like RISC computing, FORTRAN, floppy disks, and memory chips got their start in IBM's labs, there also were accomplishments that did not directly affect the bottomline, such as Benoit Mandelbrot's (1924-) discovery of fractals and Nobel-honored work by others in high temperature superconductivity and scanning tunneling microscopy. Those scientists who are particularly accomplished may be designated IBM Fellows and given time and resources to pursue research of their own choosing.

Despite its contributions, business accountants usually classify research as a cost. Because of this, support for commercial research, especially in basic science, waxes and wanes. Many commercial laboratories downsized during the late 1980s and early 1990s. AT&T took it a step further. In 1996, AT&T divested themselves of Bell Labs when they spun off their systems and technology divisions to create Lucent Technologies. Since then, one current and two former Bell Labs scientists were awarded the 1998 Nobel Prize in physics, making a total of 11 members of the lab so honored.

Perhaps the most important effect of the development of commercial research labs has been the creation of a worldwide, technical economy. Though it isn't always acknowledged, today's business strategies are driven by technology. The Internet, which leverages many key inventions of commercial labs, has disrupted many industries, especially in the distribution of books, software, and music. Laboratories are expected not just to rapidly improve on today's leading edge technologies like wireless communications and intelligent agents, but to push into new areas like nanomachines, gene therapy, and business analytics. Venture capitalists and large corporations invest in new technologies that come out of laboratories with the confidence that these are where the next opportunities will come from. But the frontier for commercial laboratories isn't only in new technologies. It's also in new methods for adopting emerging technologies and getting them to market quickly. This includes understanding how technologies converge, how firms can acquire as well as invent new technologies, and how new inventions embed themselves within the social framework for the biggest economic advantage.

PETER J. ANDREWS

Further Reading

Books

Asimov, Isaac. *Isaac Asimov's Biographical Encyclopedia of Science & Technology.* New York: Doubleday and Co., 1976.

Pursell, Carroll W. *The Machine in America: A Social History of Technology.* Baltimore: Johns Hopkins University Press, 1995.

Rosenbloom, Richard S. and William J. Spencer. *Engines of Innovation: U.S. Industrial Research at the End of an Era.* Cambridge: Harvard Business School Publishing, 1996.

Einstein's Theories of Relativity

Overview

At the dawn of the twentieth century the classical laws of physics put forth by Sir Isaac Newton (1642-1727) in the late seventeenth century stood venerated and triumphant. The laws described with great accuracy the phenomena of everyday existence. A key assumption of Newtonian laws was a reliance upon an absolute frame of reference for natural phenomena. As a consequence of this assumption, scientists searched for an elusive "ether" through which light waves could pass. In one grand and sweeping "theory of special relativity," Albert Einstein (1879-1955) was able to account for the seemingly conflicting and counter-intuitive predictions stemming from work in electromagnetic radiation, experimental determinations of the constancy of the speed of light, length contraction, time dilation, and mass enlargements. A decade later, Einstein once again revolutionized concepts of space and time with the publication of his "general theory of relativity."

Background

In the eighteenth and nineteenth centuries predominant philosophical and religious thought led many scientists to accept the argument that seemingly separate forces of nature shared a common source or absolute reference frame. Against this backdrop, nineteenth-century experimental work resulted in the unification of concepts regarding electricity, magnetism, and light by James Clerk Maxwell (1831-1879) with his four famous equations describing electromagnetic waves. Prior to Maxwell's equations, it was thought that all waves required a medium of propagation (i.e., an absolute reference frame). Maxwell's equations, however, established that electromagnetic waves do not require such a medium. Maxwell was, however, not convinced of this and he worked toward establishing the existence and properties of an "ether" or transmission medium.

The absence of a need for an ether for the propagation of electromagnetic radiation (e.g., light) was subsequently demonstrated by the ingenious experiments of Albert Michelson (1852-1931) and Edward Morley (1838-1923). The importance and implications of the Michelson-Morley experiment was lost on much of the scientific world. In many cases, the lack of determination of an ether was thought simply a problem of experimental design or accuracy. In contrast to this general dismissal, Einstein, then a clerk in the Swiss patent office developed a theory of light that incorporated implications of Maxwell's equations and demonstrated the lack of need for an ether.

In formulating his special theory of relativity, Einstein assumed that the laws of physics are the same in all inertial (moving) reference frames and that the speed of light was constant regardless of the direction of propagation and independent of the velocity of the observer.

Important components of Einstein's special theory involved length contraction and time dilation for bodies moving near the speed of light. In separate papers published in 1889, both Irish physicist George Francis Fitzgerald (1851-1901) and Dutch physicist Hendrik Antoon Lorentz (1853-1928) pointed out that the length of an object would change as it moved through he ether, the amount of contraction related to the square of the ratio of the object's velocity to the speed of light. Subsequently, this was known as a Fitzgerald-Lorentz contraction. Near the same time, French mathematician Jules-Henri Poincaré (1854-1912) pointed out problems with concepts of simultaneity and, just a year before Einstein published the special theory of relativity, Poincaré pointed out that observers in different reference frames would measure time differently.

The special theory of relativity also contained Einstein's work relating to mass and energy. Accordingly, in 1905 Einstein also published an elegantly brief but seminal paper in which he asserted and proved his most famous formula relating mass and energy (Energy = mass times the speed of light squared). Einstein's equation $E = mc^2$ is the best known—and arguably least understood—scientific formula. With ominous overtones to a nuclear age that would not dawn until the end World War II, Einstein's equation asserted that tremendous energies were contained in small masses.

A decade later, in 1915, Einstein published his general theory of relativity, which supplanted long-cherished and well-understood Newtonian concepts of gravity that had held prominence since the publication of Newton's *Principia* in 1687.

Impact

Einstein's theories of relativity brought about a revolution in science and philosophy, rivaled only by the contributions of Newton. In fact, although Newtonian physics still enjoys widespread utility, relativistic physics supplanted Newtonian cosmological concepts.

Einstein's papers on special relativity were revolutionary for both physics and science in general. Notable scientists, including German physicist Max Planck (1858-1947), who were in the process of developing quantum theory took notice, others set out to unsuccessfully reconcile relativity theory with Newtonian theories of gravitation. The special theory of relativity was quickly accepted by the general scientific community, and its implications for general philosophical thought were profound.

Despite this general acceptance, it is important to note that Einstein did not receive the Nobel Prize for relativity. Ironically, Einstein's 1921 Nobel Prize for Physics was awarded for his important contributions to quantum theory (a theory ultimately irreconcilable with relativity theory) via his explanation of the photoelectric effect.

The special theory of relativity challenged and eventually overturned classical concepts such as French chemist Antoine-Laurent Lavoisier's (1743-1794) conservation of mass and Prussian physicist Hermann von Helmholtz's (1821-1894) conservation of energy. According to special relativity theory, mass itself was not conserved except as part of a fusion of mass and energy. Mass was nothing more than a manifestation of energy.

Although Einstein's special theory was limited to special cases dealing with systems in uniform non-accelerated motion, it reverberated with philosophical consequence because it dispensed with absolutes (i.e., "absolute" rest or motion). Special relativity gave rise to a plethora of paradoxes dealing with the passage of time (e.g., the twin paradox) and with problems dependent upon assumptions of simultaneity. For example, under certain conditions, it was impossible to determine when one event happened in relation to another event. Classically separate concepts of three dimensions of geometrical space and a fourth dimension of time were fused into space-time.

As a result, those unfamiliar with the mathematical underpinnings of relativity theory often classified it as bizarre and in opposition to common sense. Einstein patiently pointed out that our perceptions of "common sense" were derived only from experiences with objects of an intermediate size, between the atomic and cosmic scale, that moved at velocities far below the speed of light.

It is not possible to understate the impact of Einstein's general theory upon the scientific community. Newtonian formulations of gravity were, prior to the publication of general relativity theory, considered proven. Only the mechanism, not the effects of gravity, were still a subject of inquiry (e.g., Poincaré's gravity waves). Although a quantum theory of gravity remains elusive, Einstein's fusion and curvature of space-time provided a predictive explanation of the mechanisms by which bodies could attract each other over a distance.

In publishing his general theory of relativity, Einstein carefully listed three potential proofs where his theory would predict phenomena differently than classical Newtonian theory. Unlike the esoteric proofs of special relativity, the proofs of general relativity could be measured by conventional experimentation.

General relativity made sensible predictions regarding the subtle shift in the position of the perihelion of a planet, not foreseen by Newtonian theory. Because of it's proximity to the strength of the Sun's gravitational field, Einstein pointed out that it was possible to test the predications of general relativity against the orbit of Mercury. Small discrepancies in the orbit of Mercury (explained by some astronomers as evidence, perhaps, for the presence of another planetoid closer to the Sun) were immediately resolved.

Einstein also asserted that light subjected to an intense gravitational field would show a red shift. Observations of the red-shift of light between Sirus and its binary star companion lent further support to general relativity.

Most importantly, general relativity demanded that light would be deflected by a gravitational field. For more than four years, through the conflicts of World War I, the scientific world awaited the opportunity to test this important prediction. Following the war, the opportunity came with the solar eclipse of 1919. The irony of the fact that the British Royal Astronomical Society sent two expeditions to confirm the work of a German physicist was not lost on a world preoccupied with war reparations. Regardless, the expeditions measured the positions of the brightest stars near the sun as it was eclipsed, and shifts in the positions of other stars established that light was bent by its passage near the Sun's intense gravitational field.

This confirmation of general relativity earned Einstein widespread fame outside the scientific world and he quickly became the most influential scientist in the world. This fame would be most profoundly manifest two decades later when he was urged to write United States President Franklin D. Roosevelt in an attempt to make Roosevelt aware of the potential uses of atomic fission. Einstein's letter influenced the establishment of the Manhattan Project, which enabled the United States to first develop atomic weapons.

Upon the rise of Adolf Hitler in Germany, Einstein, a Jew, was permitted to remain in the United States. In addition to personal fame, Einstein's work found itself transcending academic, political, and cultural borders in much the same manner as the evolutionary theory of Charles Darwin (1809-1882). Many social commentators were quick to adopt selected postulates from relativity theory to support their respective causes. The rise of relativity theory that asserted no preferred reference frames gained in popular esteem as the single perspectives of traditional political empires crumbled and worries about totalitarian rule grew.

General relativity, essentially a geometrization of physical theory, sparked a grand revision of cosmology that continues today. The fusion of space-time under the theory of general relativity dispensed with detached and measurable absolutes and made observers integral to measurement in a much broader sense than had the special theory. The general theory also described non-uniform, or accelerated, motion. The motion of bodies under general relativity is explained by the assertion that in the vicinity of mass, space-time curves. The more massive the body the greater is the curvature or attraction.

The most stunning philosophical consequence of general relativity was that space-time was not an external grid by which to measure the universe. Space-time was a creation of the universe itself. This concept was critical during debates regarding the expansion of the universe argued by Edwin Hubble (1889-1953) and others. Under general relativity, the universe was not expanding into reexisting space and time, but rather creating it as a consequence of expansion. In this regard, general relativity theory set the stage for the subsequent development of big bang theory.

General relativity sparked great excitement among astronomers who immediately grasped the significance of the theory. Shortly before his death, German physicist Karl Schwarzschild (1873-1916) proposed equations that describe the gravitational field of massive compact objects that prepared the way for prediction and—as technology improved during the course of the twentieth century—discoveries related to the evolutionary stages of stars (e.g., neutron stars, pulsars, black holes, etc.). Along with quantum theory, special and general relativity theory remain among the most influential theories in science.

K. LEE LERNER

Further Reading

Books

Howard, Don, and John Stachel, eds. *Einstein and the History of General Relativity.* Boston: Birkhäuser, 1989.

Miller, A. I. "The Special Relativity Theory: Einstein's Response to the Physics of 1905." In *Albert Einstein: Historical and Cultural Perspectives,* edited by Gerald Holton and Yehuda Elkana, 3-26. Princeton, NJ: Princeton University Press, 1982.

Pais, A. *Subtle is the Lord: The Science and the Life of Albert Einstein.* Oxford: Oxford University Press, 1982.

Stachel, J. J. "How Einstein Discovered General Relativity: A Historical Tale With Some Contemporary Morals." In *General Relativity and Gravitation: Proceedings of the 11th International Conference on General Relativity and Gravitation, Stockholm, July 6-12, 1986,* edited by M.A.H. MacCallum, 200-8. Cambridge: Cambridge University Press, 1987.

Periodical Articles

Earman, J., and C. Glymour. "The Gravitational Red Shift as a Test of General Relativity: History and Analysis." *Studies in the History and Philosophy of Science* 11, no. 3 (1980): 175-214.

Farwell, R., and C. Knee. "The End of the Absolute: A Nineteenth-Century Contribution to General Relativity." *Studies in the History and Philosophy of Science* 21, no. 1 (1990): 91-121.

The Development of Quantum Mechanics

Overview

Quantum mechanics describes the relationship between energy and matter on an atomic and subatomic scale. At the beginning of the twentieth century, German physicist Maxwell Planck (1858-1947) proposed that atoms absorb or emit electromagnetic radiation in bundles of energy termed quanta. This quantum concept seemed counter-intuitive to well-established Newtonian physics. Advancements associated with quantum mechanics (e.g., the uncertainty principle) also had profound implications for philosophical and scientific arguments concerning the limitations of human knowledge.

Background

The classical model of the atom that emerged during the last decade of the nineteenth century and the early years of the twentieth century was similar to the Copernican model of the solar system, where, just as planets orbit the sun, electrically negative electrons move in orbits about a relatively massive, positively charged nucleus. Most importantly, in accordance with Newtonian theory, the classical models allowed electrons to orbit at any distance from the nucleus. Problems with these models, however, continued to vex the leading physicist of the nineteenth century. The classical models predicted that when, for example, a hydrogen atom was heated, it should produce a continuous spectrum of colors as it cooled. Nineteenth-century spectroscopic experiments, however, showed that hydrogen atoms produced only a portion of the spectrum. Moreover, studies on electromagnetic radiation by physicist James Clark Maxwell (1831-1879) predicted that an electron orbiting around the nucleus, according to Newton's laws, would continuously lose energy and eventually fall into the nucleus.

Planck proposed that atoms absorb or emit electromagnetic radiation only in certain units or bundles of energy termed quanta. The concept that energy existed only in discrete and defined units seemed counter-intuitive, that is, outside of the human experience with nature. Regardless, Planck's quantum theory, which also asserted that the energy of light was directly proportional to its frequency, proved a powerful theory that accounted for a wide range of physical phenomena. Planck's constant relates the energy of a photon with the frequency of light. Along with constant for the speed of light, Planck's constant (h = 6.626 x 10^{-34} Joule-second) is a fundamental constant of nature.

Prior to Planck's work, electromagnetic radiation (light) was thought to travel in waves with an infinite number of available frequencies and wavelengths. Planck's work focused on attempting to explain the limited spectrum of light emitted by hot objects and to explain the absence of what was termed the "violet catastrophe," predicted by nineteenth-century theories developed by physicists Wilhelm Wien (1864-1928) and John William Strutt Rayleigh (1842-1919).

Danish physicist Niels Bohr (1885-1962) studied Planck's quantum theory of radiation and worked in England with physicists J. J. Thomson (1856-1940) and Ernest Rutherford (1871-1937), improving their classical models of the atom by incorporating quantum theory. During this time, Bohr developed his model of atomic structure. To account for the observed properties of hydrogen, Bohr proposed that electrons existed only in certain orbits and that, instead of traveling between orbits, electrons made instantaneous quantum leaps or jumps between allowed orbits. According to the Bohr model, when an electron is excited by energy it jumps from its ground state to an excited state (i.e., a higher energy orbital). The excited atom can then emit energy only in certain (quantized) amounts as its electrons jump back to lower energy orbits located closer to the nucleus. This excess energy is emitted in quanta of electromagnetic radiation (photons of light) that have exactly the same energy as the difference in energy between the orbits jumped by the electron.

The electron quantum leaps between orbits proposed by the Bohr model accounted for Plank's observations that atoms emit or absorb electromagnetic radiation in quanta. Bohr's model also explained many important properties of the photoelectric effect described by Albert Einstein (1879-1955).

Impact

The development of quantum mechanics during the first half of the twentieth century replaced classical Copernican-like atomic models of the atom. Using probability theory, and allowing for a wave-particle duality, quantum mechanics also replaced classical mechanics as the method by which to describe interactions between subatomic particles. Quantum mechanics replaced electron "orbitals" of classical atomic models with allowable values for angular momentum (angular velocity multiplied by mass) and depicted electron position in terms of probability "clouds" and regions.

When Planck started his studies in physics, Newtonian or classical physics seemed fully explained. In fact, Planck's graduate advisor once claimed that there was essentially nothing new to discover in physics. By 1918, however, the importance of quantum mechanics was recognized and Planck received the Nobel Prize for Physics. The philosophical implications of quantum theory seemed so staggering, however, that Planck himself admitted that he did not fully understand the theory. In fact, Planck initially regarded the development of quantum mechanics as a mathematical aberration or temporary answer to be used only until a more intuitive or commonsense model was developed.

Despite Planck's reservations, however, Einstein's subsequent Nobel prize-winning work on the photoelectric effect was heavily based on Planck's theory. Expanding on Planck's explanation of blackbody radiation, Einstein assumed that light was transmitted as a stream of particles termed photons. By extending the well-known wave properties of light to include a treatment of light as a stream of photons, Einstein was able to explain the photoelectric effect.

The Bohr model of atomic structure was published in 1913 and Bohr's work earned a Nobel Prize in 1922. Bohr's model of the hydrogen atom proved to be insufficiently complex to account for the fine detail of the observed spectral lines. However, Prussian physicist Arnold Sommerfeld (1868-1951) provided refinements (e.g., the application of elliptical, multi-angular orbits) that explained the fine-structure of the observed spectral lines.

Later in the 1920s, the concept of quantization and its application to physical phenomena was further advanced by more mathematically complex models, based on the work of French physicist Louis Victor de Broglie (1892-1987) and Austrian physicist Erwin Schrödinger (1887-1961), that depicted the particle and wave nature of electrons. De Broglie showed that the electron was not merely a particle but a wave form. This proposal led Schrödinger to publish his wave equation in 1926. Schrödinger's work described electrons as a "standing wave" surrounding the nucleus, and his system of quantum mechanics is called wave mechanics. German physicist Max Born (1882-1970) and English physicist P.A.M Dirac (1902-1984) made further advances in defining subatomic particles (principally the electron) as a wave rather than a particle, and reconciled portions of quantum theory with relativity theory.

Working at about the same time, German physicist Werner Heisenberg (1901-1976) formulated the first complete and self-consistent theory of quantum mechanics. Matrix mathematics was well-established by the 1920s, and Heisenberg applied this powerful tool to quantum mechanics. In 1926 Heisenberg put forward his uncertainty principle, which states that two complementary properties of a system, such as position and momentum, can never both be known exactly. This proposition helped cement the dual nature of particles (e.g., light can be described as having both wave and particle characteristics). Electromagnetic radiation—one region of the spectrum that comprises visible light—is now understood as having both particle and wave-like properties.

In 1925 Austrian-born physicist Wolfgang Pauli (1900-1958) published the Pauli exclusion principle, which states that no two electrons in an atom can simultaneously occupy the same quantum state (i.e., energy state). Pauli's specification of spin (+1/2 or -1/2) established that two electrons in any suborbital have differing quantum numbers (a system used to describe the quantum state). This insight made completely understandable the structure of the periodic table in terms of electron configurations (i.e., the energy related arrangement of electrons in energy shells and suborbitals).

In 1931 American chemist Linus Pauling (1901-1994) published a paper that used quantum mechanics to explain how two electrons, from two different atoms, are shared to make a covalent bond between the two atoms. Pauling's work provided the connection needed in order to fully apply the new quantum theory to chemical reactions.

Quantum mechanics posed profound questions for scientists and philosophers. The concept that particles such as electrons make quantum leaps from one orbit to another, as opposed to simply moving between orbits, seemed counter-intuitive. Like much of quantum theory, the proofs of how nature works at the atomic level are mathematical. Bohr himself remarked,

"Anyone who is not shocked by quantum theory has not understood it."

The rise of the importance and power of quantum mechanics carried important philosophical consequences. When misapplied to larger systems—as in the famous paradox of Schrödinger's cat—quantum mechanics could be misinterpreted to make bizarre predictions (i.e., a cat that is simultaneously dead and alive). On the other hand, quantum mechanics made possible important advances in cosmological theory.

Quantum and relativity theories strengthened philosophical concepts of complementarity, wherein phenomenon can be looked upon in mutually exclusive yet equally valid perspectives. In addition, because of the complexity of quantum relationships, the rise of quantum mechanics fueled a holistic approach to explanations of physical phenomena. Following the advent of quantum mechanics, the universe could no longer be explained in terms of Newtonian causality, but only in terms of statistical, mathematical constructs.

In particular, Heisenberg's uncertainty principle asserts that knowledge of natural phenomena is fundamentally limited—to know one part allows another to move beyond recognition. Quantum mechanics, particularly in the work of Heisenberg and Schrödinger, also asserted an indeterminist (no preferred frame of reference) epistemology, suggesting that human knowledge itself is limited by inescapable aspects of incompleteness and randomness.

Fundamental contradictions with long accepted Newtonian causal and deterministic theories made even the leading scientists of the day resistant to the philosophical implications of quantum theory. Einstein argued against the seeming randomness of quantum mechanics by asserting, "God does not play dice!" Bohr and others defended quantum theory with the gentle rebuttal that one should not "prescribe to God how He should run the world."

K. LEE LERNER

Further Reading

Bohr, Niels. *The Unity of Knowledge.* New York: Doubleday, 1955.

Feynman, Richard P. *The Character of Physical Law.* Cambridge: MIT Press, 1965.

Feynman, Richard P. *QED: The Strange Theory of Light and Matter.* Princeton, NJ: Princeton University Press, 1985.

From an Expanding Universe to the Big Bang

Overview

How did physical nature on the cosmic scale begin? What are the dimensions of the universe? These remain the two most persistent and expansive scientific questions in the collective mind of humanity. Logically, they seem to be related to each other, as they also lead to "Why are we here?" and other introspective queries. In the first half of the twentieth century, the probing for the answer to these questions reached a high point of investigative conclusions. Larger telescopes, advances in detailed photography and experimentally sophisticated spectroscopic methods, and mathematical technique provided concrete means to explore the greatest challenge to human understanding, the universe. Through about mid-nineteenth century, astronomers had catalogued stars, methodically studied the solar system, and wondered at the unexplained sights of deep space. From there on the structure of the universe and its extent became the major thrusts of specialized astronomical disciplines, cosmology and astrophysics. The result would be momentous: not only the discovery that the universe was many times larger than traditionally thought but also expanding outward at astonishing speeds approaching that of light.

Background

Nineteenth-century English astronomers father and son William (1738-1822) and John (1792-1871) Herschel theorized beyond the acceptance of a uniform distribution of stars around the Milky Way, often simply called the Galaxy. They instituted a methodical study of the "structure of the heavens." Telescopes were reaching the size at which the cosmic fuzzy patches of light, the "star clouds and nebulae," some known since ancient times, were demanding investigation. Both the philosopher Immanuel Kant (1724-1804) and astronomer Pierre Laplace (1749-1827) had suggested these were "island universes," independent star systems beyond the Milky Way.

These anomalies prompted the senior Herschel, who favored the "island universes" idea, to focus on revamping cosmic distances by recording relative distances to the stars based on observation of apparent brightness (apparent magnitude) of starlight. His observation showed that star distributions were more concentrated in the plane of the Milky Way—that is, the Galaxy was disk-shaped, perhaps like the nebulae he saw through his large telescopes. The younger Herschel concentrated on the study of what he called—anticipating the future—"extragalactic nebulae." Catalogues of stars gave way toward the end of the century to catalogues of nebulae much advanced on that of pioneer cataloguer Charles Messier (1730-1817). New spectrographic techniques were applied, an important realization being that the visible universe had the same chemical composition as analyzed on Earth. Many nebulae catalogued were gas clouds rather than galaxies and were spectrographically analyzed as such, discrediting the "island universe" concept. But intuition still kept the idea alive with observation and photography, important in providing a permanent record of astronomical observation.

By the turn of the twentieth century the mistake of erecting big reflectors with bad design in worse locations had been learned. The best operational reflectors in 1895, the Common telescope in England and the Crossley telescope of the University of California's Lick Observatory, were modest in size (about 36-in or 91.44-cm aperture) but doing important work. The Lick launched the first detailed photographic program of the so-called spiral nebulae. They appeared to be like the close Andromeda Nebula; their smaller size logically being a matter of much greater distance from the Milky Way. Into the new century two more pieces to the galactic puzzle of the Milky Way appeared: Johannes Hartmann (1865-1936) identified interstellar gas (1904), and Edward Barnard (1857-1923) did the same for interstellar dust (1909), the latter explaining areas that seemed without stars. And about 1910 the research on star brightness/temperature relationship of Ejnar Hertzsprung (1873-1967) and Henry Russell

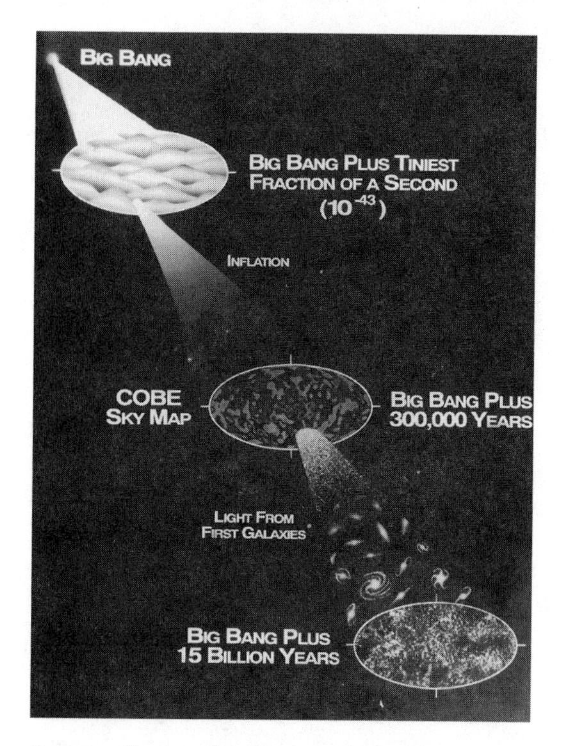

Conceptualization of the Big Bang Theory by NASA. *(AP/Wide World. Reproduced with permission.)*

(1877-1957) resulted in the Main Sequence diagram of star types, which brought a tempering order to stellar astronomy.

That the Lick telescope was in California was prophetic for the future of galactic astronomy. George Ellery Hale (1868-1938) of the Yerkes Observatory came west to California and chose Mt. Wilson above Los Angeles as the sight of a new solar observatory (1905). As would become clear, he was inaugurating the most important observatory complex for the investigation of the cosmos for half a century. With him came astronomer and telescope designer George Ritchey (1864-1945), who was able to put Hale's efforts for a 60-inch (152.4-cm) reflector, the largest working telescope in the world, into service by 1908 with himself nearly resolving individual stars in the Andromeda Nebula with it. Larger diameter mirrors meant seeing further into the origins of the stars—deep space—which became Hale's crusade for the future of astronomy. By late 1917 his persistence paid off with yet another reflector of unprecedented size, the 100-inch (254-cm) Hooker reflector, which would remain the largest in the world until 1948.

Here were hardware stepping stones toward solving the mysteries of the universe, but the mental ones were also at hand. The limited parallax and trigonometric methods for calculating the distances of stars from Earth gave way to the revolutionary technique based on the study of Cepheid variable, or pulsing, stars, the study by Henrietta Leavitt (1868-1921) in 1912, and the technique by Hertzsprung's determination of these stars as "standard candles" by their intensity-period relationship. Albert Einstein (1879-1955) had already published his theory of special relativity (1905), explaining the odd effects of relative motion with constant velocities approaching the speed of light. The cosmic scale was the only laboratory to validate the implications of relativity. By 1913, Einstein wrote Hale about proving an aspect of Einstein's theory of general relativity, which extended the physical parameters of relativity theory to acceleration and gravity on curved paths. That effect was the theory that strong gravitational fields could bend light—again, something only to be imagined in the massiveness of cosmic space. This was proved at Mt Wilson in 1915.

Einstein went on to theorize that the universe might have physical curvature and came up with three mathematical possibilities: positive curvature (as a circle) or a closed universe, negative curvature (as a saddle or hyperbolic surface) or an open universe, and uncurved. Einstein's ideas were much food for cosmic thought. Mt. Wilson had the telescopes to take the initial step toward determining the origins, structure, and extension of the universe. First, was the Milky Way the whole of the universe?—that could be found by solving the puzzle of whether the dense (globular) and open (galactic) clusters of stars and, more so, the diffuse spiral nebulae were internal or external to the Milky Way. Hale found the two astronomers to embark on this research—unlikely as it might seem, both from Missouri. Harlow Shapely (1885-1972) began observation with the 60-inch (152.4-cm) reflector in 1914 to investigate the globular clusters (Cepheids being found there). He was the first astronomer to apply the Cepheid variable yardstick to determine the distance to these clusters. Edwin Hubble (1889-1953) came in 1919 to be the first to work extensively with the Hooker 100-inch (254-cm) on investigation of the nebulae, also using variables as distance indicators.

Shapely completed his work by 1917. He was able to find the distances of variables in a dozen globular clusters, discovering a uniformity in large cluster star magnitudes and the size of the clusters. Finding the distances of 69 clusters, he was able to deduce that all the clusters were within the Milky Way and that the center of the Milky Way was in the proximity of these clusters in the constellation Sagittarius. Previously, the center was assumed to be our Sun and its solar system; now that was determined to be on the edge of the galaxy. This proved that the Milky Way was 10 times more extensive than long thought. Having shown this, Shapely did not wish to extend the reasoning to the nebulae and without further investigation assumed these were comparable in distance to the clusters and within the Milky Way, which was still the unique Galaxy. But data to the contrary began emerging. Ritchey had found (1917) a nova (erupting star) in a nebula in the constellation Cepheus, but instead of a brightness 10,000 times the Sun, it was so dim that he reasoned it must be very far off. He studied the Mt. Wilson photographic plates and found two more such novae, while Heber Curtis (1872-1942) did the same at Lick, finding three—others were also found. Curtis, whose research centered on spiral nebulae, became convinced these were independent galaxies to the Milky Way. A collision other than cosmic was coming.

Impact

Heading for the high profile Harvard Observatory, the sardonic Shapely left Mt. Wilson to the urbane and cosmopolitan Hubble, there being a strain of theoretical and professional attitude between them. Hubble's meticulous cataloguing of the shape of nebulae preceded inspection of the Andromeda Nebula for Cepheid indicators, becoming much anticipated by astronomers siding with Curtis that the spiral nebulae were independent galaxies. From about 1917 to 1923 the so-called "Great Debate" ensued with Curtis and other California astronomers opposing those lined up with Shapely, mostly in the East, whether there was one galaxy or many. A debate over the scale of the universe took place between Curtis and Shapely, still insistent that the Milky Way was the whole of the universe, in April 1920.

With the most powerful telescope in the world, Hubble was able to resolve stars including Cepheids not only in Andromeda but also in two seemingly smaller spirals in late 1923. In dramatic fashion at an astronomy conference on the first of the following year, a message from Hubble arrived stating that he had proved by the faintness and period of these Cepheids that the Andromeda Nebula and the two other nebula were far beyond the Milky Way—these were independent galaxies. This was the momentous birth of galactic astronomy with the boundaries of the universe dropping away and modern cosmology ceasing to be a matter of theory intrigued by relativity. Hubble himself was attuned to new cosmological theory, being acquainted with the cosmic models being expounded in the aftermath of relativity theory.

Einstein's conclusion (1917) that the universe was static and unchanging in curved space, a spherical universe, had stimulated many scientists to seek other solutions to his relativity field equations. About this same time Dutch astronomer Willem de Sitter (1872-1934) demonstrated other solutions, one implying an expanding universe of constantly decreasing curvature in time. Russian mathematician Alexander Friedmann (1888-1925) decided the universe would be the same everywhere with its size and density varying as functions of time (1922). In the meantime Hubble continued to advance outward into intergalactic space. Having to retire the limited Cepheids as distance indicators, he used the blue supergiant stars that inhabited the arms of Andromeda and other near galaxies for comparison with those in the further galaxies he was examining. When he could no longer distinguish individual stars, Hubble had to resort to even more approximate averaging of innate light brilliance in whole galaxies compared with those catalogued at 30 million light-years. In this manner, he reached a visible cosmic horizon he calculated as 1 billion light-years.

This tier upon tier of ever more remote galaxies had velocities of movement away from the Milky Way. Back in 1912 Lowell Observatory astronomer Vesto Slipher (1875-1969) had observed the radial (rotating) velocities of spiral nebulae with spectral analysis revealing a shift toward the red of the visible spectrum that seemed to indicate a Doppler shift in wavelength, as would be observed for an object moving rapidly away. Hubble noted this with all the galaxies being observed at Mt. Wilson, and he also observed that the further away the galaxy was the more the spectra was shifted to the red—that is, the further away the galaxy the faster it was receding from Earth. There had been theoretical talk about the redshift already, particularly from Belgian astronomer Georges Lemaitre (1894-1966), who interpreted the redshift phenomena as indicating expansion of the universe (1927). When Hubble published (1929) his findings that the entire universe was uniformly expanding in every direction, his was the observational validation of a contemporary theory. The concept relation of redshift with distance, which Hubble did not interpret as Doppler shift, or velocity with distance—something not proved—was linked to him as Hubble's law and through the scientific writings of, among others, his English friends astronomers Arthur Eddington (1882-1944) and James Jeans (1877-1946).

Of course, the idea of an expanding universe led back to a fundamental question—how did it start? Eddington seemed to be the first (1932) to tackle the thought of some initial compressed cosmic state and to coin a later famous word for it: "As a scientist I simply do not believe that the present order of things started off with a bang...." He influenced Lemaitre, who advanced the idea of an unstable, exploding "primeval atom" (after 1945) that started what he called "The Big Bang," an idea still questioned. Hubble's calculations of the galaxies' initial point were not far enough back in time (today the accepted 15-20 billion years is being questioned as too much). Ultimately, the Shapely/Hubble measuring technique was inaccurate, as came to light when colleague Walter Baade (1893-1960) discovered Cepheids varied in brightness (1941-42) with implications carrying over with research using the new (1948) Hale 200-inch (508-cm) reflector at the Mt. Palomar Observatory. Hubble judged his variables dimmer than reality, resulting in a universe only half as big as that found by Bade. Still the galaxy/redshift phenomenon was valid, and some galaxies detected are moving away at near the speed of light.

The general acceptance of a curved, expanding universe was modified in 1948 by astronomer Fred Hoyle (1915-) and colleagues with a "steady-state" universe of uniform space and unchanging in time that has constant spontaneous creation of matter. The disposal of the concentration of matter and time requirements of a Big Bang theory attracted some astronomers, but the Big Bang came back with an intriguing hypothesis from physicist George Gamow (1904-1968) and associates in the 1940s that its initial concentration of matter would infer intense heat as well and, therefore, a low temperature blackbody heat should still survive. In 1965 the first traces of that, a 3°K blackbody temperature, was discovered pervading the background of space. And new telescopes and sensors on Earth and on space platforms continue to probe the universe for proofs of several models beyond the Big Bang origin and state of the universe.

WILLIAM J. MCPEAK

Further Reading

Eddington, Arthur. *The Expanding Universe*. Reprint edition. London: University Press, 1958.

Gamow, George. *The Creation of the Universe*. New York: Compass, 1960.

Hoyle, Fred. *The Nature of the Universe*. New York: Harper & Brothers, 1960.

Hubble, Edwin. *The Realm of the Nebulae*. Reprint edition. New York: Dover, 1958.

Lankford, John, ed. *History of Astronomy: An Encyclopedia*. New York: Garland, 1997.

Shapely, Harlow. *Galaxies*. Cambridge, MA: Harvard University Press, 1961.

Life Cycles of the Stars

Overview

Until the last half of the nineteenth century, astronomy was principally concerned with the accurate description of the movements of planets and stars. However, developments in electromagnetic theories of light, along with the articulation of quantum and relativity theories at the start of the twentieth century, allowed astronomers to probe the inner workings of the stars. Of primary concern was an attempt to coherently explain the life cycle of the stars and to reconcile the predictions of newly advanced physical theories with astronomical observation. Profound questions regarding the birth and death of stars led to the stunning conclusion that, in a very real sense, humans are a product of stellar evolution.

Background

It is now known that the mass of a star determines the ultimate fate of a star. Stars that are more massive burn their fuel quicker and lead shorter lives. These facts would have astonished astronomers working at the dawn of the twentieth century. At that time, understanding of the source of the heat and light generated by the Sun and stars was hindered by a lack of understanding of nuclear processes.

Based on Newtonian concepts of gravity, many astronomers understood that stars were formed in clouds of gas and dust, termed nebulae, that were measured to be light years across. These great molecular clouds, so thick they are often opaque in parts, teased astronomers. Decades later, after development of quantum theory, astronomers were able to understand the energetics behind the "reddening" of light leaving stellar nurseries (reddened because the blue light is scattered more and the red light passes more directly through to the observer). Astronomers speculated that stars formed when regions of these cloud collapsed (termed gravitational clumping). As the region collapsed it accelerated and heated up to form a protostar, heated by gravitational friction.

In the pre-atomic age, the source of heat in stars was a mystery to astronomers who sought to reconcile apparent contradictions regarding how stars were able to maintain their size (i.e., not shrink as they burned fuel) and radiate the tremendous amounts of energy measured. In accordance with dominant religious beliefs, the Sun, using conventional energy consumption and production concepts, could be calculated to be less than a few thousand years old.

In 1913 Danish astronomer Ejnar Hertzsprung (1873-1967) and English astronomer Henry Norris Russell (1877-1957) independently developed what is now known as the Hertzsprung-Russell diagram. In the Hertzsprung-Russell diagram the spectral type (or, equivalently, color index or surface temperature) is placed along the horizontal axis and the absolute magnitude (or luminosity) along the vertical axis. Accordingly, stars are assigned places top to bottom in order of increasing magnitude (decreasing brightness) and from right to left by increasing temperature and spectral class.

The relation of stellar color to brightness was a fundamental advance in modern astronomy. The correlation of color with true brightness eventually became the basis of a widely used method of deducing spectroscopic parallaxes of stars, which allowed astronomers to estimate the distance of distant stars from Earth (estimates for closer stars could be made by geometrical parallax).

In the Hertzsprung-Russell diagram, main sequence stars (those later understood to be burning hydrogen as a nuclear fuel) form a

White dwarf stars shown in a globular cluster.*(NASA. Reproduced with permission.)*

prominent band or sequence of stars, from extremely bright, hot stars in the upper left-hand corner of the diagram, to faint, relatively cooler stars in the lower right-hand corner of the diagram. Because most stars are main sequence stars, most stars fall into this band on the Hertzsprung-Russell diagram. The Sun, for example is a main sequence star that lies roughly in the middle of diagram, among what are referred to as yellow dwarfs.

Russell attempted to explain the presence of giant stars as the result of large gravitational clumps. Stars, according to Russell, would move down the chart as they lost mass burned as fuel. Stars begin their life cycle as huge, cool red bodies, and then undergo a continual shrinkage as they are heated. Although the Hertzsprung-Russell diagram was an important advance in understanding stellar evolution—and it remains highly useful to modern astronomers—Russell's reasoning behind the movements of stars on the diagram turned out to be exactly the opposite of modern understanding of stellar evolution, which is based on an understanding of the Sun and stars as thermonuclear reactors.

Advances in quantum theory and improved models of atomic structure made it clear to early twentieth-century astronomers that deeper understanding of the life cycle of stars and of cosmological theories explaining the vastness of space was linked to advances in the understanding of the inner workings of the universe on an atomic scale. A complete understanding of the energetics of mass conversion in stars was provided by Albert Einstein's (1879-1955) special theory of relativity and his relation of mass to energy (Energy = mass times the speed of light squared, or $E = mc^2$).

During the 1920s, based on the principles of quantum mechanics, British physicist Ralph H. Fowler determined that, in contrast to the predictions of Russell, a white dwarf would become smaller as its mass increased.

Indian-born American astrophysicist Subrahmanyan Chandrasekhar (1910-1995) first classified the evolution of stars into supernova, white dwarfs, and neutron stars and predicted the conditions required for the formation of black holes, subsequently found in the later half of the twentieth century. Prior to World War II, American physicist J. Robert Oppenheimer (1904-1967), who ultimately supervised project Trinity (the making of the first atomic bombs), made detailed calculations reconciling Chandrasekhar's predictions with general relativity theory.

Over time, as the mechanisms of atomic fission and fusion worked their way into astronomical theory, it became apparent that stars spend approximately 90% of their lives as main sequence stars before the fate dictated by their mass becomes manifest.

Astronomers refined concepts regarding stellar birth. Eventually, as a protostar contracts enough, the increase in its temperature triggers nuclear fusion, and the star becomes visible as it vaporizes the surrounding cocoon. Stars then lead the majority of their life as main sequence stars, by definition burning hydrogen as their nuclear fuel.

It was the death of the stars, however, that provided the most captivating consequences.

Throughout the life of a star, a tense tug-of-war exists between the compressing force of the star's own gravity and the expanding pressures generated by nuclear reactions at its core. After cycles of swelling and contraction associated with the burning of progressively heavier nuclear fuels, eventually the star runs out of useable nuclear fuel. The spent star then contracts under the pull of its own gravity. The ultimate fate of any individual star is determined by the mass of the star left after blowing away its outer layers during its paroxysmal death spasms.

Low mass stars could fuse only hydrogen, and when the hydrogen was used up, fusion stopped. The expended star shrank to become a white dwarf.

Medium mass stars swell to become red giants, blowing off planetary nebulae in massive explosions before shrinking to white dwarfs. A star remnant less than 1.44 times the mass of the Sun (termed the Chandrasekhar limit) collapses until the pressure in the increasingly compacted electron clouds exerts enough pressure to balance the collapsing gravitational force. Such stars become "white dwarfs," contracted to a radius of only a few thousand kilometers—roughly the size of a planet. This is the fate of our Sun.

High mass stars can either undergo carbon detonation or additional fusion cycles that create and then use increasingly heavier elements as nuclear fuel. Regardless, the fusion cycles can only use heavier elements up to iron (the main product of silicon fusion). Eventually, as iron accumulates in the core, the core can exceed the Chandrasekhar limit of 1.44 times the mass of the Sun and collapse. This preliminary theoretical understanding paved the way for many of the discoveries in the second half of the twentieth century, when it was more fully understood that as electrons are driven into protons, neutrons are formed and energy is released as gamma rays and neutrinos. After blowing off its outer layers in a supernova explosion (type II), the remnants of the star formed a neutron star and/or pulsar (discovered in the late 1960s).

Although he did not completely understand the nuclear mechanisms (nor, of course, the more modern terminology applied to those concepts), Chandrasekhar's work allowed for the prediction that such neutron stars would be only a few kilometers in radius and that within such a neutron star the nuclear forces and the repulsion of the compressed atomic nuclei balanced the crushing force of gravity. With more massive stars, however, there was no known force in the universe that could withstand the gravitational collapse. Such extraordinary stars would continue their collapse to form a singularity—a star collapsed to a point of infinite density. According to general relativity, as such a star collapses its gravitational field warps space-time so intensely that not even light can escape, and a "black hole" forms.

Although the modern terminology presented here was not the language of early twentieth-century astronomers, German astronomer Karl Schwarzschild (1873-1916) made important early contributions toward understanding of the properties of geometric space around a singularity when warped, according to Einstein's general relativity theory.

Impact

There are several important concepts stemming from the evolution of stars that have had enormous impact on science and philosophy in general. Most importantly, the articulation of the stellar evolutionary cycle had a profound effect on the cosmological theories developed during the first half of the twentieth century, culminating in the Big Bang theory, first proposed by Russian physicist Alexander Friedmann (1888-1925) and Belgian astronomer Georges Lemaître (1894-1966) in the 1920s and subsequently modified by Russian-born American physicist George Gamow (1904-1968) in the 1940s.

The observations and theories regarding the evolution of stars meant that only hydrogen, helium, and a perhaps a smattering of lithium were produced in the big bang. The heavier elements, including carbon, oxygen, and iron, were determined to have their genesis in the cores of increasingly massive dying stars. The energy released in the supernova explosions surrounding stellar death created shock waves that gave birth via fusion to still heavier elements and allowed the creation of radioactive isotopes.

The philosophical implications of this were as startling as the quantum and relativity theories underpinning the model. Essentially all of the elements heavier than hydrogen that comprise everyday human existence were literally cooked in a process termed nucleosynthesis that took place during the paroxysms of stellar death. Great supernova explosions scattered these elements across the cosmos.

By the mid-twentieth century, man could look into the night sky and realize that he is made of the dust of stars.

K. LEE LERNER

Further Reading

Hawking, Stephen. *A Brief History of Time*. New York: Bantam Books, 1988.

Hoyle, Fred. *Astronomy*. Garden City, New York: Doubleday, 1962.

Sagan, Carl. *Cosmos*. New York: Random House, 1980.

Trefil, James. *Space, Time, Infinity*. New York: Pantheon Books, 1985.

The Development of Radio Astronomy

Overview

In 1932, while attempting to determine the source of static interference in radio communication systems, American engineer Karl Jansky (1905-1945) discovered the existence of radio waves emanating from beyond the Earth. Jansky traced all but one of the types of interference he encountered to terrestrial phenomena (e.g., thunderstorms). Jansky concluded that the elusive third source of interference was extraterrestrial in origin and that it emanated from a region of the sky in the direction of the center of the Milky Way. Although Jansky's findings were initially ignored, eventually his papers gave rise to the science of radio astronomy. Since Jansky's discovery, radio astronomy has advanced to become one of the most important and productive means of astronomical observation, providing astronomers with a second set of eyes with which to view the Cosmos.

Background

In the nineteenth century Scottish physicist James Clerk Maxwell (1831-1879) developed a set of equations that described the propagation of electromagnetic radiation. Other scientists soon realized that the electromagnetic radiation producing visible light was but a part of a much larger electromagnetic spectrum. In 1888, German physicist Henrich Rudolph Hertz (1857-1894) demonstrated the existence of a portion of that spectrum known as radio waves.

The earliest attempts at detecting radio signals from outer space date to the 1890s with the efforts of American inventor Thomas Edison (1847-1931) and British physicist Sir Oliver Lodge (1851-1940), who, in separate experiments, unsuccessfully attempted to detect solar radio activity (Lodge failed principally because he was listening at the wrong wavelengths). In 1921, Guglielmo Marconi (1874-1937), who, in 1901 made the first intelligible transmission of a radio signal across the Atlantic, claimed to have detected extraterrestrial signals while experimenting on his yacht.

In the1920s, communication between continents was achieved through the use of powerful short-wave radio transmitters. Scientists and engineers working for the communications companies noticed that, at various times during the year, excessive amounts of static interference interfered with communications. Bell Telephone Laboratories assigned Jansky to investigate the possible sources of this periodic interference phenomena.

GEORGE HALE'S FIRST NIGHT WITH THE 100-INCH HOOKER TELESCOPE

No early twentieth-century astronomer equaled the drive to discover the origins of the stars and the depths of the universe by building larger telescopes than American George Hale (1868-1938). His solar observatory on Mt. Wilson above Los Angeles would evolve into an observatory complex providing two of the largest telescopes in the world. After successfully constructing the first of these (a 60-inch or 152.4-cm diameter reflector, 1908), Hale found support to construct the second, a 100-inch (254-cm) reflector, the idea of his friend and benefactor Los Angeles businessman John D. Hooker. With Hooker's seed money, Hale contracted the making of the 100-inch glass blank. On cooling, the 5 tons of glass was found to have air bubbles. Two other castings were not as good.

But Hale kept faith that the first blank would not expand and contract unevenly because of the bubbles and had it mirrored and polished. Carnegie Institution money provided for constructing the lattice telescope tube, the huge polar yoke mounting, and the dome. However, World War I and other setbacks held up this work and put Hale near nervous collapse. Finally, on a November night in 1917 all was ready to take the first look through the 100-ton Hooker Telescope. The telescope was trained on Jupiter, but the result was Hale's worst nightmare. Rather than a sharp image, there was a smudge of light and half a dozen distorted images of the planet. Once again Hale had faith, suggesting that the heat of the Sun on the open dome during the day had expanded the mirror. They waited. At 2:30 AM, they looked again—now at the star Vega. Hale was right; the image was a perfect pinpoint of light. The Hooker was operational—the largest telescope in the world until Hale's last dream, the 200-inch (508-cm) Palomar reflector of 30 years later. Guided by Edwin Hubble, the Hooker would reveal a universe of galaxies like our own, a universe not only many times larger than first thought but also expanding.

WILLIAM J. MCPEAK

Jansky constructed a radio antenna to seek the source of the interference. After accounting for interference from thunderstorms, Jansky focused his investigation on a small but steady static radiation level. Intriguing to Jansky was the fact that the static varied with both direction and time. After eliminating the Sun and his own equipment as possible sources of interference, Jansky noted that the time difference in the daily reception of maximum static shifted by about four minutes each day—just as do the positions of stars on the celestial sphere. Accordingly, Jansky correctly concluded that the static emanated from a source outside the solar system. Jansky found that the static, similar to that created when electric current flows through a resistor, was strongest in the direction of the constellation Sagittarius and the center of our galaxy. Jansky suggested that the source of the static was extraterrestrial radiation of radio waves from hot charged particles in interstellar space.

In 1937, another American radio engineer, Grote Reber (1911-), attempted to duplicate Jansky's findings. Reber constructed a parabolic reflector dish receiver in his backyard and set out to search for cosmic radio waves. Early in the 1940s Reber began to systematically study the sky at varying radio wavelengths. Reber's work in radio astronomy received wide attention when his observations were published in both scholarly and popular science periodicals. In 1944, Reber published the first celestial maps marked with radio frequencies.

The first observation of radio emissions from solar flares was made in 1942, by British scientist J.S. Hey, who was tracking the source of radar jamming signals. Wartime secrecy prohibited Hey from publishing his work. After the war, however, Hey's discovery (by then repeated in the civilian sector) and other great advances in radio wave technologies were declassified. In that immediate post-WWII period, scientists who had in many cases set aside research to aid in the war effort turned their attention and talents to the development of radio astronomy. There was a literal building boom in the construction of larger and more sophisticated radio telescopes.

Engineers found that lessons learned with regard to solving problems associated with the construction of optical telescopes were, in many cases, applicable to the development of radio telescopes. Although at first surprising, the similarities between radio and optical astronomy are grounded in the fact that both deal with manifestations of electromagnetic radiation. The physics of radio waves is, therefore, exactly the same as that of visible light waves, except for differences in the wavelength and frequencies of the electromagnetic radiation received. There were, of course, important differences; instead of ground glass lens used in optical instruments, radio telescopes used parabolic-shaped metal dishes. In a similar fashion, however, radio waves can be reflected to converge at a focal point where they can be amplified and measured.

Radio telescopes grew in precision and power. Advances in radar technology improved, and by mid-century, radio astronomers had identified thousands of sources of extraterrestrial radio emissions.

Impact

Radio waves can be produced from objects that are much cooler than the temperatures required to produce visible light. In a sense, radio astronomy extends the human senses as astronomers probe the Cosmos. In addition to an ability to detect cooler objects, radio astronomy allowed astronomers to probe obscuring clouds of interstellar dust.

Although they eventually diverged as separate sub-disciplines, radio astronomers also began to experiment with the use of RADAR. In 1946, astronomers discovered that they were able to bounce radar signals off of the lunar surface. Scientists began to use radar to probe and map a number of extraterrestrial objects, including meteors not visible to the naked eye.

In a quest for better resolution of extraterrestrial radio signals, astronomers and engineers ultimately realized that instead of simply building bigger radio telescopes, they could electrically separate telescopes in a process termed radio interferometry. Today, the Very Large Array (VLA) located in New Mexico utilizes these principles of radio interferometry in linking 27 radio telescopes to achieve resolution exceeding the largest ground-based optical telescopes.

Radio astronomy also extended the established science of optical spectroscopy. Light from receding objects becomes shifted (e.g., red-shifted) so that what was once visible light appears in different parts of the electromagnetic spectrum, including radio frequencies. Radio astronomy also allowed new insights into the mechanisms operating in solar flares and sunspots, both strong radio sources.

Just as interstellar spirals can be observed in visible light to possess varying quantities of interstellar dust, radio telescope observations reveal that these spirals can also contain tremendous amounts of interstellar gas (principally in the form of molecular hydrogen). Radio astronomers have been able to identify regions of hydrogen by measuring the emission of 21-cm radiation. These determinations are critical in predicting whether the universe will continue expanding.

The information accumulated by radio astronomers has also fostered some of the greatest scientific enigmas of the modern era, including the origin of very strong radio waves from very dim stars and other star-like objects. In 1949, the Crab Nebula (Messier 1), a supernova remnant, and two other galaxies were identified as emitters of radio signals. Radio astronomers were eventually able to confirm many of the postulates of stellar evolution with the subsequent identification in the late 1960s of radio pulsars (rapidly spinning neutron stars).

By the end of 1949 an increasing number of astronomers using radio telescopes sought explanations for these unexplained radio emissions. Puzzled astronomers suspected that a variety of star-like objects might be the source of radio waves. Later known as quasars (quasi-stellar radio sources), these puzzling objects appear to be the most distant yet most energetic objects observed by astronomers. In 1963, Arno Penzias (1933-) and Robert Wilson (1936-) found that, no matter where in the sky they pointed their antenna, they found radio emissions—including emissions from parts of the sky that were visibly empty. The noise turned out to be cosmic background radiation left over from the Big Bang.

Radio astronomy also enabled man to seek scientific evidence of extraterrestrial life. Ideas regarding the mechanisms of presumed interplanetary or interstellar communication were often bizarre schemes (e.g., building mirrors to burn symbols into the sands of Mars) designed to attract the attention of alien civilizations With radio astronomy, man could, for the first time, realistically and systematically listen for evidence of extraterrestrial life and intelligence. In a serious continuation of the pioneering work of Jansky and Reber who tuned in to listen to the Cosmos, scientists continue listening for radio signals that would confirm the existence of extraterrestrial intelligence. Radio astronomy gave rise to several twentieth-century projects now commonly referred to as the Search for Extraterrestrial Intelligence (SETI). Despite the reception of some interesting signals, none of which has passed the rigorous repeatability standards of modern science, no signal has ever been identified as evidence of extraterrestrial intelligence.

K. LEE LERNER

Further Reading

Books

Abell, G. and D. Morrison. *Explorations of the Universe.* New York: Saunders College Publishing, 1987.

Trefil, J. *Space, Time, Infinity.* Pantheon Books, 1985.

Periodical Articles

Wearner, R. "The Birth of Radio Astronomy." *Astronomy* (June 1992).

"The Coldest Spot on Earth": Low Temperature Physics, Superfluidity, and the Discovery of Superconductivity

Overview

The Dutch experimental physicist and Nobel Prize laureate Heike Kamerlingh Onnes (1853-1926) worked for more than four decades in low temperature physics, a discipline he helped to establish over the years as a complete and independent field of study. When in 1908 Kamerlingh Onnes succeeded in liquefying helium, he became the very first experimentalist to reach a temperature as low as 4.2 Kelvin (or -451.84°F). His discovery of superconductivity three years later opened whole new vistas of theoretical and experimental researches that are still today of the utmost importance to the progress of science and technology.

Background

Low temperature physics really began in the second half of the nineteenth century with the discovery in 1852 of the Joule-Thomson effect, attributed to two British physicists, James Prescott Joule (1818-1889) and Sir William Thomson, Lord Kelvin (1824-1907). That year Thomson, based on his and Joule's thermodynamical studies, observed that when a gas expands in a vacuum its temperature decreases. Indeed if gases were allowed to expand, then compressed under conditions which did not allow them to regain the lost heat, and expanded once more, and so on over and over in cascade, then very low temperatures could be achieved. This Joule-Thomson effect—which gave rise to a whole new refrigeration industry aimed at the long-term conservation of perishable foodstuffs, dominated by industrialists such as the German Karl Ritter von Linde (1842-1934) and the French Georges Claude (1870-1960)—was utilized to reach temperatures never before obtained.

In 1883, Zygmunt Florenty Wroblewski (1845-1888) and Karol Stanislav Olszewski (1846-1915) were, however, the first to maintain a temperature so cold that it liquefied a substantial quantity of nitrogen and oxygen, said until then to be "permanent" gases. Fifteen years later, the Scottish physicist James Dewar (1842-1923) was able to liquefy hydrogen by first cooling the gas with liquid oxygen—kept at its low temperature with a Dewar flask, the first vacuum, or thermos, bottle ever made—then applying the aforementioned cascade method. At the turn of the twentieth century only the last so-called permanent gas, helium, still eluded liquefaction.

This achievement was to be the work of the Dutch experimental physicist Heike Kamerlingh Onnes. After studying physics in the Netherlands and Germany, he started his academic career as an assistant in a polytechnic school at Delft. It took only a few years before he received a call from Leiden University, which resulted in his appointment to the very first chair of experimental physics in the Netherlands. Kamerlingh Onnes's inaugural address leaves no doubt regarding the impetus he wanted to give to his laboratory: "In my opinion it is necessary that in the experimental study of physics the striving for quantitative research, which means for the tracing of measure relations in the phenomena, must be in the foreground. I should like to write *Door meten tot weten* (knowledge through measurement) as a motto above each physics laboratory." He always remained loyal to this declaration of principle.

It took more than 20 years for Kamerlingh Onnes to build and establish on a firm ground a cryogenic laboratory of international renown. The laboratory workshops were organized as a school, the Leidsche Instrumentmakers School; they were to have a tremendous importance in the training of qualified instrument makers, glassblowers, and glass polishers in the Netherlands. Even though he confided in his measurement aphorism, Kamerlingh Onnes's research was nevertheless upheld by a solid theoretical background ascribed to a couple of brilliant Dutch contemporaries, Johannes Diderik van der Waals (1837-1923) and Hendrik Antoon Lorentz (1853-1928). Their theories helped him understand the physics involved in the liquefaction of gases.

In 1908 Kamerlingh Onnes's efforts resulted in the liquefaction of helium, obtained at the very low temperature of 4.2 Kelvin or -451.84°F (the Kelvin absolute temperature scale, as you have probably guessed by now, was named after William Thomson, Lord Kelvin, who was the first to propose it in 1848). From then on and until his retirement in 1923, Kamerlingh Onnes would remain the world's absolute monarch of low temperature physics.

Impact

The coldest spot on Earth was now found in Leiden. By attaining this new level of temperature Kamerlingh Onnes set the stage for his next big, and probably most important, discovery. Studying the electrical resistance of metals submitted to low temperature, the Dutch physicist expected that after reaching a minimum value, the resistance would increase to infinity as electrons condensed on the metal atoms, thus impinging their movement. Experimental results, though, contradicted his claim.

Kamerlingh Onnes supposed next—based on Max Planck's (1858-1947) hypothesized vibrators used to theoretically explain the blackbody, giving birth to the quantum concept—that the resistance would decrease to zero. Using purified mercury, he found out what he had anticipated: at very low temperature electrical resistance showed a continuous decrease to zero. Superconductivity was discovered. The year was 1911. When Kamerlingh Onnes received the Nobel Prize two years later, it was for his *œuvre complète* in low temperature physics, which of course led to the production of liquid helium. But what about superconductivity? Was it ignored? In a sense, yes.

In the early 1910s this phenomenon was considered to be some sort of "peculiar oddity," for it could not yet be theoretically understood, much less used to practical ends. The reason is really quite easy to grasp when you look back at history from our modern point of view: the theoretical foundation of superconductivity is quantum mechanics, still at an embryonic stage of development when Kamerlingh Onnes discovered the empirical properties of superconductors.

From then on, the quest for absolute zero began. Large electromagnets were built in order to reach that temperature where every molecular movement stops. It became a matter of national pride to be able to say that the coldest spot on Earth was on your territory. The successor of Kamerlingh Onnes used such an electromagnet, in 1935, to achieve a temperature of only a few thousandths of a degree Kelvin. Leiden's star shined again. But astonishingly new phenomena did not always require temperatures so extreme. Since the 1920s it was shown that at 2.17K (achieved by applying moderate pressure) liquid helium (He I) changed into an unusual form, named He II.

In 1938 Pjotr Leonidovich Kapitza (1894-1984) demonstrated that He II had such great internal mobility and near vanishing viscosity, that it could better be characterized as a "superfluid." Kapitza's experiments indicated that He II is in a *macroscopic* quantum state, and that it is therefore a "quantum fluid." It now was indisputable to ascertain that low temperature physics rested on the principles of quantum mechanics. Superconductivity thus had to be tackled with this understanding in mind.

It took, however, no less than 46 years before John Bardeen (1908-1991), Leon N. Cooper (1930-), and J. Robert Schrieffer (1931-) finally found the underlying mechanism to Kamerlingh Onnes's discovery. Nicknamed the BCS theory, it can be theoretically outlined as the coupling of electrons (called *Cooper pairs*) attuned to the inner vibrations of the superconductor's crystal lattice. As the first electron in the pair flows through the lattice, it attracts toward the positively charged nuclei of the superconductor's atoms. The second electron is then "pulled" forward because it feels the attraction engendered by those same nuclei in front. The Cooper pair of electrons thus stay together as they flow through the superconductor, an unbroken interaction which helps them progress without resistance through the superconductive material.

One of the things that the BCS theory predicted was the superfluidity of the helium-3 isotope. Lev Davidovic Landau (1908-1968) theoretically explained the superfluidity of helium-4 (He II) already in the 1940s. Helium-4 is said to be a boson since each atom has an even number of particles (two protons, two neutrons, and two electrons). Helium-4, as Landau showed, must then follow *Bose-Einstein statistics*, which, among other things, means that under certain circumstances the bosons condense in the state that possesses the least energy.

Helium-3, however, having one neutron less than helium-4 (and therefore an odd number of particles), is not a boson but a fermion. Since fermions follow *Fermi-Dirac statistics* they cannot according to this theory be condensed to the lowest energy state. For this reason superfluidity should not be possible in helium-3—which, like helium-4, can be liquefied at a temperature of some degrees above absolute zero. Three Americans discovered at the beginning of the 1970s, in the low temperature laboratory at Cornell University, the superfluidity of helium-3, something that occurs at a temperature of only about two thousandths of a degree above absolute zero.

Where do all these theories and experimental facts lead? Up until 1986 the highest temperature superconductors could operate was 23.2K. Since liquid helium (expensive and inefficient) is the only gas usable for cooling to that range of temperature, superconductors were just not practical. New superconductors were found after 1986 that are operated at 77K. This higher temperature allows the use of liquid nitrogen as a coolant, far less expensive and far more efficient than liquid helium. As electronics, the designs for superconductors went from refrigerators weighing hundreds of pounds, running at several kilowatts, to far smaller units that can weigh as little as a few ounces and run on just a few watts of electricity.

This breakthrough lead to a wider use of superconductors: they are now found in hospitals as magnetic resonance imaging (or MRI) machines, in the fields of high-energy physics and nuclear fusion, and finally in the study of new means of transportation, in the form of levitating trains. Furthermore, fuel cell vehicles, run by liquid hydrogen, could one day replace the petroleum motorized cars of today. Also, by studying the phase transitions to superfluidity in helium-3, scientists may have found a theoretical explanation of how cosmic strings are formed in the universe. In light of all this we may conclude, as the 1996 Nobel Prize laureate Robert C. Richardson (1937-) did 20 years ago, that the end of physics—viewed from the lenses of low temperature physics—is yet to be at our doors.

JEAN-FRANÇOIS GAUVIN

Further Reading

Books

Mendelssohn, Kurt. *The Quest for Absolute Zero.* 2nd ed. London: Taylor & Francis; New York: Wiley, 1977.

Schechter, Bruce. *The Path of No Resistance: The Story of the Revolution in Superconductivity.* New York: Simon & Schuster, 1989.

Van den Handel, J. "Heike Kamerlingh Onnes" in *Dictionary of Scientific Biography,* edited by Charles C. Gillispie. New York: Scribner, 1973, 7: 220-22.

Vidali, Gianfranco. *Superconductivity: The Next Revolution?* New York: Cambridge University Press, 1993.

Periodical Articles

Richardson, Robert C. "Low Temperature Science—What Remains for the Physicist?" *Physics Today* 34 (August 1981): 46-51.

Models of the Atom

Overview

Until the early 1900s, the laws of classical physics, established in the seventeenth century by Isaac Newton (1642-1727) and in the nineteenth by James Clerk Maxwell (1831-1879), described the behavior of objects in the everyday world well. But investigations into the structure of the atom began to turn up strange phenomena that the Newtonian picture of the universe could not explain. The succession of atomic models put forth in the first part of the twentieth century reflect the attempts of scientists to understand and predict the weird behavior observed at tiny scales.

Background

It was the Greek philosopher Democritus who in 460 B.C. wondered what the smallest possible particles of matter might be and called them "atoms." At the beginning of the twentieth century, however, no one had yet proved that atoms really existed. All of that changed when a Swiss patent clerk named Albert Einstein (1879-1955) published a theoretical paper that established the existence of atoms once and for all.

In 1827, peering into his microscope, the Scottish botanist Robert Brown (1773-1858) noticed that a grain of pollen floating in a drop of water jiggled about in a random way. In 1905 Einstein developed a mathematical formula for this so-called "Brownian motion," arguing conclusively that it was due to the pollen grains colliding with unseen atoms. His conclusions were confirmed several years later by French physicist Jean Perrin (1870-1942).

Even before Einstein published his landmark paper, an English physicist named Joseph John (J. J.) Thomson (1856-1940) had proposed a model of what the atom might look like. While studying electric discharges in gases, Thomson discovered that rays emanating from a negative source of electricity (a cathode) were composed of negatively charged particles. Thomson concluded from measurements he made that the particles had to be fundamental constituents of the basic building blocks of matter. He called the particles "corpuscles"—later to be known as "electrons." Because atoms are electrically neutral, they had to have parts that were positively charged to balance the negative charge of the electrons. In 1904 Thomson envisaged the atom as a relatively large sphere with the positive charge spread throughout and tiny electrons embedded here and there, like raisins in a pudding.

The next development in unlocking the structure of the atom was an offshoot of French physicist Henri Becquerel's (1852-1908) discovery in 1896 that a uranium coating on a photographic plate emitted rays spontaneously, even in the absence of light. In 1902 Ernest Rutherford (1871-1937), a New Zealander who had worked with Thomson in the 1890s, showed that radioactivity results when an unstable element such as uranium is transformed into another element. Rutherford discovered that this transformation, called "radioactive decay," produced three kinds of particles that he called "alpha," "beta," and "gamma." The beta particles were soon shown to be electrons. Alpha particles, however, turned out to have a mass about four times that of a hydrogen atom and an electric charge twice that of the electron, only positive. Rutherford used the alpha particles to probe the structure of matter. He fired alpha particles at gold beaten into a thin foil. Most of the particles passed through the foil, but some bounced back. This result suggested to Rutherford that the particles were encountering bits of positively charged matter.

Ernest Rutherford. *(Library of Congress. Reproduced with permission.)*

More importantly, his findings meant that Thomson's model of diffuse positive charge throughout an atom needed revising (otherwise all the particles would have gone through the foil, not just some). Rutherford concluded that the positive charge in the atom was not spread out as Thomson thought, but concentrated in a tiny part that Rutherford called the nucleus. He imagined that negatively charged electrons occupied a spherical region surrounding the nucleus, orbiting the center as the planets do the sun. The electron sphere is 10,000 times larger than the nucleus, yet almost all the mass of an atom is concentrated in the nucleus.

No sooner had Rutherford proposed his model than it raised a question. Since opposite charges attract, according to Maxwell's equations, orbiting electrons should continuously lose energy, eventually spiraling into the nucleus. Moreover, they should give off a rainbow of colors. Yet neither of these predictions was confirmed by experiments. The answer, when it came, was totally unexpected.

In 1900 the German theoretical physicist Max Planck (1858-1947) was trying to solve a puzzle in physics called the "blackbody problem." The problem stemmed from the inability of Maxwell's theory to explain fully how color changes in objects as they get hot (for example, the way a poker stuck into a fire glows red- or white-hot). Acting out of desperation, Planck proposed that hot bodies do not emit energy in a continuous wave, as classical physics predicts, but instead in discrete lumps or packets called "quanta." Planck announced his solution at a meeting of the Berlin Physical Society in October that same year, but the reception to it was lukewarm. The theory gained some respectability five years later when Einstein used it to explain why, under certain conditions, light shining onto a metal surface in a vacuum can make electrons jump out of the metal.

The young Danish physicist Niels Bohr (1885-1962) introduced Planck's quantum theory into the theory of the atom in 1912. Bohr kept the image of the atom as a miniature solar system but explained that electrons did not spiral out of orbit into the nucleus because, first, they could only orbit at certain allowed distances from the nucleus and, second, they could jump up or down from one energy level to another but could not radiate energy continuously. (Picture the difference between a stone resting on a staircase and one poised on a ramp.) This "quantized" model of the atom also solved the riddle of why chemical elements emit light of only a few colors, a spectrum that is unique for each element. The appeal of Bohr's model was its simplicity, but it could not attack all the problems posed by quantum physics. Bohr continued to tinker with his model to bring it into closer agreement with experimental observations.

In 1927 German theoretical physicist Werner Heisenberg's (1901-1976) well-known contribution, known as the "uncertainty principle," recognized a fixed limit as to how precisely we might hope to know the world of tiny particles. Because any measurement affects what is being measured, we cannot know at any one time both the position and the velocity of a particle. Erwin Schrödinger (1887-1961), an Austrian physicist, further explained that the visual image conjured up by the description of particles as occupying positions and moving with velocities is misleading, since the distinction between wave and particle that we make from our experience in the ordinary world is itself illusory in the realm of subatomic phenomena. The implication of Heisenberg's and Schrödinger's physics to how we might picture the atom was not to try.

It was not until 1932 that the English physicist James Chadwick (1891-1974) discovered the neutron, a neutrally charged particle, proving correct an earlier speculation by Rutherford. Together, neutrons and protons make up the nucleus of an atom. As the middle of the century approached, subsequent investigations into the nature of the atom concentrated on opposite types of matter (for example, anti-electrons, antineutrons, and antiprotons, known collectively as "antimatter") and the forces at work within the nucleus.

Impact

Of such moment was the work, both experimental and theoretical, that answered the question "What are the parts of the atom and how do they fit together?" that every person mentioned in this article—with the exception of Democritus, Newton, Maxwell, and Brown—won a Nobel Prize. (The Nobel Prize began in 1901 and is not awarded posthumously.) For centuries before Thomson discovered the electron, atoms were believed to be the basic building blocks of matter, yet Thomson found a substructure that was 2,000 times lighter than the atom. The discovery of the nucleus was likewise revolutionary because it asked why negatively charged electrons do not fall into the positively charged nucleus, one of many questions about the atom that classical physics could not answer.

In science, theory and observation have to fit. When we observe things happen that we did not expect, the result is a kind of crisis. Quantum mechanics (the name given to the laws that followed from quantum theory) was a solution to the crisis brought on by the failure of classical physics to explain atomic phenomena. The planetary model of the atom provides a sense of mastery and of continuity from the very large to the very small. Imagining atoms as solar systems or as billiard balls smacking into one another has the comfortable feel of the familiar. In providing a model of the atom that is almost impossible to visualize, quantum mechanics takes away that sense of the familiar. The world is thus not so manageable as we had thought.

GISELLE WEISS

Further Reading

Baeyer, Hans Christian von. *Taming the Atom: The Emergence of the Visible Microworld.* New York: Random House, 1992.

Gonick, Larry. *The Cartoon Guide to Physics.* New York: HarperPerennial, 1990.

Gribbin, John. *In Search of Schrödinger's Cat: Quantum Physics and Reality.* London: Corgi, 1984.

Models of the Atom. http://library.thinkquest.org/28582/models/

Schrödinger, Erwin. *Science, Theory, and Man.* London: Dover, 1957.

Enrico Fermi Builds the First Nuclear Reactor

Overview

On December 2, 1942, a group led by Enrico Fermi (1901-1954) built and started up the world's first man-made nuclear reactor. This was a test to prove that a nuclear reactor could be built, and it paved the way initially for nuclear reactors that would produce plutonium for the atomic weapon to be dropped on Nagasaki, Japan, just three years later. In the following decades, more nuclear reactors would follow, producing plutonium and tritium for more advanced nuclear weapons as well as electrical energy for tens of millions of people, radioactive pharmaceuticals for medical diagnoses and treatment, and much more.

Background

The first known nuclear reactor actually goes much further back than 1942. In fact, nearly two billion years ago in what was to become the nation of Gabon in West Africa, a lucky set of circumstances led to the formation of a natural nuclear reactor in a bed of rich uranium ore. Far from the precisely engineered machines that are built today, the Gabon reactor (also called the Oklo reactor for the part of Gabon in which it lies) was a legitimate nuclear reactor that seems to have operated intermittently for several tens of millions of years.

Prior to the twentieth century atoms were thought to be indivisible. In fact, the word itself comes from the Greek *a* (meaning not) and *tomos* (meaning divisible). The latter word is also the root for the word "tome," as in microtome (a device to cut thin slices for microscopy). In investigations beginning with Henri Becquerel's (1852-1908) discovery of radioactive decay in 1896, many researchers eventually found that atoms were not indivisible, but consisted of smaller parts. James Chadwick's (1891-1974) discovery of the neutron in 1932 prepared the way for Fermi's 1934 proposal that neutrons could be used to split atoms. In 1938, the German physicists Otto Hahn (1879-1968) and Lise Meitner (1878-1968) became the first to knowingly cause uranium atoms to fission (Fermi had accomplished this a few years earlier, but thought he had instead created transuranic elements). In the process, scientists began to realize that an enormous amount of energy lay stored in the atomic nucleus. The secret to this was what may be the most famous equation in history: $E = mc^2$.

What researchers realized in the 1930s is that, when you split a uranium atom, the mass of the fission fragments is a little less than the mass of the original atom. The extra mass, when put into Albert Einstein's (1879-1955) equation, turned out to be exactly equal to the amount of energy released in a nuclear fission reaction. In fact, converting one atomic mass unit (about the mass of a neutron or proton) releases nearly 1000 MeV (or million electron volts), a tremendous amount of energy on the atomic scale. In effect, "burning" this very small amount of mass into energy gives us nuclear power.

In 1933, in the midst of these exciting discoveries, the Hungarian expatriate physicist Leo Szilard (1898-1964) realized in a flash of inspiration that it could be possible to create a self-sustaining nuclear chain reaction. As Szilard described it, he was in the process of starting to cross a street when he suddenly realized how a chain reaction would work. Realizing that a neutron is required to cause an atom to fission, or split apart, Szilard understood that any nuclear reaction would eventually die out because not all neutrons would cause fission. However, if an isotope could be found that released more than one neutron, it might be possible to have fissions go on indefinitely. This is precisely the principle behind a nuclear reactor—in a "critical" configuration, exactly the same number of neutrons are produced by fission as are lost, so the total number of neutrons in the nuclear reactor remains constant. Thus, all nuclear reactors are "critical" as a matter of course. A nuclear reactor that is not critical is simply not operating. Such a chain reaction was first shown experimentally in 1939 by French physicists Irène and Frédéric Joliot-Curie (1897-1956 and 1900-1958).

In 1941, Albert Einstein, at the behest of several concerned physicists, helped write and sign a letter to President Franklin Roosevelt, urging him to initiate a project to develop the atomic bomb before Nazi Germany could do so. At that time, the Nazi bomb was considered a very real possibility because of the large number of truly outstanding physicists and engineers in Germany at that time. This letter helped convince Roosevelt to begin the Manhattan Project, resulting in Fermi's reactor, several plutonium production reactors at Oak Ridge, Tennessee, and Hanford, Washington, and, of course, the first atomic weapons. Fermi's reactor, a pile of graphite blocks with precisely placed uranium spheres, was built in a squash court beneath the football stands at the University of Chicago. Assembled by hand, crude, dirty, with manual controls, it generated about enough energy to have lit a very small electric light bulb, if any way had existed to extract that energy. The reactor was more or less egg-shaped, 22 feet (6.7 m) high, 26 feet (7.9 m) across, and contained 6 tons of uranium oxide encased in 250 tons of graphite.

Impact

The most immediate and obvious impacts of Fermi's nuclear reactor was that it allowed the Manhattan Project to go forward and build a working atomic bomb. It made possible the plutonium production reactors and verified the physics calculations that bomb designers were using. Had the pile not gone critical at the predicted configuration, some minor recalculations would have been necessary. Had it not gone critical at all, it is likely that no nuclear weapon would have been built. Although this is potentially the most significant impact of all, it is discussed in greater detail in another essay and will not be discussed further here.

The next most apparent impact of this first artificial nuclear reactor is, of course, the construction of large civilian and military nuclear reactors, used for a variety of purposes. Military nuclear reactors are primarily used for propelling submarines and large surface ships, although several early attempts were made to develop small, portable nuclear reactors that could be used by Army units to produce power or desalinate seawater. There has also been a limited use of nuclear reactors in spacecraft. This should be clearly differentiated from the use of radioactive materials, including plutonium, in radioisotopic thermal generators (RTGs), such as those used in the *Galileo* and *Cassini* spacecraft. RTGs make use of heat generated via radioactive decay to produce electricity, but no nuclear chain reactions ever occur.

The civilian use of nuclear power is more varied, encompassing nuclear power reactors, research reactors, and isotope production reactors. Nuclear power reactors, of course, generate electrical energy for use by communities and industry alike. While the use of nuclear power for electrical energy generation is hotly debated in most nations with such reactors, several nations rely quite heavily on them for energy. These nations include Japan, France, the United States, and several republics that were formerly part of the Soviet Union. China began a massive program of nuclear reactor construction in the late 1990s in an effort to bring electrical power to most of its population, as have some other countries.

Nuclear reactors are also used extensively to manufacture radioactive isotopes for use in research and medical treatment. A very large fraction of genetic sequencing studies depend on radioactive labels attached to DNA in the lab, for example, and without radioactive materials, many of which are manufactured in nuclear reactors, much research in the medical, biological, and biotech arenas would stop. Nuclear medicine and radiation therapy also depend heavily on isotopes produced in nuclear reactors to help diagnose and treat a variety of illnesses, including many forms of cancer.

Finally, nuclear reactors are used for a variety of other research projects. For example, by bombarding a geologic specimen with neutrons in the core of a small research reactor, one can determine its chemical composition using what is called neutron activation analysis. This makes use of the fact that, by absorbing a neutron, a stable atom can be transformed into one that is radioactive. Identifying the new radioactive isotopes provides enough information to determine which elements gave rise to them. In other types of geologic work, nuclear reactors can be used to help determine the age of rocks. These dating methods, called Argon-Argon dating and fission track dating, use very different properties of atoms subjected to a neutron flux, and are both powerful and accurate ways to gain valuable geologic information.

Finally, the impact of nuclear reactors on the environmental movement has been profound. Or, perhaps a better way to put it is that the nuclear power industry has generated a long-lasting and often acrimonious debate about its potential environmental impacts, a debate that was fueled by the accidents at Three Mile Island and Chernobyl.

In a sense, the impact of these accidents is out of proportion compared to the risk actually posed by the nuclear power industry. At Three Mile Island, although the nuclear reactor was destroyed, it released only a relatively small amount of radioactivity to the environment and, in fact, nobody offsite received any more radiation than they would have from a typical series of x rays. Chernobyl, while a serious accident (although NOT a nuclear explosion, which is physically impossible in a civilian nuclear reactor), happened in a reactor plant design that is not allowed to be built anywhere except in Russia and other former Soviet Union states, making a recurrence unlikely. What contributed to the severity of the Chernobyl accident was the absence of an outer "containment" structure to keep radioactive materials from entering the environment. Such structures are required on all nuclear reactors in other nations. Therefore, holding Chernobyl up as an example of what could occur in a Canadian or French nuclear power plant is not a valid comparison. This is not to minimize these accidents, just to point out that the debate over the environmental effects of nuclear power may be largely built on misunderstanding or misinterpretation of the facts that are available.

P. ANDREW KARAM

Further Reading

Books

Rhodes, Richard. *The Making of the Atomic Bomb.* New York: Simon & Schuster, 1986.

Nero. *The Nuclear Reactor Guidebook.* University of California Press, 1976.

Internet Sites

The American Nuclear Society. http://www.ans.org

The Department of Energy. http://doe.gov

The Nuclear Regulatory Commission. http://www.nrc.gov

Frederick Kipping Develops Silicones

Overview

Silicones are a class of mixed inorganic/organic polymers developed in the early twentieth century. Since their initial discovery, they have been investigated and their uses expanded greatly because of their relative chemical inertness and their tolerance to a wide range of temperatures and environmental conditions. This versatility has made them useful in lubricants, as synthetic rubber, as water repellents, and a number of other uses.

Background

Polymers are large molecules that consist of a large number of connected smaller molecules called monomers. Monomers link together in long chains of tens or hundreds of thousands of units, forming the polymers. Polymers are common in nature. Cellulose, lignin, proteins, and other important biological molecules are all polymers. Perhaps the best known and most important biological polymer is DNA, which consists of a very long sequence of bases (the monomers) that, chemically, are very similar to one another. Adolf Spittler created the first artificial polymer in 1897, probably by accident. The first commercially successful polymer, Bakelite, was developed in 1909 by Leo Baekeland (1863-1944).

Most polymers contain carbon as their chemical "backbone," making them organic molecules. Some, however, contain silicon as the primary atom, linked to oxygen with carbon groups attached to this backbone. Such organic/inorganic polymers are called silicones. In the early twentieth century, English chemist Frederic Kipping (1863-1949) developed and investigated silicones. Kipping, under the misapprehension that he had developed ketones (a type of organic, biological molecule) with silicone substituting for one of the primary carbon atoms, called his discovery "silicone" (as opposed to silicon, the element).

Over the next 45 years, Kipping continued his researches into the properties of what became an entire class of polymers, publishing over 50 papers on the subject. During World War II, silicones became important to the war effort, used as lubricants, synthetic substitutes for rubber (much of which was produced on plantations now controlled by the Japanese), and as a water-proofing agent.

In many ways, silicones were nearly ideal as polymers. Like plastics, they could be molded into nearly any shape, making them ideal for a wide variety of uses. Unlike most plastics, they were very resistant to the effects of heat and cold, giving them a far greater versatility and value to industry. In addition, being largely inert chemically, silicones were ideal for many applications in both the chemical industry and for use in surgical procedures. Unfortunately, the chemical bond between the silicon and oxygen atoms that comprises the polymer's backbone is susceptible to some chemical attacks, including hydrolysis, acids, and bases, keeping these compounds from being universally used. However, even with these few minor weaknesses, silicones remain supremely useful in many facets of industry and modern society.

Impact

It is difficult to overstate the impact that the family of silicones has had on our society. Silicone rubber, for example, provides a soft cushion for one's eyeglasses on the bridge of the nose and, in the insole of many athletic shoes, provides an equally soft cushion between a runner's heel and the pavement. Silicone lubricants, including the ubiquitous WD-40, keep bicycles running smoothly, keep doors opening quietly, and keep engines and gears operating efficiently for years at a time. As a hydraulic fluid, silicone oils provide reliable service under the incredible variety of conditions found in military and commercial aircraft, nuclear submarines, and the family automobile. Still other compounds are used as caulking or other sealants to make showers, windows, piping systems, and boats watertight. Other uses include electrical insulation, flexible molds, heat-resistant seals, and as surgical implants (most notoriously as silicone breast implants that resulted in numerous legal suits in the 1990s, but also for plastic surgery and other surgical uses).

All of these uses take advantage of one or more of the silicones' chief chemical advantages: resiliency over a wide range of temperatures, physical and chemical versatility, and chemical inertness. The impact of these compounds, then, will be discussed in terms of each of these characteristics and the fields in which they occur.

As mentioned above, silicone compounds retain their physical properties over a wide range of temperatures. This makes them ideal for low- and high-temperature applications because, unlike conventional polymers, they remain flexible at low temperatures and they refuse to soften or melt in the heat. Take, for example, a high-performance aircraft. High-altitude air temperatures are typically far below freezing. Unlike conventional compounds, silicone grease doesn't congeal at these temperatures and silicone hydraulic fluids become only marginally more viscous. So, at relatively low speeds at high altitudes, the aircraft's control surfaces work properly. At high, supersonic speeds, the airplane will heat up considerably because of friction against the air. While most aircraft do not become red-hot, skin temperatures can rise to a few hundred degrees. As temperatures rise, the silicone grease stays semi-solid, remaining on the components it is lubricating, instead of melting and dripping off as would be the case with conventional greases. The hydraulic fluid, too, retains its viscosity, giving a uniform feel and performance to the hydraulic system (which operates the control surfaces and landing gear) under a wide range of operating conditions. Silicone compounds are also used as grease to pack pump bearings on ships that might have to operate in the sweltering equatorial waters or the below-freezing waters of polar seas.

Silicone rubbers share this relative insensitivity to temperature. Obviously, at sufficiently high temperatures any compound will melt, just as any liquid will eventually solidify. However, silicone rubber will remain useable far longer than will conventional plastics, not melting, softening, or losing its strength at temperatures of a few hundred degrees. Interestingly, this same resiliency has helped make silicones ideal for some commercial applications. For example, silicone rubber shoe inserts not only cushion the impact of one's heel against the ground, but it does so reliably, for years, without needing replacement, under conditions ranging from a midsummer's casual walk to a runner's January run. These same properties, plus their resistance to the chemicals found in perspiration and skin oils, make silicone rubber one of the favorite compounds for the nosepieces on eyeglasses.

Silicones are a chemically and physically diverse group of materials, giving them even more versatility. This family of compounds can be formulated into liquids, solids, or gels. Within each of these families, compositions can vary as well, allowing a specific product to be fine-tuned for its proposed use. Part of the reason for this versatility lies with the chemical properties of silicon, which shares many chemical properties with carbon because it lies directly beneath carbon in the Periodic Table. Like carbon, which can also assume many chemical guises, silicon can be persuaded to take on a variety of chemical bonds of different strengths, depending on which atoms it is bonded with. This means that, in some circumstances, it can be strongly linked to oxygen, making a solid material with great strength. In other circumstances, it is more loosely bonded with oxygen or even more loosely bonded with both oxygen and carbon groups, yielding a silicone grease or oil that is liquid at room temperatures. By controlling the nature of these silicon bonds and the number of silicon-based monomers that link together (or polymerize), a chemist can have a great deal of control over the final physical and chemical properties of the resulting compound.

Finally, silicone is relatively chemically inert, allowing it to be used in a wide variety of situations in which this is desirable. For example, regardless of lubricating abilities, one would not pack ball bearings in an acid because the acid would etch or corrode the bearings. The ideal lubricants are not only slippery, but are chemically unlikely to attack the surfaces they lubricate. Similarly, bodily fluids are a hostile environment for many materials since they are typically salty and replete with chemicals. Ordinary steels corrode quickly, and even many stainless steels corrode over time. Chemically reactive polymers or other compounds used in the body are also likely to either corrode or to initiate undesirable chemical reactions, leading to a breakdown of the implant, generation of unwanted chemical reaction products, or both. Many silicones manage to avoid these perils because they are not attacked by the body's chemistry and they are remarkably resistant to breakdown. This gives them widespread use in some surgical procedures in which such properties, combined with the physical properties noted above, are wanted. The use of silicone gel breast implants gained a high degree of notoriety in the 1990s because they were apparently linked with breakage and infection. However, scientific studies performed after most of these suits were settled indicated with a high degree of confidence that the silicone gel did not, in fact, cause the problems noted, even when the implants did rupture. It must also be noted that not all plastic surgeries are for the sake of vanity or aesthetics; in many cases, plastic surgery is performed to correct physical injuries from accidents or to repair birth defects.

It seems, then, that Kipping's discovery of silicones in 1904 was the impetus for a great many innovations that have had an impact on technology and society. Some of these impacts are directly noticed, such as silicone rubber shoe inserts, caulking for a leaking shower, and silicone water repellents for shoes and camping gear. Others, such as silicone-based greases and hydraulic fluids, materials for surgical implants, and the like, are not as obvious, but are equally important to society because of their role in our transportation systems, our self-image, and our commerce.

P. ANDREW KARAM

Further Reading

Books

Rochow, Eugene G. *Silicon and Silicones: About Stone-Age Tools, Antique Pottery, Modern Ceramics, Computers, Space Materials, and How They All Got That Way.* New York: Springer-Verlag, 1987.

The Advent and Use of Chlorination to Purify Water in Great Britain and the United States

Overview

Of all the conveniences of modern life, the availability of fresh, clean drinking water is perhaps the one taken most for granted. This luxury, however, was only realized around the turn of the twentieth century. In the nineteenth century, the average person in London might find tiny shrimp-like animals or putrid deposits in the drinking water. In Belgium, the water was often yellow and had an unpleasant odor. In 1844, only about 10% of the water from city fountains in Paris was potable. In Germany, the drinking water was brown and foul. In New York, residents complained about paying too much for their impure water. People all around the world were dying of diseases such as cholera and typhoid fever from the filthy water. By the early 1900s, however, this deplorable situation had been rectified. The chlorine industry had been born, and it revolutionized the world's water systems.

Background

Although the modern chlorine industry did not exist before 1900, chlorine's history dates back to 77 A.D., when the Roman scholar Pliny the

Elder's (23-79 A.D.) experiments produced hydrogen chloride. The Arabs, about 800 years later, added water to the gas to generate hydrochloric acid. Centuries later, in 1774, Swedish chemist Carl Wilhelm Scheele (1742-1786) was able to make and collect chlorine, and he accidentally discovered its bleaching capability. Five years later, in France, Claude Berthollet (1748-1822) made the first liquid chlorine bleach with potassium hypochlorite; it was called Javelle water, or *Eau de Javelle*, because it was produced in Javelle, France. In 1810, Sir Humphry Davy (1778-1829) proved that chlorine was an element, naming it after the Greek word *khloros*, for green.

Wars and revolutions in Europe postponed further advancement of chlorine research for several years. In 1820, French scientists recognized the disinfectant properties of sodium hypochlorite, and seven years later, Thomas Alcock wrote a paper that recommended using chlorites to disinfect hospitals, sewers, and water supplies. In 1894, German chemist Isidor Traube (1860-1943) demonstrated the ability of hypochlorites to disinfect water.

Water supply problems surfaced in the nineteenth century due to rapid industrialization and the concurrent population growth in major cities across Europe and the eastern United States. Suddenly, innumerable factories were in operation, using and polluting water supplies. Additionally, burgeoning cities depended on vastly inadequate sewer systems. Drinking water sources mingled with sewage and industrial waste. Unfortunately, it was not until after 1900 that experts agreed that the germs that caused serious infectious diseases like cholera and typhus flourished in water.

Before the turn of the century, people had often boiled water and then filtered it through substances such as gravel, sand, charcoal, and, later, special carbon filters. The ancient Greeks and Romans knew that boiling their drinking water made it safer. Over the centuries, people also used various metals, such as silver and copper, as disinfectants, but these elements did not consistently reduce bacteria count. Only chlorine proved to be an effective germicide. Consequently, chlorine eventually became the universal choice as a disinfectant.

By the mid-nineteenth century, a few water companies had turned to chlorine to disinfect sewage. In the United States and England, some water supplies were treated with chlorine. In 1888, Professor Albert Leeds obtained the first United States patent for water chlorination; his method combined electrolysis and hydrochloric acid. (Electrolysis refers to the use of an electric current to break up a compound by chemical conversion.) It was thought at the time that the electricity was disinfecting the water, but American William Jewell was convinced that chlorine alone could be a disinfectant agent. In 1896, Jewell used chlorine gas for the first time at a testing station in Kentucky.

Soon, permanent water chlorination plants opened around the world. The first was in 1902 in Middlekerke, Belgium, under the direction of Dr. Maurice Duyk. Polluted water was treated with chloride of lime and then filtered. American and English water experts followed the events in Belgium with enthusiasm, and chlorination soon spread to other countries.

In 1897, in Maidstone, England, Dr. Sims Woodhead had treated the water supply with bleaching powder after a typhoid epidemic there. However, influential men in England still refused to believe that routine chemical treatment of water was necessary to curb disease. An eminent British philosopher even considered compulsory sewage facilities to be an infringement of personal liberties! Finally, scientists were able to convince British medical authorities that chlorine disinfection would indeed curb the spread of serious waterborne diseases. In 1905, the world's second permanent chlorination plant opened in Lincoln, England. There, Dr. Alexander Houston and Dr. McGowan directed an operation that added sodium hypochlorite to water that was then filtered through sand. Eleven years later, the city of London added bleaching powder to the Thames River before its waters flowed into the main aqueduct and were filtered for consumption. Great Britain had at last accepted chlorination as the best way to disinfect its drinking water and sewage waste.

The path to permanent chlorination of water in America began in 1908. At that time, chlorine, in the form of hypochlorites, was used at two separate sites: one was a reservoir in New Jersey, and the other was at the water filters of the stockyards in Chicago. The first American city to install permanent chlorination was Poughkeepsie, New York, in 1909. There, George C. Whipple advised using chloride of lime to treat the water of the Hudson River, after filtration alone did not render the water potable.

Americans gradually turned to liquid chlorine as an alternative to hypochlorites. In 1908, Philadelphia had also decided to use sodium hypochlorite to disinfect its water. Five years later, in 1913, the first permanent liquid chlorination plant was opened in Philadelphia at the city's Belmont filters. Afterwards, chlorination plants spread quickly throughout the industrialized cities on the East coast. Wherever drinking water was chlorinated, typhoid death rates fell by up to 75%. By 1918, more than 1,000 cities in the United States were using chlorine to disinfect their water.

France soon also adopted chlorination methods. The French called the process *javellization* after the town of Javelle, where liquid chlorine bleach was first made in 1779. Some French cities preferred chlorine gas, while others chose hypochlorite to treat their water supplies. France, along with Britain and America, finally succeeded in obtaining drinking water that was clean, clear, and disease-free.

Impact

The role of chlorine as a disinfectant continued to grow throughout the twentieth century. Medical experts learned that the chemical was a powerful weapon that killed both bacteria and viruses by attacking the nucleic acid of bacteria and the protein coat of viruses. Typhus and cholera gradually disappeared as drinking water was chlorinated, vaccinations were developed, and cleanliness standards improved. Nevertheless, the battle to control waterborne diseases continues to the present day. In the overcrowded conditions that arise from wars and natural disasters, especially in third-world countries, people are threatened by outbreaks of cholera and typhus. Wherever and whenever sanitation levels drop, international agencies turn to chlorine to disinfect the water and thus prevent the spread of deadly diseases.

Chlorine has many other uses besides water purification. Of the 15 million tons of chlorine produced in North America alone, only 5% is used to disinfect water. More than 50% of the chlorine produced is used to manufacture other chemicals and plastics, and in products such as anesthetics, perfumes, detergents, paints, and aerospace products. In fact, NASA (National Aeronautics and Space Administration) used chlorine during its Apollo program to disinfect surfaces that might have come in contact with possibly dangerous organisms from space.

It is as a disinfectant, however, that chlorine is most commonly known. Chlorination has the added benefit of being both effective and inexpensive. About one United States dollar will pay for the treatment of one million gallons of water. In emergency situations, six drops of bleach will disinfect a gallon of dirty water in 30 minutes.

In addition to its use in purifying drinking water around the world, chlorine is also used to keep the water in swimming pools clean. Millions of households worldwide choose chlorine products to clean and disinfect their homes. Because sodium hypochlorite is scientifically recognized as perhaps the most effective weapon against dangerous bacteria and viruses, hospitals around the world use chlorine to disinfect instruments and working surfaces, and medical authorities acknowledge that bleach can protect against HIV (Human Immune Deficiency Virus) and Hepatitis B.

Although chlorine has many useful applications, it does have some limitations and drawbacks. For example, chlorine has not proven effective against Legionella, the bacteria that cause Legionnaire's Disease, a serious pneumonia-like infection. And chlorine cannot remove lead or pesticides from water. The most recent worry

CLEANING UP THE THAMES RIVER

By the mid-nineteenth century, London's Thames River was dangerously polluted by industrial waste and raw sewage. The filthy river water caused thousands to die annually from cholera, typhoid fever, and diphtheria. Summertime was especially miserable because the weather made the river's odor worse, and insects swarmed everywhere. Some members of Parliament actually fainted during the summer from the Thames' awful smell. The royal family also suffered. The windows of Buckingham Palace were kept closed because of the horrid stench and the flies. The drains of Windsor Castle were labeled "...more dangerous than a tropical jungle" by one biographer (Cecil Woodham-Smith, *Queen Victoria*, 1972, p. 430). Prince Albert, Queen Victoria's husband, had campaigned tirelessly but to no avail to have the Thames cleaned up. It is indeed ironic but not surprising that the Prince's untimely death in 1861 was attributed to typhoid fever. A few months after he died, the clean-up effort began. Fifteen years later, there was a sewage system in operation that allowed waste material to enter the Thames downstream of London.

ELLEN ELGHOBASHI

about chlorinated water is its link to cancer. When it was discovered in the 1990s that chlorinated water could form carcinogenic particles (cancer-causing particles) called THMs, or trihalomethanes, experts began searching for other ways to sterilize water, such as treating the water with ozone, ultraviolet radiation, and iodine. To date, however, these methods are either too complicated or too expensive.

Certain filtration devices can remove traces of chlorine from water. Activated carbon filters, although they do not effectively reduce bacteria, do remove carcinogenic matter. These carbon filters can also be installed below the sink or attached to the faucet. Reverse osmosis is a more expensive filtering system that uses a semipermeable membrane to remove undesirable particles. Many homes use a special filtration device to make their chlorinated drinking water safe.

Aware of the dangers posed by chlorinated water, scientists continue to search for an alternative. A Russian-developed method is an electrolytic water sterilization process that uses electricity to produce certain chemicals that purify the water. Perhaps this method, or another method of water disinfection, such as radiation, iodination, or ozonation, will one day replace chlorination. The only certain thing is that at the beginning of the twenty-first century, man is again facing the same challenge that confronted him 100 years ago: the search for entirely safe and pure drinking water.

ELLEN ELGHOBASHI

Further Reading

Books

Baker, M. N. *The Quest for Pure Water*, Vol. I. American Water Works Assoc., 1981.

Bennett, Daphne. *King without a Crown*. London: Heinmann, 1977.

Goubert, Jean-Pierre. *The Conquest of Water.* Oxford: Polity Press, 1986.

Race, Joseph. *Chlorination of Water.* London: John Wiley & Sons, Inc., 1918.

Taras, Michael J., ed. *The Quest for Pure Water*, Vol. II. American Water Works Assoc., 1981.

Woodham-Smith, Cecil. *Queen Victoria*. New York: Alfred Knopf, 1972.

Internet Sites

http://www.sgn.com/invent/extra/042395uk.html

http://www.esemag.com/0596/bleach.html

http://ourworld.compuserve.com/homepages/ottaway/disinf.htm

http://ourworld.compuserve.com/homepages/ottaway/homepage.htm

http://members.aol.com/manbio999/chlorine.htm

http://www.cxychem.com/whatwedo/chlorine/Chlorine.html

http://www.cxychem.com/whatwedo/chlorine/EndUse.html

http://edis.ifas.ufl.edu/BODY_AE009

Wolfgang Pauli's Exclusion Principle

Overview

The Pauli exclusion principle enabled the quantum structure of the atom to be understood. It provided a mechanism to explain the variety and behavior of the chemical elements, and it gave a theoretical basis for the periodic table. It was one of the key pieces of the quantum puzzle and had far-reaching implications, many of which have yet to be fully understood.

Background

Wolfgang Pauli (1900-1958) was a child prodigy, excelling in classical history as well as mathematics and science. At the Döbling Gymnasium (a high school in Vienna) he was taught classical physics just when classical notions were under attack from new quantum ideas. He read widely on general relativity and became something of an authority on the subject, publishing three academic papers within a year of graduating high school.

Pauli went on to study at the University of Munich, under Arnold Sommerfield (1868-1951), who introduced him to the radical new quantum theory. As Pauli recalled years later, "I was spared the shock which every physicist accustomed to the classical way of thinking experienced when he came to know Niels Bohr's basic postulate of quantum theory for the first time." Quantum theory seemed to contradict common sense and was difficult to visualize. However, Pauli found himself devoted to the new field, particularly the partially understood structure of the quantum atom.

Ernest Rutherford's (1871-1937) experiments had determined that the structure of the atom consisted mainly of empty space. Using classical ideas, Rutherford's atom could be described as a small core, with the nucleus (imagine an orange in the middle of a football stadium) surrounded by tiny orbiting electrons (very fast insects somewhere outside the stadium in the parking lot!) Further work by others helped define the structure of the atom, giving positions for the orbits, and so on, using classical ideas such as treating the electrons as planets orbiting a central sun.

However, classical notions of orbits and particles became increasingly unable to explain a multitude of experimental results. Niels Bohr (1885-1962) combined the ideas of Albert Einstein (1879-1955) and Max Planck (1858-1947) and others to offer a new way of defining the atom. Bohr stated that an atom could exist only in a discrete set of stable energy states; that is, atomic energy was "quantized," not continuous. Quantum objects could have only certain energy values and could jump between these values, but the energy levels between these states could not be occupied. For atomic structure this implied that electrons could only exist in certain orbits, and while they might jump between these orbits they could not exist in the "space" between.

Bohr's ideas were expanded upon, and three values were postulated to define the energy state of electrons in an atom. These three numbers roughly correspond to the electron orbit (n), the angular momentum (l), and the magnetic quantum number (m or ml) of the electron. However, these numbers did not explain the atomic structure as completely as physicists would have liked.

In 1922 Bohr gave guest lectures in Göttingen, Germany, in which he dealt with the incomplete ideas of electron structure. Bohr was searching for a general explanation to the experimentally calculated electron orbits, and this quest was to deeply inspire at least one member of his audience, Wolfgang Pauli. "A new phase of my scientific life began when I first met Niels Bohr personally," Pauli noted later.

Pauli went to Copenhagen, Denmark, to work with Bohr, and found himself frustrated by an unexplained experimental result known as the Zeeman effect. In classical physics an electric current moving in a loop acts like a magnet, producing easily calculated magnetic properties. Therefore, it was assumed that electrons in orbit around the nucleus would produce the same classical effect. However, the experimental results did not conform to theoretical predictions.

Wolfgang Pauli. *(Library of Congress. Reproduced with permission.)*

Various experimental refinements were tried, but always the Zeeman effect baffled theorists. Pauli became obsessed with the problem. "A colleague who met me strolling rather aimlessly in the beautiful streets of Copenhagen said to me in a friendly manner, 'you look very unhappy'; whereupon I answered fiercely, 'How can one look happy when he is thinking about the anomalous Zeeman effect?'"

Leaving the puzzle unsolved, Pauli moved to Hamburg, Germany, where he gave his inaugural lecture on the periodic system of elements. The periodic nature of the elements had been recognized by Dmitry Mendeleyev (1834-1907) and published in 1869. Mendeleyev realized that the 63 elements then known had a repeating pattern—they could be grouped into families with similar chemical properties. For example, helium, neon, and argon are all inert gases; that is, they do not react readily with any other elements. He arranged the 63 elements into a table showing their periodic repetition, the periodic table. There were many gaps in Mendeleyev's table, but he was able to predict what type of elements would fit into these gaps. In his own lifetime he saw three new elements added into the spaces he had left, each showing properties he had foretold. However, while the periodic table was a success, there was no underlying explanation for it, which got Pauli thinking.

In 1924 Pauli read a paper discussing quantum numbers, and this gave the final piece to his fractured thoughts. He realized that four quantum numbers, rather than three, were needed to specify the exclusive state of each electron in the atom. This new number (ms) would be unusual, having a two-valuedness not describable in classical terms. With the fourth quantum number Pauli could now make a bold statement, the exclusion principle, that in a multi-electron atom there can never be more than one electron in the same quantum state. That is, a multi-electron atom cannot have the same values for the four quantum numbers.

Impact

Pauli announced his exclusion principle in 1925, but it was given a mixed reception. The fourth quantum number was an abstract concept; it did not seem to refer to any actual property of the electrons. The exclusion principle is a created rule based on the analysis of other peoples' experiments, not a derived property. As Pauli himself noted in 1945, he "was unable to give a logical reason for the exclusion principle or deduce it from more general assumptions." However, it worked, fitting both quantum theory and experimental results. Pauli responded to his critics by writing "my nonsense is conjugate to the nonsense which has been customary so far."

Eventually, the fourth quantum number was found to represent the "spin" of the electron, with a value of 1/2 for clockwise and -1/2 for anticlockwise spin. However, spin is a classical concept, and our common conception of it does not translate exactly into quantum reality. Bohr later showed that electron spin cannot be measure in classical terms. It is a quantum property and so is a little hard to comprehend.

In some ways it is ironic that the development and success of the exclusion principle relied so heavily on the experimental work of other scientists, as the theorist Pauli was actually feared by European experimenters. A legend built up around Pauli that his mere presence in a laboratory could ruin any experiment, often in bizarre and dangerous ways. Otto Stern (1888-1969) went so far as to ban Pauli outright from any of his labs. Pauli himself came to believe in what was dubbed "The Pauli Effect," even justifying it in terms of Jungian philosophy.

The exclusion principle applies not just to atomic structure but also to any system containing electrons. Electrons in a current obey the principle, as do electrons in chemical bonds. Not only electrons, but protons, neutrons, and a whole range of other particles that are collectively called fermions (named after Enrico Fermi [1901-1954]) obey the principle. There are other particles that do not follow the exclusion principle, called bosons (named after Satyenda Bose [1894-1974]).

Pauli's work had major implications for chemistry. Mendeleyev's periodic table could be shown to be a result of the internal structure of the atom, with the families of elements he defined each relating to the same arrangement of electrons within the outer orbit. Because they had the same outer electron arrangement, the elements within a family behaved the same way chemically. This explained why some elements were more reactive than others. The behavior of electrons in chemical bonding could now be defined. The exclusion principle also allows insight into the electrical conductivity of different substances: why, for instance, metals conduct better than semi-conductors, which in turn conduct better than insulators. Everything was explained by the extremely regulated "building" plan of quantum structure as defined by the exclusion principle, which resulted in Pauli being nicknamed the "atomic housing officer."

In 1945 Pauli was awarded the Nobel Prize in physics "for the discovery of the Exclusion Principle, also called the Pauli Principle." The lack of a valid passport and other planning difficulties stopped Pauli from collecting it personally at the awards banquet in Stockholm, Sweden. Instead, a dinner was given for him at Princeton University's Institute for Advanced Study, where to everyone's amazement, Einstein offered a toast in which he designated Pauli as his successor and spiritual son. Pauli was held in high regard in the scientific community, with some physicists going so far as to call him "the living conscience of theoretical physics."

Pauli's exclusion principle was announced in a year that also saw the announcements of Maurice De Broglie's (1875-1960) idea of matter waves and Werner Heisenberg's (1901-1976) matrix-mechanics, from which followed Erwin Schrödinger's (1887-1961) wave mechanics. It was an exciting time for physics, and Pauli's exclusion principle was one of the many important concepts that helped create modern quantum physics.

DAVID TULLOCH

Further Reading

Eisberg, Robert Martin. *Fundamentals of Modern Physics*. New York: Wiley, 1961.

Hey, Tony, and Patrick Walters. *The Quantum Universe*. London: Cambridge University Press, 1987.

Pauli, Wolfgang. *Writings on Physics and Philosophy*. Edited by Charles P. Enz and Karl von Meyenn. Berlin: Springer-Verlag, 1994.

Achievements by Indian Physical Scientists

Overview

In 1871 the Indian Association for the Cultivation of Science, the first of two centers for modern science in India, was founded to give young Indians the opportunity, discouraged by the colonial British government, to conduct laboratory research. The second center was a college of science established at the University of Calcutta by an amateur mathematician whose fundraising was so effective that he was able to endow two professorial chairs, in physics and chemistry, for qualified Indian scientists. In 1930 the first occupant of the chair in physics became the first Indian as well as the first Asian to win the Nobel Prize.

Background

Chandrasekhara Venkata Raman (1888-1970) received a master's degree in science and joined the Indian Finance Department in 1907. He took up a post in Calcutta and, discovering a local scientific association, began to conduct research outside of his working hours at the finance department. By the time he was granted a university position in physics ten years later, he had already published 25 papers in journals such as *Nature* and *Physical Review.*

As a boy, Raman had acquired a love of music, and his early research concentrated on vibrations and sound and the theory of musical instruments. But in the early 1920s, in an effort to find out why the sea is blue, Raman showed that its color is caused by molecular scattering of light by water molecules. He then embarked on a serious study of how light is scattered by liquids, solids, and gases.

Raman's best-known contribution to science was his discovery of what is now known as the "Raman effect." When a beam of light passes through a transparent material, a portion of the light emerges at right angles to the original direction. Some of this scattered light shows changes in wavelength. The reason, Raman explained, is that sometimes energy is exchanged between the light and the material it is traversing. Because "Raman scattering" provides information about the energy states of scattering material, it is used as a standard laboratory tool for investigating the chemical composition of substances. Raman's work on the scattering of light won him the Nobel Prize in physics in 1930.

Satyendra Nath Bose (1894-1974) began a mathematics career in Calcutta. In 1921 Bose left Calcutta to take up a position at the University of Dacca. Three years later, he sent a brief manuscript entitled "Planck's Law and the Hypothesis of Light Quanta" to Albert Einstein (1879-1955) for comment. The paper proposed a solution to a problem having to do with German physicist Max Planck's (1858-1947) theory of how energy is emitted by a hot object, the so-called "blackbody radiation." In 1900 Planck had proposed that energy exists in the form of little lumps, or "quanta," rather than in the form of a wave. Five years later, Einstein applied Planck's theory to the study of light. At the time, Europeans were slow to accept either idea, but Indian physicists took to them quickly. In a paper published in 1919 in an astrophysical journal, Bose's colleague Meghnad Saha (see below) used the light quantum to describe radiation pressure. That same year, Bose and Saha translated Einstein's papers on the general theory of relativity, in which Einstein questioned some of Planck's assumptions. Starting from the notion that particles of light obey statistical laws different from those that describe the everyday world, Bose was able to show that Einstein's model and Planck's law were consistent, in the bargain putting Planck's law on firm mathematical footing.

Einstein himself translated Bose's manuscript for publication in the *Zeitschrift für Physik*. Louis de Broglie (1892-1987), a French physicist, used Bose's new statistics—now known as "Bose-Einstein statistics"—to show that, just as light could behave as particles, particles (molecules) sometimes behaved as waves. The Austrian physicist Erwin Schrödinger (1887-1961) based his description of the quantum world on Bose's statistics.

Einstein's support for his work made it possible for Bose to obtain a two-year fellowship to Europe, where he studied radioactivity under Paul Langevin (1872-1946) in France. On his return to India, he turned to a variety of studies, for example, x-ray crystallography (a technique for determining molecular structure) and thermoluminescence, which refers to the light emitted when stored radiation energy is heated. But his major contribution to physics would remain his work on quantum statistics.

Meghnad Saha (1893-1956) was a classmate of Bose in Calcutta. An astrophysicist, he began his research in electromagnetic theory and the theory of stellar spectra. In 1919 Saha was awarded a two-year fellowship that took him first to the Royal College of Science in London and then to Berlin. He returned to India in 1921, where he assumed a succession of university positions.

The work for which Saha is best known is his thermal ionization equation. Stars emit a range of colors (spectra) that depend on the chemical composition of the light source. By linking information about a star's spectrum with the temperature of the light source, Saha's equation could be used to determine the temperature of the star and its chemical makeup. For example, cool stars typically show the presence of familiar metals such as iron and magnesium. The spectra of hotter stars, on the other hand, are consistent with elements that require more energy to produce, such as oxygen and carbon. Saha's theory has been called the starting point of modern astrophysics.

Subrahmanyan Chandrasekhar (1911-1995) was the nephew of Chandrasekhara Venkata Raman. Chandrasekhar left India at 19 to study astronomy and physics at Cambridge University. On the boat taking him to Europe, it occurred to him that the immense pressures on ordinary matter at the core of dying stars called white dwarfs would cause the stars to collapse in on themselves. At the time, it was believed that, after exhausting their fuel, these stars contracted under the influence of gravity to dense but stable remnants about the size of Earth. Chandrasekhar published his idea about collapsing stars in an astrophysical journal and in 1930 developed it more fully, using Einstein's special theory of relativity and the principles of quantum physics. He defined a certain maximum mass possible (1.44 times the mass of the Sun) for a stable white dwarf that became known as the "Chandrasekhar limit." Stars with a mass above this limit continue to collapse into enormously dense objects now known as neutron stars and black holes.

Chandrasekhara Raman. *(Library of Congress. Reproduced with permission.)*

Chandrasekhar presented his results at a meeting of the Royal Astronomical Society in January 1935, but his announcement was greeted with great skepticism by his English colleagues. In 1936 a disappointed Chandrasekhar left England for the United States, where he accepted a position at the University of Chicago. Today, the structures implied by his early work are a central part of the field of astrophysics. Chandrasekhar began his career working on the structure and evolution of stars. But his investigations ranged from the transfer of energy by radiation in stellar atmospheres to the mathematical theory of black holes to ruminations on truth and beauty in science. For his discovery of the Chandrasekhar limit, he was awarded a Nobel Prize in 1983.

Impact

Indian physical scientists in the first half of the twentieth century did more than make important new discoveries in science. They challenged an entire climate of opinion. Beginning in the eighteenth century and until its independence in 1947, India was a colony of Great Britain. One of the more unfortunate legacies of colonialism was the establishment of a social order under which colonials were treated as superior and native Indians inferior. Applied to education, this hierarchy meant that the intent of the colonial government in building schools, colleges, and universities was not to provide Indians a liberal education but to train them for subordinate civil-service positions. Occasionally, however, an exceptional student opted to pursue a scientific career. Raman, Bose, Saha, and Chandrasekhar were all products of India's determination to develop world-class scientists. The magnitude of the contributions of these scientific leaders upended colonial assumptions about Indians' capabilities and helped India move toward self-reliance.

In addition to carrying out pioneering studies, these four scientists helped establish the infrastructure for conducting basic scientific research in India. Saha created an institute of nuclear physics at the University of Calcutta. Raman, whose influence on the growth of science in India was so profound that he became a political and cultural hero, founded the Indian Academy of Science. All were ardent teachers. Late in his life, Raman stated that the principal function of older scientists was to recognize and encourage talent and genius in younger ones. One year, Chandrasekhar drove over a hundred miles each week to teach just two students, both of whom later won the Nobel Prize. Raman's simple light-scattering technique became a routine laboratory tool, Bose laid the foundation for quantum statistics, Saha's theory of ionization became integral to work on stellar atmospheres, and Chandrasekhar's theory on the evolution of stars furthered our understanding of the cosmos. By their achievements, these researchers both advanced international science and made a place for India in it.

GISELLE WEISS

Further Reading

Blanpied, William A. "Pioneer Scientists in Pre-Independence India." *Physics Today* (May 1986): 36-44.

Chandrasekhar, Subrahmanyan. *An Introduction to the Study of Stellar Evolution.* Chicago: University of Chicago Press, 1939.

Home, Dipankar, and John Gribbin. "The Man Who Chopped up Light." *New Scientist* (January 8, 1994): 26-29.

Raman, Chandrasekhara Venkata, and K. S. Krishnan. "A New Type of Secondary Radiation." *Nature* 121 (1928): 501.

The Development of Artificial Radioactivity

Overview

In 1934, the French chemists Frédéric and Irène Joliot-Curie produced radioactivity artificially. In other words, they produced radioactivity in elements that are not naturally radioactive. One important use of artificial radioactivity is in the diagnosis and treatment of disease. In addition, the Joliot-Curies' breakthrough helped lead to the discovery of nuclear fission.

Background

All atoms of an element have the same number of protons. The number of neutrons in the atoms of an element, however, may vary. Atoms of the same element that have different numbers of neutrons are called isotopes. For example, carbon-12 and carbon-14 are isotopes of carbon. (Carbon-12 has six protons and six neutrons, and carbon-14 has six protons and eight neutrons.)

Radioactivity is a property of certain unstable isotopes. The nuclei of radioactive isotopes emit matter and/or energy. This emission is called radioactive decay. The atoms that result from radioactive decay may also be unstable. If so, the decay process will continue until a stable isotope is reached. Radioactivity occurs naturally in some elements such as uranium and polonium.

In the early 1900s, scientists began attempts to produce radioactivity in elements that are not normally radioactive. These efforts proved unsuccessful. Later in the century, the husband and wife team of Frédéric (1900-1958) and Irène (1897-1956) Joliot-Curie began researching radioactivity. Irène was the daughter of Pierre (1859-1906) and Marie Curie (1867-1934). Pierre and Marie had won the Nobel Prize for Physics in 1903 for their discovery of radioactivity. Marie won a second Nobel Prize in 1911 for her isolation of the radioactive element radium. In 1918, Irène became her mother's assistant at the Institut du Radium at the University of Paris, and she began to investigate the alpha rays emitted by polonium. (Alpha rays consist of two protons and two neutrons; they are one type of radiation.) In 1925, Frédéric Joliot was hired at the lab as an assistant, and the following year, he and Irène were married.

The Joliot-Curies continued to study the radioactive properties of polonium. At the time, only very small amounts of radioactive materials were available for use in experiments. Only about 300 grams (10.6 ounces) of radium were available worldwide, for instance. For this reason, many of the properties of newly discovered radioactive elements had yet to be examined. One piece of equipment Frédéric and Irène used in their research was a Wilson cloud chamber. This device contains gas that is saturated with moisture. When a charged particle passes though the chamber, droplets of liquid condense along its path. The cloud chamber allowed the Joliot-Curies to see the paths made by the particles emitted from radioactive elements. Although these pathways lasted for only fractions of a second, Frédéric developed a way to photograph them. Another tool the Joliot-Curies used to detect radiation was the Geiger counter. A Geiger counter emits a click when a radioactive particle passes through it. It can be used to determine how much radiation a sample is emitting.

Because they worked in Marie Curie's lab, Frédéric and Irène had access to a relatively large supply of polonium as compared to many other researchers. However, even this amount proved to be insufficient. Because some types of radioactive phenomena occurred very rarely or extremely quickly, experiments often had to be repeated over and over again. When the Joliot-Curies finally obtained enough pure polonium, they began using the alpha particles given off by its decay to bombard metals such as beryllium. They found that hitting beryllium with alpha particles caused the beryllium to emit particles as well. As soon as the polonium was removed, however, the emission from beryllium stopped. After they bombarded a sample of aluminum, however, they noticed that a nearby Geiger counter continued to click even after the polonium had been removed; the sample was continuing to emit radioactive particles on its own. In other words, the Joliot-Curies had produced radioactivity artificially.

The radioactive material that resulted from the bombardment of aluminum was an isotope of phosphorus not found in nature. This short-lived isotope gave off gamma rays (radiation emitted as energy) and beta particles (radiation in the form of high-speed electrons), and it decayed into a stable form of silicon. The Joliot-Curies also bombarded atoms of boron and magnesium with alpha particles. These collisions resulted in the production of radioactive isotopes of nitrogen and aluminum, respectively. The Joliot-Curies won the Nobel Prize for chemistry in 1935 for their discovery of artificial radioactivity.

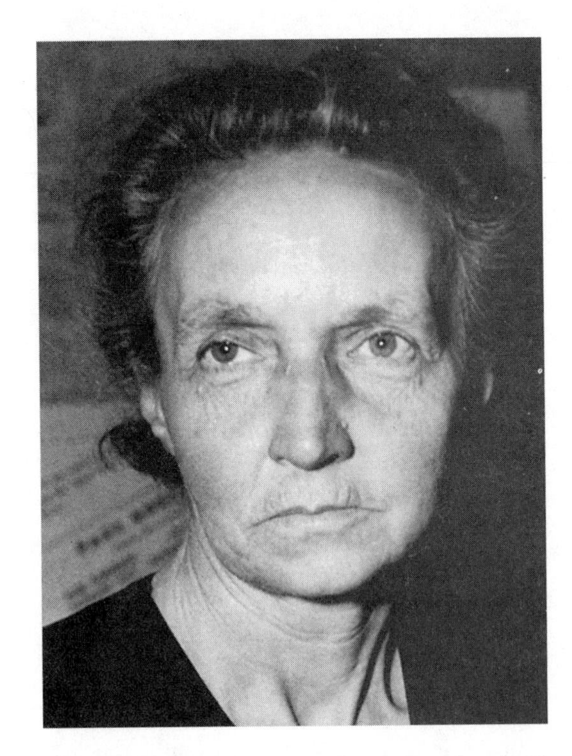

Irène Joliot-Curie. *(Library of Congress. Reproduced with permission.)*

Impact

One way in which scientists use artificially radioactive isotopes is to monitor chemical reactions either in laboratory experiments or inside organisms. Radioactive isotopes used in this way are called tracers. Early work with tracers included that by Frédéric Joliot-Curie and Antoine Lacassagne (1884-1971) to show that radioactive iodine is taken up by the thyroid gland. The thyroid gland is located at the base of the throat and produces hormones that help control metabolism—the chemical reactions that take place in an organism. Thyroid hormones contain the element iodine, and the thyroid gland takes up this element from the bloodstream in order to produce these hormones.

Today, small amounts of radioactive iodine are sometimes given to patients suspected of having certain thyroid problems. Normally, this tracer would soon make its way to the thyroid gland, where it could be detected on x-ray film. If the tracer fails to accumulate in the thyroid, however, doctors know that the gland is not working properly.

Artificially radioactive phosphorus is also used as a tracer. Small amounts can be given to plants and animals. When these organisms are released into the wild, scientists can find them again by detecting their radioactive "tag." Tracers have been used to study the flying patterns of insects, for example.

In addition, short-lived phosphorus tracers are sometimes injected into the bloodstream of a patient who has had a heart attack. Cells containing radioactive phosphorus become attached to damaged muscle in the heart. The radioactive cells then show up on x rays that doctors can use to determine the location and severity of the attack. Radioactive phosphorus is also attracted to brain tumors. By passing a special type of Geiger counter over a patient's head, doctors can locate both the radioactive phosphorus and the tumor.

Artificially radioactive isotopes can be used in the treatment of disease as well as in its diagnosis. For example, an artificially radioactive isotope of cobalt is used to treat some types of cancer patients. The radiation from the cobalt damages cancerous cells and may prevent the spread of the disease. However, the radiation also damages healthy cells, so this isotope must be used in precisely controlled amounts.

Besides its medical applications, artificial radioactivity also provided scientists with additional opportunities for studying radioactive decay and the changes that occur in nuclei. The discoveries of the Joliot-Curies inspired the Italian physicist Enrico Fermi (1901-1954) to hypothesize that artificial radioactivity could be produced by using neutrons rather than alpha particles. Unlike alpha particles, which have a positive charge, neutrons are electrically neutral. Fermi believed neutrons might be more effective at creating artificial radioactivity because they would not be repelled by the positively charged nuclei of the bombarded atoms. Through trial and error, he found that he obtained the best results by slowing the neutrons down before they hit their target atoms. Fermi found that these slow neutrons were quite effective in causing radioactivity, and he produced numerous new isotopes as a result.

When Fermi bombarded uranium atoms with slow neutrons, he got a radioactive, yet somewhat mystifying, result. Some scientists incorrectly concluded that a previously unknown element with an atomic number greater than uranium's had been produced. However, in 1938, the German physicists Otto Hahn (1879-1968), Fritz Strassmann (1902-1980), and Lise Meitner (1878-1968) realized that when a uranium atom is bombarded with slow neutrons, it breaks apart into an atom of barium and an atom of krypton. This type of reaction, in which one atom splits into two lighter atoms, is called nuclear fission. Meitner realized that some of the mass of the uranium atom was converted to energy during the process of fission. Even though the amount of mass involved was very small, the amount of energy released was enormous. It was later determined that the fission of one kilogram (2.2 pounds) of uranium will produce as much energy as the burning of three mil-

THE HIDDEN DANGERS OF SCIENTIFIC DISCOVERY

Today, researchers who work with radioactive materials take special precautions not to be exposed to radiation, which can seriously damage cells and tissues over time. When radioactivity was first discovered in 1896, however, scientists were not aware that radiation could be harmful. Marie and Pierre Curie's initial investigations of radioactivity were conducted in a poorly equipped lab. Because it had no fume hoods to draw away dangerous gases, they breathed in the radioactive radon gas given off by the radium with which they worked. Whenever a spill occurred, they merely mopped it up with a cloth. Their fingers were constantly burned and cracked from contact with radioactive materials. Their daughter Irène was probably exposed to radiation from childhood since her parents wore their work clothes home from the lab. Marie Curie and Frédéric and Irène Joliot-Curie all died as a result of complications arising from long-term radiation exposure.

Radiation safety began to be a concern about 20 years after its initial discovery. In 1915, a British scientific society proposed the use of lead or wooden shields around radioactive materials and limits to the number of hours that researchers could work with such chemicals. Scientists soon developed methods for measuring the amount of radiation to which a person was being exposed, and by 1923, maximum daily limits of radiation exposure were being suggested.

STACEY R. MURRAY

lion kilograms (3,307 tons) of coal. Scientists soon recognized that this energy could be a powerful tool if it could be harnessed.

Fermi believed that if neutrons were among the products of nuclear fission, they could be used to begin a chain reaction. In other words, the neutrons released by the fission of one atom of uranium could cause the fission of other atoms of uranium, which would release more neutrons, and so on. In 1939, Leo Szilard (1898-1964), Herbert L. Anderson, and Frédéric Joliot-Curie showed that this was true. Experiments indicated that 2.5 neutrons per uranium atom are produced during fission.

Advances in nuclear chemistry followed in quick succession, in large part because of World War II (1939-1945). The Joliot-Curies had openly published their work on artificial radioactivity. As the Nazis came to power in Europe, however, Frédéric and Irène began to worry about the implications of their recent research should it fall into the wrong hands. For this reason, when they wrote a paper in 1939 that discussed how a nuclear reactor could be built, they placed it in a sealed envelope at the Académie des Sciences. This work remained secret until 1949, after the war was over. Despite the concerns of some scientists, both sides in the war raced to be the first to develop nuclear fission as a weapon. In 1942, scientists achieved the first self-sustaining chain reaction of nuclear fission. Later that same year, the first atomic bomb was tested. (The explosive energy of such bombs comes from nuclear fission.) In 1945, the United Stated dropped the first atomic bombs used in warfare on the Japanese cities of Hiroshima and Nagasaki.

STACEY R. MURRAY

Further Reading

Books

Cooper, Dan. *Enrico Fermi and the Revolutions of Modern Physics.* Oxford: Oxford University Press, 1999.

Goldsmith, Maurice. *Frédéric Joliot-Curie: A Biography.* London: Lawrence and Wishart, 1976.

Pflaum, Rosalynd. *Marie Curie and Her Daughter Irène.* Minneapolis: Lerner Publications Company, 1993.

Rayner-Canham, Marelene F. and Geoffrey W. Rayner-Canham. *A Devotion to Their Science: Pioneer Women of Radioactivity.* Philadelphia: Chemical Heritage Society, 1997.

The Use of Poison Gases in the First World War

Overview

It is estimated that there were a million casualties from the use of poison gases in the First World War. Official figures list 180,983 British soldiers as being gassed, of whom 6,062 died in the trenches, but these figures are often considered to be an under-estimate. Apart from the difficulties of accurate reporting from the trenches, there developed a debate towards the end of the First World War and in the years before the Second World War about the ethics of using these materials as a weapon.

Background

At the outbreak of the First World War in 1914 Fritz Haber (1868-1934) was director of the newly established Kaiser Wilhelm Institute for Physical Chemistry at Berlin-Dahlem, having held the professorship of physical chemistry and electro-chemistry at Karlsruhe from 1906 to 1911. The son of a dye manufacturer, Haber was born in Bresslau, Silesia (now in Poland). He married Clara Immerwahr, the thirty-year-old daughter of another respected Jewish family in 1901. Although the Hague Conventions of 1899 and 1907 stipulated that warring countries would "abstain from all projectiles whose sole object is the diffusions of asphyxiating or deleterious gases," German scientists developed poison gases and used them by 1915. The earliest gases were derived from the work Fritz Haber did, which was designed to synthesize ammonia from the nitrogen in the air. This work also led to the development of synthetic fertilizers which greatly helped the agricultural expansion of the twentieth century. Haber received the Nobel Prize for this work on ammonia in 1918. But by that time he had proceeded to the development of chemical weapons for use by the German Army against its opponents in the First World War.

Haber's first "war gas" was chlorine, which he installed in the German trenches in 1915 and

American troops advancing on a World War I battlefield while wearing gas masks (1918). The soldier at left, unable to don his mask, clutches his throat as he breathes in the poisonous gas. *(Corbis Corporation. Reproduced with permission.)*

released when the wind was blowing towards the enemy trenches, thus getting around the prohibition on "projectiles" or gases that were thrown at the enemy in containers. In April 1915 almost 6,000 cylinders of chlorine gas were simultaneously released, and 150 tons (136,200 kg) of the poison spread along 4.3 miles (7,000 m) of the front line within about 10 minutes. Soldiers from France and Algeria breathed the gas and began to choke. This first gas attack caused 15,000 Allied casualties, of whom 5,000 died. Haber returned to his family in Berlin but on the night of May 1, 1915, his wife shot herself because of her opposition to his work. Within weeks the Allied forces were also using chlorine gas. The French developed a phosgene gas based on carbonyl chloride which was found to be 18 times as powerful as the chlorine gases. Phosgene gases were first used at Ypres on December 19, 1915.

By 1917 the Germans under Haber's direction had developed mustard gas (dichloroethyl sulphide) and used it on the Ypres front in July 1917, where it was delivered from 75- and 105-mm shells against the British 15th and 55th Divisions. The British and Americans followed soon after. Some prominent scientists such as J.B.S. Haldane (1982-1964) argued that these weapons were more humane because unlike bullets they disabled more men than they killed. His father, J.S. Haldane (1860-1936), was a physiologist whose study of oxygen and carbon monoxide in the lungs was an important part of the development of the gases. The Haldanes believed in their duty to experiment on themselves and often asked their children and friends to participate in investigations into the toxicity of different chemicals and their gases. Some observers have suggested that the argument that chemical warfare was more "humane" than bullets and bombs was based on weak statistics, because the number of deaths attributable to the poison gases was not accurately recorded during the First World War, and because there were no long-term studies of what happened to the men who had been gassed once they returned home.

Impact

Over the years poison gases have been classified by the way in which they enter the body—through the skin, causing blisters (vesicant gases); through the eye, causing tears and irritation (lacrimators); into the stomach, causing vomiting (sternulators); and those that affect the nervous system (nerve gases) or attack the lungs (respiratory gases). Some gases combine more than one effect. The effect on the victim can be acute and result in rapid death, or chronic, resulting in long-term disabilities such as eye problems, skin sensitiveness, or breathing difficulties. The gases tend to settle in the moist parts of the body, and penetrate clothing even to the genital areas. They can cause huge blisters and intense itching. If they get in the eye they cause relatively minor conditions such as conjunctivitis but also severe burns and loss of fluids. Victims who survive the first impact of the gases might be out of action for a month or more, which was considered to be the main purpose of using the poison gas—to disable the troops. Many men would recover from their first exposure after a few weeks away from the trenches only to be sent back and subjected to another attack. The sensitivity of the human to these poisons increases with exposure. Nurse Harap made the following diary entry on November 8, 1918. It described what happened to the men who were gassed in the First World War:

"Gas cases are terrible. They cannot breathe lying down or sitting up. They just struggle for breath, but nothing can be done. Their lungs are gone — literally burnt out. Some with their eyes and faces entirely eaten away by gas and their bodies covered with first-degree burns. We must try to relieve them by pouring oil on them. They cannot be bandaged or even touched. We cover them with a tent of propped-up sheets. Gas burns must be agonizing because usually the other cases do not complain even with the worst of wounds. But gas cases invariably are beyond endurance and they cannot help crying out. One boy today, screaming to die. The entire top layer of his skin burnt from his face and body. I gave him an injection of morphine. He was wheeled out just before I came off duty. When will it end?"

Mustard gas was so named because it smelled to the men of mustard or garlic, but some of the other gases were odorless. Often the first awareness of being subject to attack was to see a brown or white cloud coming over the trenches. Some gases had a slightly delayed effect so that it was several hours later before the victims began to suffer nausea, blindness, or skin burns. Some men who did not seem to be seriously injured at the time of the attack suffered for the rest of their lives from chronic chest conditions or other health problems. There was no real attempt to study the long-term effects of being exposed to these gases after the end of the First World War. The global influenza pandemic, which broke out in the early post-war years, confused the picture, and it was never certain whether the effects of the poison gases complicated these other illnesses. Similarly the relationship between exposure to poison gases and psychiatric problems suffered by soldiers has not been closely investigated.

Scientists during the First World War were more efficient at developing poison gases than in devising means to protect the soldiers from them. The earliest gas masks could not be mass-produced quickly enough to be supplied to the hundreds of thousands of soldiers on both sides of the trenches, and protective clothing had not yet been thought of. The gas masks purify and filter the air being breathed in, often through very simple filters made from charcoal. The apparatus covers the nose and eyes and is attached to a canister containing the filtering material, through which the air is passed before reaching the face.

By the end of the First World War all the major countries had well-established chemical warfare agencies such as Porton in the U.K. and the Chemical Warfare Service at Edgewood Arsenal, Maryland, in the U.S. The years leading up to the Second World War would see the development of new gases and techniques. Some of these new techniques of mass production of chemicals were applied in useful spheres, such as the development of fertilizers, pesticides, and pharmaceuticals. Haber became National Commissioner for Pest Control in Germany and developed new pesticides combining some of the chemical war agents with sweeteners to attract their victims. But one of the ironies of the early development of chemical war agents was that some of Fritz Haber's discoveries led to the manufacture of Zyklon B, which was used to gas fellow Jews in the German concentration camps of the Second World War.

SUE RABBITT ROFF

Further Reading

Books

Haber, L.F. *The Poisonous Cloud: Chemical Warfare in the First World War.* Oxford: Clarendon Press, 1986.

Harris, R. and J. Paxman. *A Higher Form of Killing: The Secret Story of Gas and Germ Warfare.* London: Chatto & Windus, 1982.

The Bergen School of Dynamic Meteorology and Its Dissemination

Overview

Meteorology as a modern science evolved after the beginning of the twentieth century, from nineteenth-century origins as a subdiscipline of geography, unsystematically defined by subjective rules of weather forecasting. Defining the motion and phenomena of the atmosphere by the mathematics of hydro- and thermodynamics and capable of being predicted by systematic methods of data analysis and forecasting was the groundwork of Norwegian theoretical physicist Vilhelm Bjerknes, assisted by his young colleagues in founding the Bergen School of meteorology.

Background

By the middle of the nineteenth century the study of meteorology, the motion and phenomena of the Earth's atmosphere, had progressed from isolated attempts at practical forecasting of the direction of weather patterns by unsystematic observational techniques and dependence on climatological data to focus on low pressure centers. Any mathematically grounded theory of atmospheric motion was limited to basic understanding of the general flow of the atmosphere on a rotating Earth, as in the work of American William Ferrel (1817-1891), followed by frictional considerations of Norwegian Henrik Mohn (1835-1916), and some others. Of fundamental importance to future atmospheric understanding was the contemporary hydrodynamic study of vortex or circular motion and its remarkable stability and conservation of its rotational flow, particularly worked out mathematically by Hermann von Helmholtz (1821-1894).

Intrigued by the possible applications of vortex theory by way of his own research with his father in hydrodynamic theory, a young Norwegian theoretical physicist, Vilhelm Bjerknes (1862-1951), had integrated his recent research in mechanics with the vortex theory by 1898. His modifications of the theory took in consideration the Sun's heating energy and frictional energy on the two largest systems on Earth behaving as fluids: the atmosphere and the oceans. These factors brought thermodynamic as well as hydrodynamic considerations into interpreting these fluid systems, and, as a result, Bjerknes developed the theory of "physical hydrodynamics," integrating relevant aspects of these sciences to explain atmospheric and oceanic motion. Most importantly, since atmospheric motion brought about weather changes, he considered the rigorous mathematical definition of atmospheric dynamics as promising a more fundamentally accurate means of predicting weather changes, given detailed analysis of atmospheric conditions.

With both theoretical and practical goals in hand, Bjerknes obtained financial support from the Carnegie Institute in Washington, D.C. (1905) to begin his research. He attracted promising students at the University of Kristiania (Oslo) from about 1907, with the basic task to develop comprehensive observational techniques entailing the three-dimensional atmosphere (particularly, using the relatively new technique of sounding balloons) to obtain an accurate meteorological database. Further opportunities offered at the University of Leipzig in Germany between 1912 and 1917 led back to Norway with the progression of World War I. At the new university at Bergen, Bjerknes was able to found a geophysical institute, where his core of researchers included his son Jacob (1897-1975), Halvor Solberg (1895-1974), one of his earlier important collaborators Harald Sverdrup (1888-1957), and soon the Swedes Tor Bergeron (1891-1977), and Carl Gustav Rossby (1898 -1957).

Bjerknes set up the first cramped location of the geophysical institute in the Bergen Museum, where, nonetheless, he continued an intensive program based on his dynamic meteorology theory. This involved systematic daily observations of and map plotting of meteorological parameters (temperature, pressure, humidity, winds, etc.), applications of both a graphical and simple mathematical formula analysis, followed by progressively more timely and extended forecasts. With the Germans cutting off Norway and the rest of Scandinavia from continental European weather communications, Bjerknes was able to convince the Norwegian government of the importance of his synoptic meteorology and forecasting to Norway's interests, thus obtaining the cooperation he needed to organize a countrywide network of observation stations for practical forecasting, as well as for continuing his research. In July of 1918 the experimental West Norwegian Meteorological

Service at Bergen (in the attic of the Bjerknes home, actually) was inaugurated.

As other contemporary meteorological investigators, the Bergen researchers had been concentrating on analysis of storm movements, posing the most practical application of accurate forecasting. German research under military authority had been emphasizing this application, especially in regard to new air warfare. Large-scale storm movement usually entailed a developing low pressure center (counterclockwise or cyclonic-upward-converging movement of air in the northern hemisphere, opposite in the southern), but a large cloud band, labeled the "squall line," brought the precipitation. A lack of comprehensive data analysis kept other researchers in Germany and England from regarding these two phenomena as separate.

Bjerknes' son Jacob had been carrying out extensive surface analysis of the atmospheric parameters of wind, temperature, and precipitation along the squall line from 1917 and had discovered that temperature gradient (change of temperature) was large across the line and that wind converged along it to coincide with the clouds and weather patterns. Jacob realized that here was a key to predicting a weather pattern's movement by analyzing wind convergence with its direction and speed and precipitation. This new development was adopted by his father as a means of validating the Bergen team's forecasts of the future movement of such storms. The density of observations obtained had enabled the researchers to gain a much more detailed picture of the mechanics of storm movement. And Jacob also discovered from his surface analyses of convergence that a low pressure center developed out of the converging squall line, along with other lines of convergence. This development of squall line and low pressure into a cyclone was termed "cyclogenesis." Jacob called this atmospheric formation the "extratropical cyclone" and published a paper on it as a model of storm formation, *On the Structure of Moving Cyclones* (1919). The squall line and developing low pressure were integrated phenomena. Here was a recurring, large-scale system of the atmosphere, revealing a definite progression of weather patterns and an efficient means upon which to base a weather forecast. Using this model Jacob was able to extend forecasts of its movement to as much as a week in advance.

The other Bergen researchers were also finding things of note in the more detailed surface and upper level data being recorded and analyzed. Halvor Solberg traced squall line patterns over time and discovered that as a squall line developed into a cyclonic wave and evolved into a low pressure center and then dissipated, a new-like formation began to the rear, and so on in sets of four, labeled the "cyclone family." Usually, the last of the four waves in the cyclone family broke loose as an outbreak of cold air southward. Tor Bergeron studied the three-dimensional structure of the new extratropical cyclone and found that in its dissipation stages it could ride over colder air in front of it or be pushed up by colder air behind. This new step in its cycle was called seclusion and "occlusion."

The basic convergent squall line separated cold air to the north from warmer air to the south, and it was actually of hemispheric proportions. Jacob and his colleagues likened this to a battle line, and lifting the term "front" from the recent world war, they renamed the squall line the "polar front" in late 1919. Solberg made exhaustive analysis of the basic lines of convergence formed as the cyclone developed, a "steering line" pushing warm air was renamed the "warm front," while the squall line proper leading the cold air from the rear was called the "cold front." The formation of these frontal boundaries was called "frontogenesis."

Impact

About mid-1918 Vilhelm Bjerknes was already in the process of spreading the Bergen theory and practice of modern meteorology. Son Jacob, who became chief of the West Norwegian Weather Service and Solberg, who was sent to Oslo to set up an eastern branch as government demand for forecasts continued, was the first to inaugurate the dissemination of the practical application of the work accomplished by going to Sweden. The Swedish scientific community, as that in Germany and England, remained unimpressed with the Bergen researchers' claimed success in weather forecasting, unfairly calling it "humbug." There was a strong sense of nationalistic pride complicating weather research and forecasting theory at this time. Both German (M. Margules, 1856-1920) and English meteorologists on similar tracks of defining airmass boundaries, especially British meteorologist William Napier Shaw (1854-1945), criticized the Norwegians for not giving credit to others as precursors to the Bergen theory. Part of Vilhelm Bjerknes's dissemination strategy was to placate the competitors, the influential Shaw in particular, and urged acceptance of the Bergen model and forecasting techniques.

Into the 1920s the Bergen group's forecasting success could hardly be ignored, and converted British meteorologists began to credit the Norwegians with breakthrough theory and practice in analyzing and forecasting the weather with validation of the Bergen techniques. The Norwegian weather forecasters showed the way to further practical applications by issuing special forecasts tailored to such commercial interests as Norwegian fishermen and farmers. Vilhelm Bjerknes left the geophysical institute at Bergen, which he had founded, to return to theoretical physics pursuits at the University of Oslo in 1926 but was a avid lecturer on the dynamic and synoptic meteorology he had initiated until his death.

Jacob and Solberg completed the mature efforts of their research with the publication of the 1922 paper "The Life Cycle of Cyclones and the Polar Front Theory of Atmospheric Circulation." They also applied results on local vertical development of instability in the atmosphere and changes in other parameters indicating weather changes (such as humidity) to the new aviation industry. Jacob remained head of the weather service until 1931, then, as a professor of meteorology at the geophysical institute, continued his research into the influences of what appeared in upper air wind data as an upper wave above frontal development . Rossby and Solberg independently studied this phenomenon on a mathematical basis. Solberg looked at a more general mathematics of atmospheric waves out of phase but superimposed on one another at different layers in the atmosphere as reflecting Jacob's observed data. But Rossby would go further.

In 1928 Rossby came to the United States to lecture at MIT on the Norwegian meteorology theory and was given the opportunity to found the first modern department of meteorology in the U.S. Rossby was a forceful proponent of the theory and its potential, and in this gained the attention of the United States Weather Bureau, which he roundly characterized as more given to bureaucracy than to a scientific approach to meteorology. He became the head of the Weather Bureau and was responsible for its complete revamping on the Norwegian weather service model. Rossby would also develop a meteorology program at the University of Chicago in the 1930s and continued to pursue data gathering and analysis of the upper troposphere. Using an automatic recording instrument package called a meteorograph which could be raised by balloon, Rossby was able to compile much data to a height of 65,000 ft (19,812 m), particularly noting a band of unusually strong winds with core speeds as much as 300 mi (483 km) per hour. It was not until late in World War II that B-29 bomber pilots verified Rossby's tracking of these winds, flying at altitudes above 20,000 ft (6,096 m). These winds were to be called the "jet stream." Rossby also defined the upper wave (first noted by Jacob and Solberg and later called

THE JET STREAM CURRENT

The conservation of momentum analogy that increasing speed of a weight on the end of a string being wound on a pencil brought up the question of why the same thing did not happen in Earth's atmosphere. Since Earth rotates and air rising over the equator tends toward the poles, the air at the poles should be moving at speeds far greater than at the equator. American meteorological theorist William Ferrel (1817-1891) noted that surface friction kept such winds from developing. But how such a building of momentum was balanced continued to be questioned. With the development of frontal theory and emphasis on atmospheric instrument sounding, the dynamic structure of the upper troposphere became an important area of research in the 1920s.

Winds of high velocity above about 20,000 feet (6,096 m) were recorded and particularly researched by Swedish meteorologist Carl-Gustav Rossby (1898-1957). Using balloon-borne automatic recording instruments (meteorographs) in the 1930s, he predicted a high wind belt based on his theory (1939-40) that such wind was the balance conserving the angular momentum of the atmosphere. The validation of this came in 1944 toward the end of World War II when American B-29 Superfortresses, the first planes to fly at 30,000 feet (9,144 m) with speeds of up to 350 miles (563 km) per hour, encountered headwinds of 200 or more miles (322 km) per hour slowing their bombing runs westward toward Japanese targets. Planes flying eastward found the opposite—flying at assisted speeds totaling 500 to 600 miles (805 to 965 km) per hour—arriving at destinations hours earlier. These maximum wind belts were coined "jet stream" winds, and meteorologists would find that they influenced the intensity and direction of lower level weather system development, such as frontal cyclones, tornadoes, and the Indian monsoon bursts.

WILLIAM J. MCPEAK

the Rossby Wave) above a surface cyclone wave as integrated with this band of strong winds. Together they developed and steered the surface cyclone and its frontal development.

Jacob came to lecture at MIT in 1939 but with the German invasion of Norway in early 1940, he decided to accept an offer from Joseph Kaplan of the UCLA physics department to head a meteorology annex. By the beginning of the war it was realized that weather forecasters would be essential to a worldwide conflict, particularly in regard to the high technical level of aerial warfare. Kaplan wanted UCLA to be one of the universities (MIT was one) chosen for military forecaster training, and enticing Jacob was an important selling point. Before war's end Jacob would be asked to found a full department of meteorology at UCLA, rivaling MIT as a center of modern synoptic meteorology and forecasting. With another of Vilhelm Bjerknes collaborators Jorgen Holmboe (1902-1979), Jacob finalized research on the integration of the upper and lower atmosphere in the development of the extratropical cyclone in their joint paper, "On the Theory of Cyclones" (1944). By the end of the war, the polar front theory of the Bergen School was widely accepted.

The immensity and changeableness of the atmosphere had long foiled attempts at realistic (time-valued), mathematically driven forecasts of weather simply because the number of equations could not be solved fast enough to be of value. But progress with the high speed digital computer resulted in, among late 1940 applications, a program of research applied to weather forecasting headed by mathematician John von Neumann in 1946. Intrigued by the challenges of weather forecasting posed by U.S. Weather Bureau meteorologists, von Neumann focused his computer's (Mathematical Analyzer, Numerical Integrator and Computer or MANIAC) initial tasks on solving equations of atmospheric flow over time. In putting together a team of scientists for the program, he turned to a young MIT scientist for important meteorological expertise. This was Jacob Bjerknes's first doctoral recipient in meteorology, Jule Gregory Charney (1917-1981). Charney had worked out mathematical equations of instability in the atmosphere in association with the upper level wave both under Bjerknes and Rossby and integrated his results into the program for MANIAC. The first successful forecast from MANIAC came in 1952—forecast of a snowstorm of 1950. By 1955 the Weather Bureau incorporated the computer into the process of analyzing and forecasting the weather. The foundation of that and a modern meteorology of airmass movement and major weather phenomena boundaries via the frontal theory was the legacy of the Bergen School initiated by Vilhelm Bjerknes.

WILLIAM J. MCPEAK

Further Reading

Books

Editors of Time-Life. *Weather.* New York: Time-Life Books, 1968.

Friedman, Robert Marc. *Appropriating the Weather: Vilhelm Bjerknes and the Construction of a Modern Meteorology.* Ithaca, New York: Cornell University Press, 1989.

Holmboe, Jorgen, George Forsythe, and William Gustin. *Dynamic Meteorology.* New York: Wiley, 1945.

Wurtele, Morton G., ed. *Selected Papers of Jacob Aall Bonnevie Bjerknes.* Los Angeles: Western Periodicals Co., 1975.

The Great Barringer Meteor Crater

Overview

The Great Barringer Meteor Crater in Arizona was the first recognized terrestrial impact crater. The confirmation of a meteor impact (subsequently identified as the Canyon Diablo meteorite) at the site proved to be an important stepping stone for advances in geology and astronomy. In solving the mystery surrounding the genesis of the Barringer crater, geologists and astronomers made substantial progress in understanding the dynamic interplay of gradual and cataclysmic geologic processes both on Earth and on extraterrestrial bodies. In addition, the story behind the early-twentieth-century controversy surrounding the origin of the crater highlights the dangers of prejudice and selective use of evidence in scientific methodology.

The Meteor Crater near Winslow, Arizona. *(Roger Ressemeyer/Corbis. Reproduced with permission.)*

Background

The Barringer Meteor Crater (originally named Coon Butte or Coon Mountain) rises 150 ft (46 m) above the floor of the surrounding Arizona desert. The impact crater itself is almost 1 mile (1.6 km) wide and 570 feet (174 m) deep. Among geologists, two competing theories were most often asserted to explain the geologic phenomena. Before the nature of hot spots or plate tectonic theory would have convinced them otherwise, many geologists hypothesized that the crater resulted from volcanic activity. A minority of geologists asserted that the crater must have resulted from a meteor impact.

In the last decade of the nineteenth century, American geologist Grove Karl Gilbert (1843-1918), then the respected head of the U.S. Geological Survey, set out to determine the origin of the crater. Gilbert assumed that for a meteor to have created such a large crater, it would had to have remained intact through its fiery plunge through the Earth's protective atmosphere. Moreover, Gilbert assumed that most of the meteor must have survived its impact with Earth. Gilbert therefore assumed that if a meteor collision was responsible for the crater, then substantial pieces of the meteor should still exist.

Armed with these assumptions Gilbert began his research. He found no substantial mass inside the crater, so he assumed that the meteor was buried. However, Gilbert was unable to find the elusive meteor, and he concluded that, in the absence of the evidence he assumed would be associated with a meteor impact, the crater had resulted from subterranean activity.

In 1902, Daniel Moreau Barringer (1860-1929), an American entrepreneur and mining engineer, began a study of the Arizona crater. After discovering that small meteors made of iron were found at or near the rim of the crater, Barringer was convinced that only a large iron meteor could be the cause of such a geologic phenomena. Acting more like a businessman or miner trying to stake a claim, Barringer seized the opportunity to form a company with the intent of mining the iron from the presumed meteor. Without actually visiting the crater, Barringer formed the Standard Iron Company and sought mining permits.

Over the next decades, Barringer, a self-confident, self-made, and wealthy man, invested his fortunes in proving the meteor impact hypothesis, and in reaping the potential profits from the mining of such a meteor. Barringer collected thousands of investment dollars from people expecting large returns. What had started out as a scientific question now became clouded by profit motive. Evidence was not subjected to scientific scrutiny as much as it was selected to bolster investor "confidence."

For the next 30 years or so, Barringer became the sword and shield of the often rancorous scientific warfare regarding the origin of the crater. In bitter irony, Barringer won the scientific battle—that the crater resulted from a meteor impact—but lost his financial gamble. In the end, the meteor that had caused the impact proved to be much smaller than hypothesized by either Gilbert or Barringer. On the heels of these findings in 1929, Barringer died of a heart attack. His lasting legacy was in the attachment of his name to the impact crater.

Impact

The debate over the origin of the Great Barringer Meteor Crater came at a time when geology itself was reassessing its methodologies. Within the geologic community there was often vigorous debate over how to interpret geologic data. In particular, debates centered on whether gradualism (similar to evolutionary gradualism) of geologic processes was significantly affected by catastrophic events.

Barringer confidently asserted that the Coon Butte crater supported evidence of catastrophic process. Although he argued with selective evidence, Barringer turned out to be correct when he asserted that the finely pulverized silica surrounding the crater could have only been created in a cataclysmic impact. Beyond the absence of volcanic rocks, Barringer argued that there were too many of the iron fragments around the crater to have come from gradually accumulated, separate meteor impacts. Moreover, Barringer noticed that instead of defined strata (layers), there was a randomized mixture of the fragments and ejecta (native rock presumably thrown out of the crater at the time of impact). Such a random mixture could only have resulted from a cataclysmic impact.

Barringer's theory gained support from mainstream geologists when American geologist George P. Merrill tested rocks taken from the rim and floor of the crater. Merrill concluded that the quartz-like glass found in abundance in the presumed eject could only have been created by subjecting the native sands to intense heat. More importantly, Merrill concluded that the absence of sub-surface fusions proved that the heat could not have come from below the surface.

The evidence collected by Barringer also influenced astronomers seeking an explanation for the large, round craters on the Moon. Once again the debate moved between those championing extraterrestrial volcanic activity (gradualism) versus those who favored an impact hypothesis (cataclysm). The outcome of these debates had enormous impact in both geology and astronomy.

One fact that perplexed astronomers was the particular shape of the lunar impact craters: they were generally round. If meteors struck the Moon at varying angles, it was argued, then the craters should have assumed a variety of oblique shapes. Barringer and his 12-year-old son set out to explain this phenomenon by conducting an experiment: they fired bullets into clumps of rock and mud. Regardless of the firing angle the Barringers demonstrated that the resulting craters were substantially round. More definitive proof was subsequently provided in 1924 by astronomers who determined that forces of impact at astronomical speeds likely resulted in the explosive destruction of the impacting body. Importantly, regardless of the angle of impact, the result of such explosions would leave rounded craters.

The confirmation that a meteor weighing about 300,000 tons (less than a tenth of what Barringer had estimated) and traveling in excess of 35,000 mph (56,315 kph) at impact proved to be a double-edged sword for Barringer. In one stroke his hypothesis that the crater was caused by a meteor impact gained widespread support, while, at the same time, Barringer's hopes of profitably mining the meteor were dashed.

In the 1960s American astronomer and geologist Eugene Shoemaker found distinct similarities between the fused rocks found at the Barringer crater and those found at atomic test sites, attesting to the power of the impact. In addition, unique geologic features termed "shatter-cones" created by immense pressure pointed to a tremendous explosion at or above the impact crater. Once scientists became aware of the tremendous energies involved in astronomical impacts, large terrestrial impacts, often hidden by erosive effects, became a focus of study. Having identified more than 150 such impact sites, scientists are researching these sites in hopes of better understanding the Earth's geologic history.

It appears that a catastrophic astronomical collision occurred at the end of the Cretaceous Period 66 million years ago. The effects of this collision are thought to have precipitated the widespread extinction of large species, including the dinosaurs. The enigmatic Tunguska explosion of 1908, which devastated a vast area of Siberian forest, may have been Earth's most recent significant encounter with an impacting object.

K. LEE LERNER

Further Reading

Books

Mark, K. *Meteorite Craters.* University of Arizona Press, 1987.

Periodical Articles

Marvin, U. "Meteorites, the Moon, and the History of Geology." *Journal of Geological Education* 34 (1986): 140.

Rampino, M.R., B.M. Haggerty, and T.C. Pagano. "A Unified Theory of Impact Crises and Mass Extinction: Quantitative Tests." *Annals of the New York Academy* 822 (1977): 403-31.

Geologist Richard Oldham's 1906 Paper on Seismic Wave Transmission Establishes the Existence of Earth's Core and Demonstrates the Value of Seismology for Studying the Structure of Earth's Deep Interior

Overview

In 1906 Richard Dixon Oldham (1858-1936) established the existence of Earth's core and the importance of seismic data for studying the structure of the planet's deep interior. By observing ways in which seismic waves were reflected and refracted, different boundary layers were later identified. Differences in wave speeds also provided information about density and rigidity as a function of radius. Seismic analyses allowed Andrija Mohorovicic (1857-1936) to identify the core-mantle boundary (1910), Beno Gutenberg (1889-1960) to deduce the core boundary (1912), Harold Jeffreys (1891-1989) to demonstrate that the core is liquid (1926), and Inge Lehmann (1888-1993) to discover the solid inner-core (1936).

Background

Magnetism provided one of the first clues to the structure of Earth's interior. Using a spherical piece of magnetized iron to simulate Earth, William Gilbert (1544-1603) found that the pattern of magnetic lines around the ore matched compass-needle patterns observed over Earth's surface (1600). This suggested that Earth contains substantial amounts of magnetized iron. However, as compass data accumulated it appeared Earth's magnetic field was slowly drifting westward—an effect that seemed impossible if Earth's interior were solid iron.

In 1692 Edmond Halley (1656-1742) proposed a core-fluid-crust model to explain this effect. According to Halley, Earth's magnetic field is produced by a solid iron core. Earth's outer shell or crust is separated from the core by a fluid region. The westward drift of the magnetic field was the result of the core rotating eastward slightly slower than the crust.

Evidence increasingly mounted indicating that Earth had a completely molten interior. Volcanoes were observed to spew forth hot material; many surface rocks appeared to have formed by crystallization from extremely hot material; and measurements indicated temperature increases with depth that extrapolated to temperatures well beyond the melting point of all known rocks at a depth of about 50 miles (80.5 km). The idea that Earth had a relatively thin crust and molten interior thus gained ascendancy after the late seventeenth century.

This view came under heavy attack beginning in the 1830s. André-Marie Ampère (1775-1836) argued that the tidal forces exerted on such an enormous volume of liquid by the Moon would render the surface unstable (1833). William Hopkins (1793-1866) echoed Ampère's critique and also noted that since the melting temperature of most substances increases with pressure, the central core might well be solid. Hopkins also adduced astronomical evidence that indicated the crust must be at least 621 miles (1,000 km) thick (1842). Lord Kelvin (1824-1907) produced similar arguments and suggested that experiments be conducted to determine the vertical motions of Earth's surface with respect to lunar and solar positions. These measurements seemed to support Kelvin's claim that Earth was "as rigid as steel," and by the end of the nineteenth century most geologists accepted that Earth was completely solid.

A new source for analyzing the structure of Earth's interior emerged at the end of the nineteenth century when British seismologist John Milne (1850-1913) suggested that with sufficiently sensitive instruments one could detect earthquakes anywhere in the world (1883). This was realized in 1889 when Ernst von Rebeur-Paschwitz (1861-1895) demonstrated that seismometers in Germany had detected vibrations that had passed through Earth's interior from an earthquake that had occurred in Tokyo. These seismic waves could be analyzed using the mathematical theory of wave motion to extract information about the physical constitution of Earth's interior.

Wave theory as applied to seismic analysis predicts three types of waves: primary (P-waves), secondary (S-waves), and surface waves. P-waves can be transmitted though solids, fluids, and gases. S-waves, however, can be transmitted only through solids, not fluids.

In his landmark report on the 1897 Assam, India, earthquake, Richard Oldham provided the first clear evidence for P-waves, S-waves, and surface waves (1899). Because seismometers located on the other side of Earth from the earthquake detected P-waves later than expected, Oldham concluded Earth must have a central core through which P-waves travel with a substantially lower velocity than the surrounding material. Also, since Earth was believed to be completely solid, he concluded the slower velocity could only be due to a much denser core, which he suggested was probably iron.

The details of Oldham's initial arguments were unconvincing. After collecting further data and refining his ideas he reasserted his initial conclusion in "The Constitution of the Interior of the Earth as Revealed by Earthquakes" (1906). In this work he conclusively established the existence of Earth's core. He did not however, as is widely believed, discover the core to be liquid. He maintained his original position—later shown to be incorrect—that the core is extremely dense and probably iron.

Impact

Oldham's analyses stimulated further seismic research on Earth's interior. While studying 1909 earthquake data, Andrija Mohorovicic noticed changes in seismic wave velocities. Mohorovicic attributed the effect to transmission through two layers of different density. He identified the boundary between these two layers as a sharp discontinuity now known to vary from 6.21 miles (10 km) under the oceans up to 31 miles (50 km) under continents (1910). Thus the Mohorovicic or "Moho" discontinuity between Earth's crust and mantle was established.

Another discontinuity was discovered by Beno Gutenberg in 1912. Gutenberg was then a student of Emil Wiechert (1861-1928) at Göttingen, Germany—a center for seismic research. Wiechert's theories provided a fairly reliable guide to seismic wave paths through Earth's interior, and mathematicians at Göttingen had developed practical computational methods based on his work for calculating variations in wave velocity with depth. Gutenberg applied these methods to determine P- and S-wave velocities in Earth's interior. His results indicated a major discontinuity at a depth of 1,802 miles (2,900 km). This is now called the Gutenberg discontinuity and marks the boundary between the mantle and core.

By 1913 Oldham became increasingly convinced that his data showed no indication of S-wave transmission through the core. He thus began to move toward the idea that Earth's core was fluid. He was anticipated in this by Leonid Leybenzon (1879-1951), whose 1911 Russian publication attracted little attention. In 1924 Wiechert suggested the sudden velocity drop below the Gutenberg discontinuity was indirect evidence that the core was fluid.

Gutenberg, however, remained firmly convinced that Earth's core was solid. He reviewed six methods for calculating core rigidity and concluded that all but one clearly indicated a solid core. The sixth method only required a fluid core if S-waves were not transmitted. Gutenberg dealt with this possibility by explaining why S-wave transmission had not yet been observed.

The fluidity of Earth's core was finally established in 1926 by Sir Harold Jeffreys. Jeffreys was able to show that each of the methods Gutenberg referred to could be suitably interpreted to support the fluid-core theory. His most conclusive demonstration was to show that the average mantle rigidity was much greater than the average rigidity of the entire Earth, thus requiring a compensating region of lower rigidity below the mantle. This obviously had to be within the core. The low core rigidity of Jeffrey's fluid-model precluded S-waves transmission.

In 1936 Inge Lehmann further refined our understanding of Earth's interior. Analyzing data from the 1929 Buller, New Zealand, earthquake, she showed that P-waves were being reflected from a sharp boundary within the core. She calculated the boundary radius to be 870 miles (1,400 km) at what is now called the "Lehmann discontinuity." Lehmann went on to argue for the solidity of the inner core, but this was not conclusively established until the early 1970s.

This model of Earth's interior—of a central iron core surrounded by a fluid then solid crust—is similar to Halley's 1692 model, but there are significant differences. Halley failed to specify the core radius and the depth of the fluid, and he had no idea how Earth's rigidity varied with depth. Halley's account of terrestrial magnetism also differs from the modern explanation.

Iron loses its magnetism at extremely high temperatures such as those believed to exist at the inner core. Consequently, Earth's iron core cannot generate terrestrial magnetism. However, it was discovered in the nineteenth century that electric currents produce magnetic fields. Ampère suggested that Earth's magnetic field and its variations might be the result of currents flowing in the planet's interior.

Walter Elsasser (1904-1991) developed this idea and in 1946 proposed a geomagnetic dynamo theory. According to Elsasser's theory, electric currents are generated within the outer-core by the induction effects of the moving fluid much as a dynamo generates electricity. Edward Ballard developed a similar though more detailed theory in 1948. Research has continued, but only in the last few years of the twentieth century have scientists been able to produce computer models of the geomagnetic dynamo capable of accurately simulating the westward magnetic drift, spontaneous pole reversals, and other secular magnetic variations.

STEPHEN D. NORTON

Further Reading

Books

Brush, Stephen G. *Nebulous Earth*. Cambridge: Cambridge University Press, 1996.

Jeffreys, Harold. *The Earth: Its Origin, History and Physical Constitution*. 6th edition. Cambridge: Cambridge University Press, 1976.

Oldroyd, David. *Thinking about the Earth: A History of Ideas in Geology*. Cambridge, MA: Harvard University Press, 1996.

Periodical Articles

Brush, Stephen G. "19th-Century Debates about the Inside of the Earth: Solid, Liquid, or Gas?" *Annals of Science* 26 (1979): 225-254.

Brush, Stephen G. "Discovery of the Earth's Core." *American Journal of Physics* 48 (1980): 705-724.

Jeffreys, Harold. "The Rigidity of Earth's Central Core." *M. N. Royal Astronomical Society* 77 (1926): 371-383.

Oldham, Richard. D. "The Constitution of the Interior of the Earth as Revealed by Earthquakes." *Quarterly Journal of the Geological Society of London* 62 (1906): 456-472.

Finding Earth's Age and Other Developments in Geochronology

Overview

The discovery of radioactivity and radioactive decay has helped to solve many problems that have plagued geologist for centuries, especially the question of absolute ages of rocks, fossils, and Earth itself. These answers have provided a more rigorous science of geology and have given scientists in a large variety of fields firm data upon which to base their studies and hypotheses. In addition, by giving a scientifically unassailable age for Earth, isotopic methods have been used by many to argue against a literal interpretation of the Bible, and it is not uncommon for practitioners of this science to be called upon to testify in legal cases involving the teaching of evolution or variants of creationism.

Background

For uncounted centuries, man either had no idea of the age of Earth or, based on a literal reading of religious works, felt Earth to have been created not more than a few thousand years ago. During the nineteenth century, as geologists gained a better understanding of geologic processes, most scientists became certain of Earth's antiquity, but still lacked any real knowledge as to what that meant. Estimates of Earth's age ranged from a few million years to many billions of years, all based on different methods of age determination.

One of the driving factors behind efforts to determine this age was the introduction of (and controversy surrounding) evolutionary theory. Evolution required vast amounts of time for species to gradually form, die off, or transform one into another. The incredible variety of life found in the fossil record simply could not occur in an Earth of only a few million, or even a few tens of millions of years old. If Earth could be shown to be old, but "only" a few million years old, evolution might yet be shown false, and man might retain a special place in creation.

The single most influential estimate of Earth's age was put forth by Lord Kelvin, William Thomson (1824-1907), the preeminent physicist of his day. Kelvin's estimates were all based to some extent on the amount of time it would take Earth to cool from an initially molten state to its current temperature. They ranged from a few tens of millions of years to nearly half a billion years. Because of Kelvin's prestige, few dared to challenge his calculations or the premise upon which they were based, even when it became apparent that Earth was, instead, much older.

In 1895, Wilhelm Röntgen (1845-1923) discovered x rays, and, in the following year, Henri Becquerel (1852-1908) discovered radioactivity in uranium. In the next few years, uranium was discovered to be present in trace amounts in virtually all rocks and soils on Earth. It was also quickly discovered by Marie and Pierre Curie (1867-1934 and 1859-1906) and others that uranium decays through a long series of radioactive elements to finally become lead, which is not radioactive. These radioactive intermediary nuclides include radium, radon, and thorium, all of which occur naturally. Work by, among others, Ernest Rutherford (1871-1937) and Frederick Soddy (1877-1956) showed that heat was released during radioactive decay while Bertram Boltwood (1870-1927) noticed that all minerals containing uranium also contained lead and helium.

These last two discoveries were of particular importance. If radioactive decay released heat, then this meant that the premise upon which all of Kelvin's calculations were based was incorrect because Earth would be cooling at a slower rate than otherwise. In addition, the invariable correlation between uranium, lead, and helium meant that uranium likely turned into helium and lead through radioactive decay. Since Rutherford had shown that the rate of radioactive decay changes predictably over time, this gave a way to construct a "clock" for determining the age of rocks. One of the first such age estimates, about 500 million years, was made by Rutherford and was based on the ratio of helium to uranium. Helium is given off during the decay of heavy elements in the form of alpha particles, which are simply the nuclei of helium atoms emitted from heavy, radioactive elements. However, Rutherford soon found his calculations to be in error because of the many alpha-emitting nuclides present in the uranium decay chains and because helium atoms escape mineral crystals with relative ease. It turned out that lead was a far better "end-point" to use for this dating. By the 1940s age estimates were converging on the current figure of 4.6 billion years of age for Earth. Later work dating meteorites indicated the solar system to be slightly older and, when the Apollo program returned with lunar rocks, we found that the Moon is a few hundred years younger.

Over the next few decades, increasingly sophisticated isotopic dating methods were developed that used a variety of radioactive elements. Some of the more widely used of these are the rubidium-strontium method, the potassium-argon and argon-argon methods, but a number of other geochronometers have been developed for specific purposes. For example, examination of isotopes of iodine and xenon in meteorites tells us about the conditions leading to the formation of the solar system, while analyzing the ratio of neodymium and samarium isotopes can help us trace the geochemical history of mountain ranges.

Impact

Today, the field of isotope geology and geochronology is far more advanced than in 1907. Geochemists routinely use mass spectroscopy equipment, including the latest advance, the tandem accelerator mass spectrometer (TAMS), to analyze isotope with an amazing degree of precision and accuracy. Since its inception, isotopic methods have had a profound impact in the scientific fields of geology, paleontology, evolutionary theory, biology, botany, and (using carbon-14 methods) in the fields of anthropology, archaeology, and history. In addition, by providing a solid and scientific basis for determining the age of Earth and its inhabitants, isotope geology has also resolved the debate over the origin of Earth for all but a handful of biblical literalists.

The chief scientific impact of isotopic dating methods has been to give an absolute timetable for events on Earth. The importance of this can scarcely be overstated. Do you want to know how long dinosaurs dominated Earth? Find the rocks with the first dinosaur fossils and the last fossils and date them. In a few days, you'll have a firm date telling you that they reigned for over 150 million years. How about determining when oxygen first appeared in the atmosphere? In this case, find the oldest rocks that can form only under conditions of low oxygen—their age tells you the last date the atmosphere could have been oxygen-deficient. Geologic dating has told us how quickly life can evolve, exactly when the dinosaurs, trilobites, ammonites, and other fossil species went extinct, when life first colonized the land, when the Gondwana supercontinent last broke up, when the Moon formed, and much more.

Having this information is interesting from the standpoint of sheer intellectual curiosity. However, it is also important because it can give us some sort of framework upon which to hang our concept of geologic and evolutionary time. Imagine trying to go to class if you can only say with certainty that social studies comes sometime after home room but before gym and that you go home sometime after track practice, which comes after lunch and before dinner. Without a clock, we might know what happened on the early Earth, but we have no idea of how fast it might have happened or what might have happened at the same time in various parts of Earth. We can construct a history of Earth based only on relative dates (that is, what happened before what); it just isn't very interesting or very informative. Virtually every historical science depends to some greater or lesser extent on isotopic methods of dating past events, and those dating methods have given us, in effect, Earth's clock and calendar.

In the social realm, isotopic dating methods have proven to be of some intrinsic interest as well as providing scientists with an outstanding tool to use in debates against those who believe in a literal interpretation of religious documents.

The intrinsic interest of non-scientists in geologic dating has a great deal to do with the general interest that most people have in trying to better understand our world. As shown by the continuing popularity of natural history museums, newspaper and other media articles on scientific topics, and the popularity of books explaining science, a large portion of the population has some interest in learning more about the world in which they live. A large part of that understanding, as with scientists, is in finding out when significant events took place, even if the time scale is almost incomprehensible in magnitude. This is especially true regarding research on our own origins.

However, it is likely that the most significant social impact of isotopic dating lies in its utility in the perennial debate over the origin of Earth and its inhabitants. So potent an argument, in fact, that virtually every court case involving the teaching of evolution versus creationism (or variants such as "scientific" creationism) at some point sees testimony by a prominent isotope geologist who explains the science behind isotopic dating methods and their results.

P. ANDREW KARAM

Further Reading

Books

Dalrymple, G. Brent. *The Age of the Earth.* Stanford University Press, 1991.

Faure, Gunter. *Principles of Isotope Geology.* John Wiley & Sons, 1986.

Hallam, A. *Great Geological Controversies.* Oxford Science Publications, 1989.

Hellman, Hal. *Great Feuds in Science.* John Wiley & Sons, 1998.

Alfred Wegener Introduces the Concept of Continental Drift

Overview

The theory of continental drift holds that the great landmasses are slowly moving, and have done so continually over the long span of geologic time. According to the theory, a single landmass called Pangaea split up about 200 million years ago, and the resulting continents eventually drifted to their present locations. The German meteorologist Alfred Wegener (1880-

1930) introduced the theory of continental drift in 1912.

Background

Looking at a world map, it is easy to see that the edges of Europe and North America, Africa and South America, would fit together nicely, almost like pieces in a jigsaw puzzle. This was noticed almost as soon as Europeans began traveling to the Americas; Francis Bacon wrote of it in 1620. But the early mapmakers had no real understanding of how the Earth formed and changed over time, and made no attempt to explain the shapes of the continents.

Nineteenth-century naturalists began to find other intriguing evidence linking opposite sides of the Atlantic Ocean. Certain rock formations stopped at the western coast of Africa and picked up again on the eastern coast of South America. In 1858, Antonio Snider, an American living in Paris, cited similarities between fossil plants and proposed that Europe and North America had once been part of the same landmass. He suggested that the continents had been separated by the biblical flood.

On December 29, 1908, the American geologist Frank B. Taylor first suggested the slow drifting of continents in a presentation to the Geological Society of America. He described a scenario in which two large sheets of the Earth's crust, originally centered at the North and South poles respectively, crept inexorably forward like the glaciers of the Ice Age, leaving behind great tears that became the South Atlantic and Indian Oceans and the Arctic Basin. His ideas went largely unnoticed.

The man who is remembered as the "father of continental drift" is the German professor Alfred Wegener. Wegener was an adventurer who enjoyed exploring and going aloft in balloons. His field of study was meteorology, but it is for his contributions to geology that he is best remembered. He put forth a theory of moving continents "not for the first time," commented a Mr. F. Debenham, attending a 1923 presentation by Wegener at the Royal Geographical Society in London, "but for the first time boldly." He also marshaled a great deal of evidence in its defense.

Impact

Wegener was intrigued by the fit of the Atlantic coastlines, and had read about the similarities in the American and European fossils. In 1912, he proposed the theory that the continents had drifted into place, and elaborated upon it in his book *The Origins of Continents and Oceans* (1915).

The theory of continental drift contradicted the way most geologists of the time believed the crust of the Earth had taken shape. The standard view was that the Earth had formed in a molten state and, as it cooled, it contracted. This contraction was thought to have resulted in the sinking of the ocean basins and the crumbled crust of mountain ranges, just as the skin of an apple becomes wrinkled and pitted as the fruit dries up and shrinks.

Some scientists were willing to listen to Wegener, however, in part because the contraction theory was itself crumbling. They now knew that recently discovered radioactive elements in the Earth's crust, such as radium, would decay and produce heat that would build up below the surface. This continual source of heat meant that cooling, and therefore contraction, was unlikely.

Wegener began accumulating evidence to support his theory. First, there were the fossils. For example, skeletons of the small reptile *Mesosaurus*, about 270 million years old, had been found in only two places, Brazil and South Africa. It strained credulity to think that this reptile could swim the Atlantic, but if its range was indeed this large, it should have been found in many places.

Living animals were also distributed in ways that were suggestive. Lemurs, for instance, the most primitive primates, are found only in southeast Africa, the nearby island of Madagascar, and across the Indian Ocean in south Asia. The usual explanation for these oddities in animal distribution were ancient land bridges that had since sunk beneath the ocean. However no traces of such bridges were found, nor was a convincing reason advanced for them to sink if their material had been lighter than the crust of the ocean floor to begin with. Very old rock formations, too, matched up on both sides of the ocean, and they certainly did not wander across any land bridge.

Finally, Wegener was able to employ his own discipline, meteorology, by seeking evidence of climatic conditions millions of years ago. The remains of tropical plants found in coal samples from Alaska provided evidence of a warm climate in the past. Large boulders deposited by glaciers were evidence of a polar climate in what is now Africa. In 1915, Wegener proposed that the continents had split off from a single landmass he called *Pangaea*, or "All Land." Surrounding Pangaea was a great ocean, which would later become the Pacific.

Wegener believed that Pangaea began breaking up about 40 million years ago because of forces created by slight irregularities in the Earth's rotation around its axis. Many scientists rejected this theory because his explanation could not begin to account for the movement of continents—the forces were far too weak. A mistaken claim that Greenland had moved 120 feet (36 m) in a single year, based on erroneous data, led to widespread ridicule, and the fact that Wegener was not a geologist did not help his cause. When he died in 1930 on a dogsled expedition to Greenland, the idea of continental drift all but died with him. It was revived after World War II, when new scientific techniques provided not only new data but, finally, a plausible explanation.

In the 1950s, *paleomagnetic* evidence supporting continental drift became available. When rock is in a molten state, the magnetic particles it contains are free to move around, and they orient themselves toward the Earth's magnetic poles. Once the rock cools and hardens, the orientation of the individual magnetic particles becomes fixed. When the position of the rock changes, the magnetic polarity changes with it. By studying the magnetic properties of ancient rocks, the British geophysicist S.K. Runcorn was able to determine that Europe and North America were at one time connected.

At about the same time, oceanographers had mapped features such as deep trenches, arcs of islands, and ridges or mountain ranges on the ocean floor. Seismologists noted that volcanoes tended to occur along the ocean ridges, and that earthquakes were common beneath the ocean trenches.

In 1961, the American scientists H.H. Hess and Robert Dietz explained continental drift in terms of *sea-floor spreading*. Their hypothesis was confirmed in 1963 by the British geophysicists F.J. Vine and D.H. Mathews using paleomagnetic measurements. In sea-floor spreading, molten rock from a layer of the Earth's interior called the *mantle* rises up through *convection currents* and breaks through the crust at the ocean ridges. It pushes out from the ridges as it hardens, spreading the ocean floor and forcing the continents farther apart.

While the rocks on the Earth's surface are billions of years old, the rocks on the ocean floor have existed for no more than 200 million years. This figure has thus replaced Wegener's 40 million year estimate for the date of Pangaea's breakup. The first split created a northern continent called *Laurasia*, and the southern landmass scientists named *Gondwanaland*. Further breaks resulted in the familiar continents of today.

The upper surface of the Earth's crust, or *lithosphere*, moves like a number of rigid plates floating on a soft zone of the mantle called the *asthenosphere*. As the plates collide, their edges crumble, pushing up to form mountain ranges or down into the mantle to create ocean trenches, island arcs—and earthquakes. The study of these changes in the Earth's surface is called *plate tectonics*.

Plate tectonics allows scientists to predict what the continents might look like millions of years in the future. The California coast may separate from the mainland and head toward Alaska. Australia may also move north until it runs into Asia. The size of the Atlantic Ocean might increase, with Africa and South America moving farther apart. We shouldn't expect to see any drastic changes in our lifetimes, however, as the continents move at a rate of only a few centimeters per year.

SHERRI CHASIN CALVO

Further Reading

Books

Kidd, J.S. and Renee A. Kidd. *On Shifting Ground: The Story of Continental Drift.* New York: Facts on File, 1997.

Kiefer, Irene. *Global Jigsaw Puzzle: The Story of Continental Drift.* New York: Atheneum, 1978.

Marvin, Ursula B. *Continental Drift: The Evolution of a Concept.* Washington D.C.: Smithsonian Institution Press, 1973.

Radar Mapping of the Solar System

Overview

Radar stands for radio detection and ranging. It is a technology that generates radio waves, reflects them from an object, and detects the reflected waves to determine where the object is located in space. An outgrowth of the tremendous advances in radar technology made during World War II, radar astronomy debuted in 1946 with the detection of a radar signals reflected from the Moon. Since that modest start, radar has been used to map the Moon, Venus, Mercury, several asteroids, and to detect numerous other bodies in space. Recently, the *Magellan* space probe took radar to, literally, new heights, mapping the surface of Venus with unprecedented accuracy during a multi-year orbital mission. Orbital radar has also been used to map the Earth's surface, including the seafloor. Radar techniques have become increasingly sophisticated over the past half century, giving astronomers yet another tool with which to explore.

Background

The history of radar actually dates back to the 1880s, when Heinrich Hertz (1857-1894) showed that radio waves exist and could be both generated and detected. Hertz also showed they could be reflected, just as light is by a mirror. Somewhat later, in 1904, German engineer Christian Hülsmeyer developed a device using radio echoes as a navigational aid. This device did not work well, however, and it was left to Guglielmo Marconi (1874-1937; the inventor of wireless radio broadcasts) to suggest a way to make radar useful in 1922. American physicists Gregory Breit and Merle Tuve developed useable radar in 1925, but its use remained limited until shortly before World War II.

During the Second World War, technological advances by Germany, England, and the United States resulted in significant improvements to radar in terms of technology, reliability, and power. Shortly after the war ended, Walter McAfee (1914-1995), an American scientist, was able to determine the strength of a radar signal that would reflect from the Moon and still be strong enough to detect with earth-bound equipment. McAfee's theory was successfully tested in 1946, ushering in the era of radar astronomy. In following years, radar was used to map portions of the Moon, to detect otherwise invisible meteor trails, and, later, to map other objects in our solar system.

Radar mapping is similar to radar detection, but differs in some significant ways. In normal detection, pulsed radio signals are sent towards an object. If the object is made of metal or something that will reflect radio waves, some are reflected back towards the antenna. These signals are picked up and the direction from which they returned shows the direction to the object. The amount of time it takes for a radar signal to return can be used to calculate the distance to the object. Radio waves, traveling at the speed of light, travel almost exactly 300 meters (about 1,000 feet) in a microsecond. By counting the number of microseconds that elapse between sending a signal and detecting its return, a computer can determine its distance.

In radar mapping, a few more steps must occur. It may help to visualize a radar signal as an expanding balloon, expanding into space at the speed of light. As this balloon comes into contact with a planet, it hits the highest points first, and they push into the balloon first, marring the balloon's surface. In quick succession, the rest of the expanding sphere presses into other high points on the planet's surface, followed by steadily lower points until an entire half of the planet is embedded in the balloon's surface. This is where the balloon analogy breaks down for, in radar astronomy, the points that were first touched by radar waves are the first to reflect those waves back at the antenna from which they came. Those first waves are soon followed by reflections from the rest of the planet, separated in time by their relative distance from the antenna. For example, a mountain that is 3,000 meters in height is, at the speed of light, 10 microseconds above the surrounding plain. It takes radar waves 10 microseconds longer to reach the plain than the mountaintop, and 10 more seconds to climb back to the elevation of the mountain. So radar waves reflected from the plain will return to Earth 20 seconds later than those from the mountaintop, because they had to travel the extra distance twice, once going to the planet and once returning. This helps to make radar mapping possible because differences in elevation are exaggerated by this phenomenon. This technique has been used to make radar maps of the Moon, Venus, Mercury, and many asteroids.

The moon as seen from *Apollo 17. (NASA. Reproduced with permission.)*

Two other points must be mentioned with regards to mapping planets with radar. First, the strength of the radar signal drops off quite rapidly with distance. In fact, signal strength declines as the square of the distance. In other words, sending a signal twice as far requires four times the power if it is to be detected at the receiving end. Simply reaching another planet with a radar signal requires one fourth the power as reaching the other planet *and* receiving a detectable return signal. This level of power is required because the signal has to travel from Earth to the other planet *and back*. In addition, it should be pointed out that different surfaces reflect radar differently, based on their geometry and composition. A hard, flat plain, for example, will return radar extremely well because there is nothing to break up, or scatter, the signal. Plains will show up as "radar-bright" objects on radar maps. On the other hand, a jumble of rocks, loose soil, mountains, and similar features will scatter the beam, sometimes greatly, leaving a "radar-dark" area. Similarly, some materials are more likely to absorb radar than others and will show up as darker areas on a map. These factors are all considered when interpreting a radar map.

Impact

The impact of radar astronomy was initially huge, then dropped off as space probes were sent to return with photographic maps of these same bodies. However, in recent years, radar astronomy and radar mapping has experienced something of a renaissance with the triumphant Magellan mission and several exceptionally successful terrestrial mapping satellite missions. At the same time, the term "radar astronomy" may no longer be strictly accurate, especially when we turn orbital radar onto our home planet. Radar mapping is a more accurate and more descriptive term that will be used for the rest of this essay.

The first impact of this field came in 1946 when the technology was first demonstrated. Simply bouncing a signal off of the Moon was a triumph because no radar signal had ever traveled so far before. Mapping the Moon with radar was more an exercise in calibrating the technology since radar maps could be compared with what we could actually see through a telescope. The first real test came when radar telescopes were turned towards Venus and, later, Mercury.

The surface of Venus is perpetually covered by thick clouds. For the first half of the twentieth century, considerable debate surrounded the question of Venus's period of rotation, in other words, how long the Venusian "day" was. The matter was finally settled in the 1950s when a series of mapping projects identified three large, distinct areas on Venus: Alpha, Beta, and

Maxwell (after the physicist who was instrumental in describing electromagnetic waves). By tracking these areas, scientists were able to show that Venus's "day" was actually longer than its "year." In fact, a single rotation of Venus takes 256 days and, unique in the solar system, Venus was found to rotate in a retrograde (i.e., backwards) direction compared to the Earth. Another interesting finding of these early radar and radio studies was that the surface temperature was over 450 degrees C, hotter than virtually any part of Mercury, in spite of Mercury's closeness to the Sun. These findings were sufficient to permanently dash any illusions of Venus as the home to life or as Earth's sister planet in the solar system.

Mercury was mapped in the 1950s and 1960s in a manner similar to Venus, but the *Mariner 10* flyby missions in 1973 and 1974 rendered these radar maps obsolete. However, radar did tell astronomers that, like Venus, Mercury has a very long "day," in this case, 59 days. Recently, radar has been used to map several asteroids with some degree of success.

The primary advantage of radar mapping is that the equipment stays on Earth. This makes maintenance possible, reduces costs enormously, and allows construction of a much larger and more powerful antenna. On the other hand, terrestrial equipment cannot achieve nearly the accuracy, precision, and resolution of spacecraft in orbit around the planet being mapped, making such craft indispensable for mapping Venus and other cloud-shrouded bodies.

The most recent advances in this field utilize such orbital craft. *Magellan*'s mission to Venus in the early 1990s resulted in mapping 98% of the planet's surface with a resolution of about 100 meters (328 feet), meaning the radar could detect objects 100 meters in size or larger. More area has been mapped on Venus than on the Earth, Moon, Mars, and Mercury combined, thanks to *Magellan*. This same technology has been turned Earthward, too. The Radar Ocean Satellite (ROSAT) launched by the United States and the French TOPEX craft have mapped the ocean surface with unprecedented accuracy. This, in turn, has allowed oceanographers to construct a very accurate map of the seafloor because, as it turns out, water mounds up slightly over top of underwater mountains and is slightly lower over subsea valleys and trenches.

P. ANDREW KARAM

Further Reading

Books

Morrison, David. *Exploring Planetary Worlds*. Scientific American Library, 1993.

Biographical Sketches

Vilhelm Frimann Koren Bjerknes
1862-1951

Norwegian Geophysicist, Meteorologist, and Physicist

Vilhelm Bjerknes made seminal contributions to the foundation of dynamic meteorology as a mathematically exact modern science with his theory of "physical hydrodynamics," while expanding practical meteorology with the development of synoptic meteorology and formula-based weather forecasting techniques. He also made important contributions to electrodynamic theory.

Bjerknes showed early signs of true genius as a youth already assisting his father, Carl Bjerknes (1825-1903), a physics teacher, with his theories on hydrodynamic forces and their similarities to electrodynamic forces. Vilhelm's formal training started with studies in mathematics and physics at the University of Kristiania (later Oslo) in 1880. Completing his masters degree (1887), he studied at Paris under physicist Jules-Henri Poincaré (1854-1912) on a state fellowship and went on to Bonn in 1890, becoming an assistant to and collaborator with Heinrich Hertz (1857-1894) in electrodynamic research. Bjerknes contributed resonance experimental verification of Hertz's electromagnetic wave theory along with adding to oscillatory circuit theory. Returning to Oslo he completed his Ph.D. with a focus on electrodynamics in 1892. Bjerknes settled into a professorship of applied mechanics and mathematical physics at the University of Stockholm in 1895, during which time he decided to return to hydrodynamics and his father's research but with limited success in the latter regard.

Bjerknes was much more open-minded in his research aspirations than his father's unformalized and eccentric goals. The study of circular or vortex motion and its stable characteristics in hydrodynamics had been advanced late in the century, and Bjerknes became interested in its applications to understanding the motion of the atmosphere and oceans. He realized that these—the two largest fluid systems on Earth—were energized to motion by the radiation of the Sun, requiring that both hydrodynamic as well as thermodynamic considerations be integrated. This he did in his theory of "physical hydrodynamics." Bjerknes particularly focused on application to the atmosphere by theorizing a rigorous mathematical interpretation of its dynamics with a systematic approach to analysis of atmospheric conditions, making possible more accurate weather forecasting.

While lecturing at MIT in 1905 and hoping to obtain American funding for his research, Bjerknes received a generous stipend (which continued until 1941) from the Carnegie Institute in Washington, D.C. By 1909 he began soliciting the meteorological community at large on his analysis techniques, stressing upper air wind data and other innovative observing methods conducive to better weather forecasting. Continuing his physics professorship at the University of Kristiania, Bjerknes began to draw bright post-graduate students to his call for meteorological research. These students included Harald Sverdrup (1888-1957), his son Jacob (1897-1975), Halvor Solberg (1895-1974), Carl-Gustav Rossby (1898-1957), and Tor Bergeron (1891-1977). Bjerknes was offered the chair of a new geophysical institute at the University of Leipzig in 1912, and several of his young collaborators followed him there to do research on storm movement forecasting, which was the most practical application in contemporary meteorology. By 1917, the war and the offer to found a geophysical institute at Bergen by his friend zoologist/explorer Fridtjof Nansen (1861-1930) brought him home to Norway.

Bjerknes intensified his ambitious program of systematic weather data gathering, analysis by mathematical and graphical techniques, and perfecting formulas for weather forecasting. The importance of timely weather forecasts brought support from the Norwegian government, which enabled him to start the West Norwegian Weather Service as part of the Bergen Geophysical Institute in mid 1918, which was expanded to a countrywide level. Bjerknes's dynamic meteorology theory was enhanced by his collaborators' theories and verification of the extratropical cyclone model, the polar front, and the airmass conception of analysis.

By 1926 the Bergen group of researchers had assumed the tasks of the weather service and the dissemination of Bjerknes's dynamic meteorology. Bjerknes left to resume his teaching of mechanics and mathematical physics at the University of Oslo. He began a series of textbooks on theoretical physics, and temporarily renewed research into his father's theories of hydrodynamic forces centering on "hydromagnetics" during this period. He retired from teaching in 1932 but continued to advocate the Bergen meteorology with great enthusiasm. Bjerknes published important seminal works supporting his dynamic meteorology theory and received many awards for his contributions to theoretical physics and modern meteorology.

WILLIAM J. MCPEAK

Niels Henrik David Bohr

1885-1962

Danish Physicist

Niels Bohr was the first to apply quantum theory in a consistent model to explain the arrangement of electrons in the atom. Bohr's model accounted for the chemical properties of the elements and for the main features in their spectra. He was awarded the Nobel Prize in Physics for this achievement in 1922.

Bohr was born in Copenhagen on October 7, 1885. After receiving his doctorate from the University of Copenhagen in 1911, he continued his education at Cambridge University under J.J. Thomson (1856-1940), the discoverer of the electron. Next he spent a year at the University of Manchester, working with Ernest Rutherford (1871-1937) just as the British physicist was discovering the atomic nucleus.

While the importance of Rutherford's work was enormous, there were a few major problems with his model of negatively charged electrons orbiting a positively charged nucleus. According to classical physics, charges orbiting in an electrostatic field should continually give off electromagnetic radiation, thus losing energy. Eventually they would spiral inwards, lacking sufficient kinetic energy to counter their attraction to the nucleus. In addition, the spectrum of light emitted from or absorbed by the atoms of an element should be a smooth continuum in this model; however, in reality the spectra showed distinct lines.

The quantum theory of the German physicist Max Planck (1858-1947) illuminated Bohr's thoughts on these problems. In 1900, Planck had put forth the idea that oscillating charges emitted and absorbed energy in discrete units called *quanta*. The energy of the quantum of light, or *photon*, was proportional to the frequency of the radiation.

In 1913, Bohr published a series of papers in which he applied quantum theory to Rutherford's atomic model. He assumed that electrons moved around atomic nuclei in certain stable orbits in which no energy was lost. Radiation was emitted or absorbed only when an electron moved from one orbit to another. The energy difference between two orbits corresponded to the energy, and thus the frequency, of the photon that was emitted or absorbed in the transition. Such transitions explained the emission and absorption lines that were seen in atomic spectra.

Bohr returned to the University of Copenhagen as a professor in 1916. Under his influence, an Institute for Theoretical Physics was established there in 1920, becoming one of the world's premier research centers. Bohr served as director of the Institute, as well as president of the Royal Danish Academy of Science. In 1922 he was awarded the Nobel Prize, and in 1932 took up residence in the "House of Honor," a mansion Denmark reserved for its most esteemed citizen. He continued his research on quantum mechanics and atomic structure throughout the 1930s, as the specter of Nazism began to cast its shadow over Europe.

Denmark was overrun by the Germans in 1940. Like many Danish intellectuals, Bohr, who was of partial Jewish descent, was involved in the resistance movement. He wrote openly about his views and attempted to protect Jewish scientists in his Institute. The situation became increasingly dangerous for him. As an atomic physicist, he risked being interned to work on Germany's weapons efforts, and he viewed with horror the prospect of the megalomaniac Hitler armed with an atomic bomb. Finally, after repeated urgings by colleagues and diplomats, he fled Denmark in 1943. In Los Alamos, New Mexico, he served as an advisor to the Manhattan Project, assisting in the U.S. atomic weapons effort. He understood it as a hedge against the Nazis and a way to deter future wars.

Niels Bohr. *(Library of Congress. Reproduced with permission.)*

Bohr returned to Copenhagen at the end of the war in Europe. After the U.S. forced the Japanese to surrender by dropping atomic bombs on Hiroshima and Nagasaki, he devoted much of the rest of his life to promoting international cooperation and peaceful uses for atomic energy. Bohr was awarded the first Atoms for Peace award in 1957. He died in Copenhagen on November 16,1962.

SHERRI CHASIN CALVO

Sir William Henry Bragg
1862-1942

English Physicist

Sir William Lawrence Bragg
1890-1971

English Physicist

The Braggs, William Lawrence and William Henry, studied the diffraction of the then newly discovered x rays by crystalline solids. Their studies confirmed that interatomic distances could be accurately determined by this technique. X-ray diffraction has come to be accepted as the most accurate method of determining the structures of molecules and complex crystals.

William Henry Bragg attended Cambridge University, graduating with honors in 1885 and moving to Australia to accept a teaching post at the University of Adelaide. In Australia he married, his wife giving birth to William Lawrence in 1890. At Adelaide he developed a reputation as an outstanding teacher and public speaker, and busied himself with the affairs of the Australian Association for the Advancement of Science. His research output was meager, however, not at all consistent with the eminence he would eventually achieve.

A turning point in William Henry Bragg's career occurred in 1904, when Bragg, serving a second term as the president of the astronomy, mathematics, and physics section of the Australian Association, was called on to make a presidential address. Perhaps concerned that his audience would be comparing him with New Zealander Ernest Rutherford, who had achieved substantial fame for research into radioactivity as a much younger man, Bragg chose to present a highly critical review of the current understanding of alpha, beta, and gamma rays. He also began publishing papers on this area every few months. In 1907 he was elected a fellow of the Royal Society and in 1908 was recalled to England as Cavendish Professor at the University of Leeds.

William Lawrence Bragg was quickly recognized as a child prodigy and entered the University of Adelaide at 15, graduating three years later with an honors degree. He then entered Trinity College in Cambridge to work under C.T.R. Wilson (1869-1959), the Scottish physicist who would become known as the inventor of the cloud chamber.

The discovery in 1912 by German physicist Max von Laue (1879-1960) and his assistants that x rays scattered by a thin crystal formed a characteristic pattern of spots on photographic film naturally attracted the senior Bragg's attention. While William Henry had believed that x rays were some sort of material particle, he and his son concluded that the experiments could only be explained as the scattering of a wave. They then derived what has come to be known as the Bragg relation. It was for this work that father and son shared the Nobel Prize for physics in 1915. Both Braggs remind active in research throughout their careers, both being knighted for their contributions. William Henry served for five years (1935-1940) as president of the Royal Society. William Lawrence became director of the Cavendish Laboratory in 1938, serving until 1953.

The technique of x-ray diffraction made it possible to determine the arrangement of atoms within a crystal. With a bit of mathematical elaboration, to take into account the different efficiencies with which different atoms would scatter x rays, it became possible to determine the arrangement of atoms even in very complicated molecules, provided that the substance could be crystallized. Even with a partially ordered sample, some useful information can be gained. X-ray diffraction data obtained by Rosalyn Franklin (1920-1958) at the Cavendish would prove critical in the determination of the structure of DNA. Today the technique, with computer controlled data collection and analysis, has become one of the mainstays of biomolecular research.

William Henry Bragg. *(Library of Congress. Reproduced with permission.)*

DONALD R. FRANCESCHETTI

Annie Jump Cannon

1863-1941

American Astronomer

Annie Jump Cannon was the first astronomer to develop a simple spectral classification system. She classified 400,000 stars—more than anyone else had achieved previously—and discovered 300 variable stars, five novas, and a double star. Cannon was the most famous female astronomer of her lifetime and was called the "Census Taker of the Sky." Cannon's successes inspired other women to pursue astronomical investigations, despite gender biases demonstrated by many male astronomers.

Born on December 11, 1863, in Dover, Delaware, Cannon was the daughter of Wilson Lee and Mary Elizabeth Cannon. Her father served in the state senate. Cannon's mother transformed their attic into an observatory for Cannon to stargaze. At Wellesley College, Cannon studied with astronomer Sarah F. Whiting (1846-1927), who taught her new research methods. After graduating in 1884, Cannon returned home, where she focused on social activities, and traveled, photographing a solar eclipse in Spain in 1892.

After her mother's sudden death, Cannon dealt with her grief by resuming her astronomical observations. She began postgraduate studies at Wellesley in 1894, assisting Whiting in the physics laboratory. Cannon also studied astronomy at Radcliffe College with Edward C. Pickering (1846-1919), director of the Harvard Observatory. He encouraged female astronomers and hired Cannon in 1896 for a position at the observatory. She examined photographic plates after her classes and observed with telescopes at night. Initially, Cannon classified 5,000 stars monthly; eventually, she was able to analyze 300 stars per hour.

Cannon's most significant achievement was improving the stellar classification system that astronomers used to survey the universe. Previous astronomers had realized that when starlight is photographed refracting through prisms, it creates a spectrum. Researchers studied spectral patterns to identify star characteristics. Early classification systems arranged spectra alphabetically from A to Q, based on composition, or by Roman numeral designations, used to indicate stars' temperatures. Pickering asked Cannon to develop a better method to record star information. She determined that the elements that compose stars create different radiant wavelengths, which form various colors in spectra. By 1901 Cannon had outlined ten star categories, based on color and brightness, that were designated by letters (O,B,A,F,G,K,M,R,N,S). The first three stars were hot white or bluish, F and G were yellow, K was orange, and the final four categories were cooler red stars. Arabic numerals were used to identify subdivisions. Cannon classified the few spectra that did not fit into this system as "peculiar" and described them in detail. The International Solar Union adopted Cannon's classification system for use in observatories worldwide. Astronomers flocked to Harvard to learn her methodology. Although other classification systems were later developed, Cannon's techniques formed their framework.

Cannon also photographed and described variable stars, compiling an extensive database for other astronomers. She served as curator of the Henry Draper Memorial Collection and, during this work, compiled more astronomical data than had any other individual. She made sure that both the northern and southern hemispheres had been completely photographed, insisting that even the faintest stars be included. Ten volumes and two supplementary editions of her work were ultimately issued, listing each star with a number and description, including its position in the sky, its brightness, its visual and photographic magnitudes, and comments on any peculiarities. Because Cannon was the sole classifier, her observations and information were consistent. This central repository helped transform astronomy from a hobby into a scientific profession with a theoretical basis.

In 1931 the National Academy of Science awarded Cannon the first Henry Draper Gold Medal given to a woman. She created the American Astronomical Society's Annie Jump Cannon Prize to fund female astronomers' research. Cannon was named Harvard's William Cranch Bond Astronomer in 1938 and given the rank of professor. She died in Cambridge, Massachusetts, on April 13, 1941. Rooms at the Harvard Observatory and the Delaware State Museum as well as a memorial volume of the *Draper Catalogue* were dedicated to Cannon.

ELIZABETH D. SCHAFER

James Chadwick
1891-1974
English Physicist

The English physicist James Chadwick is primarily remembered for his discovery of the neutron, which other physicists referred to as the historical beginning of nuclear physics. Using the neutron as a tool for investigating the atom, physicists were able to create a wide variety of new radioisotopes, split atoms and molecules, and initiate the nuclear chain reactions that led to the atomic bomb.

James Chadwick, the son of John Joseph Chadwick and Anne Mary Knowles, was born in Cheshire, an English mill town, and later moved with his family to Manchester. While enrolling at the University of Manchester, Chadwick accidentally found himself in the line for those hoping to major in physics. Chadwick, who had intended to be a mathematician, was too shy to seek out the proper line. He graduated from the Honours School of Physics in 1911 and spent the next two years at Manchester working with Ernest Rutherford (1871-1937), winner of the 1908 Nobel Prize for his theoretical work on the radioactive transformation of atoms. In 1913 Chadwick earned his Master in Science degree and a scholarship to work in Berlin with Hans Geiger (1882-1945), creator of the radiation counter. When World War I broke out, Chadwick was imprisoned as an enemy alien and confined to a Berlin stable. There he attempted to continue his research, improvising apparatus from teacups and beer glasses and a popular brand of radioactive toothpaste as his radiation source.

As soon as the war ended, Chadwick returned to England to continue his research in Rutherford's laboratory. In 1919, he accepted the Wollaston Studentship at Gonville and Caius College, Cambridge, in order to continue work with Rutherford, who had moved from Manchester to the Cavendish Laboratory, Cambridge. In 1923 Chadwick became Assistant Director of Research in the Cavendish Laboratory. Two years later he married Aileen Stewart-Brown of Liverpool. They had twin daughters.

In the 1920s Rutherford carried out the first artificial nuclear transformation. Chadwick and Rutherford were able to induce the transmutation of other light elements. Their studies of the nature of the atomic nucleus led to the identification of the proton. Rutherford had postulated the existence of the *neutron*, a subatomic particle with no charge. Chadwick began a series of experiments to demonstrate the existence of such a particle, but the neutron was elusive. Then, in 1930, Walther Bothe (1891-1957) and Herbert Becker described an unusual type of gamma ray produced by bombarding beryllium with alpha particles. Chadwick recognized that the properties of this radiation were more consistent with Rutherford's hypothetical neutron. Chadwick finally discovered the neutron after a famous three-week, round-the-clock, research marathon in February 1932. When Chadwick bombarded beryllium atoms with alpha particles, they released an unknown radiation that in turn ejected protons from the nuclei of various substances. Chadwick concluded that this radiation was composed of neutral particles that had a mass approximately equal to the proton. In 1934 in collaboration with Maurice Goldhaber, Chadwick discovered the nuclear photoelectric effect. These investigations provided evidence that the neutron is heavier than the proton. While working with slow neutrons, Chadwick and Goldhaber discovered that neutron could induce disintegration of lithium, boron, and nitrogen nuclei.

The neutron is one of the particles found in every atomic nucleus except for ordinary hydrogen. Its mass is about 1,840 times that of the electron. Because the electrically uncharged neutron is not deflected by charged atomic constituents, it could penetrate the atomic nucleus. Thus, the neutron could be used as a tool to induce atomic disintegration. Subsequently, many subatomic particles were discovered. Most of these are short-lived particles that were created in accelerators that produce high-energy collisions between particles. Chadwick's discovery pointed the way towards the fission of uranium 235 and the creation of the atomic bomb. The first self-sustaining chain reaction was achieved in a nuclear reactor in 1942. Three years later, American scientists produced the first atomic bomb.

In 1935 Chadwick was elected to the Lyon Jones Chair of Physics at the University of Liverpool. From 1943 to 1946 he worked in the United States as Head of the British Mission attached to the Manhattan Project for the development of the atomic bomb. In 1945, he witnessed the first nuclear explosion at Trinity. In 1946 he served on the U.N.'s Atomic Energy Commission, before he was able to retreat back to the sanctuary of his Liverpool laboratory. After World War II, Chadwick suffered from episodes of depression. Unlike some of his colleagues, he avoided politics and debates about the consequences of nuclear physics. In 1948, he retired from his professorship at Liverpool on his election as Master of Gonville and Caius College, Cambridge. He retired from this Mastership in 1959. From 1957 to 1962 he served as a member of the United Kingdom Atomic Energy Authority.

Chadwick was knighted in 1945. In addition to the 1935 Nobel Prize for Physics, he received many honors, including the Hughes Medal, the Copley Medal, the Franklin Medal, and many honorary doctorate degrees.

LOIS N. MAGNER

Marie Sklodowska Curie

1867-1934

Polish-born French Chemist

One of the pioneering figures in early nuclear physics and radiochemistry, Marie Curie was the first woman to win a Nobel Prize, the first person to win two Nobel Prizes (in 1903 and 1911), and remains the only person to win both prizes in the same field, and to be the mother of another Nobel Prize winner—Irène Joilot-Curie (1897-1956). Her extraordinary accomplishments served to break down many social and cultural barriers that had previously prevented women from pursuing careers as research scientists.

Maria Sklodowska was the daughter of Polish intellectuals suffering repression under the Russian occupation of their homeland. Despite such circumstances, her parents provided her with a thorough education. As a young woman, she worked as a governess before leaving for Paris in 1891 to live with relatives and pursue further studies, changing her first name to its French form, Marie. In 1893 she received her diploma in physics with high honors—an extraordinary distinction for a woman at that time—and in 1894 a second diploma in mathematics with honors. In April 1894 she met the promising young physicist Pierre Curie (1859-1906); they married in July 1895 and their daughter Irène was born in 1897.

The course of the Curies's scientific researches was set by the 1895 discovery of "x-rays" from uranium by Wilhelm Roentgen (1845-1923). Pierre and Marie undertook the laborious task of purifying several tons of pitchblende (impure uranium ore) to obtain only a few pounds of concentrated radioactive material. Early in 1898 Marie noted that the samples thereby obtained appeared to contain other substances besides uranium. In July this suspicion was confirmed when measurements of rates of radioactive emissions revealed the presence of another element, named "polonium" in honor of Marie's homeland. In November, spectroscopic analysis performed for them by Eugène Demarcay (1852-1903) disclosed the existence of a second element, named "radium," though it was not successfully isolated until 1902.

Meanwhile, in 1900 Marie assumed a position as the first female faculty member at the Ecole Normale Supérieure, an all-women's school in Sèvres. In 1903 she defended her research thesis on analysis of radiation from uranium at the Sorbonne. That same year, Marie and Pierre shared the prestigious Humphrey Davy Medal from England, and then the Nobel Prize with Henri Becquerel (1852-1908) for research on radioactivity. In 1904 a second daughter, Eve, was born, and Marie was finally allowed to join her husband, who was a member of the Faculty of Sciences at the Sorbonne, as his research assistant. This seemingly idyllic picture was shattered when Pierre was struck and killed by a horse-drawn carriage in April 1906.

The faculty of the Sorbonne thereupon offered Marie the professorship held by her late husband. This appointment was officially confirmed in 1909, making her the first female faculty member of one of the world's most prestigious universities. However, Madame Curie's life was not free from controversy; in 1910 she declined the award of the Legion of Honor, as Pierre had done in 1903, and in 1911 she was rejected for membership in the Académie des Sciences after public controversy surrounding her involvement in an extramarital affair. The award of a second Nobel Prize later that year for her discoveries of polonium and radium, along with her belated election to the Académie in 1914 and role as a founding member of the Conseil de l'Institut du Radium, represented, however, a vindication of her scientific work and reputation.

Marie continued her researches into radioactivity together with her daughter and son-in-law, Irène and Frederic Joilot-Curie, up until her death from leukemia, due to many years of unprotected exposure to radiation at a time when its hazards were not yet known.

JAMES A. ALTENA

Paul Adrien Maurice Dirac

1902-1984

British Physicist and Mathematician

Paul Dirac applied Albert Einstein's (1879-19555) theory of special relativity to quantum mechanics, the mathematical framework describing the motion of atomic particles. In doing this he represented the behavior of the electron by means of *wave functions*. Dirac's *wave mechanics* predicted the electron spin and the existence of the positron, an "anti-electron" with the same mass but a positive rather than negative charge. In 1933 he shared the Nobel Prize in Physics with the Austrian physicist Erwin Schrödinger (1887-1961).

Dirac was born in Bristol, England, on August 8, 1902. Although his mother was English, his Swiss father, a teacher at his son's school, refused to speak to the boy unless he was addressed in French. Perhaps as a result, Dirac was known for speaking as little as possible. He was good at mathematics, so he decided to study

Paul Dirac. *(Library of Congress. Reproduced with permission.)*

electrical engineering, and received his B.Sc. from the University of Bristol in 1921.

Dirac never found employment as an engineer. Instead he entered St. John's College of Cambridge University, where he received his Ph.D. in mathematics in 1926. By this time he had already made important contributions to quantum mechanics. He traveled extensively over the next few years, serving as a visiting lecturer at the University of Wisconsin, the University of Michigan, and Princeton University. In 1929 he took a trip around the world, visiting Japan with the German physicist Werner Heisenberg (1901-1976) and traveling through Siberia on his way back to Europe.

In his book *The Principles of Quantum Mechanics*, published in 1930, Dirac elucidated his *transformation theory*, a means of understanding the properties of atomic particles using statistical probability distributions. His presentation was solely mathematical, because he believed that to create a mental or visual picture was to "introduce irrelevancies." It was this work for which he was awarded the Nobel Prize.

In 1932, Dirac joined the Cambridge faculty as Lucasian Professor of Mathematics, the prestigious chair once held by Isaac Newton (1642-1727). The same year, Carl D. Anderson (1905-1991) first observed the positron by bombarding aluminum and lead with gamma rays and photographing its tracks in a cloud chamber. This provided experimental confirmation of Dirac's theory. The validation might not have excited Dirac as much as one would expect. He later wrote, "[I]t is more important to have beauty in one's equations than to have them fit experiment."

Dirac married Margit Wigner of Budapest in 1937. When he was 64 years old, he moved to the United States and became professor emeritus at Florida State University in Tallahassee. He died there on October 20, 1984.

SHERRI CHASIN CALVO

Albert Einstein
1879-1955
German-American Physicist

Albert Einstein was among the greatest scientists and most creative thinkers who ever lived. His theories changed the way physicists think about the universe, treating energy and matter as interchangeable and linking space with time and gravitation. His status as an icon of genius, coupled with his spiritual and compassionate nature, made him an important voice in world affairs as well.

Einstein was born on March 14, 1879, in Ulm, Germany. His family soon moved to Munich, where Einstein's early schooling took place. The regimentation of nineteenth-century German education held little appeal for the imaginative, curious boy, and his teachers were not particularly impressed by his promise. However, two of his uncles encouraged his fascination with mathematics and science, while his mother arranged for him to study the violin, an interest that would remain with him all his life. Einstein completed his formal education at the Federal Polytechnic Institute in Zurich, Switzerland, studying physics and mathematics. After graduation he obtained employment in the Swiss patent office at Bern and married Mileva Maric, whom he had met at the university. In 1905 he became a Swiss citizen.

The patent office was a quiet place, affording Einstein plenty of time to think. He occupied himself with theoretical physics, and in 1905 he published several papers in the prestigious journal *Annalen der Physik*. The first, entitled "A New Determination of Molecular Dimensions," prompted the University of Zurich to award a Ph.D. to the patent clerk. With the remaining papers Einstein revolutionized three branches of physics.

One of his papers explained Brownian motion, the irregular movements of microscopic particles in a liquid or gas, in terms of the atomic theory of matter. Another modeled light as a stream of tiny massless particles, or quanta, thereby explaining the photoelectric effect, in which light hitting metal causes it to release electrons. Understanding the photoelectric effect made possible the electric eye, television, and many other inventions. Einstein was awarded the Nobel Prize in physics for this work in 1921.

Einstein's third major development of 1905 was the special theory of relativity. He showed that time was not absolute; rather, it depended on the speed at which one's frame of reference moved in relation to the speed of light, *c*. For example, a space traveler making a long journey at an appreciable fraction of *c* would return to Earth noticeably younger than a homebound twin sibling. Finally, another paper that year put forth the famous equation $E = mc^2$, relating matter and energy.

In 1909 the University of Zurich offered Einstein a professorship. He also taught at Prague and at his alma mater, the Federal Polytechnic Institute. In 1914 he was lured to the University of Berlin by Max Planck (1858-1947) and resumed his German citizenship. Mileva Einstein, vacationing with the couple's two sons in Switzerland, was unable to return to Berlin because of the outbreak of World War I, and eventually this separation led to divorce. Einstein later married his second cousin, Elsa.

Einstein published his general theory of relativity, describing gravity in terms of a curvature in spacetime, in 1916. He remained in Berlin until 1933, holding positions as director of the Kaiser Wilhelm Physical Institute and member of the Prussian Academy of Sciences. This allowed him to concentrate on research, along with giving occasional lectures at the University of Berlin. With the rise of Nazism, Einstein fled to the United States, where he had been invited to join the newly formed Institute for Advanced Study at Princeton University. He remained there for the rest of his life, becoming a U.S. citizen in 1940. He died on April 18, 1955.

The great scientist lived modestly in Princeton, being completely indifferent to money. Internationally famous and greatly respected, his opinion was often sought on world affairs. He had been forced to compromise his lifelong pacifism when he thought Adolf Hitler might obtain

Albert Einstein. *(Library of Congress. Reproduced with permission.)*

an atomic bomb, and he urged President Roosevelt to launch the Manhattan Project to build a bomb as a deterrent. He was profoundly saddened when the United States dropped two atomic bombs on Japan in 1945 at the end of World War II, and thereafter advocated international law as the only way to prevent aggression between nations. Einstein supported Zionism and was offered the presidency of Israel in 1952, but he did not consider himself well-suited for the job and thus declined the offer.

In his scientific work, Einstein spent his later life on a quest for an ultimate unified theory of physics. This effort was handicapped by his discomfort with quantum mechanics, in which physical quantities such as the position and velocity of atomic particles are considered as statistical probabilities that cannot be absolutely determined. Although he was not conventionally pious or observant in any orthodox sense, Einstein was deeply religious in his nature, compassionate, and a champion of the underprivileged. Many who had the good fortune to meet him felt that simply to be in his presence was a spiritual experience, as wisdom and kindness seemed to radiate from him. Quantum mechanics offended his sense of the order and beauty in the universe. "Subtle is the Lord," he once remarked, "but malicious He is not."

SHERRI CHASIN CALVO

Charles Fabry
1867-1945
French Physicist

Fabry's career was devoted to the design of precise optical devices based on the principles of interference of light. The Fabry-Pérot interferometer would become one of the mainstays of modern optical instrumentation. Fabry was also involved in one of the first measurements of the Doppler shifts caused by the random thermal motions of atoms and in establishing the existence of the ozone layer in the earth's atmosphere.

Fabry attended the Ecole Polytechnique and the University of Paris, receiving his doctoral degree in 1889. After holding some minor teaching posts, he joined the faculty at the University of Marseilles in 1894, moving to the Sorbonne in 1920. Fabry's doctoral thesis dealt with the interference of light waves and in 1896 he devised, with Alfred Pérot, a chamber, using partially reflecting mirrors, that allowed a light beam to interfere with itself numerous times. As a result of this interference only very precisely defined wavelengths could pass through this chamber. The Fabry- Pérot interferometer, as it came to be known, could be used in very accurate distance measurements, or to produce or select light beams of a very precisely defined frequency.

Throughout his career, Fabry was interested in the applications of interferometry, the study of optical interference, to fundamental scientific questions. The kinetic theory of gases, as developed in nineteenth century by the English mathematical physicist James Clerk Maxwell (1831-1879) and the Austrian Ludwig Boltzman (1844-1906), made it possible to predict the number of molecules of any gas moving with a given velocity at each temperature. The Austrian physicist Christian Johann Doppler (1803-1853) had explained somewhat earlier how the motion of a source, while emitting a wave, would be reflected in a slight change of the frequency of the emitted wave. By 1912 Fabry and Henri Buisson were able to confirm that the frequency shifts found for helium, neon, and krypton were exactly as predicted by the Maxwell-Boltzmann theory.

Fabry also was interested astronomy—his brother Louis had become an astronomer. He conducted a number of studies of the spectra of the stars using his interferometer. In the course of this work he was the first to recognize the absorption of ultraviolet light by ozone in the upper atmosphere. Fabry was elected a member of the French Academy of Sciences in 1927 and received medals from the National Academy of Sciences in the United States and the Royal Society in England. He also served as an advisor to the International Committee on Weights and Measures.

The precise determination of atomic spectra was to provide one of the bases for the establishment of quantum theory. The combination principle developed by Lord Rayleigh (1842-1919) and Walter Ritz (1978-1909) showed that the numerous frequencies in the visible spectrum of a given atom could each be expressed as differences of a smaller set of frequencies or "terms." The terms would eventually be identified with the energy levels of the atom concerned. The terms for hydrogen were particularly simple in form and could be fit to a simple formula, which the Danish physicist Neils Bohr (1885-1962) was able to derive in 1913 from classical physics by placing a single quantum requirement on the allowed orbits of the single electron. Explaining the spectral terms of atoms with more than one electron required the development of the full quantum theory a decade later.

DONALD R. FRANCESCHETTI

Enrico Fermi
1901-1954
American Physicist

Enrico Fermi discovered a way to induce artificial radiation in heavy elements by shooting neutrons into their atomic nuclei. Using this technique he produced the first transuranium elements; that is, heavy elements that appear after uranium in the periodic table. For these accomplishments, Fermi was awarded the Nobel Prize in Physics in 1938. He was also instrumental in the design and construction of the world's first nuclear reactor at the University of Chicago. December 2, 1942, when a self-sustaining chain reaction was first achieved there, is often considered to be the beginning of the Atomic Age.

Fermi was born on September 29, 1901, in Rome, the son of railroad employee Alberto Fermi and his wife, Ida de Gattis. A precocious student, Fermi had decided to become a physicist before he had graduated from high school. He received his doctorate in physics at the University of Pisa in 1922. After postdoctoral studies with Max Born (1882-1970) at Göttingen and Paul Ehrenfest (1880-1933) in Leiden, he lectured in mathematics at the University of Florence. There, he published his first important

Enrico Fermi. *(Library of Congress. Reproduced with permission.)*

theoretical papers. By the age of 25, he had obtained a professorship at the University of Rome and begun his experimental work. Fermi married Laura Capon in 1928. Their daughter Nella was born in 1931, and their son Giulio followed in 1936.

In 1934, the French physicists Frédéric (1900-1958) and Irène Joliot-Curie (1897-1956) had first produced artificial radioactivity by bombarding atoms with helium nuclei, or *alpha particles*. However, this technique ran into difficulties with heavy elements, which had many positively charged protons in their nucleus to repel the positively charged alpha particles. The newly discovered neutron, with no charge at all, seemed to Fermi to be a good projectile to throw at these heavy atoms. The neutrons were even more effective when slowed down by collisions with the atoms in hydrogen-rich materials such as water or paraffin.

Fermi began announcing transuranium elements in 1934. It was not until 1939, after Otto Hahn (1879-1968) and Fritz Strassmann (1902-1980) had performed the same experiment, that Lise Meitner (1878-1968) and her nephew Otto Frisch (1904-1979) realized that Fermi had also been the first to split the atom. "We did not have enough imagination," he later wrote. Meitner realized that nuclear fission was accompanied by the release of enormous amounts of energy, in accordance with the famous equation of Albert Einstein (1879-1955), $E = mc^2$.

While Fermi had been working on his neutron bombardment experiments, he and his wife, who was Jewish, had grown increasingly concerned about the Fascist regime in Italy. Having received permission to leave the country to pick up Fermi's Nobel Prize in Sweden in 1938, the family fled to the United States. After three years at Columbia University, he went on to the University of Chicago.

Fermi was alarmed by the possibility of Hitler's Germany harnessing the massive destructive power of nuclear fission. With this in mind, he helped compose a letter signed by several eminent physicists and delivered to President Roosevelt by Einstein, urging him to establish what came to be called the Manhattan Project. Fermi's role was to develop the nuclear reactor, or "atomic pile" as he called it. The reactor was built in an unused squash court at the University of Chicago and went into operation in 1942. Fermi became an American citizen in 1944. He continued with the Manhattan Project in Los Alamos, New Mexico, where the first atomic bomb was tested on July 16, 1945. A few weeks later, bombs were dropped on Hiroshima and Nagasaki.

After the war, Fermi returned to the University of Chicago. His later research concentrated on nuclear particles, especially mesons. A meson is a packet of energy, or quantum, corresponding to the strong force that holds the nucleus together, just as the photon is a quantum of light. Fermi died in Chicago on November 28, 1954.

SHERRI CHASIN CALVO

Otto Hahn
1879-1968
German Chemist

Often regarded as the leading nuclear and radiochemical experimentalist of the twentieth century, Otto Hahn won the 1944 Nobel Prize for his discovery of nuclear fission. Widely respected for both his scientific research and personal integrity, he also played a leading role in reestablishing scientific research in Germany following the destruction of World War II.

Hahn was one of four children born to a professional glazier, and was initially attracted to organic chemistry in college, taking his doctorate at Marburg University in 1901 under Theodor Zincke. After a year of infantry service, he returned to work as Zincke's assistant in 1903. Dissuading Hahn from his intention to work in industry, in 1904 Zincke obtained a position for him as a research assistant in London with William Ramsay (1852-1916), where he

Otto Hahn. *(Library of Congress. Reproduced with permission.)*

isolated radiothorium, a radioactive isotope of thorium, by chemical analysis of a radioactive mineral blend. In 1905 Hahn left to spend a year with Ernest Rutherford (1871-1937) at McGill University in Montreal, where he repeated his success under Ramsay by discovering radioactinium.

In 1906 Hahn returned to Germany to work in the Berlin Chemical Institute headed by Emil Fischer (1852-1919), where in 1907 he was promoted to Privatdozent (non-stipendiary lecturer) and in 1910 to a professorship. Two other significant developments also occurred in 1907: Hahn identified mesothorium, an intermediate radioactive isotope between thorium and radiothorium, and he began a 30-year collaboration with the brilliant female physicist Lise Meitner (1878-1968). In 1912 Hahn and Meitner moved to positions at the newly established Kaiser Wilhelm Institute for Chemistry, for which Hahn later served as president in 1928.

At the Institute, Hahn was initially engaged in supporting the German military effort in World War I by research under Fritz Haber (1868-1934) on poisonous gases, a role he later greatly regretted. During the 1920s he studied emissions of beta particles (electrons ejected from a nuclear proton that changes into a neutron) by extremely weak radioactive substances, particularly radioisotopes of potassium and rubidium. His method of determining the production of strontium by the rate of radioactive decay and half-life of rubidium was subsequently utilized as a new method for dating geological strata and artifacts. In 1921 he also discovered "uranium-Z," the first nuclear isomer, though its nature remained unexplained for more than a decade until the discovery of the neutron and of artificially induced radioactivity by neutron bombardment in the 1930s.

Another important area of research was Hahn's development of the "emanation method" to study the character of and changes in the surfaces of finely divided solution precipitates. A radioactive "tracer" element was added to the precipitate, and the rate of diffusion was then measured by tracking the path and rate of emanation of a rare gas due to radioactive decay. The method proved particularly useful for working with minute quantities of matter insufficient for measurement by other techniques, and also provided information on temperature changes and crystal lattice structures. Hahn showed a direct correlation between the surface/volume ratio of the precipitate and the emanation rate.

During the 1930s, with the discovery of the neutron by James Chadwick (1891-1974), the development of neutron bombardment techniques by Enrico Fermi (1901-1954), and the creation of artificial radioactivity using Fermi's techniques by Frédéric Joilot-Curie (1900-1958) and Irène Joilot-Curie (1897-1956), Hahn's research interests shifted to the study of decay patterns and products of nuclear isotopes, particularly the transuranium elements, using electrolysis and precipitation techniques. His refusal to cooperate with the new Nazi regime made his position at the Institute increasingly difficult, especially after Lise Meitner, an Austrian Jew, was forced to flee to Sweden in 1938. Later that year, Hahn and his colleague Fritz Strassmann sent a letter to Meitner with the baffling report that neutron bombardment of a uranium sample had not produced radium as expected, but barium instead. Meitner supplied the correct interpretation of the result as the first observed example of nuclear fission.

During World War II, Hahn concentrated on research of fission products, compiling a table of over 100 nuclear isotopes by 1945. Captured by Allied troops and interred with other leading German scientists in England, Hahn learned of the belated award of the 1944 Nobel Prize for his work on fission, but heard with disbelief and despair the news of the dropping of atomic bombs on Hiroshima and Nagasaki. As Germany's most prestigious physical scientist who had not been involved in atomic weapons research during the war, Hahn was asked to take direction of the newly re-founded Kaiser Wilhelm Institute, renamed the Max Planck Institute in 1948. Despite his advanced years, Hahn worked energetically to reestablish scientific research in Germany, and became an outspoken opponent of nuclear weapons and a cautionary critic of nuclear power.

JAMES A. ALTENA

Werner Karl Heisenberg

1901-1976

German Physicist

Werner Heisenberg was a pioneer of quantum mechanics, a theory that deals with atomic particles in terms of probability functions. He was awarded the Nobel Prize in physics in 1932 for his formulation of quantum mechanics using matrices, or arrays of mathematical expressions. He is best known for his uncertainty principle, which states that it is impossible to simultaneously and precisely measure the position and velocity of a particle.

Heisenberg was born on December 5, 1901, in Würzburg into a well-to-do academic family. He was a bright and ambitious student, always at the top of his class at the *gymnasium* (secondary school), and also enjoyed playing classical piano. World War I disrupted this comfortable life, bringing shortages of food and fuel. With most men serving in the military, students were expected to be tough and independent and were indoctrinated with nationalistic views. Heisenberg belonged to a military training unit at his school, and was elected leader of a group associated with the New Boy Scouts, a right-wing youth movement.

At the University of Munich, Heisenberg studied under Arnold Sommerfeld (1868-1951); Wolfgang Pauli (1900-1958) was among his classmates. He obtained his doctorate in 1923 with the first mediocre grade of his life, derived from averaging an outstanding performance in theoretical physics with a hopeless showing in laboratory work. After a postdoctoral year in Göttingen with Max Born (1882-1970), he went to the Institute for Theoretical Physics in Copenhagen, Denmark, to work on quantum theory with Niels Bohr (1885-1962).

Werner Heisenberg. *(Library of Congress. Reproduced with permission.)*

The quantum theory described light in terms of particle-like photons, with energy related to their wavelength. Heisenberg realized that this model implied a limit to the accuracy of certain measurements. An object is typically observed by bouncing electromagnetic radiation off it. Ordinary visible light is not useful for looking at something the size of an electron, because such light's wavelength is too long. For sufficient resolution, the wavelength must be comparable to or smaller than the size of the object being observed. Photons with very short wavelengths, such as gamma rays, have very high energy. When they strike an electron, they change its velocity. The attempt to observe the situation thus affects it.

Heisenberg published his uncertainty principle in 1927. The same year, he was appointed professor at the University of Leipzig. As the Nazi regime rose to power, many other physicists urged him to leave Germany. He refused, despite increasingly shrill attacks by the Nazis upon those working in theoretical physics, which they viewed as a Jewish enterprise. In 1937, deemed guilty by association with his colleagues, he was labeled a "White Jew" and threatened with imprisonment in a concentration camp. Shaken, he pleaded his case to Nazi chief Heinrich Himmler and was "cleared."

With the outbreak of World War II, Heisenberg received orders to report to the Army Weapons Bureau, where he headed the German effort to build a nuclear weapon. Together with Otto Hahn (1879-1968), who had discovered nuclear fission, he worked on a reactor, but failed to develop a bomb, to the great relief of the rest of the world. Controversy remains to this day on whether Heisenberg's project was unsuccessful because he was provided with insufficient resources, was as a theorist a poor choice to head up what was essentially an engineering project, or because he was reluctant to arm Adolf Hitler with nuclear weapons. The latter, although an appealing theory, is dismissed by most historians. Heisenberg wasn't a Nazi, but he was steeped in German nationalism.

After the war Heisenberg was briefly interned in England with other German nuclear scientists but was released in 1946. He was appointed director of the Max Planck Institute for Physics, first at Göttingen and, after 1958, in Munich. His goals were to rebuild not only his institute but also German science as a whole, working with international authorities and the West German government. He died in Munich on February 1, 1976.

SHERRI CHASIN CALVO

Ejnar Hertzsprung
1873-1967
Danish Astronomer

Ejnar Hertzsprung introduced the concept of absolute magnitude, the intrinsic brightness of a star. He worked out the relationship between a star's brightness and its color (which indicates its surface temperature) at different stages in its evolution. The American astronomer Henry Russell (1877-1957) independently arrived at this relationship a few years later, and was the first to publish it in the form of a diagram. *Hertzsprung-Russell diagrams,* as they came to be called, are still among the standard tools of astronomy. They are used to provide insights into the changes in individual stars over time, as well as to characterize populations of stars.

Hertzsprung was born in Frederiksberg, Denmark, near Copenhagen, on October 8, 1873. He was trained as a chemical engineer at the Frederiksberg Polytechnic, and spent a few years working in Russia. He came to astronomy indirectly, through his interest in photography. In 1902 he began making photographic measurements of starlight at the Copenhagen Observatory. He published two papers, one in 1905 and the other two years later, in which he described the relationship between the color of stars and their absolute magnitude.

When we look up at the sky and notice how bright a star is, we are thinking of the apparent magnitude; that is, the brightness as seen from Earth. Naturally, this is dependent upon how far away it is. Hertzsprung defined absolute magnitude as the brightness of a star at the distance of 10 parsecs (32.6 light years), without regard to how it appears from our vantage point. His work provided a means to approximate the absolute magnitude of a star by virtue of its color. Comparing the absolute and apparent magnitudes enabled astronomers to estimate the star's distance.

This important accomplishment led to Hertzsprung's recognition as an astronomer. In 1909 Karl Schwarzschild (1873-1916), director of the Potsdam Observatory, offered him a position as senior astronomer there. While at Potsdam, Hertzsprung studied Cepheid variable stars. These stars are another important distance gauge, because their period is related to their absolute magnitude.

The chart now known as the Hertzsprung-Russell diagram was first published in 1913. It shows temperatures increasing from right to left, and brightness increasing from bottom to top. The main sequence of stars extends diagonally from the hottest, brightest, shortest-lived stars in the upper left to the dim, cool, long-lived stars in the lower right. Our Sun is an "average" star, around the middle of the main sequence.

Stars "fall off" the main sequence in the later stages of their life cycle. They might be found with the large, bright, cool red giants in the upper right, or the small, dim, hot white dwarfs in the lower left. Plotting a diagram for an entire population of stars, such as a globular cluster, gives an indication of the population's age. For example, if it is fairly old, the upper left-hand corner will be empty, because the hottest stars burn themselves out relatively quickly.

Hertzsprung became assistant director of the Leiden Observatory in the Netherlands in 1919, and published a catalog of color measurements for almost 750 stars in 1922. He was appointed director of the observatory in 1935. After he retired in 1945, he returned to Denmark. Free of administrative responsibilities, he continued his astronomical research until 1966, when he was 93 years old. He died on October 21, 1967.

SHERRI CHASIN CALVO

Edwin Powell Hubble

1889-1953

American Astronomer

Edwin Hubble's contributions opened the way to modern cosmology, the study of the universe. He first realized that many of the nebulae, often seen only as faint fuzzy patches in the sky, were actually separate galaxies like our own Milky Way. He also discovered that the universe is expanding. This eventually led to the theory that it began in a massive explosion known as the Big Bang.

Hubble was born on November 20, 1889, in Marshfield, Missouri. As an undergraduate at the University of Chicago, he earned degrees in both mathematics and astronomy. Despite his interest in science, he had promised his dying father he would study law. He went to Oxford University as a Rhodes Scholar, graduating in 1912. Having found that legal work did not hold his interest, he dissolved his Kentucky practice soon after he established it. He did retain from his Oxford experience a rather affected English accent and a taste for Savile Row tailoring—clothes made by the finest tailors in London.

Resuming his interrupted study of astronomy, Hubble returned to the University of Chicago. He received his Ph.D. in 1917. After serving in World War I, he obtained a staff position at the Mt. Wilson Observatory in California, where he worked with the 100-inch telescope there. At the time it was the most powerful telescope in the world.

It was in studying a type of variable stars called *Cepheids* that Hubble founded the discipline of extragalactic astronomy. Cepheids are particularly useful objects to observe because the period of their variability is related to their absolute magnitude, or intrinsic brightness. By comparing their absolute magnitude to their brightness as seen from Earth, it is possible to determine their distance. Hubble was able to observe Cepheids within some of the nebulae. These fuzzy objects were assumed to be within our Milky Way galaxy, which was the entirety of the known cosmos. To his surprise, his calculations indicated they were millions of light years away. In fact, he announced in 1924, those nebulae were "island universes," galaxies in their own right.

This was a major paradigm shift for astronomers, but Hubble quickly went about classifying and studying the galaxies he could observe. In doing so, he made another spectacular discovery. Like the falling pitch of a whistle on a

Edwin Hubble. *(Library of Congress. Reproduced with permission.)*

train passing by, the light from the galaxies was Doppler-shifted toward the red end of the spectrum. This indicated that the galaxies were speeding away from Earth. What's more, the farther away they were, they faster they were receding. The ratio of their speed to their distance was a number now called *Hubble's constant,* or H_0.

The importance of this work to cosmology can hardly be overstated. The assumption had always been that the universe was static and eternal. Suddenly the galaxies were found to be flying off in every direction. There was no reason that they should be fleeing the vicinity of Earth in particular. The only logical answer was that everything was receding from everything else. In other words, the universe was expanding, with the galaxies moving apart like raisins in a rising loaf of bread. Working backward from Hubble's constant, it was possible to estimate an age for the universe. Although the value for H_0 has been repeatedly revised and remains in dispute, the universe is generally believed to have been expanding for 10 to 15 billion years.

Hubble continued to study galaxies for the rest of his life, except for a hiatus during World War II, when he enlisted as head of ballistics at the Aberdeen Proving Ground in Maryland. His accomplishments brought him fame; he was popular with tourists as well as Hollywood celebrities, and he received many honors and awards. Only the Nobel Prize eluded him, despite his hiring a publicity agent in the 1940s, because there was no prize in astronomy at the time. Eventually the Nobel Committee decided that astronomy was a branch of physics, but it was too late for Hubble. He died on September 28, 1953, in San Marino, California.

SHERRI CHASIN CALVO

Heike Kamerlingh Onnes
1853-1926
Dutch Experimental Physicist

Heike Kamerlingh Onnes is best known for his life-long investigations into the properties of matter at very low temperatures. The very first experimental physicist to liquefy helium in 1908, Kamerlingh Onnes went a step further in the field of cryogenics when, in 1911, he discovered what is called today superconductivity.

Born in Groningen, The Netherlands, in 1853, Kamerlingh Onnes entered the city's university at age 17. After receiving his "candidaats" degree (approx. B.Sc.) in 1871, he went to Heidelberg, Germany, as a student of Robert Wilhelm Bunsen (1811-1899) and Gustav Robert Kirchhoff (1824-1887) from October 1871 to April 1873. He then returned to Groningen to complete his studies, where he passed his "doctoraal" examination (approx. M.Sc.) in 1878; a year later he was awarded the doctorate *magna cum laude.* Kamerlingh Onnes's academic career began when he was appointed assistant at Delft Polytechnic School, a position he held until 1882 when, at the young age of 29, he became the new professor of physics at Leiden University. He would stay in Leiden for the next 42 years.

The first thing Kamerlingh Onnes did, as new professor of physics in Leiden, was to reorganize the physical laboratory in such a way that it would suit his own program of experimental physics. This he asserted right from the start in the motto taken from his inaugural address: "Door meten tot weten," or "Knowledge through measurement." Based on the solid ground of theory developed by two eminent Dutch contemporaries—Johannes Diderik van der Waals (1837-1923) and Hendrik Antoon Lorentz (1853-1928)—Kamerlingh Onnes undertook his low-temperature studies while simultaneously establishing one of the first cryogenic laboratories in the world. Years of effort culminated in the liquefaction of helium in 1908 (something that happens at the very low temperature of 4.2

Kelvin or -451.84°F). Leiden's laboratory, at the time, was justly nicknamed "the coldest spot on Earth." From then until his retirement in 1923, Kamerlingh Onnes would remain the undisputed monarch of low-temperature physics.

Just three years later, in 1911, while doing experiments with extremely cold mercury, he discovered an entirely new phenomenon, which he called *supraconductivity* (later *superconductivity*). Maintained at very low temperatures, some materials become superconductors, which means that, as Kamerlingh Onnes discovered, they display virtually no resistance to electric currents. Even though the experiment was easily reproduced afterwards, it took 46 years before three American physicists explained theoretically the underlying mechanism for superconductivity. Yet another very important phenomenon tied to very low temperature physics was the discovery of superfluidity. A superfluid is a liquid that, when cooled to no more than a few degrees above absolute zero (-459.4°F), lacks all inner friction.

When, in 1913, Kamerlingh Onnes received the Nobel Prize for physics "for his investigations on the properties of matter at low temperatures which led, *inter alia,* to the production of liquid helium," he opened up new vistas of experimental and theoretical researches. In fact, besides his own, seven other Nobel prizes (totaling 14 people) were awarded for work on low-temperature physics, the latest being in 1998. Superconductors are now widely used in hospitals in the form of magnetic resonance imaging (MRI) machines, and in the fields of high-energy physics and nuclear fusion. Furthermore, materials with superconductivity properties are investigated so that some day they can be used in levitating trains.

Kamerlingh Onnes is one of the most prominent scientists the Netherlands has ever produced. In Leiden today, the town's old university physics laboratory (also called the Kamerlingh Onnes Laboratory) is a tourist site, as is the Boerhaave Museum, which displays the thermos flask in which liquid helium was collected for the first time and the helium liquefier that made Kamerlingh Onnes's original experiment possible.

JEAN-FRANÇOIS GAUVIN

Frederic Stanley Kipping

1863-1949

English Chemist

Frederic Stanley Kipping is best known for his pioneering work on the organic chemistry of silicon. Kipping's success in the preparation of silicone in 1904 led to 40 years of work at the fascinating frontier between organic and inorganic chemistry. This work provided the basis for the development of silicone compounds that fulfill a wide variety of industrial applications.

Kipping was one of seven children born into the family of a Manchester bank official. After attending Manchester Grammar School, he enrolled in Owens College, Manchester (now Manchester University) in 1879. He received a London University external degree in 1882 and took a position as a chemist for the Manchester Gas Department. He went to Munich in 1886 to work with William Henry Perkin, Jr. (1860-1929), who was known for his studies of the structure and synthesis of natural products. This work was carried out in the laboratory of the eminent organic chemist Johann Friedrich Adolf von Baeyer (1835-1917).

When he completed his work in Munich, Kipping took a position as demonstrator for Perkin, who had been appointed professor at the Heriot-Watt College, Edinburgh. Perkin and Kipping collaborated in writing a very successful textbook of *Organic Chemistry*, which went through several editions after it was first published in 1899. It remained the standard organic textbook for 50 years.

In 1890, Kipping was appointed chief demonstrator in chemistry for the City and Guilds of London Institute, where he worked for the chemist Henry Edward Armstrong (1848-1937). In 1897 Kipping was elected a Fellow of the Royal Society and moved to University College, Nottingham (now Nottingham University) as Professor of Chemistry. He remained at the University until his retirement in 1936.

The study of optically active compounds was Kipping's first research interest. While working with William Jackson Pope (1870-1939), an organic chemist who specialized in stereochemistry, Kipping investigated camphor derivatives (1890-1896) and nitrogen containing compounds (1900-1905). In 1899, Kipping also began working on organic compounds containing silicon. At first he was primarily interested in preparing optically active silicon compounds. Silicon is one of the most abundant elements in the Earth's crust, but research on this rather intractable element had been very difficult. The method of synthesis that had just been developed by François Auguste Victor Grignard (1871-1935) facilitated this work. Using the newly available Grignard reagents, Kipping was able to synthesize many organic compounds containing one or more atoms of silicon. He also showed that it was possible to create long chains made up of alternating silicon and oxygen atoms. Kipping's studies of organic silicon compounds from 1900 were published in a series of 51 papers. In 1927 he first characterized polymers of alternating silicon and oxygen atoms as macromolecules, but he thought that the repeating unit was essentially a ketone. He therefore called the polymers silicones. This name has persisted, although the term "siloxanes" would be more appropriate.

Silicones exhibit exceptional high temperature stability and water resistance that make them valuable substitutes for greases and oils. Silicones can be prepared in forms ranging from free-flowing liquids to heavy greases. During World War II silicones were used as synthetic rubbers, water repellents, hydraulic fluids, greases, and so forth.

Many interesting and useful silicon compounds have been synthesized, but thermodynamic data indicate that many potential silicon analogs of carbon compounds cannot be prepared because they would be too unstable or too reactive. Therefore, it does not seem possible for silicon-based life forms to exist. Several science fiction writers have proposed the existence of silicon-based life forms, but scientists generally agree that its chemistry makes it an unlikely basis for alien life forms. In honor of Kipping, the most prestigious prize for research on silicon chemistry is called the Frederic Stanley Kipping Award in Silicon Chemistry.

LOIS N. MAGNER

Max von Laue

1879-1960

German Physicist

Max von Laue is best known for his work with the diffraction of x rays by crystalline solids, including the determination of the molecular structures of crystalline materials. He received the Nobel Prize for Physics in 1914 in recognition of his work in this field.

A precocious youngster, Laue became interested in science at the age of 12. He studied at the Universities of Strassburg, Göttingen, Munich, and Berlin, where he worked with Max Planck (1859-1947) and received his doctorate. Subsequently, Laue worked at Göttingen, Berlin, and

Max von Laue. *(Library of Congress. Reproduced with permission.)*

Munich. In 1909 he wrote the first published monograph on the special theory of relativity.

X rays were discovered in 1895 by W. C. Roentgen (1845-1923). Some scientists regarded them as particles or pulses, while others considered them to be electromagnetic radiation like light. In 1912 Laue proposed a way to resolve the disagreement. When a beam of light, whose wavelength is similar to the spacing of the parallel slits in a grating, passes through these slits, the beam is diffracted, and the light is observed to travel in various directions, not just in the direction of the impinging radiation. This phenomenon is the result of constructive interference of the beams of light exiting the various slits. It was accepted that the atoms making up a crystal are arranged in an orderly geometric fashion, with each atom located at exactly the same distance from its neighbors. This spacing is similar in magnitude to the supposed wavelength of x rays. Laue proposed that if a beam of x rays is indeed made up of radiation waves, in analogy with light shining through a grating, a beam of x rays passing through a crystal should be diffracted.

In 1912 Walter Friedrich (1883-1968) and Paul Knipping, under Laue's direction, confirmed this postulate, demonstrating that a beam of x rays, passing through a crystal, is diffracted in very precise directions. This proved that x rays are a form of electromagnetic radiation. Laue further established that the pattern of the diffracted beams is a property of the individual crystal and is related to the spacing of the atoms in the crystal. Thus, the pattern of the diffracted x rays could be used to determine the crystal's internal structure.

Laue was awarded the 1914 Nobel Prize for Physics in recognition of these discoveries. His basic approach has been developed over the years to provide methods for the determination of hundreds of molecular structures, ranging from simple diatomic molecules to proteins. These methods based on Laue's discovery have provided molecular information that could not have been obtained otherwise and have dramatically facilitated the development of molecular biology and chemistry. Laue is regarded as the founder of crystallography and as one of the founders of solid-state physics.

Laue also did research in quantum theory, relativity theory, radioactivity, and superconductivity. In 1919 he became the director of the Institute for Theoretical Physics at the University of Berlin. He remained with the Institute throughout the Second World War. Although he remained in Germany during the war, he did not support the Nazi regime and assisted others to escape. After the war, he was instrumental in rebuilding the German scientific enterprise in cooperation with the occupation forces. In 1951, at the age of 71, he accepted the position of director of the Max Planck Institute for Research in Physical Chemistry.

Laue enjoyed adventure throughout his life. He was an avid motorcycle rider and mountain climber. He died in an automobile accident at the age of 81. His autobiography, published posthumously in 1961, contains numerous personal anecdotes, including his experiences during the war years, and adds a humanizing touch to the German scientific community during the first half of the twentieth century.

J. WILLIAM MONCRIEF

Walter Samuel McAfee

1914-1995

African American Astrophysicist

As mathematician for the U.S. Army's Project Diana in 1946, Walter S. McAfee helped advance lunar exploration in its earliest stages. His calculations made it possible to bounce a radar signal off the Moon, thus confirming that radio contact would be possible across space. An African American, he did not receive credit for his contributions until many years later.

Born in Ore City, Texas, on September 2, 1914, McAfee was the second of nine children to Luther (a carpenter) and Susie McAfee. He earned his bachelor's degree at Wiley College in 1934 and went on to earn an M.S. at Ohio State University in 1937. He would have continued his graduate studies at that point but, lacking funds, taught physics at a Columbus, Ohio, junior high school from 1939 to 1942. During this time, he met French teacher Viola Winston, whom he married in 1941. They had two daughters.

In 1942 McAfee went to work for the theoretical studies unit of the Electronics Research Command under the U.S. Army Signal Corps at Fort Monmouth, New Jersey. He soon became involved with Project Diana. The project's purpose was to bounce a radar signal off the Moon's surface, which would confirm that it was possible for a high-frequency signal to penetrate the Earth's dense ionosphere. Earlier efforts with low- and medium-frequency signals had failed.

One challenge the researchers faced was the discrepancy between the Earth's speed of rotation and that of the Moon. The Moon's speed varies by 750 mph (1,200 kph) from the Earth's speed of rotation. McAfee performed the calculations necessary to ensure that the radar pulse could bounce off the Moon, and on January 10, 1946, the team sent out the signal from a special antenna. Two and a half seconds later, they received a faint radio echo, thus establishing that communication with a spacecraft on the Moon was possible.

Initial reports on Project Diana failed to mention McAfee, and he did not receive any credit until the project's twenty-fifth anniversary in 1971. In the meantime, he had gone on to a doctorate at Cornell University, where he studied under theoretical physicist Hans Bethe (1906-). He remained associated with the Electronics Research Command for over forty years and from 1958 to 1975 also taught nuclear physics and electronics at nearby Monmouth College.

McAfee's honors included induction into Wiley College's Science Hall of Fame in 1982. McAfee later founded a math and physical science fellowship at Wiley to encourage minority interest in those subjects. He retired in 1985 and died on February 18, 1995, at his home in South Belmar, New Jersey.

JUDSON KNIGHT

Lise Meitner

1878-1968

Austrian-Swedish Physicist and Radiochemist

Lise Meitner is best known for her role in the discovery that heavy unstable nuclei such as uranium-235 could decay by a fission process in which the nucleus could split into two pieces of nearly equal size, releasing additional neutrons and an immense amount of energy. When physicists realized that the neutrons could be used to initiate additional fission events, it became apparent that a chain reaction could realized, leading eventually to the first atomic bombs.

Lisa Meitner entered the University of Vienna in 1901. Although she was encouraged by a number of the physics faculty, including the eminent Ludwig Boltzmann (1844-1906), she received a rather unfriendly reception from the largely male student body. In 1905 she became the second woman to receive a doctorate from the University. Following graduation she moved to Berlin where she planned to study under Max Planck (1858-1947). There she began a lasting collaboration with Otto Hahn (1879-1968), a chemist interested in discovering new elements. As a physicist, Meitner was interested in understanding the alpha, beta, and gamma rays that made the creation of the new elements possible. In 1912 Meitner joined the newly established Kaiser Wilhelm Institute in Berlin, and in 1918 she was made head of its physics department. In 1926 she received a professorial appointment at the University of Berlin, but did not teach any courses.

In 1898 Marie Curie (1867-1934), assisted by her husband Pierre (1859-1906), had announced the discovery of two previously unknown elements, polonium and radium, products of the radioactive decay of uranium. In 1919, the English physicist Ernest Rutherford (1871-1937) discovered that the bombardment of nitrogen atoms by alpha particles resulted in the transmutation of some of them into oxygen atoms. Interest in atomic transmutations increased dramatically following the discovery of the neutron by Sir James Chadwick (1891-1974) in 1932. Since the neutron, unlike the alpha particle, had no charge, it could penetrate the nucleus readily. By 1934 Meitner and Hahn were trying to understand the result of experiments in which Italian physicist Enrico Fermi (1901-1954) had bombarded uranium with neutrons. In particular, they were looking for so-called transuranic element, elements heavier than uranium.

Lise Meitner. *(Library of Congress. Reproduced with permission.)*

It was during the research on uranium that Meitner was forced to leave Germany. Of Jewish ancestry, but initially protected from the Nazi racial policies by her Austrian citizenship, her situation changed abruptly when Germany annexed Austria. Colleagues arranged safe passage for her to the Netherlands, and from there she accepted an invitation to work at the new Nobel Institute in Stockholm. Continuing the research on uranium, Hahn and Fritz Strassman (1902-1980) concluded that the neutron bombardment produced isotopes of radium, because they shared the same chemical properties as barium, as had the radium isolated by the Curies. Meitner doubted this result and asked Hahn and Strassman to prove that they had in fact produced radium. When the chemical tests were run, they found that the radioactive products could not be separated from barium at all. After discussing the new results with her nephew, Otto Frisch (1904-1979), also a physicist, Meitner came to the conclusion that what had occurred was the fission, or splitting, of the uranium nucleus into two nearly equal fragments, one of them a barium isotope, with the release of additional neutrons. Meitner calculated the energy released by the process, which was quite large, and communicated the results to Hahn.

Hahn and Strassman mentioned Meitner's work in their 1939 publication on nuclear fission, but perhaps for political reasons, they understated her contribution. As physicists learned of nuclear fission, many began to think of the possibility of a nuclear explosive, leading in the United States to the Manhattan Project and the explosion of atomic bombs over Japan in 1945. Meitner had been invited to join the project, but unlike the vast majority of refugee scientists, she refused, strongly objecting to the military use of her work. Meitner continued to work in physics and to oppose the military use of atomic energy until nearly the end of her very long and productive life.

DONALD R. FRANCESCHETTI

Robert Andrews Millikan
1868-1953
American Physicist

Robert Andrews Millikan, in his famous oil-drop experiment of 1910, measured the charge of the electron. In another important experiment, he verified Einstein's equation for the photoelectric effect and determined the exact value of Planck's constant. For these accomplishments, Millikan was awarded the Nobel Prize in physics in 1923. He also provided a role model for the American-born scientist in the early twentieth century, at a time when most of the great physicists were European.

Millikan was born in Morrison, Illinois, on March 22, 1868, the second son of the Reverend Silas Millikan and his wife, Mary Jane. Nothing about his childhood, spent in the rural Midwest, particularly hinted at a future career as an eminent scientist. After graduating from high school, he worked briefly as a court reporter. His involvement in science began after his sophomore year at Oberlin College in Ohio, when his Greek professor recruited him to teach an elementary physics course. Millikan objected that he had no real knowledge of the subject, but the professor countered that it could not be any harder than Greek. He received his Ph.D. from Columbia University in 1895 and was the only physics graduate student at the time.

After a brief stint in Germany studying with Max Planck (1858-1947), Albert Michelson (1852-1931) invited him to join the physics faculty at the University of Chicago. He remained there for 25 years, attaining the status of full professor in 1910. Millikan had little patience with lecturing, believing that students learned more effectively through solving problems and performing experiments. He attempted to put this philosophy into practice by writing better physics textbooks. It was so important to him that he was checking page proofs on the morning of his wedding to Greta Erwin Blanchard in 1902.

Robert Millikan. *(Library of Congress. Reproduced with permission.)*

Millikan began attempting to measure the charge of the electron by measuring how tiny charged water droplets moved in an electric field. The results suggested that the charge on the water droplets was always a multiple of a single elementary value. However, since water evaporates quickly, he had trouble getting sufficiently reliable data. He solved this problem by using oil instead. Spraying oil droplets into his small experimental chamber, he measured how the charge they carried affected their fall.

In the photoelectric effect, light hitting metal causes electrons to be emitted. Millikan showed that their energy varied with the frequency of the incident light. From this data he was able to obtain an experimental value for Planck's constant of proportionality relating frequency to energy. During World War I Millikan played a role in developing anti-submarine devices as Vice-Chairman of the National Research Council.

In 1921 Millikan became head of the California Institute of Technology (Caltech), where he remained until 1945. He was a skilled administrator as well as an experimentalist and was known for regularly putting in 12-hour work days. Under his leadership, the school became one of the United State's leading institutions for scientific research. He also served as professor and chairman of the Norman Bridge Laboratory of Physics there.

At Caltech, Millikan's research interests involved the radiation that Victor Hess (1883-1964) had detected impinging upon Earth from outer space. Millikan confirmed their extraterrestrial origin, and coined the term *cosmic rays*. By comparing their intensity at various latitudes on Earth, and at different altitudes, he gathered data that helped in the understanding of Earth's magnetic field and the way in which it deflects the incoming radiation.

Millikan was a prolific author. In addition to publishing more than a half-dozen physics textbooks, he wrote and lectured on the reconciliation of science and religion and the proper relationship of science and society. His autobiography was published in 1950. He was active in the larger scientific community, serving as president of the American Physical Society, Vice-President of the American Association for the Advancement of Science, and member of the Committee on Intellectual Cooperation of the League of Nations. He also enjoyed playing golf and tennis. Millikan died on December 19, 1953, in San Marino, California.

SHERRI CHASIN CALVO

J. Robert Oppenheimer
1904-1967
American Physicist

J. Robert Oppenheimer led the team that developed the first atomic bomb. Between 1943 and 1945, he directed the laboratory in Los Alamos, New Mexico, where the bomb was designed and built.

Oppenheimer was born in New York City on April 22, 1904, the indulged first child of a prosperous family. At Harvard, he majored in chemistry, but an advanced course on thermodynamics attracted him to physics. He began his postgraduate work in atomic physics at the Cavendish Laboratory at Cambridge University in England. Graduate students at Cavendish were assigned much of the work involved in setting up experiments. Although Oppenheimer excelled at theoretical analysis, he was almost entirely helpless when it came to handling mechanical equipment. Being a perfectionist, these difficulties bothered him a great deal. He was therefore delighted to accept a 1926 offer from Max Born (1882-1970) to join the Institute for Theoretical Physics at Gottingen, Germany. He received his Ph.D. there the next year.

In 1929, Oppenheimer began teaching physics at the University of California at Berkeley and at the California Institute of Technology. He liked the idea of splitting his time between the two institutions, the dry air of the West was good for his rather delicate health, and he enjoyed being closer to the ranch his family eventually purchased in New Mexico.

Events of the 1930s caused Oppenheimer to take an interest in politics for the first time. The Nazis took power in Germany, and Oppenheimer attempted to orchestrate escape plans for his relatives there as well as for Jewish scientists and teachers. He also donated money to the Spanish Loyalists fighting against the Fascist military leader Francisco Franco. Closer to home, he was troubled by the poverty and suffering brought on by the Depression, and became attracted to the ideas of socialism. However, unlike many other intellectuals of the time, he did not join the Communist Party, being afraid that the organization might attempt to take advantage of his work in nuclear physics research. In 1940 he married Katherine "Kitty" Puening, a Caltech graduate student. They soon started a family, and Oppenheimer began to lose interest in radical causes.

By the end of 1941, the United States was involved in World War II. President Roosevelt was persuaded by Albert Einstein (1879-1955) and other physicists to begin the Manhattan Project to build an atomic bomb, lest Hitler obtain such a bomb and be unopposed. At first the project was decentralized, and Oppenheimer participated by making calculations in his Berkeley office. Eventually more organization was required, and Oppenheimer was chosen to head the project's main laboratory. He suggested it be located in Los Alamos, New Mexico, near the White Sands missile range and only 60 mi (95.6 km) from his beloved desert ranch. He recruited many eminent scientists, including Enrico Fermi (1901-1954) and Niels Bohr (1885-1962).

The first atomic bomb was exploded in a test on July 16, 1945. Germany had surrendered two months before. The scientists were relieved that the project had succeeded, but many had deep reservations about the power that had been unleashed. Oppenheimer was moved to quote from the Hindu scriptures, "I am become Death, the Destroyer of Worlds." Three weeks later an atomic bomb was dropped on the Japanese city of Hiroshima and, three days later, on Nagasaki. The ancient cultural center of Kyoto had been the first intended target, but was spared when Oppenheimer objected.

After the war, Oppenheimer became the director of the Institute for Advanced Study in Princeton. He also remained an important government advisor on nuclear issues, and helped draft the policies of the U.S. Atomic Energy Commission (AEC). However, in the early 1950s, Oppenheimer's patriotism was questioned, in part because of his suggestions about international cooperation in the control of nuclear energy. In the toxic atmosphere of the "McCarthy Era," old friendships from the 1930s were cited as evidence of Communist associations. A rival scientist, Edward Teller (1908-), testified against him. An AEC security panel failed to find any evidence of disloyalty, but decided nonetheless to bar him from further access to classified information. This injustice was very hard on Oppenheimer and his family.

In 1963, with the old enmities fading, the AEC, including Teller, voted unanimously to bestow upon Oppenheimer the Enrico Fermi Award for "outstanding contributions to the development, use or control of atomic energy." Oppenheimer remained at the Institute for Advanced Study until he retired in 1966. He died of throat cancer on February 18, 1967, at his New Jersey home.

SHERRI CHASIN CALVO

Linus Carl Pauling

1901-1994

American Chemist

Linus Pauling is the only person to win two unshared Nobel prizes. His award in chemistry honored his work in elucidating protein structure, in particular the alpha helix. He was also honored for his leadership in opposing the development and use of nuclear weapons, which helped to assure the establishment of the 1963 Nuclear Test Ban Treaty. Pauling is considered by many to be the greatest chemist of the twentieth century, credited with seminal work on the chemical bond, artificial plasma, an explanation of sickle cell anemia, and many other achievements.

Linus Pauling was born in Portland, Oregon, the son of a druggist. He was a curious and imaginative child. He loved to visit his uncle's house where he would spend hours lying on the floor reading encyclopedia articles. Pauling attended public elementary and high schools, but didn't get his high school diploma until 1963 because he lacked a civics course. He received his undergraduate degree in chemical engineering in 1922. He went on to study chemistry at the California Institute of Technology, where he engaged in experimental work on crystal structures and began theoretical investigations into the nature of chemical bonds. Pauling was awarded his Ph.D., *summa cum laude*, in 1925.

Thanks to a series of grants and awards, Pauling was able to continue his investigations with some of the most accomplished scientists in the world, including Arnold Sommerfeld (1868-1951), Erwin Schrödinger (1887-1961), and Niels Bohr (1885-1962). He returned to CalTech as an Assistant Professor in 1927.

Pauling was the first to apply quantum mechanics to chemistry. In 1932, Pauling published *The Nature of the Chemical Bond*, an elegant and highly successful treatise on what he was learning. He developed the concept of electronegativity, and, in 1935, he described his valence bond method, which provided chemists with a practical way to approximate the structures and reactivities of materials. Over time, he began applying his theories to biological systems, investigating fragments of proteins called peptides and using what he learned to understand larger molecules. While lecturing at Cornell University in 1937, he became embroiled in a controversy regarding the structure of proteins. Pauling's proposal, that alpha helices were an important structural component, contradicted the "cyclo" theory of Dorothy Wrinch. Ultimately, a combination of theoretical work and experiments, including x-ray diffractions studies, established the alpha helix. It is for this work that Pauling won the 1954 Nobel Prize in Chemistry. He is considered to be the father of molecular biology.

During World War II and shortly thereafter, Pauling was involved in defense work, contributing to many developments, including better rocket propellants, a device for measuring oxygen deficiencies (in aircraft and submarines), and a substitute for human blood serum. However, in 1946, he became associated with peace organizations and began a lifelong, vigorous opposition to atomic weapons. In particular, he pointed out the dangers of nuclear fallout, calculating estimates of radiation-induced genetic deformities and publicizing his results. In the face of government opposition, which included high-level investigations and restrictions on travel, he campaigned against nuclear testing. In 1958, he presented the United Nations with a petition signed by over 9,000 scientists protesting nuclear testing. This pressure contributed to the signing of the 1963 Nuclear Test Ban Treaty, which ended above-ground testing by both the United States and the Soviet Union. In recognition of these efforts, Pauling was awarded the Nobel Peace Prize in 1962.

Pauling continued his contributions to science and peace throughout his life. He was esteemed as a teacher and an humanitarian, and is also known by many for his advocacy of vitamin C. In 1994, Pauling died at the age of 93.

PETER ANDREWS

Wolfgang Pauli
1900-1958
Austrian-American Physicist

Wolfgang Pauli developed one of the most important ideas in modern physics, which came to be known as the *Pauli exclusion principle.* It holds that no two electrons in an atom can exist in exactly the same quantum state. Pauli put forth the principle in an effort to explain a feature in atomic spectra called the *anomalous Zeeman effect.* He was awarded the 1945 Nobel Prize in physics for this work.

Pauli was born on April 25, 1900, in Vienna, Austria. His father was a professor of physical chemistry. Pauli received his Ph.D. in 1921 from the University of Munich, where he studied under Arnold Sommerfeld (1868-1951) and wrote a 200-page encyclopedia article on the theory of relativity.

In 1916 Sommerfeld had extended the atomic model of Niels Bohr (1885-1962), in which electrons orbited the atomic nucleus in specified circular paths, to include elliptical orbits. The Bohr-Sommerfeld model, with three "quantum numbers" specifying the state of the electron, was able to account for most of the features seen in atomic spectra. One of the few exceptions was the anomalous Zeeman effect. In the normal Zeeman effect, a weak magnetic field caused a single spectral line to split into a triplet. Sometimes, however, it split into many lines.

In 1924 Pauli, then a lecturer at the University of Hamburg, was investigating this phenomenon. He concluded that an additional quantum number was required to completely specify the state of an electron. If a state was so specified, only one electron could occupy it. This "exclusion principle" solved the remaining problems in the atomic spectra. In 1925 George Uhlenbeck (1900-1988) and Samuel Goudsmit (1902-1978) provided a physical model for the fourth quantum number, with a value of either $+\frac{1}{2}$ or $-\frac{1}{2}$, in their hypothesis of electron spin.

In 1928 Pauli moved to Zurich, Switzerland, in order to accept an appointment as professor of theoretical physics at the Federal Institute of Technology. He soon began attempting to uncover the reason for the wide variation in the energies of particles emitted during beta decay. "Missing" energy was a major problem, because this would contradict the conservation laws of physics. In 1931 Pauli proposed that a previously unknown particle, neutral in charge and with negligible mass, carried away the energy that was not otherwise accounted for. Enrico Fermi (1901-1954) coined the term *neutrinos* for these particles. Evidence of their existence was first observed by Frederick Reines (1918-1998) and C. L. Cowan (1919-1974) in 1953. The neutrino is difficult to detect because it rarely interacts with other matter.

Pauli was known for the clarity of his scientific publications and the influence of his extensive correspondence. During his time in Zurich, he helped build the Federal Institute of Technology into a major research center in theoretical physics. In 1940 he was appointed to the Institute for Advanced Study in Princeton, New Jersey, and began the process of obtaining United States citizenship, which was finalized in 1946. However, with World War II over, he returned to Zurich, where he died on December 15, 1958.

SHERRI CHASIN CALVO

Max Karl Ernst Ludwig Planck
1858-1947
German Physicist

Max Planck revolutionized physics by originating the quantum theory, which states that all radiant energy, such as light, is made up of irreducible packets called *quanta.* A quantum of electromagnetic radiation is called a *photon.* The energy a photon carries is proportional to its frequency. So, for example, an x-ray photon is more energetic than a photon of visible light, because x rays are of higher frequency than visible light. The constant of proportionality, h, has come to be called Planck's constant. Planck's particle-like model of light provided a way to explain the emission and absorption of energy by matter. He was awarded the Nobel Prize in physics in 1918.

Planck was born on April 23, 1858, in Kiel, Germany, to a distinguished family that valued scholarship. Planck had many interests; he did well in all his school subjects from science and mathematics to the classics, was a skilled musician with perfect pitch, and enjoyed the outdoors. Upon graduation from secondary school, he found it difficult to choose a career from among all his enthusiasms, but he finally decided that his greatest strength was in physics. He received his doctorate at the University of Munich when he was only 21, and continued there as a lecturer for six years. In 1885 he became an associate professor at the University of Kiel, where his father taught law. Four years later, he was appointed to the University of Berlin, and he remained there for the rest of his professional life.

Planck developed his radiation law in an attempt to reconcile high- and low-frequency experimental evidence regarding the emission of energy by matter at various temperatures. However, when he went back to derive it from first principles, in the process originating the concept of the quantum, he found that he was forced to accept the statistical view of thermodynamics that had been put forth by Ludwig Boltzmann (1844-1906). This was rather distasteful to Planck, who preferred natural laws to be absolutely defined. He was later to join with Albert Einstein (1879-1955) and Erwin Schrödinger (1887-1961) in opposing the indeterminate nature of quantum mechanics, in which the position of a particle is defined as a probability distribution.

Planck was a greatly respected leader in the German physics community, highly influential even after his retirement from the university in 1928. He made Berlin into a major center for theoretical physics, bringing in luminaries such as Einstein, Schrödinger and Max von Laue (1879-1960), and it remained so until the rise of the Nazis threatened many of its members. Planck bravely made his protests directly to Adolf Hitler, and as his Jewish colleagues fled for their lives, he stayed in Germany to try to preserve what was left.

Planck had already endured a great deal of tragedy; his 1918 Nobel Prize was one of the few bright spots in his later life. His wife, to whom he was greatly devoted, died in 1909 after 22 years of marriage. Between 1916 and 1919 both of his daughters died in childbirth and one of his sons was killed in World War I. During World War II the house where Planck lived with his second wife, with whom he had had one son, was completely destroyed in a bombing raid. The only remaining child of Planck's first marriage became involved in a foiled plot to assassinate Hitler and was tortured and murdered by the German police in 1945. Although Planck had stoically persevered through all the previous sadness, the horror of his son's death at the hands of the Nazis was too much for the elderly physicist, and he lost his will to live. He died on October 4, 1947, in Göttingen at the age of 89.

Max Planck. *(Library of Congress. Reproduced with permission.)*

SHERRI CHASIN CALVO

Ernest Rutherford

1871-1937

New Zealand-born British Physicist

Ernest Rutherford identified alpha and beta radiation, showed that an element could transmute by radioactivity, probed the interior structure of matter, and split the atom. He has been called the father of modern physics, but was a somewhat reluctant parent, disliking many mathematical and philosophical aspects of the "new science." Rutherford preferred simple, descriptive interpretations, once saying that if a piece of physics could not be explained to a barmaid, then it was not a very good piece of physics.

Ernest was born the fourth child of Martha and James Rutherford, in the Nelson district of the South Island of New Zealand. His education had a number of fortunate coincidences. His local primary school opened the year he was eligible to start schooling, so he did not miss years like his elder brothers and sisters. He won a scholarship to Nelson College (a high school), where his mathematical abilities were encouraged. From there he won a scholarship to attend Canterbury College (a university), where he became only the 338th graduate, and performed his first important experiments in physics, using a cloakroom as his laboratory. Rutherford missed a scholarship for further study in England, but as runner-up received the award when the winner turned down the prize. He was lucky enough to arrive at Cambridge University, England, just after major reforms allowing non-Cambridge graduates to study there.

In 1895 Rutherford joined Cambridge's respected Cavendish laboratory, the same year x rays were discovered. He established his reputation in this revolutionary field before turning to another new area, radioactivity. In 1898 he identified two different types of radiation, alpha and beta. Rutherford then became professor at McGill University in Montreal, Canada. With Frederick Soddy (1877-1956) he determined that alpha radiation consisted of helium atoms (without electrons). This work culminated in their "Disintegration Theory" (1903), which stated that one element could transform into another by radiation, which some critics called alchemy.

In 1907 Rutherford moved to Manchester, England, to become one of the highest paid physics professors in the world. He was awarded the 1908 Nobel Prize in Chemistry for his work on the transmutation of elements by radiation, and joked about his own sudden "transmutation" from a physicist to a chemist. Rutherford continued looking at atomic structure, firing alpha particles at thin sheets of gold foil, and unexpectedly finding that some were greatly deflected. Rutherford commented it was "as if you fired a 15-inch naval shell at a piece of tissue paper and the shell came right back and hit you." This led Rutherford to propose that the atom had a very small positively charged nucleus with distant orbiting negative electrons.

Rutherford was knighted in 1914. During World War One he worked in submarine detection, but still found time for some inspired research. When late for a military meeting he explained that the importance of his science experiment was "far greater than that of the war." He was in fact busy splitting the atom!

After the war Rutherford moved to the Cavendish laboratory as its new head, leading it to more ground-breaking research. In one celebrated year,1932, Cavendish researchers discovered the neutron, and purposefully split lithium atoms, both experiments resulting in Nobel prizes. As his fame grew he became a popular public lecturer around the world. Rutherford preferred simple experiments, using basic equipment, with common sense interpretations of the results, saying of mathematical theorists that "they play their games with symbols ... but we in the Cavendish turn out the real solid facts of Nature." However, he gladly used theories that supported his own experimental results.

Rutherford's loud voice and heavy build could be intimidating, but his booming laugh was legendary. He guided the careers of many talented experimenters and encouraged women scientists. From 1925-30 he was President of the Royal Society. In 1931 he was made Baron Rutherford of Nelson. He died in 1937, and was buried at Westminster Abbey.

DAVID TULLOCH

Erwin Schrödinger
1887-1961
Austrian Physicist

Best known for his wave equation and a thought experiment involving an imaginary cat, Erwin Schrödinger contributed to a variety of disciplines from biology to philosophy. Like so many of his generation, his life and career were shaped heavily by two world wars and the political divisions of Europe.

Born and raised in Vienna, Austria, Schrödinger was the son of a frustrated scientist who was determined that his son would have every possible opportunity. Until the age of 11 Erwin had a private tutor, then he entered the Akademisches Gymnasium (approximately a high school) gaining a broad education, excelling at chemistry and mathematics, but also enjoying poetry, drama, and grammar, and becoming fluent in German, English, French, and Spanish. He concentrated first on chemistry, then on Italian painting, and published several papers in botany.

In 1906 he entered the University of Vienna, where he excelled in theoretical and experimental physics. Upon graduating Schrödinger did a year of military service. He then continued research in physics until the outbreak of World War I, in which he served as an artillery officer on the Italian Front.

During the war Schrödinger's favorite teacher, Fritz Hasenöhrl (1874-1916) died. Schrödinger decided to teach at Hasenöhrl's old school in Czernowitz to honor his memory. However, the city of Czernowitz (later Chernovtsy) was taken from Austria as war reparations by Russia, making this impossible. Looking back, Schrödinger wrote, "I had to stick to theoretical physics, and, to my astonishment, something occasionally emerged from it."

In 1920 he married and received a good research position in Germany. He did important work on color theory and on the statistical thermodynamics of ideal gases. However, he was unhappy in his marriage, and his research dried up. In 1925 he wrote a very introspective account of his philosophy of life.

1926 was a turning point for Schrödinger. He was inspired by Louis de Broglie (1892-1987), whose work suggested that waves and particles could be considered interchangeable (wave-particle duality). An affair with a young woman, the first of several, provided another inspiration. Secluded in a holiday villa with his new companion, he would place pearls in his ears to block out noise and work for hours uninterrupted. The six major papers that came from this year had far-reaching implications for both physics and chemistry.

Schrödinger proposed a wave interpretation of the distribution of electrons in an atom. His wave equation was entirely theoretically based, yet gave results that agreed with experimental observations. Schrödinger felt his wave mechanics offered a more appealing approach to quantum physics, providing a way of visualizing the atom, rather than the abstract statistics of other methods.

In 1933 he received the Nobel Prize in physics, shared with Paul Dirac (1902-1984). That same year the Nazi party came to power in Germany, and Schrödinger left for Oxford, England. He was lured back to Austria, but in 1938 it was annexed by Nazi Germany, and he was forced to flee, ending up in Dublin, Ireland, where he stayed until 1955.

Never entirely happy in the direction quantum physics developed, Schrödinger often found himself at odds with other scientists. He offered many criticisms of quantum interpretations, the most famous being a thought experiment known as Schrödinger's Cat, in which a quantum event produces strange results in the observable world.

Erwin Schrödinger. *(Library of Congress. Reproduced with permission.)*

He published numerous works in physics, but also a short volume of poetry, and an influential book on biology, *What is Life* (1944). Schrödinger wrote about his career, "In my scientific work (and moreover also in my life) I have never followed one main line, one program defining a direction for a long time."

DAVID TULLOCH

Frederick Soddy
1877-1956
British Chemist and Political Economist

One of the early pioneers of radiochemistry, Frederick Soddy won the 1921 Nobel Prize for Chemistry for his theory of chemical isotopes. An iconoclastic individualist, he subsequently turned away from scientific research to utopian theories of scientific economics, foreshadowing current interest in the "social responsibility of science" and the "Green" movement in eco-politics.

Soddy came from a middle-class family, and was primarily raised by an older sister after the early death of his mother. An interest in chemistry surfaced only when he completed his secondary education. He entered Oxford University in 1895 and received his degree with first-class honors in 1898, with William Ramsay (1852-1916) as his external examiner. After two years of independent research at Oxford, Soddy impulsively applied for a lectureship in Toronto, and without waiting for a reply left for Canada. Upon arriving, he discovered that there was no hope of securing the position. He then traveled to McGill University in Montreal, where physicist Ernst Rutherford (1871-1937) accepted him as a research assistant.

Between 1901 and 1903 Rutherford and Soddy produced nine groundbreaking research papers in the nascent field of radiochemistry. Studying radioactive thorium, they proved by spectral analysis that it spontaneously disintegrates into radium by emitting an alpha particle (the nucleus of a helium atom, consisting of two protons and two neutrons)—the first recorded instance of elemental transmutation. They also showed that the rate of radioactive decay is an exponential law of a mono-molecular chemical reaction, and that this rate of decay is spontaneously counterbalanced by an equal rate of increase of radioactivity in the initial active materials, according to a cyclic law of "delay and recovery."

In addition, they further discovered that: a) the initial product of radioactive decay, dubbed "thorium X," produces argon as an emanation from a radioactive intermediate product they called "thoron"; b) radon is derived from radium as a product of further alpha particle decay; and c) uranium and thorium both produce helium by radioactive disintegration. Ironically, despite Soddy's contributions, it was Rutherford alone who won the 1908 Nobel Prize for Chemistry, not Physics, for their research.

In 1903 Soddy returned to London for a year of research with Ramsay, where he showed that radium spontaneously decays to produce helium. In 1904 he finally secured a lectureship at the University of Glasgow in Scotland, where he remained for 10 years, marrying Winifred Beilby in 1908. There Soddy performed his fundamental research on alpha- and beta- particle decay of radioactive elements. In 1910 he declared that chemically inseparable radioactive and non-radioactive materials are different species of the same element, bestowing on them the name "isotope" in 1913. In 1911 he announced the "alpha ray rule," which established that an element that undergoes radioactive decay by alpha particle emission transmutes into the element two places before it in the periodic table. In 1912-13 he es-

Frederick Soddy. *(Library of Congress. Reproduced with permission.)*

tablished that beta decay (emission of an electron when a proton decays into a neutron) is a nuclear and not a chemical change, and that the new element thus formed is the next lower one in the periodic table. Remarkably, all this was done 20 years before James Chadwick (1891-1974) discovered the existence of the neutron. In 1921, nominated by Rutherford, Soddy won the Nobel Prize for his work.

In 1914 Soddy accepted a professorship at Aberdeen University, where he stayed until 1919. His time there was spent primarily on unproductive research to support the British military effort during World War I, though in 1918 he was one of several independent co-discoverers of element 91, protactinium. In 1919 he accepted a call to Oxford University, where he was expected to establish a major research program in radiochemistry. These hopes went unfulfilled; inept at university politics and social life, Soddy quickly alienated other faculty members with his intemperate sarcasm and demands for reorganization of the university curriculum, and soon found himself without significant influence or funding. He ceased to do serious research and concentrated on teaching until his wife's death in 1936, whereupon he retired.

Already, during his Glasgow years, Soddy was attracted to socialist economic theories, and soon after moving to Oxford devoted much of his time to developing a "scientific economics" for the improvement of mankind. Influenced by art critic John Ruskin, he sought to apply laws of physics to economics, arguing that the flow of money in the economy, and of savings and investment, follows the same principles governing the flow, conservation, and conversion of energy. He became associated with various small organizations generally viewed as eccentric, serving as president of one, the "New Europe Group," from 1933 to 1954. After Word War II he became a critic of nuclear weapons, but saw his 1906 prediction that nuclear energy would one day be used to generate electrical power fulfilled in 1956, when Britain opened its first nuclear power plant.

JAMES A. ALTENA

Arne Wilhelm Kaurin Tiselius
1902-1971
Swedish Biochemist

Arne Tiselius was a Nobel Prize-winning biochemist who helped to develop techniques for separating mixtures of proteins into their individual components. These methods allowed researchers to obtain specific proteins for further study.

Arne Tiselius was born in Stockholm, Sweden, in 1902, and he became interested in chemistry during childhood. He went on to study this subject at Uppsala University. A major factor in his decision to attend this school was the presence of Theodor Svedberg (1884-1971), a well-known chemist. By 1924, Tiselius had earned his master's degree in chemistry, physics, and mathematics. He remained at the university in order to work as a research assistant to Svedberg.

Svedberg was interested in isolating individual proteins from organisms. One of the techniques he considered using was electrophoresis. In this technique, a mixture of electrically charged molecules (such as proteins) is placed in an electric field. Molecules with a negative charge move toward the positive end of the field, and vice versa. Molecules with more charge move faster than do those with less charge. In addition, smaller molecules move faster than larger molecules. As a result, electrophoresis can separate a mixture of proteins into individual bands based on the proteins' charge and size.

Tiselius began working with Svedberg on the use of electrophoresis to separate proteins called albumins. Albumins are a group of similar proteins that help to transport substances in the bloodstream. By 1930, Tiselius had earned his doctorate degree and had become an assistant professor in chemistry. At the time, the field of biochemistry—which includes the study of proteins—was not officially recognized at Uppsala University. Therefore, Tiselius decided to turn to more traditional fields of chemistry in order to become a full professor.

For several years, he studied zeolite minerals. The crystals of these minerals adsorb water; that is, they attract water to their surfaces. These crystals can also adsorb other chemicals such as ethyl alcohol, bromine, and mercury. While working with these minerals at Princeton University in the United States, Tiselius met biologists and chemists who knew of his work with Svedberg. They encouraged him to continue with his investigation into electrophoresis because they felt it would be of use to their own projects in protein research.

When Tiselius returned to Uppsala, he once again began to experiment with electrophoresis. He developed new equipment that allowed him to obtain more accurate data, and he was eventually able to separate proteins from the blood serum of horses. Serum is the liquid portion of the blood from which certain molecules have been removed. Electrophoresis separated the serum proteins into four bands. One band consisted of albumins; the other three were unknown. Tiselius named them alpha, beta, and gamma globulins. It is now known that alpha and beta globulins have several important functions such as transporting other substances and aiding in blood clotting. Gamma globulins are also known as antibodies; they play an important role in the immune system. Tiselius's work revealed the complexity of blood proteins, and as a result of this research, he was made the first professor of biochemistry at Uppsala University.

Tiselius soon realized that some biological chemicals are so similar that electrophoresis cannot separate them. In the 1940s, he began to study how adsorption chromatography could be used to isolate proteins. Adsorption chromatography relies on the fact that some chemicals adsorb, or adhere, more strongly to certain materials than do other chemicals. Tiselius developed a method of adsorption chromatography that involves a column packed with zeolite crystals, which act as an adsorbent. When a mixture of proteins is poured into the column, the proteins least attracted to the zeolite crystals pass through the bottom of the column first, and those most attracted to the crystals pass through the column last. Tiselius and Frederick Sanger (1918-1982) used this technique to separate the hormone insulin into its four components.

In 1948, Tiselius was awarded the Nobel Prize for Chemistry for his work on the separation of proteins using electrophoresis and adsorption chromatography. Between 1960 and 1964, he served as president of the Nobel Foundation. He received 11 honorary degrees from various universities and won numerous other awards.

STACEY R. MURRAY

Harold Clayton Urey
1893-1981
American Chemist

Harold Clayton Urey is best known for his work with isotopes. He received the Nobel Prize for Chemistry in 1934 for the discovery and isolation of deuterium, a heavy isotope of hydrogen. He also played a major role in the development of the atomic bomb and made important contributions to the study of isotopes of a number of elements and to the theory of the origin of the planets.

After receiving his doctorate at the University of California at Berkeley, Urey worked with Niels Bohr (1855-1962) in Denmark. Returning to the United States, he held positions at Johns Hopkins University, Columbia University, the Institute of Nuclear Study, the University of Chicago, and the University of California.

Isotopes of an element have the same number of protons (i.e., atomic number) but a different number of neutrons—and, therefore, different mass numbers. Much of Urey's research concerned the heavy isotope of hydrogen known as deuterium. The most common form of hydrogen is the simplest atom that can exist, having one proton in its nucleus and one associated electron. Its atomic number is 1, and its mass number is 1. The deuterium atom contains a neutron in its nucleus in addition to a proton. Its atomic number is 1, but its mass number is 2. Since both hydrogen isotopes have one electron and one proton, they are virtually identical in their chemical reactivity, but properties that are dependent on mass are significantly different. Urey noticed that the mass of hydrogen gas isolated from naturally occurring substances is minutely larger than would be expected if only hydrogen with mass number 1 was present. He proposed the existence of deuterium to explain this observation. In 1931 he announced the discovery of heavy water, which contains two atoms of deuterium instead of two "normal" hydrogen atoms. In 1931 Urey announced the production and isolation of pure deuterium, which he obtained by successive distillations of liquid hydrogen.

Urey was awarded the Nobel Prize for Chemistry in 1934 for his work on deuterium. He subsequently applied the methods that he had developed for the study of the isotopes of hydrogen to the separation and study of the isotopes of carbon, nitrogen, oxygen, and sulfur.

During World War II, Urey's expertise in the separation of isotopes resulted in his playing an important role in the Manhattan Project, the successful American effort to develop an atomic bomb. He directed the program at Columbia University that developed a method, employing the gaseous diffusion of uranium hexafluoride, for separating the radioactive isotope of uranium (U-235), which could be used in the bomb's fission reaction, from its more abundant isotope (U-238). After the war, he spoke out against the

ELECTROPHORESIS CRACKS THE CASE

Every person (except for identical twins) has a unique set of genetic material in the form of DNA. A technique called DNA fingerprinting takes advantage of this individuality. To create a DNA fingerprint, scientists must first acquire a sample of a person's cells, such as from blood, hair, or skin. After the DNA is removed from the cells, special proteins are added to it that cut the extremely long DNA molecules into fragments. The next step is to separate the fragments from one another based on their size. This is done by using electrophoresis—the technique Arne Tiselius helped to develop. Electrophoresis separates the fragments into bands. The pattern formed by the bands is different for each person.

The technique of DNA fingerprinting is sometimes used in criminal investigations. If a blood or skin sample is found at a crime scene, for example, a DNA fingerprint can be made from these cells. This fingerprint can then be compared to that of a suspect. If the patterns of the bands are identical, the results indicate that the suspect was present at the crime scene. If the patterns of the bands differ, the results prove that someone other than the suspect was the source of the crime-scene sample.

STACEY R. MURRAY

danger inherent in the use of nuclear power, especially nuclear war.

His work with isotopes led to a study of the abundance of naturally occurring isotopes on earth, resulting in a theory of the origin of the elements and their relative abundance in the sun and on other planets in the solar system. His research led him to propose that the earth's atmosphere was once made up of ammonia, methane, and hydrogen, and that reactions among these molecules could lead to the production of living entities. One of his students, Stanley Miller (1930-), successfully demonstrated that when energy is supplied to a mixture of these gases, biological molecules are indeed produced. Theoretically, at least, these molecules could then interact to build living systems. Urey proposed further theories that explain the origin of the solar system as a condensation of gasses around the sun. He published the results of this theoretical work in *The Planets: Their Origin and Development* (1952).

J. WILLIAM MONCRIEF

Ernest Thomas Sinton Walton
1903-1995
Irish Physicist

Ernest Thomas Sinton Walton shared the 1951 Nobel Prize for Physics with John Douglas Cockcroft (1897-1967) for their pioneering studies of the transmutation of atomic nuclei by artificially accelerated atomic particles. Cockcroft and Walton had developed the first nuclear particle accelerator, which became known as the Cockcroft-Walton generator.

Ernest Walton, son of a Methodist minister, was born at Dungarvan, County Waterford, Ireland. After attending day schools in Banbridge and Cookstown, Walton was sent as a boarder to the Methodist College, Belfast, in 1915. Having excelled in mathematics and science, he was able to enter Trinity College, Dublin, on a scholarship in 1922. He specialized in physics, and graduated with first-class honors in both mathematics and experimental science. In 1927, he received his M.Sc. degree and a Research Scholarship, and went to Cambridge University to work in the Cavendish Laboratory under Ernest Rutherford (1871-1937). Walton earned his Ph.D. in 1931. From 1932 to 1934, he was a Clerk Maxwell Scholar. He returned to Trinity College in 1934 as a Fellow. He was Erasmus Smith Professor of Natural and Experimental Philosophy from 1946 to 1974. In 1952 he became chairman of the School of Cosmic Physics at the Dublin Institute for Advanced Studies. He was elected Senior Fellow of Trinity College in 1960.

Walton's early research involved hydrodynamics and the production of fast particles by means of a linear accelerator. His first attempts at high-energy particle acceleration failed, but his methods were later developed and used in the betatron and the linear accelerator. A particle accelerator is a device that produces a beam of fast moving, electrically charged atomic or subatomic particles. Walton and Cockcroft joined forces to produce an instrument that could accelerate protons. The high voltage source for their initial experiments was a four-stage voltage multiplier assembled from four large rectifiers and high-voltage capacitors. The acceleration achieved by their apparatus was large enough to cause a reaction with lithium nuclei. Cockcroft and Walton's experimental system for building up high direct voltages could be expanded to levels many times greater than that of the original apparatus.

In 1932, using their new particle accelerator, they disintegrated lithium nuclei by bombarding them with accelerated protons. Cockcroft and Walton observed that helium nuclei were emitted from the lithium nuclei and concluded that when a proton penetrated a lithium nucleus, the lithium nucleus split into two helium nuclei, which were emitted in nearly opposite directions. Their interpretation was later fully confirmed. This was the first time that a nuclear transmutation had been produced by techniques totally under human control. Using their basic technique and apparatus, Cockcroft and Walton investigated the transmutations of many other atomic nuclei and established the importance of accelerators as a tool for nuclear research. When heavy hydrogen was discovered, they carried out experiments using heavy hydrogen nuclei as projectiles. Their experiments produced some atomic nuclei that were previously unknown and provided a valuable method for comparing masses of atomic nuclei.

Within the limits of experimental error, Cockcroft and Walton's analysis of the energy relations in the lithium transmutation they had achieved provided verification of Einstein's law concerning the equivalence of mass and energy. Eventually, more exact investigations based on the same principles gave a complete verification of Albert Einstein's law. In addition to the Nobel Prize in Physics, Walton received many other honors, including the Hughes Medal, and honorary degrees. Walton married Freda Wilson, daughter of a

Methodist minister and a student at Methodist College, in 1934. They had two sons and two daughters. Walton died in Belfast in 1995.

LOIS N. MAGNER

Alfred Lothar Wegener
1880-1930
German Meteorologist and Geophysicist

Alfred Lothar Wegener was a meteorologist and geophysicist best known for his theory of continental drift, according to which Earth's continents once formed a single landmass and over time drifted to their present positions. Largely rejected during his lifetime, Wegener's idea of continental motion is now universally accepted, although the details of his work have been superseded by plate tectonics.

Wegener was born in Berlin, Germany, on November 1, 1880, to Richard and Anna Schwarz Wegener. He attended the universities at Heidelberg and Innsbruck and received his doctorate in astronomy from Berlin in 1905. In 1906 he accompanied a Danish expedition to Greenland as their meteorologist. Upon returning Wegener lectured in meteorology at the Physical Institute in Marburg (1908-12). Wegener helped lead a second expedition to Greenland in 1912-13 to study glaciology and climatology. In 1924 a special chair of meteorology and geophysics was created for Wegener at the University of Graz. He again led expeditions to Greenland in 1929-30 and 1930-31. He died during the latter trip while attempting to resupply his party.

The idea of continental drift occurred to Wegener in 1910 when he noted the correspondence between the Atlantic shores of Africa and South America. He initially dismissed the idea as improbable. His interest was rekindled in 1911 when he learned paleontological evidence was being used to argue that a land bridge once connected Africa with Brazil.

Wegener became convinced the paleontological and geological similarities required explanation and presented a provisional account of his continental drift theory in 1912. An extended version appeared in 1915 under the title *Die Entstehung der Kontinente und Ozeane* ("The Origins of the Continents and Oceans") which was not widely read until the third edition appeared in 1924. Wegener was not the first to conceive of continental drift, but his account was the most fully developed and supported by extensive evidence.

Wegener proposed the existence of a supercontinent, which he named Pangaea, surrounded by a supersea, Panthalassa. He believed Pangaea began to split and move apart about 200 hundred million years ago. His strongest supports for the theory were the rocks and the flora and fauna on both sides of the Atlantic. Also, geodetic measurements indicated that Greenland was moving away from Europe. Finally, the separation of Earth's crust into a lighter granite floating on a heavier basalt suggested the possibility of continental horizontal transport.

The most serious problem for Wegener's theory was the lack of suitable mechanism. It was difficult to imagine a force strong enough to displace the continents through the solid mantle and oceanic crust. Wegener suggested two mechanisms that were later shown to be only one millionth as powerful as required. By 1928 Wegener's theory had been generally discounted.

Arthur Holmes suggested a viable driving mechanism in 1929. He argued that radioactive heating created convective currents within Earth's mantle, resulting in an internal zone of slippage. The motions thus generated were small but sufficient to account for the displacement of continental landmasses. Holmes' work drew little attention due to the disrepute of continental drift theory.

The discovery of mid-oceanic ridges after World War II renewed interest in Wegener's theory. Paleomagnetic studies in the early 1950s indicated that rocks have magnetic orientations that vary from continent to continent—consistent with continental drift. Wegener's work gained further support in the 1960s with the acceptance of sea-floor spreading and the discovery of subduction zones. Holmes's driving mechanism was then revived and is still widely accepted.

It was eventually recognized that Earth's crust is composed of plates moving relative to each other. Today, plate tectonics is the principal theory of the genesis, structure, and dynamics of Earth's continents.

STEPHEN D. NORTON

Biographical Mentions

Walter Sydney Adams
1876-1956

American astronomer who applied spectroscopic techniques in classifying stars and determining stellar motions. Adams established spectroscopic parallax. This method for measuring stellar distances led to his discovery of white dwarfs (1915). In 1925 he provided support for Albert Einstein's general theory of relativity by observing the predicted red-shift in heavy star spectra due to intense gravitational fields. Most of Adams's career was spent at Mount Wilson Observatory, where he served as director from 1923 until 1946.

Sir Edward Victor Appleton
1892-1965

English physicist who was awarded the 1947 Nobel Prize for Physics for investigations of the ionosphere. His experiments on radio waves confirmed the existence of the Heaviside-Kennelly or E-layer originally proposed to explain Marconi's 1901 transatlantic radio transmission. In the first ever radio distance measurement, Appleton determined the layer's height to be approximately 100 km (62 miles). Further research revealed another layer above this—the Appleton or F-layer. Appleton's discoveries made possible many later advances in radio, shortwave, and radar.

Francis William Aston
1877-1945

British chemist and physicist who was awarded the 1922 Nobel Prize for Chemistry for discoveries and research in mass spectrography. Aston invented the mass spectrograph in 1919. His spectrographic observations led him to put forward the whole-number, or Aston rule, according to which atomic weights are always whole numbers. He correctly attributed apparent deviations from this rule to the presence of isotopes. His later measurements of isotopic masses allowed more accurate estimates of nuclear binding energies.

Hertha Marks Ayrton
1854-1923

British engineer-physicist who was awarded the 1906 Royal Society Hughes Medal for work on the electric arc and sand ripples. Ayrton's 1902 volume *The Electric Arc* became the standard text on the subject. She provided searchlight carbon specifications for the British admiralty and, during World War I, invented the Ayrton Fan for dissipating poison gases. She was the first female member of the Institution of Electrical Engineers (1899) and first woman to read one of her papers before the Royal Society (1904).

Wilhelm Heinrich Walter Baade
1893-1960

German-American astronomer who classified stars into two different types: Population-I (hot, younger blue stars) and Population-II (cool, older red stars). His observations showed the Hertzsprung period-luminosity curve for Cepheid variables—used to calculate stellar distances—only applied to Population-I stars. Baade established a new curve, showing that estimates of the size and age of the universe had to be doubled. Baade also worked with Fritz Zwicky on explaining supernovae and how neutron stars form.

Charles Glover Barkla
1877-1944

English physicist noted for his research and important discoveries concerning x-rays. He was awarded the 1917 Nobel Prize for Physics for discovering that x-ray scattering depends on the molecular weight of gases through which it passes (1903). In 1904 Barkla observed x-ray polarization, which suggested x-rays were a form of electromagnetic radiation. He also noted that secondary radiation produced by x-ray absorption was of two types—later referred to as K-radiation and L-radiation.

Joseph Barrell
1869-1919

American geologist best known for his work on the strength of Earth's crust. He argued Earth's rigid outer crust rests on a semi-plastic layer, proposing the names lithosphere for the former and aesthenosphere for the latter. Barrell also developed a revolutionary interpretation of sedimentary strata. Prior to his work all sediments were believed to be produced by the oceans. Barrell showed that at least one fifth of all land surfaces are composed of other sediments.

Daniel Moreau Barringer
1860-1929

American mining engineer and geologist best known for investigating the Diablo Crater in Arizona. Approximately 600 feet (183 m) deep and over 4,000 feet (1,219 m) in diameter, the crater's origin was a matter of much speculation, most believing it volcanic in nature. However, based on his discovery of numerous nickel-iron rocks at the site, Barringer believed it was in fact a meteor impact crater. He drilled for, but never found, meteorite remains. The site is now known as the Barringer Meteor Crater.

Friedrich Johann Karl Becke
1855-1931

Austro-Hungarian mineralogist best known for his fundamental work on metamorphism—alterations in rock mineralogy, texture, and internal structure by heat, pressure, and chemical reactions. Using his graphical method for representing rock components, he differentiated between metamorphic and sedimentary rock transformations. He put forth the Becke volume rule, according to which mineral formation with the greatest molecular density is favored under isothermal conditions and increasing pressure (1896). Becke also did considerable work on crystalline schists.

Alexander Karl Friedrich Frans Behm
1880-1952

German physicist and engineer who developed acoustic methods of detection and distance measurement for ships and aircraft. Originally a teacher of technical physics, Behm was motivated by the *Titanic* disaster in 1912 to invent an early form of sonar, which was of wide scientific and military interest after the use of submarines in World War I. In 1920 he began manufacturing the first practical "echo-location" systems in Kiel, where he also founded a related scientific society.

Friedrich Bergius
1884-1949

German chemist who won the Nobel Prize for chemistry in 1931 for his research into the effects of high temperature and pressure on chemical reactions. One of the processes he pioneered involved using coal and hydrogen gas to produce oil. This method, called coal liquefaction, was used to produce gasoline in Germany during World War II. Bergius also developed a method of converting cellulose, a compound found in wood, to sugar.

Hans Albrecht Bethe
1906-

German-born American physicist who was awarded the 1967 Nobel Prize for Physics for contributions to nuclear reactions, especially for his 1939 theory describing stellar energy production. One of his many contributions to quantum electrodynamics was the theoretical explanation of the "Lamb shift" (1947). Bethe also co-authored (with Ralph Alpher and George Gamow) the alpha-beta-gamma "big bang" theory of the universe's origin. During World War II, Bethe's research was crucial for developing the atomic bomb, though he later actively opposed its use.

Jacob Aall Bonnevie Bjerknes
1897-1975

Norwegian-American meteorologist who developed the extratropical cyclone model (1919) of modern meteorology while conducting research on weather systems and forecasting techniques with his father Vilhelm Bjerknes and others at Bergen, Norway. Bjerknes and collaborator Halvor Solberg went on to develop frontal theory in terms of warm and cold fronts defined by airmass temperature boundaries. Bjerknes also identified the upper atmospheric wave that generated the surface extratropical cyclone/frontal dynamics. He founded the UCLA department of meteorology and was an early pioneer defining the El Niño mechanism.

Bart Jan Bok
1906-1983

Dutch-American astronomer who studied the structure of the Milky Way Galaxy. In 1947, he discovered small dark clouds that can be seen against the light of stars. He proposed that these clouds, now known as Bok globules, are made up of gas and dust that will eventually condense to form new stars. Bok, along with his wife, Priscilla, wrote an influential astronomy book titled *The Milky Way* in 1941.

Max Born
1882-1970

German-born English physicist who was awarded the 1954 Nobel Prize for Physics for contributions to quantum mechanics, especially his statistical interpretation of the wave function (1926). Born collaborated with Werner Heisenberg and Pascual Jordan in developing matrix mechanics (1925). His collaboration with Theodore Kármán on the heat capacity of solids resulted in the Born-Kármán theory of specific heats (1913). Born also contributed to the theory of crystal lattice vibrations and the statistical theory of fluids or BBGKY-equations.

Jagadis Chandra Bose
1858-1937

Indian physicist who successfully conducted wireless signaling experiments two years before Guglielmo Marconi's experiments on Salisbury Plain. Ahead of his time, Bose was the first to use semiconductor junctions to detect radio waves and invented many now familiar microwave components such as wave-guides, horn antennas, polarisers, dialectric prisms, and semiconductors for detecting electromagnetic radiation. Shortly after presenting his work before the Royal Society in 1897, Bose directed his research efforts towards response phenomena in plants.

Percy Williams Bridgman
1882-1961

American physicist who was awarded the 1946 Nobel Prize for Physics for work in high-pressure physics. Bridgman's self-tightening joint allowed him to extend the range of pressures under which substances could be studied from 3,000 to 100,000 atmospheres. His work later served as the basis for General Electric's development of synthetic diamonds. Bridgman is also known for his "operational" philosophy of scientific methodology, according to which science should restrict itself to concepts definable by specific physical operations.

Prince Louis-Victor-Pierre-Raymond, 7th Duc de Broglie
1892-1987

French physicist awarded 1929 Nobel Prize in physics for discovering the wave nature of matter. By the 1920s it was known that electromagnetic waves sometimes behave like particles. In his doctoral dissertation de Broglie proposed that particles sometimes behave like waves (1924). The wave properties of matter (de Broglie waves) only manifest themselves at atomic levels and had never before been observed. The existence of de Broglie waves was confirmed by diffraction experiments in 1927.

Subrahmanyan Chandrasekhar
1901-1995

Indian-born American astrophysicist who was awarded the 1983 Nobel Prize for Physics for theoretical studies of the structure and evolution of stars. Chandrasekhar showed that as stars exhaust their nuclear fuel they reach a stage where they no longer generate sufficient pressure to sustain their size and gravitational collapse occurs. Chandrasekhar calculated that stars of 1.4 solar mass (the "Chandrasekhar limit") or less collapse to white dwarfs, while stars of greater mass become supernovas, typically leaving neutron stars.

Frank Wigglesworth Clarke
1847-1931

American geochemist best known for his meticulous analyses and compilations of chemical and physical data. His *Data of Geochemistry* (1908) was a standard source on minerals in Earth's crust, and he proposed a theory of evolution of the heavier elements from the lighter. Clarke was simultaneously chief chemist of the U.S. Geological Survey and honorary curator of minerals of the U.S. National Museum (1883-1924). He was the first president of the national American Chemical Society (1901).

John Douglas Cockcroft
1897-1967

British physicist who was awarded the 1951 Nobel Prize for Physics with Ernest Walton for pioneering work on the transmutation of atomic nuclei by accelerated particles. Cockcroft played an integral role in the development of nuclear energy as an atomic advisor to the British government and as the leader of various nuclear research projects and facilities, including head of Britain's first atomic energy research laboratory at Harwell. During World War II he was responsible for developing radar equipment for anti-aircraft guns.

Arthur Holly Compton
1892-1962

American physicist who was awarded the 1927 Nobel Prize for Physics for discovering the Compton effect—shifts in x-ray wavelength due to scattering. His quantum theoretical explanation of this phenomenon and derivation of equations for predicting scattered wavelengths confirmed the particle-wave duality of electromagnetic radiation. Throughout the 1930s Compton conducted extensive research on cosmic radiation, which helped establish that at least some cosmic rays are charged particles. He was one of the leaders in developing the atomic bomb during World War II.

Sir Arthur Stanley Eddington
1882-1944

English astronomer who was one of the founders of modern astrophysics. Eddington showed that, to avoid collapse, the outward gas and radiation pressure of a star must equal its inward gravitational pull. This placed an upper limit of 50 solar masses on stable stars. Cepheid variable pulsation, he argued, was due to a star's instability. Eddington also established the mass luminosity law and led the 1919 solar eclipse expedition, which confirmed the gravitational bending of light.

Julius Elster
1854-1920

German physicist who helped to develop a simple thermionic device that converts heat energy to electrical energy. Elster's apparatus consisted of a sealed tube that contained an electrode, or electrical conductor, at either end. One electrode could be heated while the other was cooled. The heated metal emitted electrons, which passed through the tube to the cool electrode. In other words, an electric current flowed through the tube as a result of temperature differences between the electrodes.

Kasimir Fajans
1887-1975

Polish-American chemist who helped to formulate the radioactive displacement law. This law states that when a radioactive atom emits an alpha particle (two protons and two neutrons), its atomic number decreases by two. By contrast, when an atom emits a beta particle (an electron), its atomic number increases by one. In addition, Fajans developed rules that describe the conditions under which covalent bonding is more likely to occur than ionic bonding, and vice versa.

Clarence Norman Fenner
1870-1949

American geologist who analyzed igneous rocks. Educated at Columbia University, Fenner briefly worked as an assayer and surveyor for mining companies. He accepted a position with the Carnegie Institution's geophysical laboratory in Washington, D.C. Fenner studied how minerals form and crystallize, especially volcanic rocks in South and North America. He also examined radioactive rocks and how they represent geological age. Fenner assessed geological composition and steam pressure in core samples from Yellowstone National Park's geyser.

Aleksandr Evgenievich Fersman
1883-1945

Russian geochemist and mineralogist who made significant contributions to geology and crystallography. Fersman was elected to the Russian Academy of Sciences and made director of the society's Geological Museum in 1919. During the 1920s he determined the distribution of chemical elements in the earth's crust and provided reasons for the observed differences in terrestrial and cosmic distributions. During the last 20 years of his life he was responsible for reassessing the Soviet Union's mineral resources.

Aleksandr Alexandrovich Friedmann
1888-1925

Russian mathematician and astrophysicist best known for his mathematical model of an expanding universe. Albert Einstein's 1917 static model of the universe required an arbitrary "cosmological constant" to explain why the universe does not collapse. Friedmann argued that Einstein's model was inconsistent with general relativity. He discarded the cosmological constant and demonstrated that expansion or contraction of the universe was due solely to gravity. Friedmann models or universes laid the foundation for "big bang" theories.

George Gamow
1904-1968

Russian-born American physicist who made important contributions to cosmology, nuclear physics, and molecular biology. Gamow was co-developer of the Alpher-Bethe-Gamow "big bang" theory of the universe's origin, which explained the observed abundance of lighter elements. He later concluded that heavier elements formed in the center of stars. He independently developed alpha decay theory and suggested how DNA governed protein synthesis. Gamow was a great popularizer of physics and widely known for his Mr. Tompkins books.

Sir David Gill
1843-1914

Scottish astronomer and pioneer in applying photography to astronomy. As director of South Africa's Cape Observatory (1879-1907), he conducted a photographic survey of the southern skies. Gill and J. C. Kapteyn measured the plates (1886-98) and published the positions and magnitudes of over 400,000 stars. Their survey was the basis for later work on stellar distribution. In 1888 Gill redetermined the distance between the Sun and Earth to within one part in a thousand.

Victor Moritz Goldschmidt
1888-1947

Swiss-born Norwegian chemist and mineralogist known as the father of modern geochemistry. Goldschmidt's research was largely devoted to the earth's chemical composition. He developed a model of Earth in which elemental distribution was a function of charge and size. Using x-ray crystallography he determined the structures of over 200 compounds and developed the first tables of ionic radii. Goldschmidt's Law (1929), based on these findings, allowed him to predict the presence of elements in particular minerals.

Moses Gomberg
1866-1947

Ukrainian-born American chemist noted for preparing the first stable free radical triphenylmethyl (1900). His discovery met with initial skepticism since it was generally believed that long-lived radicals could not exist. Further findings convinced most chemists by 1911. Gomberg devoted the remainder of his career to pioneering researches in free-radical chemistry. He also discovered ethylene glycol, which is commonly used in antifreeze for automobile radiators. Gomberg was president of the American Chemical Society in 1931.

François Auguste Victor Grignard
1871-1935

French chemist who shared the 1912 Nobel Prize for his discovery of the Grignard reaction process. By treating magnesium filings with methyl iodide under anhydrous ether to avoid spontaneous atmospheric combustion, compounds of the general formula RMgX (R = an ether molecular group, Mg = magnesium, X = an anion) could be formed and then used as reagents for synthesizing numerous basic organic compounds. The reaction remains one of the most important synthetic tools employed in organic chemistry today.

Beno Gutenberg
1889-1960

German-American geophysicist who demonstrated Earth's outer core to be liquid and calculated the boundary between it and the solid mantle above to be at a depth of 2,900 km (1,802 miles), now called the Gutenberg discontinuity. In collaboration with Charles Richter, Gutenberg worked to derive more accurate earthquake travel time curves, redetermine major earthquake epicenters, and to quantify interrelationships between magnitude, intensity, energy, and acceleration of vibrations within the earth. Gutenberg also extended the Richter scale to deep-focus shocks.

George Ellery Hale
1868-1938

American astronomer who was a pioneer in astrophysics. A Massachusetts Institute of Technology graduate, Hale designed and built scientific instruments. Hale's spectroheliograph took the first photographs of solar gas clouds. As director of the Mount Wilson Observatory, Hale procured large telescopes. His research revealed that sunspots are cooler than the surface around them and that solar magnetic fields exist. He discovered that the polarity of sunspots reverses every eleven years. A prolific author, Hale founded and edited the *Astrophysical Journal*

Johannes Franz Hartmann
1865-1936

German astronomer whose observations provided the first clear evidence for interstellar matter. While observing the spectrum of the binary system Delta Orion, Hartmann noticed that the calcium lines did not exhibit the periodic Doppler-shifting expected from stars orbiting each other. Since these stationary lines could not be produced by calcium in the binary stars' atmospheres, Hartmann concluded they were the result of a mass of stationary gas between Delta Orion and Earth.

Walter Heinrich Heitler
1904-1981

German-Irish-Swiss physicist best known for his work with Fritz London in providing the first quantum mechanical treatment of the hydrogen molecule (1927). The Heitler-London theory, as it is known, successfully deployed wave mechanics in accounting for covalent bonding in hydrogen molecules. Their work was later generalized by Linus Pauling and is known as the valence bond method in quantum chemistry. Heitler's later years were devoted to attempts at reconciling science and religion.

James Stanley Hey
1909-

British radio astronomer and author who was the first to detect radio emissions from the Sun. In 1942, during the Second World War, Hey was responsible for investigating the jamming of British radar by German forces. He studied certain "rushing noises," thought to be jamming signals, and discovered instead that the noise came from a sunspot. This discovery led to further study of the Sun as a source of radio emission, advancing the science of radio astronomy.

Christopher Kelk Ingold
1893-1970

British chemist who applied quantum theory to explain organic chemical reactions as the breaking and formation of bonds resulting from the rearrangement of bonding electrons. Ingold proposed that dipoles (separation of charge centers) in molecules may be either permanent (due to the structure of the molecules) or temporary (caused by an external polarizing agent). He also defined reactants as nucleophilic if they give up electrons in a reaction, and electrophilic if they accept electrons.

Vladimir Nikolaevich Ipatieff
1867-1953

Russian-born American chemist who was one of the first to investigate high-pressure catalytic reactions of hydrocarbons. He also developed a process for manufacturing high-octane gasoline. Ipatieff was an officer in the Imperial Russian Army and later attended the Mikhail Artillery Academy in St. Petersburg, where he was a professor of chemistry. In 1930, he took a position with the Universal Oil Products Company in Chicago where he developed his catalytic processes for industrial use.

Karl Guthe Jansky
1905-1950

American radio engineer who discovered the first extraterrestrial radio source. While investigating noise in short-wave radio communications, Jansky identified a steady hiss-type static and demonstrated its source to be the center of the Milky Way (1932). Since interstellar dust obscures many astronomical objects of interest, such as the galactic core, from optical instruments, Jansky's discovery established the utility of radio observations in astronomy. Jansky also helped to refine the design of transmitting and receiving antennas.

Sir James Hopwood Jeans
1877-1946

English astrophysicist best known for his popular books on astronomy and physics. Jeans made significant contributions toward the understanding of stability in rotating masses and was the first to propose matter's continual creation throughout the universe (1928). He also put forward the now defunct "tidal" theory of the solar systems' origin. According to this theory, the gravitational pull of a passing star drew out a filament of solar debris that subsequently condensed into the planets.

Sir Harold Jeffreys
1891-1989

British geophysicist and astronomer known for his work on Earth's internal structure and earthquake travel time tables compiled with Keith Bullen—commonly known as the JB Tables. After studying tidal effects, Jeffreys proposed that Earth's core was molten rock. He opposed continental drift theory and estimated the solar system's age at several billion years. He calculated the surface temperatures of the outer planets and modeled the internal formations of Jupiter, Saturn, Uranus, and Neptune.

Frédéric and Irène Joliot-Curie
1900-1958
1897-1956

French nuclear physicists who were awarded the 1935 Nobel Prize for Chemistry for their discovery of artificial radioactivity. Frédéric Joliot (1900-1958) and Irène Curie (1897-1956), daughter of renowned scientist Marie Curie, married in 1926 and subsequently collaborated under the joint surname Joliot-Curie. Through alpha-particle bombardment they were able to transform elements into radioactive isotopes. Such isotopes are now widely used for medical diagnosis. The Joliot-Curie's work also made nuclear fission and development of nuclear power plants possible. Irène was named director of the Radium Institute in 1932, and Frédéric was instrumental in establishing the French Atomic Energy Commission after World War II.

Sir Harold Spencer Jones
1890-1960

British astronomer best known for his precise determination of the astronomical unit (distance between Earth and Sun), which thus established the scale of the solar system. Jones photographed the asteroid Eros over 1,200 times during its 1930-31 close approach, and after 10 years of data analysis announced the astronomical unit to be 93 million miles (149,668,992 km). Jones also showed that Earth's rate of rotation was not constant. He served as the tenth astronomer royal (1933-55).

Ernst Pascual Jordan
1902-1980

German physicist instrumental in the evolution of quantum mechanics. Jordan collaborated with Max Born and Werner Heisenberg in elucidating the principles of quantum mechanics through the development of matrix mechanics (1926). Jordan then did pioneering work in quantum electrodynamics. He also proposed an alternate to Einstein's general theory of relativity. Independently and concurrently developed by Robert Dicke and Carl Brans (1961), it is known as the Brans-Dicke-Jordan theory. Precise measurements have not supported it.

Paul Karrer
1889-1971

Swiss chemist who won the Nobel Prize for chemistry in 1937 for determining the three-dimensional structures of numerous organic (carbon-containing) compounds. Karrer was the first to determine the structure of a vitamin, that of vitamin A. He also investigated the structures of carotenoids, yellow-to-red pigments found, for example, in carrots and lobster shells. The structure of carotenoids is closely related to that of vitamin A, and Karrer showed how the body can use carotenoids to produce this vitamin.

Hendrik Anthony Kramers
1894-1952

Dutch theoretical physicist known for his important contributions to quantum mechanics. He demonstrated that quantization of classical orbits in helium atoms produces instabilities, indicating inadequacies in provisional quantum theory (1923). Kramers also developed a quantum theory of dispersion allowing him to quantitatively explain the effect later experimentally found by Raman. He later established the Kramers-Kronig relations connecting the real and imaginary components of polarizability and contributed to the W(entzel)-K(ramers)-B(rillouin) method (1926) for obtaining approximate Schrödinger equation solutions.

Gerard Peter Kuiper
1905-1973

Dutch-born American astronomer best known for discovering Uranus's fifth satellite, Miranda (1948), and Neptune's second satellite, Nereid (1949). Kuiper confirmed the existence of an atmosphere around Saturn's satellite Titan. He discovered methane bands on Uranus and Neptune, and deduced the existence of frozen water at Mars's polar caps. Kuiper was active in the United States space program, especially the Ranger and Mariner programs. He also helped select lunar landing sites for the Apollo missions.

Henrietta Swan Leavitt
1868-1921

American astronomer who contributed to stellar measuring methodology. Leavitt analyzed photographic plates at the Harvard College Observatory. She recognized that a relationship exists between the period in which a star's brightness varies and its magnitude. She composed tables of data that astronomers used to measure stellar distances. Later astronomers realized that Leavitt's theory applied only under specific conditions, but her research aided in determining that the Magellanic Clouds were actually two galaxies. Leavitt located 2,400 variable stars and discovered four novae.

Pyotr Nicolayevich Lebedev
1866-1912

Russian physicist best known for his research into the effects of electromagnetic radiation on matter. Nineteenth-century physicist James Maxwell earlier predicted that electromagnetic waves, such as light, would exert pressure on matter. In 1901 Lebedev designed a refined version of William Crookes's radiometer and successfully measured this light pressure. Lebedev later argued that radiation pressure caused comet tails to point away from the Sun. This effect was later shown to be insignificant compared to that of solar winds.

Inge Lehmann
1888-1993

Danish seismologist who was awarded the 1971 American Geophysical Union's William Bowie Medal for "outstanding contributions to fundamental geophysics and unselfish cooperation in research." Lehmann's seismic wave observations led her to conclude in 1936 that Earth has a solid inner-core surrounded by a liquid outer-core. She calculated the inner-core radius to be 1,400 km (870) and noted that seismic waves were reflected at its outer-core boundary. This boundary has been named the "Lehmann discontinuity" in her honor.

Georges Edouard (Abbé) Lemaître
1894-1966

Belgian mathematical physicist who formulated an early version of the "big bang" theory in which the universe was conceived to have evolved from a cataclysmic explosion of an immeasurably dense "primeval atom." Five years after Aleksandr Friedmann's 1922 work suggested the possibility of a contracting or expanding universe, Lemaître independently showed Albert Einstein's general field equations permitted an expanding universe solution. This expansion was confirmed with Erwin Hubble's 1929 announcement that galaxies recede from Earth with velocities directly proportional to their distance from Earth.

Willard Frank Libby
1908-1980

American chemist who won the Nobel Prize for chemistry in 1960 for his work on the development of radiocarbon dating. He invented techniques for measuring the amount of radioactive carbon in the remains of organisms. This amount then could be used to determine the age of the remains (up to 50,000 years old). Libby also developed a method of separating different forms of uranium. This work proved important in the development of the atomic bomb.

Fritz Wolfgang London
1900-1954

German-born American physicist who, with Walter Heitler, devised the first quantum mechanical treatment of covalent bonding in hydrogen molecules. London and R. Eisenschitz employed quantum mechanics to calculate the force between atoms, known as the "dispersion force." During the 1930s, London and his brother, Heinz, developed the phenomenological theory of superconductivity (the phenomena in which super-cooled substances lose all electrical resistance). London's research from 1938 on focused on superfluidity (fluid flow without resistance).

Hendrik Antoon Lorentz
1853-1928

Dutch theoretical physicist best known for co-developing the Lorentz-Fitzgerald transformations. Derived in 1892 from his theory of electrons, these equations allowed Lorentz to account for the Michelson-Morely null-result. Albert Einstein later derived the transformations from his theory of relativity. Another consequence of Lorentz's electron theory was the splitting of spectral lines in magnetic fields—the Zeeman effect, confirmed by Pieter Zeeman in 1896. For these discoveries, Lorentz and Zeeman shared the 1902 Nobel Prize for Physics.

Adriaan van Maanen
1884-1946

Dutch-American astronomer known for his careful measurement of distant astronomical objects. Beginning in 1916 he claimed to have detected motion in certain spiral nebulae. If true, these spiral nebulae were thought to be within our own Milky Way and not galaxies distinct from it. But in 1932 these results were shown to be in error, and in 1935 he admitted his mistake. It was then shown that those nebulae were in fact independent galaxies.

Edwin Mattison McMillan
1907-1991

American physicist who co-discovered the first transuranium element, neptunium. McMillan was director of the radiation facility at the University of California, Berkeley. His research, along with colleague Glenn Theodore Seaborg's, led to the discovery of plutonium. He and Seaborg share the 1951 Nobel Prize in Chemistry for their work in transuranium elements. Transuranium elements are elements with an atomic number greater than 92.

Albert Abraham Michelson
1852-1931

American physicist best known for inventing the interferometer (1881) and for his 1887 experiments with Edward Morley that failed to detect the motion of Earth relative to the ether—a hypothetical substance most nineteenth-century physicists believed necessary for light propagation. This "null-result" was later explained by Albert Einstein's special theory of relativity. Michelson was awarded the 1907 Nobel Prize for Physics for his high-precision optical instruments and measurements carried out with their aid.

Thomas Midgley Jr.
1889-1944

American industrial chemist best known for discovering chlorofluorocarbons, which are widely used for producing plastics and cleaning electronic components. Midgley introduced the chlorofluorocarbon Freon-12 in 1930 as a noncorrosive, nontoxic, and nonflammable refrigerant. Midgley also discovered the anti-knock fuel additive tetraethyl lead (1921). He then developed a method for extracting the very scarce bromine from seawater when it was discovered that it was required to prevent tetraethyl lead from fouling engine valves and spark plugs.

Paul Alwin Mittasch
1869-1953

German chemist who pioneered numerous industrial applications of chemical catalysis during a 33-year career at the German chemical firm BASF. Mittasch's work focused on catalytic activators (which accelerate reaction speeds), "poisons" (which decelerate reaction speeds), and catalytic mixtures (which simultaneously accelerate and decelerate different stages of a multi-step reaction process). After retiring in 1934 he wrote numerous works devoted to chemistry and philosophy, focusing upon Friedrich Nietzsche, Arthur Schopenhauer, and Robert Julius Mayer.

Andrija Mohorovicic
1857-1936

Croatian geologist who discovered Earth's crust-mantle boundary—the Mohorovicic discontinuity. Mohorovicic's study of seismic records from a 1909 earthquake revealed two distinct sets of waves. This indicated that the seismic waves had changed velocity, which Mohorovicic attributed to transmission through two layers of different density. He identified the boundary between the two layers as a sharp discontinuity, now known to vary from 10 km (6.2 miles) under the oceans to as much as 50 km (31 miles) under continents.

Henry Gwyn Jeffreys Moseley
1887-1915

English physicist who derived the relationship between an element's x-ray wavelengths and atomic number—Moseley's Law. This allowed him to assign characteristic numbers to each element and order them into a periodic table. Moseley then predicted the existence of several undiscovered elements. He was able to predict their spectral patterns, which proved invaluable in the search and later discovery of these elements. Moseley was killed at the battle of Sulva Bay during World War I.

Paul Hermann Mueller
1899-1965

Swiss-born chemist who discovered that the chemical compound dichlorodiphenyltrichloroethane (DDT) killed common crop-eating pests. He was granted a Swiss patent for his work in synthesizing the compound in 1940. During World War II, DDT proved to be of enormous value in controlling the mosquitoes that spread typhus and malaria. Despite DDT's success, it did not degrade safely into the environment and caused widespread health concerns in the 1960s. By the 1970s, many countries banned its use.

Robert Sanderson Mulliken
1896-1986

American chemical physicist who was awarded the 1966 Nobel Prize for Chemistry for applying quantum theory to explain molecular structure. In 1928 he published his molecular orbital method of chemical bonding, according to which the energy levels of bonding electrons in molecules are determined by quantum mechanical properties of the molecules, not independent atomic characteristics. Mulliken's orbital method was adopted by quantum chemists in the 1950s, having competing with Linus Pauling's valence bond method for two decades.

Hermann Walther Nernst
1864-1941

German chemist who won the Nobel Prize in 1920 for his formulation of the third law of thermodynamics, also known as the Nernst heat theorem (1905). This law states that no substance can be cooled to absolute zero (no atomic motion), but as this limit is approached, the difference between the changes in the heat content and the free energy of a chemical system become zero. Nernst's law allowed for exact measurement of free energy changes and related chemical thermodynamics to quantum theory in physics.

Julius Arthur Nieuwland
1878-1936

American chemist best known for his vinyl derivatives of acetylene. Nieuwland produced divinylacetylene, which he showed could be polymerized into a rubber-like solid (1925). In 1929 he catalytically converted acetylene into monovinylacetylene, which when treated with acid forms chloroprene. Working with DuPont chemists, Nieuwland polymerized chloroprene into the first commercially successful synthetic rubber neoprene. Neoprene is more durable and useful than natural rubber. First marketed in 1932 as Duprene, Neoprene is widely used in industry.

Richard Dixon Oldham
1858-1936

Irish geologist and seismologist who established the existence of Earth's core. Based on his landmark survey of the 1897 Assam earthquake, Oldham identified three types of seismic waves—primary (P-waves), secondary (S-waves), and surface waves—predicted earlier by Siméon Poisson. Oldham noted that during an earthquake, seismographs located on the opposite side of Earth detect P-waves later than expected, leading him to conclude that Earth has a core less dense and rigid than its mantle.

Cecilia Payne-Gaposchkin
1900-1979

English-American astronomer who analyzed the universe's composition. After Payne-Gaposchkin graduated from Cambridge University, she moved to the United States to study at the Harvard College Observatory. Earning a doctorate from Radcliffe College, she accepted an appointment at the observatory. Payne-Gaposchkin studied galactic structures, variable stars, and stellar atmospheres. Her research revealed that most celestial bodies are formed from hydrogen and helium. Payne-Gaposchkin's classification of stars by varying brightness advanced the study of stellar evolution.

Marguerite Catherine Perey
1909-1975

French nuclear chemist who discovered francium, the last naturally occurring element to be found (1939). Perey's discovery is especially remarkable because the total quantity of francium in the earth's crust is less than 24.5 grams. Perey was the first woman elected to the Académie des Sciences (1962), which had been closed to women since its founding in 1662. In 1949 she was appointed to the University of Strasbourg's new chair of nuclear chemistry.

Jean Baptiste Perrin
1870-1942

French physicist awarded the 1926 Nobel Prize in physics for his work on Brownian motion. In 1895 he settled the cathode ray particle-wave debate by demonstrating that cathode rays carry negative charges and are thus particles. Perrin's classic studies of Brownian motion (1908) provided the first demonstration of the existence of atoms. He confirmed Einstein's mathematical analysis of Brownian motion. This enabled Perrin to accurately calculate Avogadro's number and derive the size of water molecules.

Charles Snowden Piggot
1892-1973

American chemist and geophysicist best known for developing a coring technique (1936) for sampling material from the ocean floors, providing the first reliable dates for sediments and sedimentary rates. His work on determining the mass spectrum for lead played a significant role in unraveling the mysteries of radioactive decay and initiating the science of geochronology. Piggot was awarded the Order of the British Empire for his mine disposal work during World War II.

Ludwig Prandtl
1875-1953

German engineer considered the father of aerodynamics. His explanation of the boundary layer in fluid flow helped bridge the gap between empirical hydraulics and modern hydrodynamics and aerodynamics. Prandtl's studies of turbulent airflow laid the foundation for later work on chaotic motions and had important applications for aircraft and rocket design. He helped establish and served as director of the Kaiser Wilhelm Institute for Fluid Dynamics (now the Max Planck Institute) in 1925.

Sir Chandrasekhara Venkata Raman
1888-1970

Indian physicist who was awarded the 1930 Nobel Prize for Physics for discovering the effect that bears his name. Announced in 1928, the Raman effect refers to the frequency shifts of light due to scattering by molecules. The effect has been effectively exploited as a tool in analyzing molecular structure. Raman was instrumental in promoting the development of Indian science—creating the *Indian Journal of Physics* (1926) and instigating establishment of the Indian Academy of Sciences (1934).

Sir Willans Owen Richardson
1879-1959

British physicist awarded the 1928 Nobel Prize in physics for work on thermionic emission of electrons from hot surfaces. In 1901 he discovered what is now known as Richardson's law, which states that in thermionic emission electron current increases exponentially with increasing temperature. His gyromagnetic researches led him to conclude that electron currents are responsible for magnetism—the Richardson-Einstein-de Haas effect. Richardson was instrumental in the application of physics to radio, telephony, television, and x-ray technologies.

Charles Francis Richter
1900-1985

American geophysicist who developed the earthquake magnitude scale. Earning a physics doctorate from the California Institute of Technology, Richter conducted seismology research at the Pasadena Geophysical Laboratory. Together with Beno Gutenberg, he devised a scale to gauge an earthquake's magnitude instead of using assessments of structural damage. The new scale measured an earthquake's entire power, not just its energy at specific sites. Richter wrote two books about seismology, promoting quantitative studies of the geophysical forces that cause earthquakes.

Carl-Gustaf Arvid Rossby
1898-1957

Swedish meteorologist who defined the upper atmospheric wave dynamics (later called the Rossby Wave), which generates surface frontal systems, and later discovered the associated jet stream upper level winds, which he further theorized as steering the frontal mechanism. An early member of the Bergen School of Meteorology under Vilhelm Bjerknes, he began the first meteorology department in the United States at MIT (1928) and became head of and revamped the U.S. Weather Bureau. His research in large-scale atmospheric instability would be the foundation for the important baroclinic instability theory of his younger associate, Jule Gregory Charney at MIT.

Henry Norris Russell
1877-1957

American astronomer who preferred developing theories to collecting data. Unlike many astronomers, Russell amassed information only to analyze specific problems. As a result of his methodology, Russell often was one of the first astronomers to announce a discovery, such as that two classes of red stars—dwarfs and giants—exist. He also theorized about stellar evolution. During annual visits to the Mount Wilson Observatory, Russell used data collected there for his research. Russell wrote a *Scientific American* column to popularize astronomy.

Sir Martin Ryle
1918-1984

British radio astronomer who shared the 1974 Nobel Prize in physics with Antony Hewish for pioneering research in radio astrophysics. Ryle produced the first detailed map of radio-emitting objects in the sky. His extensive cataloguing of radio sources was made possible in part by his development of the aperture synthesis technique for improving telescope resolving power. Ryle worked on radar during World War II, was knighted in 1966, and was honored as the Astronomer Royal (1972-74).

Meghnad Saha
1893-1956

Indian astrophysicist who gave birth to the development of accelerators in India when he developed a 37-in (94-cm) cyclotron in 1940. Before Saha's interest in research with particle accelerators peaked in the 1930s, he studied the thermal ionization that occurs in the extremely hot atmosphere of stars. Saha demonstrated that elements in stars are ionized in proportion to their temperature, as defined by the equation that now bears his name. In 1938 he was appointed professor of physics at Calcutta, where he established the first Indian Institute of Nuclear Physics.

Bruno Hermann Max Sander
1884-1979

Austrian mineralogist and geologist who used microscopic and x-ray studies of the fine structures of rocks and minerals to determine the large-scale geological processes that formed them. He obtained a Ph.D. from Innsbruck University in 1907 and became professor of mineralogy and petrography there in 1922. Sander illuminated the mechanical forces that have shaped Earth over time and developed mathematical methods for studying the structures of materials in general.

Charles Schuchert
1858-1942

American paleontologist who described the changes that have occurred over time in the positions of the oceans and continents. Originally a self-educated fossil illustrator, he became a curator at the U.S. National Museum and finally a professor at Yale. Schuchert collected enormous amounts of data on the worldwide distribution of fossils of different types and ages, enabling him to reconstruct past geography. His textbook on North American historical geology was very influential.

Karl Schwarzschild
1873-1916

German astronomer who predicted the existence of black holes (1916). Schwarzschild showed that a star undergoing gravitational collapse that shrinks below a certain radius—the Schwarzschild radius—develops a gravitational pull so great that not even light can escape. Such objects are called black holes. Schwarzschild also produced the first exact solution to Einstein's field equations (1916). During the 1890s he developed a method for accurately determining the apparent magnitude of stars from photographic plates.

Glenn Theodore Seaborg
1912-1999

American physicist who discovered ten atomic elements. Seaborg conducted research at his alma mater, the University of California at Berkeley. Working with colleagues, he isolated new isotopes. Seaborg was credited with identifying ten elements, including seaborgium, which was named for him. During World War II, he worked on the Manhattan Project. He won the 1951 Nobel Prize in chemistry with Edwin McMillan for their investigations of transuranium elements. Seaborg served as chairman of the Atomic Energy Commission from 1961 to 1971.

Harlow Shapley
1885-1972

American astronomer who improved astronomical education. Studying at Princeton under Henry Norris Russell, Shapley began his career at the Mount Wilson Observatory, where he researched globular star clusters. He stated that the Milky Way's diameter was 300,000 light years, which was later proven incorrect. As director of the Harvard Observatory, Shapley promoted astronomy to the public with frequent news releases. He established a graduate program in astronomy at Harvard. Shapley catalogued galaxies and his research aided other astronomers' discoveries.

John Clarke Slater
1900-1976

American physicist who pioneered the development and application of quantum mechanics to atoms, molecules, and the solid state. Slater, Niels Bohr, and Hendrik Kramers collaborated on the idea that light consists of probability waves (1924). This became a central tenet of quantum mechanics. Slater determinants are well known, as is the Slater orbital—an extremely useful atomic orbit approximation. Slater was an outstanding teacher and administrator while at Harvard and Massachusetts Institute of Technology and published many influential physics textbooks.

Vesto Melvin Slipher
1875-1969

American astronomer who contributed to spiral nebulae theory. Earning a doctorate from Indiana University, Slipher measured planets' rotation periods at the Lowell Observatory. Slipher's most important research concerned the spectrographic study of spiral nebulae. He examined data to determine if spiral nebulae resembled the Earth's solar system. Slipher measured spiral nebulae's velocities, noting that they were moving away from Earth. This information supported other astronomers' theory of an expanding universe. Slipher also served as director of the Lowell Observatory.

Arnold Johannes Wilhelm Sommerfeld
1868-1951

German physicist whose investigations of atomic spectra led him to revise Niels Bohr's atomic theory (1916). Sommerfeld replaced Bohr's circular orbits with elliptical ones and introduced the azimuthal quantum number. These changes made it possible to explain the fine structure of the hydrogen spectrum. The Bohr-Sommerfeld model was fruitful but failed to explain the Zeeman effect—splitting of spectral lines due to strong magnetic fields. Sommerfeld introduced the magnetic quantum number to account for these effects.

Søren Peter Lauritz Sørensen
1868-1939

Danish biochemist best known for introducing the pH (potential of hydrogen) scale for measuring the acidity or alkalinity of solutions (1909). Sørensen did extensive work on enzymes and proteins. His laboratory successfully crystallized egg albumen in 1917, an important step in attempts to characterize proteins. Sørensen also contributed to medical research on epilepsy and diabetes as well as being involved in chemical application in industry, working on explosives and with spirits and yeast.

Hermann Staudinger
1881-1965

German chemist who won the 1953 Nobel Prize as the major founder of macromolecular chemistry during the 1920s. Staudinger proved that macromolecules are not aggregates of smaller molecules joined by weak inter-molecular bonds, but rather are long chains of single molecular units connected by regular organic (Kekule) bonds. Staudinger also founded the journal *Makromolekulare Chemie* and pioneered research into biological macromolecules such as cellulose and amino-acid sequences.

Otto Struve
1897-1963

Russian-born American astronomer who with B. Gerasimovic demonstrated the pervasiveness of interstellar matter (1929). This discovery was significant for attempts to map stellar distribution, because intersellar matter distorts starlight by absorption. Struve later discovered the presence of interstellar hydrogen, a result crucial for radio astronomy. He was director of both Yerkes Observatory (1932-1947) and McDonald Observatory (1939-1947), which he was instrumental in founding. Struve was awarded the Royal Astronomical Society's Gold Medal in 1944.

Harald Ulrik Sverdrup
1888-1957

Norwegian meteorologist and oceanographer who did extensive research on ocean physics and chemistry, explained the equatorial countercurrents, and developed a method of predicting tides and surf heights. Sverdrup was one of Vilhelm Bjerknes's original collaborators in the Bergen School of Meteorology and then focused on oceanography. As professor of geophysics at the University of Bergen (from 1926), he was involved with magnetic and oceanographic theory and the Bergen weather service. In 1935 he became head of Scripps Institute of Oceanography (La Jolla, California). He wrote *Oceanography for Meteorologists* and co-authored *The Oceans*, both in 1942. Sverdrup returned to Norway, heading the Norwegian Polar Institute at Oslo from 1948.

Leo Szilard
1898-1964

Hungarian-born American physicist and a leading contributor to the development of nuclear energy and atomic weapons. One of the first to realize the significance of nuclear fission, Szilard helped Enrico Fermi organize efforts behind the first controlled nuclear chain reaction (1942).

His post-war efforts to control nuclear weapons include helping establish the *Bulletin of Atomic Scientists* (1945) and The Council for a Livable World (1962). After World War II Szilard began research in molecular biology, working on aging and memory.

Joseph John Thomson
1856-1940

British physicist who discovered the electron (1897). Thomson improved upon Jean Perrin's results by accurately determining the charge-to-mass ratio of cathode ray particles. These new particles were later named "electrons" by George Stoney. Thomson's device for measuring charge-to-mass ratios developed into the cathode ray oscilloscope, which is widely used as a research tool and in television receivers. Thomson helped establish the preeminence of the Cavendish Laboratory in experimental physics and was awarded the 1906 Nobel Prize in physics.

Clyde William Tombaugh
1906-1998

American astronomer who discovered the planet Pluto. Tombaugh helped Lowell Observatory director Vesto Slipher search for a new planet. The previous director, Percival Lowell, had suggested the possible location of a planet affecting the orbits of Uranus and Neptune. Tombaugh studied photographs taken of that site at different times, looking for light and movement varying from that of stars. On February 18, 1930, he detected such movement in the Gemini constellation, and the discovery of Pluto was publicized the next month.

Mikhail Tsvet
1872-1919

Russian botanist who developed chromatography, the technique for separating the pigments that made up plant dyes. Tsvet's method has been extensively used for identifying many biologically important materials, especially after it was adapted to use paper as the support. Paper chromatography has been widely used to study colorless amino acids, steroids, carbohydrates, and other complex materials of natural origin. Tsvet is also noted for his research on chlorophyll, of which he discovered several new forms, and the carotenoids, a term he first coined.

John Hasbrouck Van Vleck
1899-1980

American physicist best known for his quantum explanation of magnetic phenomena. He introduced the concept of temperature-independent susceptibility known as Van Vleck paramagnetism and used quantum theory to evaluate the behavior of an atom or an ion in crystals. His work on electron correlation was exploited by others in developing the laser. Van Vleck shared the 1977 Nobel Prize in physics with Philip Anderson and Neville Mott for "investigations of the electronic structure of magnetic and disordered systems."

Paul Villard
1860-1934

French physicist who discovered gamma radiation. During his study of nuclear reactions in 1900, Villard was the first to observe an emission, which he named gamma rays. He proposed that a gamma ray is radiation similar to light, rather than made up of particles like the previously observed nuclear reaction emissions, alpha and beta rays. Ernest Rutherford confirmed Villard's conclusion in 1914, showing that gamma rays are electromagnetic radiation with a much shorter wavelength than X-rays.

Carl Friedrich, Freiherr (Baron) von Weizsäcker
1912-

German physicist who, independent of Hans Bethe, proposed a series of nuclear-fusion reactions as the mechanism whereby stars produce energy (1938). Weizsäcker also modified Pierre Simon Laplace's nebular hypothesis. According to Laplace, the planets formed from a gaseous nebula rotating uniformly—like a solid disk—around the Sun. Weizsäcker noted there would be differential rotation resulting in vortices from which the planets coalesced. Though failing to consider the solar system's angular momentum, Weizsäcker's improvements advanced our understanding of planetary formation.

Chaim Weizmann
1874-1952

Russian-Israeli chemist who pioneered microbiological applications in industrial chemistry, such as using bacteria to break down starches for production of ethanol, butanol, and acetone. A politically active Zionist, Weizmann played a key role in securing the 1917 Balfour Declaration establishing the British mandate of Palestine, which ultimately became the state of Israel in 1947. Weizmann became the new nation's first president in 1948. He established an Institute of Sciences (now named after him) in the city of Rehovot in 1934.

Sarah F. Whiting
1846-1927

American astronomer who took the first American x-ray photographs. Whiting was a physics professor at Wellesley College. She established the first physics laboratory for female students in the United States, initiating astronomical study. Whiting was director of the college's Whiting Observatory. Whiting studied at major European laboratories to share current research with her classes. In 1895 she experimented with x-ray photography. Whiting's book *Daytime and Evening Exercises in Astronomy* (1912) outlined her teaching methodology. She inspired other women also to become scientists.

Bibliography of Primary Sources

Books

Bohr, Niels. *The Theory of Spectra and Atomic Constitution* (1922). This book represents Bohr's attempt during the 1920s to explain the structure and properties of the atoms of all chemical elements, including the spectra emitted by atoms.

Bohr, Niels. *Atomic Theory and the Description of Nature* (1934). An important book by Bohr offering his insights into atomic theory.

Clarke, Frank. *Data of Geochemistry* (1908). This work was for many years standard source on minerals in Earth's crust.

Pauling, Linus. *The Nature of the Chemical Bond* (1932). Pauling was the first to apply quantum mechanics to chemistry, and in this book he offered an elegant and highly successful treatise on what he was learning.

Schrödinger, Erwin. *What is Life?* (1944). An influential work, this book inspired a generation of scientists in regards to molecular biology, including Francis Crick.

Wegener, Alfred. *Die Entstehung der Kontinente und Ozeane* (1915). This book (whose title means "The Origins of the Continents and Oceans") contained Wegener's explanation of his pioneering continental drift theory, which he conceived of in 1912. Wegener was not the first to conceive of continental drift, but his account was the most fully developed and supported by extensive evidence.

Periodical Articles

Bjerknes, Jacob. "On the Structure of Moving Cyclones" (1919). In this paper Bjerknes described the atmospheric formation he called the "extratropical cyclone" and offered it as a model of storm formation The squall line and developing low pressure were integrated phenomena. Here was a recurring, large-scale system of the atmosphere, revealing a definite progression of weather patterns and an efficient means upon which to base a weather forecast.

Bjerknes, Jacob and Halvor Solberg. "The Life Cycle of Cyclones and the Polar Front Theory of Atmospheric Circulation" (1922). This paper represents the mature efforts of the authors' research on cyclones and atmospheric circulation.

Einstein, Albert. "A New Determination of Molecular Dimensions" (1905). This was the first of five landmark papers published in the German periodical by Einstein in 1905. Einstein later earned a Nobel Prize for this paper.

Einstein, Albert. "On the Motion—Required by the Molecular Kinetic Theory of Heat—of Small Particles Suspended in a Stationary Liquid" (1905). In this paper Einstein offered a theoretical explanation of Brownian motion, the irregular movements of microscopic particles in a liquid or gas.

Einstein, Albert. "On a Heuristic Viewpoint Concerning the Production and Transformation of Light" (1905). Here Einstein modeled light as a stream of tiny massless particles, or quanta, thereby explaining the photoelectric effect, in which light hitting metal causes it to release electrons.

Einstein, Albert. "On the Electrodynamics of Moving Bodies" (1905). Einstein's fourth influential paper of 1905 introduced his special theory of relativity, in which he showed that time was not absolute; rather, it depended on the speed at which one's frame of reference moved in relation to the speed of light.

Einstein, Albert. "Does the Inertia of a Body Depend upon Its Energy Content?" (1905). This paper was a mathematical explanation of the special theory of relativity presented in "On the Electrodynamics of Moving Bodies." Here, Einstein put forth his famous equation, $E = mc^2$.

Jeffreys, Harold. "The Rigidity of Earth's Central Core" (1926). The fluidity of Earth's core was finally established in this paper by Jeffreys.

Meitner, Lise and Otto Frisch. "Disintegration of Uranium by Neutrons: A New Type of Nuclear Reaction" (1939). In this classic paper Meitner, along with her nephew Otto Frisch, first proposed the term "fission" for the process of uranium fission.

Oldham, Richard. "The Constitution of the Interior of the Earth as Revealed by Earthquakes" (1906). In this work Oldham conclusively established the existence of Earth's core. He did not however, as is widely believed, discover the core to be liquid. He maintained his original position—later shown to be incorrect—that the core is extremely dense and probably iron.

NEIL SCHLAGER

Technology and Invention

Chronology

1903 Using a craft he designed with his brother Wilbur, Orville Wright makes the first controlled and powered flight in history, at Kitty Hawk, North Carolina.

1907 Leo Hendrik Baekeland, a Belgian-American chemist, creates the first modern plastic, Bakelite.

1908 Henry Ford designs his Model T, which will become the first automobile produced by assembly-line techniques; by the time of its discontinuation 20 years later, Ford will have sold more than 15 million Model Ts.

1920 Building on a quarter-century of technological development that includes the creation of wireless telegraphy by Guglielmo Marconi (1896), and Lee De Forest's triode amplifying tube (1907), KDKA in Pittsburgh begins the world's first regular radio broadcasts.

1926 American physicist Robert Goddard designs, builds, and launches the first modern rocket, which burns liquid oxygen and fuel.

1927 Talking motion pictures make their debut with *The Jazz Singer,* starring Al Jolson.

1930 Frank Whittle, an English inventor, patents his design for the world's first jet engine.

1935 The first commercial tape recorder, using magnetic recording tape (invented by J. A. O'Neill in 1927), appears on the market.

1935 English physicist Robert Alexander Watson-Watt develops and patents the first practical radar equipment for the detection of aircraft.

1935 DuPont chemist Wallace Hume Carothers patents the first successful synthetic fiber, nylon.

1936 More than a decade after pivotal inventions by Vladimir Zworykin made TV technology possible, the British Broadcasting Corporation (BBC) begins the world's first regular television broadcasts.

1945 Years of atomic research culminate in the destruction of two Japanese cities, Hiroshima and Nagasaki, by U.S. bombs—the first and only use of nuclear power as a weapon—which leads to the Japanese surrender and the end of World War II.

1946 The first large, general-purpose electronic computer, ENIAC (Electronic Numerical Integrator and Computer), becomes fully operational.

1947 American physicists William Bradford Shockley, John Bardeen, and Walter House Brattain build the first transistor, beginning the semiconductor revolution.

Overview: Technology and Invention 1900-1949

Overview

The industrial revolution, which transformed technology in the nineteenth century, entered a second or mature phase by the beginning of the twentieth century. The widespread use of iron, coal, and steam in the 1800s provided a foundation for developments in chemistry, electricity, steel, and increases in mass production and consumption as the century changed. With the spread of industrialism in the Western world, many early-twentieth-century people viewed technology and science as venues for human progress and anticipated a near-future world free of want and war.

No figure of the first half of the twentieth century represents the promise of technology better than Henry Ford (1863-1947). His innovation of the moving assembly line to manufacture the Model T, a complex technological system, with standardized, interchangeable parts, transformed industrial production and made the products of industrialism affordable for a larger population. The resulting mass consumption and materialism became a defining characteristic of a mature industrial culture. That culture also developed institutions dedicated to deliberate invention and innovation with the goal of expanding the marketplace and material comfort. By mid-century industrialized nations relied on modern technology for the postwar affluence and prosperity characteristic of the second half of the twentieth century.

Technology Transformed

The developed industrial age provided mankind with a host of new technological devices. From Henry Ford's Model T to transistors, technology became more dominant in people's lives. Inexpensive motor cars created an automobile culture with its demand for highways, petroleum products, and repair and support services. Research and development in industrial laboratories, government agencies, private workshops, and universities produced myriad devices from airplanes to plastics, from rockets to ranges. Diesel engines and electric motors supplemented, then surpassed, steam power as the means of locomotion. Domestic technology provided many households with conveniences such as washing machines, vacuum cleaners, gas and electric ranges, electric irons, refrigerators, dishwashers, and garbage disposals. These laborsaving devices created as much work as freedom for women, who took on the mantle of household manager with increasing responsibilities for home hygiene and family care. Entertainment within and without the house relied increasingly on electric devices from radios to motion pictures. In countless ways, the new technology of industrialism transformed the way people lived in the first half of the century.

The Age of Electricity

Electrical technology developed fully in the first half of the twentieth century. The successful advancement of incandescent lighting resulted in electric lights replacing the widely used gas lighting in both interior and exterior applications. The discovery and use of detection and amplification devices such as diodes and triodes ushered in an era of mass communications. Inventors and innovators such as Lee De Forest (1873-1961) and Edwin Armstrong (1890-1954) made significant contributions to radio broadcasting with the increased use of vacuum tubes to send and detect radio signals. Vladimir Zworykin (1889-1982) and Philo Farnsworth (1906-1971) in the United States, and John Logie Baird (1888-1946) in Britain, developed television technology in the 1930s and 1940s. Because depression and war kept video from the commercial marketplace, it did not become a viable communications technology until the late 1940s, with its widespread use further delayed until the postwar prosperity of the 1950s and 1960s.

An equally important result of using electrical technology to detect and amplify signals was the development of calculators and computers, first for rather simple arithmetic and, later, for more sophisticated computations and analyses. During World War II, the need for rapid ballistics calculations drove the development of devices such as the ENIAC computer by John Presper Eckert Jr. (1919-1995) and John W. Mauchly (1907-1980) in the mid-1940s. The appearance of the transistor as a substitute for vacuum tubes had an impact on computer design and a host of other electronic devices that dominated electronic communications in the second half of the twentieth century. By mid-century the age of electrical technology was firmly in place with phonographs, motion pictures, electric lighting, motors, and railroads, while the dawn of the electronic era had begun with radio, tape recorders, television, and computers.

Monumental Technology

The increased use of steel as a structural element and the enhancement of steam, electrical, and internal combustion power sources provided the foundation for technology on a massive scale. With projects such as the construction of the Panama Canal, Soviet efforts to mechanize collective agriculture, the drilling of offshore petroleum reserves, pioneering work on rocketry, and the erection of soaring steel skeleton skyscrapers and towering suspension bridges, technology in the first half of the twentieth century became larger and more massive in scale, more monumental in its impact, and more dominant in affecting the lives of those living in industrialized societies. Deeply rooted in research and development efforts and increasingly reliant on a scientific base, technological change accelerated so that new products and new processes became hallmarks of technology of the time. Consumer credit and advertising aided mass consumption of the fruits of industrialism.

This expansion of technology and widespread technological change were accompanied by more government involvement in the support of technological development from war efforts in weaponry, submarines, tanks, aircraft, and medicine to large-scale projects such as the Tennessee Valley Authority and Rural Electrification Authority in the United States in the 1930s. During a time of economic depression, technological opportunity and change were the means to improve living conditions in less affluent and remote rural areas of an industrial nation. Increasingly, people sought to extend the materialism of industrialism to larger and larger segments of the population.

Conclusion

As the culmination of the industrial era, the twentieth century experienced the promise and the problems of technology. Greater materialism, faster communication and transportation, man-made products, mass production and consumption, and widespread and large-scale technology marked the era. At the same time, the first half of that century endured two world wars and a major economic depression that deeply affected people's lives and attitudes. Technology was both friend and foe to humanity; the prospect of plenty and material comfort coexisted with horrors of large-scale war and the potential for massive destruction. Clearly industrialism transformed everyday life in urban, technological societies. Henry Ford had begun a revolution of production with consequences he could not foresee.

Yet, even with its failings, technology emerged as the source of progress. Partnered with science, it provided the means for improving living conditions, widening the world, accelerating change, and imagining the future. The public embrace of technology's value and the maturing of industrialism throughout the first part of the twentieth century provided a foundation for the half-century of relative peace and prosperity that followed World War II. The achievements and promise of technology during the first half of the twentieth century were tempered by large-scale war, the potential for nuclear annihilation, and the environmental costs of rapid scientific and technological change. As with much of technological development, the benefits realized were accompanied by long-range costs that threatened the world's resources and increased people's dependence on technology for their existence. By and large, though, industrial societies welcomed the major changes and achievements created by this mature industrial era.

H. J. EISENMAN

The Development of Airships

Overview

Lighter-than-air airships and heavier-than-air airplanes both became practical around 1900, and coexisted for the next 40 years. Airships reached technological maturity by 1920 and were, for the next two decades, superior to airplanes for long flights with heavy payloads. Germany, Great Britain, and the United States all built airships after World War I, and all saw airships as a key part of aviation's future. The promise of airships dimmed rapidly in the 1930s, however, after a series of spectacular crashes tarnished their reputation. World War II smothered any hope of a quick revival, and produced a new generation of airplanes superior to any foreseeable airship. Airships survived, after 1940, in the significant but narrowly defined niche of aerial observation platform.

Background

The two basic elements of an airship—a balloon for lift and an engine-driven propeller for power—were readily available to nineteenth-century designers. The problem, as many discovered, was finding a light, powerful engine. The first successful airship was built and flown in 1852 by Henri Giffard (1825-1888), then Europe's leading steam engine designer. His 350-lb (159 kg), 3-horsepower engine drove the 44-ft (13.5 m) airship at little more than 5 mph (8 kph), making operation in even a gentle breeze impossible. The gasoline engine, developed in the 1890s, provided the kind of power-to-weight ratio airships needed. In 1898 Alberto Santos-Dumont (1873-1932) flew a gasoline-powered airship, *Number 1,* that produced the same 3 horsepower as Giffard's steam engine, but weighed 80% less. The growing power of gasoline engines gave *Number 1*'s successors greater speed and maneuverability, making them the first truly practical airships.

Santos-Dumont's airships, like virtually all of those built before them, were non-rigid: the pressure of the lifting gas maintained their shape. Beginning in 1900, however, airships of a radically different design began to emerge from Germany. Designed by (and named for) Count Ferdinand von Zeppelin (1838-1917), the new airships had rigid, cigar-shaped hulls built of metal frames braced by wires and covered by fabric. Tall, narrow gas bags, separated by frames and maintenance catwalks, filled the interior space. The engines, crew, and passengers occupied small pods suspended from the underside of the hull. The Zeppelin's most striking feature, however, was its size. The first was 420 ft (128 m) long with a gas capacity of 400,000 cu ft (11,200 cu m)—10 times the size of Giffard's pioneering airship, and 60 times the volume of Santos-Dumont's *Number 1.*

Zeppelins were powerful enough to do useful work in both military and civilian roles. They inaugurated the golden age of the airship, roughly 1906-1937, and ensured that Germany would dominate its first dozen years. They also defined the form that the airships of the golden age would take. The big airships built by Britain, France, Italy, and the United States would all be variations on Count Zeppelin's technological themes.

Impact

Zeppelin and his partner, Hugo Eckener (1868-1954), used the new airships to found the world's first airline in November 1909. Known as DELAG, the line began operations in 1910 and quickly grew to serve most of Germany's major cities. It made more than 1,500 flights in the four years leading up to World War I, carrying more than 10,000 passengers and traveling more than 100,000 mi (160,934 km) without even a minor accident or injury. The prewar success of DELAG raised hopes of a postwar era in which a new generation of larger airships would serve international or even intercontinental passenger routes. The 1917 flight of a naval Zeppelin, *L59,* from Bulgaria to East Africa reinforced these hopes. *L59,* the largest airship yet built, carried 13 tons (11.8 metric tons) of cargo 4,200 mi (6,759 km) in 96 hours on an aborted mission to resupply besieged German troops. Her mission, though a failure, provided what many airship watchers saw as a glimpse into the future.

The airship's military record in the 1910s was more mixed. Their greatest success came in the essential but unglamorous role of coast defense, where they served as platforms for observers looking for enemy naval vessels. Their range and endurance gave them substantial advantages over the still-primitive airplanes of the time, and their proven reliability made them well suited to long flights over water. The airships used by the German army and navy as strategic bombers were more visible, but less

The *Graf Zeppelin* airship (1929). *(Library of Congress. Reproduced with permission.)*

successful. Three-and-a-half years of raids on London killed only 557 people and did only localized property damage. The difficulty of hitting specific targets at night limited the raids' tactical value, and the growing effectiveness of air defenses steadily eroded their impact on Londoners' morale.

Interest in airships as strategic bombers evaporated after the end of World War I, but interest in them as airliners and coastal sentinels remained high. Having lost the war, however, Germany also began to lose its lead in the airship business. Britain, France, and Italy seized Germany's four surviving military airships, along with two new ones built by the Zeppelin Company for the reborn DELAG, as spoils of war. The United States, alone among the allies, paid for its German airship, contracting with the Zeppelin Company to build the *Los Angeles.* Using their appropriated German technology, the allied nations entered the airship business in earnest. Britain's *R34,* based on wartime studies of Zeppelins that crashed on British soil, made the first round-trip aerial crossing of the Atlantic in 1919. Seven years later, a joint Italian-American-Norwegian expedition flew the Italian-built airship *Norge* over the North Pole, from King's Bay, Spitzbergen, to Teller, Alaska.

Both *R34* and *Norge* followed in the footsteps of airplanes. A converted World War I bomber had made the first one-way, non-stop Atlantic crossing a few weeks before *R34,* and a Fokker transport commanded by Admiral Richard Byrd (1888-1957) had flown from King's Bay to the North Pole and back hours before *Norge* departed. The airplanes, however, had been operating at the limits of their capabilities, while the airships had made more ambitious flights with ease. The ultimate demonstration of the airship's range and endurance came, however, from Germany. The *Graf Zeppelin,* launched in 1928, inaugurated commercial air travel across the Atlantic in the fall of that year, carrying both mail and fare-paying passengers. She went on to circle the world in the summer of 1929 before beginning scheduled service flights from Europe to South America.

REPORTING THE *HINDENBURG* DISASTER

The May 6, 1937, landing of the giant airship *Hindenburg* was a routine event but still a newsworthy one. Reporters waited and watched. A newsreel crew crouched behind their movie camera as the biggest airship in the world edged toward its mooring mast just before 7:30 PM. Speaking into a portable tape recorder, reporter Herb Morrison described the arrival for an upcoming radio broadcast. Suddenly, the giant airship burst into flame and sank toward the ground, a smoldering wreck. Newsreel cameras silently recorded her destruction. Morrison sobbed into his recorder: "It's burning, bursting into flames, and it's falling on the mooring mast and all the folks. This is one of the worst catastrophes in the world.... Oh, the humanity and all the passengers!" The newsreel images and Morrison's words had a visceral impact no written account or still photograph could match. They demonstrated, for the first time, the unique power of radio and film to make audiences feel like eyewitnesses to real-world events. They also showed the new technologies' power to shape viewer perceptions. Most viewers, then and now, assume that all aboard the airship died; in fact, nearly two-thirds escaped alive.

A. BOWDOIN VAN RIPER

The success of the *Graf Zeppelin,* already the largest airship built up to that time, encouraged the construction of even larger craft. The United States Navy commissioned *Akron* and *Macon* for coastal patrol work in the early 1930s, while the British government backed the building of two passenger airships: *R100* and *R101.* Germany's own super-airship appeared in 1935, built by the Zeppelin Company with the backing of the Nazi government and named for Germany's last pre-Nazi leader—*Hindenburg.*

The immense size of these super-airships gave them capabilities that no airplane, and no earlier airship, could match. *Akron* and *Macon* could carry up to five fighter planes each in hangars built into their bellies, launching and recovering them high over the ocean. *R100* and *R101,* designed as the prototypes of an air fleet that would link Britain to its distant colonies, could carry 100 passengers apiece—five times the capacity of the *Graf Zeppelin* and four times that of the largest passenger airplanes then flying. *Hindenburg* was, in its way, the most spectacular of all—designed to transport 50 passengers in the luxurious style of the finest ocean liners. Its passengers could enjoy gourmet meals, a lounge with a grand piano (made of aluminum to save weight), and even a carefully engineered smoking room—in close proximity to 7 million cu ft (196,000 cu m) of explosive hydrogen.

The careers of four of the five super-airships ended in catastrophe. *R100* performed flawlessly on her first and only flight (to Canada in 1930), but months later *R101* crashed on a French hillside only hours after departing on her inaugural flight to India. *Akron* crashed in the Atlantic during a 1933 storm, and *Macon* went down in the Pacific two years later. *Hindenburg* had the most spectacular end, bursting into flames as she landed at the Lakehurst, New Jersey, airship terminal on May 6, 1937. None of the crashes, in retrospect, reflected badly on the airship concept: *Akron* and *Macon* were lost to weather and pilot error, *R101* doomed by grossly incompetent engineering, and *Hindenburg* most likely destroyed by a freak accident. The death toll in all four crashes was only 159: 74 aboard *Akron,* 2 aboard *Macon,* 48 aboard *R101,* and 35 aboard *Hindenburg.* Nevertheless, the crashes led first Britain, then the United States, and finally Germany to abandon support for large rigid-hulled airships. The last rigid airship, *Graf Zeppelin,* was grounded after the *Hindenburg* explosion and never flew again. She, and her never-flown successor, *Graf Zeppelin II,* were broken up for the aluminum in their frames on the eve of World War II.

The end of the passenger airship coincided with the airplane's rise to technological maturity. World War II produced a new generation of long-range bomber and cargo planes that, after the war, became the basis for the first land-based airliners capable of crossing oceans. Jet engines and pressurized cabins, also products of World War II, allowed the airliners of the 1950s to match the smooth, quiet ride of prewar airships at speeds many times greater. Airships thus lost any hope of reclaiming their prewar share of the passenger-carrying market. The small, non-rigid airships (blimps) built since 1940 have been built for the few, limited roles where their unique attributes are still valuable. The most significant of those roles, coastal patrol, evaporated in the 1950s as ballistic missiles became the principal threat to the security of Western nations. Airships now serve only as camera platforms and flying billboards—as fixtures in the sky rather than a means of traveling through it.

A. BOWDOIN VAN RIPER

Further Reading

Botting, Douglas. *The Giant Airships.* Alexandria, VA: Time-Life Books, 1981.

Collier, Basil. *The Airship: A History.* New York: G. P. Putnam's Sons, 1974.

Eckener, Hugo. *My Zeppelins.* London: Putnam, 1958.

Horton, Edward. *The Age of the Airship.* New York: Henry Regnery, 1973.

Robinson, Douglas H. *Giants in the Sky: A History of the Rigid Airship.* University of Washington Press, 1973.

Santos-Dumont, Alberto. *My Airships.* New York: Dover, 1973.

Shute, Neville. *Slide Rule: The Autobiography of an Engineer.* London: Heron Books, 1953.

Toland, John. *The Great Dirigibles: Their Triumphs and Disasters.* New York: Dover Books, 1972.

Drilling for Offshore Oil

Overview

Although the first "offshore" oil drilling took place in the mid-1890s, the first true offshore oil rig, out of sight of land, was not built until 1947 off the Louisiana coast. Since that time, offshore and deep-water drilling has become increasingly sophisticated, and oil recovery is taking place in ever-deeper waters on the continental shelf throughout the world. In addition to the Gulf of Mexico, offshore oil rigs have sprouted off the California coast, between Indonesia and Australia, the Caspian Sea, the Mediterranean Sea, and in the North Sea. Near-shore drilling takes place in Venezuela's Lake Maracaibo and, north of Indonesia, in waters claimed by China, Vietnam, and Indonesia lies what may be the world's largest natural gas field. The world has come to depend heavily on offshore oil, in spite of the risk of large-scale marine oil spills, the expense of constructing an oil platform, and the risky nature of the work.

Background

Seeps of oil have been known since antiquity, including some places just off the coast of California. In the mid-1890s, a few oil wells were drilled off a pier near Santa Barbara, but they produced only a few barrels of oil per day. Other oil platforms were built in the next few decades, primarily in lakes in Louisiana and Venezuela, as well as in the shallow waters off the Texas coast. Although more productive than the early Santa Barbara wells, they were still a far cry from true offshore oil rigs.

In 1947, the Kerr-McGee company, an independent Oklahoma-based oil firm, purchased the rights to drill for oil about 10 mi (16 km) off the Louisiana shore. They were fairly sure that the oil was there; what they didn't know was how to build an oil rig that would operate out of sight of land. Hurricanes, possible ship collisions, high waves, uncertain bottom conditions, and more were all important and all unknown. Kerr-McGee solved these problems, and others, and went on to build the first true offshore oil rig, striking oil in 1947. Other companies soon followed suit, although at a deliberate pace because of the high cost of offshore oil recovery (about five times the cost of similar wells on land). Offshore drilling really started to take off in the 1960s, in the Gulf of Mexico and off the California coast. In the 1970s, oil was discovered in the North Sea and, over the next 10 years, in many other locations. By 1980 it was thought that up to a quarter of all recoverable oil could be located at sea, worth over $5 trillion in year 2000 dollars.

Several technological innovations made offshore oil production possible. Among these were methods of finding oil away from any visible geological indicators, oil platform design and construction, getting the oil to shore, and making the platform able to withstand wind, weather, waves, and other possible disasters. All of these challenges were met, of course, as indicated by the presence of numerous oil platforms today.

North Sea oil pushed oil rig technology as never before. Forced to drill in deeper, often stormy waters that experienced a full four seasons, engineers developed a number of innovative solutions. Where the first offshore rigs rested on the bottom, supporting their weight and the weight of long supports, some later rigs were supported by huge hollow floats, tethered to the seafloor with the platform riding high above the waves. Other rigs were constructed ashore and towed to their drilling sites, where the legs were cranked down through the platform to rest on the seafloor. To maximize production, some huge platforms were constructed with 50 or more separate wells beneath them. These, and other technological advances, made it possible to drill for oil in ever-deeper ocean waters.

With increased offshore drilling came greater chances for pollution if something went wrong. The first such "blowout" occurred off the coast of California in 1969, spilling a few thousand barrels of oil into the sea. A more serious blowout occurred in 1979 in the Gulf of Mexico. These events captured the public's attention at the time, but interest seemed to wane quickly, mainly because such events are not common.

Impact

Offshore oil is a significant part of the world's energy picture and shows signs of becoming even more important as the large oil fields on land become depleted. However, their discovery and exploitation has already affected society in some interesting and important ways. These are: 1) Pushing technological innovations that have benefited both the oil industry and science. 2) The addition of an important new source of energy. 3) The economic and political impacts on countries with offshore oil reserves. 4) Influences on the mass media and popular culture.

One of the most obvious issues about drilling for oil at sea is that it's difficult. One person compared it to trying to drill a hole in the sidewalk by dangling a piece of spaghetti from the top of the Empire State Building. As oil rigs moved into deeper water, the engineering problems became ever more challenging, requiring an array of innovative engineering solutions to overcome them. For example, wave stress on deep-water oil platforms in the North Sea has forced engineers to find better ways of monitoring stress-induced cracks in the platform supports. Pulling a drill string to change the bit requires lowering it again to successfully hit a 12-in (30-cm) hole at the bottom of 100 m (328 ft) of water, without being able to see the hole or the end of the drill string. Even prospecting for oil underwater is difficult, requiring whole new systems for conducting geophysical investigations in deep water. Among these systems was the development of new methods of seismic studies, setting off explosions to map underground structures by timing the return of echoes from different rock and sediment layers. All of these innovations have found uses elsewhere, many of them in science.

In 1968, the *Glomar Challenger* set out on the first leg of the Deep Sea Drilling Project. Using technology largely developed by oil companies, its goal was to conduct drilling operations in the deep ocean to better understand the history of the oceans and the Earth. Other scientific vessels had conducted ocean drilling before, but none had been specially designed for the task. The *Glomar Challenger* and its successor vessels (currently the *JOIDES Resolution*) have made a number of extremely important discoveries, including confirmation of many aspects of plate tectonic theory, a better understanding of the origins of ice ages, the recent passage of the solar system through a relatively dense interstellar dust cloud, and the discovery of apparent supernova debris in deep-sea sediments. All of these discoveries have served to give us a far better understanding of the Earth, and factors that can affect it (and us). Would we have made these discoveries without offshore oil platforms? Almost certainly. However, it is likely that the technology to make such discoveries would have lagged without the impetus provided by the desire and the need to recover oil from the depths.

Probably the best-known impact of developing offshore oil rigs is the tremendous impact it has had on the world's energy picture for the short-term and intermediate-term future. It is now thought that at least 25% of the world's oil is to be found at sea. This is a tremendous amount of oil, especially when we consider that most of the large fields ashore are thought to have been located already. Petroleum geologists have covered the Earth's surface in search of promising geologic formation that might house oil and, since the discovery of Alaskan oil, only one large oil field has been added to the picture, the oil near the Caspian Sea. And this is not a new oil field, just one that was underutilized until recently. Many geologists feel that, if there are any huge new oil fields yet to be found, they underlie the oceans. As of this writing, the biggest offshore oil fields are in the Gulf of Mexico, between New Guinea and Australia, and in the North Sea. With probable lifespans of several decades, these major oil fields are an important source of energy for the world.

Offshore oil is also an important source of income for many nations and, since many nations with large amounts of offshore oil are not OPEC members, it is politically important, too. While petrodollars have not had the same impact on Britain, Norway, and the Netherlands that they had on the oil-rich nations of the Middle East, they have proven to be a welcome addition to national coffers. At the same time, it must be remembered that these nations were already established as first-world countries when they discovered their offshore oil, as opposed to the small and then-backward kingdoms of the Middle East. Other nations, particularly the resource-rich but poverty-stricken nations of Venezuela and Indonesia, have failed to capitalize on their oil resources, primarily because oil revenues have been intercepted by the ruling elite where they serve to help maintain the power structure. In addition, since most of these nations are not OPEC members, the development of offshore oil has served to dilute OPEC's power to impose oil prices on the rest of the world. In fact, since North Sea oil recovery became important in the 1980s, OPEC's power dwindled markedly with only a few brief resurgences in the 1990s. This, in turn, has helped keep oil prices down for a remarkably long time and is one of the factors behind the record-setting U.S. economy during the 1990s.

Finally, oil platforms have entered the public arena to some extent because of their size, sophistication, and use of high technology. At least one James Bond movie was set on an oil rig, as were two thrillers written by the late Alistair MacLean. Other books and movies have made use of oil platforms, too, and a number of documentary shows have been devoted in full or in part to these behemoths.

P. ANDREW KARAM

Further Reading

Books

Yergin, Daniel. *The Prize*. New York: Simon & Schuster, 1991.

The Invention of the Airplane and the Rise of the Airplane Industry for Military and Civilian Purposes

Overview

The development of the airplane is a twentieth-century phenomenon. From the first powered aircraft to the creation of the supersonic transport, airplanes improved quickly. This was aided by the innovations of World War I and World War II. Demand for air travel led to the creation of an industry including aircraft construction companies, engine and equipment makers, as well as firms that built and operated airports. When military leaders recognized its value, the airplane became central to defense as well as in the strategy and tactics of wars.

Background

Men have dreamed of being able to fly for centuries. Leonardo da Vinci (1452-1519) drew pictures of machines that could fly. In 1782 the Montgolfier brothers invented a hot air balloon that floated over Paris for 25 minutes. The development of powered balloons, however, did not lead to practical aircraft. In the nineteenth century a glider did fly, and men tried to get airborne with steam and other engines. In 1896 a powered model airplane actually flew. Various configurations of wings and propellants were tried unsuccessfully, but the airplane was on the verge of being born.

Brothers Orville and Wilbur Wright (1871-1948 and 1867-1912, respectively), bicycle makers from Dayton Ohio, worked out problems of how to control aircraft in the air. They built a biplane fitted with a 12-horsepower gasoline engine and took it to North Carolina. On December 17, 1903, with Orville at the controls, the plane flew for 12 seconds and traveled 120 feet (36.5 m). Four men and a boy witnessed this flight, and one took a picture. A few newspapers mentioned the event, but little attention was paid to the airplane for five years, because at the time they were still impractical. When World War I began in 1914, the design, capability, and production of airplanes advanced rapidly. Manufacturers built airplanes in England, France, and Germany. The United States, however, had only 110 airplanes available when it entered the war in 1917. To rectify this deficiency, airplane manufacturers cooperated with automakers to produce 15,000 additional planes by the end of the war in 1918. Two aircraft companies, established in time to make planes for the war, became leading aircraft producers. Boeing in Seattle, Washington, and Lockheed in Santa Barbara, California, are both still in business.

CLIPPER SHIPS OF THE SKY: TRANSOCEANIC PASSENGER SERVICE ON FLYING BOATS

Passenger air travel began shortly after the first flights, and the first airline began flying passengers on a schedule in 1914 with a scheduled flight (in a Benoist Flying Boat) from St. Petersburg, Florida, to Tampa, 18 miles away. Twenty-two years later, commercial air travel was to make another huge leap, this time crossing oceans, using more sophisticated flying boats.

This next step was taken in 1926 by Ralph O'Neill, who inaugurated flying boat service from New York to Buenos Aires, Argentina, via Rio de Janeiro, Brazil, and Havana, Cuba. O'Neill's airline was put out of business in 1930 by Juan Trippe and his Pan American Airlines. Trippe realized that air travel between continents could be highly profitable because of the time savings over ships. Over the next decade, Trippe commissioned the Pan Am Clippers, flying boats that carried up to 50 passengers across the Pacific Ocean to destinations in Asia in less than a week. These planes set a standard of luxury and romance that has never been equaled.

To make this route work, Trippe first had to have an airplane capable of making long journeys over water, and then he had to have places to land them. This involved designing new aircraft capable of flying 2,400 miles (3,862 km) safely and reliably over water, developing new navigation systems, and constructing facilities and hotels in Hawaii, Midway Island, Guam, Wake Island, and Manila. Service across the Atlantic followed a few years later, and the era of intercontinental air travel was born.

P. ANDREW KARAM

After the war, advances continued at a rapid pace. By 1920 airlines were carrying passengers across the United States as well as Europe. The U.S. government used airplanes to speed up delivery of the mail. In 1924 beacons were established at airports, and night flying became feasible. By then, passenger airlines were operating in South America, Africa, and Australia. KLM, Lufthansa, and Qantas were early airlines that are still in operation today.

Europeans led in building airplanes as well as in running airlines. DeHaviland and Vickers built planes in England in the 1920s, and in France several companies arose in the 1930s. In Germany, banned from manufacturing military aircraft by the peace treaty at the end of World War I, Messerschmitt, Heinkel, Fokker, and Junkers created an aviation industry. They also built military aircraft in secret after Adolf Hitler came to power in 1933. Japan also built military aircraft in the 1930s. Their biggest company was Mitsubishi, which produced the Zero, a famous World War II fighter plane.

In the United States several more aircraft companies were established. Donald Douglas located a company in Santa Monica, California, and Consolidated Aircraft in Rhode Island took over the designs of the company the Wright brothers had begun in New York in 1909.

During this period air transportation was becoming safer, and by 1935 four U.S. airlines were operating: American, Eastern, United, and TWA. Other airlines, like Northwest, Delta, and Braniff, established regional and restricted schedules. Pan American Airlines was the only American international airline at the time, flying to Latin America, Asia, and the Pacific.

To accommodate more passengers and make longer trips without refueling, larger planes were needed. To this end, Boeing produced the two-engine Boeing 247, Lockheed built the single engine Vega, and Ford Motor Company built the Trimotor, all of which were successful planes. Douglas Aircraft created a twin-engine transport, the Douglas DC-3, which became the most widely used plane in the world. In fact, some are still flying today.

In the mid-1930s several companies produced flying boats, or seaplanes. These four-engine planes, designed to land and take off from

The first flight: Wilbur and Orville Wright at Kitty Hawk (1903). *(The Granger Collection. Reproduced with permission.)*

the water, could cross oceans and sustain long flights. Engine makers kept up with this rapid growth. One of the most widely used engines was the Pratt and Whitney, manufactured by several companies. Allison also made a popular airplane engine. By 1938 there were more than three and a half million passengers flying around the world, half of them traveling on U.S. airlines.

When World War II began in 1939, air travel was severely restricted. After December 7, 1941, when the United States entered the war, 40 companies began producing aircraft for the United States and its allies. Production of American bombers, transports, and fighters reached 100,000 a year by 1944. Without doubt the airplane was a crucial element for all parties during World War II.

A new propulsion system was on the horizon that would make these planes obsolete and lead the aircraft industry into the twenty-first century. Though jet propulsion was understood for centuries, the Germans were first to put it in an airplane in 1939. Jet engines became the engine of choice in the last half of the twentieth century and would become the engine that enabled airplanes to break the sound barrier and fly supersonically.

Impact

The impact of the airplane and the airline industry on the people of the world has been enormous and rapid. From its first successful flight to its ability to fly faster than the speed of sound, the airplane has made the world accessible to everyone. Speaking only of the United States, in its early years the airplane became a tool that brought this huge country together. By the end of the first half of the century, it was a preferred method of travel and was affordable for ordinary people. The availability of pilots trained during the war helped the build-up of airline operations after the war and aided the manufacture of airplanes. Airports and support services also blossomed, and travel became easier, safer, and more accessible. A major impact of airplanes on modern life is the speed of delivery of mail. In addition, as communications advanced by the development of radios, airlines and airports quickly saw the value of fitting airplanes with radios for safety and convenience.

Government became interested and involved in air travel from the first. Regulations covering routes, airports, and flying over cities were common. An agency was established that regulated aircraft safety as well as one overseeing pilot licenses. In the United States the Civil Aeronautics Board, established in 1938, was responsible for the rates charged by the airlines, as well as their routes. Every state established its own agency to control airports, but from the earliest flights between cities in the United States, there were safety concerns. Control towers to oversee the takeoff and landing of airplanes were built in every airport. They have steadily improved to include systems to aid planes landing in bad weather and provide radio beacons to bring a plane into an airport accurately. The Federal Aviation Administration (FAA) was established in 1958, combining the regulatory tasks of the CAA and local agencies.

As mentioned above, airplanes were a major tool for all parties during World War II. Able to deliver material and men quickly to a battle site, they also carried bombs to enemy cities. It was the aerial bombing of Pearl Harbor in Hawaii that brought the United States into the war against Japan. Aerial activity greatly advanced the development of aircraft carriers, which became the major naval weapon of the Pacific theater of war. Aerial bombing of enemy cities was an important tactic in this war, and the nation that could supply huge numbers of bomber and fighter planes had an edge. The United States was the one nation that had the industrial might to overwhelm Germany and Japan and win the war. It also could also supply ample men and material without being touched itself from the air. The event that ended the war, the dropping of two atomic bombs that forced Japan to surrender, could not have been carried out in any vehicle but an airplane. In 1948 and 1949, in one of the first maneuvers of the Cold War, Russia attempted to cut off Berlin, then in the center of occupied Germany. United States airplanes foiled this attempt by sending a constant stream of food and supplies to Berlin by air. The Berlin airlift was a success and Berlin remained a western city in East Germany.

In the 1950s the United States took the flying units of all military services except the navy and combined them into a new unit called the Air Force. This is an indication of how important air operations had become. They were very necessary in the ensuing 40 years of Cold War with Russia. In the last half of the century, the airplane was the also the center of several extremely destructive trends. The terrorist highjacking of airplanes, with terrorists holding the hostages in return for demands, became a way for small groups of terrorists to make a political statement. Because of it, the security of airports was increased and tightened. In the latter part of the twentieth century airplanes became the primary way to reach people involved in earthquakes, fires, hurricanes, typhoons, and floods. While the airplane has been beneficial in many ways and destructive in others, there is no denying its enormous impact on modern life.

LYNDALL B. LANDAUER

Further Reading

Baker, David. *Flight and Flying: A Chronology*. New York: Facts on File, 1993.

Bilstein, Roger E. *Flight in America: From the Wrights to the Astronauts*. Baltimore, MD: Johns Hopkins University Press, 1987.

Hall, Richard R. *Rise of Fighter Aircraft: 1914-1918*. Annapolis, MD: Nautical and Aviation Publishing Co. of America, 1984.

Kirk, Stephen. *First in Flight: The Wright Brothers of North Carolina*. Winston-Salem, NC: J.F. Blair, 1995.

Leary, William E., ed. *The Airline Industry*. New York: Facts on File, 1992.

The Soviet Union Promotes Rapid Technological Development in the Communist Ideology

Overview

In the early twentieth century, Russia and her republics experienced tremendous political and cultural upheaval with the installation of socialism and the rise to power of the Communist Party. The ideology of Communism greatly affected the scientific community, and technology in particular, as Lenin and then Stalin attempted to shape most scientific disciplines into applicable ideas that would further Soviet society. Overall, the era of the Soviet Union had a detrimental effect on the whole scientific community, depriving Russia of some of its top scientists and systematically censoring certain scientific disciplines deemed too "esoteric" for application. Nevertheless, some sciences thrived in the socialist regime. Russian technology, in particular, was especially progressive and made great strides during World War II and the Cold War, initially relying on American styles of manufacturing and industrial design. Eventually though, as the Stalin-led Soviet Union closed itself to the rest of the world, science and technology was handicapped by a ban on the free exchange of ideas. Even today, Russia struggles with a bank of talented scientists working in ill-funded, poorly equipped research institutions.

Background

At the turn of the century, Russia suffered severe political unrest in a society that was marked by extreme privilege and terrible poverty. Only a small percentage of Russians were educated in the elite state-sponsored schools, and those few intellectuals were, in turn, society's few professionals. The average income of a member of the Imperial Russian Academy of Sciences, for example, was almost 30 times higher than that of an industrial worker. And while many scientists enjoyed their privileged status in society, most still supported the idea of bringing down the oppressive tsarist monarchy. Of the many different democratic and socialist groups competing for power just before the Bolshevik Revolution in 1917, most scientists supported the more moderate parties who favored reform without extreme measures.

With the onset of World War I the provisional government that took power after the Bolshevik Revolution established several research institutes aimed at studying technology for military needs and the extraction of mineral resources necessary to manufacture and operate this technology. The Bolshevik administration vowed "knowledge and education for the masses" and then criticized the "bourgeois" or elite scientists and experts who were seen as anti-Soviet supporters merely because of their previous association with the monarchy. Many of these scientists found themselves defenseless in a confused political Russia, and were frequently arrested and even executed. A great exodus of terrified scientists followed, including the departure of some of Russia's "star" scientists, like aircraft engineer Igor Sikorsky (1889-1972), who fled to the United States, biologist V. Korenchevsky, who emigrated to Britain, and chemist G.B. Kistiakovsky, who came to the U.S. and served as scientific advisor to President Eisenhower.

Soon, Lenin and the leaders of the Bolsheviks realized that they were losing their greatest strength and that rapid industrialization of the country depended on scientists trained in the much-hated tsarist system of education. Communist ideology relied heavily on the plan that technology would bring the country to a greater social and economic success. Lenin recognized that the prolonged Russian Civil War from 1918-1921 depleted existing Russian technology, and demanded widespread industrial and economic development. Lenin admitted the need to keep scientists well-funded and in the political loop, "in spite of the fact that they are inevitably impregnated with bourgeois ideas and customs." Life for technical experts changed dramatically, with new scientific institutions opening up around the country, and a special decree granting scientists better living conditions in the new technocratic Soviet society. Lenin's hope was to prevent the exodus of more of Russia's experts, train a new group of revolutionary "red" scientists, and eventually demote the "bourgeois" elite.

Several governmental agencies were formed to oversee the burgeoning scientific community, and by 1922 at least 40 separate institutes were organized. The free exchange of ideas with the West was encouraged, and Lenin, especially, argued that in order to move successfully into a technocratic economy, the USSR would need America. "We will need America industrial goods," he wrote, "locomotives, automobiles, etc., more than any other country." Visits abroad were frequent, Russian scientific journals were in abundance, and Russian science was seen as an international leader, led by scientists such as Ivan Pavlov (1849-1936) and S.S. Chetverikov. The period was referred to as "The Golden Years" of Soviet science, but this intellectual freedom would soon come to an abrupt, bloody halt.

Impact

By 1928, when Joseph Stalin came to control the USSR, Soviet science was regarded with high esteem by the international scientific community. The natural sciences, in particular, excelled with groundbreaking Soviet discoveries in genetics and behavioral psychology. Nevertheless, Stalin was set on purging Communist Russia of any remnants of the "bourgeois" days of the tsar, and put science on a course for superindustrialization. In the Stalinist regime, science was expected to be rational, applicable, and practical and, once again, all "bourgeois" scientists were regarded as enemies of the state. Stalin went on to eliminate all those who fell under anti-Soviet suspicion by conducting mock trials and interrogations. Many well-known scientific experts were arrested and shot during this period, while the lesser-qualified party scientists took their places.

In order to jump start this move toward industrialization, Stalin imported large quantities of Western machinery and technical equipment. The United States, in particular, was much admired by the Soviet leaders, especially the assembly line and standardization labor methods used in steel mills and in Henry Ford's automotive factories. "Fordism" became the model with which Soviet leaders built up their vast industrial manufacturing centers. Tractor plants modeled after Ford's Detroit factories were built and Ford tractors from abroad were imported by the thousands. Foreign technical experts were brought in to live in and guide the construction of industrial sites, while Soviet officials further condemned the "impractical" sciences such as biology and theoretical physics. The Dnieper Dam, the largest hydroelectric dam of that era, was built in consultation with Colonel Hugh Hooper, the engineer who designed the Coolidge Dam in Washington. This pressure for purely pragmatic science almost eliminated branches of natural science, or forced these natural scientists to compromise their research to create practical, immediate applications in their field.

But while most natural sciences were foundering, the metallurgic sciences flourished. Once again, the U.S. design in ore-dressing plants, zinc and lead production, and smelter technology was frequently copied. An immense iron and steel complex "city" called Magnitogorsk especially reflected the extremes of this period of industrialization: a city modeled after a U.S. Steel plant in Gary, Indiana, the complex became the symbol for "gigantomania" that defined the industrial boom of the late 1920s and early 30s. Stalin erected huge skyscrapers, enormous dams, sprawling industrial complexes, and began construction on the largest blast furnace in the world at that time.

But many of these complexes suffered a high rate of industrial accidents despite the reliance on foreign experts to fill the holes left by the persecution of the bourgeois experts. Rather than admit his mistake and release the qualified, persecuted scientists, Stalin opened scientific prison camps, called *sharshki* and drew on the expertise of the "criminal" scientists. The influence of these bourgeois scientists was tolerated, briefly, while their knowledge was used for the explosive industrial growth. It was believed, in the Communist ideology, that after this initial reliance, the socialist science would lead the world in technological progress with its own experts.

A period referred to as the Great Terror began in 1936, when several million people were arrested and more than half a million executed in a purge of suspected "anti-Communists." While the technology established before this period—tractor plants, smelters, power stations—were unaffected by the disappearance of the thousands of experts in these fields, the development of all new technology halted. International connections were completely severed. Foreign experts left the country. Technology suffered, especially in areas where innovation is critical, like aviation and military technology.

These technical weaknesses soon became critical as World War II started in 1941. The USSR, thanks to its mass production capabilities, had large numbers of tanks and airplanes, but they were not equal to the 1941 standards of Western countries. Stalin once again turned to his prisons and the experts he had locked up in the Great Terror. Many aviation and engineering experts were released on temporary work leave and prison "institutes" were once again revived. In one of these prison research centers the leading aircraft experts developed high-tech military aircraft and jet-mounted weaponry. From 1941 to 1945 the average speed of a Russian fighter plane increased by 62 mi (100 km) an hour.

Lenin, and the Communist ideology, taught that capitalism limited the development of technology, science, and the productivity of labor. But when America unleashed the atomic bomb, Stalin was clearly embarrassed. In 1946 the beginning of the Cold War started, with Soviet science striving not just to compete with Western technology, but soundly surpass it. Soon, all branches of military-oriented science received the highest state priority. Financial resources for science skyrocketed, the average salary for scientists doubled, and living conditions were improved for specialists in these disciplines, even though most of the country was slogging in poverty.

The number of students attending higher technical schools and universities grew from 817,000 to 1,500,000 in one year. By 1949 the newest mission of Stalin's was clear: the Soviet Union would be a socialist leader through the prowess of its science. But, each area of research was expected to have a "socialist" focus, meaning that research should still have purpose and immediate application. A research center for atomic energy was built in Moscow, and the first nuclear explosion in the USSR was made in 1949, when Stalin celebrated his seventieth birthday. A special rocketry center was built in 1946 to test military rockets, large industrial plutonium reactors were constructed, and German engineers were brought to Russia to help develop new aircraft.

Russia certainly achieved great strides in aircraft, space, and nuclear sciences during the Cold War, but so many important "idealistic" sciences suffered that the whole of the Soviet science community was permanently weakened. Cybernetics, for example, was declared a "pseudo-science" and is blamed for Russia's lagging electronics and computer industry today. Genetics, once the example of the genius of the Russian scientific community, was taken out of the hands of Russia's famed genetics experts and reworked to become more applicable, eventually failing in its theories and research.

Even Russia's once admired, impressive industrial technology lost its innovative edge, as computerization took over complex industrial sites and progressive, advanced metals replaced steel as a stronger, more resilient material. While the Communist ideology had periods of scientific excellence in limited fields, namely technology, the short-sighted approach of focusing on only technology and only on applicable research caused the Soviet Union to lose what was once its strongest resource: science.

LOLLY MERRELL

Further Reading

Books

Krementsov, Nikolai. *Stalinist Science.* New Jersey: Princeton University Press, 1997.

Medvedev, Zhores A. *Soviet Science.* New York: Norton, 1978.

Parrott, Bruce. *Politics and Technology in the Soviet Union.* Massachusetts: The MIT Press, 1983.

Popovsky, Mark. *Manipulated Science.* New York: Doubleday, 1979.

Periodical Articles

Josephson, Paul R. "Soviet Scientists and the State." *Social Research* vol. 59, Issue 3 (Fall 1992): 589.

The Development of Plastics

Overview

At the beginning of the twentieth century manufacturers made consumer goods and electrical insulation from natural materials like shellac, rubber, cellulose, and camphor. Leo Hendrik Baekeland's (1863-1944) invention of Bakelite in 1907 ushered in the era of chemically synthesized moldable materials called plastics.

Background

The word plastic comes from the Greek adjective *plastikos*, which means moldable. The first citation of plastic as a material dates from 1910, when a dictionary defined it as a substance that could be molded into shape when soft.

Alexander Parkes (1813-1890) presented the first artificial molding compound at London's Great International Exhibit of 1862. A mixture of nitrocellulose softened with camphor and castor oil, it could be molded after heating and held its shape when cooled. Parkes believed his invention would one day be a substitute for rubber, but its raw materials were expensive and its manufacturing process inefficient.

John Wesley Hyatt (1837-1920) invented celluloid in 1869, after reading about a $10,000 prize for the discovery of a substitute for ivory billiard balls. Celluloid, also made of nitrocellulose and camphor, became a cheap and durable substitute not only for ivory, but also for marble, pearl, tortoiseshell, linen, and other natural substances. It was popular for decorative items and for detachable shirt collars that could be wiped clean with a sponge, a convenience that appealed to the new class of clerical and service workers. By the end of the 1920s celluloid's own properties were exploited to make automobile parts, household goods, and photographic film.

The importance of celluloid diminished once synthetic plastics entered the market. The first was Bakelite. Leo Baekeland was familiar with experiments in which phenol and formaldehyde were heated in the presence of hydrochloric acid. The result was either a sticky mass or a rubber-like solid, often accompanied by an explosion. The growing market for shellac and rubber insulators encouraged Baekeland, a Belgian-born immigrant to the United States with a doctorate in chemistry, to find a more reliable way of combining phenol and formaldehyde. He found that heating them under pressure in an apparatus he called the Bakelizer produced a material that would not melt, burn, dissolve, crack, or break. Although Baekeland did not understand the new chemical composition of this material, he did foresee its commercial significance and patented both his process and several potential applications. He built a small pilot plant where he experimented with materials and molding techniques before licensing them to other manufacturers. After describing Bakelite in a speech to the American Chemical Society in 1909, Baekeland was deluged with inquiries and founded the General Bakelite Company.

New companies sprang up to manufacture similar compounds. After years of fighting patent infringements, Baekeland joined with his two main competitors to form the Bakelite Corporation. During the 1920s, Bakelite's sales skyrocketed to meet the needs of the new automobile and radio industries. An advertising campaign called Bakelite "the material of a thousand uses" in an effort to convince manufacturers to try it and consumers to demand it. Although the shiny black of Bakelite radios epitomized modernity in the 1920s, black was also the only color that could hide the fillers that gave Bakelite its strength. As Bakelite's patents expired, new plastics were invented.

Baekeland and other chemists created synthetic plastics through trial and error, before their molecular structure was understood. In the 1920s, Hermann Staudinger (1881-1965) showed that plastics consist of giant molecules strung together in chains called polymers. These chains slide apart when heated, allowing the material to soften but not melt, then snap back together when cooled. The interwoven polymers, like a chain link fence, give plastic its hardness and ductility. Bakelite and other plastics derived from cellulose and coal tar are thermosetting. This means that once heated and shaped, they cannot resoften. This makes them hard, heat resistant, and practically indestructible. The second wave of plastics, derived from petroleum and natural gas, are thermoplastic. This means they can be reheated and shaped in their final chemical state, making them easier to mold, more economical (waste is reused), but less durable than thermosetting plastics. Once the fundamental structure of plastics was known, research into new plastics progressed rapidly.

Wallace Hume Carothers (1896-1937) was hired from Harvard by the DuPont chemical company in 1928 to direct basic research in polymerization. His goal was to create new synthetics by building polymer chains. Not only did he further the knowledge of polymerization, he also created the first synthetic fiber, nylon. Nylon stockings, introduced with fanfare in 1939, were greeted with immediate enthusiasm. Within three years, however, nylon production was diverted to the military. The demand for nylon stockings after the war was so frenzied that riots broke out wherever they were sold.

Several other thermoplastics were invented in the 1930s. Cellulose acetate was popular because of its strength and visual appeal. Polyvinyl chloride was used for everything from shower curtains to shoes. Polystyrene had the advantages of transparency and the ability to capture the tiniest details of a mold. They were all shaped in a speedy process called injection molding. Acrylic (polymethyl methacrylate), manufactured in sheets, was sold under the trade names Plexiglas and Lucite. Despite the vastly different characteristics of the hundreds of synthetic materials available by 1940, the public tended to lump them all together under one name—plastic.

World War II transformed the plastic industry and changed its public image. As *Life* magazine proclaimed, "War Makes Gimcrack Industry

Baekeland (left) holding the first tube of Bakelite. *(Bettmann/Corbis. Reproduced with permission.)*

into Sober Producer of Prime Materials." The urgent need for military supplies forced the industry to develop new materials and processes. In the interest of patriotism manufacturers worked together, and the U.S. output of synthetic resins almost quadrupled between 1939 and 1945.

Many wartime materials and fabrication techniques found peacetime uses. The resin-bonded plywood that had formed curved airplane wings and fuselages was shaped into furniture and prefabricated housing. Fiberglass, developed for radar housings because it did not interfere with transmissions, was molded into small boats. Polyethylene, a thermoplastic used for electrical insulation during the war, was injection-molded into cups by Tupper Plastics and into fishing lures by Loma Plastics, two companies that became giants in the 1950s.

Some people believed the return to peacetime would be a return to a totally plastic world, with a plastic house and a plastic car for every family. Predictions about the potential of plastics became so unrealistic that the Society of the Plastics Industry began a campaign to "deglamorize" it by offering more balanced information to the public, manufacturers, and retailers. The risk to the industry was that disappointed customers would reject all plastics.

Manufacturers tried to transform plastic's modernistic image by designing traditional, wood-grained furniture made of Formica with leather-like upholstery made of vinyl. *House Beautiful* magazine devoted an entire issue to decorating with plastic. The characteristic of plastic that received the highest praise was neither appearance nor cost, but ease of cleaning.

As another publication promised, all homemakers had to do was "swoosh and smile."

Impact

One hundred years after the invention of Bakelite, life without plastics is unimaginable. Almost every manufactured object in homes, schools, offices, and shopping mall would either cease to exist or look entirely different. At the beginning of the twenty-first century, the word "plastic" resonates with meaning, both positive and negative. It conjures up both control—the dominance of science over nature—and loss of control—the fear of plastics burying the planet. People depend on the convenience, strength, versatility, and low cost of plastics, but at the same time worry about dependence on a material they see as phony, polluting, and even dangerous.

Early in its history, plastics were promoted as a social equalizer. No longer would only the rich be able to own brightly colored jewelry, silk stockings, and ivory comb and brush sets. Cast phenol, nylon, and celluloid made these luxuries available to everyone. Plastics, created from seemingly endless supplies of coal and petroleum, bridged the gap between the demands of an increasing population and the fixed supply of natural materials.

In doing so, the plastics industry transformed the economy of the industrialized world. Not only did the chemical companies that made plastic resins flourish, so did the large and small manufacturers that produced finished products. The number of molders and fabricators jumped from 60 in 1929 to 170 in 1939 to 370 in 1946, most of them small businesses. After the war, although the major chemical companies continued to control the supply of resins, manufacturing entrepreneurs flourished. In fact, plastics became synonymous with opportunity, as ironically illustrated in the movie *The Graduate,* when a family friend advises the recent college graduate, "I just want to say one word to you...Plastics....There's a great future in plastics."

Twentieth-century design is inextricably linked with plastics, alternately capitalizing on and rejecting their unnaturalness. The physical characteristics of early plastics elicited a style called modernism characterized by streamlined shapes and simple lines. Rounded molds were cheaper to create than molds with sharp edges, and molten plastic flowed through them more easily, making certain styles economically as well as esthetically appealing. Industrial designers in the 1930s and again in the 1960s recommended that new materials be treated "on their own terms" instead of as imitations of natural materials.

The first seeds of anxiety about plastics were planted as early as the 1920s when people started to question the wisdom of defying the natural order by creating new materials. Twenty years later, rumors spread that nylon stockings were made of decayed corpses and caused rashes or even cancer. Although these particular concerns evaporated over time, the underlying anxiety about plastics did not. It reappeared when babies were suffocated by dry-cleaning bags in the 1950s, when fears spread about the safety of Teflon coatings in the 1960s, and when statistics showed higher-than-average cancer rates among workers in the plastic industry in the 1970s.

The discovery of plastic was motivated by the need to save scarce natural resources. By the time of the first Earth Day in 1970, plastics were accused of harming the environment by overloading landfills, creating pollution, and depleting fossil fuels. The plastic industry has responded by promoting resource conservation through recycling, reuse, and the efficient use of resources.

LINDSAY EVANS

Further Reading

Books

Fenichell, Stephen. *Plastic: The Making of a Synthetic Century.* New York: HarperBusiness, 1996.

Meikle, Jeffrey L. *American Plastic: A Cultural History.* New Brunswick, N.J.: Rutgers University Press, 1995.

Sparke, Penny, ed. *The Plastics Age: From Bakelite to Beanbags and Beyond.* Woodstock, NY: Overlook Press, 1993.

Other

http://www.plasticsresource.com/

Model T: The Car for the Masses

Overview

Henry Ford (1863-1947) revolutionized the automobile industry with the Model T. Instead of seeing the car as a plaything of the rich, Ford envisioned a vehicle "for the great multitude." The Model T, Ford decreed, would be both dependable and affordable. The success of the Model T had a profound impact on the United States, as it made the automobile a common household possession. The automobile freed people to move, and with millions of vehicles clamoring for space, an infrastructure had to be created to support them. Roads, gas stations, and other amenities became part of an ever-expanding culture built around the automobile. Average people were able to buy the Model T, and as a result the car became an everyday necessity.

Background

Contrary to popular belief, Henry Ford did not invent the automobile or mass production. In fact, the honors for inventing the automobile are usually given to two Germans, Gottlieb Daimler (1834-1900) and Karl Benz (1844-1929), who expanded upon German technology to produce the first car in the mid-1880s. In the United States, Charles and Frank Duryea (1861-1938 and 1869-1967, respectively) from Springfield, Massachusetts, demonstrated their gasoline-powered engine in 1893.

The idea of developing automobiles spread like wildfire. By 1900 there were more than 50 car manufacturers in America. Although the machines were highly publicized and caught the nation's imagination, automobiles had to be hand-crafted and thus were expensive and mainly produced for the rich.

Henry Ford's simple, yet revolutionary, idea of allowing everyday people to benefit from his innovations stands him apart from other early automobile makers. After a lifetime of tinkering with machinery, Ford realized that mass production would allow him to lower the price of his models. The other key factor would be standardization. Each car that rolled off the assembly line would be exactly like the one that preceded it.

Much of Ford's thinking derived from his own background as part of a farming family. As biographer Robert Lacey explains it, Ford had an "almost didactic impulse to share the joy of machines with the world." Ford also epitomized the American rags-to-riches story, and his ascent became part of the culture of the early-twentieth-century United States.

Ford explained that "to make them all alike, to make them come through the factory just alike," the company could conquer the economy of scale that kept auto prices high. The uniformity of his cars even carried through to the color. Ford is famous for saying that buyers could have "any color you choose, so long as it's black."

Ford held strong beliefs about his low-priced automobile. However, his financial backers were happy to continue selling high-end cars. Ford's work with the Model N in 1906-07, both with mass production and pricing, convinced him that his idea was sound. By 1908, he bought out his opponents within the company and owned 58% of the company.

In the fall of 1908 Ford unveiled the first Model T. The car had several features separating it from other vehicles being produced, such as changes making the car more negotiable on primitive country roads and an engine encased for protection. Ford set the price at $825, which he knew made it too expensive for most people, but he believed the price would fall in the future due to assembly line technology. With Ford in control, efficiency became the keystone of his operations. For the next 20 years Ford produced black Model Ts, and only Ts (often called the "Tin Lizzie" or "flivver").

The Model T itself was a sturdy looking automobile with a high roof, giving it a refined look. It had a 4-cylinder, 20-horsepower engine. It was a powerful and dependable vehicle—exactly what Ford desired. Another innovation introduced by Ford was the use of lighter, stronger vanadium steel, which European carmakers used.

The company sold 11,000 cars from 1908-09, which raised $9 million, a 60% increase over the previous year. Ford then outdid himself with the 1910-11 model, selling 34,528. Sales skyrocketed, reaching 248,000 in 1914, or nearly half the U.S. market. The heavy demand for cars forced Ford to pioneer new methods of production. He built the largest and most modern factory in America on a 60-acre tract in Highland Park in Detroit, Michigan. Ford's net income soared from $25 million in 1914 to $78 million by 1921.

The Ford Model T (1908). *(American Automobile Manufacturers Association. Reproduced with permission.)*

Ford, preaching modern ideas of efficiency, introduced the continuously moving assembly line. He tinkered with the process until finding the exact pace his workers could handle. Chassis production dropped from 6 hours to 90 minutes. The one millionth Model T was built in 1916. The Highland Park plant churned out 2,000 cars a day.

As he had predicted, Ford lowered the price of Model T's, eventually down to $300. The company controlled 96% of the inexpensive car market. By the end of World War I Ford Model Ts represented nearly half the cars on Earth.

Impact

The impact of the Model T was enormous, influencing both the local and national economy and ushering in a dramatically different lifestyle for many American citizens.

Farmers took to the Model T in droves because the car ran well over rut-filled roads and gravel. This was especially important given that the United States had little in the way of paved roads through the 1920s.

The popularity of Ford automobiles led to a constant labor shortage. Carrying his innovative thinking into the labor arena, Ford instituted a profit-sharing and bonus system. However, it was his introduction of the "five-dollar day" in 1914 that really took the manufacturing world by storm. Ford's plan outlined an eight-hour workday, shorter than the industry average. More importantly, Ford would pay workers a basic wage of $5 a day, eclipsing the industry's standard $1.80 to $2.50. The program made Ford a national hero and his legend approached cult status. In addition, the high wages had a profound impact on the thousands of workers who made their way to Detroit to work for Ford.

By 1921, nearly 5.5 million Ford automobiles had been built. When he halted production of the Model T seven years later, more than 15 million had been produced.

The Model T also helped grow and define the world of car dealerships. By 1913 Ford had 7,000 affiliated dealers selling his car in all parts of the country. In fact, there was at least one Ford dealership in every town with a population that exceeded 2,000 people. This is an overlooked but vastly underappreciated aspect of the Model T's impact on the nation. It would be impossible to calculate the money generated for local economies by just this small portion of the Ford empire.

After the high-water mark of the early 1920s, the Ford Company began to slip. A new 1,100-acre factory complex, the River Rouge in Dearborn, Michigan, opened and marked Ford's attempt at vertical integration. The size and sprawl of "The Rouge" proved too much for Ford. Personality clashes with subordinates left the company a virtual one-man operation, which proved dreadful.

The Model T also looked outdated by the late 1920s. Stylish models from General Motors and Chrysler forced Ford to drop the car and replace it with the Model A. Just when Ford began regaining market share, the Depression hit and spelled doom.

What Ford's Model T did for the nation is nearly incomprehensible. Not only did the automobile become an everyday necessity, it allowed formerly isolated people to explore the world around them. Driving became a national pastime.

Over the years, America transformed into one large neighborhood. Local, state, and federal governments fed the car culture by building the infrastructure that encouraged people to travel. The business community soon followed suit. A thriving service industry sprouted up to assist drivers, whether going across town or across the country.

Although the Ford Motor Company suffered in its leader's later years, at one point even losing up to $1 million a day in the 1940s, the public's fascination with the Model T set in motion a love for cars that exists to this day.

BOB BATCHELOR

Further Reading

Ford, Henry. *My Life and Work*. Garden City, NY: Doubleday, 1922.

Lacey, Robert. *Ford: The Men and the Machine*. Boston: Little, Brown, 1986.

Nevins, Allan. *Ford: The Times, the Man, the Company*. New York: Scribners, 1954.

Building the Panama Canal

Overview

In the early years of oceanic commerce, ships carrying goods between Europe and the Far East had to travel a long, circuitous 12,000-mile (19,308 km) route around the continent of South America. As early as the 1500s, Spanish rulers first explored the idea of creating a canal through the Isthmus of Panama to drastically reduce travel time. In 1903, a treaty between Panama and the United States finally paved the way for the construction of the Panama Canal, a massive feat of engineering which not only united two oceans, the Atlantic and the Pacific, but opened an invaluable artery for international trade.

Background

While the construction of the Panama Canal was not undertaken until the early twentieth century, it was first envisioned several centuries earlier. Christopher Columbus (1451-1506) searched in vain for a passage between the North American and South American Continents. In 1534, Charles I of Spain recognized the value of cutting a route through Panama in order to gain greater accessibility to the riches of Peru, Ecuador, and Asia. He ordered a survey of a proposed canal route through the Isthmus of Panama drawn up, but unrest in Europe put the project on permanent hold.

The first construction did not get underway for another three centuries, when a Frenchman named Ferdinand Marie de Lesseps (1805-1894) decided that his country should take responsibility for building an interoceanic canal. Lesseps, who had completed Egypt's Suez Canal just 10 years earlier, felt this new canal would be equally successful. His crew began work on a sea-level canal in 1882, but their equipment proved insufficient to cut through the rocky terrain, money was growing short, and diseases such as yellow fever and malaria decimated the French work force. Lesseps's company was left bankrupt, and the operation was forced to shut down with the canal still incomplete. In 1894, the French created the New Panama Canal Company to finish the task, and set out to find a buyer.

Meanwhile, America too had been eyeing the isthmus. President Theodore Roosevelt, newly at the country's helm after the assassination of President McKinley, began building up America's military might and revitalizing its navy. Strategically, the canal took on new importance, as the U.S. empire stretched from the Caribbean across the Pacific. At the time, it took a full 67 days for the U.S. battleship, the *Oregon*, to travel from San Francisco to the Caribbean. Roosevelt was determined to prove America's military superiority and show that his country could bridge the distance between the Atlantic and Pacific oceans.

A few years earlier, the U.S. Congress had chartered the Maritime Canal Co., under the leadership of millionaire J.P. Morgan, to build a canal in either Nicaragua or Panama. Nicaragua was chosen, but a financial panic in the U.S. caused operations to shut down within five years. Now, a battle waged in Congress between completing the Nicaragua project or taking on construction of the Panama canal. In 1902, Iowa Senator William Hepburn introduced a bill to initiate construction of the Nicaraguan canal, to which Senator John Spooner of Wisconsin attached an amendment that literally overrode the bill, providing for a canal at Panama instead.

The Spooner Act gave President Roosevelt $40 million to purchase the New Panama Canal Company from the French, but there was one more obstacle to overcome. Panama wanted to sell the land to America, but Colombia refused. Roosevelt predicted there would soon be a revolution in Panama, and he was right. A local revolt put a new government in Panama, which the U.S. firmly supported.

In 1903, the United States and Panama signed the Hay-Bunau-Varilla treaty by which the United States guaranteed Panama's independence, and secured the right to construct an interoceanic canal through the Isthmus of Panama. America was given a perpetual lease on the 10-mile (16 km) wide canal zone.

Impact

There were several major obstacles standing in the way of successfully completing the canal. It was, first of all, a major feat of engineering. The crews had to dig through the Continental Divide, a ridge of land separating two oppositely flowing bodies of water. The canal was to be the largest earth dam ever built at the time, with the most massive canal locks and gates ever constructed.

Sanitation was also a potential problem. The warm Panamanian climate was a breeding ground for mosquitoes, which carried malaria and yellow fever. Early French crews lost an estimated 10,000 to 20,000 workers to yellow fever outbreaks between 1882 and 1888. The American government vowed this would not happen to their own crews and to this end, sent in doctor William Gorgas (1854-1920) to examine the area. In 1900, U.S. Army tropical disease expert Walter Reed (1851-1902) had proven that yellow fever was spread by the female mosquito. Gorgas vowed to remove the mosquito population from the canal zone, a task easier said than done. His first job was to treat all standing or slow-moving bodies of water with a combination of oil and insecticide. He covered windows with wire screens and sent health workers door-to-door to search for mosquitoes and their eggs. He fumigated houses and quarantined sick workers. By December 1905, yellow fever was virtually eliminated from the Canal Zone.

Before the actual construction of the canal could begin, the administrative infrastructure had to be put in place. Congress set up a commission to control the canal area, overseeing all transactions. Locals were hired to pave roads, repair the ailing French buildings, lay a new railroad track to accommodate American train cars, and put in a working water and sanitation system.

Next, the commission hired its chief engineer, John F. Wallace, to oversee the project. Wallace began his excavation at Culebra Mountain. Culebra Cut was a 10-mile (16 km) stretch through the rockiest terrain along the canal route. Wallace quickly tired of the commission's layers of red tape, which were draining his time and patience, and petitioned Congress for a new governing body. Roosevelt quickly fired the commission, replacing them with seven new members. Wallace returned to the canal site, but ended up resigning a short time later.

In 1905, Wallace was replaced by John Stevens, and building of the canal began in earnest. Stevens first had to decide what type of canal to build—at sea level, as the French had begun, or by using locks—closed off sections of the water used to raise and lower levels as ships moved through. He decided on a lock canal, a decision which was seconded by President Roosevelt.

But just when work seemed to be finally progressing, Stevens sent a letter to the president, indicating that he was not "anxious to continue in service." His resignation was accepted, and yet another replacement engineer was sought. Roosevelt appointed Army Lieutenant George Washington Goethals, whom he knew would exercise tight control and keep the project within the watchful eye of the U.S. Government. Quickly, Goethals organized his team, introducing a system to cut costs and keep everyone on track.

Panama Canal lock gates under construction (1913). *(Archive Photos. Reproduced with permission.)*

By 1907, more than 39,000 people were hard at work digging through the rock of the Culebra Cut. But there were problems. When the French had excavated Culebra Cut years earlier, their method involved chopping the tops off of hills and piling the dirt on either side, which led to mudslides. Indeed, the area was plagued by landslides. The American engineers were forced to increase the amount of rock and clay they excavated and adjust the angle of their cut in order to overcome the problem. In 1908, changes were made to the initial design of the canal because of unforeseen problems. The width was increased from 200 feet (61 m) to 300 feet (91 m) and the size of the locks was increased from 95 (29 m) to 110 feet (33 m).

When completed, the canal stretched 50 miles (80 km), through three sets of locks. At its Atlantic entrance, a ship would pass through a 7-mile (11 km) dredged channel in Limón Bay. It would then travel 11.5 miles (18 km) to two parallel sets of locks at the town of Gatun, which were fueled by an enormous man-made lake. Each of the locks would raise or lower ships 85 feet (26 m). Then it was another 32 miles (51 km) through a channel in Gatun Lake to Camboa, where the Culebra Cut began. Another set of two locks located at Pedro Miguel and at Miraflores on the Pacific side would then lower the ship to sea level. Water flowed in and out of the locks through huge culverts, or drains, in the walls of the locks. A railroad-like locomotive served as a tow to pull the ships through the locks.

Ten years after construction had begun, the canal was finally completed in August of 1914 at a cost of approximately $387 million. The first ship to cross through was the concrete vessel *Cristobal*, however the first publicized voyage was that of the freighter *Ancon*. Unfortunately, Europe and the United States had just entered into World War I, so traffic across the canal was initially light. At first, around 2,000 ships per year passed through but by the end of the war, that number rose to 5,000. Soon, the Panama Canal was accommodating nearly all of the world's interoceanic trade vessels.

In 1977, the United States and Panama signed two treaties granting the United States control of the canal until the end of the twentieth century. The U.S. was to oversee management, operation, and defense for the canal, and American ships would be free to travel back and forth across it. On December 31, 1999, just as the new millennium was about to dawn, the U.S. handed control of the Canal back over to Panama. At the ceremony, President Jimmy Carter, who oversaw the initial transfer treaty in 1977 told Panama's President Mireya Moscoso, "It's yours."

The Panama Canal would not only prove a modern marvel of engineering, it was to open up a whole new world in international trade. Roosevelt once said of his achievement, "The canal was by far the most important action I took in foreign affairs during the time I was President. When nobody could or would exercise efficient authority, I exercised it."

STEPHANIE WATSON

Further Reading

Books

Bennett, Ira. *History of the Panama Canal*. Washington, D.C.: Historical Publishing Co., 1915.

Chidsey, Donald. *The Panama Canal—An Informal History of its Concept, Building, and Present Status*. New York: Crown Publishers Inc., 1970.

McCullough, David. *The Path Between the Seas: The Creation of the Panama Canal*. New York: Simon and Schuster, 1977.

The Development of RADAR and SONAR

Overview

Although they rely on two fundamentally different types of wave transmission, Radio Detection And Ranging (RADAR) and Sound Navigation and Ranging (SONAR) both are remote sensing systems with important military, scientific, and commercial applications. RADAR sends out electromagnetic waves, while active SONAR transmits acoustic (i.e., sound) waves. In both systems these waves return echoes from certain features or targets that allow the determination of important properties and attributes of the target (i.e., shape, size, speed, distance, etc.). Because electromagnetic waves are strongly attenuated (diminished) in water, RADAR signals are mostly used for ground or atmospheric observations. Because SONAR signals easily penetrate water, they are ideal for navigation and measurement under water.

Background

For hundreds of years, non-mechanical underwater listening devices (listening tubes) had been used to detect sound in water. As early as 1882, the Swiss physicist Daviel Colladen attempted to calculate the speed of sound in the known depths of Lake Geneva.

Based upon the physics of sound transmission articulated by nineteenth-century English physicist Lord Rayleigh (1842-1914) and the piezoelectric effect discovered by French scientist Pierre Curie (1509-1906) in 1915, French physicist Paul Langevin (1872-1946) invented the first system designed to utilize sound waves and acoustical echoes in an underwater detection device. In the wake of the *Titanic* disaster, Langevin and his colleague Constantin Chilowsky, a Russian engineer then living in Switzerland, developed what they termed a "hydrophone" as a mechanism for ships to more readily detect icebergs (the vast majority of any iceberg remains below the ocean surface). Similar systems were put to immediate use as an aid to underwater navigation by submarines.

Improved electronics allowed the production of greatly improved listening and recording devices. Because passive SONAR is essentially nothing more than an elaborate recording and sound amplification device, these systems suffered because they were dependent upon the strength of the sound signal coming from the target. The signals or waves received could be typed (i.e., related to specific targets) for identifying characteristics. Although quite good results could be had in the hands of a skilled and experienced operator, estimates of range, bearing, and relative motion of targets were far less precise and accurate than results obtained from active systems, unless the targets were very close—or made a great deal of noise.

The threat of submarine warfare during WWI made urgent the development of SONAR and other means of echo detection. The development of the acoustic transducer that converted electrical energy to sound waves enabled the rapid advances in SONAR design and technology during the last years of the war. Although active SONAR was developed too late to be put to much of a test during WWI, the push for its development reaped enormous technological dividends. Not all of the advances, however, were restricted to military use. After the war, echo-sounding devices were placed aboard many large, French ocean-liners.

During the early battles of WWII, the British Anti-Submarine Detection and Investigation Committee (its acronym, ASDIC, became a name commonly applied to British SONAR systems) made efforts to outfit every ship in the British fleet with advanced detection devices. The use of ASDIC proved pivotal in the British effort to repel damaging attacks by German submarines upon both British warships and merchant ships keeping the island nation supplied with munitions and food.

While early twentieth-century SONAR developments proceeded, another system of remote sensing was developed based upon the improved understanding of the nature and propagation of electromagnetic radiation achieved by Scottish physicist James Clerk Maxwell (1831-1879) during the nineteenth century.

In the 1920s and early 1930s, Scottish physicist and meteorologist Sir Robert Alexander Watson-Watt (1892-1973) successfully used short-wave radio transmissions to detect the direction of approaching thunderstorms. Another technique used by Watson-Watt and his colleagues at the British Radio Research Station measured the altitude of the ionosphere (a layer in the upper atmosphere that can act as a radio reflector) by sending brief pulses of radio waves upward and then measuring the time it took for the signals to return to the station. Because the speed of radio waves was well established, the measurements provided very accurate determinations of the height of the reflective layer.

In 1935, Watson-Watt had the ingenious idea of combining these direction- and range-finding techniques, and, in so doing, he invented RADAR. Watson-Watt built his first practical RADAR device at Ditton Park.

Almost immediately, officials at the Royal Air Ministry asked Watson-Watt whether his apparatus might have the potential of damaging or downing enemy aircraft. Watson-Watt responded that radio wave transmissions were far too weak to achieve this end. Regardless, he suggested to Ministry officials that radio detection was feasible. In 1935, Watson-Watt wrote a letter titled "Detection and Location of Aircraft by Radio Methods." Watson-Watt carefully set forth that reading the weak return signal from an aircraft would pose a far greater engineering challenge than encountered in his meteorological experiments. The signal sent out needed to be more than a hundred times more energetic. In addition, a more sensitive receiver and antenna would need to be fabricated.

Shortly thereafter, without benefit of a test run, Watson-Watt and Ministry scientists conducted an experiment to test the viability of RADAR. The Watson-Watts apparatus was found able to illuminate (i.e., detect) aircraft at a distance of up to 8 mi (13 km). Within a year, Watson-Watt improved his RADAR systems so that it could detect aircraft at distances up to 70 mi (113 km). Pre-war Britain quickly put Watson-Watt's invention to military use and by the end of 1938 primitive RADAR systems dotted the English coast. These stations, able to detect aircraft regardless of ground fogs or clouds, were to play an important role in the detection of approaching Nazi aircraft during WWII.

The development of RADAR was not the exclusive province of the British. By the outbreak of WWII all of the major combatants had developed some form of RADAR system. On many fronts battles were often to be influenced by dramatic games of scientific and technical one-upsmanship in what British war-time Prime Minister Sir Winston Churchill called the "Wizard War." During the war, Watson-Watt became one of those wizards as he took up the post of scientific advisor to the Royal Air Ministry.

By the end of the war the British and American forces had developed a number of RADAR types and applications, including air interception (AI), air-to-surface vessel (ASV), Ground

Controlled Interception (GCI), and various gun sighting and tracking RADARs.

Regardless of their application, both RADAR and SONAR targets scatter, deflect, and reflect incoming waves. This scattering is, however, not uniform—and in most cases a strong echo of the image is propagated back to the signal transmitter in much the same way as a smooth mirror can reflect light back in the specular direction. The strength of the return signal is also characteristic of the target and the environment in which the systems are operating. Because they are electromagnetic radiations, RADAR waves travel through the atmosphere at the speed of light (in air). SONAR waves (compression waves) travel through water at much slower pace—the speed of sound. By measuring the time it takes for the signals to travel to the target and return with echoes, both RADAR and SONAR systems are capable of accurately determining the distance to their targets.

Within their respective domains, both RADAR and SONAR can operate reliably under a wide variety of adverse conditions to extend human sensing capabilities.

Impact

As a result of the wartime success of RADAR, scientists and engineers quickly sought new applications for such systems. The benefits to meteorological science were obvious.

RADAR technology developed during WWI also had a dramatic impact on the fledgling science of radio astronomy. During the war British officer J.S. Hey correctly determined that the Sun was a powerful source of radio transmissions. Hey discoed this while investigating the causes of systemwide jamming of the British RADAR net that could not be attributed to enemy activity (Hey attributed the radio emission to increased solar flare activity). Although kept secret during the war, British RADAR installations and technology became the forerunners of modern radio telescopes as they recorded celestial background noise while listening for the telltale signs of enemy activity (e.g., V-2 rocket attacks).

The historical credit given to the decisiveness and impact of Churchill's Wizard War remains hotly debated. Churchill himself described the Battle of Britain as largely a battle decisively fought and won with "eyes and ears." Regardless, it is unarguable that the remote sensing devices and RADAR networks (named Chain High and Chain Low) employed by the British allowed British commanders to more effectively concentrate their out-gunned and out-manned forces against the Nazi air onslaught.

Ironically, WWII induced design improvements in SONAR technology that laid the foundation for the development of non-invasive medical procedures such as ultrasound in the last half of the twentieth century. Sound- and electromagnetic signal-based remote sensing technologies and techniques became powerful medical tools that allowed physicians to make accurate diagnosis with a minimum of invasion to the patient.

Remote sensing tools such as RADAR and SONAR also allow scientists, geologists, and archaeologists to map topography and subsurface features on Earth and on objects within the solar system. SONAR readings led to advances in underwater seismography that allowed the mapping of the ocean floors and the identification of mineral and energy resources.

RADAR systems are critical components of the modern commercial air navigation system. One British wartime invention, Identification Friend or Foe (IFF) RADAR, used to identify and uniquely label aircraft, remains an important component in the air traffic control system.

K. LEE LERNER

Further Reading

Books

Cox, A.W. *Sonar and Underwater Sound.* Lexington, MA: Lexington Books, 1974.

Heppenheimer, T.A. *Anti-Submarine Warfare: The Threat, The Strategy, The Solution.* Arlington, VA: Pasha Publications Inc., 1989.

Holmes, J. *Diagnostic Ultrasound: Historical Perspective.* Mosby, 1974.

National Defense Research Committee. *Principles and Applications of Underwater Sound.* Washington, D.C., 1976.

Rowe, A. *One Story of Radar.* Cambridge, England: Cambridge University Press, 1948.

Watson-Watt, R.A. *Three Steps to Victory.* Odhams Press, 1957.

Rocketing into Space: The Beginnings of the Space Age

Overview

In the first half of the twentieth century, scientists and engineers in Russia, Germany, and the United States engaged in theoretical and experimental investigations of travel beyond Earth's atmosphere using rockets. Stimulated by science fiction fantasies of space travel and funded by meteorological and military organizations, pioneering rocket scientists of the time developed missiles that traversed distances ranging from thousands of feet to hundreds of miles. Their efforts laid the foundation for the successful missile and space programs of the United States and the Soviet Union in the post-World War II era and helped to launch the Space Age.

Background

The promise and products of science and technology in the industrial age of the nineteenth century fed the aspirations of those who longed to conquer space with rockets. The close alliance of science and technology, especially in the emergence of the research and development laboratories which institutionalized change, provided a new landscape for those individuals who desired to improve the understanding of propulsion and reactive motion so critical to rocket travel within and beyond Earth's atmosphere. Understanding the physics of motion was a necessary first step to devising a successful rocket, and transforming theoretical designs into workable mechanisms required painstaking experimental tests. Typical of so much of the technology of the time, rocketry required an extensive period of trial and error to transform theoretical constructs into workable missiles. Institutional funding, experimental acumen, devotion, and patience combined to transform the dreams of rocket flight into reality in just one generation.

Three men hold a prominent place in the history of rocket technology: Russian engineer Konstantin Tsiolkovsky (1857-1935), American physicist Robert H. Goddard (1882-1945), and German researcher and instructor Hermann Oberth (1894-1989). The Russian scientist Tsiolkovsky was the first person to translate Newton's law of action-reaction into a theoretical analysis of rocket motion. (English physicist and mathematician Isaac Newton [1642-1727] proposed three basic laws of motion, the third of which is that for every action, there is an equal and opposite reaction.) Tsiolkovsky also was the first to propose liquid fuels and to devise a multi-stage design to provide travel beyond Earth's atmosphere. Although Tsiolkovsky is considered the father of rocketry by Russians, little of his extensive work was known outside of the Soviet Union, and his failure to perform experimental work limited the impact of his pioneering analyses.

Likewise, the extensive theoretical and experimental work of American physicist Robert Goddard had limited influence. Captivated as a young man by the science fiction accounts of space travel written by Jules Verne (1828-1905) and H. G. Wells (1866-1946), Goddard used his training as an academic scientist to analyze rockets and create an elaborate experimental program to test high-altitude designs. First with solid fuel and then with liquid fuels he tested various designs, and by 1914 he held patents covering key principles such as using a combustion chamber with a nozzle to exhaust burning gases, creating a system to feed fuel into that chamber, and using a multi-stage device. During World War I, he received modest support from the Smithsonian Institution for meteorological research and from the United States Army for weaponry research. After the war, U.S. government interest in rockets waned, and Goddard returned to his private experimental work.

In 1919 Goddard published *A Method of Reaching High Altitudes*, which, much like Tsiolkovsky's works, promoted space travel. The public reacted to Goddard's treatise with sensationalism, labeling him a mad scientist for proposing travel to the Moon. This misunderstanding of Goddard's work increased his desire to work secretly and to share his results with only a small, inner circle of associates. By March of 1916 Goddard was successful in launching a liquid-fueled rocket that reached a height of over 400 feet (122 m) and traveled almost 200 feet (61 m) from the launch site. This pioneering feat encouraged him to continue his experimental research efforts, which eventually captured the attention of American aviator Charles Lindbergh (1902-1974), who persuaded the Guggenheim Foundation to fund Goddard's rocket investigations. Throughout the 1930s and early 1940s, Goddard, who established a research facility in Roswell, New Mexico, made many incremental improvements to his rocket design with these resultant developments: using a nozzle for thrust in a rocket; demonstrating that rockets would function in a vacuum; using liquid fuel; using gyroscopes (a body that spins on a movable axis) for controlling a rocket with an inertial guidance system; using a multi-stage rocket design; and, using deflector vanes for stabilization and guidance. He also developed several rocket motors and pumps during this era of extensive experimentation. These achievements gave Goddard nearly every part of a workable rocket for space travel—so much so that he is considered the father of American rocketry. But, because Goddard kept much of this technology secret from other Americans working in rocket design during his lifetime and resisted invitations from rocket societies to share these findings, he influenced a relatively small number of people.

The most successful theoretical and experimental rocket researches centered on the work of Hermann Oberth and his German associates. Like Tsiolkovsky and Goddard, Oberth was influenced by science fiction literature, especially the work of Jules Verne, and had an early interest in space travel. He spent his early years speculating about and devising the mathematical analysis for various aspects of rocketry. Oberth continued this interest in his university studies; the result was his publishing *The Rocket into Planetary Space* in 1923. This influential book treated manned and unmanned rocket flights, various aerodynamic and thermodynamic analyses, proposals for liquid fuel designs, and a host of other technical aspects regarding space technology. Unlike Goddard, Oberth did not experiment with his designs; instead, he produced mainly theoretical accounts and analyses, yet his book was well known, and, as a result, much more influential among his contemporaries than Tsiolkovsky's lesser-known publications and Goddard's relatively secretive works.

Impact

The possibilities of space travel using rocket technology articulated by Tsiolkovsky, Goddard, and Oberth spread interest in the subject during the 1920s and 1930s. Each independently concluded that space travel was possible with rockets, that liquid-fueled rockets and multi-stage designs were the best for space travel, and that mathematics could be created to describe the behavior of rocket technology. As a result, rocket societies emerged in the Soviet Union, Germany, and the United States. These groups encouraged and engaged in rocket research and experiments so that by the 1930s several successful tests of amateur rockets drew the attention of the military, especially in Germany. The emerging Nazi regime in Germany embraced work on this potential weapon, which was not banned by the Treaty of Versailles, and absorbed most civilian rocket research and development under the umbrella of military control.

That control centered on the work of Walter R. Dornberger (1895-1980) and his young assistant, Wernher von Braun (1912-1977), who headed a research and development program in Germany in the 1930s. Their efforts, particularly at Peenemunde, an isolated island in the Baltic Sea, eventually produced the V-2 (Vengeance Weapon 2) used against Britain during the closing months of World War II. Although this new weapon wreaked havoc on parts of Britain, it also demonstrated the potential for space flight by reaching altitudes of 60 miles (97 km) on the edges of space. Government support of rocket research, especially in Germany, had produced both a new weapon and a new opportunity for the development of travel beyond Earth's atmosphere. Having ignored the potential of long-range rockets during World War II, the United States military took advantage of German rocket achievements by capturing the devices, records, and documents, and even personnel central to the work of the V-2. In the immediate post-war period, the United States began a military rocket research program in earnest. By 1949, modified V-2 rocket designs allowed the American Army to launch a rocket which reached a record altitude of almost 250 miles (402 km). German V-2 rocket technology became the basis of much of America's missile and rocket development, for scientific and military purposes, in the post-war era.

In 1945, at the war's end, the Soviets also recognized the importance of German rocket advances during World War II. They removed what artifacts and documents they found remaining at the abandoned Peenemunde site after von Braun and many of his associates vacated the facility in order to surrender to the Americans rather than the Russians. Some of the staff at Peenemunde chose to work for the Soviets, and in the post-war era they recruited others to help with a high profile, well-funded government program in the Soviet Union to improve the V-2 and create a new generation of powerful

Robert Goddard with the first successful liquid fuel rocket (1926). *(NASA. Reproduced with permission.)*

rockets. The Soviets recognized the importance of long-range missiles for their post-war strategy and, once they had exploited the German staff, turned to their own engineers and scientists to develop intra- and inter-continental missiles for military use and space exploration.

By 1950, both the United States and the Soviet Union had active rocket technology programs in place designed to serve the military in the Cold War and the scientific community in learning more about the heavens. The origins of that rocket technology lay with the pioneering theoretical and experimental work of Tsiolkovsky, Goddard, and Oberth. Those achievements that made space travel sensible and possible generated amateur societies, private and government research programs, and military efforts that provided the foundation for the successful space age in the second half of the twentieth century. German success with the V-2 rocket made the space age possible because the United States and the Soviet Union based much of their post-World War II developments on the war work of the Germans. With strong government support, both countries developed rockets capable of traveling distances across oceans or into the atmosphere, thereby turning the science fiction musings of the nineteenth century into the technological reality of roaring rockets reaching for the Moon, Mars, and beyond.

H. J. EISENMAN

Further Reading

Bainbridge, William S. *The Spaceflight Revolution*. New York: Wiley, 1976.

Braun, Wernher von, and Frederick I. Ordway III. *Space Travel: A History*. New York: Harper & Row, 1985.

Lehman, Milton. *This High Man*. New York: Farrar, Straus & Company, 1963.

Pisziewicz, Dennis. *Wernher von Braun: The Man Who Sold the Moon*. Westport, Connecticut: Praeger, 1998.

Winter, Frank H. *Rockets into Space*. Cambridge, MA: Harvard University Press, 1990.

Wulforst, Harry. *The Rocketmakers*. New York: Orion Books, 1990.

The Empire State Building: Skyscraper Symbol of America's Power

Overview

The skyscraper is one of the most impressive tributes to the twentieth century. These buildings are a celebration of modern technology and innovation. On the other hand, skyscrapers have a pedestrian role—part of zoning and tax law, political squabbles, and real estate battles for control of prime locales. Despite all this, skyscrapers are awe-inspiring. After 1900 in the United States, many cities were transformed as numerous tall buildings appeared on the landscape. Of these, none is more famous than the Empire State Building in New York City, built between 1929 and 1931. Since its construction, the Empire State Building has symbolized the technological prowess and economic strength of the United States.

Background

Tall buildings have captured the imagination of people throughout history, dating back to the obelisks of ancient Egypt. However, it wasn't until the development of iron and steel as structural materials in the nineteenth century that the tall building became a reality. These materials allowed architects to design buildings beyond the limitations of masonry and brick.

Steel allowed architects to move skyward with a minimum of bulk, thus allowing larger windows and flexible interior spaces. The work of bridge builders inspired architects to apply metal technology to buildings. The development of the passenger elevator in 1857 was the final obstacle to erecting tall structures. Before the elevator, the traditional building height limit was five stories.

Two early skyscrapers appeared in New York City in 1875: the 9-story Tribune Building and the 10-story Western Union building. New York architects continued building early skyscrapers, culminating in the Woolworth Building (1913), designed by Cass Gilbert (1859-1934). It was the tallest building in the world, reaching a height of 792 feet (241 m).

Many of the early advances in skyscrapers can be directly attributed to the devastating fire that wiped out most of Chicago in 1871. City planners and architects turned to fireproof iron and steel instead of wood and masonry. Modern business also demanded large working spaces, and this combined with high real estate costs helped the skyscraper take shape in Chicago. William Le Baron Jenney (1832-1907) is considered the founder of the "Chicago School" of architecture. His firm built an entirely metal 9-story structure in 1885. He used Bessemer steel, which brought down the price, allowed mass production, and increased the use of metal framework.

The next phase of evolution took place in New York City. By the mid-1890s New York skyscrapers pushed past 20 stories. The soaring price of land pushed architects in the city to build taller and taller buildings. From 1900 until the Great Depression hit in 1929, one or more new skyscrapers appeared every year.

Impact

In 1931 New York City celebrated the opening of a skyscraper labeled the "eighth wonder of the world." Built at the corner of Fifth Avenue and Thirty-fourth Street, the Empire State Building was the tallest building in the world. It weighed in at a whopping 600 million pounds (272,160,000 kg) but was placed on 220 columns, which gave it the impact on the earth of only a 45-foot (14 m) high pile of rocks. In the nation's capital, President Herbert Hoover threw the switch that symbolically lit the building.

Opening in the midst of the Great Depression, the building lifted the spirits of American citizens, even though its owners had rented only 25% of the 2 million feet (609,600 m) of office space available. In fact, they wouldn't achieve full occupancy until the late 1940s. Regardless, the nearly $41 million building gave a psychological boost to the nation at one of its darkest times.

At the gala opening, a carnival atmosphere prevailed. People fought for the chance to inspect the marble lobby and buy tickets to the observation deck on the eighty-sixth floor. Those lucky enough to make it up were awestruck by what they saw—it was as if the world had changed forever.

Designed by the architecture firm of Shreve, Lamb, and Harmon, and engineer H. G. Balcom, the Empire State Building symbolized the development of skyscrapers in the interwar years. The building embodied the refinement of steel construction that had been maturing over the previous decades. William Frederick Lamb, who studied at Columbia University and the Ecole des Beaux-Arts in Paris, was the chief designer on the structure.

Although the building is not the most stunning visual work in the development of skyscrapers, its sheer size makes it an amazing success. Hundreds of logistical problems had to be overcome in putting up the 85-story monster, which reaches a height of 1,239 feet (377.6 m) at the tip of its spire. None of its individual parts (ziggurat base, interlocked setbacks of the crown, or Art Deco spire) are innovative separately, but taken as a whole the building holds a special place in architecture history.

The Empire State Building was the inspiration of John Jacob Raskob, a self-made man who rose to become an officer and shareholder in General Motors. Raskob wanted the building to represent the ideal executive office building. It would achieve maximum efficiency, but with artistry and a tastefully modern style. The Empire State Building would also feature top-notch engineering. Rather than name the structure after himself (like the Woolworth, Chrysler, and Chanin buildings in New York), Raskob had more grandiose thoughts. It would embody New York itself in all its power and glory. The building's height and stature as the tallest in the world added to its majesty.

Lamb preferred functional architecture and designed the Empire State Building with simplicity in mind. He planned the structure with the practical aspects of budget, time, and zoning regulations in mind.

The grandeur of the Empire State Building immediately captured the hearts of Americans. The building's steel skeleton went up in just over eight months and was a constant news story around the nation. New Yorkers followed the construction with an almost cult-like obsession. Telescopes were installed in Madison Square Park for average citizens to gaze at the work in progress. At the time, it was called the world's greatest monument to man's ingenuity, skill, mind, and muscle.

Workers built Empire State Building in mind-numbing speed. The walls of the upper floors averaged one story a day, and the top 14 stories were laid in brick in just 10 days. The building was constructed with assembly line technology in mind. In fact, the building contractor later wrote that the pace led him to suffer "a rather severe nervous breakdown."

The Empire State Building, serving as a symbol for both New York and the United States, plays an important role in popular culture. The building has played a starring role in movies ranging from *King Kong* to *Sleepless in Seattle*. Millions of pictures, postcards, and paintings of the skyscraper have been sold to tourists around the world.

Americans, unlike their contemporaries in Europe, built great skyscrapers as symbols of power and splendor. In serving as headquarters for the nation's business elite, tall buildings served a variety of functions, not least of which was as pawns in competitive displays of wealth.

Over the years, skyscrapers marked a rite of passage for cities around the world. On one hand, they were perpetual advertisements for their owners. On the other, skyscrapers catered to the romanticism of the masses. They reflected the technological and economic power of the United States and the righteousness of the modern technological age.

BOB BATCHELOR

Further Reading

James, Theodore Jr. *The Empire State Building*. New York: Harper & Row, 1975.

Landau, Sarah Bradford and Carl W. Condit. *Rise of the New York Skyscraper, 1865-1913*. New Haven, CT: Yale University Press, 1996.

Nye, David E. *American Technological Sublime*. Cambridge, MA: MIT Press, 1994.

Reynolds, Donald Martin. *The Architecture of New York City: Histories and Views of Important Structures, Sites, and Symbols*. New York: Macmillan, 1984.

Roth, Leland M. *A Concise History of American Architecture*. New York: Harper & Row, 1979.

Wiseman, Carter. *Shaping a Nation: Twentieth-Century American Architecture and Its Makers*. New York: W.W. Norton, 1998.

The Development of Jet Engines

Overview

Jet engines, invented in 1930 by Frank Whittle (1907-1996), have become the dominant form of propulsion for the multimillion-dollar commercial air transportation industry. Jet aircraft's ability to deliver products and services at fast speeds has changed the way business is conducted, and its affordability has enabled more people to travel by air.

Background

Before the development of jet engines, the aviation industry had an absolute limit on how fast, how far, and how high their planes could fly, and how much they could carry. Frank Whittle had a dream to eliminate these boundaries.

In 1923, at the age of 16, Whittle entered RAF College at Cranwell. When he was selected for officer and pilot training in 1926, he wrote his thesis on *Future Developments in Aircraft Design*. Whittle explored new possibilities for propulsion, which, in 1929, led to his idea of using a gas turbine for jet propulsion. Whittle applied for a patent in 1930, but interest (and funding) from the government was meager. The only report on file regarding the idea of jet propulsion was discouraging, and, even though the analysis was based on outdated materials, the Air Ministry developed an attitude of skepticism toward Whittle's research, which lasted for years.

In fact, the British government thought so little of Whittle's patent on the jet engine that they allowed its publication when it was approved in 1932. Within a year, Germany had its own jet research program under way. Whittle continued to work on his project with little official encouragement. In fact, when his patent ran out in 1935, Whittle did not have the five pounds to renew it, so he never received royalties for his invention.

To properly fund experiments, Whittle brought investors and colleagues together to form Power Jets Ltd. in 1936. Ironically (and too late), by then the government had decided to classify his research, putting Whittle in a position where he could not tell investors what they were investing in. At the same time, the government was not willing to provide sufficient funding for Whittle to continue his research.

Whittle faced daunting technical challenges as well. The three basic elements of a jet engine are the compressor, the combustion chamber, and the turbine. A jet engine sucks in air, compresses it by three- to 12-fold, mixes it with fuel (burned to superheat the air, with a small amount used to turn the turbine for more air compression), and forces air and combustion products out the end to create thrust. Though gas turbines existed, Whittle had to rethink them entirely. The goal of contemporary turbines was to harness as much of the energy of combustion as possible to drive machinery. Whittle's jet engine took most of the combustion products and used them for thrust, using only a small portion to drive the turbine. In addition,

Whittle needed to develop materials that could stand the enormous forces the engine generated, and he needed to find the optimum way to mix fuel and air in his system.

Despite many obstacles, Whittle was able to test the first jet engine, the WU (Whittle Unit) turbojet, in 1937. (By that time, Whittle had also patented his idea for the turbofan engine, but the conditions of his funding did not allow a test of this new idea.) That same year, aircraft designer Hans Pabst von Ohain (1911-) secretly tested a jet engine at Germany's Heinkel Aircraft Works.

With a working prototype, Whittle continued to develop his engine, working to make it more durable, more powerful, and more efficient. The work often involved physical courage, as the test engines roared, fan blades broke, and machinery seized at thousands of rpm. On some occasions, everyone fled from the machinery but Whittle. He not only had to solve technical problems, but he also had to continue his fight against official resistance. Even as work had shifted to a jet aircraft, the National Academy of Sciences Committee on Gas Turbines said the goal "seems beyond the realm of possibility."

By August of 1939, however, there was little room for disagreement. The Germans tested the first operational jet aircraft, the Heinkel He 178. By 1941, the Germans had a production model aircraft that could go at 100 mph (161 kph) faster than the fastest Allied fighter.

Impact

The first British jet aircraft did not fly until 1941, weeks after the German production model had had its maiden flight. The Gloster E.28/39 (Pioneer) was piloted by Gerry Sayer. It was powered by Whittle's W1A engine and had a top speed for level flight of 370 mph (595 kph) at 25,000 ft (7,620 m). Plans for the next generation W2 engine and the Meteor aircraft were taken out of the hands of Power Jets Ltd. and given to competitors—first to Rover, then Rolls-Royce—and the British jet did not see action until 1944, when it shot down a V-1 rocket. (With Whittle's plans, the Americans built their own version of the Meteor, the P-59, which was secretly tested as early as 1942, but never used in combat.)

Luckily for the Allies, the Germans did not exploit their lead in jet aviation. The Germans concentrated on rocketry, and the Messerschmitt Me-262, first used in combat in 1942, was used as a ground attack aircraft rather than a fighter.

Frank Whittle with an early jet engine (1948). *(Archive Photos, Inc. Reproduced with permission.)*

There were no dogfights between the Meteor and the Me-262.

While the impact of jet aviation during World War II was minimal, it has been crucial in most major conflicts since. The 1950s, with its test pilot heroes, has been called the Golden Age of Aviation. Jet engines allowed aircraft to fly higher and faster than was possible for propeller-driven craft. Though the sound barrier was broken with a rocket-powered vehicle, all production models of supersonic aircraft were powered by jet engines. Jet fighters, capable of traveling at Mach speeds, are components of the arsenals of most industrialized nations today. The manufacture of military jets has had an economic impact as well. As an example, a single order of 50 F-16s by Greece in the year 2000 was valued at $2.1 billion.

Jet engines themselves continued to develop. The first major commercial application was the turboprop. These engines used most of their power to turn the turbine rather than to create thrust. The turbine was used to drive propellers, and took advantage of the high power to weight ratio of jet engines. The turboprop could be used with traditional airframes and became popular in Europe. It did not compete well in the U.S., where longer flights and greater fuel economy were demanded. The first pure jet was the Boeing 707, which began operations in 1958.

Whittle's turbofan, which forces more air through the jet, increasing thrust without increasing fuel consumption, has assumed a prominent place in aviation and is the engine for the popular Boeing 757. There are also several versions of the jet engine, including the ramjet and the scramjet, designed to push performance for aircraft beyond the limits of Whittle's turbojet.

The effect of the jet engine on commercial aviation is incalculable and came as a surprise. Introduced at first to shorten travel time for passengers, jet engines soon became a means of opening up a much wider market for commercial flight. The carrying capacity of a jet engine far exceeds that of a propeller plane, meaning that more passengers and freight can be carried with each trip. Jets require less maintenance than propeller planes, and they last longer. Economies of scale came into effect and, over time, drove consumer costs for air transportation down to less than half (in constant dollars). As a result air travel is competitive with alternatives such as driving or taking a train. U.S. airlines carried over 600 million passengers in 1998, 10 times as many as they did in 1960. Airfreight also has become popular, with a five-fold increase in tonnage carried from 1970 to 1998. In 1998, profits for all U.S. airlines were $9 billion. In 1955, the peak year for non-jet aircraft, the profits were $140 million. The commercial aerospace manufacturing industry has had important economic impacts as well. As an example, in the year 2000 Kenya Airways ordered five jets from Boeing at a cost of half a billion dollars.

The broad availability of economical air transportation has made long distance travel common. It has facilitated the development of international businesses and global trade. The speed of jet aircraft has made overnight delivery of mail and packages commonplace. Economical air transportation has also increased the speed at which ideas are shared across borders, but it also has accelerated the spread of diseases, as passengers unwittingly carry bacteria across borders.

Whittle saw his dream come true, often in the face of almost unexplainable resistance, but, personally, he only benefited modestly from his contribution to aviation. When the British government nationalized Power Jets Ltd., Whittle resigned. His genius was more appreciated in America, and he became a research professor at the U.S. Naval Academy in 1953.

PETER J. ANDREWS

Further Reading

Books

Chaikin, Andrew L. *Air and Space: The National Air and Space Museum Story of Flight.* New York: Bullfinch Press, 1997.

Golley, John. *Genesis of the Jet—Frank Whittle and the Invention of the Jet Engine.* Shrewsbury: Airlife Publishing, 1996.

Internet Sites

"Midland Air Museum: The Jet Engine." http://www.jetman.dircon.co.uk/mam/thejet.htm

"Whittle's Machine." http://people.aero.und.edu/~draper/whittle.html

The Invention of Nylon

Overview

The invention of nylon by Wallace Carothers (1896-1937) in 1935 launched the age of artificial fabrics and established basic principles of polymer chemistry that made plastics an ubiquitous part of civilization. Nylon itself has an unparalleled range of advantageous properties, including high strength, flexibility, and scratch resistance. Nylon has been overtaken in popularity by polyester, but it is still widely used in clothing, carpeting, toothbrushes, and furnishings. Artificial fibers, which make up a multi-billion dollar industry, offer the ability to control characteristics in ways that are impossible with natural fibers. In fact, today's polymers have replaced natural materials in many applications, including most textiles in the U.S. They have provided new materials, such as lightweight, shock-resistant body armor, that have characteristics that are impossible to reproduce by natural methods. Nylon and other polymers have also created environmental concerns and disposal problems, leading to widespread efforts at recycling.

Background

In 1931, U.S. access to silk was in jeopardy, due to political and trading tensions with Japan.

Magnified nylon fibers. *(Chael Siegel/Phototake NYC. Reproduced with permission.)*

There was a great interest in finding a substitute, an artificial fiber. Wallace Hume Carothers made the breakthrough in 1934, thanks to a combination of a systematic approach to research and his deep understanding of polymer chemistry.

Carothers's lab at DuPont was an exception within the world of industrial research. It was dedicated to basic science and allowed top scientists to pursue experiments that were driven by their curiosities, rather than by market demands. This was new for DuPont, which had lured the young chemistry professor from Harvard University.

DuPont was a successful company, founded in 1802, that had built its wealth on explosives. In 1902, competition in gunpowder forced the company to look for new sources of revenue, and its research laboratory was founded. This was the beginning of the company's interest in high molecular weight molecules (macromolecules), and its research laboratory began investigating fibers in 1909. By then, DuPont had become a diversified chemical company with expertise in solvents, acids, and stabilizers. That year, a notable example of success in the field of macromolecules was introduced, Bakelite. Named after Leo Baekeland (1863-1944), Bakelite was the first thermosetting plastic. It was hard, resistant to solvents, and worth a fortune to his company.

So it was that Carothers arrived in 1928 to direct research in organic chemistry. Carothers was an expert in polymers, molecules that are made up of long chains of repeating units. It was an arcane field at the time, with few principles and without even a standard way of naming and classifying compounds. Carothers's approach was both fundamental and practical, and, by 1931, his lab had produced a synthetic form of rubber called neoprene. They had also developed an understanding of radical polymerization, using and creating charged organic molecules in chain reactions that led to large molecules. They determined the compositions of condensation polymers, which were created by the splitting off of water as bonds were formed, and established basic principles for driving these reactions.

When the challenge came to create a new fiber, they were ready. There was precedent for their attempt. Louis Chardonnet (1839-1924) had created a hit at the Paris Exposition of 1891 with rayon, so-called because it was so shiny it seemed to be sending out the rays of the Sun. He first made the material in 1889 by extruding dissolved nitrocellulose through tiny holes and letting the solvent dry. Rayon was the first artificial fiber to be widely used, though the fibers were sometimes weak and could be highly inflammable.

Carothers aimed to create the first fully synthetic fiber. When he combined amine, hexamethyldiamine, and adipic acid in 1934, he was able to produce fibers in a test tube. These were the result of a condensation reaction, and it was a pivotal insight by Carothers that changed this laboratory curiosity into the basis for a new industry. Condensation reactions produce water as a byproduct, and Carothers realized that this water was interfering with further reactions, limiting the size of the fibers. By distilling off the water as it was formed, he was able to produce molecules that were long, strong, and elastic. The molecule was called Nylon 66 because each of the two constituent molecules contained six carbon atoms. DuPont patented nylon in 1935 and brought it to market in 1939.

Impact

Nylon was an immediate success. It found dozens of uses, including in toothbrushes, as fishing lines, surgical thread, and especially stockings (which came to be called nylons). It is the physical properties of nylon that make it so attractive. Nylon, a polyamine, has a high strength to weight ratio, it resists changes to its shape, and doesn't scratch easily. It is resistant to moisture and has flow properties that make it perfect for injection molding.

Nylon lends itself to a wider variety of fabrics than any natural fiber. It can be woven into tricot, reversible knot, taffeta, crepe, satin, velvet fleece, brocade, lace, organza, and seersucker. Each weave takes advantage of the natural properties of the nylon in a different way, making it silky or coarse, shiny or dull, sheer or bulky. This means that anything from lingerie and swimwear to sweaters and gloves can be made from nylon. Beside clothing, nylon has found uses in parachutes, ropes, screening, body armor, and cords for automobile tires.

While replicating the qualities of natural fibers, chemists and chemical engineers use what they have learned about polymers to change the properties of fibers, tuning them to specific tasks by, for example, making them more insulating, lighter weight, or fire resistant. This is true for plastics as well as for textiles, and many of the artificial materials that surround us in everyday life owe their origins to the discoveries of Carothers and his colleagues.

The synthetic textile industry is a huge and growing enterprise. In 1900, manufactured fibers accounted for, at most, 1% of the American fiber market. By 1998, manufactured fibers accounted for 70% of the fiber used. The phenomenon is worldwide, with 16 million metric tons (17,636,684 tons) of synthetic fiber produced annually in Asia, and about a third as much each in Europe and North America. Polyester is the king of textiles, with a market size over three times that of nylon and a worldwide annual market value in the tens of billions of dollars. The impact of artificial fibers on national economies is much higher when mark-ups for finished apparel, marketing, and distribution are factored in.

The new materials that began with nylon also changed the culture and the language. Polyester will always be associated with disco and the fashions of the 70s. The word "plastic" suggests more than just synthetic materials; it also refers to anything that's false in culture and society. One reaction to the ubiquity of synthetic materials has been a market for "natural" foods, fabrics, and furnishings. Two recent developments have made synthetics more popular with the public. First, blends have been used to get the look and feel of natural fibers while getting advantages, such as durability and permanent press, of synthetics. Second, finer threads of polyesters (microfibers 100 times thinner than a human hair) have been developed to improve moisture handling and the feel of the fabric. The features of microfiber polyester are so appealing that it is now frequently used for high fashion. At every level of society, synthetic fibers are clothing the world.

An unintended consequence of the success of nylon is the waste it generates. Synthetic polymers, in contrast to natural polymers like wood, cotton, and silk, are not biodegradable. They represent a significant contribution to dumps and landfills and may persist in the environment for hundreds of years. The recognition of the problem has led to two initiatives: First, many municipalities have started recycling programs, which collect and reuse old plastics in new products. Nylon is an especially good target for recycling because of its high melting point. Manufacturers have made a specific commitment to recycle carpeting (for which 2 billion pounds [907,200,000 kg] of nylon are used each year), but as recently as 1996 only 1% of discarded carpet was finding its way into new carpeting. Second, researchers have begun to develop biodegradable polymers that have the beneficial physical characteristics of traditional synthetic polymers while having a route to biological breakdown, usually by microorganisms. This often involves using a biological material, such as chitin, as a starting material, rather than petroleum products. Biodegradable polymers may have the additional benefit of being good choices for medical uses, such as being the basis for dissolving stitches or providing scaffolding for growing replacement organs.

After his discovery of nylon, Carothers's reputation grew among his peers, and he was the first organic chemist elected to the National Academy of Sciences, but he did not live to see the world that he had created. He committed suicide in 1937, before nylon was commercialized. DuPont, on the other hand, benefited enormously from the products. Not only did nylon add to DuPont's wealth, but the laboratory that Carothers had established went on to create non-stick coatings, spandex fiber, Kevlar, and many other polymers of commercial importance.

PETER J. ANDREWS

Further Reading

Books

Asimov, Isaac. *Isaac Asimov's Biographical Encyclopedia of Science & Technology.* New York: Doubleday and Co., 1976.

Hermes, Matthew E. *Enough for One Lifetime: Wallace Carothers, Inventor of Nylon.* Philadelphia: Chemical Heritage Foundation, 1996.

Internet Sites

Biodegradable Polymer Research Center. http://www.eng.uml.edu/Dept/BPRC

DuPont. http://www.dupont.com

The Birth of Television

Overview

The invention of television has exerted a profound and wide-reaching effect on the nature and quality of modern everyday life. More vivid than radio, more intimate than film, television became one of the central and most significant technologies of the twentieth century. Television took a long time to reach maturity, as it required the technology to broadcast as well as receive images, along with the cooperation of government and commercial interests to coordinate the supply of programming. But once television broadcasting became a reality and television sets were for sale to the average home, it quickly became the primary source for entertainment and information, first in the United States and England, and eventually throughout the world.

Background

Television—a term first used in 1900—lived in the imaginations of inventors and writers long before it became a technological reality. The desire to transmit images was piqued by the ability to carry sound on the telephone, invented in 1876. Advances in photography, cinematography, and facsimile transmission also stimulated interest in television. The earliest patented design for a "television" system came in 1884 from a German engineer named Paul Nipkow (1860-1940). Nipkow's system consisted of a disk with 24 holes through it in the shape of a spiral. Spinning the disk in a light source produced a pattern of images on a photosensitive cell, which would then produce a varying electrical current. Reception would be achieved by reassembling the dissected image using another rotating disk and a magneto-optical light modulator. Nipkow never actually constructed the system he proposed, but mechanical systems based on the Nipkow disk became the first reliable televisions. Mechanical television systems were, however, challenged nearly from the beginning by an alternative strategy using cathode-ray tubes. For several decades, mechanical and electronic television struggled alongside one another as inventors and engineers tried to make real the dream of transmitting images.

The chief proponents of mechanical television were John Logie Baird (1888-1946), a Scotsman, and an American, Charles Francis Jenkins (1867-1934). Baird began his television experiments in 1923. His invention was introduced to the public in 1925, during a publicity affair at London's Selfridge's department store. This early apparatus, based directly on the principle of the Nipkow disk, was extremely crude and could transmit only rough patterns of light and dark to represent images. By 1929, Baird had advanced his system of "low-definition" television significantly, and undertook broadcasting experiments with British Broadcasting Corporation (BBC) radio transmitters. Jenkins also began his investigations into television using a Nipkow disk system in 1923. He was issued the first American television broadcasting license by the Federal Radio Commission in 1928, for experimental transmissions from a station in the suburbs of Washington, D.C. The first commercial receivers for the Jenkins system went on sale in 1930, with a viewing screen of 2 x 1.5 inches (5.08 x 3.81 cm)! These early television signals had no more than 48 lines per screen; in comparison, televisions at the end of the twentieth century typically used 525 or 625 lines, giving much finer definition of images suitable for larger screens. Most images produced were black and white (or "monochrome"), although experiments using color filters began as early as 1929.

But lone inventors like Baird and Jenkins did not have television research to themselves for long. During the mid-1920s, large companies including General Electric (GE), American Telephone & Telegraph (AT&T), and Radio Corporation of American (RCA) joined in the efforts to produce a commercially viable mechanical television. Between 1928 and 1932 both the United States and England had some limited amount of television broadcasting (entertainment programs of various kinds as well as occasional news shows) available to the public; television broadcasting had begun in Germany as well. American and British efforts focused primarily on bringing television signals into homes; in Germany television was treated as an alternative to cinema and was usually viewed in large theaters.

While broadcasting began with mechanical television systems, a technological competitor was emerging that would leave mechanical television stillborn. The idea of using a cathode ray tube as the foundation for "distant electric vision," as it was called by A. A. Campbell Swinton (1863-1930) in his own proposal in 1908, was first raised by Russian engineer Boris Rosing in 1907. An assistant to Rosing, Vladimir Zworykin (1889-1982), became the driving force behind electronic television. Zworykin emigrated to the United States and joined the Westinghouse company in 1919. While at Westinghouse (and later at RCA), Zworykin designed the "iconoscope," a transmitting tube where an image could be projected onto a photosensitive array. These photosensitive elements would then be scanned by a beam of electrons, which in turn released charged elements to form a picture signal that could be received and displayed by a modified cathode-ray tube. In-house field tests of television using the iconoscope began at RCA in 1932.

Zworykin and his mighty corporate backers did have strong competition. Philo Farnsworth (1906-1971), an independent inventor, demonstrated publicly the first complete electronic television system in 1934, following more than eight years of development. His system was based upon an invention of his own called the image-dissector tube. In this electronic version of the Nipkow disk, a moving electronic image moved passed a stationary aperture through which passed electrons that in turn stimulated an electrode to collect and reassemble picture elements. While Farnsworth is largely given credit for the invention of electronic television, his system failed to achieve the commercial success of the system backed by RCA.

Of course, for television to fulfill the dreams of its creators required a system for broadcasting. In England and Germany television broadcasting was run by companies closely affiliated with the government. British broadcasting switched from Baird's mechanical system to an all-electronic system in 1936, but stopped broadcasting entirely in 1939 after the outbreak of World War II. For a time during the war, the German government made use of television for propaganda purposes, including television broadcasts of the 1936 Olympics from Berlin. In the United States, television was promoted by commercial interests—companies including RCA and Columbia Broadcasting System (CBS) that had been enormously successful in the radio industry. Although experimental broadcasts continued, the public was hesitant to invest in receivers for their homes out of fear that they would become rapidly obsolete as several companies competed to dominate the television market. The Federal Communications Commission (FCC) finally endorsed a set of standards in 1941 that would insure compatibility of all receivers with all broadcasts, but the United States entered World War II just a few months later, effectively bringing about a hiatus in the development of television.

Beginning in 1946, the television industry started to grow again. Wartime research produced improved equipment, and the number of stations exploded so rapidly that in 1948 the FCC was forced to place a four-year freeze on the assignment of new stations while problems of interference could be worked out. Another area of contention was color broadcasting, where rival companies CBS and RCA championed two completely different and incompatible systems. While CBS's "field sequential system," which utilized colored disks rotating in the signal in the transmitter and receivers, was developed more quickly, RCA's "dot sequential" system, which sent three separate signals carrying different colors, ultimately triumphed following four years of FCC hearings beginning in 1949. Thanks to this controversy, color television got off to a slow start—it was not until the mid-1960s that sales finally eclipsed those of monochrome sets, and color broadcasting became routine.

Impact

In 1947 there were 16,000 television sets in American homes; in 1949 there were four million, and by the end of 1950, 11 million. This number continued to increase until eventually, virtually every home in the United States had at least one television. Initially, most homes used large rooftop antennas to receive signals broadcast through the air like radio, but alternative methods of delivering signals appeared early and have continued to grow in importance. The concept of "cable" television, where signals are brought to each home via coaxial cable from a central antenna, began in remote areas far from broadcasting centers in the 1940s; in the 1980s, it became a multi-billion dollar industry that competed successfully with broadcast television throughout the country. The distribution of television programs directly into homes from satellites became possible in 1979 and, like cable television, it continued to increase in popularity and influence.

Television programming was initially controlled by a small number of "networks" of stations broadcasting in major cities and affiliated with existing radio stations. Three of these networks—the National Broadcasting Company (NBC, the broadcasting arm of RCA), CBS, and the American Broadcasting Company (ABC)—dominated television broadcasting for more than 30 years. While they remained influential, by the 1980s the growth of cable television provided viewers with new alternatives to the broadcast networks. Forms of entertainment programming, including game shows, variety shows, serial dramas, and situation comedies, appeared very early in television history and continued in popularity. Television also became a vital source of news and information, initially following formats established on radio but soon taking great advantage of the visual medium by incorporating images of events like those formerly shown on cinema newsreels.

Television is truly a dream made real. It has influenced nearly every aspect of life, providing images profound and profane that people share and even experience simultaneously. While the technology of television continues to improve and spread to even the most remote places in the world, the impact of even the earliest television broadcasts was very powerful. Although the television may be superseded someday by newer technologies, those technologies will undoubtedly have been themselves shaped by television.

LOREN BUTLER FEFFER

Further Reading

Abramson, Albert. *The History of Television, 1880-1941.* London: McFarland, 1987.

Burns, R. W. *Television: An International History of the Formative Years.* London: IEE Press, 1998.

Inglis, Andrew F. *Behind the Tube: A History of Broadcast Technology and Business.* Stoneham, MA: Butterworth, 1990.

Ritchie, Michael. *Please Stand By: A Prehistory of Television.* Woodstock, NY: Overlook Press, 1994.

Udelson, Joseph H. *The Great Television Race: A History of the American Television Industry, 1925-1942.* University of Alabama Press, 1982.

Computers: The Dawn of a Revolution

Overview

By the end of the twentieth century, computers could be found in devices from wristwatches to automobiles, from medical equipment to children's toys. But while scientists and philosophers had dreamed of the possibility of automating calculation nearly one hundred years earlier, very little progress was made toward modern computers before 1940. As scientists and engineers worked to face the challenges of World War II—including cracking codes and calculating the physics equations to produce atomic weapons—they finally made computers a reality. In a few short years, the theoretical vision of computing was brought together with existing technologies such as office machines and vacuum tubes to make the first generation of electronic computers.

Background

Mechanical calculating machines had their origins in the mid-nineteenth century with the work of Charles Babbage (1792-1871), whose "Analytical Engine" was intended to use gears and punched cards to perform arithmetical operations. Babbage's design anticipated many of the concepts central to modern computers, such as programming and data storage. Interestingly, a fragment of Babbage's machine survived to the 1940s, when it was discovered in an attic at Harvard and helped inspire the computer innovator Howard Aiken (1900-1973). Punched cards were also used to collect and store data beginning with the 1890 census. The man responsible for this innovation, Herman Hollerith (1860-1929), founded the company that became International Business Machines (IBM), the largest and perhaps the most important computer company of the twentieth century. Another nineteenth-century development that proved most vital to the evolution of computers was the mathematical work of George Boole (1815-1864). Boole showed that the binary number system, which has only two symbols 0 and 1, could be used to perform logical operations by letting 0 stand for false and 1 for true. Later, this "Boolean Algebra" was used to represent the on and off states of electronic switches and formed the framework for electronic computers.

As with many technological achievements, ideas and visions have often spurred ahead practical plans for computers. One of the most important visionaries who helped to bring about the invention of digital computers was Alan M. Turing (1912-1954), a British philosopher and mathematician who in the 1930s wrote a number of important papers proposing ideas about theoretical machines designed to test mathematical or logical statements. Turing's imaginary machines helped to establish the actual potential and limitations of computers. Turing also helped to design and build some of the very first electronic computing machines for use in deciphering encrypted transmissions during World War II, including in 1943 a device called the "Colossus" made from 1500 vacuum tubes.

There is much dispute about who really built the first modern computer. World War II was a great impetus for computer research in Germany as well as among the Allied nations. While Konrad Zuse (1910-1995), a German, did design and construct several computing machines in the 1930s and 1940s, it was in Britain and the United States that computing research advanced furthest and fastest. Following his work with Colossus during the war, Turing joined the staff at Britain's National Physical Laboratory (NPL) to assist in the development of electronic digital computers. Turing produced an important program report that included one of the first designs for an electronic stored-program device. The project based on Turing's ideas, the Automatic Computing Engine (ACE) took some years to advance, but a prototype machine was finally produced in 1948. Computer research proceeded elsewhere in Britain as well; advanced machines were developed in the 1940s at the universities of Cambridge and Manchester.

In the United States, early computer projects flourished at industrial research laboratories and university campuses, driven ahead by commercial, scientific, and wartime interests. During the 1930s, Vannevar Bush (1890-1974) at MIT built a "Differential Analyzer" machine that used an array of gears, shafts, wheels, wires, and pulleys driven by electric motors to solve differential equations. At the Bell Telephone Laboratories of AT&T, a binary calculator that used telephone relays was built in 1939, the same year that engineers at IBM, with the help of Harvard professor Howard Aiken, started to build the electromechanical Mark I computer, which when finished consisted of 750,000 moving parts. In 1940, John Atanasoff (1903-1995) and Clifford Berry (1918-1963) of Iowa State College built the first computer to use vacuum tubes, a technology that would dominate the next decade.

War efforts were the impetus behind a major computer project at the Moore School of the University of Pennsylvania. There, beginning in 1942, engineers built the vacuum-tube based Electronic Numerator, Integrator, and Computer (ENIAC) to calculate ballistic tables for the Army Ordnance Corps. The 30-ton (27 metric ton) ENIAC was housed in 48-foot-tall (15-meter) cabinets, and was made out of 18,000 vacuum tubes and miles of wires. It performed calculations one thousand times faster than its electromechanical predecessors. John von Neumann (1903-1957), a Hungarian immigrant mathematician working on the atomic bomb project at

Los Alamos, collaborated with Herman Goldstine of the Moore School to produce an important report that described the storage of programs in a digital computer. Two engineers who worked at the Moore School, J. Presper Eckert (1913-1995) and John Mauchly (1907-1980), subsequently conceived of a computer they called the Universal Automatic Computer (UNIVAC). They built the first UNIVAC for the U.S. Census in 1951; eventually, nearly 50 UNIVAC computers were constructed. The corporation founded by Eckert and Mauchly became one of the important early computer companies in the U.S., helping to move computers into the business world.

All of these very early computers were unreliable behemoths, filling up entire rooms with cases lined with fragile vacuum tubes that generated vast amounts of heat. While the conceptual limits of computing as described by Turing and others seemed boundless, the physical and practical limits of these machines were quite obvious to all who used them. But in 1948, three scientists at Bell Laboratories—John Bardeen (1908-1992), William Shockley (1910-1989), and Walter Brattain (1902-1987)—published a paper on solid-state electronic "transistors." The transistor took advantage of a discovery that had been made by physicists of substances, such as silicon, that have the property of conducting electricity in one direction but not the other—called "semiconductors." Small, cool transistors soon replaced the large, hot vacuum tubes as "switches" in the electronic circuits that make up computers. This discovery started the evolution of ever-smaller computer devices that continued to the end of the century and beyond. Another 1948 discovery by a Bell Labs scientist, Claude Shannon (1916-), provided a vital theoretical framework to explain the electronic transmission of information. By 1949 all the elements for the rapid development of computer technology and a computer industry were finally in place.

Impact

During the 1950s, the computer became an important force in business as well as science. The large computer systems, known as mainframes, cost anywhere from several hundred thousand to several million dollars. Most were made, sold, and maintained by IBM, along with a handful of much smaller competitors that came to be known as the "seven dwarves." All computers required special air-conditioned rooms and highly trained staff to operate them. Information entered the computer through paper tape or punched cards; output would be delivered (sometimes days later) in the form of more cards, or on large paper printouts. As the storage capacity of computers improved, the first software programs were developed: John Backus of IBM produced the earliest version of FORTRAN in 1954. Programming languages enabled scientists and engineers to use computers much more effectively. As computers became more accessible, new applications were found for them. The business functions performed by typewriters, adding machines, and punched-card tabulators were gradually taken over by computers. Eventually, sophisticated software programs allowed people without technical training, even children, to use computers for work, study, and entertainment.

In the two decades that followed the introduction of the transistor, computers became smaller, cheaper, faster, simpler, and much more common. No branch of science was left untouched by the rapid development of the computer, and many activities of everyday life were transformed as well. The replacement of the transistor by the microprocessor in the 1970s made far smaller machines practical, and new uses for computers helped maintain a growing market for small machines even as mainframes remained essential to business, education, and government. But by the 1980s, micro, or "personal" computers (PCs), began to challenge the supremacy of mainframes in the workplace and in fact networks of inexpensive PCs replaced mainframes for many applications.

Large computers remained essential, however, especially to the evolution of computer communication including the Internet. It is through the Internet, a large system of computers linked together to transmit messages and images quickly throughout the world, that most individuals have come to experience and understand the computer and its contributions. Computers have provided easy access to large volumes of information and have made difficult calculations and investigations feasible. They have made it possible to travel into outer space, and to communicate instantaneously with people across the globe. While some have raised concerns about the effect of computers on personal privacy and independence, the computer has brought much of great value to commerce, science, and humanity—far more even than Babbage, Turing, or von Neumann could have imagined.

LOREN BUTLER FEFFER

Further Reading

Aspray, William, ed. *Computing Before Computers.* Ames: Iowa State University Press, 1990.

Burks, Alice, and Arthur W. Burks. *The First Electronic Computer: The Atanasoff Story.* Ann Arbor: University of Michigan Press, 1988.

Campbell-Kelly, Martin, and William Aspray. *Computer: A History of the Information Machine.* New York: Basic Books, 1996.

Ceruzzi, Paul. *Reckoners: The Prehistory of the Digital Computer from Relays to the Stored Program Concept, 1935-1945.* Westport, CT: Greenwood Press, 1983.

Cohen, I. B., and Gregory W. Welch, eds. *Makin' Numbers: Howard Aiken and the Computer.* Cambridge, MA: MIT Press, 1999.

McCartney, Scott. *ENIAC: The Triumphs and Tragedies of the World's First Computer.* Walker Publishers, 1999.

Williams, Michael. *A History of Computing Technology.* Englewood Cliffs, NJ: Prentice Hall, 1985.

The Development of Mass Production Has a Dramatic Impact on Industry and Society

Overview

Mass production is the modern system of manufacturing that uses principles such as interchangeability and the use of the assembly line. Although notions regarding mass production existed in many industrialized nations, the concept wasn't fully realized until Henry Ford (1863-1947) put it to use in 1914. Ford's success in producing the Model T automobile set the early standard for what mass production could achieve. As a result, mass production quickly became the dominant form of manufacturing around the world. The idea of mass production also took hold in popular culture. Numerous artists, writers, and filmmakers used the image of the assembly line to symbolize either the good or the evil of modern industrial society.

Background

Notions of mass production date back to the 1800s and the development of machine tools. However, the nineteenth century witnessed the birth of a true machine tool industry. The earliest machine tool pioneers were in Britain. Henry Maudslay (1771-1831) built precision tool works necessary for mass production. Many of England's early machine tool artisans worked for Maudslay as apprentices. Later, these individuals crafted precision lathes, plane surfaces, and measuring instruments.

Even with the early successes in Europe, technology scholars attribute the widespread adoption of mass production to trailblazers in the United States. With its abundant waterpower, coal, and raw material but shortage of workers, America was the ideal place for building skill into the machine. From its earliest industrial beginnings, American leaders attempted to mechanize production of barrels, nails, and other goods. In the early 1800s American inventor Thomas Blanchard used mechanized production to make rifles and muskets for the U.S. Armory in Springfield, Massachusetts. Blanchard's efforts were supported by the War Department, which then supported other mass production applications.

The distinct system developed in the United States became known as the "American system of manufacturing." In the nineteenth century, the nation witnessed the rise of innovators such as Eli Whitney (1765-1825), Samuel Colt (1814-1862), and Cyrus McCormick (1809-1884). These leaders were committed to interchangeability and mechanized production. By 1883 the Singer Manufacturing Company sold over 500,000 sewing machines, while McCormick produced thousands of reapers.

Many factors came together in the early twentieth century to make mass production possible. In fact, Henry Ford's decision to produce an inexpensive automobile that working people could afford was a gamble. He succeeded in convincing his financial backers only through sheer determination. Also, Detroit had a history of mechanical innovation, which provided the skilled engineers and designers that could build Ford's factories. Another important component was that the immigration boom in the area provided the company with workers to man the lines.

Ford's determination to make Model Ts and only Model Ts helped in the development of mass production techniques. Each process was broken down into its smallest part, and as the components moved down the line, the pieces soon formed the whole. Throughout the process Ford emphasized accuracy, and experts noted the durability and soundness of the automobile.

As the early production process peaked, Ford introduced the assembly line to the mix. The assembly line gave Ford factories a fluid appearance and dramatically increased productivity. Without the assembly line, Ford would not have been able to keep pace with consumer demand for the Model T. More important for buyers, the increased efficiency brought with it a reduced cost. Model T prices quickly dropped from more than $800 to $300.

"Every piece of work in the shop moves," Ford explained. "It may move on hooks on overhead chains going to assembly in the exact order in which the parts are required; it may travel on a moving platform, or it may go by gravity, but the point is that there is no lifting or trucking of anything other than materials." Soon, Model Ts were being produced once every two minutes.

The company sold 11,000 cars from 1908 to 1909, which raised $9 million, a 60% increase over the previous year. Ford then outdid himself with the 1910 to 1911 model, selling 34,528. Sales skyrocketed, reaching 248,000 in 1914, or nearly half the U.S. market. The heavy demand for cars forced Ford to pioneer new methods of production. He built the largest and most modern factory in America on a 60-acre tract at Highland Park in north-central Detroit. Ford's net income soared from $25 million in 1914 to $78 million by 1921.

Another essential facet of Ford's mass production system was his willingness to adopt resourceful means of finding labor to man the assembly lines. The sheer size of the workforce Ford needed to keep pace combined with the monotony of the assembly line led to great turnover in the factories.

Early in 1914 Ford introduced the "Five Dollar Day" to deal with labor shortage. Ford decided that he would pay workers the then-outrageous sum of $5 for an 8-hour workday, much shorter than the industry average. The new wage far surpassed the industry's standard of $1.80 to $2.50 per day. The program made Ford a hero and extended his growing legend.

Because of mass production and Ford's high wages, company workers were given the ability to elevate themselves above working-class means. With the extra pay, they participated in the accumulation of material items previously out of their reach. In turn, other mass producers, especially of middle-class luxuries, were given another outlet for goods. The Five Dollar Day ensured the company that it would always have the workers needed to produce, while at the same time allowing working-class families a means to participate in America's consumer culture.

Impact

Ford's use of mass production techniques was a landmark step for American industry. Within the automobile industry, Ford's beloved Model T eventually declined in popularity, but mass production became a permanent part of the business. Mass production techniques spread to other car makers, and Alfred P. Sloan (1875-1966) of General Motors introduced the annual model change in the 1920s. The changing look of automobiles, made affordable by mass production, mirrored the changing national landscape. A sweeping car craze prompted the desire for material abundance that would mark the genesis of modern America after World War I.

Advertisers, artists, and writers used the factory and assembly line to symbolize life in the United States. Often, they associated manliness with technology and engineering. Many looked upon the factories that linked American cities with romanticism. Corporate marketing, advertising, and public relations staffs and outside agencies developed to massage this message into the public's subconscious.

Many factories even began offering tours to show off production capabilities. Ford's Highland Park factory received more than 3,000 visitors a day before 1920. General Electric, National Cash Register, and Hershey Chocolate established tours as well. They were a new form of public relations and left visitors with a deep impression of the company.

Over the next several decades, the influence and dominance of mass production solidified around the world. In preparing for both World War I and World War II, nations intensified mass production of arms and ammunition. The efficiencies of mass production allowed American businesses to switch from consumer goods to war stuffs quickly. The amount of armaments brought to the war effort by the United States turned the tide in both wars.

After World War II American industry shifted back to consumer goods but did not slow its pace of production. The rise of suburban living and the subsequent "baby boom"—a huge increase in babies born in the post-war era—kept assembly lines producing at phenomenal rates. The growth of the middle class, in both wages and desire for material goods, can be traced to the development and dominance of mass production.

Mass production also bears responsibility for the negative outcomes associated with unskilled labor. The process made workers dispensable and increased the power of the foremen, managers, and department heads that wielded power over them. These influences were mocked in the famous 1936 film by Charlie Chaplin, *Modern Times*. In real life, mass production led to worker unrest, turnover, and social conflict. Unionization efforts intensified as workers became more alienated in the factory setting. Thus, the advent of mass production had both positive and negative effects on society.

BOB BATCHELOR

Further Reading

Cowan, Ruth Schwartz. *A Social History of American Technology*. New York: Oxford University Press, 1997.

Hounshell, David A. *From the American System to Mass Production, 1800-1932*. Baltimore: Johns Hopkins University Press, 1984.

Kranzberg, Melvin, and Joseph Gies. *By the Sweat of thy Brow: Work in the Western World*. New York: Putnam's, 1975.

Nye, David E. *American Technological Sublime*. Cambridge, MA: MIT Press, 1994.

Helicopters: The Long Journey

Overview

The invention of the helicopter was a long and difficult process. Although the first practical helicopter was developed years after the first successful flight of an airplane, the concept of rotary flight predates that of the fixed-wing plane and goes back to the fourth century A.D. However, the helicopter as we now know it is an invention of the twentieth century and required the perseverance of many inventors to make this machine a reality.

Background

The helicopter is described as a VTOL or vertical take-off and lift machine, which means it ascends vertically without benefit of a runway. The whirling action of the rotor, to which the blades are attached, gives the helicopter its lift. The blade is more curved on the top than the bottom, and the air is forced to flow faster over the top of the blade. A difference in air speed creates a difference in air pressure with less being on the upper surface of the blade, and in this way lift is created. Rotor tilt controls the direction of flight and in whatever direction the rotor is tilted the helicopter moves. It can move forward, backward, side to side, and up and down at very low speeds, and can hover in the air.

Written in the "Pau Phu Tau," the first mention of rotary flight dates back to the fourth century A.D. In this document, the alchemist Ko Hung proposed a type of flying top that could be made "with wood from the inner part of the jujube tree, using ox leather (straps), fastened to returning blades so as to set the machine in motion." Something like this later emerged in Europe as a "Chinese toy" or "Chinese top," and was a children's toy. Leonardo da Vinci (1452-1519) described a rotary machine late in the fifteenth century, a screw made of "starched linen" to be turned "swiftly" by a single passenger. The term helicopter comes from his use of Greek to describe this craft, *helix*, meaning "spiral" and *pteron*, meaning "wing." He never attempted to build the machine, however.

As theories of rotary flight were being developed, inventors began to create helicopter models to test these ideas. In 1754, the Russian Mikhail Vasilyevich Lomonosov (1711-1765) proposed to the Russian Academy of Science, a co-axial, contra-rotating model to be operated by clockwork for the lifting of meteorological instruments. Contra-rotating means there are two sets of blades, one set below the other, on the same rotor. This counteracts the effect of torque, which causes the fuselage to spin in the opposite direction of the rotor. It is thought his model may have flown, but Frenchmen Launoy and Bienvenu are generally credited with the first flying helicopter models. They demonstrated their model to the Académie des Sciences on April 28, 1784. The model consisted of co-axial rotors powered by the spring loading action of a whalebone bow. In Britain in 1796, Sir George Cayley (1773-1857), also known as the "father of aerial navigation" due to his contributions to fixed-wing flight, constructed and flew a model helicopter. Although his model was not unlike that of Launoy and Bienvenu, Cayley probably did not know of the French models.

Recognizing the need for a substantial power source to lift a man-operated helicopter, inventors experimented with steam-powered helicopter designs during the 1800s, but they were heavy and awkward. In 1876 the German Nikolaus August Otto (1832-1891) developed the four-stroke internal combustion engine, leading the way to smaller, lighter and more powerful engines. The American Thomas Edison (1847-1931) had experimented with model helicopters driven by electricity and even gunpowder during this time. Although unsuccessful, he felt a practical helicopter would one day be built.

Other great inventors like Wilbur (1867-1912) and Orville (1871-1948) Wright did not agree. They flew the first airplane in 1903, the *Flyer II*, at Kitty Hawk, North Carolina. The success of the airplane in many ways made continuing exploration into rotary machines appear foolish and backwards, and the Wright brothers condemned helicopter development by saying, "Like all novices, we began with the helicopter (in childhood) but soon saw it had no future and dropped it." Work on rotary designs did continue, as the piston engines of the early 1900s became lighter and more efficient.

Impact

Full-sized helicopters did not achieve any success until 1907, when Frenchmen Louis Breguet and Charles Richet designed the first. It was a massive machine weighing 1,272 pounds (577 kg) with the pilot, powered by a 50-hp Antoinette eight-cylinder inline engine that drove four rotors. On September 29, 1907, at Douai, France, it lifted for two minutes but was stabilized by four men with poles. Next, on November 13, 1907, fellow Frenchman Paul Cornu (1881-1914) piloted his design, which remained aloft for a few seconds at six feet (1.8 m). It did not need to be steadied and although tethered, it never reached the end of the lines and is now recognized as the first free flight of a piloted rotary-winged vehicle. Their successes inspired further development.

One of the most important figures in the history of rotary flight was the Spaniard Juan de la Cierva (1895-1936). He was responsible for the development of the autogiro, differing from the helicopter in that the rotors are not powered by an engine, but rather depend on auto-rotation for sustained flight. The success of autogiros aided the development of the helicopter. After the First World War in 1919, he began to work on rotary machines. In 1922, he designed and built an autogiro that had flapping hinges where the blades attach to the rotor, and is credited with solving the problem of dissymmetry of lift, which causes the machines to flip over and crash. The machine was flown on January 9, 1923, about 200 yards (183 m) off the ground and executed some maneuvers. In 1926 he launched the Cierva Autogiro Co., Ltd. and built the C. 6C, the first rotary vehicle for the British Royal Air Force, as an experimental machine. Cierva flew the English Channel in 1928, in his autogiro and toured 3,000 miles (4,827 km) in short flights across Europe. Harold Pictarin, in the United States, secured the rights to manufacture autogiros in 1928, and in the late thirties he and Cierva licensed the Kellett Autogiro Co., developing the Kellet YG-1, which became the U.S. Army's first rotary-winged craft.

As rotary flight was shown to have practical applications, interest in helicopters ignited. Heinrich Focke (1890-1979), gaining experience with rotary flight when Germany built C.19s and C.30s under license, built the Focke-Wulf 61. This machine is considered the first practical helicopter ever built, and it flew to 11,000 feet (3,353 m) for 80 minutes in 1937. That year Hannah Reitsch flew the FW 61 for a crowd at the Berlin Sports Hall, which proved a great success and showed off the machine's maneuverability. Hannah Reitsch was to later become Hitler's personal pilot.

After experimenting with rotary flight for years Igor Sikorsky (1890-1972), a Russian-American, built the VS 300 in 1939. The VS 300 had a single main rotor with a tail and tail prop similar to many of the designs today, and marked the first flight of a helicopter in the U.S., flying untethered in 1940. Sikorsky improved upon this design, and the production of the VS 300 as the R.4 is heralded as the beginning of helicopter production in the United States. The R.4, based on the VS 300, flew in 1942 and the U.S. military bought hundreds of models during WWII. This design became a direct link between early designs and the helicopter of today.

During the mid-1900s military use of helicopters increased, leading to improvements in design. Able to hover in one place, move in every direction slowly, and land in an area little more than the size of the craft itself, the maneuverability of the helicopter made it valuable to the American military during the wars in Korea and Vietnam. Initially, it was not used for combat, but was used to carry troops, supplies, guns, and ammunition into remote and difficult areas. Helicopters were used to rescue downed pilots behind enemy lines, and airlifted more than 23,000 causalities to field hospitals. The Bell H-13 was responsible for 18,000 of those rescued and received the nickname "Korean Angel," for that reason. It was during the Vietnam War that helicopters became true gunships. Beginning in 1964, they were used in combat and the Huey UH-13 was fitted with anti-tank missiles and turret mounted grenade launchers. While the helicopter still carried on its important role of transporting and rescuing troops, it was also being used to suppress enemy fire. During the war, helicopters were modified to function better under attack and were made narrower and faster. These wars prompted much advancement in the development of the helicopter, and now there are more helicopters in the world's airforce than airplanes.

While helicopters do not enjoy a glamorous reputation, they are indispensable pieces of equipment. They have developed into more economical and practical machines powered by lighter, more powerful engines. Today, helicopters are used to perform important services for cities, industry, and government. Rescue missions and operations depend on the versatility of the helicopter for disaster relief efforts at sea and on land. The Coast Guard uses them regularly, and the ability of the helicopter to hover allows for harnesses to be extended to victims on the ground or at sea, who can then be transported to safety. Helicopters are also useful when rescuing lost or injured hikers or skiers in mountainous terrain. Hospitals now have helipads so accident victims can be transported faster for emergency treatment. Police use them for aerial observation, tracking fleeing criminals, searching for escaped prisoners, or patrolling borders for illegal activity. Police and news agencies even use the helicopter to watch for traffic problems in major cities.

Wildlife and forestry employees rely on the helicopter to conduct aerial surveys of animal populations and to track animal movements. Forestry personnel use the helicopter to observe the condition of tree stands and to fight fires. Helicopters transport personnel and equipment to base camps, and spray fires. The agricultural industry engages helicopters to spray fields and to check on and round up cattle.

Helicopters are of particular interest to industry, performing a variety of jobs that require strength and maneuverability, such as hoisting heavy building materials to the upper levels of a skyrise and hauling awkward and large objects under their fuselage. They have also been used to erect hydro towers and other tall structures. Petroleum industries rely on the helicopter to observe pipelines for damage and to transport personnel to offshore drilling operations.

Able to perform a number of passenger services, helicopters are frequently employed by businesses to transport clients and employees. Although expensive, it is a convenient way to beat the traffic for those with time constraints, and downtown businesses in large cities will often have heliports on top of their buildings. Helicopters transport passengers from the airports and are enjoyed recreationally by site seers and helicopter enthusiasts.

While helicopters have improved greatly since the first piloted rotary machines of 1907, they are much slower than airplanes and cannot reach the same altitudes. Expensive and difficult to fly, helicopters are not always economical, but they are highly versatile and maneuverable, and can move in ways that are impossible for fixed-wing craft. This maneuverability makes the helicopter an essential tool for industrial, military, and civil service.

KYLA MASLANIEC

Further Reading

Books

Bryan, Leslie A., et. al. *Fundamentals of Aviation and Space Technology.* Urbana, Illinois: Institute of Aviation, 1968.

Skjenna, Olaf. *Cause Factor: Human, A Treatise on Rotary Wing Human Factors.* Ottawa, Canada: Canadian Government Publishing Centre, 1986.

Taylor, Michael J. *History of Helicopters.* Toronto, Ontario: Royce Publications, 1984.

Other

Kimmet, Katie and Amanda Nash. "History of the Helicopter." http://165.29.91.7/classes/humanities/amstud/97-98/helicptr

"Mr. Carmody, We Want Lights": The Tennessee Valley Authority and Rural Electrification Under the New Deal

Overview

Any discussion of the 1930s conjures images of the Dust Bowl, unemployment, and economic deprivation across the United States. Perhaps more disadvantaged than most in the nation were the residents of the Tennessee River Valley. Generations of farmers had overworked and deforested marginal land. The region had failed to attract industry, and the general standard of living for valley inhabitants was poor. Already plagued by frequent epidemics, inadequate housing, and meager incomes, the Great Depression only exacerbated problems in the region.

The Tennessee River Valley was not the only region whose people lived in conditions that seemed primitive when compared to contemporary urban standards. Many rural dwellers across the nation faced similar problems. Recognizing the importance of aiding family farms across the United States, and the potential for revitalizing and developing the Tennessee River Valley, the Rural Electrification Administration and the Tennessee Valley Authority were created by President Franklin D. Roosevelt and his administration as part of his "New Deal" to get American citizens out of the choke hold of the Depression.

Background

Roosevelt appealed to Congress for help in establishing "a corporation clothed with the power of government but possessed of the flexibility and initiative of a private enterprise." On May 18, 1933, Congress passed the Tennessee Valley Authority Act. The TVA Act covered parts of seven states—Alabama, Georgia, Kentucky, Mississippi, North Carolina, Tennessee, and Virginia—an area that extended the geographic limits of the Tennessee Valley, but shared many of the same economic woes. The historic act sought not only to alleviate some of the economic effects of the Depression, but, in its unprecedented Section 23, provided a mandate to improve the "social well-being of the people living in [the] river basin." Thus, the Tennessee Valley Authority, or TVA, was one of the most ambitious components of Roosevelt's New Deal.

From the beginning, TVA pioneered an integrated approach to resource management. The mission of TVA was comprehensive—whether it was power production, rural poverty, flood and erosion control, navigation, agricultural reform, disease prevention, reforestation, or cultural resource management—and each project was studied to determine all attendant conditions. Special attention was paid to concerns of long-term impact upon man and the environment.

The TVA began its first project the summer after the TVA Act was signed. Norris Dam, named after Senator George Norris of Nebraska—dubbed "the father of TVA," was built along the Clinch River. The project was one of the most ambitious and controversial of the TVA's early years. Norris Dam was constructed to flood the entire Norris Basin area, one of the poorest in all of the TVA region. The thousands of residents who lived in the basin were forced to move. Those who owned their property outright were compensated financially, but many basin residents were sharecroppers and tenant farmers and thus received nothing. The TVA, with the assistance of the Works Progress Administration (WPA), constructed model communities with modern amenities, schools, and stores, to which they hoped the basin residents would move. However, the construction of the dam required the influx of a great number of workers; these workers moved into the planned communities, which then became in effect "company towns." Citizens displaced by TVA dam and reservoir projects often found themselves relocating to areas that had faced the same endemic problems as the places they had left.

By June 1934 the TVA employed 9,173 workers. Several thousand more were employed by the WPA to assist with the construction of TVA projects. Sixteen dams were built between 1933 and 1944. In regions previously prone to catastrophic flooding, the construction of a series of dams enabled better control of excess water and almost eliminated the threat of serious floods on the stretch of river between Chattanooga, Tennessee, and Muscle Shoals, Alabama. The dams also provided reservoirs, hydroelectric power, and locks that eased shipping difficulties on the Tennessee River. The construction of the Wheeler and Wilson dams near Muscle Shoals and Florence, Alabama, improved the navigability of the river so greatly that the tonnage of river trade increased from 32 million ton-miles in 1933 to 161 million ton-miles in 1942. This ease of shipping meant that local farmers had greater access to markets and also made the region more attractive to industrial interests.

The TVA was also charged with helping the Valley's farmers become more productive and cause less damage to already fragile farmlands. Outdated farm practices, over fertilization of land with chemicals, deforestation, and the need to produce more and more crops in order to meet rising costs took its toll on much of the region's arable land. Erosion, flooding, and poor harvests were endemic problems. The TVA established model farms to teach Valley farmers about crop rotation, responsible fertilization, and counter-erosion techniques. The organization worked closely with local land grant colleges in an effort to further educate farmers and aid them in becoming more productive and to raise their standard of living.

The main component of the social and economic plan for the region was the generation and distribution of electricity. Working through direct commissions and private distributors, the TVA quickly began the process of electrification in rural areas and provided a greater quantity of power, at a lower cost, to the Valley's larger cities. Tupelo, Mississippi, became the first city to purchase its power wholesale from the Tennessee Valley Authority.

Unlike the majority of residents in the Tennessee River Valley, most private power companies in the region were critical of the government's intervention. Utility companies saw the TVA as a threat to competitive business and regarded the government as not being capable of adequately generating, selling, and distributing electricity. The TVA was challenged in court several times during the 1930s; almost all of the cases were brought forth by private power companies. In 1939 the constitutionality of the TVA Act was upheld in the Supreme Court. By 1941, just eight years after its inception, TVA had become the largest producer of electrical power in the United States.

Striving to implement some of the ideas of the TVA program across the nation, the Roosevelt administration created the Rural Electrification Administration (REA) in 1935. The main goal of the REA was to provide electricity to all rural areas, most especially family farms. Private power companies in many regions did give farmers the opportunity to purchase electricity, but in order to run rural lines, most demanded all or a majority of the construction costs up front. Furthermore, many companies proposed to charge higher rates for the power they supplied the countryside. These factors made electricity a prohibitive cost to many farmers and rural citizens.

The head of the REA, John Carmody (1881-1963), visited areas adjacent to the TVA region in order to assess the people's reaction to the projects that the government was undertaking nearby. Unlike that of the private utility companies, the response from citizens was overwhelmingly positive. In areas that had been denied electricity based upon claims by power companies that rural lines were too expensive to construct and that the residents could not afford to pay for power, the residents were generally in favor of government intervention in the region. During his visit, Carmody spotted a sign in north Georgia that read: "Mr. Carmody, we want lights!" The plea of the region's residents to be included in New Deal programs was heard, and Georgia became one of the pilot regions for REA sponsored rural electric cooperatives. Five years later, under REA guidance, Georgia was one of the most widely electrified states.

By 1939 the percent of rural households nationwide that had electricity had risen from just under 10% to 25%. The REA aided in the establishment of over 400 electric cooperatives that served 288,000 individual households. While the push for rural electrification was largely completed during the New Deal and immediately after the end of World War II, the Tennessee Valley Authority remained active. In the postwar era, TVA shifted from its New Deal broad economic and social mission to a more streamlined organization focusing on the production and sale of power and the maintenance of its dams and reservoirs along the Tennessee River and its tributaries.

In 1959 the TVA petitioned for the authority to issue bonds, and later that year Congress enacted legislation that made the TVA power system self-financing. Over the course of the next three decades, the TVA continued to develop its innovative approach to regional planning. In the 1960s TVA began using nuclear power in some areas; in the 1970s the organization was one of the first to implement pollutant emission standards. Though TVA has followed a general trend toward becoming more similar to other power companies, in the last decade of the twentieth century TVA began to reincorporate updated versions of their New Deal initiatives.

The TVA is one of the chief moderators in the ongoing social dialogue between various business, cultural resource management, environmental, and industrial groups in the region.

Impact

Rural electrification brought many changes beyond electric lighting to farmers and residents of small towns. As power cooperatives were established, the REA instituted programs to help members purchase modern appliances like stoves, ovens, and refrigerators. Such appliances were not only meant to make life more convenient and comfortable for rural residents, but also to reduce waste and help the farms to become more productive. For example, REA studies conducted in 1936 estimated that American farmers who did not have electricity or a means to refrigerate food wasted one-third of all of their meat stuffs. Starting programs that helped collectives and individual farmers purchase walk-in coolers was supposed to alleviate not only problems of spoilage, but also improve the daily diet of farm families.

Similar initiatives provided low-interest financing for tractors, combines, and other farm equipment. Such programs were intended to make small family farms more productive and able to compete with larger farms. Though the productivity of small farms did increase as a result of rural electrification efforts, the same programs and technology were available to the large farms. While the REA did help to improve the quality of life and the income of many American farmers, it failed to offset the general trend of families leaving rural areas to move to the city. The number of small family farms continued to decline through out the New Deal era.

Despite the progressive nature of most TVA and REA projects, some citizens were not included in the government's plans for improvement in rural areas. During the course of the 1930s, the National Association for the Advancement of Colored People (NAACP) investigated the TVA on three separate occasions for racial discrimination in the housing, hiring, and training of blacks. TVA planned communities excluded black families and African-American land grant colleges were not permitted to take part in the agricultural training or extension programs. In an attempt to justify policies of racial segregation, the organization stated that such exclusions reflected the traditional social structure of the older Valley communities. However, black leaders noted that the region's white and black sharecroppers and farmers had worked together and lived as neighbors for years.

From its inception, the TVA was concerned not only with the impact of its projects upon man and the environment, but also upon relics of the past as well. With the assistance of workers from the WPA, large archaeological surveys and excavations were carried out on all land that was marked for TVA development. Some of the most famous archaeological discoveries in the southeastern United States were made in conjunction with TVA projects. Working closely with the WPA, and later the Civilian Conservation Corps (CCC), TVA archaeological projects included the identification, cataloging, and excavation of sites as well as the establishment of local museums. In more recent decades, the TVA has helped to pioneer cultural resource management legislation that is more sensitive to the interests of Native American groups as well as scientific interests. TVA continues to be one of the most prominent sponsors of archaeological research in all seven states that it spans.

ADRIENNE WILMOTH LERNER

Further Reading

Badger, Anthony J. *The New Deal: The Depression Years, 1933-1949.* New York: Farrar, Straus & Giroux, 1991.

Creese, Walter L. *TVA's Public Planning: The Vision, the Reality.* Knoxville: University of Tennessee Press, 1990.

Grant, Nancy L. *TVA and Black Americans: Planning for the Status Quo.* Philadelphia: Temple University Press, 1990.

Household Work Is Transformed by Technological Developments

Overview

Household work was transformed by new technologies during the first half of the twentieth century. Some machines were intended to automate work that servants used to do, while others took advantage of new sources of energy such as electricity and gas to make homes cleaner, safer, and more efficient. Technological innovations such as refrigerators, washing machines, and ranges were first available only to the wealthy, but after the end of World War II household appliances became widely affordable. Household appliances have not eliminated time or effort spent on domestic chores, but they have helped dramatically raise the general standard of household comfort and convenience.

Background

While many of the activities that took place in homes at the turn of the twentieth century would still be familiar one hundred years later, how those tasks were performed changed dramatically. Before 1900, nearly all household chores were performed manually. Affluent homes employed servants to assist in cooking, cleaning, and other tasks; other chores, such as laundry, were sometimes performed outside the home. But the first decades of the twentieth century saw a great decline in the number of household servants, and women in the upper classes were faced with the challenge of performing laborious household chores themselves. This created a marketplace for new automated appliances for the kitchen and throughout the house. Poorer women, on the other hand, had always been responsible for their own housework, but, as electricity became available and prices for appliances fell, they too were able to welcome gadgets large and small into their homes.

One of the most important home technologies of the early part of the twentieth century had nearly disappeared by its end: the sewing machine. Early sewing machines were powered by pedals or trestles that the operator pumped with her feet. This was tiring work, and made machine sewing often seem a greater chore than simply sewing by hand. Singer, the world's largest sewing machine manufacturer, introduced a fully electric model in 1921, and the popularity of sewing machines for the home soared. These machines made basic sewing chores much easier, and also performed elaborate specialized stitching previously beyond the reach of domestic sewing. The popularity of sewing machines began to decline after World War II, however, as inexpensive clothes and household linens became widely available for purchase and made home sewing unnecessary.

The chores of cleaning homes and clothes were changed perhaps more than any other by the use of electricity. Prior to the invention of electric-powered cleaners, homes were swept by hand or with the aid of mechanical carpet sweepers. By all accounts, these manual techniques left a great deal of dust and dirt behind! The first electrical carpet cleaners were not portable machines for the home, but large commercial machines that were brought to the home for periodic cleaning sessions. These were followed by permanently installed centralized vacuuming systems, where light hoses could be attached to suction outlets on the wall in order to easily vacuum any room. These centralized systems were popular in apartment buildings, where the large, expensive units were most practical. The earliest version of the modern style portable vacuum cleaner appeared in 1908, manufactured by the W. H. Hoover Company and soon to be known simply as "Hoovers." Hoover and his competitors continued to improve the vacuum cleaner, though the basic design—a small motor, suction, a dust bag, and perhaps a rotating brush—remained the same for decades.

Machines to help do laundry were especially important to households no longer able to rely on domestic help. Many women were reluctant to make use of commercial laundries because of their persistent bad reputation for abusing clothes and linens, so even manually operated laundry aids were desirable. Early machines consisted of a cylindrical tank that filled with water, a gas heating element, and a hand crank to turn the cylinder. Clothes could be washed, rinsed, and (slowly) dried in these machines. With the addition of electric motors to washing machines, their popularity skyrocketed; by the late 1920s, millions were sold annually in the United States. Washing machines became fully automatic in the 1930s with the introduction of electro-mechani-

A woman stocks an electric household refrigerator. *(Hulton-Deutsch Collection/Corbis. Reproduced with permission.)*

cal controls and the ability to "spin-dry" clothes. Another laundry aid that became electrified in this era was the iron. Electric irons for the home were first introduced in 1904.

Kitchens were completely transformed in the first decades of the century with the introduction of gas power. Stoves powered by coal, wood, or other solid fuels had to be lit well in advance of use. The heat could not be easily controlled, and they were often smoky and messy—homes were routinely covered with a layer of grime produced by stove exhaust. Gas stoves came into use rather slowly beginning in the 1890s, but were quite popular by 1914. Manufacturers made their gas stoves more and more elaborate while at the same time improving their safety and convenience. Gas stoves continued to outpace their electric competitors in popularity for some time, although the opposite was true for refrigerators. There, electrical machines were introduced early (they were widely available in the United States by the 1920s, replacing older non-electric "iceboxes") and took over the market. Gas powered refrigerators, although quiet, safe, and efficient, failed ever to become commercially successful.

While large appliances were important to reducing manual household work, numerous small electrical gadgets were introduced at the same time. Some, like toasters and fans, came to enjoy perpetual popularity. Others were curious hybrids, like an iron that also made tea, that failed. Still others, like machines designed to mix or mash food, went through decades of evolution and modest success without ever becoming indispensable kitchen aids.

Of course, not all technologies in the home were intended to assist in household work. Some were for personal comfort, others for entertainment. Bathrooms saw significant technological improvements in this era. Effective flush toilets with elevated cisterns were perfected late in the nineteenth century, and were gradually improved, becoming more compact and decorative. Mass-produced enameled bathroom fixtures—toilet, bathtub, and sink—began to be produced around 1916, and this helped bring fully appointed bathrooms into even modest homes. Entertainment in the home was transformed by a series of electrical devices, including the phonograph, the radio, and the television, as well as more specialized instruments such as the very popular electric organ.

Impact

These new technologies changed life and work at home, but not always in the manner they promised. Many, if not most, new devices for the home were marketed to prospective consumers with the promise of saving effort and of making housework more efficient. While some manual labor was saved (for example, it was physically easier to wash laundry in an electric machine than in one with a hand crank), little time was saved by the new inventions. Household technology did not give women the freedom to do other things besides housework; in fact, by providing machines to do work such as laundry in the home rather than outside of it, many sociological opportunities to reduce housework were lost. The net effect of fewer servants, fewer services like home deliveries, and more household machines left middle and upper class women doing more housework in the middle of the twentieth century than they had been doing at its start.

The affluent years after World War II saw a flood of new home construction and new household technologies as well. By this time, nearly all homes in the United States had running water and electricity service. Improved and less expensive versions of washing machines, refrigerators, electric stoves, and small appliances proliferated, and standards of cleanliness and comfort that had been unattainable for all but the wealthy came within the reach of almost everyone. Brand new appliances appeared, such as air conditioners, tumble dryers, freezers, and dishwashers, and became widely available in the 1950s. New home entertainment technologies such as television sets, "high-fidelity" stereo systems, and portable transistor radios also attracted postwar consumers.

While the home technologies of the years 1900-1949 were mostly designed to save labor, in the aftermath of the energy crisis of the 1970s engineers worked to design machines that would save energy. High divorce rates and other demographic changes in the 1980s found most women working outside the home. This created a powerful demand for technologies that saved not only labor but—finally—time. Microwave ovens, wrinkle-free fabrics, frost-free refrigerators, and self-cleaning ovens all helped save significant time performing daily household tasks. For home entertainment, video tape recorders gave people more control over their recreational television viewing time, by letting them tape programs or watch films at their own convenience, and personal cassette players (such as the Sony "Walkman") made music listening more flexible as well.

In general, household technologies have made homes safer, cleaner and more comfortable; they have also made the preservation and preparation of food more convenient. But as new technologies have been developed to perform traditional tasks in the home, they have actually changed how we live. Freezers have encouraged the consumption of prepared food, air conditioners have made awnings and front porches uncommon, vacuum cleaners helped make wall-to-wall carpeting popular. While household technologies have greatly improved the general quality of life in homes of all kinds, there have also been unintended consequences. Lifestyles have evolved alongside technologies, each influencing the other, as we have become increasingly dependent on machines to perform even the most basic functions in our homes.

LOREN BUTLER FEFFER

Further Reading

Cowan, Ruth Schwartz. *More Work for Mother.* New York: Basic Books, 1983.

Du Vall, Nell. *Domestic Technology: A Chronology of Developments.* Boston: G. K. Hall, 1988.

Hardyment, Christina. *From Mangle to Microwave: The Mechanization of Household Work.* Cambridge: Polity Press, 1988.

American Physicists William B. Shockley, Walter H. Brattain, and John Bardeen Produce the First Transistor, Initiating the Semiconductor Revolution

Overview

In 1947 Bell Laboratories scientists invented the transistor—a semiconductor device that could amplify electrical signals transmitted through it. Those in the "know" recognized the significance of the transistor as a compact, reliable replacement for the inefficient vacuum tube; but the development of what many now consider the twentieth century's most important invention, was not prominently reported. Not even the team responsible for the transistor, John Bardeen (1908-1987), Walter Houser Brattain (1902-1987), and William Bradford Shockley (1910-1989), were aware of the singular role their discovery was about to play in initiating the information age and making possible everything from miniature hearing aids to high-speed computers.

Background

In 1904 John A. Fleming (1849-1945) developed a vacuum tube diode—known as a "valve" because it forced current within the tube to flow in one direction. This was essential for converting alternating currents to direct current. In 1907 Lee De Forest (1873-1961) patented the Audion vacuum tube, which functioned as a valve as well as amplifying current. De Forest achieved amplification by inserting a metal grid into the tube. Varying grid input current allowed him to control the flow of a secondary current in the tube such that weak grid inputs resulted in strong secondary currents. De Forest conceived of cascading Audions to provide the necessary amplification for long-distance broadcasts and signal reception.

About this time Alexander Graham Bell's (1847-1922) telephone patents began expiring. AT&T sought an advantage over the competition. Their solution was to develop a transcontinental telephone service. They bought De Forest's Audion patent hoping it could be adapted for amplifying signals along telephone lines. AT&T's design improvements yielded an acceptable device that was quickly deployed and thus formed the basis not only for early long-distance telephony but also radio, radar, and computers.

AT&T's improved Audion was far from ideal, however. It proved extremely unreliable, consumed too much power, and generated a great deal of heat. By the 1930s Mervin Kelly, then Bell Labs' research director, believed a better solid-state device could be produced. He pushed research on semiconductors, which he thought might provide the answer.

Semiconductors possess the unusual property that their conductivity varies. They are composed mostly of atoms that do not conduct electricity. However, they do possess small numbers of conducting atoms. Depending on how they are handled, semiconductors can be made to conduct more or less electricity. It was hoped that this property could be exploited to produce an alternative to vacuum tubes.

After World War II Kelley assembled a team of Bell Lab scientists to develop a solid-state semiconductor switch to replace vacuum tubes. Shockley was the team leader. The team also included the experimentalist Walter Brattain and theorist John Bardeen.

In 1945 Shockley designed a semiconductor amplifier based on the field effect. The underlying principle here is that an electric field applied through a semiconductor surface should alter the charge density within and thus change the semiconductor conductivity. Shockley's field-effect mechanism, however, failed to amplify electric currents. He assigned Bardeen and Brattain the task of determining why.

Bardeen and Brattain developed a close working relationship. Bardeen suggested experiments and interpreted results while Brattain built the equipment and ran experiments. Their work went largely unsupervised, but by 1947 they were becoming increasingly frustrated by their inability to produce positive results. It was then that Bardeen had his historic insight that surface charges were hindering electric field penetration of semiconductor material.

Without notifying Shockley, they produced a new device having closely spaced contacts lightly touching the semiconductor surface. According to Bardeen's calculations, current coming through one contact should produce a stronger current at the other contact. On December 16, 1947, they successfully tested their device—the first point-contact transistor (from transfer resistor).

Chagrined at having had no direct role in the crucial breakthrough, Shockley conceived the junction transistor shortly thereafter. Special semiconductor crystals were necessary to make this device work. These were produced by Gordon Teal, and the first junction transistor was successfully tested on April 12, 1950. Shockley's device was more durable and practical than the point-contact transistor. It also proved much easier to manufacture and became the most commonly used transistor until the late 1960s.

Impact

On June 30, 1948, Bell Labs held a press conference to announce the invention, but the story gained only limited coverage and generated little excitement. This initial indifference quickly dissipated as appreciation of transistor applications grew. However, even after Bardeen, Brattain, and Shockley were jointly awarded the 1956 Noble Prize in physics for their work, no one fully grasped how profound and pervasive the transistor's influence would be.

Transistors were initially used as telephone amplifiers, in operator headsets, and for hearing aids. Widespread commercial use in the early 1950s was limited, though, because transistors were too expensive, unreliable, and difficult to manufacture. Considerable effort was devoted to developing more functional and efficient transistors that could be cheaply and reliably mass produced, but most U.S. companies concentrated on defense applications for which miniaturization was desirable and high costs were not prohibitive.

As transistors became faster, more reliable, and cheaper, the commercial market began to look more attractive. Texas Instruments decided to probe this market by producing a hand-held radio. Released in October 1954, the 5-inch (12.7-cm) Regency was the first transistorized radio and popular culture's first highly visible exposure to the new technology. When the Regency was discontinued the next year, Sony Electronics was poised to capture the market. Their mass-produced miniature, transistorized radios made it possible for information to be quickly disseminated across the globe. Sony also developed the first transistorized TV and went

William Shockley. *(Library of Congress. Reproduced with permission.)*

onto manufacture VCRs, stereo equipment, and computer games.

In 1955 Shockley left Bell Labs to found Shockley Transistor Corporation in Palo Alto, California. This was the first semiconductor company established in what is now called Silicon Valley. The research team Shockley assembled consisted of some of the brightest people working in the field, but within a year Shockley's domineering managerial style drove eight of his best employees away. Known as the "traitorous eight," they went on to found Fairchild Semiconductor (1957).

In 1959 Fairchild Semiconductor and Texas Instruments independently developed integrated circuits—single "chips" of semiconducting material that combine the functions of transistors and other electronic components. Instead of making separate bulky components and inefficiently wiring them together, it was now possible to produce complete miniaturized circuits from single crystals. Texas Instruments exploited these chips to produce the first hand-held calculator in 1967, and Intel was founded in 1968 to produce computer memory chips. Today centimeter-by-centimeter chips contain millions of transistors.

Integrated circuits were a significant advance but limited precisely because they were wired to perform specific functions. This was overcome by Intel's microprocessor.

In 1969 the Japanese calculator firm Busicom contracted Intel to produce chips for a new calculator. Intel assigned Ted Hoff (1937-) to work on the contract. After reviewing Busicom's 12-chip calculator design, Hoff proposed an alternate single-chip architecture that combined the separate functions into a 2300-transistor CPU. The resulting 4004-chip was announced to the industry in November 1971.

The 4004 was more than a complicated integrated circuit; it was programmable. Calculator chips perform specific functions that are hard-wired in during the manufacturing stage. Such chips can perform these and only these functions. The functions wired into the 4004 were different. They could be combined and controlled in various ways by user-implemented programs, thus making the microprocessor more versatile.

Microprocessors were quickly incorporated into computers. Measuring only ⅛ inch by ¹⁄₁₆ inch (0.3 x 0.16 cm), the 4004 was as powerful as the 30-ton ENIAC computer built in 1946. Such a reduction of size made possible super computers and initiated the PC revolution. Today Intel is the world's leading manufacturer of computer chips, with 1999 revenues over $21 billion.

Often referred to as the "invisible technology," transistors are at the core of every major information age innovation. Computer networks and the internet as well as satellite communication would all be impossible without transistors. Miniaturization has also made possible pacemakers, cellular phones, remote controls, fuel injection, compact disc players, video games, smoke detectors, photocopiers, global positioning, and literally thousands more devices and applications.

As miniaturization continues further, applications will be found, but miniaturization cannot continue indefinitely. Fundamental physical principles—the size of the atom and electron—place ultimate constraints on the size of transistors. Nevertheless, much progress is to be expected, and it is likely that by 2010 transistors will be twice as small as they are now, opening up new applications possibilities.

STEPHEN D. NORTON

Further Reading

Books

Eckert, Michael, and Helmut Schubert. *Crystals, Electrons, Transistors: From Scholar's Study to Industrial Research*. New York: American Institute of Physics, 1990.

Hoddeson, Lillian, E. Braun, J. Teichmann, and S. Weart. *Out of the Crystal Maze: Chapters from the History of Solid State Physics*. New York: Oxford University Press, 1992.

Queisser, H. *The Conquest of the Microchip: Science and Business in the Silicon Age*. Cambridge: Harvard University Press, 1988.

Riordan, Michael, and Lillian Hoddeson. *Crystal Fire: The Birth of the Information Age*. New York: W. W. Norton & Company, 1997.

Seitz, Frederick, and Norman G. Einspruch. *Electronic Genie: The Tangled History of Silicon*. Urbana, IL: University of Illinois Press, 1998.

Periodical Articles

Bardeen, John, and Walter H. Brattain. "The Transistor: A Semiconductor Triode." *Physical Review* 74 (1948): 230-231.

Bardeen, John, and Walter H. Brattain. "The Physical Principles Involved in Transistor Action." *Physical Review* 13 (1949): 1208-1225.

Brattain, Walter H., and John Bardeen. "Nature of the Forward Current in Germanium Point Contacts." *Physical Review* 74 (1948): 231-232.

Kilby, J. "Invention of the Integrated Circuit." *IEEE Transactions on Electron Devices* ED-23 (1976): 648-654.

Shockley, William. "The Theory of P-N Junctions in Semiconductors and P-N Junction Transistors." *Bell System Technical Journal* 28 (1949): 435-489.

The Technology of War

Overview

World War I and World War II can be respectively described as the first mechanized, and first electronic, wars. Though warfare throughout the centuries had seen the development and improvement of weapons and other technologies of warfare, truly mechanized warfare, where advanced mechanical technology began to play a major role in how wars were fought, was only possible after the Industrial Revolution and the establishment of an advanced industrial base that included advanced refining, machining, and assembly-line techniques. In the First World War, such techniques were still in their infancy, while in the Second World War they reached their peak and were augmented by the new world of electronic warfare. For this reason, while the Second World War is often seen as a continuation of the First World War in political terms, it can also be seen as a continuation and refinement of First World War technologies, and their marriage to increasingly sophisticated electronic information systems.

Background

When war broke out in Europe in 1914, military strategy and tactics were still based in the nineteenth century, and the very basic idea that the overwhelming and breaking of an enemy line could be accomplished through the sheer onslaught of bodies. What changed this was the use of the Maxim machine gun by the German military, then the adoption of other machine gun types by the other combatants. The machine gun, in combination with the use of barbed wire as a defensive line, meant that those troops who survived the charge across no-man's land would then have to face becoming entangled in the wire, where they were easy targets for machine gunners and riflemen in the defending trenches. Artillery could be used in an attempt to weaken the enemies' defenses, and open holes in the wire, but this proved to be largely ineffective. What was needed was a vehicle that could afford protection to the men inside as they crossed no-man's land, and which could then crush its way through the barbed wire and cross the enemy's trenches to reach the artillery and supplies in the rear area. In other words, what was needed was a tank.

The tank's earliest ancestor might be the war chariot, an armored vehicle carrying offensive weapons and capable of crossing the territory between the attackers and the defending line and breaking through. The problem faced by all such vehicles, however, is terrain; on soft, rugged, or uneven terrain, wheeled vehicles quickly bog down or break. In the First World War, this problem was compounded by the legendary mud of no-man's land, which had been churned and softened by years of small arms and artillery fire. What was needed to cross such terrain was a vehicle that could lay down its own track, a path-laying vehicle. At the same time, the vehicle needed to be self-propelled; if it needed to be drawn forward by horses, for example, then it would be a simple question of killing the horses to halt the vehicle's advance.

Throughout the eighteenth century, with the development of the steam engine and vehicles that used various systems to lay their own "railroad track" or similar paths, the basic construction of the tank came ever closer to being realized. From about 1900 until the outbreak of the war, various experiments were carried out with so-called "endless chain" track-laying vehicles, which featured a track-and-roller system similar to what can be seen on bulldozers and other tracked vehicles today. These vehicles also benefited from the development of the internal combustion engine. In the late nineteenth/early twentieth century in America, the Holt company developed the Holt tractor, was the design that would be studied and utilized for British First World War tanks.

The development of an armored fighting vehicle was first proposed in 1914 by Lt. Col. E.D. Swinton and Capt. T.G. Tulloch of the British army, and was supported by Winston Churchill, then first lord of the admiralty, but it wasn't until early 1916 that a vehicle had been developed that was ready for tactical deployment. In September of 1916, during the battle of the Somme, 49 of the new British tanks went into battle for the first time. These 49 tanks were of two types, one designated "male," the other "female"; while both models were nearly identical in weight (31 and 30 tons, respectively), and size (26 feet [8 m] long, 13 feet [4 m] wide, and 8 feet [2.4 m] high), the male was armed with a 6-pounder naval gun and four machine guns, while the female was armed with six machine guns only. Of these 49 tanks, however, the majority were afflicted with mechanical or other problems, so that only 9 actually fulfilled their combat missions. Despite this disappointing first "trial by combat," the usefulness of the tank became apparent to all combatants engaged in the conflict, and by the end of the war the British, French, and German forces had all developed and deployed tanks in combat, with only the United States lagging behind and having to use French tanks when American forces entered the conflict in 1917.

The actual impact of the tank on warfare in WWI is not as significant as it was in WWII, due to both the mechanical and operational unreliability of the vehicles in combat, and a lack of real tactical theory to support their combat use. The development of the tank, however, intro-

American B-17 bombers in action over Europe during World War II. *(Bettmann/Corbis. Reproduced with permission.)*

duced a new age of European warfare, in which the onslaught of human bodies was replaced by the onslaught of armored, heavily armed, mechanized vehicles with the capacity to make deep drives into enemy territory. In WWII, the *blitzkrieg* or "lightning war" tactics of the German military would rely heavily upon the use of armored vehicles, and would represent the most advanced tactical thinking in the deployment of this new land warfare technology.

If the introduction of the machine gun and the tank began to change the face of land warfare in WW I, the introduction of the airplane during the conflict opened the theater of air combat. Airships and balloons had been used in late-nineteenth-century combat to fulfill reconnaissance roles, and when war broke out in 1914, the airplane, still in an early form of development, was used in the same way. Aircraft were small and unarmed, with a pilot and an observer, and usually flew on solo missions to observe enemy positions and maneuvers. If an enemy aircraft was encountered on the mission, combat between the two usually consisted of the two aircraft closing to a short range, then the pilot and the observer blazing away with whatever weapons they had on them at the enemy aircraft.

By 1915, the combatants on both sides began to arm their aircraft with rear-firing machine guns. This meant, however, that the weapons could really only be used for defensive purposes, in the case that an observation aircraft found itself under attack from an enemy. In the autumn of 1915, the German military introduced the Fokker E.I., a fighter plane with a forward-firing machine gun. A synchronization device allowed the machine gun to fire only when the propeller blade was not opposite the muzzle of the gun. This gave the Fokker an offensive, attack capability, and all combatants began to realize the need for specialized types of planes such as observation aircraft, fighters and bombers, and suitable tactics for the use of each type.

This led, in 1916, to the concept of formation flying, and what would later be called "the flying circus." Instead of lone aircraft flying out on missions, a formation of aircraft, flying together and cooperating with one another, assured mutual protection in combat. This meant that fighters would be put in formation with observation or bombing aircraft to provide those planes with protection. It was also realized that fighter aircraft could be used in an offensive capacity, and formations were sent out with the direct intent of engaging enemy air and land units in support of other land and air operations.

The principle developments in aircraft technology involved the design of the pilot and observer seating and armament—with the final designs incorporating a pilot who sat in the front of the plane with a forward-firing machine gun, and an observer who sat in the rear and utilized a rear-firing machine gun mounted on a rail that allowed for an arc of fire—and improvements in overall aircraft design that led to increased airspeed, cruising altitude, and diving capabilities. Unlike tank tactics, air tactics evolved rapidly, with the principal notion of formation flying carrying over into Second World War air tactics. The principal impact of the airplane on WW I combat was its use as an offensive weapon that could engage the enemy deep in its own territory. While bombers were used to a limited extent, due to problems with designing an aircraft that could carry a significant bombing load, fighters enjoyed an extensive combat role that ushered in the era of air warfare.

Finally, the First World War saw the development of an entirely new system of warfare at sea. While naval warfare for centuries was essentially an affair that took place between large, heavily armed, massed naval formations, WW I saw the introduction of a lone hunter that could not only inflict serious damage on enemy shipping, but could also function as a weapon of terror: the U-boat.

Like the tank and the airplane, the concept of an underwater vessel is centuries old, but only with the development of the electric engine, the internal combustion engine, and the screw propeller did the submarine actually become a feasible naval vessel. The internal combustion engine and the screw propeller gave all types of naval vessels improved speed and propulsion, but it was the electric engine that made it possible for submarines to run submerged without generating fumes that would eventually have poisoned the crew.

Though the first true submarines were developed and deployed in the late nineteenth century, their role in overall naval warfare was extremely limited; as they had no real ocean-going capacity, they were essentially limited to patrolling coastal waters and lacked the capacity to truly engage other naval vessels. During the First World War, however, the German military adopted a policy of "unrestricted submarine warfare," in which they actively sought out and engaged enemy merchant ships. This was generally considered to be in violation of international law, since it involved the torpedoing of what were normally considered to be unarmed civilian ships without any warning or attempt to provide capture and rescue operations. That passenger ships were also considered to be enemy ships led to a considerable outcry against the German policy, and with the sinking of the passenger ship *Lusitania* on May 7, 1915, resulting in the deaths of 1,198 men, women, and children, international outrage reached such a pitch that the German government was forced to suspend its operations against passenger ships.

The damage inflicted by the German U-boats, however, was so extensive that it threatened to carry the war in favor of the Germans. By the end of the war, the Germans had sunk 5,408 allied merchant ships, and had lost only 203 submarines. The German threat was countered only by the introduction of the convoy system, which gave armed naval escort to merchant ships. The introduction of American naval forces in 1917, provided a considerable naval fleet at the hands of the Allies for this very purpose.

The introduction of the submarine in naval combat demonstrated that a small, underwater vessel that could be cheaply manufactured could not only impose considerable damage on enemy shipping, it could also be used as a weapon of psychological warfare to terrorize civilian populations. The advent of the submarine, then, may not be as significant for its tactical and strategic value as for its significance in opening the theater of warfare to include civilian populations as a target, a strategy that would be extensively employed by both sides during World War Two.

If World War I can be described as the first mechanized war, then World War II might be thought of as the first electronic war. While improved manufacturing processes and a stronger industrial base allowed for the machines of World War I to be manufactured in greater numbers, and with greater degrees of structural integrity and strength, the major change in the quality of warfare in the Second World War was brought about largely by the development of electronic technology such as radio, radar, and sonar, which in turn meant that the primary ammunition of the war was no longer bullets, but information.

During the First World War, the coordination of units was still largely dependent upon communication through hand signals, voice, carrier pigeon, signaling apparatus, and, occasionally, telegraph. However, no real means existed for tanks or airplanes in a squadron to communicate with each other in tactical situations, and once an attack was launched, it was extremely difficult for commanders in the rear to change strategy based on prevailing battle conditions. The ability of forward units to communicate information to the rear was extremely limited, and thus the most aggressive attack could be blunted simply because there was no way to relay orders or information from the front to the rear or vice-versa.

The adoption of wireless radio communications by the world's major military forces, and the development of small, mobile wireless sets, enabled two major changes in the conduct of military operations. First, it allowed for greater communication and coordination between individual tactical units (squad, individual tanks, or planes), which in turn allowed for a more rapid ability for the units themselves to adapt to changing battle conditions—command and control at the tactical level, in other words, became more decentralized. At the same time, however, forward units could more quickly communicate intelligence to the rear, providing commanders with up-to-the-minute reports on changing battle conditions. Radio, in other words, made possible the type of coordinated shock attack that the German military would perfect in the *blitzkrieg* attack, as well as the communication of important strategic information from forward units. World War One may have seen the development of the tank and the airplane as devices, but it was the radio that enabled these devices to reach their full tactical and strategic potential.

On the tactical level, radio provided a greater degree of coordination between individual units, and on the strategic level, it can be said to have given commanders a larger degree of remote control. A commander could unleash his units in a combat situation and expect that they would function in a coordinated tactical way among themselves, but would maintain communication that would allow for large-scale strategic decisions to be made in response to the fortunes of battle. Thus, for example, squadrons of hundreds of fighters and bombers could be dispatched on a bombing run, and while radio would allow the individual planes to communicate with one another during combat situations, radio would also allow air commanders in the rear to re-direct planes to targets as new intelligence information was gathered. Such technology was also critically

LAUNCHING AIRCRAFT FROM SHIPS: THE FIRST EXPERIMENTS

Aircraft carriers have been in combat use for well over a half century, and the technology that makes them such effective weapons seems fairly well perfected. The technology and techniques that now seem almost routine, however, were developed laboriously over time, and the first attempts to launch aircraft from ships were awkward or amusing at best and disastrous at worst.

On October 7, 1903, a few months before the Wright brothers' first flight, Samuel Langley's aerodrome was launched by catapult from a ramp atop a houseboat in the Potomac River. According to the *Washington Post*, "It simply slid into the water like a handful of mortar...." Langley's second attempt, two months later, ended as poorly, as did all subsequent attempts to marry ship and aircraft for the next seven years.

On November 14, 1910, Eugene Ely took off from a wooden ramp on the USS *Birmingham* in a Curtis biplane, beginning the era of naval aviation, with a few caveats. The pilot did not land on the ship, he and his plane were civilian, and the *Birmingham* was stationary. But an airplane had successfully made it into the air from the deck of a ship. Two months later, Ely successfully landed on a wooden ramp built onto the USS *Pennsylvania*, marking the first successful landing on a ship. This landing is noteworthy for another reason—to stop the brakeless aircraft, Ely attached hooks to the bottom of the plane and strung 22 ropes, each tied to two 50-pound sandbags—the forerunner to today's arresting gear on modern ships.

P. ANDREW KARAM

important in communicating with naval units, and it was in the ciphers and decoding of German radio transmissions to U-boats that the true Atlantic naval war was fought.

Because radio allowed for remote control of large-scale operations, and because massive troop movements were carried out through radio signals, the interception of radio signals and decoding them became a major concern, as was the development of ciphers and codes. In many ways, information became the real medium of warfare in World War Two, as can be seen in the development of ever-more complex electronic systems, such as radar and sonar, for gathering information, and the attempt to develop other electronic systems, such as the German Enigma coding machine, to frustrate the gathering of electronic information. World War Two was a war that was fought in the conventional way with bullets and grenades, but it was also the first war which was fought with electromagnetic waves.

Because of the way it allowed for large-scale coordination and rapid communication, radio changed the scale and speed of warfare. It also changed the perception of war, as radio became the chief propaganda and psychological warfare weapon. Edward R. Murrow's live broadcasts from London during the Blitz, as well as the use of radio by the German propaganda ministry, are examples of the ways in which radio was used to solidify and win public opinion, and to manipulate the public's perception of the war. Both the Allies and Axis powers used radio broadcasts, such as those of Tokyo Rose, to undermine the morale of enemy forces.

Finally, World War Two also saw the deployment of weapons of mass destruction against civilian populations on a scale that had never before been realized. German rocket technology, coupled with radio control, enabled the German military to deploy a continuous rain of V-1 and V-2 rocket weapons against the civilian population of London, while the Allies responded with saturation bombing attacks against such cities as Dresden, where one overnight firebombing raid led to 35,000 civilian casualties. The most significant use of such weapons, however, would have to be the use of atomic bombs on the Japanese cities of Nagasaki and Hiroshima, where casualties continue to mount as Japanese exposed to the blast succumb to radiation-induced diseases like leukemia.

Impact

World Wars One and Two might be said to have given us both the technology and the infrastructure to support that technology, that are at the heart of modern warfare. But World War Two also showed that war, in the future, would be fought in a different way, that "total war" against entire enemy populations would now be the order of the day, and that the war of opinion, information, and misinformation would be as important as the war of bullets and grenades.

PHIL GOCHENOUR

Further Reading

Books

Brown, Louis. *A Radar History of World War II: Technical and Military Imperatives.* Institute of Physics, 2000.

Cross, Robin. *Technology of War.* Raintree Steck-Vaughn, 1994.

Keegan, John. *The First World War.* New York: Knopf, 1999.

McFarland, Stephen L., and Richard P. Hallion. *America's Pursuit of Precision Bombing, 1910-1945.* Washington, DC: Smithsonian Institution Press, 1997.

Hydroelectricity and the "Big Dam Era"

Overview

Dams have been used to control river flow since the time of the ancient Egyptians, but it was not until the twentieth century that the full potential of harnessing this natural resource was realized. Hydroelectric power has garnered a great deal of attention since the early 1900s as a clean and renewable source of energy. Traditional fossil fuels are expensive, have a finite supply, and produce a great deal of pollution. The extraction of substances like oil and coal is often damaging to the environment. During the early to mid-1900s hydro projects provided as much as 40% of the electricity consumed in the United States. Today, waterpower supplies about one-tenth of the electricity consumed in the U.S. and about one-quarter of the world's electricity. The leading producers of hydroelectricity are Canada, United States, and Brazil.

Background

A dam is defined as a barrier placed across a river to stop the flow of water. It can be a small earth or rock structure or a huge concrete dam. Controlling or stopping the flow of water, a dam creates a reservoir behind it. The stored water can be made available for irrigation, flood control, municipal or city water supplies, recreational activities, and hydroelectric development. Hydropower can be created by releasing the higher water in the reservoir through penstocks in the dam that can be delivered to turbines, which use the gravitational force of the falling water to power generators, which create electricity. Transmission lines carry the electricity to different locations for a variety of uses.

People have always opted to settle near water sources and have felt the need to control and manipulate water for their own use. The first recorded dam was made by Egyptians about 2800 B.C. and was located on the Nile River. Ancient Romans built dams out of stone and connected water wheels to grinding stones to mill grain in the fourth century B.C. Early in North American history, dams provided power for gristmills with water wheels that operated by water falling onto the wheels much like those used by the Romans. They created small mills and water wheels to grind flour and to operate forges in the 1700s. These mills were essential in rural areas and were later used to make paper, powder, textiles, and lumber. However, early settlers found themselves at the mercy of the hydrocycle. They could not operate their mills during times of flooding or drought and soon found it necessary to further control their water supply. Rudimentary dams were constructed to store water for use during times of drought and for flood protection. These dams were made of locally available materials, such as trees, rocks, and soil.

As rural areas began to enter the market place, larger mills were being used on a larger scale, particularly for lumber and textile factories, and towns began to form around these operations. America began to shift from a rural lifestyle to the industrialized town, but the increasing use of the steam engine caused a decline in the use of waterpower, and some of the villages around the mills began to die. Waterpower was to have its rebirth through the growth of the electric lighting industry. In 1842, the Fourneyron turbine had been introduced to the United States, a highly upgraded version of the water wheel. Later the development of electric generators in the 1870s made it possible to produce a constant and reliable electrical current in commercial quantities. Parallel wiring allowed people to use the lights they wanted, when they wanted, unlike lights that had previously been wired in a series. Incandescent lamps along with an alternate current system began to be used instead of a direct current system. These advancements laid the basis for the giant regional station.

Hoover Dam. *(Roy Morsch/Stock Market. Reproduced with permission.)*

The use of electricity increased in the late 1800s, and the first dam built specifically to create hydroelectric power was built in Appleton, Wisconsin, in 1882. It powered two paper mills and one home. The larger Niagara Falls project in New York was a culmination of European and American technology and went into operation in 1895. The 5000 hp station supplied power to both Ontario and New York, and spurred many ambitious projects after it.

Impact

The beginning of the twentieth century led to bigger stations servicing larger areas. The Reclamation Act of 1902 authorized a series of federal irrigation projects. The focus was on irrigation but some produced small amounts of hydro for use around the dams. Excess amounts were sold, and the revenue was applied to the projects. The first of many hydro projects in the U.S. in the early 1900s was the Keokuk Canal and Hydro Plant on the Mississippi River at Keokuk, Iowa. The canal was constructed in 1877, but the hydro plant was not built until 1913. At the time it was the largest hydro plant in the world. It was 4,649 feet (1,418 m) long, and had a head of 34.5 feet (10.5 m).

During the same period, the American Southwest was being hit with floods and drought, and the Bureau of Reclamation began to work on ways to solve these problems. In 1928, President Herbert Hoover (1874-1964) signed into law the Boulder Canyon Project. Dam construction began on the Colorado River on the Arizona-Nevada border in 1931, and was completed in 1935. The dam's purpose was to improve navigation, regulate the flow of the Colorado River, and to store and deliver water to cities and towns. Hydroelectricity was a secondary function. The dam was a huge engineering marvel. The Boulder Dam, later to be named the Hoover Dam after President Hoover, was 727 feet (222 m) high, and 1,244 feet (379 m) long. The hydro plant capacity was 1.5 million kilowatts. The reservoir created behind the dam, called Lake Mead, after the Reclamation Commissioner Dr. Elwood Mead, remains one of the world's largest artificial lakes at 110 miles (177 km) long and 500 feet (152 m) deep. The Hoover Dam is a National Historic Landmark and a National Historic Civil Engineering Landmark.

In 1933, the national Industrial Recovery Act authorized the construction of dams and other public works as an antidote to the dour economic effects of the Depression. The development of the Colombia River was a part of this plan. Some of the dams constructed include the Grand Coulee, Bonneville, and Foster Creek. In 1933, construction began on the first of these, the Bonneville Dam. That same year construction was authorized on the Grand Coulee Dam, the third largest hydro plant in the world and the largest in the United States.

The "big dam era" of the early 1900s had a profound impact on the twentieth century. The shift to large central electric stations and large hydro projects encouraged urbanization and economic growth in many areas. Populations congregated around the large plants and encouraged industrialization. Initial construction and operation of the plants created jobs, and later businesses and industries were attracted to areas like Tennessee and New York by the lure of excess economical hydroelectricity. These large projects provided much-needed jobs during the Depression of the 1930s and helped revive local economies.

Towns and cities benefited directly from the cheap hydro, as in the case of the TVA (Tennessee Valley Authority) customers who pay approximately half of the average cost of electricity. Farms and ranches in many areas depend on hydropower to run their farms and also use the water stored by the reservoirs for irrigation and flood control, like areas along the Colorado River. Large plants like these also encourage tourism by providing reservoirs for recreation. Lake Mead is a popular vacation and recreation spot, and the Hoover Dam itself is a tourist attraction because of its size and grandeur. Government revenues from hydro plants are applied to other projects, and decrease the amount that has to be taxed.

Hydropower is a flexible, cheap, reliable, and renewable source of energy and could be of great benefit to areas like South America that have low coal resources but an abundance of water resources. The largest hydro plant in the world is located in Brazil on the Parana River. The plant began operation in 1983 and has an installed capacity of 12,600 megawatts. Hydropower does not produce pollution in its creation the way coal does, and in a time when nonrenewables like coal and oil are a concern, a renewable and clean source of energy could become even more important.

Hydropower can still have a huge impact on the environment. Areas behind the dam are flooded and habitat and land are destroyed. Substances, such as mercury, trapped in the soils, accumulate in the reservoir and poison fish stocks. Dams interrupt fish spawning by blocking their movement, and the increased siltation destroys spawning beds. People living in areas behind the dams are displaced by flooding, sometimes losing their homes and whole way of life. Some aboriginal communities in the Canadian north have lost their land and traditional way of life because of flooding caused by hydro projects. Negative impacts like these have made large hydro projects the subject of hot debate.

An effort has been made to lessen the impacts of dams and hydro projects. Some dams have fish ladders, stepped pools that allow fish movement through the dam areas, so they can continue to move up the river. However, these are not always successful. In 1968, the Wild and Scenic Rivers Act was passed to protect rivers in a natural state from hydroelectric projects. Other efforts include the push for smaller developments that serve smaller areas. These projects will not have the same profound impacts on the environment that large projects do, and could make use of many smaller dams in the country that currently do not have hydroelectric operations.

KYLA MASLANIEC

Further Reading

Books

Berlow, Lawrence H. *The Reference Guide to Famous Engineering Landmarks of the World.* Phoenix, Arizona: Oryx Press, 1998.

Dales, John E. *Hydroelectricity and Industrial Development in Quebec 1898-1940.* Cambridge, Massachusetts: Harvard Press, 1957.

Hunter, Louis C. *Water Power, A History of Industrial Power in the United States, 1780-1930, Vol. 1.* Charlottesville, Virginia: University Press of Virginia, 1979.

Vennard, Edwin. *Government in the Power Business.* New York: McGraw-Hill Book Company, 1968.

Other

The History of Hydropower Development, Bureau of Reclamation Power Program. http://www.usbr.gov/power/edu/history.htm

Birth of the "Talkies": The Development of Synchronized Sound for Motion Pictures

Overview

The development and institution of synchronized sound brought about a total revolution in the artistic potential of motion pictures, but the history of sound film's development is also a testament to the forces of economics in the film industry.

Background

Films were never truly silent; by 1900, major theaters provided some form of musical accompaniment to motion pictures, whether through scores written for films and played out on large organs, or through the improvised accompaniment of a pianist or other musicians. Attempts were made to bring recorded sound to the filmgoing experience, but the only available technology were cylinders or discs of recordings, such as were used in the early Edison phonographs. (Thomas Alva Edison's [1847-1931] original phonograph used a tin-foil covered cylinder that was hand-cranked while a needle traced a groove on it.) These had substantial drawbacks in that they could only hold about four minutes of sound, the sound itself was difficult to amplify for a large audience, and synchronization with the action on the screen was almost impossible. Musical accompaniment, then, was limited to special performances in large theaters that could afford to hire live musicians. For these reasons, it was in the economic interest of film producers to find inexpensive ways to bring sound to all films, in the hope that musical accompaniment would increase audience interest in the art form and, subsequently, increase film attendance. While the original intent was to provide synchronized musical accompaniment, it was the potential of synchronized sound systems to play back synchronized speech and dialogue that finally captured the audience's attention.

In 1919 three Germans—Josef Engl, Joseph Masserole, and Hans Vogt—invented the Tri-Egon System, which allowed sound to be recorded directly on film. In this system, a photo-electric cell was used to convert sound waves to electrical impulses, which were then converted to light waves and recorded directly on the strip of film as the soundtrack. A projector equipped with a reader reconverted the light waves to sound for playback, while a special flywheel regulated the speed of the playback. This made it possible to have synchronized sound that ran for the entire length of the film.

In America Dr. Lee De Forest (1873-1961) was at work on a synchronized sound system based on the Audion 3-Electrode Amplifier Tube, developed in 1923 to solve the problem of amplification for playback in a large auditorium. By 1924, 34 theaters in the East had been wired for the De Forest system, and another 50 were planned in the United States, Britain, and Canada. De Forest began producing films under the name of the De Forest Phonofilm Company, releasing short one- or two-reel films that featured scenes from musicals and operas, famous vaudeville acts, speeches from celebrities and politicians, pieces performed by famous musicians, and an occasional narrative film. The emphasis, however, was on reproducing music.

Neither of these systems, however, was adopted by a major Hollywood studio because the studios feared that the conversion to sound film would be an extremely expensive undertaking for what was really a fad. The development of the Vitaphone system and its adoption by the fledgling Warner Brothers Studio would, however, force the studios to reconsider both the expense and necessity of converting to synchronized sound, as would the success of the Fox Movietone news.

The Vitaphone system developed by Western Electric and Bell Telephone Laboratories was a sound-on-disc system that Warner Brothers intended to use to provide musical accompaniment to films. In 1926 Warner Brothers premiered the system with the screening of the film *Don Juan* in New York City. By 1927 Warner Brothers had wired 150 theaters across America for sound, a huge capital investment considering it sometimes cost as much as $25,000 a theater to make the conversion. Other studios, again fearing the cost that the conversion to sound would entail, as well as anticipating the loss of revenue from silent films that had already been produced, banded together to resist the move to sound films, or to create a competing sound system of their own.

This was the avenue pursued by the Fox Film Corporation. In 1927 Fox acquired the rights to the Tri-Egon system in America, and had been developing, since 1926, a sound-on-film system with Theodore W. Case and Earl I. Sponable. On January 21, 1927, Fox premiered its system with a series of performances from a Spanish singer. In May of the same year Fox presented another series of shorts, including a performance by the comedian Chic Sale, but it was the program of June 14, 1927, that captured the audience's imagination. During this program, Fox presented the reception of American aviator Charles Lindbergh (1902-1974) at the White House, and a speech by Italian dictator Benito Mussolini (1883-1945). The audience response to seeing these celebrities and hearing them speak was so enthusiastic that Fox created the Fox Movietone News, and began to screen three or four newsreels, featuring clips of celebrities or special events, at every Fox theater. Convinced that sound was the wave of the future, Fox president William Fox signed a reciprocal contract with Warner Brothers that allowed for the exchange of equipment and technicians, effectively covering both studios if one system became more popular than the other, or if rival studios attempted to develop a competing system. In this way, their huge financial investment in the future of sound films would be protected.

The breakthrough came with the Warner Brothers film *The Jazz Singer* (1927), in which the actor Al Jolson (1886-1950) ad-libbed a few lines of synchronized dialogue. The effect was sensational, as audiences heard an actor speak lines for the first time as though they were natural and spontaneous. As a result, *The Jazz Singer* grossed over $3,000,000 internationally, and the "talkies" were born. In 1928 Warner Brothers produced the first "100% talkie," *Lights of New York*, and the era of the sound film was fully underway. By 1929 fully three-fourths of all Hollywood films had some form of sound accompaniment, and by 1930 silent films were no longer being produced.

Impact

The impact of sound film on the film industry was monumental. First, the very form of films changed, due at first to the difficulties of recording and editing sound film. Because the microphones in use at the time could only pick up nearby sound, and were extremely sensitive within their limited range, actors had to stand very still and very close to the microphone. Camera noise could also be picked up by the microphones, so the cameras and their operators were enclosed within glass booths. Movement of the actors within the frame, and movement of the camera itself, became near impossibilities, so that films once again came to resemble the filmed stage plays that were typical of the early days of motion pictures. Also, because the sound was recorded directly on the film at the time of filming, the film could not be edited after it was shot, except for cuts made during scene transitions. The whole art of editing and montage, developed in America by such directors as D.W. Griffith, and refined to a high art under Soviet directors like Sergei Eisenstien, was simply no longer possible. Additionally, many great silent film stars, such as the German actor Emil Jannings (1887-1950), who had a heavy accent, and John Gilbert, whose voice did not match his screen image, discovered that they could no longer find work in sound films. Because of sound film, acting for film began to concentrate less on the expressiveness of the body or face to carry the meaning of the scene, and to concentrate more on the expressiveness of the voice.

Since sound now seemed to limit the potential of film, instead of expanding it, a major theoretical debate among filmmakers developed. Many felt that sound should be used to record exactly what was seen on the screen, such as dialogue and sound effects relating to the on-screen action—what was referred to as synchronous sound. Others, such as Eisenstein, felt that sound should be used to provide non-related elements that could interact in meaningful ways with the on-screen action, which was referred to as contrapuntal, or asynchronous, sound. This approach would have also freed film from some of the constraints forced upon it by the crude sound technology, but as improvements were made to microphones and camera equipment, many of these constraints were lifted. Finally, in 1929, post-synchronized sound systems were developed that enabled sound to be recorded and synchronized with the film after the film was shot; this allowed for the editing and montage effects that had been impossible with early sound film. *Hallelujah*, directed by King Vidor in 1929, took full advantage of post-synchronous sound techniques, and is generally regarded as the first film of the full-sound era. In one critical scene, as the film's characters are running through a swamp, the camera moves with them, and rapid cutting takes place as the audience hears the sound of birds, branches breaking, and dialogue, all sounds that were recorded after the scene was shot and were later added to the soundtrack.

The coming of sound film not only impacted the movies as art, but the movies as an industry as well. Despite the high costs associated with the conversion to sound, sound films saved Hollywood from the Great Depression. Throughout the 1920s film attendance began to fall off as audiences discovered new technologies like the radio, and had sound film not been developed and adopted during the short period between 1926 and 1930, many Hollywood studios would have been forced into bankruptcy during the Great Depression, as audiences tired of the limited expressive capabilities of the silent film. Though resisted at first because of economic factors, sound opened up entirely new dimensions in film as art, dimensions that audiences were anxious to explore, and provided Hollywood with an economic base that sustained it through the worst economy in American history.

PHIL GOCHENOUR

Further Reading

Cook, David A. *The History of Narrative Film*. Third edition. New York: W. W. Norton & Co., 1996.

Crafton, Donald. *Talkies: America's Transition to Sound Film, 1926-31* (History of American Cinema 4). Los Angeles: University of California Press, 1999.

Women Inventors between 1900-1949: Setting the Stage for Equal Opportunity

Overview

New inventions are absolutely vital to a strong nation and a prosperous economy, so it is not surprising that the patent system is one of the oldest parts of the United States government. Anybody can legally patent an invention, even children. However, until very recently, most inventions were patented by men. Where were all the women inventors? The answer has to do with education, beliefs about women, and the nature of the patent itself. Historically, women were usually excluded from scientific and technical education, so they learned fewer technical skills than men. To make matters worse, many people in the eighteenth and nineteenth centuries argued that women lacked the ability to solve mechanical problems, and these ideas discouraged women from pursuing patent applications. Perhaps most importantly, women found invention less profitable than men did. The patent system rewards successful inventors with a "patent," or limited monopoly on an invention. An inventor can sometimes make a lot of money from a patent, because he or she has the exclusive right to sell or use the invention for a certain number of years. However, because women faced discrimination in business, they found it hard to turn patents into profit, and they therefore had less motivation to invent. Although women inventors continued to face these obstacles through much of the twentieth century, changing times gradually improved the educational, social, and economic conditions for inventive women and encouraged them to patent more of their ideas.

Background

By 1900, women held only 1% of all U.S. patents. One reason is that throughout the 1800s women faced pervasive discrimination in education. Women therefore had fewer opportunities to gain the technical skills and knowledge that often lead to a successful invention. In the nineteenth century it was rare for girls to learn about chemistry, engineering, metallurgy, or electricity. In fact, during the entire nineteenth century, only about a dozen women earned doctorates in chemistry, and only three earned a Ph.D. in physics. Until the mid-twentieth century, many universities did not admit women at all. At home, boys often had opportunities to learn how to use tools and fix machines, but their sisters learned how to sew and cook. As late as the 1970s many high schools in America still provided separate cooking classes for girls and "shop" classes for boys.

Lack of technical education only partially explains why so few women held patents. There are many kinds of things, after all, that women could invent without technical knowledge. Lena Wilson Sittig, for example, never went to school, but she filed more than five patents for outdoor garments that she sewed herself. Although Harriet Carter had no training in chemistry, she invented a formula made from ground coal, wax, and other common household ingredients, which reduced the cost of fuel in winter. Other women patented feather dusters, toys, fire escapes, and other inventions that required imagination but little education. Why didn't women patent more inventions like these?

One reason is that during the nineteenth century and well into the first half of the twentieth, women experienced constant exposure to ideas about their physical and mental inferiority that, without question, discouraged their creativity and confidence. Since women's bodies were different than men's bodies, in 1900 people assumed that women's brains were physically different from men's brains. Unfortunately, even by 1920 many people still thought that women were not equipped for logical thinking and problem solving, and some people actually believed that women would become ill if they attempted too much intellectual activity. From their earliest girlhood, women were therefore bombarded with the message that they were not as smart or as inventive as men. These notions were completely wrong, of course, but after being taught that they were not mentally capable of invention, it is not surprising that so few women thought of trying to patent their ideas.

At the dawn of the twentieth century, there was another reason why women did not often patent. It takes money and effort to patent, so most inventors hope to earn a profit from their invention. A patent can indeed be valuable, but it is worthless until the inventor can manufacture and sell her invention, or unless she can sell the patent rights to someone else. Patent law allows for both these options, but women inventors could not necessarily use them. Women, particularly married women, have historically faced restrictions on their ability to own property, form partnerships, and sign business contracts. For example, in 1900 there were still states that gave a woman's husband control of everything she owned, and in other states a woman had to register property before it was considered hers. Significantly, one of the largest increases in women's patenting was 1885-1900, a period when this country saw the most dramatic change in women's property rights.

These gains were nonetheless very limited because women inventors often needed loans or business partners in order to manufacture their inventions. Even when women legally owned their patents, many states restricted a woman's ability to sign contracts, and it was perfectly legal for a bank to refuse a woman a business loan simply because she was a woman. Moreover, business was considered "men's work," so few women leaned much about accounting and the other day-to-day skills needed to run a company. Until the 1920s many people argued that business and politics would corrupt women's superior virtue and morality. Ridiculous as it sounds today, this argument is one reason why women had to fight so hard not only for equal opportunity in business, but for the right to vote and participate in politics.

Impact

During the first half of the twentieth century, reforms in education, transformation of the workplace, and a gradual change in beliefs about women's abilities encouraged more women to invent. From 1905 to 1922, for example, women received about 500 new patents a year, double the rate in 1895. While still a small percentage of patents over all (about 2%), it is significant that women's application rates increased faster than men's rates.

Part of this improvement in women's inventiveness had to do with changes in education. Increasing educational opportunities made it easier for inventive women to learn the skills that they needed. Although even in the 1950s and 1960s many schools still refused to admit women, during the first 50 years of the century, more women attended public universities or received scientific training at women's colleges than ever before. The number of women Ph.D.s is one measure of women's advance in technical education. This figure quadrupled from about 50 in 1920 to more than 200 in 1949. However, from 1940 to 1949, the increase in women Ph.D.s slowed down, showing that women still had a long way to go. While women made gains in certain sciences like biology, where they accounted for 11% of all biologists, a government study in 1948 revealed that only eight women—about 0.01% of the total—were employed as engineers.

However, one hopeful trend in this period was new opportunities for women in industry and business. When men went off to fight in World War I and World War II, women took their places in factories, where they gained new knowledge of industrial technology. These women not only learned new technical skills, they had a chance to observe mechanical and logistical problems in factories and patent ways to solve them. Women also participated more in the business work place. Women clerks had been rare in the nineteenth century, but growing reliance on the typewriter and telephone in the twentieth century brought more women into the office as typists, secretaries, and telephone operators. Although these women faced unfair discrimination in pay and promotion, they did gain exposure to the business world in greater numbers than ever before. Change in corporate structure unfortunately did create one new problem for inventive women: As technology became more complex, more patent applications came not from individuals but from teams of workers in companies like General Electric. These companies often blatantly excluded women from technical jobs, so women did not have the opportunity to contribute to many exciting new inventions.

During the period 1900-1949, views of women also began to change for the better. Feminists and suffragists confronted social ideas about feminine mental inferiority, and in 1919 the U.S. Congress passed the Nineteenth Amendment, which gave women the right to vote. Meanwhile, the public became aware of the accomplishments of women such as Nobel laureate Marie Curie (1867-1934), and these achievements challenged the idea that women were not capable of technical and scientific work. Although ideas about women were changing, the pace of change was very slow. One government study of women inventors published in 1923 still identified "lack of faith" in women's abilities as one of the main reason for women's failure to patent.

Although there was no time during the first half of the century when women held more than 4% of all U.S. patents, gradual improvement in conditions for women during this period laid the groundwork for the future. However, it was not until the second half of the twentieth century that legal changes truly began to level the playing field for women inventors. It was only in 1974, for example, that the courts amended civil rights laws regulating lending to prohibit discrimination based on sex. During the 1970s and 1980s, amendments to the Equal Credit Opportunity Act and other legislation made it illegal to refuse a loan on the basis of sex and made it illegal to consider marital status when assessing a loan application. Discrimination against women in university admissions faced legal challenges, even at all-male educational institutions. Significantly, women's patents, which were 3.5% overall by 1977, surged to nearly 6% by 1988, and to 10.3% in 1998. Women excelled in chemistry patents in the 1980s, a trend that reflected greater acceptance for women in chemistry than in engineering. However, from 1994 to 1998, the employer with the most prolific women inventors was IBM, a computer hardware and software engineering firm.

Why are changes in women's patenting rates important? Invention of new technology is absolutely critical to advances in medicine, production of food, care of the environment, better computers and other electronics, safer cars and buildings, and improvements in everything we use each day. Since women account for about

50% of the population, women have 50% of the world's inventiveness and brainpower. However, social discrimination against women historically has dramatically reduced women's ability to contribute their half of technological improvements. (Similar discrimination against African Americans and ethnic minority groups has further degraded our nation's potential for technical achievement.) As technical education continues to open to women, as women advance in corporate technical roles, and as women entrepreneurs market wonderful new inventions, women's inventive contribution will rise, and advances in our quality of life will rise along with it.

DEVORAH SLAVIN

Further Reading

Blashfield, Jean F. *Women Inventors*. Minneapolis: Capstone Press, 1996.

Casey, Susan. *Women Invent: Two Centuries of Discoveries That Have Shaped Our World*. Chicago: Chicago Review Press, Inc., 1997.

Macdonald, Anne L. *Feminine Ingenuity*. New York: Ballantine Books, 1993.

Rossiter, Margaret. *Women Scientists in America*. Baltimore: Johns Hopkins University Press, 1982.

Showell, Ellen H., and Fred M. B. Amram. *From Indian Corn to Outer Space: Women Invent in America*. Cobblestone Publishing, Inc., 1995.

Stanley, Autumn. *Mothers and Daughters of Invention: Notes for a Revised History of Technology*. Metuchen, NJ: Scarecrow Press, Inc, 1993.

The Development of the Tape Recorder

Overview

A number of experimental sound recording devices were designed and developed in the early 1900s in Europe and the United States. Recording and reproducing the sounds were two separate processes, but both were necessary in the development of the tape recorder. Up until the 1920s, especially in the United States, a type of tape recorder using steel tape was designed and produced. In 1928 a coated magnetic tape was invented in Germany. The German engineers refined the magnetic tape in the 1930s and 1940s, developing a recorder called the Magnetophon. This type of machine was introduced to the United States after World War II and contributed to the eventual widespread use of the tape recorder. This unique ability to record sound and play it back would have implications politically, aesthetically, and commercially throughout Europe and the United States during World War II and after. Sound recording and reproduction formed the foundation of many new industries that included radio and film.

Background

Sound recording and reproduction began to interest inventors in the late nineteenth century, when several key technological innovations became available. Recordings required the following: a way to pick up sound via a microphone, a way to store information, and a playing device to access the stored data. As early as 1876 American inventor Alexander Graham Bell (1847-1922) invented the telephone, which incorporated many of the principles used in sound recording. The next year, American inventor Thomas Alva Edison (1847-1931) patented the phonograph, and German-American Emil Berliner (1851-1929) invented the flat-disc recording in 1887. The missing piece was a device to play back the recorded sounds.

The history of the tape recorder officially begins in 1878, when American mechanic Oberlin Smith visited Thomas Edison's laboratory. Smith was curious about the feasibility of recording telephone signals with a steel wire. He published his work in *Electrical World* outlining the process: "acoustic cycles are transferred to electric cycles and the moving sonic medium is magnetized with them. During the playing the medium generates electric cycles which have the identical frequency as during the recording." Smith's outline provided the theoretical framework used by others in the quest for a device that would both record and play the sound back.

In 1898 a Danish inventor, Valdemar Poulsen (1869-1942), patented the first device with the ability to play back the recorded sounds from steel wire. He reworked Smith's design and for several years actually manufactured the first "sonic recorders." This invention, patented in Denmark and the United States, was called the telegraphon, as it was to be used as an early kind of the telephone answering machine.

The recording medium was a steel chisel and an electromagnet. He used steel wire coiled around a cylinder reminiscent of Thomas Edison's phonograph. Poulsen's telegraphon was shown at the 1900 International Exhibition in Paris and was praised by the scientific and technical press as a revolutionary innovation.

In the early 1920s Kurt Stille and Karl Bauer, German inventors, redesigned the telegraphon in order for the sound to be amplified electronically. They called their invention the Dailygraph, and it had the distinction of being able to accommodate a cassette. In the late 1920s the British Ludwig Blattner Picture Corporation bought Stille and Bauer's patent and attempted to produce films using synchronized sound. The British Marconi Wireless Telegraph company also bought the Stille and Bauer design and for a number of years made tape machines for the British Broadcasting Corporation. The Marconi-Stilles recording machines were used until the 1940s by the BBC radio service in Canada, Australia, France, Egypt, Sweden, and Poland.

In 1927 and 1928 Russian Boris Rtcheouloff and German chemist Fritz Pfleumer both patented an "improved" means of recording sound using magnetized tape. These ideas incorporated a way to record sound or pictures by causing a strip, disc, or cylinder of iron or other magnetic material to be magnetized. Pfleumer's patent had an interesting ingredient list. The "recipe" included using a powder of soft iron mixed with an organic binding medium such as dissolved sugar or molasses. This substance was dried and carbonized, and then the carbon was chemically combined into the iron by heating. The resulting steel powder, while still heated, was cooled by being placed in water, dried, and powered for the second time. This allowed the recording of sound onto varieties of "tape" made from soft paper, or films of cellulose derivatives.

In 1932 a large German electrical manufacturer purchased the patent rights of Pfleumer. German engineers made refinements to the magnetic tape as well as designing a device to play back the tape. By 1935 a machine known as the Magnetophon was marketed by the German Company AEG. In 1935 AEG debuted its Magnetophon with a recording of the London Philharmonic Orchestra.

Impact

By the beginning of World War II the development of the tape recorder continued to be in a state of flux. Experiments using different types and materials for recording tapes continued, as well as research into devices to play back the recorded sounds. Sound recording on coated-plastic tape manufactured by AEG was improved to the point that it became impossible to distinguish Adolf Hitler's radio addresses as a live or a recorded audio transmission. Engineers and inventors in the United States and Britain were unable to reproduce this quality of sound until several of the Magnetophons left Germany as war reparations in 1945. The German version combined a magnetic tape and a device to play back the recording. Another interesting feature, previously unknown, was that the replay head could be rotated against the direction of the tape transport. This enabled a recording to be played back slowly without lowering the frequency of the voice. These aspects were not available on the steel wire machines then available in the United States.

The most common U.S. version used a special steel tape that was made only in Sweden, and supplies were threatened at the onset of World War II. However, when patent rights on the German invention were seized by the United States Alien Property Custodian Act, there were no longer any licensing problems for U.S. companies to contend with, and the German innovations began to be incorporated into the United States designs.

In 1945 Alexander Poniatoff, an American manufacturer of electric motors, attended a Magnetophon demonstration given by John T. Mullen to the Institute of Radio Engineers. Poniatoff owned a company called Ampex that manufactured audio amplifiers and loudspeakers, and he recognized the commercial potential of the German design and desired to move forward with manufacturing and distributing the Magnetophon. In the following year he was given the opportunity to promote and manufacture the machine in a commercially viable way through an unusual set of circumstances.

A popular singer, Bing Crosby, had experienced a significant drop in his radio popularity. Crosby attributed his poor ratings to the inferior quality of sound recording used in taping his programs. Crosby, familiar with the Magnetophon machine, requested that it be used to tape record a sample program. He went on to record 26 radio shows for delayed broadcast using the German design. In 1947 Bing Crosby Enterprises, enthusiastic about the improved quality and listener satisfaction, decided to contract with Ampex to design and develop the Magnetophon recording device. Ampex agreed to build 20 professional recording units priced at $40,000 each, and Bing Crosby Enterprises then sold the units to the American Broadcasting Company.

In the film world, Walt Disney Studios released the animated film *Fantasia.* This film used a sound process called Fantasound, incorporating technological advances made in the field of sound recording and sound playback. These commercial uses of the magnetic tape recording devices allowed innovations and expansion in the movie and television-broadcasting field.

By 1950, two-channel tape recorders allowing recording in stereo and the first catalog of recorded music appeared in the United States. These continued advancements in tape recorder technology allowed people to play their favorite music, even if it had been recorded many years prior. Radio networks used sound recording for news broadcasts, special music programming, as well as for archival purposes. The fledgling television and motion pictures industries began experimenting with combining images with music, speech, and sound effects. New research into the development and use of three and four-track tape recorders and one-inch tape was in the works as well as portable tape cassette players and the video recorder. These new innovations were possible and viable because of the groundwork laid by many individuals and companies in the first half of the twentieth century.

LESLIE HUTCHINSON

Further Reading

Eargle, John. *Handbook of Recording Engineering.* New York: Chapman & Hall, 1996.

Millard, A. J. *America on Record: A History of Recorded Sound.* London: Cambridge University Press, 1995.

Moylan, William D. *The Art of Recording: The Creative Resources of Music Production and Audio.* Lowell, MA: University of Massachusetts Press, 1993.

Biographical Sketches

John Vincent Atanasoff
1903-1995

American Inventor, Physicist and Mathematician

It is impossible to imagine a world without computers. Computers control nearly every facet of our lives, from car air bags to airplanes, schools, businesses, and space shuttles. None of these modern marvels would have been possible without the work of John Vincent Atanasoff. He was responsible for creating the first digital computer, and with his invention, launching the modern era of computing.

Atanasoff was born on October 4, 1903, in Hamilton, New York. Early on, he became fascinated with mathematics, and was especially interested in a Dietzgen slide rule his father had bought, and the mathematical principles by which it operated. (A slide rule allows the user to perform calculations mechanically by sliding one ruler against another-both rulers are marked with graduated scales.) Atanasoff's family later moved to Florida, where he completed high school as a straight-A student.

In 1921, Atanasoff entered the University of Florida, continuing his perfect grade point average. He graduated with a bachelor of science degree in electrical engineering in 1925. He continued his studies at Iowa State College (now University), where he studied mathematics, earning a master's degree in 1926. Within a few days of graduating, Atanasoff married his college sweetheart, Lura Meeks. Four years later, in 1930, he received his doctorate in theoretic physics from the University of Wisconsin, and soon thereafter returned to Iowa State University to become an assistant professor in mathematics and physics.

Ever since his early fascination with the slide rule, Atanasoff had been interested in finding new and faster ways of performing mathematical computations. He was frustrated by the slow, time-consuming devices that were available at the time—the Monroe calculator and the IBM tabulator—and knew he could create a more efficient calculating machine. He envisioned a machine that was digital, rather than the relatively slow and inaccurate analog models. While sitting in an Illinois pub in 1937, he

John Atanasoff. *(AP/Wide World Photos, Inc. Reproduced with permission.)*

came up with the principles that would form the basis of his electronic digital computer. His machine would utilize capacitors (components in an electrical circuit) to store the electrical charge that would represent numbers in the form of 0s and 1s. Rather than using a base-10 counting system, as in analog devices, it would use binary numbers (those with a numerical base of 2) and compute by direct logic. And finally, it would have condensers for memory, and a regenerative process, which could store information even when power was lost.

To assist him in his work, Atanasoff called in Iowa State graduate student Clifford Berry, a well-respected electrical engineer. The two developed their prototype, the Atanasoff Berry Computer (ABC), in December of 1939. The ABC had two rotating drums containing capacitors, which held the electrical charge for the memory. Data was entered using punched cards.

The following year, Atanasoff attended a lecture given by physicist Dr. John W. Mauchly (1907-1980), who expressed interest in the ABC prototype. Atanasoff offered to show Mauchly his invention, a decision which he would later regret, as Mauchly apparently stole many of Atanasoff's ideas when designing his own computer, called the ENIAC (Electronic Numerical Integrator and Computer). It is a general misconception that the ENIAC was the world's first digital computer. The ABC actually predated the ENIAC.

In the 1970s, a landmark court case (Honeywell Inc. vs. Sperry Rand Corp et al.) overturned ENIAC's patent, ruling that Mauchly's computer borrowed many of its design features from the ABC. While Atanasoff eventually established his own successful business, he was never able to secure a patent for his ABC and failed to gain the recognition he deserved. It was not until 1990, when he was awarded the National Medal of Technology by President George Bush, that his contributions to the field of computing were finally recognized.

Atanasoff died of a stroke on June 15, 1995, at his home in Monrovia, Maryland. He was 91 years old. Although his invention never took off on its own, the methods used in his prototype ABC computer were to form the backbone of every modern computer used today.

STEPHANIE WATSON

Leo Hendrik Baekeland

1863-1944

Belgian-American Chemist

A 1940 *Time* magazine article called Leo Hendrik Baekeland the "Father of Plastic," and with good reason. His invention of the first synthetic plastic, Bakelite, proved cheaper and more versatile than other substances, and has since been used in everything from pot handles to electronics.

Baekeland was born in Ghent, Belgium, on November 14, 1863, the son of working-class parents. As a teenager, Baekeland spent his days in high school, and his evenings at the Ghent Municipal Technical School, where he studied chemistry, physics, mathematics, and economics. He was awarded a scholarship to the University of Ghent, and in 1884 received his doctorate of science, graduating with honors.

From 1882-89, Baekeland taught chemistry at the University of Ghent. Between 1885 and 1887 he also taught chemistry and physics at the Government Higher Normal School of Science in Bruges, Belgium. In 1889, Baekeland and his wife, Celine, traveled to the United States and decided to stay permanently. His dual interest in photography and science led him to a position as a chemist at the photographic manufacturing company E. and A. T. Anthony and Company. In 1893, he left that position to found the Nepera Chemical Company in Yonkers, New York.

Leo Baekeland. *(Corbis Corporation. Reproduced with permission.)*

In the late 1800s, photographs were developed using sunlight, which made it impossible for photographers to complete their work under cloudy skies. Baekeland perfected a new photographic paper, named Velox, which allowed photographs to be developed with the use of an artificial light. In 1899, George Eastman (1854-1932) of Eastman Kodak bought the rights to Velox for the enormous sum of one million dollars. Baekeland celebrated his new wealth by purchasing an estate in an upscale community north of Yonkers.

After the success of his first invention, Baekeland shifted his focus to creating a new, cheaper, synthetic form of shellac to replace that which was naturally secreted by the Laccifer lacca beetle. The latter form was becoming more and more expensive, and harder and harder to obtain. Baekeland and his assistant began experimenting in 1904, in search of a substance that would not only dissolve in solvents to create an effective insulator, but one that would also be malleable, or pliable, like rubber.

Three years later, while working with a mixture of phenol and formaldehyde, Baekeland noticed that the two chemicals, when combined, produced a shellac-like residue which could be used to coat surfaces like a varnish. When Baekeland put this mixture into his "Bakelizer," a heavy iron pot similar to a pressure cooker, out came a moldable, yet solid, substance. The synthetic plastic, which he called Bakelite, was unique in that it would not melt under extreme heat. That made it especially useful for things like automobile engine parts and handles for kitchen utensils. Baekeland received a patent for Bakelite in 1906, and founded the General Bakelite Corporation (later the Bakelite Company) four years later to manufacture and sell his invention. Baekeland served as president of the company until it merged with the Union Carbide Corporation in 1939.

Bakelite was known as "the material of a thousand uses," and became part of everything from beads to telephones, from jewelry to electronics. In 1945, a year after Baekeland's death, annual plastic production in the United States rose to more than four hundred thousand tons. Today, it is everywhere we look, in every aisle of the local grocery store, in virtually every item used in the home. Baekeland's invention came to represent an industrial, as well as a cultural, revolution, as epitomized in the classic line from the film *The Graduate*: "I just want to say one word to you. Just one word: plastics."

STEPHANIE WATSON

John Logie Baird
1888-1946
Scottish Inventor

John Logie Baird is not credited as the inventor of the television, though he did create the first working television set in 1923. He went on to a number of "firsts," including the first color transmission. Yet his was a mechanical rather than an electronic system, and by the mid-1930s it would be rendered obsolete by a far better machine using a cathode-ray tube.

Born on August 13, 1888, in Helensburgh, Scotland, Baird was the youngest of John and Jessie Morrison Inglis Baird's four children. His father was the minister of a local parish church. Baird studied electrical engineering at the Royal Technical College in Glasgow, and went on to the University of Glasgow. He never completed his work on a bachelor of science degree, however, due to the outbreak of World War I.

Baird was not fit for military service, but spent the war years as engineer for an electrical power company. After the war, he started a series of entrepreneurial enterprises, beginning with a sock factory in 1919. A year later, he moved to the Caribbean island of Trinidad, where he set up a facility to manufacture jam and chutney. This, too, he sold after a year, moving to London to establish a soap-making company, which he sold as well. In the autumn of 1922, Baird moved to the seaside town of Hastings, where he invested his profits in experiments to develop a successful means of transmitting visual signals via wireless technology.

In order to transmit such signals, scientists had come to realize, it was necessary to develop a means for scanning the object to be transmitted. The scanned image, obtained by moving a beam of light in a series of lines from top to bottom and left to right, would create light signals which, when converted into electrical impulses, could be transmitted to a receiver. The receiver would then convert these signals into an image, which it would reproduce as a series of lines on a screen.

In 1894, German inventor Paul Nipkow had patented an idea he called the Nipkow disc, which had a series of small holes cut in a spiral. A powerful beam of light would be focused on one disc, which by rapidly rotating would create the lines of light necessary to scan the object. These would then be collected by a light-sensitive photoelectric cell that would create the electrical impulses and transmit these to a second Nipkow disc.

For the system to work, the second disc would have to be rotating rapidly, in exact synchronization with the first. Problems associated with synchronization had kept Nipkow's system from advancing past the theoretical stages until Baird in 1922 developed a way to correctly synchronize the two disks. He built his first two discs out of cardboard and other discarded objects, and in early 1923 made the world's first television transmission: the blurry images of a cross and a human hand, which he transmitted a distance of a few feet.

The decade that followed was without question the high point of Baird's career. He began demonstrating his invention to other scientists and the public, and soon became a celebrity. In 1929, under pressure from various sectors of the British government, the British Broadcasting Company (BBC) allowed Baird to begin making experimental television broadcasts from its facilities, and this only further increased his prominence. In 1931, 43-year-old Baird married Mary Albu, a concert pianist with whom he later had a son and daughter.

Forming the Baird Television Development Corporation, Baird and his associates set out to develop a number of firsts, starting with the first television transmission station, in London in 1926. In 1928, he transmitted the first television signal across the Atlantic, from London to New York; the first television pictures taken outdoors in daylight; and the first color television image. Using a system he developed called "noctovision," Baird also figured out how to use infrared rays to scan objects in total darkness.

By the early 1930s, however, events were in motion that would sound the death knell on the enterprise to which Baird had devoted so much energy: both Marconi-EMI in Great Britain and RCA in America were hard at work on developing television systems that used the cathode-ray tube. The latter, invented in Germany in 1897, produced a far better picture—as Marconi-EMI proved when, in early 1935, it broadcast via the BBC's facilities.

The BBC initially agreed to allow both Baird and Marconi-EMI to broadcast, but the superiority of the cathode-ray system became so overwhelmingly apparent that in 1937, it ended its relationship with Baird. The inventor lived another nine years, and died in Bexhill, Sussex, on June 14, 1946.

JUDSON KNIGHT

Vannevar Bush

1890-1974

American Electrical Engineer and Computer Scientist

Vannevar Bush is known as "The Godfather of Information Science." In the early 1930s he and his colleagues at the Massachusetts Institute of Technology (MIT) invented the first practical analog computer, the differential analyzer. During World War II he led the Manhattan Project to develop an atomic bomb. In 1945 he wrote a visionary essay that anticipated the Internet, hypertext, and the World Wide Web by over 40 years.

Bush was born in Everett, Massachusetts, into an old New England seafaring family. His father, Richard, was a Unitarian minister. At school in Chelsea, Massachusetts, he proved to be a genius at mathematics. He received both his B.S. and M.S. from Tufts University in 1913 and his doctorate in engineering from both MIT and Harvard in 1916. The same year he married Phoebe Davis.

After teaching mathematics and electrical engineering at Tufts from 1914 to 1917, Bush did submarine-detection research for the U.S. Navy in World War I. In 1919 he returned to MIT to teach electrical power transmission, electrical engineering, and related subjects until 1939. In 1922 he co-founded the Raytheon Company, thus setting the stage for the Route 128 technology corridor around Boston and foreshadowing what President Eisenhower called in 1961 the "military-industrial complex." In 1932 he was appointed Dean of the MIT School of Engineering.

The MIT differential analyzer was cumbersome by subsequent standards of computing, but in its day it was state-of-the-art. It was a large mechanical assemblage of gears that could solve differential equations and manage as many as eighteen variables. Throughout his career Bush worked mainly with mechanical and analog computers and never appreciated the potential of digital computing.

Bush was President of the Carnegie Institution of Washington from 1939 to 1955. President Roosevelt appointed him Chairman of the National Defense Research Committee (NDRC) in 1940 and Director of the Office of Scientific Research and Development (OSRD) in 1941. As head of OSRD, he coordinated the Manhattan Project to build the first atomic bomb.

Even if Bush had made no scientific advances, he would be remembered for writing "As We May Think," a remarkable piece of prophecy published in the *Atlantic Monthly* in July 1945. At the dawn of the third millennium it remains required reading for almost all library school students in the U.S. It described a device called the "memex" that would fit on a desktop, store vast amounts of data on microfilm, allow quick search and retrieval, and display any requested information on a screen. It would be operated with a keyboard, buttons, and levers. It would be able to index data, record "trails" of searches, and select among simultaneous "projection positions." Bush thus prefigured hypertext, the graphic user interface (GUI), windows, and multitasking. He was among the first to imagine computers as more than just glorified calculators.

Among his books are *Principles of Electrical Engineering* (1922); *Science: The Endless Frontier* (1945), which led to the creation of both the National Science Foundation (NSF) and the Advanced Research Projects Agency (ARPA); *Endless Horizons* (1949); *Modern Arms and Free Men* (1949); *Science Is Not Enough* (1967); and an autobiography, *Pieces of the Action* (1970).

Vannevar Bush. *(Library of Congress. Reproduced with permission.)*

Probably Bush's greatest contribution to the science of communication lies in the later engineers he inspired, such as Joseph Carl Robnett Licklider (1915-1990), who worked on the human-to-computer interactive interface and was one of the early developers of the Internet; Doug Engelbart (1925-), who invented the computer mouse in 1963; Ted Nelson (1937-), who coined the term "hypertext" in 1965; Tim Berners-Lee (1955-), creator of the World Wide Web; and Bill Gates (1955-), founder of Microsoft Corporation. Each of these computer scientists acknowledged "As We May Think" as a major influence, and each made significant progress toward realizing Bush's vision of fundamentally reinventing society though information technology.

ERIC V.D. LUFT

Wallace Hume Carothers
1896-1937
American Chemist

Wallace Carothers quite literally transformed the texture of human life. His assignment as director of a research team for the DuPont Company marked the culmination of a promising career, and the promise bore fruit: Carothers's experiments with polymer plastics yielded nylon, one of the most significant inventions of the twentieth century, and the first synthetic fiber. He would probably have won a Nobel Prize, as some have speculated, yet just before he turned 40, Carothers—who in addition to his career successes was a newlywed with a first child on the way—took his own life.

Carothers was born on April 27, 1896, in Burlington, Iowa, the oldest of Ira and Mary McMullin Carothers's four children. His sister, Isobel, was destined to go on to celebrity as a member of the radio musical trio "Clara, Lu, and Em." Carothers's father was a teacher at Capital City Commercial College in Des Moines, and the son enrolled there to study accountancy in 1914. A highly advanced student, he completed the accountancy program in just one year, and went on to study chemistry at Tarkio College in Missouri.

During World War I, wartime duties forced Arthur Pardee, head of the chemistry department, to leave Tarkio. Unable to replace him, the college appointed Carothers, who was ineligible for military service due to health reasons, to take his place—this despite the fact that he was only an undergraduate. Carothers earned his B.S. in chemistry in 1920, and went on to the University of Illinois, where he completed his master's degree a year later.

Pardee convinced Carothers to join him at the University of South Dakota, where he taught for a year, but by 1922 he was back at the University of Illinois to complete his Ph.D. work. He earned his doctorate in 1924, then began teaching at the university. In 1926, however, Harvard selected Carothers above numerous candidates, and he went east to teach. At Harvard, he began experimenting with high molecular weight polymers, an activity that attracted the attention of the DuPont Company. In 1928, DuPont invited him to direct a new research program at their headquarters in Wilmington, Delaware, where Carothers would head a team of scientists. It was an offer even more attractive than Harvard, and 32-year-old Carothers accepted.

At first Carothers focused on molecules in the acetylene family, and by 1931 his experiments with combinations of vinylacetylene and chlorine had yielded a synthetic rubber that the company marketed as neoprene. This alone would have been a great claim to fame, but within a few years Carothers produced something even greater, a synthetic equivalent of silk. He knew that he needed to create a chemical reaction that would produce long molecules oriented in the same direction. A series of experiments led to an understanding of how polymers were created, and this in turn yielded nylon. The latter would debut in 1938, and became a huge sensation at the World's Fair a year later. By then, however, Carothers was dead.

In 1936, he married Helen Sweetman, who worked in the patent division at DuPont. Though highly enthusiastic about his work, Carothers was always shy and withdrawn, so his marriage must have come as a surprise. He was, however, a talented singer, something he shared with his sister Isobel, his closest sibling. Her death in January 1937 sent him into a bout of depression, and on April 29, he committed suicide in Philadelphia. Seven months later, on November 27, Helen gave birth to a daughter, Jane.

JUDSON KNIGHT

Lee De Forest
1873-1961
American Physicist

Lee De Forest introduced a third element into the diode vacuum tube designed by John Ambrose Flemming (1849-1945), providing a means by which a small voltage applied to this element could have a large effect on the current of electrons flowing between cathode and anode. The triode made possible the amplification of weak electrical signals, which in turn made possible the transmission of high quality audio and video signals by radio waves. The transistor, a solid-state device with the electrical properties of a triode, eventually replaced the vacuum tube triode in most applications.

De Forest had a somewhat unusual childhood. His father was a Congregationalist minister and President of the Talledega College for Negroes in Alabama. As a result, the white community ostracized his family, and Lee made his friends among the black children of the town. De Forest attended a private college preparatory school in Massachusetts, and entered the Sheffield Scientific School at Yale University, receiving his bachelor's degree in 1896. He returned to Yale to complete his Ph.D. degree in 1899, submitting a thesis on the reflection of radio waves. De Forest was interested in the transmission of telegraph messages by radio, and improvements made by him on the earlier system were sufficient for him to obtain support from the United States government for his early work.

In order to improve the detection of radio signals, De Forest added a third electrode to the

Lee De Forest. *(Library of Congress. Reproduced with permission.)*

vacuum tube diode invented by John Ambrose Flemming in 1904. The third electrode, placed between the heated anode and the cathode, would eventually make possible for a small electrical signal to modulate, or control, a large current, a capability required for the effective transmission of sound waves by radio and to amplify the small radio signal when received. Before this could be fully developed, however, a number of advances had to be made in the theory of vacuum tube circuits and in the production of high vacuums. Historians now agree that De Forest did not at first appreciate the full range of possibilities afforded by the triode, and viewed it primarily as a technical improvement on the diode. Nonetheless De Forest was happy to take some credit for subsequent developments. He entitled his autobiography *Father of Radio* and many people regarded him as such.

De Forest spent several years in New York City working with the fledgling radio industry and then moved to California where he worked at first for the Federal Telegraph Company. He continued to make many new inventions, receiving over 300 United States patents. His later inventions included an amplifier based on a cascade of triodes and a medical diathermy machine.

De Forest is regarded as a transitional figure in the history of electronics. He was one of the last independent inventors. As a result he was often forced to sell his patents at bargain prices to better-financed corporations, and he lost large sums of money defending his patents. A number of subsequent inventions in the field involved adding additional electrodes to the triode to fine-tune its behavior. Vacuum-tube technology also gave rise to the need for better vacuums and for economical methods of producing them. As a result, large corporations like General Electric and American Telephone and Telegraph began establishing laboratories for fundamental scientific research related to their products. Industrial research became a major employer of scientific talent, increasing the number of scientists and engineers actively at work. Well-financed industrial research would lead in 1947 to the transistor, a solid-state version of triode that would replace vacuum tubes in most applications.

DONALD R. FRANCESCHETTI

Philo Taylor Farnsworth
1906-1971
American Inventor

Philo T. Farnsworth is one of several individuals who invented the television, although his role in the development of the television camera is sometimes overlooked. He was a lifelong inventor and was the holder of 165 patents during his lifetime.

Once the transmission of sound by radio was accomplished in the early twentieth century, it was not difficult to imagine transmitting a series of pictures that would be interpreted as a moving picture by the viewer. Indeed, the first steps in this direction were achieved in the eighteenth century. A primitive facsimile transmission line operated between Paris and Lyons, France, between 1865 and 1870. In 1884, Dr. Paul Nipkow (1860-1940) of Berlin received a patent for a picture transmission system based on a two spinning perforated disks. Light from the scene being transmitted would pass through one of the spinning holes and be collected by a photocell. The voltage produced would carry the information to a light source behind a second spinning disk, synchronized with the first, and fall on a screen which then could be viewed. Although this mechanical method had many flaws and did not produce a very good image, it was commercialized in England by the British engineer John Logie Baird (1888-1946).

In 1922 the sixteen-year-old Farnsworth showed one of his high school teachers his idea to replace the spinning disk with an electronic scanning system. With his teacher's encouragement he began work on his ideas. After two years as a student at Brigham Young University, he moved to California to devote full time to his invention. In 1927 he successfully transmitted several sixty-line images. He received a patent for his electronic camera tube in 1930.

Also working on an electronic picture transmission system was Vladimir K. Zworykin (1889-1982), a Russian immigrant who had filed a patent application in 1923 for a primitive electronic television camera. In 1930, he left his job at Westinghouse, where people were unimpressed with his ideas about television, to work for the Radio Corporation of America. RCA was willing to invest a substantial amount of money to develop this new technology. That same year, Zworykin visited Farnsworth and concluded that RCA was on its way to a superior system that would not infringe on Farnsworth's patents. RCA's lawyers thought differently, however, and tried to buy the rights to Farnsworth's patent. Farnsworth was unwilling to accept a lump sum payment but was willing to sell for a royalty on RCA's income from television. RCA was unwilling to do this, however, as it thought, correctly, that there was a tremendous amount of money to be made in television, and it did not want to forego a fixed percentage. RCA tried until 1939 to get Farnsworth to accept a flat sum but eventually agreed to pay a royalty, the first time the company had done so.

The same year, RCA introduced commercial electronic television at the New York World's Fair, selling sets to a few hundred wealthy New Yorkers during the year. Both Zworykin and Farnsworth had made improvements to their original design by that time. Because RCA devoted so much of its effort to publicity and marketing, Farnsworth is often omitted when the story of its invention is told. Farnsworth, nonetheless, produced numerous other inventions, earning 165 patents over his lifetime. Farnsworth, a Mormon and native of Utah, is memorialized as the Father of Television in a statue erected by the State of Utah in Washington, D.C.

DONALD R. FRANCESCHETTI

Reginald Aubrey Fessenden
1886-1932

Canadian Electrical Engineer and Inventor

A disciple of Thomas Edison (1847-1931), Fessenden made major contributions to wireless telephony and radio. In 1906 he made the first broadcast of voice and music from Brant Rock, Massachusetts. The principle of using electromagnetic waves to send detailed information over great distances to one or more receivers was central to many of the most significant technologies of the twentieth century.

"Reg" Fessenden was born in East Bolton, Quebec, where his father was a minister. Although the family moved to Niagara Falls, Ontario, Fessenden returned to Quebec, where he completed his bachelor's degree at Bishop's College while teaching high school mathematics. Fessenden changed jobs frequently during his life, alternating between academics and industry. He was a school principal in Bermuda, from 1884-86, where he met his future wife, Helen, but he returned to New York to work for Edison, from 1887-90, before moving on to further industrial positions. Fessenden then returned to teaching, as professor of electrical engineering at Purdue University (1892), and then at the University of Pittsburgh (1893-1900). In 1900, Fessenden began working for the U.S. Weather Bureau, on the new technology of wireless telegraphy, but two years later he left to form his own business, the National Electric Signaling Company.

Fessenden's inventions in the first decade of the twentieth century were originally intended to improve the transmission of telegraph signals, but they were also crucial for making a successful wireless telephone. His "electrolytic detector" received radio waves much more clearly than the crude "coherer" used by other systems. His high-frequency alternator, developed in conjunction with the Swedish engineer E. Alexanderson (1878-1975), produced smooth, continuous radio waves capable of carrying a clear sound signal, unlike the noisy, intermittent bursts generated by existing spark gap technology. Fessenden also developed the key ideas of "amplitude modulation" (AM) and the "heterodyne principle," which explained how to mix sound with radio waves for transmission, and how to decode them at the receiving end. Many wireless telegraph stations used Fessenden equipment, and thus received his first radio broadcasts of December 24 and 31, 1906, hearing voices and music through headsets that had previously only produced dots and dashes.

Fessenden soon succeeded in producing wireless telephone communication over thousands of miles. Unfortunately, relations between Fessenden and his financial backers broke down, and the company collapsed in 1911. It took Fessenden until 1928 to resolve ownership of his wireless patents, after lengthy and expensive court battles. In the meantime, he consulted for the British and American governments during World War I, developing signaling systems for submarines, antennas used to detect incoming zeppelins, and a superior depth-measuring instrument for ships. During the 1920s, when commercial radio began, Fessenden received greater recognition for his work, and won several honors. He was increasingly in poor health, however, worn out from his legal disputes, and he gradually withdrew from public life. Fessenden retired to Bermuda in 1928, to devote himself to his new interest, research on ancient civilizations. He was happier in Bermuda than in the U.S., but died just four years later at 65.

Reginald Fessenden was a talented, original, and multi-faceted inventor. Although he had many interests, his contributions to radio and wireless communication were by far the most important. Virtually every aspect of twentieth-century life was affected by technologies that his work made possible, such as radio and television broadcasting, wireless point-to-point communications, satellites, and personal mobile phones. Fessenden was also highly idealistic, and always stuck to his personal beliefs, even at his own cost. This meant that he was not as successful in business as his contemporary, Guglielmo Marconi (1874-1937), but he did manage to attract a few loyal friends and followers.

BRIAN SHIPLEY

Sir John Ambrose Fleming

1849-1945

British Physicist and Electrical Engineer

Fleming invented the thermionic valve or vacuum tube diode, making possible the rectification or conversion of alternating current into pulsating direct current. The diode was the first reliable rectifier and permitted a vast expansion of radio technology. Fleming remained active in research in electronic technology for 65 years.

Fleming, the son of a minister, attended University College, London, graduating in 1870. He taught science for a time, and then in 1877 entered Cambridge University to work under the eminent British physicist James Clerk Maxwell (1831-1879), with whom he had begun a correspondence. In 1881, Fleming joined the faculty of the newly established University of Nottingham, but left after a year to return to London, where he worked as a consultant to the London Telegraph Company and the Swan Lamp Factory. In 1885 he became a professor of electrical engineering at University College London, a position he held for the next 41 years.

In 1884 Fleming visited American inventor Thomas Edison (1847-1931) in the United States. There he learned about the Edison effect, in which a current would flow from the heated filament in an evacuated light bulb to a separate electrode sealed in the bulb if the electrode was connected to the positive terminal of the battery providing power to the bulb. No current would flow if it were connected to the negative terminal. When Joseph John Thomson (1856-1940) discovered the electron in 1897, the explanation of the effect became apparent. Electrons could escape from the heated filament into the vacuum, a process dubbed "thermionic emission" and were attracted to the positive charge on the electrode.

The first demonstration of radio waves occurred in 1887 when Heinrich Hertz (1857-1894) induced a spark in a coil of wire a short distance from his transmitter. The main obstacle to using radio waves for communication was their frequency. The signal voltage reversed so many times per second that no mechanical device could respond. In 1889 Fleming became a collaborator with Italian inventor Gugliemo Marconi (1874-1937) and began exploring the use of the Edison effect to turn the alternating radio signal to one that consisted of a series of pulses in only one direction. In 1904 Fleming was awarded a patent for a device, the vacuum tube diode or "valve," which would rectify, that is allow current to flow in only one direction. The diode was exactly what was needed by the new radio industry since it could convert a radio signal, which changed directions thousands of times a second, to a series of pulses all in the same direction, which could then register on a meter or other physical device. In 1906, American physicist Lee De Forest (1873-1961) added a third electrode, initially to improve the performance as a rectifier. In the right type of circuit, De Forest's new triode allowed for the amplification of signals, allowing both the production of stronger signals and the detection of much weaker ones.

Although Fleming did not receive the public acclaim that would go to De Forest, he enjoyed a very long and productive career in physics, and in the new field of electronic technology. In 1874 he gave the first scientific paper ever presented to the newly formed London Physical Society. In 1939, 13 years after his official "retirement," he gave his last paper to the society at the age of 90. He was knighted in 1929, and received many awards from scientific societies.

DONALD R. FRANCESCHETTI

Henry Ford
1863-1947
American Businessman

Although he did not invent the technology that made him famous, Ford's profound impact on American culture cannot be underestimated. The development of the automobile, which opened up the nation for its people, and the assembly line, which established mass production as the mechanism for economic power, were perfected by Ford. Both transformed American life. Under Ford's leadership, the automobile went from a luxury to a necessity and inaugurated the "car culture."

Born in Dearborn, Michigan, in 1863, Henry Ford had an aptitude for machinery. However, it was the sight of a coal-fired steam engine in 1876 that set in motion his later triumphs. By age 16, leaving the family farm against his father's wishes, Ford apprenticed in a machine shop in Detroit. He found the work inspirational, and was especially interested in the new gasoline-powered internal combustion engines, which had been developed by German engineer Karl Benz (1844-1929). Ford was hooked.

After three years in the machine shop, he joined the Westinghouse Engine Company as a part-time employee, spending his off hours in a machine shop of his own. His free time also allowed him to travel around Detroit asking questions of its best engineers. In 1896 he produced a two-cylinder, four-cycle engine that generated four horsepower. Ford mounted the engine on a borrowed chassis with four bicycle wheels, and called it the Quadricycle. It was a huge success. People clamored for Ford's invention and wanted to try it for themselves. He sold his first machine for $200, then built a second one, bigger and more powerful than the first. Backed by investors, Ford opened the Detroit Automobile Company (soon reorganized as the Henry Ford Company), the first car manufacturer in what would come to be called the "Motor City."

Ford entered his cars in races and soon won a reputation for speed, setting new records in the process. His success brought more investors' money into the company, beginning the close union between the auto industry and racing.

Henry Ford, who vowed to "build a motor car for the great multitude."

Over the years, this alliance led to improvements in car design and technology, benefiting the industry as a whole.

In 1902 Ford left the company he'd founded, and started the Ford Motor Company in 1903. The company produced eight different models, and within five years made 100 cars a day. During this time Ford bought many out many of his original investors, and by 1908 owned 58% of the company. Ford's dissatisfaction with the small number of cars produced was growing—he wanted to produce 1,000 a day. He announced that the company would reduce its production to one type, the Model T. "I will build a motor car for the great multitude," he proclaimed. The way to make them affordable, he said, was "to make them all alike, to make them come through the factory just alike." Stockholders were furious, but since Ford controlled 58% of the company there was little they could do.

In the fall of 1908 the first Model T rolled out. The car had several new features that made it more negotiable on country roads, and the engine was encased for protection. Ford set the price at $825, which he knew was expensive, but he believed the price would fall through assembly line technology. With Ford in control, efficiency became the keystone of his operations. For 20 years, Ford produced black Model Ts, and only Ts (often called the "Tin Lizzie" or "flivver").

Ford sold 11,000 cars from 1908 to 1909, then outdid himself with the 1910 model, selling 19,000. Sales skyrocketed, reaching 248,000 in 1914, or nearly half the U.S. market. The heavy demand for cars forced Ford to pioneer new methods of production. He built the largest and most modern factory in America on a 60-acre tract in Highland Park in Detroit.

Ford, preaching modern ideas of efficiency, introduced the continuously moving assembly line. He tinkered with the process until he found the exact pace his workers could handle. Chassis production dropped from 6 hours to 90 minutes. The one millionth Model T was built in 1916. Highland Park churned out a phenomenal 2,000 cars a day.

As he predicted, the price of Model Ts soon fell to $300, giving Ford 96% of the inexpensive car market. Ford next focused on labor problems. He instituted profit-sharing and a bonus system, but it was the "five-dollar day" in 1914 that really took hold.

Ford workers had an eight-hour workday, shorter than the industry average. This allowed the factory to run three eight-hours shifts a day. More importantly, Ford paid workers a basic wage of $5 a day, eclipsing the industry's usual $1.80 to $2.50. The move made Ford a national hero, and his legend approached cult status. He represented the all-American success story. By 1921 nearly 5.5 million Ford automobiles had been built.

A stern man, Ford changed with the outbreak of World War I. His attempts to orchestrate an end to the fight made him look like a maniac. In 1918 he won the Michigan Democratic nomination for a U.S. Senate seat but lost the general nomination by a small margin. Next, he bought the *Dearborn Independent*, a local newspaper, and used it as a mouthpiece for his racist and isolationist views.

After the high-water mark of the early 1920s, the Ford Company began to slip. A new 1,100-acre factory, the River Rouge complex in Dearborn, opened and marked Ford's attempt at vertical integration. The size and sprawl of the Rouge proved too much for Ford. Personality clashes with subordinates left the company a virtual one-man operation, which proved dreadful.

The Model T also looked outdated by the late 1920s. Stylish models from General Motors and Chrysler forced Ford to drop the car and replace it with the Model A. Just when Ford began regaining market share, the Depression hit and spelled doom.

The outbreak of World War II also hurt Ford's image. At home, he intimidated his workers through campaigns of espionage and subversion against labor unions. His fuzzy pro-Hitler remarks and turn to isolationism led many to label Ford a Nazi apologist.

Ford died in 1947, just as suburbanization and a reinvigorated car craze swept the nation. Although he did not invent the automobile, Ford is most closely linked to its glory. By allowing the masses to purchase cars, he set in motion the creation of a car culture in the United States. In a nation linked by a spiderweb of roads and highways, shopping plazas and fast-food joints, Ford's influence on popular culture is felt daily.

BOB BATCHELOR

Robert Hutchings Goddard
1882-1945
American Physicist

Goddard was an early advocate of rocketry and space travel and is one of the principal inventors of the liquid fluid rocket. Beginning in 1926 he was able to launch a series of successful rockets, eventually obtaining modest financial support for his work. Many of the concepts developed by him are used in current rocket design. Goddard's contribution to rocket science was only belatedly acknowledged by the United States government.

Robert Goddard, the son of a bookkeeper and machine shop operator, spent his childhood in Worcester, Massachusetts, during the period of rapid industrialization that followed the American Civil War. His interest in the possibility of space travel was apparently kindled in 1898, when he read the serialized version of H.G. Wells's *War of the Worlds* that appeared in the *Boston Post*. He received his college education at the Worcester Polytechnic Institute and went on to earn a doctorate from Clark University, also in Worcester. Following a year conducting research at Princeton University, he returned to Clark, an institution with which he would remain associated throughout his career.

Goddard began testing some of his ideas in the laboratory in 1912, obtaining two patents in 1914. In 1919 he published a small booklet entitled "A Method of Reaching Extremely High Altitudes." Goddard was the first to develop a liquid fuel rocket engine, launching the first liquid-fueled rocket from his aunt's farm in 1926. Goddard's initial tests were funded by the Smithsonian Institution. A second test in 1929 carried

Robert H. Goddard *(Library of Congress. Reproduced with permission.)*

a package of instruments to an even greater height, but also attracted the attention of the police. Local authorities forbade further rocket testing in Massachusetts.

At this point the famous aviator Charles A. Lindbergh (1902-1974) came to Goddard's aid. Lindbergh persuaded the philanthropist Daniel Guggenheim to make a grant of $50,000 to Goddard. Goddard began a new series of tests at Roswell, New Mexico, with his rockets attaining an altitude of over 2 km (1.25 mi) by 1935. Goddard was unable, however, to interest the American military in the development of rocketry. During World War II, he received a small government stipend to perfect a small-scale rocket weapon, the bazooka, that he had designed during the First World War, and to work on rocket-assisted aircraft launches for the Navy.

Goddard's early interest in space flight was matched in Germany by Hermann Oberth (1984-1989), and in Russia by Konstantin Tsiolkovsky (1857-1935). All three worked independently, obtaining many of the same theoretical results, although by 1925 Oberth had begun to correspond with the other rocket researchers. While Goddard had been the first to experiment with liquid fuel rockets, the German rocket enthusiasts, with the support of the Nazi regime, were the first to extensively develop rockets, the V-1 and V-2, for military use. When, at the end of the Second World War, American agents interviewed captured German rocket scientists, they were surprised to learn just how much of the German rocket technology had been based on Goddard's ideas. The American government subsequently paid Goddard's estate one million dollars for the rights to use Goddard's more than 200 patents. Goddard's contribution to space exploration is commemorated in the Goddard Spaceflight Center in Greenbelt, Maryland.

DONALD R. FRANCESCHETTI

George Washington Goethals
1858-1928
American Army Engineer

George Washington Goethals was a United States Army Engineer appointed by President Theodore Roosevelt as the chief engineer of the Panama Canal when John F. Stevens resigned in 1907. Goethals took a special interest in the employees working under him, which he was to be well known for, and created an atmosphere of cooperation on the project. His engineering and people skills helped him complete the Panama Canal six months ahead of schedule in 1914. Goethals remained at the Canal from 1914 to 1916 as governor of the Canal Zone.

Born on June 29, 1858, in Brooklyn, New York, Goethals was described as a quiet, slow-moving boy. He was a serious child who spent much of his youth planning his future. He worked after school to save money for college and attended the College of the City of New York. He later entered the Military Academy at West Point in 1876. Goethals graduated from West Point in 1880 as a second lieutenant and served in the Corps of Engineers. Four years after his graduation Goethals married Effie Rodman and they had two sons.

Before his work on the Panama Canal, Goethals gained practical experience building dams, bridges, and locks on rivers like the Ohio and the Tennessee. He served as an instructor and taught at West Point, employing his valuable field experiences. Goethals gained a reputation as a highly skilled and qualified engineer.

In 1907, President Theodore Roosevelt (1858-1919) appointed Goethals to the position of chairman and chief engineer of the Panama Canal project. This followed the resignation of two other engineers. At the beginning of the following year, he took complete control of the construction of the Canal. Goethals faced an enormous task complicated by technical problems and problems associated with coordinating the activities of 30,000 workers depleted by bouts of malaria and yellow fever. Goethals maintained a hands-on knowledge of the day-to-day construction of the Canal by visiting construction sights in person, and by holding informal sessions with his men every Sunday to work out problems with the crew. The Canal was completed six months ahead of schedule and opened for traffic in 1914. Goethals remained at the Canal to act as the governor of the Canal Zone from 1914 to 1916.

The Panama Canal is regarded as one of the world's most important and spectacular engineering feats. It links the Atlantic and the Pacific Oceans and extends 51 miles (82 km) from Limón Bay on the Atlantic to the Bay of Panama on the Pacific. The Canal shortened the ocean voyage between New York and San Francisco to less than 5,200 miles (8,367 km) from 13,000 miles (20,917 km). An important commercial and military waterway, approximately 13,500 ships pass through it a year carrying 220 million short tons (200 million metric tons) of cargo. Most of the traffic is to or from American ports, but other countries such as Canada and Japan frequently use the Canal as well. The Canal was a significant waterway during the Second World War, the Korean War, and the Vietnam War, with large amounts of equipment and troops passing through it. While the Panama Canal has often been a source of trouble between the Americans and Panamanians, it has inspired the imaginations of many and has demonstrated the amazing reach of human innovation.

George Washington Goethals enjoyed a long and distinguished career that extended past his achievements at the Panama Canal. He served as quartermaster general during the First World War, as director of purchase, storage, and traffic and as assistant chief of staff in charge of supplies. President of the engineering firm George W. Goethals and Company, from 1919 to 1928, Goethals continued to work in the field of engineering. He was also a consultant for many engineering organizations and on projects such as the Port of New York. Goethals died on January 21, 1928.

KYLA MASLANIEC

Ferdinand Anton Ernst Porsche

1909-1998

Austrian Automobile Designer

The family name of Ferdinand "Ferry" Porsche long ago became a household word, even if very few households could afford to have a Porsche in the garage. Less well-known is Porsche's role in creating a far more popular car—perhaps the most popular vehicle in history—the Volkswagen Beetle.

Born in Wiener Neustadt, Austria, Porsche was the son of Ferdinand and Aloysia Kaes Porsche. His father was an automobile designer who worked for the Austro-Daimler company, destined to become Daimler-Benz. In the year prior to his son's birth, the elder Ferdinand designed a car that traveled at 85 mi (137 km) an hour—a speed almost unbelievable in 1908. By 1923, when the family moved to Stuttgart, Germany, the Porsche's father was a board member with Daimler-Benz, and in 1930 he opened his own shop to build race cars.

When he was just 11 years old, young Ferdinand was given a two-seater car that his father had built, and later he helped his father build a lightweight race car. At age 12, he witnessed a fatal car crash during a race, and amazed investigators when he correctly traced the cause of the accident to the collapse of a wheel. As an adult, Porsche went to work with his father. He married Dorothea Reitze in 1935, and together they had four sons.

The Nazis' accession to power in 1933, an event that ended the career of so many great German scientists, actually stimulated Porsche's career. An automobile enthusiast, Hitler was determined to built a low-cost "people's car," or *Volkswagen,* and he commissioned Porsche and his father to design it. Later Porsche, like other collaborationist businesspeople, insisted that he had no choice but to work with Hitler. He also referred to the Russian prisoners of war who worked in the Volkswagen factory as "employees," though in fact they were slave-laborers.

After the war, Porsche moved the factory to Gmund, Austria. Due to the family's Nazi ties, the German government had taken away their contract to produce Volkswagens. Therefore, Porsche turned to a design he had created in 1939, and thus was born the Porsche sports car. In 1948 he introduced the first Porsche—the 356. Two years later, he moved production back to Stuttgart, and introduced a new model, the Carrera, with a new engine design. The company continued making Carreras, of which it sold nearly 80,000 models, until 1965. In 1964 Porsche introduced the 911. The Carrera RS followed in 1973, the 930 in 1974, and the 924 later in the 1970s.

Critics savaged the 924, not because it was not a good car, but because its fuel-efficiency (an outgrowth of the 1970s oil crisis) and its low cost threatened the Porsche brand's "snob appeal." Yet the car remained a symbol of prestige, highly popular and admired both by those who could afford it and those who could not. Though the company went public in 1972, Porsche's family retained a majority share of voting stock. Porsche died on March 27, 1998, in Zell-am-See, Austria.

JUDSON KNIGHT

Igor Ivanovich Sikorsky

1889-1972

Russian-American Aircraft Designer

Today remembered as the father of the helicopter, Igor Sikorsky had three distinct aviation careers. During the birth of aviation, Sikorsky designed and constructed the first successful large four-engine airplanes. After immigrating to America following the Russian revolution, Sikorsky's company built large flying boats for long-range airline service. Not until 1938 did Sikorsky embark on the third of his aviation careers, beginning design work on the helicopters for which he became famous.

Igor Sikorsky. *(Library of Congress. Reproduced with permission.)*

Igor was born in the Russian (now Ukrainian) city of Kiev, one of five children. His father taught psychology at the university and established a successful private practice. Igor's mother was also well educated. In one of Igor's earliest memories, his mother described Leonardo da Vinci's (1452-1519) designs for helicopter-like flying machines. Sikorsky spent three years studying at the Imperial Russian Naval Academy, then resigned in 1906 to pursue engineering. While visiting Germany in 1908, Sikorsky read his first account of the Wright brothers' flight. Recognizing his fascination, his older sister Olga offered Igor, then just 19, enough money to purchase an engine and some building materials. But the heavy engines of the time rendered his attempts to build a helicopter hopeless. Always persistent in the face of difficulties, Sikorsky instead designed an airplane and awaited the day when technological developments would make his helicopter dreams possible.

Sikorsky's success with airplanes was remarkable. In less than two years, by 1911, one of his planes set a world speed record. The planes were still frail, though. In one case, Sikorsky crash-landed after a mosquito was caught in his fuel tank and clogged the carburetor. But aviation progressed quickly. In 1913 Sikorsky constructed the first four-engine planes in the world. During World War I these massive planes became the first heavy bombers. Though the army initially found them almost useless, by 1917 the planes were quite successful. Several times they fought off attacks by five or more German fighter planes. No longer did mere mosquitoes endanger Sikorsky's aircraft. However, the Russian Revolution in 1917 ended Sikorsky's first career in aviation, forcing him to flee Russia for his own safety.

By March 1919 Sikorsky had arrived in New York City, ready to resume work in aviation. The end of the war led to hard times for an aircraft designer. While earning a meager living teaching math to other Russian immigrants, Sikorsky met Elizabeth Semion, and they were married in 1924. By 1923 Sikorsky was back on his wings, and 1928 marked the return of Sikorsky's success, as he became a United States citizen and sold the first of his flying boats to Pan American Airways. These designs culminated in the large "Clipper" planes that introduced long-range commercial air travel in the 1930s. However, again social turmoil overcame the technical innovations of Sikorsky's designs. By 1938 the Great Depression had dried up the market for large luxury flying boats, thus ending Sikorsky's second aviation career.

Fortunately, Sikorsky managed to keep his crack engineering team together as he entered his third aviation career, returning to his lifelong dream: building a practical helicopter. By late 1939 the prototype VS-300 was flying. An infusion of military support led to the creation of the R-4, the world's first mass-produced helicopter. Though helicopters played little role in World War II, they were rapidly adapted to military, civilian, and industrial uses after the war. In 1950 Sikorsky accepted the Collier Trophy, one of aviation's highest awards, on behalf of the helicopter industry he had founded. Igor Sikorsky retired in 1957 but remained active as a spokesman for the helicopter industry. He died in 1972 in Easton, Connecticut.

ROGER TURNER

Wernher Magnus Maximilian von Braun

1912-1977

German-American Rocket Engineer

Wernher von Braun developed the world's first guided missiles for the German military during World War II. After the war, his rockets were used to launch America's first space probes, and he supervised the development of the Saturn rockets that took astronauts to the Moon during the Apollo era.

Von Braun was born on March 23, 1912, in Wirsitz, Germany (now Wyrzysk, Poland), into a wealthy family. His boyhood interest in astronomy was encouraged by his mother, and he built an observatory at the boarding school he attended. He did not, however, do particularly well in his physics and mathematics classes, until frustration in trying to understand a book by the rocketry pioneer Hermann Oberth (1894-1989) motivated him to apply himself. In 1930 he began attending the Berlin Institute of Technology, where he joined the German Society for Space Travel and assisted Oberth in his liquid-fueled rocket experiments. He obtained his bachelor's degree in mechanical engineering in 1932.

During his studies for his Ph.D. in physics, which he received in 1934 from the University of Berlin, von Braun continued his rocket research with a grant from the German Ordnance Department. He then took a job with the department and soon launched two liquid-fueled rockets to altitudes of about 2 miles (3.2 km). Over the next three years he established a team of about 80 people working on a guided rocket that could carry a 100-lb (45 kg) payload to an altitude of 15 miles (24 km), and experimenting with a rocket-driven fighter plane. This team formed the core of the fa-

Wernher von Braun. *(The Library of Congress. Reproduced with permission).*

mous German rocket stronghold at Peenemunde, and von Braun became its director.

During World War II the Peenemunde rocket production site supplied the Nazis with the world's first guided ballistic missile, the V-2 (*vergeltungswaffen* or vengeance weapon), and the first guided anti-aircraft missile, called the Wasserfall. The V-2 was used with devastating effect on the cities of Allied nations. Von Braun later said that patriotism and the pursuit of scientific research outweighed moral issues in his involvement with the German military.

After the Nazis were defeated, von Braun and 116 other German rocket engineers surrendered to the United States Army and were promptly employed by its Ordnance Department. At first, stationed at White Sands, New Mexico, all they had to work with was captured V-2 rockets. During the early 1950s they developed the Redstone ballistic missile and a longer-range four-stage missile called the Jupiter at the Redstone Arsenal in Huntsville, Alabama. Von Braun became a U.S. citizen in 1955.

When the Russians launched the world's first artificial satellite, *Sputnik*, in 1957, the resulting space race between the superpowers changed the focus of von Braun's work. The first satellite launched by the United States, the *Explorer 1*, was placed into orbit using a modified Redstone rocket in January 1958. It was followed the next year by America's first interplanetary probe, *Pioneer IV*, launched with a Jupiter rocket and proceeding to an orbit around the Sun. Von Braun's team was transferred to the National Aeronautics and Space Administration (NASA) in 1960. The rocket facilities at the Redstone Arsenal became the nucleus of NASA's Marshall Space Flight Center, and von Braun was appointed the center's director. Marshall Space Flight Center's main task during the 1960s was to develop the huge Saturn launch vehicles for the Apollo program.

Von Braun wrote a number of popular books on space exploration, including *Across the Space Frontier* (1952), *Conquest of the Moon* (1953), *The Exploration of Mars* (1956), *First Men to the Moon* (1960), *History of Rocketry and Space Travel* (1969), and *Moon* (1970). He led Marshall Space Flight Center until 1970, when he was appointed deputy associate administrator for planning at NASA Headquarters. He resigned from NASA two years later to accept an executive position with the aerospace contractor Fairchild Industries. In 1975 he founded the National Space Institute, an organization dedicated to promoting space exploration. He died in Alexandria, Virginia, on June 16, 1977.

SHERRI CHASIN CALVO

Sir Robert Alexander Watson-Watt
1892-1973
Scottish Physicist

Historians of science regard radar, which uses radio waves to detect the positions of aircraft, and the atomic bomb as the two most important results of defense research in World War II. But whereas the names of J. Robert Oppenheimer (1904-1967), Enrico Fermi (1901-1954), and other creators of the bomb are well-known, that of Sir Robert Watson-Watt is hardly a household word. Yet Watson-Watt, who also coined the term "ionosphere," transformed the character of peacetime as well as wartime, and can justly be credited with saving many lives.

Watson-Watt was born on April 13, 1892, in Brechin, Scotland, the son of Patrick, a master carpenter, and Mary Matthew Watson Watt. Patrick had chosen to retain both his mother's and father's family names, but these were not hyphenated; Robert only began doing so after he was knighted in 1942.

After winning a scholarship to University College in Dundee, Watson-Watt studied electrical engineering, and became intrigued with wireless telegraphy. In 1912, he earned a bachelor of science degree in electrical engineering, and afterward worked briefly as an assistant professor at University College. With the beginning of World War I in 1914, he tried unsuccessfully to obtain a job with the British War Office, and instead went to work for the Meteorological Office. In this apparent setback were the seeds of his greatest triumph.

While working with the Meteorological Office, Watson-Watt suggested the use of radio wave triangulation as a means of tracking weather patterns. Others were receptive to the idea, but the necessary technology was years away. Once the war was over, Watson-Watt earned another bachelor of science degree, this time in physics, from the University of London. He then went to work at a field observing station in Ditton Park, Slough. In 1927, when Watson-Watt was appointed its director, this became the Radio Research Station, a unit whose mission was to conduct research on the atmosphere, the detection of naval signals—and the radio location of thunderstorms.

A strange request in 1935 led to the discovery of radar. An official of the British Air Ministry asked Watson-Watt if radio waves could be concentrated in such a fashion that they could destroy enemy aircraft. Watson-Watt and his assistant, A.F. Wilkins, explained the impossibility of such a Flash Gordon–style device, but they noted that by sending out radio waves, one could indeed detect the position of aircraft. On February 12, 1935, Watson-Watt explained this idea in a top-secret memo to the Air Ministry; later he would refer to this date as the birth of radar.

The Air Ministry responded with enthusiasm to the idea of radar, itself a term coined in the United States as an acronym for "radio detection and ranging." By then it was becoming increasingly clear that Germany was gearing up for war, and the Air Ministry hastily prepared radio stations throughout Britain. Historians would later credit radar with helping England survive the Nazi bombing raids in the Battle of Britain.

For his contribution to the war effort, Watson-Watt was knighted, and received a number of other honors as well. Following the war, he founded a consulting company, Sir Robert Watson-Watt and Partners. He was married three times, the last time in 1966, to Dame Katherine Jane Trefusis-Forbes, former head of the Women's Royal Air Force. After 1952, she and Watson-Watt lived primarily in Canada and the United States, where he apparently worked as a freelance consultant. She died in 1971, and Watson-Watt on December 5, 1973, in Inverness, Scotland.

JUDSON KNIGHT

Sir Frank Whittle
1907-1996
English Aviation Engineer

Sir Frank Whittle was responsible for one of the most important inventions to come out of World War II—a machine that, like the computer, arrived on the scene late and only came to prominence in the postwar years: the jet engine. It is a distinction Whittle shares with someone he never worked with, German engineer Hans von Ohain (1911-1988), who simultaneously built a jet engine for the German war effort.

Whittle was born in Coventry, England, on June 1, 1907. His father, a machinist, was also an inventor, and in 1916 the elder Whittle went into business for himself as director of the Leamington Valve and Piston Ring Company. His young son helped out at the factory, where he gained considerable experience with the mechanics of machinery.

Young Frank found little to capture his attention in school, where the subjects that most interested him—astronomy, engineering, and other sciences—were not on the curriculum. In secondary school he became intrigued with aeronautics, and after graduation this led to his joining the Royal Air Force (RAF). Because he was only five feet tall, the RAF very nearly refused to accept Whittle, but such was his record of service that after three years of rigging aircraft, he was accepted to the RAF College at Cramwell as a cadet. During his off time, he participated in the Model Aircraft Society, which he later credited with greatly expanding his knowledge of flying.

After he joined the 111 Fighter Squadron in 1928, Whittle became intrigued with the central problem facing the aircraft industry at that time: overcoming the limitations that both propellers and piston engines placed on altitude and speed. At high altitudes, the air was too thin to properly engage propellers, nor could a piston engine continue to run on the meager oxygen content of high-altitude air. After considering the problem, Whittle proposed a means of doing away both with propellers and pistons by using a turbine engine. The turbine could compress the thin oxygen of the upper stratosphere, combine it with fuel, and ignite it, and the expanding gases would result in a jet blast that would propel the craft.

Instead of being applauded, Whittle met with resistance and objections, chief among them the fact that the RAF possessed no materials that could withstand the heat and stress created by a jet engine. Whittle persevered, however, and eventually found a group of outside backers with whom he formed a corporation called Power Jets in 1936. By April 1941, more than 18 months after the beginning of World War II in Europe, Whittle was ready to test his W.1 jet engine.

On May 15, 1941, an RAF pilot took up the Gloster-Whittle E28/29 aircraft, created by Whittle, for a 17-minute flight. During that time, the aircraft reached a speed of 370 miles per hour (592 km/hour)—unheard-of at the time—at an elevation of 25,000 ft (7,620 m). Duly impressed, the British government supported Power Jets in the process of refining the engine and aircraft in time to begin production in June 1942.

In 1943 Rolls-Royce took over Power Jets, and in the following year Great Britain nationalized it. Only in 1944 did jets reach the skies over Britain, when the Meteor I helped defend the island against German dictator Adolf Hitler's last-gasp bombardment with V-1 rockets. By then it was clear that the Allies would win the war with or without the jet, and enthusiasm waned among Whittle and his colleagues. Whittle himself left the company in 1946.

During the remaining half-century of his life, Whittle accepted a number of honors and awards, among them knighthood in 1948. He had married Dorothy Lee of Coventry in 1930, and they had two sons; they were divorced in 1976. Whittle served as technical consultant for a number of firms, and following his divorce moved to the United States. There he married a second wife, Hazel, and went to work developing jets for the U.S. Navy. He died at his home in Columbia, Maryland, on August 8, 1996, aged 89.

JUDSON KNIGHT

Wilbur Wright
1867-1912
American Pilot and Engineer

Orville Wright
1871-1948
American Pilot and Engineer

Wilbur and Orville Wright built and flew the airplane that made the first controlled, powered flight in December 1903. Their design was shaped both by their extensive knowledge of others' work on aeronautics and by their own careful research on engines, propeller design, and control systems. They went on, in 1908-12, to play a key role in popularizing the new technology.

Born near Dayton, Ohio, during the last third of the nineteenth century, Wilbur and Orville Wright brought business acumen, mechanical skill, and disciplined minds to their experiments with flying machines. They closely followed the work of Otto Lilienthal (1848-1896) and Samuel Pierpont Langley (1834-1906) and, most importantly, corresponded extensively with Octave Chanute (1832-1910). Chanute, a French-born aviator living in Chicago, provided the Wrights not only with news of his own experiments with gliders, but also with a connection to aeronautical work underway in Europe.

Simultaneously, the brothers began a systematic program of experiments and glider flights designed to give them first-hand knowledge. Unable to find reliable scientific data on the aeronautical properties of wings and propellers, they built a wind tunnel and compiled their own data to establish which shapes provide maximum lift and thrust. The brothers also built a series of full-size aircraft, flying them first as kites and then as gliders from the sand dunes near Kitty Hawk, on North Carolina's Outer Banks. The dozens of glider flights they made in 1901 and 1902 turned both Wrights into expert pilots—a comparative rarity among airplane designers of that era. The gliders also allowed them to experiment with different methods of controlling an aircraft in flight.

The end-product of this research program was the powered *Flyer I,* which rose into the air on December 17, 1903. It used the same basic design as their most advanced gliders, but added a pair of propellers driven by a lightweight, 12-horsepower gasoline engine of their own design. The first flight of the day, with Orville at the controls, lasted 12 seconds and covered 120 ft (37 m); the fourth and last flight, made by Wilbur, lasted a full minute and covered nearly 600 ft (183 m). Though witnessed by five reliable observers and documented in a photograph, the landmark flights of December 17 were ignored for nearly five years. An editorial acknowledgement in the prestigious magazine *Scientific American* began to turn the tide in 1906, but the Wrights began to receive full credit only in 1908.

Orville and Wilbur Wright. *(Corbis Corporation (New York). Reproduced with permission.)*

The Wrights, methodical in business as they were in engineering, took carefully planned steps to promote and profit from their invention. Orville sold the first of many Wright-built observation planes to the United States Army in 1908, and Wilbur went to Europe to drum up military interest there. Orville started flying schools in the United States in 1910, training novice pilots such as Henry H. ("Hap") Arnold, future commander of the United States Army Air Forces, and Cal Rodgers, who made headlines in 1911 by flying a Wright machine across the United States. Several graduates of the Wrights' flying schools went on to work for the brothers as demonstration pilots, raising money and promoting the new technology by performing at fairs and exhibitions. The company remained solidly, but not spectacularly, successful until 1912, when Wilbur died of tuberculosis and a disheartened Orville sold the firm and its patents to a group of investors.

Orville continued to fly regularly until 1918. He lived to see his belief in the airplane's enormous military and commercial potential fulfilled, and to hear, in the year before his death, of the first airplane to fly faster than the speed of sound.

A. BOWDOIN VAN RIPER

Ferdinand Adolf August Heinrich, Graf von Zeppelin

1838-1917

German Military Officer and Inventor

Ferdinand Graf von Zeppelin was responsible for the development of the dirigible, a lighter-than-air vehicle known as the zeppelin. These airships were used extensively in the early 1900s until the *Hindenburg* disaster of 1937.

Ferdinand von Zeppelin was born in Konstanz, Baden, Germany, in 1838. Even at 12 years of age, Zeppelin exhibited a keen interest in technology. At the age of 17, he attended the military academy in Ludwigsburg and was also educated at the University of Tübingen, receiving a degree in civil engineering.

He entered the Prussian Army in 1858, and served in the Austro-Prussian War in 1866 and the Franco-Prussian War from 1870-1871. In 1863 he went to the United States as a military observer with the Union Army with the purpose of gaining a better understanding of the military use of hot air balloons. Zeppelin became convinced that airships could be used practically in the military. He took his first balloon ride in 1863, in St. Paul, Minnesota, increasing his interest in aeronautics. Publishing a comprehensive plan for a civil air transportation system using large, lighter-than-air ships in 1887, he tried to convince the German military of the potential value of the airships. The Germans were not interested initially, and in 1890 Ferdinand von Zeppelin retired from military service to pursue his interest in airships full-time.

Zeppelin's first concept was of an "air train," with wagons that could be connected for the transport of goods. Struggling with his plans for 10 years, he patented his design and began construction of the airship in 1899. The initial concept was gradually transformed into the zeppelin airship of 1900. First taking flight on July 2, 1900, from a floating hangar on Lake Constance, near Friedrichshafen, Germany, the LZ-1 carried five persons 3.75 mi (6 km) at an altitude of 1,300 ft (396 m).

The first zeppelin had a rigid frame made of metal, and the entire cylindrical framework of the ship was covered in a smooth-surfaced cotton cloth. Gas bubbles filled with hydrogen gas gave the craft its lift, and it was powered by two 15 hp Daimler internal combustion engines. Forward and aft rudders controlled the steering and the passengers were carried in aluminum

Ferdinand Graf von Zeppelin. *(Corbis Corporation. Reproduced with permission.)*

gondolas. The first flight experienced some problems but inspired enough interest in the public that Zeppelin was able to continue his work through public donations. The German government began to take notice, and when a zeppelin craft achieved sustained flight over a 24-hour period in 1906, the Germans commissioned an entire fleet. Before Graf von Zeppelin's death, 130 of the ships were built with nearly 100 of them being used in the First World War.

Zeppelin airships could reach speeds of 81 mi (136 km) an hour and could reach heights of 13,943 ft (4,250 m). The height achieved by the craft was much greater than that of any airplane at the time, and was used in air raids over Britain and France. They were armed with five machine guns and could carry 4,410 lbs (2,000 kg) of bombs. The airships were large and slow and proved to be vulnerable to anti-aircraft fire, but they were still used to transport supplies.

Zeppelins were also used in commercial passenger service, and were the first airships to be used for that purpose in 1910. The *Graf Zeppelin*, completed in 1928, and the *Hindenburg* of 1936 were the two most famous Zeppelins. Used in transatlantic flight service, the *Graf Zeppelin* made 590 flights before being decommissioned in 1937. In 1929, the airship completed a world trip, covering 21,550 mi (34,600 km) in 21 days. The *Hindenburg* was 804 ft (245 m) long, and achieved a maximum speed of 84 mi (135 km) per hour. While landing at Lakehurst, New Jersey, on May 6, 1937, the hydrogen-filled airship burst into flames. Thirty-six people died in the crash and soon after the airships ceased to be used.

While airships like the zeppelins are no longer used today, they had a great impact on the early 1900s and indeed the rest of the century. Ferdinand Graf von Zeppelin's airships inspired the imaginations of other aeronautic inventors and proved the potential value of air passenger service, and blimps filled with hot air are used at sporting events today. These developments helped to fuel the fire of aviation research in the early part of the century.

KYLA MASLANIEC

Vladimir Kosma Zworykin
1889-1982
Russian-American Physicist and Electrical Engineer

Recognized as the "father of television," Vladimir Zworykin created the iconoscope and the kinescope, two inventions that made that machine possible. Yet that was far from the only achievement credited to this prolific genius, who in his lifetime obtained more than 120 patents. Among his other inventions was the electron microscope, which greatly expanded scientists' knowledge by making it possible to see objects much smaller than those glimpsed by regular microscopes. As for his principal invention, Zworykin was asked in 1981 what he thought of American television programming: "Awful," was his reply.

The son of Kosma, who operated a fleet of river boats, and Elaine Zworykin was born on July 30, 1889, in Mourom, Russia. Zworykin studied electrical engineering under Boris Rosing, an early advocate of cathode ray tubes, at St. Petersburg Institute of Technology. Cathode ray tubes shot streams of charged particles, and Rosing—going against the prevailing wisdom among the few scientists then considering the possibility of television—maintained that this, and not the mechanical systems then being tested, was the most viable television technology.

After earning his degree at St. Petersburg in 1912, Zworykin went on to the Collège de France in Paris, where he studied x-ray technology under Paul Langevin (1872-1946), a renowned French physicist. He served as a radio officer in the Russian army signal corps in World War I, during which time he also married Tatiana Vasilieff. The couple later had two children, but in the face of the Communist takeover, they decided to leave Russia in 1918. By 1919 they were living in the United States.

Zworykin first took a job with Westinghouse, where in 1920 he began work on developing radio tubes and photoelectric cells, small devices whose electrical properties are modified by the action of light on them. He wrote his doctoral dissertation on photoelectric cells at the University of Pittsburgh, and in 1923 filed a patent for his iconoscope. In contrast to the mechanical systems then under development by figures such as Great Britain's John Logie Baird (1888-1946), Zworykin's iconoscope was electronic. It replicated the actions and even the structure of the human eye, and produced a far better picture than a mechanical system— without requiring nearly as much light.

In the following year, Zworykin filed a patent for the invention that, with the iconoscope, would make television possible: the kinescope, or picture tube. Zworykin's special cathode ray tube overcame problems first noted by Scottish physicist A. A. Campbell Swinton in 1908, and provided a practical means for bombarding a signal plate with electrons, thus producing an image.

Zworykin demonstrated his invention to executives at Westinghouse, and was told that he should spend his time on something "a little more useful." Thus the company missed one of the greatest business opportunities in history, and Zworykin took his talents to RCA, the Radio Corporation of America, in 1929. RCA had to invest plenty before it reaped any rewards, however. Initially Zworykin told RCA's David Sarnoff (1891-1971) that development of television would cost "about $100,000"; in fact, as Sarnoff later told the *New York Times,* "RCA spent $50 million before we ever got a penny back from TV."

In 1930 Zworykin and G. A. Morton had created an infrared image tube that made night vision technology possible, and the military adapted this as Sniperscope and Snooperscope during the war. Also during the war, Zworykin collaborated with John von Neumann (1903-1957) at Princeton's Institute for Advanced Studies on the development of one of the first computers.

With the end of World War II and the lifting of restrictions regarding the manufacture of receivers, television exploded. Soon Zworykin's invention, if not his name, was making its way into virtually every household of the industrialized world. Meanwhile, he turned his attention to a number of inventions, among them the electron microscope, the electric eye used in security systems and automatic door openers, electronic missile controls, and some of the earliest electronic technology to aid the blind in reading print.

Zworykin and his first wife divorced, and in 1951 he married Katherine Polevitsky. Among the many awards he received in his lifetime were the Edison Medal from the American Institute of Electrical Engineers (1952) and the National Medal of Science (1967). He was elected to the National Academy of Sciences in 1943, and also received the French Legion of Honor. Zworykin died one day before his 93rd birthday, on July 29, 1982.

JUDSON KNIGHT

Biographical Mentions

Howard Hathaway Aiken
1900-1973

American mathematician who invented the first automatic calculator. Howard Hathaway Aiken was born at the turn of the century in Hoboken, New Jersey. He studied engineering at the University of Wisconsin, Madison, then received his doctorate at Harvard University. In 1939, with the help of three other engineers at the International Business Machines (IBM) Laboratory, Aiken began work on the first automatic calculating machine. Their Mark I, which was completed in 1944, performed four operations—addition, subtraction, multiplication, and division—and was able to check back on previous results. The machine weighed 35 tons (32 metric tons), contained about 500 miles (805 km) of wire, and was able to complete its calculations without human intervention. Aiken went on to produce three more computers, including the electric Mark II in 1947.

Edwin H. Armstrong
1890-1954

American electrical engineer who invented the regenerative circuit, which laid the foundation for modern radio and television circuitry. (A circuit indicates the complete path followed by an electric current from an energy source, to the energy-using device, and back to the source again.) Edwin H. Armstrong was born in New York City in 1890. At the age of 14, he was thrilled by tales of Guglielmo Marconi's (1874-1937) first transmission of a wireless message across the Atlantic Ocean, and was determined to one day become an inventor. Armstrong studied electrical engineering at Columbia University, and it was there that he began work on his regenerative circuit, a device which amplified television or radio signals so that they could be heard across a room. He later invented the superheterodyne circuit, which amplified weak electromagnetic waves and today forms the basis of most radio, radar, and television reception. Arguably his greatest achievement was the creation of a wide frequency radio transmission signal, which became known as FM radio.

John Bardeen
1908-1991

American physicist who was awarded the 1956 Nobel Prize for Physics, shared with Walter Brattain and William Shockley, for research on semiconductors and discovery of the transistor effect, which revolutionized the electronics industry. In 1972 Bardeen won a second Nobel Prize with Leon Cooper and J. Robert Schrieffer for their explanation of superconductivity, known as the BSC theory. Bardeen is one of only three people to have received two Nobel Prizes, and the only two-time recipient of the Nobel Prize for Physics.

Manson Benedicks

American physicist who in 1915 discovered that the germanium crystal could be used to convert alternating current (AC) into direct current (DC). This would become the basis of the integrated circuit, a miniature electric circuit packaged as a single unit with the input, output, and power-supply connections self-contained. Integrated circuits are used in computers as memory chips.

Emile Berliner
1851-1929

German-born inventor who developed a helicopter that flew in 1919. Berliner came to the United States in 1870 to study physics. He developed a carbon microphone transmitter that was used by the Bell Telephone Company in their phones. The gramophone he designed was the first to use a flat disc rather than the cylinder shape proposed by Thomas Edison, and was bought by the RCA Company. In 1919, he designed one of the first helicopters that flew.

Clifford E. Berry
1918-1963

American computer designer who along with John V. Atanasoff designed and built the first electronic digital computer. The first design (1939) featured about 300 vacuum tubes for control and arithmetic calculations, use of binary numbers, logic operations, memory capacitors, and punched cards as input/output. This prototype influenced the direction of electronic computing technology, as it was the first to use electronic means to manipulate binary numbers.

Clarence Birdseye
1886-1956

American businessman and inventor whose name became synonymous with frozen foods. Clarence Birdseye was born in Brooklyn, New York. He was a naturalist from early on, but turned to business when he found that he lacked the funds to finish his studies at Amherst College. He left school to become a fur trader in Labrador in 1912 and again in 1916. While there, he noticed that the locals froze food in order to sustain themselves during the long harsh winter. When Birdseye returned to the United States, he formed General Seafoods Company, where he began selling his own brand of frozen foods. His process of quick freezing preserved the flavor of the food. Birdseye's frozen fish, fruits, and vegetables quickly became a huge success with the public. From 1930-34 he served as president of Birds Eye Frosted Foods, and from 1935-38 he ran Birdseye Electric Company. Birdseye held nearly 300 patents on inventions ranging from heat lamps, to a harpoon gun.

Harold Stephen Black
1898-1983

American electrical engineer who revolutionized the telecommunications industry with his method of eliminating distortion in amplification. Harold Stephen Black was born in Leominster, Massachusetts. He graduated from Worcester Polytechnic Institute in 1921, and several years later received an honorary doctorate in engineering from Worcester Tech. After graduation, he accepted a job with the Western Electric Company, the forerunner of Bell Telephone Laboratories. While there, he discovered the principles for his negative-feedback amplifier, which fed systems output back into the input, producing amplification virtually without distortion. His principles were later used in telephones, radar, weaponry, and electronics.

Henry Albert Howard Boot
1917-1983

British physicist who designed the radar. The device known as the cavity magnetron had a powerful output of centimeter-waves that allowed for the necessary preciseness. This innovative invention is thought to have influenced the outcome of World War II. The radar enabled a precise beam and small lightweight transmitters to be used especially at night on bomber planes. In 1945 Boot began working in the emerging field of nuclear physics, where he helped to design the cyclotron and the magnetron.

Andrew Donald Booth

British physicist who designed and produced an electronic stored-program computer at the University of London. During World War II, while working as a mathematical physicist, he was researching crystalline structure using x-ray diffraction data. This work required tedious complex mathematical calculations and provided an incentive to develop an automatic calculator. Booth designed and produced the Automatic Relay Computer (ARC) during 1947-49 at the Birkbeck College computation laboratory at the University of London.

Herbert Cecil Booth
1871-1955

British engineer who invented the vacuum cleaner. In 1900 he demonstrated the principle of removing dust from carpets by suction, and in 1901 he patented an electrically powered machine he called a "vacuum cleaner." The machine was mounted on a horse-drawn wagon and had a long tube to access buildings and houses. His Vacuum Cleaning Company was the first commercial cleaning company.

Robert August Bosch
1861-1942

German precision mechanic who helped design and produce the low-tension magneto, the spark plug, and the distributor for multi-cylinder engines in 1902-03. He is also known for introducing the eight-hour workday, as well as advocating industrial arbitration and free trade. In 1932 he wrote a book on the prevention of world economic crises.

Walter Houser Brattain
1902-1987

American physicist who was awarded the 1956 Nobel Prize for Physics, with John Bardeen and William Shockley, for research on semiconductors and creating the transistor. Brattain joined Bell

Telephone Laboratories in 1929. Shockley arrived in 1936 and designed various semiconductor devices that Brattain built and tested. After a 22-month war hiatus they were joined by Bardeen. Brattain and Bardeen completed the first working transistor in 1947. Brattain successfully constructed an improved transistor by 1950.

George Harold Brown
1908-1987

American electrical engineer whose efforts led to the development of radio and television broadcast antennas. In 1933 he received a Ph.D. from the University of Wisconsin and then joined the Radio Corporation of America (RCA) as a research engineer. During 1934-36 Brown developed broadcast antennas that could transmit electromagnetic waves in a certain direction, which led to the "turnstile" antenna. This latter device in turn became the industry standard for television and frequency-modulated (FM) radio broadcasting.

John Moses Browning
1855-1926

American inventor who designed and produced a wide variety of firearms. The son of a gunsmith, he produced his first gun from scrap metal when he was thirteen. In the late 1890s he developed an automatic machine-gun that used propelling gas as the driving force to drive the piston, extract the cartridge case, reload, and refire the ammunition. In 1917 the United States Army purchased nearly 57,000 of these guns for use in World War I.

Arthur Walter Burks
1915-

American philosopher-logician who was among the principal designers of the ENIAC computer, built in 1945. While at Princeton's Institute for Advanced Study, Burks, John von Neumann, and Herman Goldstine produced one of the most influential reports on computers—*Preliminary Discussion of the Logical Design of an Electronic Computing Instrument* (1946). Burks's research group at the University of Michigan (1946-1985) worked on programming, automata theory, neural net simulation, and self-reproducing cellular systems. Burks later turned his attention to adaptive computation.

Chester Floyd Carlson
1906-1968

American physicist whose work led to the development of paper-copying machines. By 1938 Carlson had devised a basic system of electrostatic copying onto plain paper, which after 12 years work gave the xerographic method that is widely used. The process uses chemical, electrical, or photographic techniques to copy printed or pictorial documents. The process was bought by the Xerox Corporation.

Willis Haviland Carrier
1876-1950

American engineer and inventor who developed the formulae and equipment that made air conditioning possible. Carrier developed an apparatus for treating air that was both safe and nontoxic. It became the first air conditioner and was used to enhance comfort and improve industrial processes and products. In 1911, he disclosed his formulae to the American Society of Mechanical Engineers, and it remains the basis for fundamental calculations used by the air conditioning industry.

Juan de la Cierva
1886-1936

Spanish aviator and aeronautical engineer who pioneered rotary flight. In 1918, after receiving his engineering degree, he built the first trimotor airplane in Spain. In 1923 he designed an aircraft called the *autogyro* that combined the capabilities of a conventional fixed wing aircraft and the rotary action of the helicopter. The autogyro was used in the United States during 1931-33. In 1936 design variations on the autogyro led to the construction of the first successful helicopter.

Georges Claude
1870-1960

French chemist and physicist known for his invention of the neon light. Vapor-tube lamps, popular at the time, produced light by striking an arc between electrodes in a tube containing certain gases. He discovered that a vapor tube filled with neon gas produces a bright orange-red light when put under low pressure. By adding other substances to the tube, additional colors could be produced. By the 1920s neon lighting became a popular form of advertising.

Paul Cornu
1881-1914

French aeronautical engineer who designed and flew a prototype of the helicopter. In 1907 he was successful in completing the first manned helicopter flight by flying one-foot (.3048 m) off the ground and hovering for 20 seconds. On later flights he acquired a height of 5 feet and (1.524 m) was timed at 6 miles per hour (9.656 km/hour). The two-rotor machine measured 40 ft (12.2 m), weighed 573 pounds (260 kg), and was powered by a 24-horsepower Antoinette engine. Its landing gear comprised 4 bicycle tires.

Jacques-Yves Cousteau
1910-1997

French diver who spent 60 years exploring the world's oceans. Jacques-Yves Cousteau was born in a small town near Bordeaux, France, in 1910. Although he was a sickly child, he loved to swim and often spent hours at the beach. His first dive was in Lake Harvey, Vermont, in the summer of 1920. From that point on, the sea truly became his passion. He joined France's Naval Academy and served in World War II, assisting the French Resistance. It was during the war that he made his first underwater films, with the help of the Aqualung, which he invented with engineer Emile Gagnan. Their invention freed divers from having to use unwieldy diving helmets, and allowed divers to stay underwater for longer periods of time by providing them with pressurized air while submerged. In 1950, with money given to him by a millionaire, Cousteau bought his now-famous boat—the *Calypso*—and turned it into a floating oceanic laboratory. He authored several books and produced numerous documentaries on the sea, which garnered him 40 Emmy nominations.

Glenn Hammond Curtiss
1878-1930

American aviator and inventor who made the first public flights in the United States and designed several aircraft, including the flying boat and the seaplane. (A seaplane is designed to take off from or land on a body of water.) Glenn Hammond Curtiss was born in Hammondsport, New York, in 1878, his middle name being chosen in honor of his town's founder. Curtiss was thrilled by speed, experimenting with bicycles, motorcycles, and flying machines. His exploits even made him the model for a series of children's books entitled *The Adventures of Tom Swift*. Curtiss designed and built engines for aircraft, then tried them out in the air. In 1908, he won the *Scientific American* trophy after his successful flight of 0.6 miles (1 km). He was the first to build seaplanes in America, earning him the title "The Father of Naval Aviation." His planes were widely used by Great Britain, Russia, and the United States during World War I.

Jacques Arsène D'Arsonval
1851-1940

French inventor known for his idea called "Ocean Thermal Energy Conversion" (OTEC). OTEC uses the temperature difference between the warm surface of seawater and the colder deeper layers of ocean water to generate electricity. His original concept used a fluid with a low boiling point that vaporized by using the heat extracted from the warm water on the surface. This heated fluid was to turn a turbine to produce electricity.

Walter Robert Dornberger
1895-1980

German rocket engineer and scientist. As an engineer and officer in the German Army, he established an experimental rocket station that successfully fired a 60-pound (27-kg) thrust rocket in 1932. During World War II he directed the development of the V-2 rockets. After spending three years as a prisoner of war, he came to the United States as a consultant to the Air Force and in 1950 joined the Bell Aircraft Corporation.

William Henry Eccles
1875-1966

English physicist and engineer best known for pioneering work in radio communication. To explain Marconi's transatlantic radio transmissions, Oliver Heaviside postulated an upper atmospheric layer capable of reflecting radio waves (1902). Eccles demonstrated how upper atmospheric ionization could produce this "Heaviside layer" and how diurnal variations in radio wave propagation could be accounted for by solar radiation (1912). He investigated effects of atmospheric disturbances on reception and developed various radio wave detectors and amplifiers.

J. Presper Eckert, Jr.
1913-1995

American electrical engineer who is credited with co-inventing the first electronic computer. J. Presper Eckert, Jr. graduated from the University of Pennsylvania's Moore School of Electrical Engineering with a degree in electrical engineering in 1941. At Moore, he became fascinated with his professor John Mauchly's (1907-1980) ideas of building a computer. In 1943, Eckert began work on an electronic digital computer for the United States government. His enormous ENIAC (Electronic Numerical Integrator and Computer) weighed 30 tons (27 metric tons) and filled an entire room. The computer performed high-speed calculations, and became the precursor to all modern-day computers. Eckert and Mauchly also designed the BINAC (Binary Automatic Computer) and UNIVAC I (Universal Automatic Computer).

Alva John Fisher
1862-1947

American inventor of the domestic electric washing machine. In 1910 he patented a washing machine powered by a small electric motor and equipped with a self-reversing gearbox to ensure that the clothes were not compacted into a solid mass. Named the "Thor" by the Hurley Machine Company of Chicago, this was the forerunner of the modern washing machine.

Heinrich Focke
1890-1979

German aeronautical engineer who designed the first successful helicopter flown in 1936. The autogyro, a hybrid airplane-helicopter, was introduced in 1920 by Spanish engineer Juan de la Cierva. However, the craft was difficult to maneuver. Focke's design, a twin-rotor aircraft, was stable and easier to maneuver. In 1924 he and George Wulf started an aircraft company building a variety of civil and military aircraft including the helicopter. After World War II he taught aeronautical engineering at the Technical University in Stuttgart, Germany.

Enrico Forlanini
1848-1930

Italian engineer who was at the forefront in the development of the hydrofoil. The hydrofoil is a cross between a ship and an airplane. It is raised above the water by small wing-like foils. At a specific speed the lift generated by foils raises the hull out of the water. Forlanini's hydrofoil reached a speed of 38 knots in 1905.

Emile Gagnan

French engineer who with Jacques-Yves Cousteau designed the Self-Contained Underwater Breathing Appartus, also known as the aqualung. In 1943 Cousteau and Gagnan developed the aqualung tank that released compressed air activated by the breath itself. Gagnan perfected the regulator valve that would automatically feed the compressed air to the mouthpiece in ratio to the diver's breath. This invention allowed for the exploration of the ocean depths for scientific observation, military use, and recreational purposes.

Hans Wilhelm Geiger
1882-1945

German nuclear physicist who developed various instruments and techniques for measuring charged nuclear particles—the most famous of which is the Geiger-Müller counter, commonly referred to as the Geiger counter. Geiger invented an alpha-particle detector in 1908 that allowed him and Ernst Rutherford to show alpha-particles to be helium nuclei. In 1909 Geiger performed his now famous alpha-particle scattering experiments with Ernest Marsden. Their results directly led to Rutherford's nuclear theory of the atom.

Kate Gleason
1865-1933

American engineer and businesswoman who became the first female president of a United States bank, as well as the first woman to join the American Society of Mechanical Engineers. Kate Gleason was born in Rochester, New York. At the age of 12, she began working in her father's machine-tool factory. In 1884, she was admitted to Cornell University, where she studied mechanical arts. After graduation, she helped her father create a machine that beveled (or cut at an angle other than 90 degrees) gears quickly and cheaply. American industrialist Henry Ford (1863-1947) called the invention "the most remarkable work ever done by a woman." From 1890-1901 she served as secretary-treasurer of her father's company, helping make Gleason Works a leading producer of gear-cutting machinery. In 1913, Gleason went out on her own and four years later became the first woman president of Rochester's First National Bank. In 1918, she also became the first woman elected to the American Society of Mechanical Engineers.

Peter Carl Goldmark
1906-1977

Hungarian-American engineer who pioneered the first color television. From 1936-40, while working for the Columbia Broadcasting System, he designed a color television system that utilized a rotating, three-color disk and a cathode ray tube. In 1948, with the invention of vinyl, he developed the long-play record that used a groove of 0.003 inch (0.008 cm). This enabled the equivalent of six 78-rpm records to be compressed into one record and led to the availability of stereo recordings.

Herman Heine Goldstine
1913-

American mathematician involved in the development of the Electronic Numerical Integrator and Computer (ENIAC), the first electronic digital computer. Designed for the U.S. Army, this first prototype contained 18,000 vacuum tubes and was used to compute ballistic tables. After leaving the Army in 1945, he joined the Institute for Advanced Study. There, with John von Neumann, he helped build the prototype of the present-day computer using a magnetic drum for data storage.

Leon Alexandre Guillet
1873-1946

French engineer whose scientific work was concerned with the theory of metal alloys. He investigated the properties of special kinds of steel: those made with nickel, manganese, chromium, and tungsten as well as bronzes and brasses. During World War I, he experimented with thermally treating metal alloys and developed a form of stainless steel. In 1905 he became director of the research department of one of the largest automobile factories in France.

Douglas Rayner Hartree
1897-1958

English physicist who developed new methods of numerical analysis. Douglas Rayner Hartree was born in Cambridge, England. He entered St. John's College, Cambridge in 1915, but was called away from his studies to join a team studying anti-aircraft gunnery during World War I. After the war, he completed his studies at Cambridge and went on to earn his doctorate in 1926. From 1929-37 he served as chair of theoretical physics at Manchester, Cambridge. As a physicist, Rayner developed both new methods of numerical analysis, as well as ways of using differential equations to calculate atomic wave functions. Before computers had been invented, he developed a differential analyzer which calculated sophisticated equations. When J. Presper Eckert, Jr. (1913-1995) developed his ENIAC computer, Hartree traveled to the United States to assist the machine with its earliest calculations.

Harold Locke Hazen
1901-1980

American engineer who with Vannevar Bush invented the electromechanical analog computer that could solve sixth-order differential equations and three simultaneous second-order differential equations. A lifelong professional colleague of Bush at Massachusetts Institute of Technology, he worked on the atomic bomb, the Rapid Arithmetic machine, and the Rockefeller differential analyzer.

Peter Cooper Hewitt
1861-1921

American electrical engineer who invented the mercury-vapor lamp. Between 1901-03 he made his mercury-vapor lamp commercially available as a reliable form of industrial lighting. His quartz-tube mercury lamp was widely used in the biological sciences. Other inventions included a device for converting alternating current into direct current and a radio receiver.

René Alphonse Higonnet
1902-1983

French inventor who, along with engineer Louis Moyroud, developed the first practical, high-speed phototypesetting machine. Higonnet began his engineering education in France, then traveled to the United States to complete his studies at several American schools, including Harvard University. Higonnet worked as an engineer with Matériel Téléphonique, a French subsidiary of ITT, from 1924 to 1948. In 1946 he teamed up with Moyroud to develop the Lumitype (also known as the Photon), the first phototypesetting machine, which debuted in the United States in 1948.

Albert Wallace Hull
1880-1966

English physicist who independently discovered the powder method for x-ray analysis of crystals. After receiving his Ph.D. at Yale in 1909, Hull joined General Electric as a research physicist in 1914, and later served as assistant director of its research laboratory. In 1916 scientists Peter Debye and Paul Scherrer devised the powder method technique of x-ray analysis. Unaware of their work, a year later Hull used this same method to examine the crystal structure of several common metals, including iron. He then turned his attention to electronics, inventing several components used in electric circuits, including the magnetron, an oscillator that generates microwaves.

Millar Hutchinson

American inventor who invented the electrical hearing aid in 1902. The hearing aid, by amplifying sounds, has allowed countless numbers of people to retain or improve their hearing.

F.W. Jordan

American inventor whose work with W.H. Eccles led to the invention of the first "flip-flop" circuit in 1919. Flip-flop circuits are electrical circuits that can have one of two "states," such as energized (or on) or de-energized (or off), and that can "flip-flop" between those two states. Such circuits were essential to designing the first logic circuits, such as those found in all computers, and this invention was crucial to the invention of digital computers.

Hugo Junkers
1859-1935

German aviation engineer who built some of the first all-metal airplanes and was responsible for innovations in heating systems and engine designs. In 1890 Junkers founded a research institute devoted to the study of engine and airflow technology. He opened an aircraft factory at Dessau in 1910, patented his flying-wing design, and, in 1915, launched the first successful all-metal airplane, the *JU-1 Blechesel* ("Sheet Metal Donkey") monoplane. During the 1920s and 1930s he developed commercial passenger aircraft and dive bombers used by the German Luftwaffe during World War II. Despite his great success, in 1933 the Nazis banned Junkers from his work and took over his plant.

Charles Franklin Kettering
1876-1958

American engineer whose more than 300 patented inventions include the first electrical ignition system for automobiles, the self-starter, and the first practical engine-driven generator. In 1909 Kettering left the National Cash Register Company (NCR) to establish the Dayton Engineering Laboratories Company (Delco) with businessman Edward A. Deeds. In 1916 he sold his company to General Motors (GM) and became the vice president of its research corporation. Under Kettering's leadership the GM lab developed the lightweight diesel engine, the refrigerant Freon, four-wheel brakes, and safety glass, among other inventions.

Arthur Korn
1870-1945

German physicist who invented the first practical photoelectric facsimile system. A professor of physics at the Berlin Institute of Technology, in 1904 Korn developed a device to transmit pictures from one place to another over electric wires. The system used the light-sensitive element selenium to convert the different tones of a scanned image into a varying electric current. He also developed a phototelegraph, which enabled aircraft to pick-up weather maps and aerial pictures and which was widely used by European military authorities.

Igor Vasilevich Kurchatov
1903-1960

Russian nuclear physicist who headed the development of Soviet nuclear technology. Kurchatov's early work on non-conductors (dielectrics) led to studies in atomic physics. After studying neutrons, Kurchatov proposed the splitting of heavy atoms in 1939, and became the head of Soviet atomic research efforts during World War II. On August 12, 1953, his team detonated the first Soviet hydrogen bomb. An advocate of peaceful atomic uses, he helped build the first atomic energy station. He also devised a method to protect shipping from magnetic mines.

Edwin Herbert Land
1909-1991

American inventor with over 500 optical and plastics patents who was the driving force behind the Polaroid Corporation. Land developed the polarizer Polaroid-J used in sunglasses and optical instruments. During World War II he developed the invaluable reconnaissance tool "vectography"—a three-dimensional photographic system. In 1947 he invented the first instant-developing film and the Polaroid-Land Camera, which produced prints in 60 seconds. Land also developed an optical system for observing living human cells in their natural colors.

Paul Langevin
1872-1946

French physicist best known for establishing the modern theory of magnetism. Langevin explained paramagnetism and diamagnetism in terms of interactions between magnetic fields induced by orbiting electrons and external magnetic forces. He showed how to slow down neutrons, which proved essential for later work on atomic reactors. He is important for introducing and explaining Albert Einstein's theory of relativity to the French scientific community. His echolocation technique for detecting submarines was the basis for sonar.

Irving Langmuir
1881-1957

American chemical physicist who was awarded the 1932 Nobel Prize for Chemistry for work on surface chemistry. From studies of surface films on liquids Langmuir deduced approximate molecular sizes and shapes, and his studies of gas absorption at solid surfaces led to the Langmuir absorption isotherm. He also developed a shared-electron theory of chemical bonds. While at General Electric (1919-1950) Langmuir developed many devices, including an improved light bulb, enhanced vacuum pump, and hydrogen welding torch.

Willard Frank Libby
1908-1980

American chemist who received the 1960 Nobel Prize for Chemistry for developing the technique of carbon-14 dating. As a member of the Manhattan Project, Libby helped develop a method of separating uranium isotopes, a critical step in the creation of the atomic bomb. After completing this project, he and a group of his students devised the carbon-14 method for dating organisms as old as 50,000 years. His technique measures small amounts of radiation in carbon-based organisms, and became widely used by archaeologists and anthropologists for dating artifacts.

Alexander Martin Lippisch
1894-1976

German aviation designer who pioneered the development of high-speed jet and rocket-powered aircraft. Lippisch initially devoted his research to tailless planes, and developed his first successful tailless glider in 1921. Seven years later he made history with the first rocket-powered glider flight. In 1940 he designed the rocket-powered *ME 163* interceptor for the German Luftwaffe. This aircraft initially flew in excess of 600 mph (966 kph), and by the next year reached speeds of 754 mph (1213 kph) and heights of 30,000 ft (9144 m). In 1946 Lippisch emigrated to the United States, where he worked with the U.S. Air Force and founded the Lippisch Research Corporation in 1965.

Guglielmo Marconi
1874-1937

Italian inventor and entrepreneur who developed wireless telegraphy and related technologies. After experimenting with electromagnetic waves in Italy, Marconi moved to Britain, where he patented his method for transmitting signals through space. In 1901 Marconi was the first to send a wireless signal across the Atlantic, from England to Newfoundland. One of the best-known inventors of his time, Marconi received the Nobel Prize for physics in 1909.

John W. Mauchley
1907-1980

American physicist who proposed and later built the world's first successful, general digital computer. Mauchley constructed the mammoth Electronic Numerical Integrator And Computer (ENIAC), with J. Presper Eckert, while at the Moore School of Engineering at the University of Pennsylvania in Philadelphia. It used cathode ray tubes as its memory. Mauchley and Eckert later went on to make UNIVAC, the first computer sold commercially by the Rand Corporation, running on Hollerith cards.

Louis Marius Moyroud
1914-

French inventor who, along with engineer René Higonnet, developed the first practical, high-speed phototypesetting machine. Moyroud earned a degree in engineering at the Ecole Nationale Supérieure des Arts et Métiers and served as an officer the French military from 1936 to 1940. He briefly worked for LMT Laboratories, a French subsidiary of ITT, but left in 1946 to devote himself to research on photocomposition. In 1948 Moyroud and Higonnet introduced their phototypesetting machine, the Lumitype (also called the Photon), in the United States. The first book to be composed with this device was *The Wonderful World of Insects* (1953).

Richard Dixon Oldham
1858-1936

Irish geologist and seismologist who established the existence of Earth's core. Based on his landmark survey of the 1897 Assam earthquake, Oldham identified three types of seismic waves—primary (P-waves), secondary (S-waves), and surface waves—predicted earlier by Siméon Poisson. Oldham noted that during an earthquake, seismographs located on the opposite side of Earth detect P-waves later than expected, leading him to conclude that Earth has a core less dense and rigid than its mantle.

Ransom Eli Olds
1864-1950

American automobile inventor who designed the first successful American-made car—the Oldsmobile—in 1896. Olds's first car was a small carriage with a single cylinder, four-cycle engine and a curved dashboard, appropriately named the "Curved Dash Olds." In 1899 he formed the Olds Motor Works with backing from businessman Samuel L. Smith, and by 1904 5,000 Oldsmobiles had been sold. In 1904 Olds parted with Smith to form the REO Motor Car Company. Olds pioneered the progressive assembly method, a precursor to the modern automotive assembly line, but after several years his new company lost ground to its competitors and he turned his attention to other enterprises, including his own brand of lawn mower.

Fritz Pfleumer

German inventor who in 1928 produced a system for recording on paper coated with a magnetizable powdered steel layer. The German manufacturer AEG bought the patent from Pfleumer and by 1935 had developed the magnetophon tape recorder. His patented design was not available until after World War II in the United States.

George A. Philbrick

American electrical engineer who helped develop the first electronic sight simulator for fighter aircraft and guided bombsight simulator for bombers during World War II. To do this, Philbrick also developed the first use for a small-scale operational amplifier, later to become a very important electronic device that was also used for automatic antiaircraft weapons during the Second World War. This led to the development of the first electronic analog computers a few years later.

Valdemar Poulsen
1869-1942

Danish engineer who invented the telegraphone, the first practical device for recording and replaying sound—the forerunner of modern audio and video recording systems. Poulsen developed the telegraphone, an electromagnetic phonograph that could record and reproduce human speech on a steel wire, while employed with the Copenhagen Telephone Company. In 1903 he founded the American Telegraphone Company to market his invention, which by then was able to record for 30 continuous minutes. That same year he patented arc wireless transmission, a modification of Englishman William Duddell's "singing arc," which could generate continuous audio waves and paved the way for radio broadcasting.

Sir John Turton Randall
1905-1984

British physicist and biophysicist who worked with Henry Albert Boot designing the radar or cavity magnetron that was influential in the Allies winning World War II. He also worked for the General Electric Company developing luminescent powders for use in discharge lamps. In 1944, as professor of natural philosophy at St. Andrews University, he pioneered the use of new types of light microscopes to be used by research biologists, physicists, and biochemists.

Louis Renault
1877-1944

French automobile manufacturer who, with his brothers, manufactured racing cars and, during World War I, tanks. Renault continued manufacturing military equipment for the French army and, during the Second World War, for the occupying German forces. This led to his imprisonment after France was liberated, and he died in jail in 1944. Subsequently, the Renault automobile company became one of the largest in Europe and was later nationalized by the French government.

Boris Lwowitsch Rosing
1869-1933

Russian engineer who invented the first cathode ray tube to reproduce television images. In 1907 Rosing was able to transmit black and white silhouettes of simple shapes using a mechanical mirror-drum as a camera and a cathode ray tube as a receiver. He was the first to suggest that electrical signals from a mechanical transmitter might be transformed into visual images when sent to a cathode ray receiver.

Erik Rotheim
1898-1938

Norwegian inventor known for the invention of the aerosol can. In 1926 he filed for a patent for an aerosal-type dispenser. He later modified his original design, specifying a spray nozzle and using hydrocarbons as the propellant gas.

David Sarnoff
1891-1971

Russian-American businessman who became a pioneer in television and radio broadcasting. David Sarnoff was born in Minsk, Russia, and traveled via steerage with his family to New York nine years later. He spoke no English, but helped his family make ends meet by selling newspapers and holding odd jobs. At the age of 15, he learned Morse code (a series of dots and dashes representing letters and numerals) and became a junior operator with the Marconi Wireless Telegraph Company of America. On April 14, 1912, he made history by picking up a fateful message relayed from a ship at sea: "*RMS Titanic* ran into iceberg, sinking fast." Sarnoff was soon promoted within the company, and in 1916 he suggested the first "radio music box." In 1921, he became general manager of the newly formed Radio Corporation of America (RCA), where he made headlines by broadcasting a fight between Jack Dempsey and George Carpentier to between 200,000 and 300,000 listeners, and at the same time making RCA a household name. In 1926, he formed the National Broadcasting Company (NBC) and helped launch television as a new broadcasting medium.

Louis Schmeisser

German inventor who developed the first efficient, hand-carried machine gun. Although the first machine gun, the Gatling gun, was invented in 1862 and other machine guns followed, these were all cumbersome and unwieldy. Other nations such as England and the U.S. developed domestic versions of the machine gun, and, by the Second World War, machine guns were found on airplanes, tanks, jeeps, and in the hands of foot soldiers.

William Bradford Shockley
1910-1989

American physicist awarded the 1956 Nobel Prize in physics with John Bardeen and Walter Brattain for developing the transistor at Bell Laboratories (1948). Shockley conceived of the junction transistor shortly thereafter, which Brattain successfully built and tested by 1950. The commercially successful Shockley Transistor Corporation was later founded to produce miniaturized circuits for radios, televisions, and computer equipment. In the 1960s Shockley advanced the controversial theory that the genetic component of intelligence was determined by racial heritage.

Frederick Simms
1863-1644

English engineer who is considered the father of the British motor industry. Impressed with Daimler's pioneering internal combustion engine, Simms purchased the British rights, intending to manufacture automobiles. He was initially unsuccessful and sold his rights. Later, Simms helped develop the tank and other vehicles important to modern, mechanized warfare. Simms is also generally recognized as inventing the words "motor car" and "petrol." Simms founded the Manufacturing Company to produce engines and Simms Motor Units, which produced ancillary equipment for cars.

Adolph Karl Heinrich Slaby
1849-1913

German mathematician and electrical engineer. As professor of electrical engineering, he was asked to institute the new Department of Electrical Engineering at Technical University of Berlin. While teaching there he was instrumental in the design and development of the wireless telegraph in Germany. Along with the design, he contracted the services of an electric company to promote the wireless telegraph. He eventually left the university and formed the commercial company Telefunken.

Perry LeBaron Spencer

American engineer known for his work in inventing the microwave oven. While working with radar equipment he noticed that candy in his pocket had melted. Radar detects objects by bouncing microwaves from them. Spencer he realized that it was the microwaves themselves that had melted the candy. In 1945 his microwave oven was commercially available, known as a radarange. The first microwave prototypes were large, cumbersome, and expensive.

Elmer Ambrose Sperr
1860-1930

American electrical engineer who invented the gyrostabilizer, used to control ships, airplanes, and torpedoes, and developed other gyroscopic compasses and devices such as autopilots. Sperry was a prolific inventor with more than 400 patents to his name. He formed numerous companies to manufacture and sell his diverse inventions, including an electric generator, mining machinery, electric streetcars, precision instruments, farm machinery, and hydraulic equipment.

Alan Archibald Campbell Swinton
1863-1930

Scottish electrical engineer and inventor known for his work with the medical applications of radiography and the cathode ray. He published the first x-ray photograph taken in Britain, and his work became popular and in demand. In 1908 he outlined the principles of "distant electric vision"—an electronic system of television that used the cathode ray. This patent incorporated the basic process on which modern television cameras and receivers work.

Sir Ernest Dunlop Swinton
1868-1951

British inventor, writer, and member of the military. One of the originators of the tank, he is responsible for the use the word *tank* to describe armored vehicles used for war purposes. He later became professor of military history at Oxford University (1925-39).

John Talafierro Thompson
1860-1940

American soldier and inventor who developed the submachine gun in 1918. The "Tommy gun," named for Thompson, weighed only 10 lb (4.5 kg), making it easy to carry. The gun saw its first military use in Nicaragua in 1925, carried by U.S. Marines. Since then, variations and improvements on this design have given rise to a plethora of weapons manufactured by nearly every major industrial nation. The submachine gun has seen action in nearly every war since its invention.

Leonardo Torres y Quevedo
1852-1936

Spanish mathematician and engineer who built an algebraic equation-solving machine, a robot that could play a chess endgame, and in 1920 demonstrated that calculations of any kind could be achieved by purely mechanical means. Due to his dislike of writing, he left behind few descriptions of his many inventions, which included innovative and practical designs for aero-ships and the cable-car line at Niagara Falls, Ontario.

Konstantin Eduardovich Tsiolkovsky
1857-1935

Russian rocket scientist who was an early pioneer of space travel research. In the 1890s Tsiolkovsky developed Russia's first wind tunnel to test aircraft aerodynamics. His passion for the stars led him to try his hand at science fiction writing, and his stories of rocket ships eventually evolved into a study of the potential for space travel. Tsiolkovsky's research in aerodynamics led him to theorize about the challenges of using rocket engines in space, including navigation and fuel supply maintenance. His "Tsiolkovsky Formula" formed the basis of contemporary astronautics, and paved the way for space exploration.

Thomas J. Watson, Sr.
1874-1956

American businessman who established the International Business Machines Corporation (IBM). In 1913 Watson left a sales manager position at the National Cash Register Company (NCR) to become president of the Computing-Tabulating-Recording Company, which made electrical punch-card computing systems. In 1924 the company changed its name to IBM, and went on to dominate the business machine industry in the 1930s and 1940s. In 1952 Watson turned the company over to his son, Thomas Jr. At the time of the elder Watson's death in 1956, IBM had 200 offices throughout the country and employed 60,000 people.

Maurice Vincent Wilkes
1913-

English physicist and computer scientist who developed the first stored-program computer—the EDSAC—and later assisted in the creation of the first computer network, called the "Cambridge Ring," in the mid-1970s. Wilkes headed the computing laboratory at Cambridge, where, after meeting computer pioneers Howard Aiken and Sam Caldwell on a trip to America in 1946, he set out to build his own computer, one that could store programs. The EDSAC (electronic delay storage automatic calculator) was completed in June 1949, followed by EDSAC 2 in 1958. The EDSAC 2 was the first computer to have a microprogrammed control unit.

Frederic Calland Williams
1911-1977

English electrical engineer who invented the first digital computer memory system, called the Williams tube store. During World War II, Williams helped to develop IFF (Identification Friend or Foe) systems to distinguish between friendly and enemy aircraft, and AI (Airborne Interception) systems, which allowed aircraft to track and intercept other planes. In the mid-1940s he visited MIT, where he was inspired to develop a cathode-ray-tube digital storage system for computers. In 1947 Williams and his team built the first working Random Access Memory (RAM) system, which they used in their Manchester Mark 1 computer. Williams was knighted in 1976.

Richard Adolf Zsigmondy
1865-1929

Austrian chemist who received the 1925 Nobel Prize for Chemistry for his work with colloids. Zsigmondy studied quantitative analysis at the Medical Faculty in Vienna, and organic chemistry at the University of Munich. He later took a position in a glassworks, where he was introduced to colloids, a suspension of tiny particles in another substance. He discovered the presence of tiny particles of gold in ruby glass, and believed he could learn more about colloids by studying the way in which particles scatter light. To this end, he and Heinrich Siedentopf invented the ultramicroscope in 1903, which was subsequently used in the fields of biochemistry and bacteriology.

Konrad Zuse
1910-1995

German engineer who built the world's first binary calculating machine—the Z1—in 1938. Zuse completed his degree in civil engineering, and began his career as a design engineer in the aircraft industry. In the 1930s he set out to build a computer that utilized a binary set of numbers, allowed for automatic arithmetical calculation, and had a high memory capacity. In 1938 he completed the Z1, and in 1941 produced the Z3, the first program-controlled electromechanical computer. His subsequent Z4 design was used at the Swiss Federal Polytechnical Institute until 1955.

Bibliography of Primary Sources

Books

Bush, Vannevar. *Science: The Endless Frontier* (1945). Bush's book was influential in prompting the creation of both the National Science Foundation (NSF) and the Advanced Research Projects Agency (ARPA).

Goddard, Robert. *A Method of Reaching High Altitudes* (1919). Contains Goddard's proposals for advanced rocket designs, which, much like Konstantin Tsiolkovsky's works, promoted space travel. The public reacted to Goddard's treatise with sensationalism, labeling him a mad scientist for proposing travel to the Moon.

Neumann, John von, Arthur W. Burks, and Herman Goldstine. *Preliminary Discussion of the Logical Design of an Electronic Computing Instrument* (1946). Also known as the "Princeton Reports," this highly influential paper outlined the basic design of nearly all future computers, including the idea of stored programs.

Norbert, Wiener. *Cybernetics* (1948). In this work, Wiener applied principles of control theory, automation, and information feedback to comparative analysis of human and machine processes, introducing ideas that later influenced the development of artificial intelligence.

Oberth, Hermann. *The Rocket into Planetary Space* (1923). An influential work containing discussion of manned and unmanned rocket flights, various aerodynamic and thermodynamic analyses, proposals for liquid fuel designs, and a host of other technical aspects regarding space technology.

Tsiolkovsky, Konstantin. *The Investigation of Outer Space by Means of Reaction Apparatus* (1903). In this visionary work, Tsiolkovsky discussed the possibility of rocket flight into space and calculated the force necessary for such a rocket to escape Earth's gravitational pull.

Periodical Articles

Bardeen, John, and Walter H. Brattain. "The Transistor: A Semiconductor Triode" (1948). One of three key papers that described the authors' invention of the transistor.

Bardeen, John, and Walter H. Brattain. "The Physical Principles Involved in Transistor Action" (1949). The third of three landmark papers by the inventors of the revolutionary transistor.

Brattain, Walter H., and John Bardeen. "Nature of the Forward Current in Germanium Point Contacts" (1948). This article appeared in the same issue of *Physical Review* as the authors' paper titled "The Transistor."

Bush, Vannevar. "As We May Think" (1945). In this prophetic article, which first appeared in *Atlantic Monthly,* Bush described a device, called the "memex," that would fit on a desktop, store vast amounts of data, allow quick search and retrieval, and display any requested information on a screen—thus predicting the future computer with startling accuracy.

Clarke, Arthur C. "Extra-Terrestrial Relays" (1945). This article, first published in *Wireless World,* contains a proposal for the use of orbiting satellites to relay radio and television transmissions around the world.

Shockley, William. "The Theory of P-N Junctions in Semiconductors and P-N Junction Transistors" (1949). Shockley in this article described an important refinement of the original transistor developed by his colleagues at Bell Labs, Walter Brattain and John Bardeen.

JOSH LAUER

General Bibliography

Books

Agassi, Joseph. *The Continuing Revolution: A History of Physics from the Greeks to Einstein.* New York: McGraw-Hill, 1968.

Allen, Garland E. *Life Science in the Twentieth Century.* New York: Cambridge University Press, 1978.

Anderson, E. W. *Man the Navigator.* London: Priory Press, 1973.

Anderson, E. W. *Man the Aviator.* London: Priory Press, 1973.

Arnold, Caroline. *Genetics: From Mendel to Gene Splicing.* New York: F. Watts, 1986.

Asimov, Isaac. *Adding a Dimension: Seventeen Essays on the History of Science.* Garden City, NY: Doubleday, 1964.

Bahn, Paul G., editor. *The Cambridge Illustrated History of Archaeology.* New York: Cambridge University Press, 1996.

Basalla, George. *The Evolution of Technology.* New York: Cambridge University Press, 1988.

Benson, Don S. *Man and the Wheel.* London: Priory Press, 1973.

Bowler, Peter J. *The Norton History of the Environmental Sciences.* New York: W. W. Norton, 1993.

Brock, W. H. *The Norton History of Chemistry.* New York: W. W. Norton, 1993.

Bruno, Leonard C. *Science and Technology Firsts.* Edited by Donna Olendorf, guest foreword by Daniel J. Boorstin. Detroit: Gale, 1997.

Bud, Robert and Deborah Jean Warner, editors. *Instruments of Science: An Historical Encyclopedia.* New York: Garland, 1998.

Bynum, W. F., et al., editors. *Dictionary of the History of Science.* Princeton, NJ: Princeton University Press, 1981.

Carnegie Library of Pittsburgh. *Science and Technology Desk Reference: 1,500 Frequently Asked or Difficult-to-Answer Questions.* Washington, D.C.: Gale, 1993.

Crone, G. R. *Man the Explorer.* London: Priory Press, 1973.

Dunn, L. C. *A Short History of Genetics: The Development of Some of the Main Lines of Thought, 1864-1939.* Ames: Iowa State University Press, 1991.

Elliott, Clark A. *History of Science in the United States: A Chronology and Research Guide.* New York: Garland, 1996.

Erlen, Jonathan. *The History of the Health Care Sciences and Health Care, 1700-1980: A Selective Annotated Bibliography.* New York: Garland, 1984.

Fearing, Franklin. *Reflex Action: A Study in the History of Physiological Psychology.* Introduction by Richard Held. Cambridge: MIT Press, 1970.

French, Roger and Andrew Wear. *British Medicine in an Age of Reform.* New York: Routledge, 1991.

Good, Gregory A., editor. *Sciences of the Earth: An Encyclopedia of Events, People, and Phenomena.* New York: Garland, 1998.

Graham, Loren R. *Science in Russia and the Soviet Union: A Short History.* New York: Cambridge University Press, 1993.

Grattan-Guiness, Ivor. *The Norton History of the Mathematical Sciences: The Rainbow of Mathematics.* New York: W. W. Norton, 1998.

Gregor, Arthur S. *A Short History of Science: Man's Conquest of Nature from Ancient Times to the Atomic Age.* New York: Macmillan, 1963.

Gullberg, Jan. *Mathematics: From the Birth of Numbers.* Technical illustrations by Pär Gullberg. New York: W. W. Norton, 1997.

Hellemans, Alexander and Bryan Bunch. *The Timetables of Science: A Chronology of the Most Important People and Events in the History of Science.* New York: Simon and Schuster, 1988.

Hellyer, Brian. *Man the Timekeeper.* London: Priory Press, 1974.

A History of Science Policy in the United States, 1940-1985: Report Prepared for the Task Force on Science Policy, Committee on Science and Technology, House of Representatives, Ninety-Ninth Congress, Second Session. Washington, D.C. U.S. Government Printing Office, 1986.

Holmes, Edward and Christopher Maynard. *Great Men of Science.* Edited by Jennifer L. Justice. New York: Warwick Press, 1979.

Hoskin, Michael. *The Cambridge Illustrated History of Astronomy.* New York: Cambridge University Press, 1997.

Lankford, John, editor. *History of Astronomy: An Encyclopedia.* New York: Garland, 1997.

Lincoln, Roger J. and G. A. Boxshall. *The Cambridge Illustrated Dictionary of Natural History.* Illustrations by Roberta Smith. New York: Cambridge University Press, 1987.

Martin, Ernest G. *The Story of Our Bodies: The Science of Physiology, Organs, and Their Functions in Human Beings.* New York: P. F. Collier & Son Company, 1930.

Porter, Roy. *The Cambridge Illustrated History of Medicine.* New York: Cambridge University Press, 1996.

Rassias, Themistocles M. and George M. Rassias, editors. *Selected Studies, Physics-Astrophysics, Mathematics, History of Science: A Volume Dedicated to the Memory of Albert Einstein.* New York: Elsevier North-Holland Publishing Company, 1982.

Reingold, Nathan, editor. *Science in America Since 1820.* New York: Science History Publications, 1976.

Reingold, Nathan and Ida H. Reingold, editors. *Science in America, A Documentary History, 1900-1939.* Chicago: University of Chicago Press, 1981.

Rothenberg, Marc. *The History of Science in the United States: An Encyclopedia.* New York: Garland, 2000.

Rudwick, M. J. S. *The Meaning of Fossils: Episodes in the History of Paleontology.* New York: American Elsevier, 1972.

Sachs, Ernest. *The History and Development of Neurological Surgery.* New York: Hoeber, 1952.

Sarton, George. *Introduction to the History of Science.* Huntington, NY: R. E. Krieger Publishing Company, 1975.

Sarton, George. *The History of Science and the New Humanism.* New Brunswick, NJ: Transaction Books, 1987.

Schneiderman, Ron. *Computers: From Babbage to the Fifth Generation.* New York: F. Watts, 1986.

Smith, Roger. *The Norton History of the Human Sciences.* New York: W. W. Norton, 1997.

Spangenburg, Ray and Diane K. Moser. *The History of Science from 1895 to 1945.* New York: Facts on File, 1994.

Stiffler, Lee Ann. *Science Rediscovered: A Daily Chronicle of Highlights in the History of Science.* Durham, NC: Carolina Academic Press, 1995.

Stwertka, Albert and Eve Stwertka. *Physics: From Newton to the Big Bang.* New York: F. Watts, 1986.

Swenson, Loyd S. *Genesis of Relativity: Einstein in Context.* New York: B. Franklin, 1979.

Travers, Bridget, editor. *The Gale Encyclopedia of Science.* Detroit: Gale, 1996.

Webster, Charles. *Biology, Medicine and Society, 1840-1940.* New York: Cambridge University Press, 1981.

Weindling, Paul, editor. *International Health Organizations and Movements, 1918-1939.* New York: Cambridge University Press, 1995.

Whitehead, Alfred North. *Science and the Modern World: Lowell Lectures, 1925.* New York: The Free Press, 1953.

Willmore, A. P. and S. R. Willmore, consultant editors. *Aerospace Research Index: A Guide to*

World Research in Aeronautics, Meteorology, Astronomy, and Space Science. Harlow, England: F. Hodgson, 1981.

World of Scientific Discovery. Detroit: Gale, 1994.

Young, Robyn V., editor. *Notable Mathematicians: From Ancient Times to the Present.* Detroit: Gale, 1998.

JUDSON KNIGHT

Index

Numbers in bold refer to main biographical entries

A

B

C

D

E

F

G

H

M

N

O

P

Q

R

S

T

U

V

W

X

Y

Z